AF829142

121

Plant Growth-Promoting Microbes for Sustainable Biotic and Abiotic Stress Management

Heba I. Mohamed
Hossam El-Din Saad El-Beltagi
Kamel A. Abd-Elsalam
Editors

Plant Growth-Promoting Microbes for Sustainable Biotic and Abiotic Stress Management

Springer

MÄNGELEXEMPLAR

Editors
Heba I. Mohamed
Ain Shams University
Cairo, Egypt

Hossam El-Din Saad El-Beltagi
King Faisal University
Al-Ahsa, Saudi Arabia

Kamel A. Abd-Elsalam
Agricultural Research Center
Giza, Egypt

ISBN 978-3-030-66586-9 ISBN 978-3-030-66587-6 (eBook)
https://doi.org/10.1007/978-3-030-66587-6

© The Editor(s) (if applicable) and The Author(s), under exclusive license to Springer Nature Switzerland AG 2021
This work is subject to copyright. All rights are solely and exclusively licensed by the Publisher, whether the whole or part of the material is concerned, specifically the rights of translation, reprinting, reuse of illustrations, recitation, broadcasting, reproduction on microfilms or in any other physical way, and transmission or information storage and retrieval, electronic adaptation, computer software, or by similar or dissimilar methodology now known or hereafter developed.
The use of general descriptive names, registered names, trademarks, service marks, etc. in this publication does not imply, even in the absence of a specific statement, that such names are exempt from the relevant protective laws and regulations and therefore free for general use.
The publisher, the authors, and the editors are safe to assume that the advice and information in this book are believed to be true and accurate at the date of publication. Neither the publisher nor the authors or the editors give a warranty, expressed or implied, with respect to the material contained herein or for any errors or omissions that may have been made. The publisher remains neutral with regard to jurisdictional claims in published maps and institutional affiliations.

This Springer imprint is published by the registered company Springer Nature Switzerland AG
The registered company address is: Gewerbestrasse 11, 6330 Cham, Switzerland

Preface

The growth rate of the global population demands increasing food production. However, the increase in agricultural productivity is in many circumstances largely dependent on the use of chemical fertilizers that many farmers around the world do not have economic provision and can have a negative effect on the climate. Environmental stresses may also impede plant growth and yield, causing low crop productivity, which can impact global food safety. There must also be less use of chemical fertilizers and an increased plant tolerance for abiotic stresses in order to increase global agricultural production economically, ecologically, and sustainably. Plant growth-promoting microorganisms (PGPM) have a potential benefit to improve crop production, food, and safety in sustainable and more environmentally friendly agricultural systems. The use of beneficial microbes like fungi, bacteria, algae, cyanobacteria, and actinomycetes, those microbes which enhance the growth of plants under abiotic stressors including drought, salinity, waterlog, temperature extremes, mineral nutrients, heavy metal, and biotic stress conditions including plant diseases, nematodes, viruses, and diseases. The achievement of sustainable agriculture while maintaining environmental, agroecosystem functions and biodiversity is a major challenge to current agricultural practices and also poses serious risks for crop productivity, soil fertility, and nutritional value of agricultural production through the conventional use of chemical inputs (fertilizer, pesticides, nutrients, etc.). Given these threats, the management of pests and diseases, the preservation of the health of the agroecosystems, and the avoidance of public and animal health problems now become key priorities. Researchers, scientists, agriculturists, farmers, and policymakers have been very aware of PGPM as biofertilizers, plant growth promoters, biopesticides, and managers of soil and plant health. PGPM are receiving increasing attention from agronomists and environmentalists as candidates to develop an effective, eco-friendly, and sustainable alternative to conventional agricultural (e.g., chemical fertilizers and pesticide) and remediation (e.g., chelators-enhanced phytoremediation) methods employed to deal with climate change-induced stresses. Using PGPM will help satisfy the demand for global agricultural productivity, which is projected to hit approximately nine billion by 2050 to feed the world's growing population. To achieve this goal, however, PGPM strains must be environmentally friendly, be compatible with useful soil rhizobacteria, give considerable plant growth promotion and biocontrol potential, and can withstand different biotic and abiotic stresses.

This book provides up-to-date knowledge on biofertilizers and the roles of microorganisms in plant health, with specific emphasis on the mitigating strategies to combat plant stresses. The application of microorganisms for quicker, more cost-effective, and precise diagnostic procedures of plant disease control and antimicrobial mechanisms has been discussed in detail.

The first chapter by Shah et al. reviews Cyanobacteria and Algae as Biocontrol Agents Against Fungal and Bacterial Plant Pathogens. Chapter 2 by Monteiro et al. highlights Plant Growth Promoting Rhizobacteria in Amelioration of Abiotic Stresses: A Functional Interplay and Prospective. In Chap. 3, Jampílek and Kráľová describe Seaweeds as Indicators and Potential Remediators of Metal Pollution. The Role of Microorganisms in Managing Soil Fertility and Plant Nutrition in Sustainable Agriculture is described by Mohamed et al. in Chap. 4. In Chap. 5, Prasher and Sharma highlight the Role of Endophytic Bacteria in the Alleviation of Heavy Metals from an Ecosystem. Chapter 6 by Silva et al. provides insights into Microbial Enzymes and Soil Health. In Chap. 7, Yasmeen et al. state *Pseudomonas* as Plant Growth-Promoting Bacteria and Its Role in Alleviation of Abiotic Stress. In Chap. 8, Basit et al. highlight Plant Growth-Promoting Rhizobacteria (PGPR) as Biocontrol Agents for Viral Protection. Chapter 9 by Lonkar and Bodade describes the Potential Role of Endophytes in Weeds and Herbicide Tolerance in Plants. Almoneafy et al. in Chap. 10 detail the Auspicious Role of Plant Growth Promoting Rhizobacteria in the Sustainable Management of Plant Diseases. Chapter 11 by Basit et al. gives an overview of Microbial Bioactive Compounds Produced by Endophytes (Bacteria and Fungi) and Their Uses in Plant Health. Biosynthesis of Nanoparticles by Microorganisms and Applications in Plant Stress Control are discussed in Chap. 12 by Ramadan and El-Beltagi. Chapter 13 by Padhi and Behera explains Nano-enabled Approaches for the Suitable Delivery of Fertilizer and Pesticide for Plant Growth. Shanab and Shalaby give information about the Production of Plant Hormones from Algae and Its Relation to Plant Growth in Chap. 14. In Chap. 15, Misra and Ansari state the Role of Trichoderma in Agriculture and Disease Management. The Production of Antibiotics from PGPR and Their Role in Biocontrol of Plant Diseases are highlighted in Chap. 16 by Hamid et al. In Chap. 17, Jilani et al. describe the Role of Phosphate-Solubilising Microorganisms in Agricultural Development. Gören-Sağlam in Chap. 18 gives an overview of Cyanobacteria as Biofertilizer and Their Effect Under Biotic Stress. Shah et al. in Chap. 19 describe Microorganism: A Potent Biological Tool to Combat Insects and Herbivores. In Chap. 20, Ahmad et al. focus on Eco-Friendly Approaches for the Alleviation of Root-Knot Nematodes. In Chap. 21, Sharma et al. write on Rhizosphere, Rhizosphere Biology, and Rhizospheric Engineering. Finally, Siddiqua et al. explain Microbial Enzymes and Their Role in Phytoremediation in Chap. 22.

We wish to thank Springer officials, particularly C. Aishwarya and Shanthini Kamaraj, for their generous support and efforts in accomplishing this volume. We are highly delighted and thankful to all our contributing authors for their vigorous support and outstanding cooperation to write altruistically these authoritative and valuable chapters. We specially thank our families for consistent support and encouragement.

With a bouquet of information on the role of plant growth-promoting microorganisms for sustainable biotic and abiotic stress management, the editors hope that this book will be a valuable resource for students of different divisions; researchers and academicians, working in the field of nanoscience, nanotechnology, plant sciences, agriculture microbiology, and fungal biology; and scholars interested in strengthening their knowledge in the area of environmental microbiology.

Cairo, Egypt — Heba I. Mohamed
Al-Ahsa, Saudi Arabia — Hossam El-Din Saad El-Beltagi
Giza, Egypt — Kamel A. Abd-Elsalam

Contents

About the Editors

Heba I. Mohamed is a Professor of Plant Physiology, Faculty of Education, Biological and Geological Sciences Department, Ain Shams University. Dr. Heba completed her M.Sc and Ph.D in Plant Physiology, Faculty of Education, Ain Shams University. Dr. Heba's research interests include biotic and abiotic stresses, plant biochemistry, use of eco-friendly compounds to alleviate plant stress, plant secondary metabolites, and genetic differences between different genotypes. She has published 8 book chapters, 1 review article, and more than 57 research articles in international peer-reviewed journals including *Agronomy*, *Plants*, *Molecules*, *Sustainability*, *Plant Physiology and Biochemistry*, *Biological Trace Element Research*, *Photosynthetica*, *European Journal of Plant Pathology*, and other journals. Dr. Heba is a reviewer of journals including *Plant Physiology and Biochemistry*, *Scientific Reports*, *The Journal of Horticultural Science & Biotechnology*, and *Journal of Soil Science and Plant Nutrition*. She also is editor of the *Microbial Biosystems* journal. Dr. Heba is a member of the project titled Genome wide association study (GWAS) of dodder tolerance in the model plant *Medicago truncatula* to detect key genetic factors and accelerate MAS in fodder legumes. Belonging to PHC IMHOTEP bilateral Egypt France cooperation in science and technology program 2019. Dr. Heba has obtained certificate of recognition in honor of achievement in international publication that supports Ain Shams University World Ranking 25/6/2012, 2017, 2018. Dr. Heba has obtained country encouragement prize in Cairo 2018.

Ain Shams University
Cairo, Egypt

Hossam El-Din Saad El-Beltagi, Ph.D. is a Professor of Agricultural Biotechnology, College of Agriculture and Food Sciences, King Faisal University, Al-Ahsa, Saudi Arabia, and a Professor of Biochemistry, Faculty of Agriculture, Cairo University, Giza, Egypt. Dr. Hossam's research interests include plant biotic and abiotic stresses, plant biochemistry, biotechnology, and plant secondary metabolites.He has published 5 book chapters, 1 review article, and more than 102 research articles in international peer-reviewed journals including *Food Chemistry*, *PLoS One*, *International Journal of Food Science and Nutrition*, *Food Biochemistry*, *Molecules*, *Sustainability*, *Animals*, *Antioxidants*, *Radiation Physics and Chemistry*, *Physiological and Molecular Plant Pathology*, *Grasas y acetes*, *Gesunde Pflanzen*, *Journal of Biotechnology*, *Notulae Botanicae Horti Agrobotanici cluj napoca*, *Fresenius Environmental Bulletin*, and *Journal of Medicinal Plants Research*. He is Associate Editor of many international journals by different publishers including *Journal of Advanced Research* from Elsevier and *Notulae Scientia Biologicae*. Dr. Hoosam attended more than 25 international conferences and workshops. He is a member of more than 8 international scientific societies. Dr. Hossam is a reviewer of more than 30 international peer-reviewed journals from many publishing houses. He received many Prizes including Ministry of Higher Education Prize for International Publishing Award in top Eigenfactor Score journal 2012; Country Encouragement Prize for Agriculture Sciences, 2013; Cairo University Encouragement Prize in advanced technological sciences 2013; and Cairo University scientific excellence Prize in the field of interdisciplinary environmental science 2015. Dr. Hossam served as visiting professor in Biochemistry Department at College of Science in Granada University, Spain 2017. He served as Consultant in Deanship of Scientific Research in King Faisal University since 2019 to date.

Agircultural Biotechnology Department
College of Agriculture and Food Sciences
King Faisal University
Al-Ahsa, Saudi Arabia

Biochemistry Department, Faculty of Agriculture
Cairo University
Giza, Egypt

Kamel A. Abd-Elsalam, Ph.D. is a Research Professor at Plant Pathology Research Institute, Agricultural Research Center, Giza, Egypt. Dr. Kamel's research interests include developing, improving, and deploying plant biosecurity diagnostic tools, understanding and exploiting fungal pathogen genomes, and developing eco-friendly hybrid nanomaterials for controlling toxicogenic fungi and plant diseases. He has published 10 book chapters, 6 review articles, 1 translated book, and more than 120 research articles in international peer-reviewed journals including *Fungal Diversity*, *Fungal Biology*, *FEMS Review Microbiology*, *PLOS One*, and *PLOS Genetics*. He is associate editor for *Mycosphere* and review editor for *Frontiers in Genomic Assay Technology* and referees for journals, including *Plant Pathology*, *Journal of Phytopathology*, and *Crop Protection*, *IET Nanotechnology*, *Fungal Diversity*, *BMC Genomics*, and *Foodborne Pathogens and Diseases* and edited or authored some books. Dr. Kamel has also served as molecular mycologist for 5 years in the Department of Botany and Microbiology, College of Science, King Saud University, Saudi Arabia. He received the Federation of Arab Scientific Research Councils Prize for distinguished scientific research in biotechnology (fungal genomics) during 2014 (first ranking). Dr. Kamel has pursued his Ph.D. in Molecular Plant Pathology from Christian Albrechts University of Kiel (Germany) and Suez Canal University (Egypt) and then been awarded postdoctoral fellowship from Christian Albrechts University of Kiel in 2008. Dr. Kamel served as visiting associate professor in Institute of Excellence in Fungal Research, Mae Fah Luang University, Thailand, Institute of Microbiology, TUM, Germany, Laboratory of Phytopathology, Wageningen University, the Netherlands, and Plant Protection Department, Sassari University, Italy.

Plant Pathology Research Institute, Agricultural Research Center (ARC)
Giza, Egypt

Chapter 1
Cyanobacteria and Algae as Biocontrol Agents Against Fungal and Bacterial Plant Pathogens

Syed Tanveer Shah, Abdul Basit, Izhar Ullah, and Heba I. Mohamed

Contents

S. T. Shah · A. Basit (✉) · I. Ullah
Department of Horticulture, Faculty of Crop Production Sciences, The University of Agriculture, Peshawar, Pakistan

H. I. Mohamed
Department of Biological and Geological Sciences, Faculty of Education, Ain Shams University, Cairo, Egypt

© The Author(s), under exclusive license to Springer Nature Switzerland AG 2021
H. I. Mohamed et al. (eds.), *Plant Growth-Promoting Microbes for Sustainable Biotic and Abiotic Stress Management*,
https://doi.org/10.1007/978-3-030-66587-6_1

1 Introduction

Present sustainable agricultural farming is severely reliant on an intensive use of tillages, excessive fertilizer use and irrigation, and chemical pesticides; with no doubt, the food requirement of the majority of countries has been fulfilled, even though this raised many health and environmental problems. Now the question of enhancing the crop production in agriculture sector without deteriorating the environment and harming water resources and land/soil fertility has arisen (Singh and Strong 2016). The need of quality environment and food crops production can be fulfilled with sustainable practices of agriculture (Singh and Strong 2016); this philosophy of sustainable agricultural production includes environmentally friendly farming with low cost through natural techniques of resources conservation, i.e. water and soil, maintaining the crop profitability and productivity and also making the agricultural ecosystem self-regulating and resilient (Koller et al. 2012). For the last few decades, green energy production of various processes of microbes has attained great attention as a sustainable technique for biofuel generation, namely, ethanol, butanol, methane (CH4), syngas and H_2. Recent studies reported remarkable growth in cyanobacterial biomass production for biofertilizers; a different supplement of foods, i.e. superfoods; and biofuels for farming of safe agricultural production (Benson et al. 2014).

A highly diversified group of various microorganisms have been found in association to different plant species in the endosphere, rhizosphere and phyllosphere. These microorganisms associated to plants and known for metabolite production may have a neutral, beneficial or harmful impact on crop productivity (Mendes et al. 2013). Ever since, the approach of sustainable crop protection can be characterized by synthetic pesticide alternatives, i.e. derived compounds and microorganisms of plants (Gwinn 2018). Amongst all the alternatives, cyanobacteria and algae are distinguished bioactive agents which have gained a remarkable consideration by scientists globally.

Phytopathogenic biocontrol agents (Fig. 1.1) in the very wide-ranging sense encompass the methods for utilization of organisms except for human. Remarkable research literature can be found as a result of this long-lasting present strategy, with an impetus principle of the enhanced activity and limitations to the application of chemical pesticides. This also included high costs of purchase, concerns to the environment and the highly increased regulations and restrictions of governments. Cyanobacteria, autotrophic blue-green algae which are known to be the most diverse, largest and abundantly distributed group of small, prokaryotic, unicellular and photosynthetic organism, found specifically in fresh and marine water, all together with eukaryotic algae 'could have been the world's largest biomass'. As a promising biocontrol agent growing in huge colonies and causing plant diseases, no great attention has been attained by cyanobacteria (Pisciotta et al. 2010). Cyanobacteria with diversified sizes and shapes have covered 150 identified genera with the features of oldest-ever fossils of 3.5 billion years, approximately, and are responsible for the oxygenic environment of the current days (Hoekman et al. 2012).

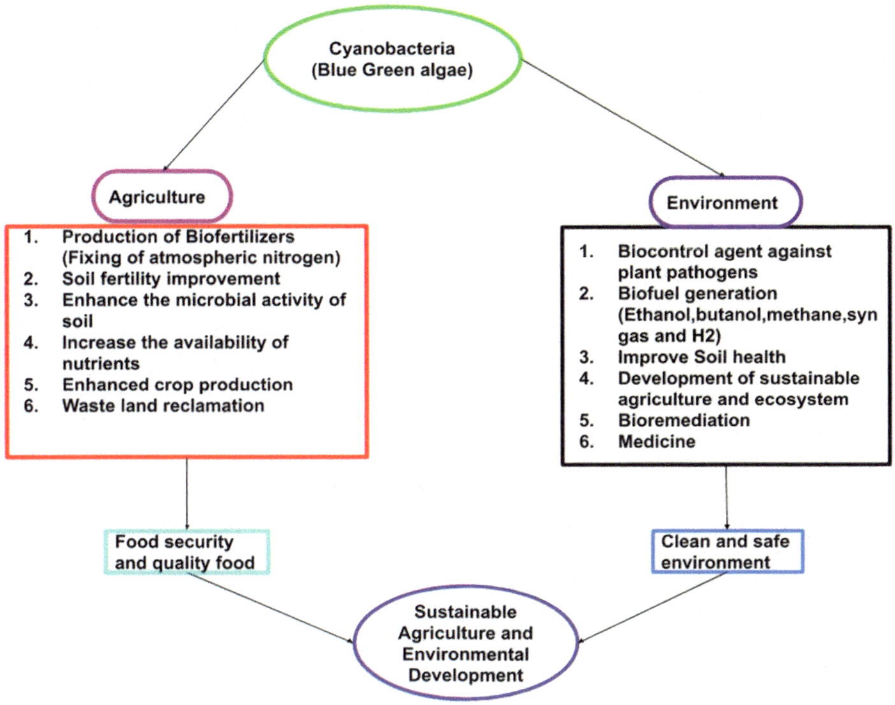

Fig. 1.1 The theoretical framework shows potential cyanobacterial functions in the environment or sustainable agriculture

A proposed classification of cyanobacteria since 1985, with four bacterial orders, has been recognized, namely, *Oscillatoriales*, *Chroococcales*, *Stigonematales* and *Nostocales*, with their phyla *Gloeobacterales*, *Pleurocapsales* and *Chroococcales*. Cyanobacteria are associated with the periods of origin of plants. The cyanobacteria are immensely important in determining the path of evolution and ecological changes all over the earth's history. In the late Proterozoic or the early Cambrian period, cyanobacteria began to take up residence within certain eukaryote cells, this event is called endosymbiosis, for the origin of the eukaryotes. They have the potential to fix atmospheric nitrogen, so that could be used as a biofertilizer for the cultivation of economically important crops such as rice and beans (El-sohaimy 2012; Meena et al. 2019; Koller 2015) (Fig. 1.1). Mostly, the eukaryotic algae have been categorized in 18 different taxonomic classes (Wainwright et al. 1993), where a majority of the algae can be found either in marine or in freshwater habitats and almost 147 genera have a large number of species found in soil. They are mostly photosynthetic; however, a great number are heterotrophic facultative species while few are non-photosynthetic (heterotrophic obligates). These are largely untapped and rich sources of a varied wide collection of naturally active products. This chapter is aimed to understand cyanobacteria as an alternative for sustainable development without the harmful effect of chemicals, synthetic fertilizers and other pesticides/insecticides, to elaborate the antifungal and antibacterial activities of cyanobacterial extract against pathogenic colonial growth and to study the role of Cyanobacteria as a source of exopolysaccharides to improve soil structure and microbial growth.

2 Cyanobacteria and Algae Against Phytopathogens

Since a long time, cyanobacteria and algae have been used scarcely as beneficial extracts against pathogenic fungus because of their stimulation effect for plant productivity and vigour. They have also been used as media substrates for microbial cultures and biofuel production (Fig. 1.1). A fungal pathogen may be responsible for causing infection in cultivated cropping systems, severe postharvest losses and fruit decay. *Schlerotinia sclerotiorum* and various other species of *Fusarium*, *Rhizoctonia*, *Verticillium*, *Pythium* and *Phytopthora* are known to be the most significant polyphagous fungi found in soil (Pastrana et al. 2016). While hindering water absorption and nutrients in the soil, they directly attack root structures and cause wilting, damping off, yellowing, root rot and color rot. Amongst leaf-related pathogens, fungus of Erysiphales order are strong mediators of the disease powdery mildew causing huge economic losses, thus the need for various applications of chemicals (Romero et al. 2007; Jarvis et al. 2002). *Colletotrichum* spp. and *Botrytis cinerea* are the vectors for anthracnose and grey mould, respectively, and many other *Rhizopus*, *Mucor* and *Penicillium* species are the fungal pathogens which are responsible for postharvest losses and fruit decay (Husaini and Neri 2016). Such fungal pathogens could be controlled normally by the fungicidal applications during the time of growing crop cycle from the time of flowering to harvest time.

2.1 Cyanobacteria: A Prevalent Evolutionary Phylum

The most successfully emerged phyla of prokaryotes which were sustained during the evolutionary course were cyanobacteria. They are known to be the most primitive forms of life on the Earth planet, and their evidence reported chloroplasts with eukaryotic photosynthesis have their origin from cyanobacteria, bringing about the evolutionary aerobic respiration as long time ago (2.22–2.45 billion years) (Dixit and Suseela 2013; Gothalwal and Bajpai 2012). Even though the autotrophic nutritional mode is more dominant while some of the species of cyanobacteria can grow well in dark and anaerobic environments including *Oscillatoria* and *Nostoc*, few cyanobacterial species, i.e. *Nostoc*, can also function in atmospheric nitrogen fixation (Yadav et al. 2011; Uzair et al. 2012). However, cyanobacteria can be included in those phyla where simple and unicellular structures have developed complex and multicellular structures during the evolutionary era (Schirrmeister et al. 2011). Extending from a range of unicellular gram-negative to colonial and multicellular filamentous forms (Singh et al. 2011), being the principal phytoplankton constituent, this provides sufficient opportunity to exploit them as a secondary metabolite producer. Along with so many applications in nutraceuticals, pharmaceuticals, food and feed industries, the ecological, morphological and genetic cyanobacterial diversity has led to the wide array of compound production (Tan 2007).

2.1.1 Antifungal Activity

Cyanobacteria have been found to be the most significant source of naturally occurring bioactive compound with antiprotozoal, antimicrobial, antibacterial, antiviral, anti-proliferative and anticancer activities (Dixit and Suseela 2013; Russo and Cesario 2012; Simmons et al. 2008). Many authors revealed efficiency against the growth of the fungal colony of various phytopathogens is among the different activities of cyanobacteria (Table 1.1). Many studies have observed numerous species of *Nostoc*, *Microcystis* and *Anabaena*. Crude source of ethanol extracted from *Anabaena laxa* reported an inhibitory effect and a counter to various fungi, namely,

Table 1.1 Antifungal activity of cyanobacterial extracts on plant pathogenic colonial growth

Cyanobacteria	Pathogen	References
Anabaena sp.	*Alternaria alternata, Botrytis cinerea, Colletotrichum gloeosporioides, Fusarium oxysporum*	Kim (2006)
Anabaena sp.	*Macrophomina phaseolina, Fusarium moniliforme, Alternaria solani, Pythium aphanidermatum, Fusarium solani*	Prasanna et al. (2008)
A. laxa	*F. moniliforme, F. oxysporum* f. sp. *lycopersici*	Prasanna et al. (2015)
Anabaena variabilis	*F. moniliforme, F. oxysporum* f. sp. *lycopersici*	Prasanna et al. (2015)
A. variabilis	*Aspergillus Niger, A. solani*	Tiwari and Kaur (2014)
Calothrix sp.	*A. alternata, B. cinerea, C. gloeosporioides, F. oxysporum, Phytophthora capsici, Pythium ultimum*	Kim (2006)
Microcystis aeruginosa	*F. Oxysporum, M. phaseolina, P. aphanidermatum, Pythium oedochilum, Rhizoctonia solani*	Khalid et al. (2010)
Microcystis aeruginosa	*Aspergillus flavus, Fusarium verticillioides, Fusarium proliferatum*	Marrez and Sultan (2016)
Nostoc sp.	*A. alternata, B. cinerea, C. gloeosporioides, F. oxysporum, P. capsici, P. ultimum, Rhizopus stolonifer*	Kim (2006)
Nostoc commune	*F. oxysporum* f. sp. *lycopersici*	Kim and Kim (2008)
Nostoc entophytum	*R. solani*	Osman et al. (2011)
N. muscorum	*R. solani*	Osman et al. (2011)
Chlorella sp. *Halopithys* sp.	*R. solani*	Righini et al. (2020)
Anabaena sp. *Ecklonia* sp. *Jania* sp.	*Botrytis cinerea*	Righini et al. (2019)

Candida albicans, *Aspergillus oryzae*, *Saccharomyces cerevisiae*, *Trichophyton mentagrophytes* and *Penicillium notatum*. The colony growth of *Fusarium moniliforme* was reduced by 23 different Anabaena strains observed in a screening (Prasanna et al. 2008), whereas *Alternaria solani* and *Nostoc muscorum* were observed as an effective counter to *R. solani*, inhibited by 17 several strains. Different *Anabaena* and *Calothrix* strains carried action countering to various species of *Rhizoctonia* and *Pythium* (Manjunath et al. 2010). *Nostoc muscorum* is known to be effective against *R. solani*, and was effective and inhibited the colony growth more than *N. entophytum* (Osman et al. 2011). Among all the compounds synthetized by cyanobacteria, chitosanase homologues, endoglucanase and benzoic acid were detected, and their presence was correlated to the activity against fungi (Gupta et al. 2011). The terpenoid noscomin, as an extract of *N. commune*, was tested to function against *Escherichia coli*, *Bacillus cereus* and *Staphylococcus epidermidis* (Jaki et al. 2000). Methanolic extract taken from *M. aeruginosa* exhibited an activity of fungus contrary to seven different pathogens of human and eight saprophytic and five phytopathogens (Khalid et al. 2010). *M. aeruginosa* strain revealed the growth inhibition of *Aspergillus flavus* and *A. niger* and *Fusarium verticillioides* (Marrez and Sultan 2016).

2.1.2 Antibacterial Activity

Recently, nosocomial infections (Enterococci and *Staphylococcus aureus* resistant to vancomycin and methyicillin, respectively, Amp C β-lactamase produced by Enterobacteriaceae) showing a big threat worldwide and a challenge to therapeutic studies are caused by bacteria resistant to multidrug (Reinert et al. 2007). Cyanobacteria having antibacterial activity and possibly energetic against bacteria are of great importance to scientists to produce new antibiotics (Biondi et al. 2008). Noscomin shows antibacterial activity against *Escherichia coli*, *Staphylococcus epidermidis* and *Bacillus cereus* taken from *Nostoc commune* (Jaki et al. 2000). Furthermore, stronger inhibition of green algea as compared to cyanobacteria was observed from isolated Nostocine A from *Nostoc spongiaeforme* (Hirata et al. 2003), while the growth of other cyanobacteria and green algae was introverted by *Nostoc*-isolated nostocarboline (Blom et al. 2006). Growth of *Salmonella typhi* MTCC3216, *E. coli* ATCC25992, *Mycobacterium tuberculosis* H37Rv, *Enterobacter aerogenes* MTCC2822, *Staphylococcus aureus* ATCC25923 and *Pseudomonas aeruginosa* ATCC27853 was inhibited by hapalindole isolated from *Fischerella* sp. and *Nostoc* CCC537, thus showing antimicrobial activity (Asthana et al. 2009).

2.1.3 Control of Diseases

Some studies are carried out on cynobacterial activity on the suppression of plant pathogen both in plant (especially leaf portion) and soil. The activity of *R. solina* was significantly reduced by *N. entophytum* and *Nostoc muscorum*, which also increased the survival rate, plant height, dry weight of root and shoot of soybean (Osman et al. 2011). Wilt disease of tomato caused by *F. oxysporum* was significantly controlled by *Nostoc linckia* when applied in soil (Alwathnani and Perveen (2012). Powdery mildew caused by *P. xanthii* in zucchini was significantly reduced by application of *Anabaena* sp. which also enhanced the defense enzyme activities (Roberti et al. 2015) which were also confirmed by Prasanna et al. (2015) in maize.

2.2 Algae

A varied assemblage of photosynthetic species mostly aquatic is known as algae. Algae mostly include green algae which are unicellular organisms (*Chlorella*) and seaweeds which are marine multicellular algae (*Sargassum*). Table 1.2 shows the

Table 1.2 Classification of algae

Phylum/class	Alga species
Chlorophyta	*Caulerpa sertularioides*
	Chlorella
	Ulva lactuca
	Zygnema czurdae
	Zygnema stellinum
	Zygnema tenue
Phaeophyceae	*Ascophyllum nodosum*
	Cystoseira myriophylloides
	Ecklonia sp.
	Ecklonia kurome
	Durvillaea potatorum
	Fucus spiralis
	Laminaria digitata
	Leathesia nana
	Padina gymnospora
	Pelvetia canaliculata
	Sargassum
	Sargassum filipendula
	Sargassum liebmannii
	Stypopodium zonale
	Undaria pinnatifida
Rhodophyta	*Corallina* sp.
	Eucheuma denticulatum
	Gelidium pusillum
	Gracilaria edulis
	Halopithys sp.
	Kappaphycus alvarezii
	Porphyra umbilicalis
	Rhodomela confervoides

different phyla that include red algae (Rhodophyta), chlorophyta (green algae) and Ochrophyta (brown algae) (Guiry 2012). Algal composition is effected by harvest season, compounds such as polysaccharides and geographic location (Schiener et al. 2015), essential nutrients (Cu, Zn, Mn, Co, Mo etc.) and plant hormones (auxins, abscisic acid and cytokinins (Craigie 2011). Furthermore, algae also show antiviral, antioxidant, antimicrobial and antifungal activities which have many applications in cosmetics, bioactive substances, pharmaceuticals and pigments production (Sharma and Sharma 2017). Due to all these properties, algal application plays an important role in soil fertility and crop productivity (Arioli et al. 2015) in different agronomic and horticultural crops (Alam et al. 2013). The increment in postharvest life, disease control and resistance to biotic and abiotic stresses was reported in fruits due to the application of various algal extracts (Esserti et al. 2017).

2.2.1 Antifungal Activity

Several studies have shown the direct importance of pathogen resistance by algal extract (brown algae) having antifungal activity. Mycelial growth of *Aspergillus* spp. and *Penicillum* spp. and *Fusarium oxysporum* was significantly reduced by cyclohexanic and aqueous extracts from *Sargassum* sp. (Khallil et al. 2015). Colonies of *Rhizoctonia* and *Fusarium solani* were inhibited by *Padina gymnospora* and *Sargassum laftifolium* which contain methanolic extract (Ibraheem et al. 2017). *Ascophyllum nodosum, stypopodium zonale, Fucus spiralis, Pelvetia canaliculata* and *Sargassum muticum* extracts contain terpenes and phenols which can suppress the growth of *Colletotrichum lagenarium* (Fernandes Peres et al. 2012). De Corato et al. (2017) reported that the mycelial growth and germination of spores of *Botrytis cinerea* were completely inhibited by *Undaria Pinnatifida* and *Laminaria digitata*. Furthermore, *Gracilaria edulis* that contains methanolic extract significantly inhibits the mycelial growth of *Macrophomina phaseolina* (Ambika and Sujatha 2015) while water extract obtained from *Gracilaria edulis* minimized infections of *Corallina* sp. and *Halopithys* in zucchini (Roberti et al. 2016). Brown algae (*Leathesia nana*) and red algae (*Rhodomela confervoides*) contain bromophenol bis(2,3 dibrom-4,5- dihydroxybenzyl) ether which is an antifungal substance which reduced *B. cinerea* growth and *Colletotrichum gloeosporioides* (Liu et al. 2014). An extract acquired from *Chlorella vulgaris* (green microalgae) by process of enzymatic digestion can reduce the growth of *B. cinerea* (El-ghanam et al. 2015) and showed antifungal activity associated with flavonoids and phenols. These antioxidant compounds are found abundant in alga (Ahmed 2016).

2.2.2 Antibacterial Activities

Active metabolites such as alkaloids, sterols, peptides and phlorotannins produced by marine macroalgae have a wide range of biocontrol activities against different pathogens in the ecosystem (Abdel-Raouf et al. 2015) which have attained much consideration due to their antibacterial, antioxidant and cytotoxic properties (Moubayed et al. 2017). For example, leaf spot disease of *Gymnema sylvestre* (a precious medicinal plant) caused by *Pseudomonas syringae* can be minimized by methanolic extract obtained by *Sargassum wightii* (Kumar et al. 2008) but the little effect has been shown by ethyl acetate extract. Several other studies have shown that other taxa (*Turbinaria conoides*, *Ulva lactuca*, *G. verrucosa*, *Chaetomorpha antennina* and *Halimeda tuna*) have less effective antibacterial activities against *P. syringae* though an evident effect was recorded by acetonic extract from *Sargassum polyceratium* (brown macroalga) against *Erwinia carotovora*, *Escherichia colia*, etc. (Kumar et al. 2008). Now ethanolic extract acquired from *Caulerpa racemosa* and *S. polyceratium* can work actively against *Staphylococcus aureus* (Arunkumar et al. 2005). Esserti et al. (2017) reported a reduction of crown gall disease of tomato caused by *Agrobacterium tumefaciens* by foliar application of aqueous macroalgal solution obtained from *Fucus spiralis* and *Cystoseira myriophylloides*.

2.2.3 Control of Diseases

The effectiveness of algal extract through soil or foliar application against different diseases has been reported in different crops in which disease control ability of brown algal extract is extensively reported (Righini et al. 2018). *E. maxima* algal extract applied through soil minimized the incidence of *Verticillium* wilt of pepper (Rekanović et al. 2010), while *Ecklonia* sp. algal extract applied as foliar spray was effective against powdery mildew in zucchini caused by *P. xanthii* (Roberti et al. 2016). This fungal activity is due to the antioxidant activities of secondary metabolites especially phenols that work both against plant and human pathogens. Moreover, Nagayama et al. (2002) reported the effectiveness of phlorotannins (algal phenols) acquired from *Ecklonia kurome* against methicillin-resistant *Staphylococcus aureus*. Athukorala et al. (2006) described the anticancer and antioxidant activity of *Ecklonia cava* phenols which was effective against murine colon cancer cell line CT-26.

In a 2-year experiment, Pugliese et al. (2018) reported minimizing powdery mildew (causal agent: *Erysiphe necator*) in grapevine by applying laminarin which also controlled powdery mildew and leaf spot in several strawberry cultivars which were caused by *B. cinerea* and *M. fragariae*, respectively (Meszka and Bielenin 2011). Furthermore, the application of laminarin can be used as an alternative against grey mould in raspberry (Krawiec et al. 2016) and disease severity in strawberry caused by (Feliziani et al. 2015). *B. cinerea* hence can reduce *B. cinerea* resistance to fungicide (Krawiec et al. 2016). Bromophenol (BDDE), a compound extracted from brown algae and red algae, can be used to treat strawberry for disease control caused by *B. cinerea* (Liu et al. 2014).

As previously mentioned, algal extracts can work directly (antifungal activity) or indirectly (induce plant resistance) against a fungal pathogen. In particular, it has been shown that cell wall and storage polysaccharides from green, brown and red seaweeds (marine macroalgae) corresponding to ulvans, alginates, fucans, laminarin and carrageenans can trigger defense responses in plants enhancing protection against pathogens (Vera et al. 2011). Extracts obtained from brown algae (*A. nodosum*) is successfully investigated on many plant species especially in carrot against *Alternaria radicina* (Jayaraj et al. 2008) and cucumber against *F. oxysporum*, *Alternaria cucumerinum* and *B. cinerea* (Jayaraman et al. 2011). This might be due to their role to enhance enzymatic activities and synthesis of pathogenesis-related proteins (PRs) by brown algal extract (Abkhoo and Sabbagh 2016). *Sargassum filipendula* extract (Ramkissoon et al. 2017) and polysaccharide-enriched extract from green algae *Ulva lactuca* (Hernández-Herrera et al. 2014) are effectively used to control disease symptoms in tomato caused by *Xanthomonas campestris* and *A. solani*, respectively.

3 Role of Algae in the Agriculture Sector

A diversified class of microbes that can perform photosynthesis is known as algae, which play a vital role in agriculture, used as a soil stabilizer and biofertilizer. Seaweeds from algal source can be grown on arid, desert and waste lands with very low demand of water and are used as a fertilizer and more effective to reduce runoff of phosphorus and nitrogen as compared to seaweed obtained from manures of livestock producing a carbon-less and healthy food. These are produced around the world which can be consumed as food supplements. These are an essential source of iodine, and levels of iodine in milk depend upon the feeding mechanism of a cow which is increased by application of seaweeds. Feed seasonings with algae enhanced the rate of egg-laying in hen (Abdel-Raouf et al. 2012).

4 Biocontrol Strategy of Phytopathogen

The protection of plants by using precise and directed components of an ecosystem is termed as biocontrol. Similarly, biological control is the regulation of pests' population by use of biotic organisms, i.e. parasites, predators, and pathogens. It is also the use of organisms that live inside the host cells or utilization of introduced organisms other than host plants resistant to the diseases for combating multiplication in their populations. They are classified based on their contribution to controlling phytopathogen attacks like facultative symbionts of plants and hyperparasites, and saprophytes. The use of environmentally friendly methods like biological control is of great importance in maintaining the natural balance of pest population and is a slow process, acquiring few quick profits, but their effect is long-lasting and environmentally friendly.

4.1 Biological Control of Phytopathogen by Cyanobacteria

Different alternative methods of synthetic pesticides by natural microbial sources and light components have been used for controlling the attack of the pathogen in different ways as agents of insecticides, fungicides, and acaricide (Hassan 2007; Safonova and Reisser 2005; Amer et al. 2000; Ibraheem and Abdel-Raouf 2007; Duke 2002). These environmentally friendly methods despite their lethal effect on pests are widely used for maintaining the health of the environment and fauna distribution without effecting beneficial organisms. Biological agents like fungi and bacteria have been used for control of soil-borne pathogens. Various concentrations of nitrogen-fixing cyanobacterium (*N. muscorum* Ag) were used against damping-off disease. It was observed that the growth of damping causal agents like *Sclerotinia sclerotiorum* and *Rhizoctonia solani growth was inhibited by the use of* extracts from cells of *N. muscorum* or by extracellular products of this cyanobacterium (Caire et al. 1997). The usage of biological control agents like different algal taxa of different habitats was found effective against different plants or animal diseases and also against some plant pests. Some other studies were reported about the residual effect of algal sources against insect pests (Nassar et al. 1999). They also reported the larval development inhibition and the development and survival delay of adult female mosquitoes by cyanobacteria- and green algae-producing substances. The production of some potential substances by blue-green algae is of great importance quantitatively which acts as an approach to integrated pest management, and also a suitable environmental system of disease control revealing the significance of biological control agent as a basic component of techniques for the management of pests (Hassan 2007).

4.2 Exopolysaccharide

Cyanobacteria can be counted as the most significant source of exopolysaccharides, the extracellular polymers with changed chemical composition improving microbial growth and soil structure, as well as exoenzymes activity (Hamed 2007; Ibraheem 2007). Sustainable agricultural farming to yield high crop production could be achieved by maintaining an adequate quantity of organic matters available in the soil. Cultivation of crops on soil with adequate nutrient changes the structural characteristics of soil and causes reduction of nitrogen amount and organic matter. Reduction in organic matter availability in the soil disturbs the soil aggregate stability. Some cyanobacteria increase the availability of nutrients and improve soil structure by excreting mucilage or slime which spread all around the organism to an extent of dissolution in the soil solution or culture medium partially. The reclamation of uncultivated soil such as saline and calcareous soil could be made suitable for agricultural operations by application of algal biofertilizers (Hedge et al. 1999). *N. muscorum is* one of the cyanobacteria that excrete exopolysaccharides and enhance the overall saline soil stability (Caire et al. 1997).

5 Role of Cyanobacterial Products on Antagonistic Activity of Fungal and Bacterial Agents

Wide-ranging mechanisms have been used as defensive agents against pathogen development and disease incidence by showing different responses, i.e. to prevent, combat or control plant disease. For achieving maximum productivity and quality products, most of the growers depend on inorganic fertilizer and pesticides. However, the excessive use of agrochemicals for controlling disease and pest attack causes environmental pollution, adversely affecting the health of the environment, and also fear-mongering by some challengers of pesticides which significantly changed the attitude of people towards chemical pesticide use in the agriculture sector. Many plant diseases could be successfully controlled with the application of biocontrol agents, i.e. antagonistic microorganisms. However, their effect is very slow and costly, but its effect is time lasting and acts as the best control method of disease in greenhouse farming. The concern of people towards pesticide use as a preventive measure against pest and disease problems has brought the increasing interest in using alternatives to inorganic pesticides. Currently, few eco-friendly biological control vectors were used to control plant pathogenic fungi causing soil-borne disease. The biologically active compound produced by cyanobacteria (blue-green algae) and eukaryotic algae exhibiting antifungal, antibiotic and toxic activity counters to phytopathogens (Schlegel et al. 1998). *Anabaena* spp., *Scytonema* spp. and *Nostoc* spp. were found active against the growth of *Cunninghamella blakesleeana* (soil-borne fungus) and damping-off (Bloor and England 1989). Seeds were treated with culture filtrates, or extracts produced by cyanobacteria and algae performed better against damping-off fungi, i.e. *Pythium* spp., *Rhizoctonia solani* and *Furarium* spp. (Kulik 1995).

6 Mechanism of Interaction Between Pathogenic and Plant Diseases Biocontrol Agent

A combination of different types of interaction between organisms results in biological control under different experimental conditions; many mechanisms are operated during the characterization of mechanism followed in the biocontrol process. Almost in all conditions, the presence and activities of other organisms encounter the antagonistic effect of pathogens. Therefore, the adaptation of different antagonistic mechanisms produced by the directional spectrum associated with the specificity of interactions and the interspecies contact quantity has been focused in this study (Fig. 1.2).

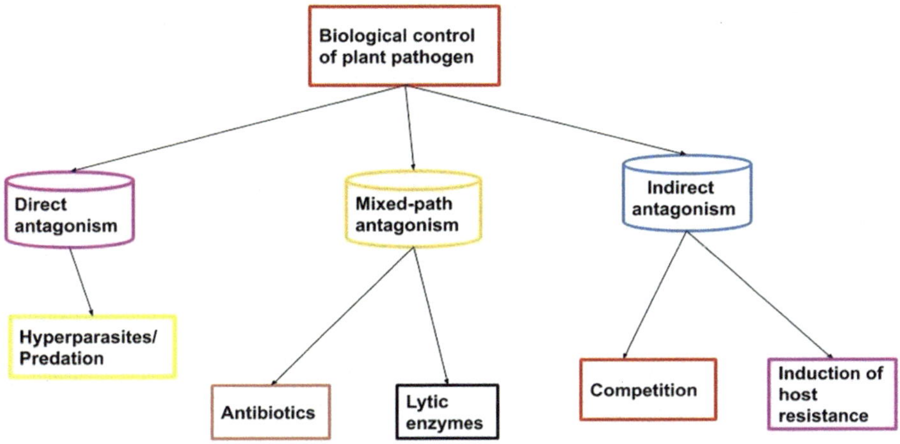

Fig. 1.2 Mechanism of interaction between pathogenic and plant diseases biocontrol agent

6.1 Hyperparasites and Predation

In hyperparasitism, biological control agents (BCA) are used to kill directly the pathogen or its propagules. Generally, hyperparasites are classified into major four classes, i.e. hypoviruses, obligate bacterial pathogens, facultative predators and parasites. An obligate bacterial pathogen of root-knot nematode, i.e. *Pasteuria penetrans*, was used as an agent of biological control. Hyperparasites such as hypoviruses (a fungus, i.e. *Cryphonectria parasitica*) that cause chest nut diseases were infected by virus that shows the effective result in reducing the disease-producing ability of the pathogen (hypovirulence) (Milgroom and Cortesi 2004). However, the success or failure of hypovirulence is dependent on the interaction of viruses, fungus, trees and the environment. A number of plant pathogenic fungal parasites have been specified where some of them like *Coniothyrium minitans* attack on sclerotia and others like *Pythium oligandrum* attack on living hyphae, whereas hyperparasites attack on the individual fungal pathogen, e.g. pathogens of powdery mildew were parasitized by a small fungal group, i.e. *Acrodontium crateriforme*, *Acremonium alternatum*, *Ampelomyces quisqualis*, *Cladosporium oxysporum* and *Gliocladium virens* (Kiss 2003). Some other attack on different developmental stages of phytopathogenic nematodes (e.g. *Dactylella oviparasitica*, and *Paecilomyces lilacinus*). Microbial predation compared to hyperparasitism gives less predictable results of disease control and is more general and pathogen nonspecific. However, some under limited availability of nutrients show predatory behaviour compared to typical growing conditions. Some species of *Trichoderma* show differential response by activating the chitinase genes that are helpful against the cell wall of fungi to parasitize *R. saloni*.

6.2 *Antibiotic-Mediated Suppression*

Antibiotics are known as microbial toxins which damage or kill different other organisms a minute quantity. Some of the microbes are considered as an important source of producing and secreting either single or many compounds with an antibiotic action which play an effective role in suppressing disease-causing plant pathogens. The growth of the target pathogens has been significantly suppressed by using antibiotics in vitro/in situ. Different biocontrol agents are important means of in situ antibiotics production (Pal and McSpadden 2006); moreover, estimated effective quantities are hard to measure because of their small quantity products as compared to other less toxic organic compounds available in the phytosphere. The suppression of diverse microbial competitors could be resolved by the production of antibiotics. Biological control could be enhanced by the production of antibiotics that inhibit the activity of different pathogens differentially. Phenazine and DAPG were produced by genetically engineered strains of *Pseudomonas putida* WCS358r and have been found effective by suppressing phyto-disease in wheat grown in the field (Glandorf et al. 2001).

6.3 *Lytic Enzymes and Other By-Products of Microbial Life*

The activity and growth of pathogen were significantly effective in the production of metabolites of a diverse group of microorganisms. Many microbes are used in suppressing plant pathogen activity directly by secreting lytic enzymes which hydrolyze several polymeric compounds, i.e. chitin, proteins, cellulose, hemicellulose and DNA. Biocontrol activities of *Lysobacter enzymogenes* strain C3 significantly show a positive response by a b-1,3-glucanase (Palumbo et al. 2005). The fungal plant pathogen is suppressed by *Lysobacter* and *Myxobacteria* that produce a significantly large number of lytic enzymes (Bull et al. 2002). In addition to these, indirect suppression of disease could be achieved by the activity of some products of the lytic enzyme. Oligosaccharides obtained from the cell wall of fungus are identified as an important source of plant host defense induction. Plant host resistance against diseases was achieved by a strain of *Lysobacter enzymogenes* (C3) (Kilic-Ekici and Yuen 2003), though induction of these activities is not clearly understood. The ratio and composition of C:N of organic matter in the soil is mostly dependent on the activity of any and above compounds in disease suppression of phytopathogens that play a major role in providing a nutrient-rich environment in the soil and rhizosphere. Sometimes maximum disease suppression could be achieved by the improvement of these activities. The use of chitosan as a postharvest biocontrol agent can stimulate the damage of microbial activity similar to that of using hyperparasites (Benhamou 2004) and also found effective against root rot caused by *Fusarium oxysporum* f. sp. *radices lycopersici* in tomato plant.

7 Role of Cyanobacteria in Crop Protection

7.1 Cyanobacteria and Allelopathy

Biologically active metabolites used by single species are effectively used in inhibiting sympatric species growth which may act as a potential competitor for controlling annual variability and resources in the communities of phytoplankton (Vardi et al. 2002). Several heterocystous cyanophycean genera such as *Anabaena sp., Nostoc sp., Cylindrospermum, Scytonema, Calothrix, Rivularia, Chlorogloea, Gloeotrichia, and Nostochopsis* have been shown to fix atmospheric nitrogen efficiently, which improve the nutritional status of soil. *Fischerella* produce fischerellins (A and B), and play role as alternative approach of allelopathy (Ganter et al. 2008) but also play a role as alternative approach of allelopathy. The pentacyclic calothrixins produced by *Calothrix* strains act in allelopathic interactions in inhibiting RNA polymerase and DNA synthesis (Doan et al. 2000). Nostoc 78-12A produced nostocarboline (a carboline alkaloid) that helps in inhibition of the toxin produced by cyanobacterium. *Microcystis aeruginosa* acts as an allelopathic agent and has significant effects on photosynthesis (Shao et al. 2009).

7.2 Application of Cyanobacterial Secondary Metabolites

Cyanobacteria are a major source of bioactive metabolites or compounds that contain a varied range of nitrogen-rich alkaloids and peptides (Gervick et al. 2001). The significance of such microbes, which are known as the source of cyano-toxins and different other newly found biologically active compounds, is accepted and recognized worldwide (Mundt et al. 2001; Kumar et al. 2005); however, their role as a chemical potential agent like biocontrol agents or in crop protection is less explored in agriculture. The attack of a disease-causing organism like bacteria, fungi, zooplankton and eukaryotic microalgae could be reduced possibly by using cyanobacteria as a potential defense option of synthesis of highly active toxins.

7.3 Biological Control Perspective of Cyanobacteria Against Diseases

Cyanobacteria produce various secondary metabolites having hormonal, toxic, antimicrobial and antineoplastic effects (Jaki et al. 2000) and are targeting prokaryotic as well as eukaryotic microorganism. Selected microorganisms show bioassays of aqueous and organic solvent extracts of antimicrobial compounds. Bioactive compounds showed fungal activity against important fungi that act as a synthesis of antibiotics by lead production and open a gateway in the agriculture sector (Nagle

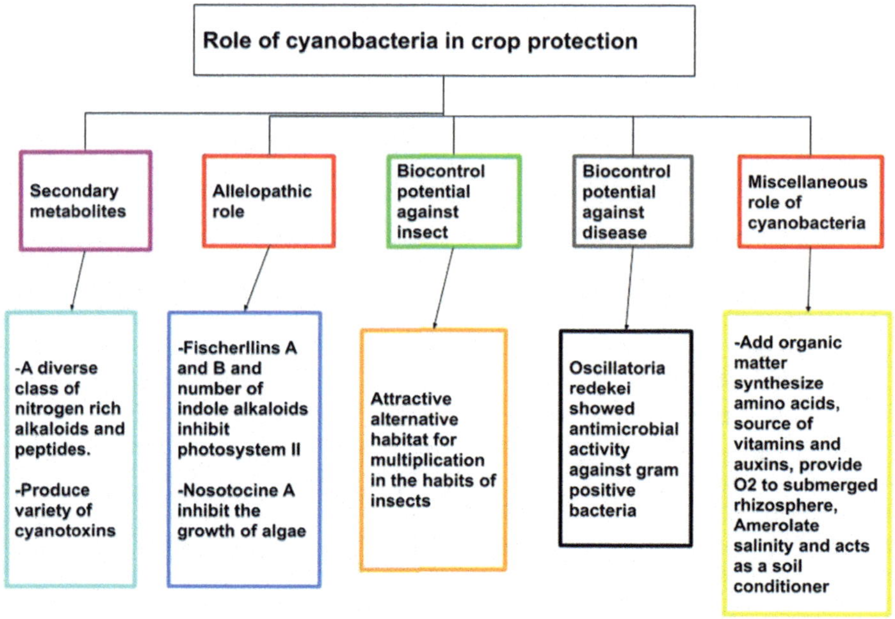

Fig. 1.3 Biological control perspective of cyanobacteria against diseases

and Wedge 2002; Volk and Furkert 2006). *Tolypothrix tjipanensis*, cyanobacterium that produced tjipanazoles, revealed an effective response of showing fungicidal activity against *Aspergillus flavus* (Ozdemir et al. 2004). Several pathogenic fungi activities were reduced by fischerellin-A from *Fischerella*. Gram-positive bacteria showed antimicrobial activity by an unsaturated mixture of fatty acids from *Oscillatoria redekei* (Sabin et al. 2003). Gram-positive bacteria showed positive activity in a sample of 22 different strains of cyanobacteria obtained either from terrestrial or freshwater environments while reported minute activity counters to gram-negative bacteria (Fig. 1.3).

7.4 Multiple Significant Roles of Cyanobacteria

Cyanobacteria can be considered as an excellent source of organic matter to the soil, synthesis of amino acid, auxins and vitamins, and oxygen supply in submerged conditions, increase phosphate solubility and enhance fertilizer efficiency in plants, while decreasing the contents of oxidizable matter and salinity (Kaushik 2004). They are also considered as an important agent of soil conditioner and nitrogen that represent renewable biomass resource which is increasing as a source of the novelty of bioactive molecules. They promote the production of plant hormones, and their antibiotics or toxic compounds have been observed in enzymes inhibiting the activities of phytopathogen (Sergeeva et al. 2002). Besides toxins, cyanobacterial biomass is also a source of a large number of active substances having fungicide, cytotoxic, algicidal, antibacterial and antiviral activities (Jaki et al. 1999). Green algae in paddy fields were controlled by the use of algicides obtained from cyanobacteria, and their better growth was also observed. The major component of Nostoc 31 was nostocyclamide that has antibiotic and algicide activities. Mundt et al. (2002) observed the response of lipophilic and hydrophilic extracts for antibiotic, immunomodulating, antiviral and inhibiting activity of various enzymes in in vitro systems reporting various interesting effects.

8 Conclusion

For sustainable environmentally friendly farming, the reduction of synthetic fertilizers application and pesticides use is an interesting topic of the recent time. There has been mproved progress in cyanobacterial and algal biofertilizer products. On the other hand, application of cyanobacteria and algae use in controlling fungal and bacterial diseases is an innovative concept in sustainable agriculture. Recent studies reported remarkable growth in cyanobacterial biomass production for biofertilizers; a different supplement of foods, i.e. superfoods; and biofuels for farming of safe agricultural production. The most successfully emerged phyla of prokaryotes which were sustained during the evolutionary course were cyanobacteria. Along with so many applications in nutraceuticals, pharmaceuticals, food and feed industries, the ecological, morphological and genetic cyanobacterial diversity has led to the wide array of compounds production. The autotrophic nutritional mode is more dominant while some of the species of cyanobacteria can grow well in dark and anaerobic environments including *Oscillatoria* and *Nostoc*, while few cyanobacterial species, i.e. *Nostoc*, can also function in atmospheric nitrogen fixation. The cyanobacterial activity against phytopathogens has been studied both through their applications on leaves and soil surfaces. Green algae being unicellular and photosynthetic organisms, i.e. seaweeds (multicellular marine algae), and *Chlorella*, such as a brown alga named *Sargassum* can reach up to a length of 1–3 m. Cyclohexane and aqueous extracts from *Sargassum* sp. inhibit mycelial growth of *Aspergillus* spp. by 37–54.5%. Marine microalgae with an enormous spectrum production of synthetically vigorous metabolites, i.e. polysaccharides, cyclic peptides, sterols, polyketides, diterpenoids, alkaloids, quinones, glycerols and lipids, have a wide-ranging capability of bacterial/biological activitity function against many other organisms. Both cyanobacteria and algae play a key role in sustainable agricultural farming as biological control agents. They also play a vital role as allelopathic sources and are eco-friendly for safe sustainable crop production.

References

Abdel-Raouf N, Al-Homaidan AA, Ibraheem IBM (2012) Agricultural importance of algae. Afr J Biotechnol 11(54):11648–11658

Abdel-Raouf N, Al-Enazi NM, Al-Homaidan AA, Ibraheem IBM, Al-Othman MR, Hatamleh AA (2015) Antibacterial b-amyrin isolated from Laurencia microcladia. Arab J Chem 8:32–37

Abkhoo J, Sabbagh SK (2016) Control of Phytophthora melonis damping-off, induction of defense responses, and gene expression of cucumber treated with commercial extract from Ascophyllum nodosum. J Appl Phycol 28:1333–1342

Ahmed EA (2016) Antimicrobial activity of microalgal extracts isolated from Baharia Oasis, Egypt. Glob Adv Res J Microbiol 5:033–041

Alam MZ, Braun G, Norrie J, Hodges DM (2013) Effect of Ascophyllum extract application on plant growth, fruit yield and soil microbial communities of strawberry. Can J Plant Sci 93:23–36

Alwathnani HA, Perveen K (2012) Biological control of Fusarium wilt of tomato by antagonist fungi and cyanobacteria. Afr J Biotechnol 11:1100–1105

Ambika S, Sujatha K (2015) Antifungal activity of brown, red and green alga seaweed extracts against Macrophomina phaseolina (Tassi) Goid., in pideonpea var. CO (Rg) 7. Int J Agric Sci 11:210–216

Amer SAA, Mohamed SM, El-Gengaihi S, Dimetry NZ (2000) Acaricidal activity of lipoidal matter of different plant extracts against the two-spotted mite *Tetranychus urticae koch*. Insect Sci Appl 20(3):191–194

Arioli T, Mattner SW, Winberg PC (2015) Applications of seaweed extracts in Australian agriculture: past, present and future. J Appl Phycol 27:2007–2015

Arunkumar K, Selvapalam N, Rengasamy R (2005) The antibacterial compound sulphoglycerolipid 1–0 palmitoyl-3-0(6-sulpho-aquinovopyranosyl)-glycerol from Sargassum wightii Greville (Phaeophyceae). Botanica Mar 40:441–445

Asthana RK, Tripathi MK, Deepali A et al (2009) Isolation and identification of a new antibacterial entity from the Antarctic cyanobacterium *Nostoc* CCC 537. J Appl Phycol 21:81–88

Athukorala Y, Kim KN, Jeon YJ (2006) Antiproliferative and antioxidant properties of an enzymatic hydrolysate from brown alga, Ecklonia cava. Food Chem Toxicol 44:1065–1074

Benhamou N (2004) Potential of the mycoparasite, *Verticillium lecanii*, to protect citrus fruit against *Penicillium digitatum*, the causal agent of green mold: a comparison with the effect of chitosan. Phytopathology 94:693–705

Benson D, Kerry K, Malin G (2014) Algal biofuels: impact significance and implications for EU multi-level governance. J Clean Prod 72:4–13. https://doi.org/10.1016/j.jclepro.2014.02.060

Biondi N, Tredici MR, Taton A et al (2008) Cyanobacteria from benthic mats of Antarctic lakes as a new source of bioactivities. J Appl Microbiol 105:105–115

Blom JF, Brutsch T, Barbaras D et al (2006) Potent algicides based on the cyanobacterial alkaloid nostocarboline. Org Lett 8:737–740

Bloor S, England RR (1989) Antibiotic production by the cyanobacterium *Nostoc muscorum*. J Appl Phycol 1(4):367–372

Bull CT, Shetty KG, Subbarao KV (2002) Interactions between Myxobacteria, plantpathogenic fungi, and biocontrol agents. Plant Dis 86:889–896

Caire G, de Cano SM, de Mulé ZMC, Palma RM, Colombo K (1997) Exopolysaccharides of *Nostoc muscorum* ag. (cyanobacteria) in the aggregation of soil particles. J Appl Phycol 9:249–253

Craigie JS (2011) Seaweed extract stimuli in plant science and agriculture. J Appl Phycol 23:371–393

De Corato U, Salimbeni R, De Pretis A, Avella N, Patruno G (2017) Antifungal activity of crude extracts from brown and red seaweeds by a supercritical carbon dioxide technique against fruit postharvest fungal diseases. Postharvest Biol Technol 131:16–30

Dixit RB, Suseela MR (2013) Cyanobacteria: potential candidates for drug discovery. Antonie Van Leeuwenhoek 103:947–961

Doan NT, Rickards RW, Rothschild JM, Smith GD (2000) Allelopathic actions of the alkaloid 12-epi-Hapalindole E Isonitrile and Calothrixin a from cyanobacteria of the genera Fischerella and Calothrix. J Appl Phycol 12:409–416
Duke SO (2002) Chemicals from nature for weed mangement. Weed Sci 50:138–151
El-ghanam AA, Farfour SA, Ragab SS (2015) Bio-suppression of strawberry fruit rot disease caused by Botrytis cinerea. J Plant Pathol Microbiol S3:005
El-Sohaimy SA (2012) Functional foods and nutraceuticals—modern approach to food science. World Appl Sci J 20(5):691–708
Esserti S, Smaili A, Rifai LA, Koussa T, Makroum K, Belfaiza M, Kabil EM, Faize L, Burgos L, Alburquerque N, Faize M (2017) Protective effect of three brown seaweed extracts against fungal and bacterial diseases of tomato. J Appl Phycol 29:1081–1093
Feliziani E, Landi L, Romanazzi G (2015) Preharvest treatments with chitosan and other alternatives to conventional fungicides to control postharvest decay of strawberry. Carbohydr Polym 132:111–117
Fernandes Peres JC, de Carvalho LR, Gonçalez E, Saggion Berian LO, D'arc Felicio J (2012) Evaluation of antifungal activity of seaweed extracts. Ciênc Agrotec Lavras 36:294–299
Ganter M, Berry JP, Thomas S, Wang M, Perez R, Rein K (2008) Allelopathic activity among cyanobacteria and microalgae isolated from Florida freshwater habitats. FEMS Microbiol Lett 64:55–64
Gervick WM, Tan LT, Siachitta N (2001) Nitrogen-containing metabolites from marine cyanobacteria. In: Cordell G (ed) The alkaloids. Academic Press, San Diego, pp 75–184
Glandorf DC, Verheggen P, Jansen T, Jorritsma JW, Smit E, Leefang P, Wernars K, Thomashow LS, Laureijs E, Thomas-Oates JE, Bakker PA, Van Loon LC (2001) Effect of genetically modified Pseudomonas putida WCS358r on the fungal rhizosphere microflora of field-grown wheat. Appl Environ Microbiol 67:3371–3378
Gothalwal R, Bajpai P (2012) Cynobacteria a comprehensive review. Int Res J Pharm 3(2):53–57
Guiry MD (2012) How many species of algae are there? J Phycol 48:1057–1063
Gupta V, Natarajan C, Kumar K, Prasanna R (2011) Identification and characterization of endoglucanases for fungicidal activity in *Anabaena laxa*. J Appl Phycol 23:73–81
Gwinn KD (2018) Bioactive natural products in plant disease control. In: Atta-ur-Rahman (ed) Studies in natural products chemistry, vol 56. Elsevier, Amsterdam, pp 229–246
Hamed SMM (2007) Studies on nitrogen fixing cyanobacteria. M.Sc. Thesis, Botany Department, Faculty of Science, Beni-Suef University, Beni-Suef, Egypt
Hassan NAM (2007) Studies on the algal flora distributed at Wadi-Sannur of the eastern- desert of Egypt. M.Sc. Thesis, Faculty of Science, Beni-Suef University, Egypt
Hedge DM, Dwivedi BS, Sudhakara-Babu SN (1999) Biofertilizers for cereal production in India—a review. Indian J Agric Sci 69(2):73–83
Hernández-Herrera RM, Virgen-Calleros G, Ruiz-López M, Zañudo-Hernández J, Délano-Frier JP, Sánchez-Hernández C (2014) Extracts from green and brown seaweeds protect tomato (*Solanum lycopersicum*) against the necrotrophic fungus *Alternaria solani*. J Appl Phycol 26:1607–1614
Hirata K, Yoshitomi S, Dwi S, Iwabe O, Mahakhant A, Polchai J, Miyamoto K (2003) Bioactivities of nostocine a produced by a freshwater cyanobacterium *Nostoc spongiaforme* TISTR 8169. J Biosci Bioeng 95:512–517
Hoekman SK, Broch A, Robbins C, Ceniceros E, Natarajan M (2012) Review of biodiesel composition, properties, and specifications, Renew Sustain Energy Rev 16(1): 143–169
Husaini AM, Neri D (2016) Strawberry growth, development and diseases. CABI, Wallingford
Ibraheem IBM (2007) Cyanobacteria as alternative biological conditioners for bioremediation of barren soil. Egypt J Phycol 8:99–116
Ibraheem IBM, Abdel-Raouf N (2007) Allelopathic activity of some local cyanobacterial extra metabolites against some pathogenic bacteria. Egypt J Phycol 8:120–129

Ibraheem IBM, Hamed SM, Abd Elrhman AA, Farag FM, Abdel-Raouf N (2017) Antimicrobial activities of some brown macroalgae against some soil borne plant pathogens and in vivo management of *Solanum melongena* root diseases. Aust J Basic Appl Sci 11:157–168

Jaki B, Orjala J, Sticher O (1999) A novel extracelular diterpenoides with antibacterial activity from the cyanobacterium, *Nostoc commune*. J Nat Prod 62:502–503

Jaki B, Orjala J, Heilmann J, Linden A, Vogler B, Sticher O (2000) Novel extracellular diterpenoids with biological activity from the cyanobacterium *Nostoc commune*. J Nat Prod 63:339–343

Jarvis WR, Gubler WD, Grove GG (2002) Epidemiology of powdery mildews in agricultural pathosystems. In: Bélanger RR, Bushnell WR, Dik AJ, Carver TLW (eds) The powdery mildews: a comprehensive treatise. APS, St. Paul, pp 169–199

Jayaraj J, Wan A, Rahman M, Punja ZK (2008) Seaweed extract reduces foliar fungal diseases on carrot. Crop Prot 27:1360–1366

Jayaraman J, Norrie J, Punja ZK (2011) Commercial extract from the brown seaweed Ascophyllum nodosum reduces fungal diseases in greenhouse cucumber. J Appl Phycol 23:353–361

Kaushik BD (2004) Use of blue-green algae and Azolla biofertilizers in rice cultivation and their influence on soil properties. In: Jain PC (ed) Microbiology and biotechnology for sustainable development. CBS Publishers & Distributors, New Delhi, pp 166–184

Khalid M, Shameel M, Ahmed V, Shahzad S, Lghani S (2010) Studies on the bioactivity and phycochemistry of Microcystis aeruginosa (Cyanophyta) from Sindh. Pak J Bot 42(4):2635–2646

Khallil AM, Daghman IM, Fady AA (2015) Antifungal potential in crude extracts of five selected brown seaweeds collected from the western Libya coast. J Micro Creat 1:103–107

Kilic-Ekici O, Yuen GY (2003) Induced resistance as mechanisms of biological control by Lysobacter enzymogenes strain C3. Phytopathology 93:1103–1110

Kim JD (2006) Screening of Cyanobacteria (Blue-Green algae) from Rice Paddy Soil for Antifungal Activity against Plant Pathogenic Fungi. Mycobiology 34(3): 138–142

Kim J, Kim JD (2008) Inhibitory effect of algal extracts on mycelia growth of the tomato-wilt pathogen, *Fusarium oxysporum* f. sp. lycopersici. Mycobiol 36: 242–248

Kiss L (2003) A review of fungal antagonists of powdery mildews and their potential as biocontrol agents. Pest Manag Sci 59:475–483

Koller M (2015) Cyanobacterial polyhydroxyalkanoate production: status quo and quo vadis? Curr Biotechnol 4(4):464–480

Koller M, Salernoa A, Tuffnerb P, Koiniggb M, Bochzeltb H, Schoberd S et al (2012) Characteristics and potential of microalgal cultivation strategies: a review. J Clean Prod 37:377–388. https://doi.org/10.1016/j.jclepro.2012.07.044

Krawiec M, Szot I, Krawiec P (2016) The use of laminarin as an effective tool for anti-resistance management in chemical control of grey mould in raspberry. Acta Hortic 1133:469–472

Kulik MM (1995) The potential for using cyanobacteria (blue-green algae) and algae in the biological control of plant pathogenic bacteria and fungi. Eur J Plant Pathol 101(6):585–599

Kumar K, Lakshmanan A, Kannaiyan S (2005) Bioregulatory and therapeutic effects of blue green algae. Indian J Microbiol 43:9–16

Kumar CS, Raju D, Sarada VL, Rengasamy R (2008) Seaweed extracts control the leaf spot disease of the medicinal plant Gymnema sylvestre. Indian J Sci Technol 1:93–94

Liu M, Wang G, Xiao L, Xu X, Liu X, Xu P, Lin X (2014) Bis (2,3-dibromo-4,5-dihydroxybenzyl) ether, a marine algae derived bromophenol, inhibits the growth of *Botrytis cinerea* and interacts with DNA molecules. Mar Drugs 12:3838–3851

Manjunath M, Prasanna R, Nain L, Dureja P, Singh R, Kumar A, Jaggi S, Kaushik BD (2010) Biocontrol potential of cyanobacterial metabolites against damping off disease caused by Pythium aphanidermatum in solanaceous vegetables. Arch Phytopathol PFL 43:666–677

Marrez DA, Sultan YY (2016) Antifungal activity of the cyanobacterium Microcystis aeruginosa against mycotoxigenic fungi. J Appl Pharm Sci 6:191–198

Meena M, Divyanshu K, Kumar S, Swapnil P, Zehra A, Shukla V, Yadav M, Upadhyay RS (2019) Regulation of L-proline biosynthesis, signal transduction, transport, accumulation and its vital role in plants during variable environmental conditions. Heliyon 5(12):e02951. https://doi.org/10.1016/j.heliyon.2019.e02952

Mendes R, Garbeva P, Raaijmakers JM (2013) The rhizosphere microbiome: significance of plant beneficial, plant pathogenic, and human pathogenic microorganisms. FEMS Microbiol Rev 37:634–663

Meszka B, Bielenin A (2011) Activity of laminarin in control of strawberry diseases. Phytopathologia 62:15–23

Milgroom MG, Cortesi P (2004) Biological control of chestnut blight with hypovirulence: a critical analysis. Annu Rev Phytopathol 42:311–338

Moubayed NMS, Al Houri HJ, AlKhulaifi MM, Al Farraj DA (2017) Antimicrobial, antioxidant properties and chemical composition of seaweeds collected from Saudi Arabia (Red Sea and Arabian gulf). Saudi J Biol Sci 24:162–169

Mundt S, Kreitlow S, Nowotny A, Effmert U (2001) Biochemical and pharmacological investigations of selected cyanobacteria. Int J Hygiene Env Health 203:327–334

Nagayama K, Iwamura Y, Shibata T, Hirayama I, Nakamura T (2002) Bactericidal activity of phlorotannins from the brown alga Ecklonia kurome. J Antimicrob Chemother 50:889–893

Nagle DG, Wedge DE (2002) Chemical ecology of plants. In: Malik B (ed) Allelopathy in aquatic and terrestrial ecosystems. Birkhauser Verlag, Basel, pp 7–32

Nassar MMI, Hafez ST, Nagaty I, Khalaf M, Samy AA (1999) Insecticidal activity of cyanobacteria against four insects, two of medical importance and two agricultural pests with reference to the action on albino mice. Egypt Soc Parasitol 29(3):939–949

Osman MEH, El-Sheekh MM, Metwally AM, AEA I, Ismail M (2011) Antagonistic activity of some fungi and cyanobacteria species against Rhizoctonia solani. Int J Plant Pathol 2(3):101–114

Ozdemir G, Karabay NU, Dalay MC, Pazarbasi B (2004) Antibacterial activity of volatile component and various extracts of Spirulina platensis. Phytother Res 18:754–757

Pal KK, McSpadden BG (2006) Biological control of plant pathogens. Plant Health Instruct. https://doi.org/10.1094/PHI-A-2006-1117-02

Palumbo JD, Yuen GY, Jochum CC, Tatum K, Kobayashi DY (2005) Mutagenesis of beta-1,3-glucanase genes in Lysobacter enzymogenes strain C3 results in reduced biological control activity toward Bipolaris leaf spot of tall fescue and Pythium damping off of sugar beet. Phytopathology 95:701–707

Pastrana AM, Basallote-Ureba MJ, Aguado A, Akdi K, Capote N (2016) Biological control of strawberry soil-borne pathogens *Macrophomina phaseolina* and *Fusarium solani*, using *Trichoderma asperellum* and *Bacillus spp*. Phytopathol Mediterr 55:109–120

Pisciotta JM, Zou Y, Baskakov IV (2010) Light-dependent electrogenic activity of cyanobacteria. PLoS One 5(5):e10821. https://doi.org/10.1371/journal.pone.0010821

Prasanna R, Nain L, Tripathi R, Gupta V, Chaudhary V, Middha S et al (2008) Evaluation of fungicidal activity of extracellular filtrates of cyanobacteria—possible role of hydrolytic enzymes. J Basic Microbiol 48:186–194

Prasanna R, Bidyarani N, Babu S, Hossain F, Shivay YS, Nain L (2015) Cyanobacterial inoculation elicits plant defense response and enhanced Zn mobilization in maize hybrids. Cogent Food Agric 1:998507. https://doi.org/10.1080/23311932.2014.998507

Pugliese M, Monchiero M, Gullino ML et al (2018) Application of laminarin and calcium oxide for the control of grape powdery mildew on Vitis vinifera cv. Moscato J Plant Dis Prot 125:477. https://doi.org/10.1007/s41348-018-0162-8

Ramkissoon A, Ramsubhag A, Jayaraman J (2017) Phytoelicitor activity of three Caribbean seaweed species on suppression of pathogenic infections in tomato plants. J Appl Phycol 29:3235–3244

Reinert R, Donald EL, Rosi FX, Watal C, Dowzicky M (2007) Antimicrobial susceptibility among organisms from the Asia/Pacific rim, Europe and Latin and North America collected as part of TEST and the in vitro activity of tigecycline. J Antimicrob Chemother 60:1018–1029

Rekanović E, Potočnik I, Milijašević-Marčić S, Stepanović M, Todorović B, Mihajlović M (2010) Efficacy of seaweed concentrate from Ecklonia maxima (Osbeck) and conventional fungicides in the control of Verticillium wilt of pepper. Pestic Phytomed 25:319–324

Righini H, Roberti R, Baraldi E (2018) Use of algae in strawberry management. J Appl Phycol 30:3551–3564

Righini H, Baraldi E, GarcíaFernández Y, Martel Quintana A, Roberti R (2019) Different antifungal activity of Anabaena sp., Ecklonia sp., and Jania sp. against Botrytis cinerea. Mar Drugs 17:299

Righini H, Roberti R, Quintana AM (2020) Biocontrol of Rhizoctoniasolani by water extracts from Chlorella sp. and Halopithys sp. Open Acc J Agric Res. OAJAR-100029

Roberti R, Galletti S, Burzi PL, Righini H, Cetrullo S, Perez C (2015) Induction of defense responses in zucchini (*Cucurbita pepo*) by Anabaena sp. water extract. Biol Control 82:61–68

Roberti R, Righini H, Pérez Reyes C (2016) Activity of seaweed and cyanobacteria water extracts against *Podosphaera xanthii* on zucchini. Ital J Mycol 45:66–77

Romero D, de Vicente A, Zeriouh H, Cazorla FM, Fernández-Ortuño D, Torés JA, Pérez-García A (2007) Evaluation of biological control agents for managing cucurbit powdery mildew on greenhouse-grown melon. Plant Pathol 56:976–986

Russo P, Cesario A (2012) New anticancer drugs from marine cyanobacteria. Curr Drug Targets 13(8):1048–1053

Sabin M, Susann K, Jansen R (2003) Fatty acids with antibacterial activity from the cyanobacterium Oscillatoria redekei HUB 051. J Appl Phycol 15:263–267

Safonova E, Reisser W (2005) Growth promoting and inhibiting effects of extracellular substances of soil microalgae and cyanobacteria on *Escherichia coli* and *Micrococcus luteus*. Phycol Res 53:189–193

Schiener P, Black KD, Stanley MS, Green DH (2015) The seasonal variation in the chemical composition of the kelp species Laminaria digitata, Laminaria hyperborea, Saccharina latissima and Alaria esculenta. J Appl Phycol 27:363–373

Schirrmeister BE, Antonelli A, Bagheri HC (2011) The origin of multicellularity in cyanobacteria. *BMC Evolution*. Biol 11(45):1–21

Schlegel I, Doan NT, De Chazal N, Smith GD (1998) Antibiotic activity of new cyanobacterial isolates from Australia and Asia against green algae and cyanobacteria. J Appl Phycol 10(5):471–479

Sergeeva E, Liaimer A, Bergman B (2002) Evidence for production of the phytohormone indole 3-acetic acid by cyanobacteria. Planta 215:229–238

Shao J, Wu Z, Yu G, Peng X, Li R (2009) Allelopathic mechanism of pyrogallol to Microcystis aeruginosa PCC7806 (Cyanobacteria): from views of gene expression and antioxidant system. Chemosphere 75: 924–928

Sharma P, Sharma N (2017) Industrial and biotechnological applications of algae: a review. J Adv Plant Biol 1:1–25

Simmons TL, Engene N, Ureña LD, Romero LI, Ortega-Barría E, Gerwick L, Gerwick WH (2008) Viridamides a and B, lipodepsipeptides with antiprotozoal activity from the marine cyanobacterium Oscillatoria nigro-viridis. J Nat Prod 71:1544–1550

Singh JS, Strong PJ (2016) Biologically derived fertilizer: a multifaceted bio-tool in methane mitigation. Ecotoxicol Environ Saf 124:267–276. https://doi.org/10.1016/j.ecoenv.2015.10.018

Singh RK, Tiwari SP, Rai AK, Mohapatra TM (2011) Cyanobacteria: an emerging source for drug discovery. J Antibiot 64:401–412

Tan LT (2007) Bioactive natural products from marine cyanobacteria for drug discovery. Phytochemistry 68:954–979

Tiwari A, Kaur A (2014) Allelopathic impact of Cyanobacteria on pathogenic fungi. Int J Pure App Biosci 2 (3): 63–70

Uzair B, Tabassum S, Rasheed M, Rehman SF (2012) Exploring marine cyanobacteria for lead compounds of pharmaceutical importance. Sci World J 179(82):1–10

Vardi A, Schatz D, Beeri K, Motro U, Sukenik A, Levine A, Kaplan A (2002) Dinoflagellate cyanobacterium communication may determine the composition of phytoplankton assemblage in a mesotrophic lake. Curr Biol 12:1767–1772

Vera J, Castro J, Gonzalez A, Moenne A (2011) Review-seaweed polysaccharides and derived oligosaccharides stimulate defense responses and protection against pathogens in plants. Mar Drugs 9:2514–2525

Volk BR, Furkert HF (2006) Antialgal, antibacterial and antifungal activity of two metabolites produced and excreted by cyanobacteria during growth. Microbiol Res 161:180–186

Wainwright PO, Hinkle G, Sogin ML, Stickel SK (1993) Monophyletic origins of the metazoa: an evolutionary link with the fungi. Science 260:340–342

Yadav S, Sinha RP, Tyagi MB, Kumar A (2011) Cyanobacterial secondary metabolites. Int J Pharma Bio Sci 2(2):144–167

Chapter 2
Plant Growth Promoting Rhizobacteria in Amelioration of Abiotic Stresses: A Functional Interplay and Prospective

Pedro Henrique Riboldi Monteiro, Franquiéle Bonilha da Silva, Caíque Menezes de Abreu, and Glacy Jaqueline da Silva

Contents

1 Introduction

Climate change alters abiotic factors, such as periods of drought, precipitation rates, temperatures, evaporation and light intensity, which, associated with the physical and chemical characteristics of the soil, trigger stresses on plants, which

P. H. R. Monteiro (✉) · C. M. de Abreu
Department of Agronomy, Federal University of Vales do Jequitinhonha and Mucuri, UFVJM, Diamantina, Minas Gerais, Brazil

F. B. da Silva
Department of Soils, Faculty of Agronomy, Federal University of Rio Grande do Sul, Porto Alegre, Rio Grande do Sul, Brazil

G. J. da Silva
Department of Biotechnology Applied to Agriculture, Universidade Paranaense — UNIPAR, Umuarama, Paraná, Brazil

© The Author(s), under exclusive license to Springer Nature Switzerland AG 2021

H. I. Mohamed et al. (eds.), *Plant Growth-Promoting Microbes for Sustainable Biotic and Abiotic Stress Management*,
https://doi.org/10.1007/978-3-030-66587-6_2

consequently can impair development and productivity (Cramer et al. 2011; Ullah et al. 2015; Maxton et al. 2018; Martins et al. 2018). Agricultural production is closely linked to environmental quality, which, when threatened, affects natural resources, of which soil plays an essential role because it is the primary source of the elements necessary for the growth and development of plant species (Akhtar et al. 2020).

2 World Agricultural Sector X Productive Challenges

This productivity is obtained when biotic and abiotic factors, including microbiota, water, light, soil and nutrients, are available and in balance, allowing cultures to express the maximum genetic potential (Cramer et al. 2011; Fan et al. 2018). Genetically improved cultivars are selected to enhance the gene expression of desired characteristics, such as rapid growth; productivity in the volume of grains, fruits, oils, seeds, biomass and wood; stem form; adaptability; quality; and resistance to pests, diseases and abiotic factors, that is, anything that adds commercial value to culture and derived products (Miguel et al. 2016; Fonseca et al. 2017).

The new challenges facing the global agricultural sector are sustainable production, that is, production without harming the environment. This takes into account the change in the global climate panorama that has undergone an intense process of change, according to the reports of the Intergovernmental Panel on Climate Change (IPCC). These reports highlight climate change in the world in recent decades, through rising temperatures and changes in rainfall patterns (IPCC 2007; Ullah et al. 2015). This modification will result in changes in the productive bioclimatic zones (Garcia et al. 2014a, b; Maxton et al. 2018).

3 The Interaction of the Soil System-Microorganism-Plant

The microorganisms and plants' interaction system has the soil as a common component since it can influence plant and microbial growth, whether in multiplication, survival and even in the metabolic activity present in ecosystems. In the soil, there is a high concentration of microorganisms, especially in the regions close to the roots of the plants. This region is defined as the rhizosphere, which is conceptually considered the portion of soil that is influenced by the roots (Hiltner 1904). Due to its complexity and diversity, the rhizosphere has no measurable shape and size, but a biological and physical-chemical gradient that changes radially and longitudinally around the root (McNear Jr 2013).

The root microbiome enables the presence of microorganisms by being enriched with root exudates, which have a high concentration of components with varying molecular weights, ions, carbon and free oxygen, mucilage and different primary and secondary metabolites. In addition to inhabiting this rich space and benefiting from it, these microorganisms establish ecological relationships with the plant. Among the most important relationships for the agricultural sector are parasitism, commensalism and mutualism (Lebeis 2015; Ullah et al. 2019).

Mutualistic associations allow microbial communities to influence the adaptation, development, health and survival of hosts, with plant-microorganism interactions. This bond may be optional, as is the case of bacteria that promote plant growth, which colonize the rhizosphere, the rhizoplane and even the (endophytic) tissues, which act positively on the plant. These microorganisms can promote plant growth and control pathogens and pests, in addition to serving as anti-stress agents. Also, there are mandatory ones such as mycorrhizae (fungus-root association), whose function is to increase the water and nutrient uptake area (Gray and Smith 2005; Farrar et al. 2014).

Plant growth-promoting rhizobacteria (PGPR) have several physiological mechanisms that contribute to the adaptation and induction of plant species resistance to abiotic stresses caused by edaphoclimatic changes (Kloepper et al. 1989; Kumar et al. 2019; Xia et al. 2020). The PGPR denomination helps to identify bacteria that are beneficial to plants and that consequently are efficient in increasing productivity. There is a range of PGPRs that are tolerant to abiotic stress conditions, which are the most suitable for use and/or applications in agricultural production (Vimal et al. 2017), thus being a sustainable production strategy to reduce the side effects of xenobiotics to the environment (Fig. 2.1).

This productive strategy aims to use the natural relationships between microorganisms and plants, calling this as the study of microbial biotechnology. The techniques employed for this research line have been widely studied, as they are natural, sustainable and economical strategies in the development of production technologies for the industrial and agribusiness sectors (Bianchi et al. 2016; Fukami et al. 2017; Ullah et al. 2019).

The use of microbial biotechnology in the handling of microbial inoculants for plant species has sought the development of new processes and methodologies to quantitatively assess the benefits induced by these microorganisms, such as increased nutritional content, productive gain (production, productivity and mass), tolerance and others. Many of these forms of assessment have been estimated by bioinformatics tools, based on molecular studies, encompassing all "omics" (Genomics, Proteomics, Metabolomics and Transcriptomics), the most used being Genomics and Proteomics (Sarim et al. 2020) (see Sect. 6).

4 PGPRs X Abiotic Stresses: The Main Mechanisms Involved in Resistance

Microbial biotechnology has been studied and applied in the screening of PGPRs that can survive in environments with extreme abiotic conditions, such as water or drought deficit, extreme temperature variations, flooding, salinity, soil acidity and the presence of heavy metals, able to quickly colonize the rhizosphere, mitigate and even induce host resistance (Xia et al. 2020).

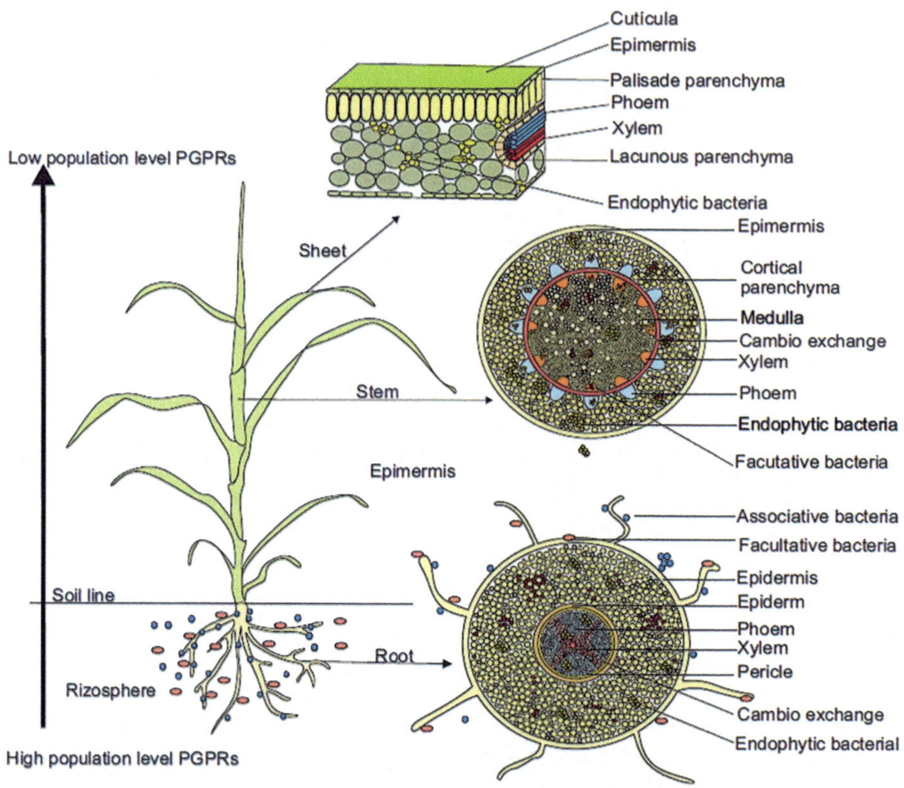

Fig. 2.1 The interactions between microorganisms (PGPR) and their availability in plant

Soil microorganisms, which are capable of directly promoting plant growth (PGP), can provide nutrients and produce phytoregulatory substances and also show how the PGP produced can be evaluated (Fig. 2.2).

Plants under prolonged exposure to stress conditions show a significant reduction in AIA (indole-acetic acid) levels. Therefore, inoculation of AIA-producing microorganisms can induce variations in the synthesis of endogenous AIA in the plant (Goswami and Deka 2020). This concentration of AIA, provided by microorganisms, allows roots to develop under stress, as well as photosynthetic production, which is not interrupted by low ethylene production.

The ethylene hormone has its biosynthesis regulated by environmental conditions, which is why it is known as the "stress hormone" (Glick 2014). When in unfavourable conditions, the plant has its ethylene levels altered, and its homeostasis is affected. ACC (aminocyclopropane carboxylic acid) is an immediate precursor to ethylene, which can be degraded by the enzyme ACC deaminase, preventing the production of the hormone. Some bacteria act positively in such situations, as they produce ACC deaminase, thus preventing excessive endogenous ethylene production. Rhizobacteria that produce high rates of ACC deaminase manage to reduce

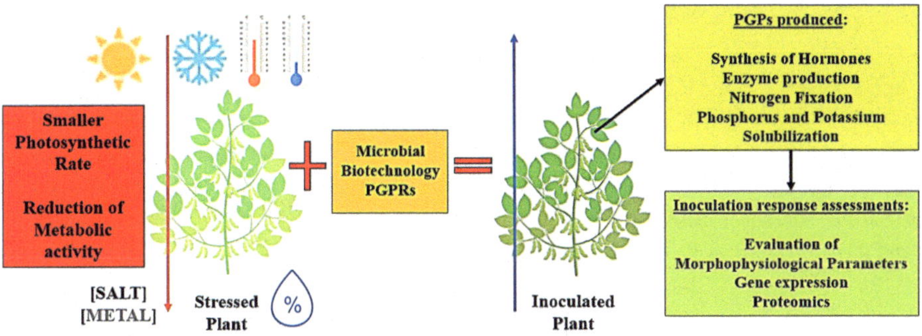

Fig. 2.2 The figure summarizes the difference between plants under abiotic stress and inoculated plants, which produce PGPs to combat or induce resistance to abiotic stress. Moreover, it shows the ways to evaluate the efficiency of the inoculation and the benefits of this. On the left is a plant under abiotic stress and the consequences of that stress. In the figure on the right, we can see an inoculated plant, accompanied by the production of PGPs beneficial to the plant, combating abiotic stress and evaluating the efficiency of these PGPs for plants

the susceptibility of plants to stressful environments, by stimulating the production of exopolysaccharides (EPS) that are on the surface and bind to cations (such as Na^+). This connection results in the formation of films called "biofilm" (Maxton et al. 2018).

There are rhizobacteria capable of producing EPS that help them to survive harsh environments. Exopolysaccharides are formed from long chains of sugar units like glucose, galactose and rhamnose in several portions. They are released by the bacterial cell during its growth and are not bound to the cell, being divided into two groups: homopolysaccharides (formed by monomers of the same species) and heteropolysaccharides (composed of different types of monosaccharides) (Etesami and Maheshwari 2018). EPS have anionic properties due to the acyl group, which also increases its lipophilicity and its interactions with other cations and polysaccharides (Kaushal and Wani 2016). Studies show that EPS production is more pronounced when bacteria are exposed to stress than under normal environmental conditions (Martínez-Gil et al. 2014; Naseem and Bano 2014). These EPS not only help in the survival of bacteria but also play an important role in the tolerance of plants to drought and salinity, as well as in plant growth (Vurukonda et al. 2016).

5 Types of Abiotic Stress

The authors of this chapter have built it by surveying the types of stress present/recurring in annual and perennial crops (fruit and woody). The most described types of abiotic stress are of a water nature, always portrayed as drought, water deficiency and flood that causes anoxia; of a thermal nature that is described through low and high temperatures, in addition to freezing and frosts; of chemical nature of the soil, the saline environment composed of high concentrations of Na^+, Ca^{+2}, Mg^{+2} and K^+; and heavy metals, reaching toxic levels and pH, correlated to the concentration of Al^{+3} present in the soil.

An example of the multifunctionality of PGPR is *Bacillus xiamenensis* (PM14), proven to be a bacterium that promoted the tolerance of sugar cane (*Saccharum officinarum*) to salt stress and heavy metals, thermotolerant up to 45 °C with a production of ACC deaminase and EPS production. It additively promotes the resistance of sugarcane to fungal diseases of the root, by presenting 12 antibiotics (Xia et al. 2020).

5.1 Water Nature

Among the types of abiotic stresses, the most worrying are those of a water nature as they are one of the main causes of growth and productivity restriction of plants around the world (Martins et al. 2018). As a result of this stress, there is a reduction in the availability of animal and human food, in addition to losses in the economy (Kumar and Verma 2018). Therefore, looking for strategies that minimize the impact of drought, so that plant production can be maintained, is of great importance to satisfy the global demand for food production (Goswami and Deka 2020).

The result of an environment with a scarcity of water, with restricted and irregular rainfall distribution, is a dry environment. This has the effect of limiting water in the soil or the excessive loss of water by plants through the transpiration process in relation to the absorption of the roots, affecting vital processes such as photosynthesis, respiration, carbohydrate metabolism and ion absorption (Ullah et al. 2018; Akhtar et al. 2020). Water stress has historically reduced the productive yield of vegetables, especially grains and cereals, with rates of up to 10%. This is the result of changes in bioclimatic zones, as a result of climate change. It is estimated that in approximately 2050, productive land will suffer from drought in more than 50% of the areas, which will consequently affect world production (Jochum et al. 2019).

The mechanisms of reaction to water stress are being understood employing the adoption of traditional technologies that are linked to genetic engineering. The area of interest is microbial biotechnology, which uses the interactions between microorganism-plant cultivation (agricultural and forest) aiming to increase the productivity of plant resistance (Card et al. 2016; Silva et al. 2016; Tang et al. 2017; Bilal et al. 2018).

When using *Bacillus subtilis*, strain GOT9, as a microbial inoculant in *Arabidopsis thaliana* and *Brassica campestris*, we observed responses to correlated saline and hydric stress. This statement is proven by the expressive gene expression of these plants. When the plant interacts with this microorganism in a dry environment or in the field, the RD29 and RAB18 genes encode the dehydrin protein, which forms a protective biomolecule during exposure to stress. RD20 is a kaleosin isoform, which, when expressed in greater quantity, increases the tolerance due to the control of the stomatal opening (Woo et al. 2020). Dehydrins are the first proteins identified in plants with low- or high-water potential (dehydration), in saline environments and even at low temperatures (Banerjee and Roychoudhury 2016).

Other responses to water stress occur by reducing the turgidity of plant cells, which induce stomatal closure and consequently cause restriction of photosynthetic rate and cell elongation, managed by the hormonal balance of plants, mainly through the NCED3 gene, which acts on the biosynthesis of abscisic acid (ABA) (Woo et al. 2020). The second response occurs in the roots, where proteins called aquaporins are activated. These proteins are present in the membranes of root cells that absorb water promptly (Gaspar 2011). However, understanding the processes of signalling the plants' response to a lack of water is very complex, as there is no single universal route considered (Martins et al. 2018).

Research shows that some PGPR can stimulate the production of phytohormones such as endogenous ABA, AIA (Belimov et al. 2015; Forni et al. 2017), gibberellic acid (AG3) (Maxton et al. 2018) and cytokinin (Peleg and Blumwald 2011). The regulation of ABA stimulates the hydraulic conductivity in the roots and also the regulation of aquaporins, being an ally for plants in tolerance to drought (Goswami and Deka 2020). The combination of the production of AIA and AG3 increases the potential for water intake due to the increase in the number of root hairs (Maxton et al. 2018).

Potato plants grown in a greenhouse in different water potentials with and without PGPR inoculation showed a 50% increase in root biomass and 40% in tuber yield when inoculated (Belimov et al. 2015). In the same study, when evaluated in a field experiment, inoculation increased tuber yield by 27%. Tahir et al. (2019) reported that PGPRs inoculated in corn plants in dry condition produced AIA, ACC deaminase and EPS, increasing the water content and chlorophyll content in the leaves. Also, inoculation caused a gain in grain yield, not only in plants under dry conditions but cultivated in ideal field capacity. Chandra et al. (2019) observed that the inoculation of bacteria producing ACC deaminase in wheat plants under water deficit promoted the accumulation of nutrients and grain productivity gain, both in the variety of wheat resistant to drought and in those typically sensitive.

Another vital hormone during drought tolerance is cytokinin, as it delays senescence and premature death of the leaves, which can increase the plant's yield (Peleg and Blumwald 2011). The origin of the cytokinin can be endogenous by plants or microbial. Alfalfa plants inoculated with *Sinorhizobium meliloti*, a cytokinin producer, experienced a delay in drought-induced senescence (Xu et al. 2012). When *Methylobacterium oryzae* was inoculated in lentil plants (*Lens culinaris*), changes in morphological and physiological patterns were correlated with water use efficiency and cytokinin levels, deferring drought tolerance (Jorge et al. 2019). Prolonged exposure to drought stimulates the production and regulation of phytohormones, osmolytes, EPS and antioxidants by these bacteria, which induce the plant to a greater tolerance to stress, mainly due to morphological changes in the roots (Yang et al. 2009), as described earlier.

Although less frequent than drought, excess water is an abiotic factor that gives stress to the plant, caused by an anoxic habitat. Flooding also influences the productivity of crops, causing physiological disturbances that result in a deficit in growth and agricultural yield (Grichko and Glick 2001; Sairam et al. 2009).

Excessive rainfall and flooding in poorly drained soils are the main situations that can lead plants to root hypoxia or anoxia (Barnawal et al. 2012). Some plants show tolerance to flooding due to the formation of aerenchyma and adventitious roots, induced by the interaction of AIA and ethylene (Ashraf 2019). Plants sensitive to flooding can suffer several damages with the reduction of photosynthesis, closure of stomata, reduction of growth, epinastia and necrosis, which result in the loss of agricultural productivity (Grichko and Glick 2001; Barnawal et al. 2012).

Excessive endogenous ethylene produced in the roots due to flooding is the main inducer of chemical signals and physiological changes that affect all plant tissues (Barnawal et al. 2012). Ethylene, previously described as “stress hormone”, when in ideal concentrations, helps in root initiation, together with AIA, but in high concentrations, there are a decrease in oxygen and an increase in ACC synthase activity that suppress formation and root elongation (Ahmed et al. 2006).

The reduction of the levels of endogenous ethylene in the plant is one of the ways to mitigate the stress caused by flooding. Therefore, the inoculation of PGPR producing ACC deaminase is an economical and ecological alternative that makes the plant more tolerant of flooding, as has been demonstrated in several studies. The ACC deaminase produced by these bacteria synthesizes ACC (immediate precursor to ethylene) in ammonia and α-ketobutyrate (Glick 2014).

Grichko and Glick (2001) evaluated the inoculation of *Pseudomonas putida* UW4 and *Enterobacter cloacae* CAL2 in tomato plants under flooding conditions. The study showed that inoculation increased the tolerance of tomatoes, resulting from the activity of bacterial ACC deaminase, which reduced the content and effect of ethylene in plants. *Ocimum sanctum* subjected to flooding conditions and inoculated with several PGPRs that present desirable growth promotion mechanisms (ACC deaminase, phosphate solubilization, production of AIA and siderophores) had an increase in root and shoot growth and also a reduction in the ethylene content (Barnawal et al. 2012). Nascimento et al. (2012) tested the inoculation with the *Mesorhizobium* strain transformed with an exogenous ACC deaminase plasmid in chickpea plants (*Cicer arietinum*) under flooding. The results showed that there was an increase of 127% in nodulation and 125% in biomass of plants inoculated with the transformed strain, showing that ACC plays an important role in the development of microbial inoculants, especially under stress conditions.

5.2 *Thermal Nature*

Current estimates of climate change suggest that the increase in temperature is a reality in the coming years (IPCC 2007). This increase in temperature will drastically reduce agricultural production on a global scale, as it alters the distribution of productive bioclimatic zones and growing seasons, leading plants to early maturity and consequently to the anticipation of harvest (Porter 2005; Garcia et al. 2014a, b). The stress caused by high temperatures is one of the main limitations during anthesis and grain filling in cereal crops from temperate regions (Ali et al. 2011). In wheat, for example, which has an ideal temperature after anthesis of 15 °C, each 1 °C above can cause a 3 to 5% reduction in grain weight (Wiegand et al. 1981).

High temperatures also cause denaturation, aggregation and inhibition of protein synthesis and inactivation of enzymes in mitochondria and chloroplasts, affecting the fluidity of membrane lipids causing loss of their integrity (Howarth 2005). All of these factors can lead to reduced growth and production of toxic compounds. Given this, the use of easily accessible, low-cost and environmentally friendly methods would be an alternative to minimize the impact of climate change, with the use of PGPRs being a promising path, as has been shown in some studies. This reflects directly on the search for wild PGPR, which inhabits regions with high temperatures to assess the potential in the production of growth promoters.

The low temperature is another limiting factor of productivity and with geographic distribution for many agricultural species. The stress caused by the cold leads to a series of molecular, biochemical, physiological and morphological changes, which are reflected in the fall in productivity (Barka et al. 2006). Cold-resistant plants tend to increase their tolerance to freezing when exposed, due to a phenomenon called cold acclimation (Thomashow 1999).

As previously reported in this chapter, inoculation with PGPR is a great ally of plants in tolerating adverse conditions, including low temperatures. PGPR species with the potential to reduce damage from this type of stress are generally isolated from regions with constant low temperatures or are found in mountainous regions. *Bacillus subtilis* and *Pseudomonas corrugata* strains, from the Himalayan region, inoculated in wheat (*Tricicum tivum*), proved efficient through the growth of plants in development and height, with greater enzymatic activity and phosphorus concentration (Trivedi and Sa 2008; Trivedi et al. 2012). Vine seedlings (*Vitis vinifera* L.) inoculated with *Burkholderia phytofirmans* PsJN, between 4 and 26 °C, showed cold tolerance in addition to the increase in seedling biomass (6 and 2.2 times more biomass than in controls at 26 and 4 °C, respectively) and root growth (11.8 and 10.7 times more than in controls at 26 and 4 °C, respectively), with an increase in the content of starch, proline and phenols, to the non-inoculated, demonstrating the growth and development stimulus of the vine by PGPR to withstand cold stress (Barka et al. 2006). In a second step, the use of the inoculant *B. phytofirmans* PsJN in grapevines increased the concentration of total soluble sugars, starch and sugars related to cold tolerance (glucose, sucrose and raffinose with its precursor, galactinol) (Fernandez et al. 2012). These results were later explained by the increase in metabolic levels and expression rates of genes related to cold stress (Theocharis et al. 2012).

5.3 Soil Chemical Nature

5.3.1 Salinity

Saline-sodium soils are considered to be those with electrical conductivity greater than 4 dS m^{-1} in the saturation extract in the rhizospheric region and 15% of exchangeable sodium percentage (Sparks 2003). Saline environments cause ionic and osmotic stress, which results in the suppression of plant growth and consequently loss of crop productivity (Baek et al. 2020). This type of stress influences almost 70% of rainfed land in the world (Goswami and Deka 2020), making it the most important factor for the abandonment of agricultural areas (Maxton et al. 2018).

Bacterial EPS shows remarkable performance for plants in dry conditions. However, they can be very favourable in salinity conditions. In these environments, EPS can bind to Na^{+} free ions due to their anionic property, thus preventing their absorption by the plant (Upadhyay et al. 2012). In a study that evaluated the effect of *Pseudomonas putida* (strain Rs-198) under conditions of salt stress in cotton plants, it was possible to verify that there was an increase in the uptake of Ca^{+2}, K^{+} and Mg^{+2}, in addition to the decrease in uptake of Na^{+} by plants, reducing the negative impact of stress (Yao et al. 2010). Tewari and Arora (2014) found an increase in the growth of sunflower plants under salinity when inoculated with *P. aeruginosa*. The authors attributed this effect to the production of EPS by the bacterium.

Several studies demonstrate the benefit of inoculating AIA-producing bacteria in plants under salinity conditions. The inoculation of wheat plants with strains of *Pseudomonas* spp., which produce AIA, increased root growth by 40% and the number of new shoots in salt stress situations by 52% (100 mmol L^{-1} of NaCl) (Egamberdieva 2009). Rabhi et al. (2018) found that inoculation of *Pseudomonas knackmussii*, producer of AIA in *Arabidopsis thaliana*, promoted growth and decreased oxidative stress caused by salinity compared to uninoculated plants.

Saline stress is also capable of inducing the supra-optimum production of ethylene by the plant, causing damage to its development. For this reason, the production of bacterial ACC deaminase acts directly in maintaining plant growth in saline environments (Bal et al. 2013). The application of *Pseudomonas fluorescens* TDK1 promoted the growth of peanut plants, as well as a greater tolerance to salinity (Saravanakumar and Samiyappan 2007). In wheat plants inoculated with *Bacillus mojavensis* k78 at different levels of salinity, a positive effect occurred, such as gain in dry mass of roots and shoots, mainly in the highest concentrations of salts (Pourbabaee et al. 2016). In pepper plants, Wang et al. (2018) found that inoculation with *Bacillus* sp. AWU5 increased the fresh and dry mass, root length and shoot in salinity conditions, compared to uninoculated plants.

In addition to helping plants under salinity conditions through hormonal regulation, rhizobacteria also produce osmolytes, which are organic molecules responsible for balancing the osmotic difference between the cytosol and the external environment. Rhizobacteria can also produce antioxidant substances, which reduce the damage caused by salinity in plants. Jha et al. (2011) observed that co-inoculation with *Pseudomonas pseudoalcaligenes* and *Bacillus pumilus* increased salinity tolerance through the production of osmoprotectors and antioxidants in the early stages of rice plant growth. Abd-Allah et al. (2018) inoculated chickpea plants with a strain of *Bacillus subtilis*, which significantly increased plant growth by modulating the antioxidant system, eliminating oxidative damage caused by salinity.

Plants, when subjected to salt stress, can express genes such as RD29, RD29A and RD29B, which are part of the dehydrin family, proteins responsible for reducing the water conductivity of the roots and inducing the autophagic degradation of aquaporins. The RD20 and NCED3 genes are expressed during water stress (see Sect. 5.1) and, when in greater quantities, increase salinity tolerance (Woo et al. 2020). Other genes expressed, but in *Brassica campestris* is the SOS1 gene, essential for the occurrence of Na^+ and K^+ ions homeostasis; and WRKY8, considered as a positive regulator in salt stress, which stimulates the increase in the levels of endogenous ABA and the ABA responsible for stress. *Bacillus oryzicola,* strain YC7007, when inoculated in *Arabidopsis thaliana,* induced the plant's resistance to salinity through the expression of the SOS1 gene, plant growth and increased number of side roots, in addition to greater fresh weight and chlorophyll rates (Baek et al. 2020).

5.3.2 Soil pH

The pH variation depends on the soil formation factors, that is, source material, relief, climate and organisms as a function of time. Another factor that influences is the use of the soil, being the management adopted crucial to maintain the physical-chemical and mainly biological characteristics, which will directly affect the productive potential of the crops, whatever they may be. Therefore, it is possible to obtain acidic soils with a pH below 5.5 and alkaline soils where the pH is above 6.5. This varies depending on the species and its pH tolerance.

Acid soils limit the growth of PGPRs groups and the root system, consequently the development of the aerial part of crops. The main responsibility for the toxicity in acidic soils is the high concentration of aluminium in the form Al^{+3}, which is soluble in water and readily available for the absorption of plants (Sparks 2003). In alkaline soils, exchangeable aluminium is neutralized, as it is converted into species unavailable for root absorption. In addition to aluminium, other nutritional elements in the soil have reduced availability for the plant, such as micronutrients, iron, manganese, cobalt, zinc, boron and selenium, and also macronutrients, nitrogen, phosphorus and sulfur, causing damage to plant development. Also, molybdenum and chlorine have increased availability, making them potentially toxic to plants.

The inoculation of maize (*Zea mays*) with *Bacillus* and *Burkholderia* in soils with pH below 5 increased the length of the roots, between 1.4 and 2 times, respectively, in soils contaminated with aluminium, which reduced by up to 65% the formation of roots in control plants, proving that these microorganisms are efficient in combating the toxic effect of aluminium. However, only plants inoculated with *Burkholderia* were able to reduce the accumulation of aluminium by 50%, with a 30% increase in phosphorus in the roots, while plants inoculated with *Bacillus* did not show the same result (Arora et al. 2017). This observation of the inoculation of *Bacillus* in maize was not observed in wild isolates of the family Bacillaceae

(*Bacillus* and *Halobacillus*), performed by Banik et al. (2018), which proved the ability of these isolates to tolerate aluminium and acid pH variations (up to 6), to alkaline environments (pH up to 10), enhancing their use in search of the benefits that these microorganisms can provide to plants when used in limited environments.

Phosphorus is one of the main elements whose availability is influenced by soil pH. It is estimated that 70 to 90% of the phosphate applied via mineral fertilization to the soil is immobilized in iron and aluminium oxides in acidic soils and precipitated as calcium phosphate in alkaline soils (Norrish and Rosser 1983; Lindsay et al. 1989).

Insoluble forms of P, such as tricalcium phosphate ($Ca^3PO^4)^2$, aluminium phosphate (Al^3PO^4) and iron phosphate (Fe^3PO^4), can be converted into soluble P by solubilizing organisms that inhabit different soil ecosystems (Song et al. 2008; Sharma et al. 2013). Among the soil bacterial communities, ectorizospheric strains such as *Pseudomonas* sp., *Bacilus* sp., *Burkholderia* sp. and symbionts such as rhizobia have been described as effective phosphate solubilizers (Igual et al. 2001; Song et al. 2008).

The ability of rhizosphere bacteria to solubilize insoluble phosphates is attributed to primary mechanisms such as the excretion of H^+ ions, the production of organic acids and the biosynthesis of acid phosphatases (Arcand and Schneider 2006; Richardson and Simpson 2011). Among the organic acids produced, we can mention acetate, lactate, malate, oxalate, succinate, citrate and gluconate, which form complexes with iron or aluminium present in iron and aluminium phosphates, thus making the phosphate present in the soil available for plants (Gyaneshwar et al. 2002). These microorganisms can grow in media with tricalcium phosphate or similar insoluble materials as the only source of phosphate and not only assimilate the element but also solubilize quantities beyond their nutritional requirements, making the surplus available to plants (Chen et al. 2006).

Countless studies have already demonstrated the effect of inoculation of phosphate-solubilizing bacteria on plants. Afzal and Bano (2008) inoculated a rhizobia strain associated with a phosphate solubilizing bacterial strain, in wheat cultivation in a phosphorus-deficient sandy soil. They found an increase in phosphorus absorption, 30 to 40% more than when compared to the treatment that received only fertilizer application. In addition, there was a significant increase in dry root weight, plant height, ear length, grain yield and sugar and leaf protein content.

5.3.3 Heavy Metals

One of the ways to decrease toxicity by heavy metals in plants is through the conversion of these bioavailable elements to inert organisms. Many rhizobacteria can perform this conversion, reducing the toxicity of the element in the rhizospheric region, so that the plant does not absorb them (Etesami and Maheshwari 2018). This is an economical and ecological alternative for the reduction of heavy metals in contaminated environments (Congeevaram et al. 2007; Khan et al. 2012). PGPRs produce organic acids such as citrate, oxalate, gluconic acid, etc., with complexing properties of heavy metals, reducing the potential toxic effect for plants (Archana et al. 2012; Xu et al. 2012). They are also capable of performing bioleaching and/or immobilization by intracellular accumulation and transformation into non-toxic forms by extracellular enzymes (Wani et al. 2009; Khan et al. 2012).

In addition to the direct effects on toxicity, these heavy metal-tolerant microorganisms can promote plant growth through their multiple mechanisms. Organic acids produced by these bacteria can increase the availability of nutrients such as P and K (Patel et al. 2010); in addition to protecting plants from pathogens by synthesizing antimicrobial, cyanogenic and siderophore compounds, they produce AIA, resulting in a better yield of plants grown in contaminated soils (Oves et al. 2013).

Oves et al. (2013) evaluated the inoculation of a chromium-resistant strain (*Pseudomonas aeruginosa* OSG41) in chickpea plants, grown in the presence of hexavalent chromium. The authors observed that inoculation with the bacteria reduced the uptake of Cr by 36% in the root, 40% in the aerial part and 40% in the grains. Also, there were an increase in dry mass of the aerial part and greater absorption of nutrients (N and P), grain production and nodulation. Moreira et al. (2014) found that inoculation with *Ralstonia eutropha* reduced cadmium toxicity in corn plants in a greenhouse. The inoculation reduced the accumulation of metal in the plant by immobilization, which led to a low translocation to the aerial part.

A study that evaluated rhizoremediation using the inoculation of *Sedum plumbizincicola* plants with *Bacillus* sp. SC2b proved to be efficient in reducing metals such as cadmium and zinc in contaminated soils (Ma et al. 2015). The results of the study showed that the inoculated plants had a significantly higher accumulation of Cd and Zn, demonstrating the effectiveness of inoculating plants with this strain in mobilizing metals for rhizoremediation in contaminated soils. Besides, the bacteria increased the root and shoot biomass and also the leaf chlorophyll content. In the same sense, Shreya et al. (2020) tested the inoculation of Cr-tolerant strains in the cultivation of chickpea plants in a greenhouse, in the presence of Cr. The authors found that inoculation promoted plant growth (increasing root and aerial part biomass and plant length), in the presence of Cr in the soil. It was also reported in the study that inoculation increased phytostabilization, accumulating more Cr both in the root and in the aerial part of the plant, demonstrating the potential as a soil bioremediation agent.

The use of *B. subtilis* and *P. putida* in two species of wheat (*Triticum aestivum*) grown in a greenhouse in two doses, 10 mg kg^{-1} (low) and 20 mg kg^{-1} (high) of Cd, presented positive responses with a higher germination rate even in the presence of two heavy metal dosages. It also showed higher growth and lower metal concentration rates in plant tissues, demonstrating the ability of these bacteria to induce resistance to stressful environments, as well as the potential for rhizorremediation (Khatri et al. 2020).

The use of PGPR tends to be successful because they colonize plants subjected to different types of stress. Two bacteria from the Bacillaceae family (*Bacillus* and *Halobacillus*) inoculated in peanut seedlings in vitro under saline stress (1% NaCl) and heavy metals (Zn, Al and Pb) showed a positive response to physiological parameters. In addition, plants inoculated with bacteria showed lower concentrations of heavy metals, demonstrating the possible proto-cooperation in supporting resistance to saline and contaminated environments (Banik et al. 2018).

6 Genetic Plant-Microorganism Interaction: An Application of Proteomics in the Detection of Phyto-Beneficial Actions Induced by PGPRs

The use of genomics, transcriptomics and proteomics in the agricultural scenario (according to topic 2) allows the researcher to understand how genes and genetic information of organisms are organized within the genome and how they interact. Through the generation and order of sequences, integration of physical maps, identification of genes and polymorphism, expression of gene products (transcribed) and the structuring of proteins, the researcher can analyze the functional data and predict possible genetic expressions and metabolic routes that are associated with the induction of phyto-beneficial actions of PGPRs in cultivated plants.

The soil-rhizosphere-rhizoplane-endophyte-plant system is a dynamic and coordinated unit that exercises a unique interaction with the components of the environment, generating several phyto-beneficial responses for plant growth (Carvalho et al. 2016). These adaptive responses are correlated with the reciprocal genetic regulation between bacteria and plants during the signaling process, plant colonization and establishment (Dhawi 2020). When we analyze the mutual influence between microorganisms and plants, the understanding of the genetic and biochemical mechanisms that regulate the type of plant-microorganism interaction has not yet been fully elucidated. One of the key points for understanding this interaction is to understand how plants use different signaling pathways to recognize and distinguish beneficial microorganisms from pathogens.

The sequence of steps that occurs during the associative process has demonstrated the existence of the formation of different amounts of transcriptomes in a single individual, which makes it difficult to read sequentially all transcripts; moreover, mRNA abundance is not always well correlated with protein abundance, since the activity of proteins encoded by mRNA is regulated at various levels after expression (Salvato and de Carvalho 2010). The development of proteomic tools has greatly facilitated the application of protein analysis to investigate plant-rhizobacterial interaction (Singh et al. 2017).

Proteins and their functional interactions form a network of connectivity that needs to be considered for a complete understanding of biological phenomena. The STRING platform (https://string-db.org/) is the main database that involves and integrates all publicly available sources on protein-protein interaction information (Szklarczyk et al. 2019). Thus, the BLAST (Basic Local Alignment Search Tool) matrix determines the average percentage similarity between the proteomes, measuring the proportion of gene families conserved and shared between the strains and the total number of gene families within each strain. The absolute number of families of genes shared and combined for each microorganism is displayed at the output of the matrix, determining the number of proteins shared between each proteome.

Transcriptional factor-type proteins function as repressors or gene activators (Poupin et al. 2016) in stressful situations such as anoxia, high temperatures, freezing, water deficit, pH and salinity, among others. PGPR inoculation improves the expression of sLTPs (non-specific lipid transfer protein), which are small basic proteins present in abundance in plants and which are involved in key processes of plant cytology, such as membrane stabilization, cell wall organization and signal transduction (Singh et al. 2017).

The expression of several proteins associated with biotic stress such as ACC deaminase, APx, ArsR, Aux_AIA, kaleosin, carbohydrate kinases (FGGY_C), cafeoil-CoA O-methyltransferase (CCoAOMT), cold shock, dehydrin, FTsZ, phosphodiesterase, glutathione S-transferase (GST), heat shock protein 70 (HSP70), MreB_Mbl, nsLTPs4, sulfatase, thioredoxin H, thaumatin, ubiquitin, ribulose carboxylase large chain, hama-glutamyl-hydrolase (GGH) and vegetative storage glycoprotein (VSP) among others (Tables 2.1 and 2.2), are increased 2 to 10 times more in the plant cell system.

Studies with *Pseudomonas* sp. inoculated in sorghum seedlings revealed the increased presence of reciprocal stimulation of proteins FtsZ, MreB, Mbl, AcrR and ArsR (Table 2.1) during and after colonization (Dhawi 2020). The use of *Enterobacter cloacae* in response to salt stress (with NaCl) resulted in the regulation of several proteins such as tubulin, profilin, retinoblastoma, CASP (casparian membrane protein) and xyloglucan endotransglycosylase (Singh et al. 2017). These act in the lignin biosynthesis and acceleration of protein synthesis, strengthening of the cell wall and in the maintenance of the cell structure to prevent damage during the salt stress condition, leading to the synthesis of storage proteins and osmoprotectors. *Paenibacillus polymyxa* inoculated in watermelon roots induced the expression of caffeoyl-CoAO-methyltransferase (CCoAOMT), glutathione S-transferase (GST) and ubiquitin, which act in the regulation of unwanted proteins, increase the plant's hardiness and resistance to compression, while it resists damage, metabolic, biotic and abiotic stress involved in growth, photosynthesis and other metabolic and physiological activities (Yaoyao et al. 2017). The use of *Bacillus* sp. and *Arthrobacter pascens* also under salt stress resulted in the production of ACC deaminase, bacteriocin and siderophores (Ullah and Bano 2015). *Pseudomonas simiae* inoculated in soybeans produces positive regulation of the large subunit of RuBisCo, showing an increased abundance of photosynthesis and proteins related to abiotic stress in the expression of VSP and GGH (Vaishnav et al. 2015).

The mRNA expressions of ACS gene transcripts are commonly found in plants located in typically stressful environments, such as nutritional imbalance, heavy metal toxicity, drought and salinity, potentiating the production of ACC oxidase (ACO). The inoculation of PGPRs in several cultures reduces the accumulation of ACO in the roots and increases the expression of ACC deaminase (typically of microbial origin) reducing the deleterious action of ethylene (ACO) (Saikia et al. 2018), inducing morphological changes and biochemical, resulting in systemic tolerance induced to abiotic stresses (Etesami and Maheshwari 2018).

Table 2.1 Role of proteins in promoting plant growth induced by PGPRs on abiotic stress

Protein[a]	Function	References
ACC deaminase[b]	Generation and regulation of phytohormones, production of siderophores, fixation, solubilization, mineralization of essential nutrients, hydrolysis of the fungal cell wall and ISR systemic induction and inducer of EPS production	Etesami and Maheshwari (2018)
Aux_AIA	Repressor of genes of early response to auxin in low concentrations, with an accumulation of RNA in roots, inflorescences and flowers	Poupin et al. (2016)
Kaleosin	Responsible for the functional regulation of the stomatal opening and closing control mechanism	Woo et al. (2020)
Carbohydrate kinase (FGGY_C)	Bacterial signaling and increased sugars and carbohydrates (>plant biomass)	Dhawi (2020)
Cafeoil-CoA O-methyltransferase (CCoAOMT)	Regulation and biosynthesis of lignin	Yaoyao et al. (2017)
Large ribulose carboxylase chain	Ribulose carboxylase bisphosphate activity (RUBISCO) in the activation of photosynthesis	Vaishnav et al. (2015)
Cold shock	Reduction of damage caused by intracellular freezing, by hydrogen bonding to the ice crystal network, modification of the structure and propagation of ice crystals	Singh et al. (2017)
Dehydrin	Reduction of water conductivity of roots and induction of autophagic degradation of aquaporins	Woo et al. (2020)
Gamma-glutamyl hydrolase (GGH)	Activation of peptidase, metabolic process of glutamine	Vaishnav et al. (2015)
Vegetative storage glycoprotein (VSP)	Activation of acid phosphatase activities inducing the nutrient reservoir	Vaishnav et al. (2015)
FTsZ	Tubulin-related protein connected to bacterial membrane	Dhawi (2020)
Phosphodiesterase	Repair of DNA-protein crosslinking in plants (>tolerance to biotic and abiotic conditions)	Dhawi (2020)
Heat shock protein 70 (HSP70)	Protein translocation, protection and maintenance of cellular homeostasis during the stress period; regarding the biogenesis of other proteins	Dhawi (2020)
MreB_Mbl	Encodes genes related to bacterial cell membrane protein FtsZ	Dhawi (2020)
nsLTPs[c]	Transfer of phospholipids and glycolipids across membranes and deposition of wax on the cell wall of expanding epidermal cells and secretory tissues	Singh et al. (2017)
Sulfatase	Hormonal regulation; cell degradation and remodeling of signaling pathways	Dhawi (2020)
Thioredoxin H	Antioxidant action on shoot development and leaf photosynthesis under stress	Singh et al. (2017)
Thaumatin	Formation of disulfide bonds, which provide protein stability under varying thermal and pH conditions	Singh et al. (2017)
Ubiquitin	Regulator of unwanted proteins in the autoimmune system by the ubiquitin-proteasome pathway	Yaoyao et al. (2017)

[a]Domain

[b]1-aminocyclopropane-1-carboxylate

[c]Non-specific lipid transfer protein

Table 2.2 Response of PGPRs inoculation on protein production and the resulting benefit in plants subjected to different types of stress

PGPR	Protein	Tolerance/resistance						References
		Anoxia	Water deficit	Heavy metals	pH	NaCl	Temp.	
Ochrobactrum pseudogrignonense, *Pseudomonas* sp. and *Bacillus subtilis*	ACC deaminase	+	+	+	+	+	+	Etesami and Maheshwari (2018) and Saikia et al. (2018)
Enterobacter cloacae	ACC deaminase					+		Singh et al. (2017)
Bacillus cereus	ACC deaminase						+	Mukhtar et al. (2020)
Bacillus xiamenensis	ACC deaminase			+	+	+	+	Xia et al. (2020)
Bacillus sp. *Arthrobacter pascens*	APx					+		Ullah and Bano (2015)
	Aux_AIA							Poupin et al. (2016)
Enterobacter cloacae	Cold shock					+	+	Singh et al. (2017)
Pseudomonas simiae	Ribulose carboxylase					+		Vaishnav et al. (2015)
Paenibacillus polymyxa	Cafeoil-CoA O-methyltransferase (CCoAOMT)			+	+		+	Yaoyao et al. (2017)
Enterobacter cloacae	Xyloglucan endotransglycosylase					+		Singh et al. (2017)
Enterobacter cloacae	Profilin					+		Singh et al. (2017)
Enterobacter cloacae	Retinoblastoma					+		Singh et al. (2017)
Enterobacter cloacae	Thaumatin				+	+	+	Singh et al. (2017)

In the ethylene biosynthetic pathway, S-adenosylmethionine is converted by ACC synthase into ACC, the immediate precursor of ethylene, which immediately after the onset of stress produces small peaks of ethylene, consuming the ACC present in plant tissues and probably activating the synthesis of defensive genes within the plant, initiating the inhibitory processes of plant growth and survival, as an example of chlorosis, abscission and senescence, caused by secondary peaks of harmful ethylene. The ACC deaminase produced by PGPRs degrades the plant's ACC under stress to provide nitrogen and energy, which reduces the harmful effect of ethylene and consequently improves the plant's stress tolerance (Etesami and Maheshwari 2018).

The bioprospecting of new proteins has revealed major discoveries such as chlorophyll a/b binding protein 40 (cab-BO3–1), known to increase in quantity during development and exposure to light, increased about twice with inoculation of *Bacillus subtilis* JS in plants of *Lactuca sativa* and *Nicotiana tabacum* (Kim et al. 2018). The combined inoculation of *Pseudomonas brassicacearum*, *Bacillus thuringiensis*, *Bacillus cereus* W6 and *Bacillus subtilis* induced thermotolerance through the production of heat-tolerant proteins of high molecular weight, in addition to increasing the activity of antioxidant enzymes in the plants of *Zea mays* L. (Ashraf et al. 2019). The identified proteins can be useful for genetic transformation and thus improve the mechanism of tolerance to adverse culture conditions. Corroborating the application of PGPR-based bioinoculant benefits the plant, inducing the expression of several proteins involved in growth, photosynthesis and other metabolic and physiological activities.

7 Conclusions and Future Perspectives

World agriculture is undergoing major changes due to climate change and human pressure. These are capable of modifying agricultural soils, depleting it by the intense exploitation of mineral nutrients that allow for the productive supply required by global food demand. These changes cause a series of abiotic stresses, which are limiting production, such as availability of water (drought, flood), temperature (heat, cold and freezing), salinity, soil acidity and heavy metals, creating an alert about the management of crops. Thus, there is a need to explore more sustainable production systems, in which the exploration of the environment must be balanced and/or more similar to the natural ecological functions, mainly with the reduction of the use of resilient synthetic xenobionts.

In this chapter, we present the success of microbial biotechnology, which through its techniques, can be used as an alternative means by rural producers to overcome the challenges of today's agriculture. The adoption of the use of biotechnological products of microbial origin in production systems has proven the potential of their use through the processes of bioprospecting of microorganisms in plants and inoculation of PGPRs that benefit plants. The review shows how it is possible to evaluate and quantify the benefits promoted by microorganisms in association with plants, such as increased production, development and plant growth and expressiveness in secondary metabolism. Numerous quali-quantitative tools can be used to evaluate the gains achieved by the use of this technology, such as analysis of gene expression and proteomics, production determination by scale and estimates.

Understanding the beneficial effects of PGPs on the interaction processes that make up the soil-microorganism-plant system points to the need for further research on biotechnological development for microorganisms present in soil and inside plants, to leverage and improve technologies developed over the last few decades. This is necessary since the adaptability of organisms to climate and soil conditions is a key point for the positive effect of microbial inoculants.

The authors believe that the new directions of microbial biotechnology will occur on the bioprospecting of new microorganisms in the production system in locu, which will make it more efficient, production of formulations and techniques for applying mixtures of strains that have distinct and complementary secondary action/metabolisms of action. The prospects for this branch of science are a reality, as technical-scientific and productive advances have proven to be efficient when applied to the most diverse agricultural environments, under different types of management and cultivated agricultural species, with economic gains. These results have leveraged the public and private sectors, stimulating the creation of public policies aimed at promoting and encouraging the research and commercial sectors to create, develop and produce biotechnological products aimed at sustainable agricultural production.

References

Abd-Allah EF, Alqarawi AA, Hashem A, Radhakrishnan R, Al-Huqail AA, Al-Otibi FON, Egamberdieva D (2018) Endophytic bacterium *Bacillus subtilis* (BERA 71) improves salt tolerance in chickpea plants by regulating the plant defense mechanisms. J Plant Interact 13(1):37–44

Afzal A, Bano A (2008) Rhizobium and phosphate solubilizing Bacteria improve the yield and phosphorus uptake in wheat (*Triticum aestivum*). Int J Agric Biol 10:85–88

Ahmed S, Nawata E, Sakuratani T (2006) Changes of endogenous ABA and ACC, and their correlations to photosynthesis and water relations in mungbean (*Vigna radiata* (L.) Wilczak cv. KPS1) during waterlogging. Environ Exp Bot 57(3):278–284

Akhtar SS, Amby DB, Hegelund JN, Fimognari L, Großkinsky DK, Westergaard JC, Roitsch T (2020) *Bacillus licheniformis* FMCH001 increases water use efficiency via growth stimulation in both normal and drought conditions. Front Plant Sci 11:297

Ali SZ, Sandhya V, Grover M, Linga VR, Bandi V (2011) Effect of inoculation with a thermotolerant plant growth promoting *Pseudomonas putida* strain AKMP7 on growth of wheat (*Triticum* spp.) under heat stress. J Plant Interact 6(4):239–246

Arcand MM, Schneider KIMD (2006) Plant- and microbial-based mechanisms to improve the agronomic effectiveness of phosphate rock: a review. Ann Braz Acad Sci 78(4):791–807

Archana G, Buch A, Kumar GN (2012) Pivotal role of organic acid secretion by rhizobacteria in plant growth promotion. In: Microorganisms in sustainable agriculture and biotechnology. Springer, Dordrecht, pp 35–53

Arora P, Singh G, Tiwari A (2017) Effect of microbial inoculation in combating the aluminium toxicity effect on growth of *Zea mays*. Cell Mol Biol 63(6):79–82

Ashraf A, Bano A, Ali SA (2019) Characterisation of plant growth-promoting rhizobacteria from rhizosphere soil of heat-stressed and unstressed wheat and their use as bio-inoculant. Plant Biol 21(4):762–769

Baek D, Rokibuzzaman M, Khan A, Kim MC, Park HJ, Yun DJ, Chung YR (2020) Plant-growth promoting *Bacillus oryzicola* YC7007 modulates stress-response gene expression and provides protection from salt stress. Front Plant Sci 10(1646):1–13

Bal HB, Nayak L, Das S, Adhya TK (2013) Isolation of ACC deaminase producing PGPR from rice rhizosphere and evaluating their plant growth promoting activity under salt stress. Plant and Soil 366(1–2):93–105

Banerjee A, Roychoudhury A (2016) Group II late embryogenesis abundant (LEA) proteins: structural and functional aspects in plant abiotic stress. Plant Growth Regul 79(1):1–17

Banik A, Pandya P, Patel B, Rathod C, Dangar M (2018) Characterization of halotolerant, pigmented, plant growth promoting bacteria of groundnut rhizosphere and its in-vitro evaluation of plant-microbe protocooperation to withstand salinity and metal stress. Sci Total Environ 630:231–242

Barka EA, Nowak J, Clément C (2006) Enhancement of chilling resistance of inoculated grapevine plantlets with a plant growth-promoting rhizobacterium, *Burkholderia phytofirmans* strain PsJN. Appl Environ Microbiol 72(11):7246–7252

Barnawal D, Bharti N, Maji D, Chanotiya CS, Kalra A (2012) 1-Aminocyclopropane-1-carboxylic acid (ACC) deaminase-containing rhizobacteria protect *Ocimum sanctum* plants during waterlogging stress via reduced ethylene generation. Plant Physiol Biochem 58:227–235

Belimov AA, Dodd IC, Safronova VI, Shaposhnikov AI, Azarova TS, Makarova NM, Tikhonovich IA (2015) Rhizobacteria that produce auxins and contain 1-amino-cyclopropane-1-carboxylic acid deaminase decrease amino acid concentrations in the rhizosphere and improve growth and yield of well-watered and water-limited potato (*Solanum tuberosum*). Ann Appl Biol 167(1):11–25

Bianchi L, Germino GH, de Almeida SM (2016) Adaptação das plantas ao déficit hídrico. Acta Iguazu 5(4):15–32

Bilal S, Shahzad R, Khan AL, Kang SK, Imran QM, Al-Harrasi A, Yun BW, Lee J (2018) Endophytic microbial consortia of phytohormones-producing fungus *Paecilomyces formosus* LHL10 and bacteria *Sphingomonas* sp.LK11 to *Glycine max* L. regulates physio-hormonal changes to attenuate aluminum and zinc stresses. Front Plant Sci 9:1–18

Card S, Johnson L, Teasdale S, Caradus J (2016) Deciphering endophyte behaviour: the link between endophyte biology and efficacious biological control agents. FEMS Microbiol Ecol 92:1–19

Carvalho TLG, Ballesteros HGF, Thiebaut F, Ferreira PCG, Hemerly AS (2016) Nice to meet you: genetic, epigenetic and metabolic controls of plant perception of beneficial associative and endophytic diazotrophic bacteria in non-leguminous plants. Plant Mol Biol 90(6):561–574

Chandra D, Srivastava R, Gupta VV, Franco CM, Paasricha N, Saifi SK, Sharma AK (2019) Field performance of bacterial inoculants to alleviate water stress effects in wheat (*Triticum aestivum* L.). Plant and Soil 441(1–2):261–281

Chen YP, Rekha PD, Arun AB, Shen FT, Lai WA, Young CC (2006) Phosphate solubilizing bacteria from subtropical soil and their tricalcium phosphate solubilizing abilities. Appl Soil Ecol 34(1):33–41

Congeevaram S, Dhanarani S, Park J, Dexilin M, Thamaraiselvi K (2007) Biosorption of chromium and nickel by heavy metal resistant fungal and bacterial isolates. J Hazard Mater 146(1–2):270–277

Cramer GR, Urano K, Delrot S, Pezzotti M, Shinozaki K (2011) Effects of abiotic stress on plants: a systems biology perspective. BMC Plant Biol 11(1):1–14

Dhawi F (2020) Plant growth promoting Rhizobacteria (PGPR) regulated Phyto and microbial beneficial protein interactions. Open Life Sci 15(1):68–78

Egamberdieva D (2009) Alleviation of salt stress by plant growth regulators and IAA producing bacteria in wheat. Acta Physiol Plant 31(4):861–864
Etesami H, Maheshwari DK (2018) Use of plant growth promoting rhizobacteria (PGPRs) with multiple plant growth promoting traits in stress agriculture: action mechanisms and future prospects. Ecotoxicol Environ Saf 156:225–246
Fan C, Yao H, Qiu Z, Ma H, Zeng B (2018) Genome-wide analysis of *Eucalyptus grandis* WRKY genes family and their expression profiling in response to hormone and abiotic stress treatment. Gene 678:38–48
Farrar K, Bryant D, Cope-Selby N (2014) Understanding and engineering beneficial plant–microbe interactions: plant growth promotion in energy crops. Plant Biotechnol J 12(9):1193–1206
Fernandez O, Theocharis A, Bordiec S, Feil R, Jacquens L, Clément C, Fontaine F, Barka EA (2012) *Burkholderia phytofirmans* PsJN acclimates grapevine to cold by modulating carbohydrate metabolism. Mol Plant Microbe Interact 25:496–504
Fonseca ES, Peixoto RS, Rosado AS, Balieiro SC, Tiedje JM, Rachid CTCC (2017) The microbiome of Eucalyptus roots under different management conditions and its potential for biological nitrogen fixation. Microb Ecol 75(1):183–191
Forni C, Duca D, Glick BR (2017) Mechanisms of plant response to salt and drought stress and their alteration by rhizobacteria. Plant and Soil 410(1–2):335–356
Fukami J, Ollero FJ, Megías M, Hungria M (2017) Phytohormones and induction of plant-stress tolerance and defense genes by seed and foliar inoculation with *Azospirillum brasilense* cells and metabolites promote maize growth. AMB Express 7(1):153
Garcia LG, Ferraz, SDB, Alvares CA, Ferraz KDB, Higa RCV (2014a) Modeling the climatic aptitude of *Eucalyptus grandis* in the face of climate change scenarios in Brazil. Embrapa Florestas-Artigo em periódico indexado (ALICE)
Garcia RA, Cabeza M, Rahbek C, Araújo MB (2014b) Multiple dimensions of climate change and their implications for biodiversity. Science 344(6183):1247579
Gaspar M (2011) Aquaporins: from water channels to multifunctional transporters in plants. Rev Bras Bot 34(4):481–491
Glick BR (2014) Bacteria with ACC deaminase can promote plant growth and help to feed the world. Microbiol Res 169(1):30–39
Goswami M, Deka S (2020) Plant growth-promoting rhizobacteria—alleviators of abiotic stresses in soil: a review. Pedosphere 30(1):40–61
Gray EJ, Smith DL (2005) Intracellular and extracellular PGPR: commonalities and distinctions in the plant–bacterium signaling processes. Soil Biol Biochem 37:395–412
Grichko VP, Glick BR (2001) Amelioration of flooding stress by ACC deaminase-containingplant growth-promoting bacteria. Plant Physiol Biochem 39(1):11–17
Gyaneshwar P, Kumar GN, Parekh LJ, Poole PS (2002) Role of soil microorganisms in improving P nutrition of plants. Plant and Soil 245(1):83–93
Hiltner L (1904) Ueber neuere Erfahrungen und Probleme auf dem Gebiete der Bodenbakteriologie und unter besonderer BerUcksichtigung der Grundungung und Brache. Arb Deut Landw Gesell 98:59–78
Howarth CJ (2005) Genetic improvements of tolerance to high temperature. In: Ashraf M, Harris PJC (eds) Abiotic stresses: plant resistance through breeding and molecular approaches. Howarth Press, New York, pp 277–300
Igual J, Valverde A, Cervantes E, Velázquez E (2001) Phosphate-solubilizing bacteria as inoculants for agriculture: use of updated molecular techniques in their study. Agronomie 21:561–568
IPCC – Intergovernamental Panel on Climate Change (2007) Climate change 2007: mitigation—contribution of working group III to the fourth assessment report of the intergovernmental panel on climate change. Cambridge University Press, Cambridge
Jha Y, Subramanian RB, Patel S (2011) Combination of endophytic and rhizospheric plant growth promoting rhizobacteria in *Oryza sativa* shows higher accumulation of osmoprotectant against saline stress. Acta Physiol Plant 33(3):797–802

Jochum M, McWilliams KM, Borrego E, Kolomiets M, Niu G, Pierson E, Jo YK (2019) Bioprospecting plant growth-promoting rhizobacteria that mitigate drought stress in grasses. Front Microbiol 10:2106

Jorge GL, Kisiala A, Morrison E, Aoki M, Nogueira APO, Emery RN (2019) Endosymbiotic *Methylobacterium oryzae* mitigates the impact of limited water availability in lentil (*Lens culinaris* Medik.) by increasing plant cytokinin levels. Environ Exp Bot 162:525–540

Kaushal M, Wani SP (2016) Rhizobacterial-plant interactions: strategies ensuring plant growth promotion under drought and salinity stress. Agric Ecosyst Environ 231:68–78

Khan N, Mishra A, Chauhan PS, Sharma YK, Nautiyal CS (2012) *Paenibacillus lentimorbus* enhances growth of chickpea (*Cicer arietinum* L.) in chromium-amended soil. Antonie Van Leeuwenhoek 101(2):453–459

Khatri S, Sharma RK, Shridhar V (2020) Influence of cadmium-tolerant and plant growth-promoting Rhizobacteria on cadmium accumulation and growth response of wheat seedlings under mountain ecosystem. Agric Res 9(1):56–65

Kim JS, Lee JE, Nie H, Lee YJ, Kim ST, Kim SH (2018) Physiological and proteomic analysis of plant growth enhancement by the rhizobacteria *Bacillus* sp. JS. Genes Genomics 40(2):129–136

Kloepper JW, Lifshitz R, Zablotowicz RM (1989) Free-living bacterial inocula for enhancing crop productivity. Trends Biotechnol 7:39–44

Kumar A, Verma JP (2018) Does plant - microbe interaction confer stresstolerance in plants: a review? Microbiol Res 207:41–52

Kumar A, Patel JS, Meena VS, Ramteke PW (2019) Plant growth-promoting rhizobacteria: strategies to improve abiotic stresses under sustainable agriculture. J Plant Nutr 42(11–12):1402–1415

Lebeis SL (2015) Greater than the sum of their parts: characterizing plant microbiomes at the community-level. Curr Opin Plant Biol 24:82–86

Lindsay WL, Vlek PLG, Chien SH (1989) Phosphate minerals. In: Dixon JB, Weed SB (eds) Minerals in soil environment, 2nd edn. Soil Science Society of America, Madison, WI, pp 1089–1130

Ma Y, Oliveira RS, Nai F, Rajkumar M, Luo Y, Rocha I, Freitas H (2015) The hyper accumulator *Sedum plumbizincicola* harbors metal-resistant endophytic bacteria that improve its phytoextraction capacity in multi-metal contaminated soil. J Environ Manage 156:62–69

Martínez-Gil M, Ramos-González MI, Espinosa-Urgel M (2014) Roles of cyclic Di-GMP and the Gac system in transcriptional control of the genes coding for the Pseudomonas putida adhesins LapA and LapF. J Bacteriol 196(8):1484–1495

Martins GS, Freitas NC, Máximo WPF, Paiva LV (2018) Gene expression in two contrasting hybrid clones of *Eucalyptus camaldulensis* x *Eucalyptus urophylla* grown under water deficit conditions. J Plant Physiol 229:122–131

Maxton A, Singh P, Masih SA (2018) ACC deaminase-producing bacteria mediated drought and salt tolerance in *Capsicum annuum*. J Plant Nutr 41(5):574–583

McNear DH Jr (2013) The rhizosphere—roots, soil and everything in between. Nat Educ Knowl 4(3):1

Miguel PSB, de Oliveira MNV, Delvaux JC, de Jesus GL, Borges AC, Tótola MR, Costa MD (2016) Diversity and distribution of the endophytic bacterial community at different stages of Eucalyptus growth. Antonie Van Leeuwenhoek 109(6):755–771

Mukhtar T, Smith D, Sultan T, Seleiman MF, Alsadon AA, Ali S, Chaudhary HJ, Solieman THI, Ibrahim AA, Saad MA (2020). Mitigation of heat stress in *Solanum lycopersicum* L. by ACC-deaminase and exopolysaccharide producing *Bacillus cereus*: Effects on biochemical profiling. Sustainability, 12(6); 2159

Moreira H, Marques AP, Franco AR, Rangel AO, Castro PM (2014) Phytomanagement of cd-contaminated soils using maize (*Zea mays* L.) assisted by plant growth-promoting rhizobacteria. Environ Sci Pollut Res 21(16):9742–9753

Nascimento F, Brígido C, Alho L, Glick BR, Oliveira S (2012) Enhanced chickpea growth-promotion ability of a *Mesorhizobium* strain expressing an exogenous ACC deaminase gene. Plant and Soil 353(1–2):221–230

Naseem H, Bano A (2014) Role of plant growth-promoting rhizobacteria and their exopolysaccharide in drought tolerance of maize. J Plant Interact 9(1):689–701

Norrish K, Rosser H (1983) Mineral phosphate. In: Soils: an Australian viewpoint. Academic Press, CSIRO, Melbourne, London, pp 335–361

Oves M, Khan MS, Zaidi A (2013) Chromium reducing and plant growth promoting novel strain *Pseudomonas aeruginosa* OSG41 enhance chickpea growth in chromium amended soils. Eur J Soil Biol 56:72–83

Patel KJ, Singh AK, Nareshkumar G, Archana G (2010) Organic-acid-producing, phytate-mineralizing rhizobacteria and their effect on growth of pigeon pea (*Cajanus cajan*). Appl Soil Ecol 44(3):252–261

Peleg Z, Blumwald E (2011) Hormone balance and abiotic stress tolerance in crop plants. Curr Opin Plant Biol 14(3):290–295

Porter JR (2005) Rising temperatures are likely to reduce crop yields. Nature 436(7048):174–174

Poupin MJ, Greve M, Carmona V, Pinedo I (2016) A complex molecular interplay of auxin and ethylene signaling pathways is involved in Arabidopsis growth promotion by *Burkholderia phytofirmans* PsJN. Front Plant Sci 7:492

Pourbabaee AA, Bahmani E, Alikhani HA, Emami S (2016) Promotion of wheat growth under salt stress by halotolerant bacteria containing ACC deaminase. J Agric Sci Technol 18(3):855–864

Rabhi NEH, Silini A, Cherif-Silin H, Yahiaoui B, Lekired A, Robineau M, Aït Barka E (2018) *Pseudomonas knackmussii* MLR6, a rhizospheric strain isolated from halophyte, enhances salt tolerance in *Arabidopsis thaliana*. J Appl Microbiol 125(6):1836–1851

Richardson AE, Simpson RJ (2011) Soil microorganisms mediating phosphorus availability. Plant Physiol 156:989–996

Saikia J, Sarma RK, Dhandia R, Yadav A, Bharali R, Gupta VK, Saikia R (2018) Alleviation of drought stress in pulse crops with ACC deaminase producing rhizobacteria isolated from acidic soil of Northeast India. Sci Rep 8(1):1–16

Sairam RK, Kumutha D, Ezhilmathi K, Chinnusamy V, Meena RC (2009) Waterlogging induced oxidative stress and antioxidant enzyme activities in pigeon pea. Biol Plant 53(3):493–504

Salvato F, de Carvalho MCCG (2010) Methods and strategies in proteomics and their applications in plants. Cienc Rural 40(3):727–734

Saravanakumar D, Samiyappan R (2007) ACC deaminase from *Pseudomonas fluorescens* mediated saline resistance in groundnut (*Arachis hypogea*) plants. J Appl Microbiol 102(5):1283–1292

Sarim KM, Srivastava R, Ramteke PW (2020) Next-generation omics technologies for exploring complex metabolic regulation during plant-microbe interaction. Elsevier, Amsterdam

Sharma SB, Sayyed RZ, Trivedi MH, Gobi TA (2013) Phosphate solubilizing microbes: sustainable approach for managing phosphorus deficiency in agricultural soils. Springerplus 2:587

Shreya D, Jinal HN, Kartik VP, Amaresan N (2020) Amelioration effect of chromium-tolerant bacteria on growth, physiological properties and chromium mobilization in chickpea (*Cicer arietinum*) under chromium stress. Arch Microbiol:1–8

Silva MCS, Mendes IR, Paula TA, Dias RS, Paula SO, Silva CC, Kasuya MCM (2016) Expression of the nifH gene in diazotrophic bacteria in *Eucalyptus* urograndis plantations. Can J For Res 46:190–199

Singh RP, Runthala A, Khan S, Jha PN (2017) Quantitative proteomics analysis reveals the tolerance of wheat to salt stress in response to *Enterobacter cloacae* SBP-8. PLoS One 12(9):e0183513

Song OR, Lee SJ, Lee YS, Lee SC, Kim KK, Choi YL (2008) Solubilization of insoluble inorganic phosphate by *Burkholderia cepacia* DA23 isolated from cultivated soil. Braz J Microbiol 39(1):151–156

Sparks DL (2003) Environmental soil chemistry. Academic Press, San Diego, p 352

Szklarczyk D, Gable AL, Lyon D, Junge A, Wyder S, Huerta-Cepas J, Jensen LJ (2019) STRING v11: protein–protein association networks with increased coverage, supporting functional discovery in genome-wide experimental datasets. Nucleic Acids Res 47(D1):D607–D613

Tahir M, Khalid U, Khan MB, Shahid M, Ahmad I, Akram M, Ahmad N (2019) Auxin and 1-Aminocyclopropane-1-carboxylate deaminase activity exhibiting rhizobacteria improved maize quality and productivity under drought conditions. Int J Agric Biol 21:943–954

Tang Q, Puri A, Padda KP, Chanway CP (2017) Biological nitrogen fixation and plant growth promotion of lodgepole pine by an endophytic diazotroph *Paenibacillus polymyxa* and its GFP-tagged derivative. Botany 95:611–619

Tewari S, Arora NK (2014) Multifunctional exopolysaccharides from *Pseudomonas aeruginosa* PF23 involved in plant growth stimulation, biocontrol and stress amelioration in sunflower under saline conditions. Curr Microbiol 69(4):484–494

Theocharis A, Bordiec S, Fernandez O, Paquis S, Dhondt-Cordelier S, Baillieul F, Clément C, Barka EA (2012) *Burkholderia phytofirmans* PsJN primes *Vitis vinifera* L. and confers a better tolerance to low nonfreezing temperatures. *Mol.* Plant Microbe Interact 25:241–249

Thomashow MF (1999) Plant cold acclimation: freezing tolerance genes and regulatory mechanisms. Annu Rev Plant Biol 50(1):571–599

Trivedi P, Sa T (2008) *Pseudomonas corrugata* (NRRL B-30409) mutants increased phosphate solubilization, organic acid production, and plant growth at lower temperatures. Curr Microbiol 56(2):140–144

Trivedi P, Pandey A, Palni LMS (2012) Bacterial inoculants for field applications under mountain ecosystem: present initiatives and future prospects. In: Maheshwari DK (ed) Bacteria in agrobiology: plant probiotics. Springer, Berlin, pp 15–44

Ullah S, Bano A (2015) Isolation of PGPRS from rhizospheric soil of halophytes and its impact on maize (*Zea mays* L.) under induced soil salinity. Can J Microbiol 61(4):307–313

Ullah A, Heng S, Munis MFH, Fahad S, Yang X (2015) Phytoremediation of heavy metals assisted by plant growth promoting (PGP) bacteria: a review. Environ Exp Bot, v. 117: 28–40.

Ullah A, Sun H, Yang X, Zhang X (2018) A novel cotton WRKY gene, GhWRKY6-like, improves salt tolerance by activating the ABA signaling pathway and scavenging of reactive oxygen species. Physiol Plant 162:439–454

Ullah A, Nisar M, Ali H, Hazrat A, Hayat K, Keerio AA, Khan A (2019) Drought tolerance improvement in plants: an endophytic bacterial approach. Appl Microbiol Biotechnol 103(18):7385–7397

Upadhyay SK, Singh JS, Saxena AK, Singh DP (2012) Impact of PGPR inoculation on growth and antioxidant status of wheat under saline conditions. Plant Biol 4(4):605–611

Vaishnav A, Kumari S, Jain S, Varma A, Choudhary DK (2015) Putative bacterial volatile-mediated growth in soybean (*Glycine max* L. Merrill) and expression of induced proteins under salt stress. J Appl Microbiol 119(2):539–551

Vimal SR, Singh JS, Arora NK, Singh S (2017) Soil-plant-microbe interactions in stressed agriculture management: a review. Pedosphere 27(2):177–192

Vurukonda SSKP, Vardharajula S, Shrivastava M, SkZ A (2016) Enhancement of drought stress tolerance in crops by plant growth promoting rhizobacteria. Microbiol Res 184:13–24

Wang W, Wu Z, He Y, Huang Y, Li X, Ye BC (2018) Plant growth promotion and alleviation of salinity stress in *Capsicum annuum* L. by *Bacillus* isolated from saline soil in Xinjiang. Ecotoxicol Environ Saf 164:520–529

Wani PA, Zaidi A, Khan MS (2009) Chromium reducing and plant growth promoting potential of *Mesorhizobium* species under chromium stress. Biorem J 13(3):121–129

Wiegand CL, Gerbermann AH, Cuellar JA (1981) Development and yield of hard red winter Wheats under semitropical conditions 1. Agron J 73(1):29–37

Woo OG, Kim H, Kim JS, Keum HL, Lee KC, Sul WJ, Lee JH (2020) *Bacillus subtilis* strain GOT9 confers enhanced tolerance to drought and salt stresses in *Arabidopsis thaliana* and *Brassica campestris*. Plant Physiol Biochem 148:359–367

Xia Y, Farooq MA, Javed MT, Kamran MA, Mukhtar T, Ali J, Chaudhary HJ (2020) Multi-stress tolerant PGPR *Bacillus xiamenensis* PM14 activating sugarcane (*Saccharum officinarum* L.) red rot disease resistance. Plant Physiol Biochem 151:640–649

Xu J, Li XL, Luo L (2012) Effects of engineered *Sinorhizobium meliloti* on cytokinin synthesis and tolerance of alfalfa to extreme drought stress. Appl Environ Microbiol 78(22):8056–8061

Yang J, Kloepper JW, Ryu CM (2009) Rhizosphere bacteria help plants tolerate abiotic stress. Trends Plant Sci 14(1):1–4

Yao L, Wu Z, Zheng Y, Kaleem I, Li C (2010) Growth promotion and protection against salt stress by *Pseudomonas putida* Rs-198 on cotton. Eur J Soil Biol 46(1):49–54

Yaoyao E, Yuan J, Yang F, Wang L, Ma J, Li J, Shen Q (2017) PGPR strain *Paenibacillus polymyxa* SQR-21 potentially benefits watermelon growth by re-shaping root protein expression. AMB Express 7(1):1–12

Chapter 3
Seaweeds as Indicators and Potential Remediators of Metal Pollution

Josef Jampílek and Katarína Kráľová

Contents

1 Introduction

Marine macroalgae or seaweeds are multi-cellular autotrophic organisms mostly inhabiting shallow waters and generally live attached to rock or other hard substrata in coastal areas using holdfast. They absorb water and nutrients in all their tissues, directly from the surrounding water and perform photosynthesis also in all their tissues (PRNSA 2013; Seaweeds 2020). Consequently, they can grow only when the penetrating light is sufficient for photosynthesis, and therefore in clear waters, macroalgae can survive and grow at depths of >200 m but in muddy water, this is only

J. Jampílek (✉)
Department of Analytical Chemistry, Faculty of Natural Sciences, Comenius University, Bratislava, Slovakia
Regional Centre of Advanced Technologies and Materials, Czech Advanced Technology and Research Institute, Palacky University, Olomouc, Czech Republic

K. Kráľová
Institute of Chemistry, Faculty of Natural Sciences, Comenius University, Bratislava, Slovakia

© The Author(s), under exclusive license to Springer Nature Switzerland AG 2021

H. I. Mohamed et al. (eds.), *Plant Growth-Promoting Microbes for Sustainable Biotic and Abiotic Stress Management*,
https://doi.org/10.1007/978-3-030-66587-6_3

a few meters (Seaweeds 2020). Macroalgae can produce even 2–14 kg of organic carbon per square meter on an annual basis, which is pronouncedly more compared to terrestrial plants in temperate climates (ca. 1 kg) (Mouritsen 2013). Moreover, seaweeds contain high mineral content which could achieve even tenfold greater values compared to that of terrestrial plants, accompanied with a wide variety of vitamins, and therefore several of them can serve as a food source for consumption by humans or as feed for animals (O'Sullivan et al. 2010; Kraan 2013; Mouritsen 2013; Cherry et al. 2019). Over 10,000 species of seaweeds are known, and based on their unique coloration affected by pigments (chlorophyll and accessory pigments) they could be classified into three different groups: brown algae (Pheophyta), red algae (Rhodophyta) and green algae (Chlorophyta) (PRNSA 2013). From seaweeds more than 1800 species belong to Chlorophyta, ca. 2000 species to Phaeophyta and over 7200 species to Rhodophyta (Seaweeds 2020). Some widespread genera of brown, red and green seaweeds are presented in Fig. 3.1.

Seaweeds inhabiting marine environment are exposed to various abiotic stresses such as increasing temperature, high light, UV light or toxic metals and eventually repeated desiccation-rehydration cycles in intertidal species adversely affecting their growth and development (Schmidt et al. 2012; dos Santos et al. 2014; Ramesh et al. 2015; Zhu et al. 2015; Farias et al. 2017b; Tala et al. 2017; Scherner et al. 2018; Capdevila et al. 2019; Quintano et al. 2019; Terada et al. 2020; Thomsen et al. 2019). Exposure of many seaweed species to strong solar irradiation results in reversible photoinhibition (Betancor et al. 2015; Tala et al. 2017). However, high light alleviated inhibitory effects of a high temperature of 30 °C on the growth and physiological parameters of *Ulva prolifera* (Jiang et al. 2020). Biomass yield and photosynthetic performance of seaweeds pronouncedly decrease at exposure to metal excess, accompanied with declined levels of photosynthetic pigments and ultrastructural alterations of chloroplasts (dos Santos, et al. 2014; Farias et al. 2017b; Scherner et al. 2018; Celekli and Bulut 2020; Dong et al. 2020). Moreover,

Fig. 3.1 Some widespread genera of brown, red and green seaweeds

metal amounts bioaccumulated in seaweeds are excellent indicators of heavy metal pollution (Okuku and Peter 2012; Reis et al. 2014; Chalkley et al. 2019; Sun et al. 2019; Bonanno et al. 2020) and are frequently in good correlation with the metal content in the surrounding seawaters (Okuku and Peter 2012; Reis et al. 2014; Bonanno et al. 2020) and sediment (Rybak et al. 2013; Malea and Kevrekidis 2014). These characteristics could be used for monitoring the extent of metal pollution. Moreover, as a response to metal stress, seaweeds evolved several defense mechanisms including cellular exclusion of metals, preparation of metal-chelating compounds and the activation of the antioxidant system (Moenne et al. 2016). Heavy metal caused oxidative stress in seaweeds is accompanied with increased generation of reactive oxygen species (ROS), and therefore for monitoring of metal-polluted aqueous environment also the levels of malondialdehyde (MDA) and non-enzymatic antioxidants as well as activities of antioxidant enzymes such as superoxide dismutase (SOD), catalase (CAT), peroxidase (POD), etc. can be used (Kumar et al. 2012; Saez et al. 2015b; Zhu et al. 2017; Costa et al. 2019; Rodriguez-Rojas et al. 2019).

Seaweeds can concentrate metals from seawater due to the presence of numerous functional groups in the cell wall that interact with metal ions, and therefore they could be applied for monitoring of metal pollution in coastal waters and estuaries as well as for bioremediation purposes (Ali et al. 2017; Chalkley et al. 2019; Sun et al. 2019). Living as well as dead biomass of seaweeds represents a cheap, easily available and effective biosorbent able to uptake and accumulate toxic metals from wastewater via metabolic process or through physicochemical pathways (Cid et al. 2015; do Nascimento et al. 2019). In living macroalgae, toxic metals are firstly adsorbed onto the cell surface (biosorption) followed by intracellular uptake. Whereas biosorption is a metabolically passive process and effectiveness of metal removal depends on kinetic equilibrium and the composition of the sorbent's cellular surface, bioaccumulation is an active metabolic process driven by energy from a living organism (Volesky 2007; Kanamarlapudi et al. 2018; Shamim 2018). For example, the photosynthesis of seaweed increased bioaccumulation and bioabsorption from Cd-containing environment (Han et al. 2020). Biosorption occurs at a faster rate than bioaccumulation, and it is a reversible process because metals bound onto the cellular surface could be removed, whereas bioaccumulation is only partially reversible. Therefore, for the elimination of toxic metals by environmental remediation, biosorption is favorable because metal biosorption occurs also on dead seaweed biomass (Vijayaraghavan et al. 2012a; Gabruk et al. 2015).

The efficiency of removing toxic metals from wastewater depends on the pH, biosorbent amount, metal ion concentration and contact time of the biosorption process (Deniz and Ersanli 2018). Phaeophyta was reported to have high heavy metal adsorption capacity, while Chlorophyta and Rhodophyta have moderate adsorption capacities, whereby dead algal biomass was found to be more effective in practical applications than living algae (Lin et al. 2019). Biosorption using seaweed can be considered as a good alternative to the existing technologies' ineffective removal of toxic metals from wastewater (Kaur et al. 2012; Arumugam et al. 2018). Alginate extraction residue from seaweeds (Moino et al. 2017; Nishikawa et al. 2018; Filote et al. 2019), as well as biochars derived from wasted marine macro-algae (Kidgell et al. 2014; Yang et al. 2018; Saravanakumar et al. 2020), were also successfully applied for heavy metal removal.

This chapter presents an overview of findings related to the photosynthetic processes in seaweeds and impact of toxic metals on these processes, the feasibility of seaweeds application as indicators of toxic metal pollution as well as the use of living and dead seaweed biomass as sorbents for phytoremediation purposes. Nutritional benefits of edible seaweeds are discussed as well.

2 Impact of Environmental Conditions on Photosynthetic Processes in Seaweeds

*S*eaweeds as photosynthesizing organisms convert sunlight into chemical energy that is then stored in carbohydrates, and under normal conditions, photosynthesis is the dominant process, allowing the seaweeds to build up their carbohydrate content. Seaweeds utilize sunlight more efficiently than terrestrial plants because the access to light in the water decreases with the increasing distance from the surface of the sea and turbidity of waters (Millar 2011). Light-harvesting chlorophylls (Chls) absorb light in the red (> 640 nm) and blue (< 440 nm) regions of the light spectrum, whereby chlorophylls in reaction centers of photosystem (PS) II and PSI absorb light at 700 and 680 nm, respectively. Chl*c* absorbs moderately in red region, i.e. around 620 nm, but approximately tenfold more strongly in 400–450 nm (Lal 2018). Accessory pigments of light-harvesting systems, namely, carotenoids, phycobilipoteins, phycocyanin and phycoerythrin, absorb light at wavelength, where Chls do not function efficiently (470–630 nm) (Chan 2003). On the other hand, seaweed species that live at the ocean surface may also contain pigments that protect them from the ultraviolet radiation. However, it could be mentioned that seaweeds can photosynthesize to a certain extent also when they are exposed to air and are partially dehydrated (Mouritsen 2013).

2.1 Chlorophyta

The green algae division is the largest, insofar as some species is concerned. It includes 500 genera and 8000 species: single-cell, multi-cell and sometimes multinuclear. Most of the species develop in freshwater and only one-tenth in the sea (Einav 2020). These algae are typically green in color due to the presence of Chl*a* and Ch*b* in their chloroplasts. Their overall colouration depends on the balance between the Chls and other pigments such as β-carotene and xanthophylls. Storage product of green algae is starch (amylose and amylopectin), and their cell wall consists of cellulose, hydroxyproline, glucosides, xylans and mannans (Ortiz-Calderon et al. 2017). Main genera include *Codium*, *Enteromorpha*, *Ulva*, *Chaetomorpha* and *Cladophora*. Green algae are common in areas where light is abundant, such as shallow waters and tide pools (Heuzé et al. 2017).

Gao et al. (2019) at studying structure and organization of the thylakoid membranes in *Ulva prolifera* found that the macroalga had few but long loosely stacked membranes, which lack the conventional grana found in vascular plants and the thylakoid membrane complexes demonstrated lateral heterogeneity. Under salt stress, a supercomplex composed of PSII, light-harvesting complex (LHC) II and PSI of ca. 720 kDa contained two important photoprotection proteins, the PSII S subunit and the light-harvesting complex stress-related protein, as well as xanthophyll cycle pigments (violaxanthin, antheraxanthin and zeaxanthin), whereby the excitation energy was efficiently transferred from PSII to PSI, even when PSII was inhibited. Xu and Gao (2016) demonstrated that green, red and brown macroalgae were capable of utilizing UV-A irradiance to drive photosynthetic carbon fixation also in the absence of photosynthetically active radiation (PAR) and the gross photosynthetic rates ranged from 6.5±0.3 to 52.3±1.8 μmol C/g (fresh weight) per hour, the highest rate being estimated in the green alga *Ulva lactuca* Linnaeus. The ratio of gross photosynthesis driven by UV-A alone to that observed by saturating PAR varied from 14% to 22%.

Ulva compressa as an intertidal species tolerates repeated desiccation-rehydration cycles in nature. Desiccation to 73% relative water content (RWC) resulted in ca. 50% decrease of the maximum quantum yield of PSII, while relative electron transport rates were stimulated. Higher desiccation up to 48 or 27% RWC resulted in a strong decrease of relative electron transport rates. The flexibility of the pectin-rich cell wall layers was assumed to be a major contribution to desiccation tolerance in *Ulva* (Holzinger et al. 2015).

In *Ulva lactuca* treated with 1 and 5 mM Cr^{6+} solution, a decrease of the F_v/F_m ratio was observed 2 h after treatment with 1 and 5 mM Cr^{6+}, the cell viability decreased, and at exposure to 1 and 5 mM Cr^{6+} also the rate of necrotic cells was found to increase by ca. 76.93 and 84.23%, respectively (Unal et al. 2010).

2.2 *Rhodophyta*

Most marine red algae species occur from low tide marks to 100 m depth. In exceptionally clear water, some of them can grow as far as 250 m below the surface of the sea, and a calcareous red alga was found even at a depth of 268 m, where only 0.0005% of the sunlight penetrates. Major red algae genera include *Pyropia*, *Porphyra*, *Chondrus* and *Palmaria* (Mouritsen 2013). From ca. 6500 red algae, the prevalent majority of which are marine algae found in the intertidal and the subtidal to depths of up to 40, or occasionally, 250 m (Kim 2011). Red macroalgae are benthic, but their distinctive combination of photosynthetic pigments allows them to survive in deeper water (Parrott 2017).

The unique coloration of red algae originates from their two accessory pigments, the bluish phycocyanin and the reddish phycoerythrin, which absorb longer bluish wavelengths of light and this allows them to grow in deep waters where longer wavelengths of light can penetrate; red algae contain also allophycocyanin. Ch*a* is the only type of chlorophyll, and zeaxanthin (and sometimes lutein) is the predominant carotenoid besides α- and β-carotene and several xanthophylls. Cell wall of red algae consists of cellulose, xylans, several sulfated polysaccharides (galactans) and calcification in some species. Storage product of red algae is amylopectin-like floridean starch (Ortiz-Calderon et al. 2017). Red algae have two types of light-harvesting antennas, the phycobilisome (PBS), which is directly connected to the reaction centers of PSII, and a LHCI complex connected to the reaction centers of PSI (Gantt et al. 2003). The main reserves of red algae are typically floridean starch, and floridoside and their walls are made of cellulose, agars and carageenans.

In *Gracilaria tenuistipitata* exposed to Cd and Cu at EC_{50} concentrations, differences in gene expression and response patterns were estimated. Cd up-regulated the expressions of SOD and the nitrate transporter (NRT) even after 6 days of exposure, and expressions of both nuclear and chloroplast-encoded proteins were affected. However, after 6 days of exposure to Cu, a slower acclimation was detected, and acclimation over time was observed also based on the analysis of the photosynthetic rate suggesting tolerance of this macroalga to tested metal ions (Tonon et al. 2018).

The investigation of responses of *Gracilaria lemaneiformis* (Gracilariales, Rhodophyta) macroalga, which was incubated at 20 °C and 24 °C and at CO_2 levels of 390 ppm and 700 ppm, respectively, showed that increased temperature affected the growth and photosynthesis of the seaweed much more than increased CO_2 concentrations. Consequently, the ongoing climate change accompanied by increasing atmospheric CO_2 concentrations and sea surface temperatures will be reflected in improved growth and carbon sequestration of this macroalga (Liu et al. 2018a). Comparison of the photosynthetic performances of PSI and PSII measured in different generations of *Pyropia yezoensis* (leafy thalli and filamentous thalli) exposed to air containing enhanced CO_2 concentrations showed that in contrast to filamentous thalli, in the leafy thalli the increasing CO_2 concentration was accompanied with an increase of the effective photochemical quantum yield of PSII (Φ_{PSII}) also under moderate stress (dehydration and salt treatment) suggesting that CO_2 in the air was utilized directly and CO_2 uptake and biomass, as well as the expression of genes involved in the Calvin cycle of *P. yezoensis* leafy thalli, was higher than that in filamentous thalli, likely due to its different carbon utilization mechanism and the adaptation to intertidal environment (Huan et al. 2018).

At studying photosynthetic electron flow during desiccation and re-hydration of the intertidal macroalgae *Porphyra haitanensis*, it was found that cyclic electron flow around PSI was still active after inactivation of linear electron flow following severe desiccation at absolute water content <24%, and after re-hydration the PSI activity was restored more rapidly than that of PSII. It was supposed that reduced nicotinamide adenine dinucleotide phosphate (NADPH) accumulated during desiccation can serve as an electron donor for $P700^+$ and stimulate its recovery during re-hydration, thereby favoring the operation of cyclic electron flow (Gao et al. 2013).

2.3 Phaeophyta

Brown macroalgae (Phaeophyta) are benthic macroalgae that form the primary element of coastal ecosystems in temperate and cold waters. At present, approx. 2000 species in approx. 300 genera are described. Main genera include *Sargassum*, *Fucus*, *Lessonia*, *Turbinaria* and *Cystoseira*. Photosynthetic pigments of brown macroalgae are Chl*a*, Chl*c*, fucoxanthin, α-, β- and γ-carotenes and several xanthophylls, and these macroalgae produce laminaran (β-1,3-glucopyranosid) as the storage polysaccharide; their cell walls consist of alginates, fucoidan (fucan) and cellulose, and is traversed by plasmodesmata (Davis et al. 2003; Kawai and Henry 2017; Ortiz-Calderon et al. 2017). The accessory pigment found in the chloroplasts of brown macroalgae, fucoxanthin ($C_{42}H_{58}O_6$), is a xantophyll significantly absorbing light primarily in the blue-green to yellow-green part of the visible spectrum (450–540 nm). It acts as antennae for light-harvesting and energy transmission in the photosystem LHCs.

Green thylakoid membrane fraction isolated from the brown seaweed *Ecklonia radiata* had Chl*a*/P700 ratio of 80 and spectral characteristics similar to those of P700-Chl*a*-protein complex of green plants. In the membrane fraction rich in Chl*c* and fucoxanthin, both these constituents transferred effectively their energy to Chl*a*, functioning similarly to the light-harvesting Chl*a*/*b*-protein complex of vascular plants (Barett and Anderson 1980). Impact of mono and divalent cations on PSII activity of phaeoplasts prepared from *Fucus* sp. was reported by Berkaloff and Duval (1980). The relative electron transport rate of PSII of five *Sargassum* species (Fucales), *S. piluliferum*, *S. patens*, *S. fusiforme*, S. *crispifolium* and *S. alternato-pinnatum*, was found to increase with increasing temperature, being greatest at 28–30 °C, while at temperatures >32 °C showed a decrease (Tsuchiya et al. 2012).

Bright light affected the rate of electron flow from the reaction center Chl to the secondary electron acceptor, redox-active quinone B (Q_B), in the pelagic phaeophyte, *Sargassum natans* (L.) Gaillon, whereby the proportion of Q_B bound to the D_1/D_2 complex in PSII decreased during the protracted periods of bright light indicating that *S. natans* tolerated high irradiances by down-regulating its quantum yield during the day, decreasing its functional absorption coefficient via the uncoupling of LHCs, and suppressing the efficiency with which absorbed light was utilized (Schofield et al. 1998). Wang et al. (2020) who investigated the toxic impact Co on *Saccharina japonica* found that the spore germination percentage pronouncedly decreased after 1 and 2 days of exposure to 1000 μg Co/L, the gametophyte growth was considerably suppressed after 9 days of exposure to 1 μg Co/L and the relative growth rate and maximum quantum yield (F_v/F_m) of PSII of the juvenile sporophytes after 2 weeks of exposure to Co concentrations ≥10 μg/L were markedly reduced.

Figueroa et al. (2019) reported that warming can increase photoprotection through yield loss and decrease the photosynthetic activity in *Fucus serratus* (Ochrophyta) macroalgae. In *Sargassum fusiforme*, a brown seaweed species inhabiting lower intertidal zones, where algae are often exposed to various stresses, the

PSII activity declined markedly under salinities of 4.5 and 6%, although the PSI activity did not change pronouncedly suggesting that PSI was much more tolerant to saline stress than PSII (Gao et al. 2016).

Adaptation to high light irradiances was found to enhance the photosynthetic Cu^{2+} resistance in Cu^{2+} tolerant and non-tolerant populations of the brown macroalgae *Fucus serratus*. Cu^{2+} did not induce dynamic photoinhibition suggesting that the xanthophyll cycle was probably not involved in the protection of photosynthetic apparatus against the toxic effect of Cu^{2+} ions. The higher photosynthetic Cu^{2+} resistance of high light algae did not result in increased growth (Nielsen and Nielsen 2010).

3 Inhibition of Photosynthetic Electron Transport in Algal PSII by Metal Ions

Many metal ions when applied in excess harm photosynthetic processes of both algae and vascular plants. ROS generated by ions of heavy metals causing lipid peroxidation strongly damage the photosynthetic apparatus of photosynthesizing organisms (Pinto et al. 2003; Kumar et al. 2012; Moenne et al. 2016; Kráľová et al. 2019). To the reduced rate of photosynthetic electron transport (PET) in metal-treated photosynthesizing organisms can contribute also decreased levels of photosynthetic pigments such as Chl*a* and pigments contained in light-harvesting systems (Chl*b* and carotenoids in Chlorophyta, phycocyanin and phycoerythrin in Rhodophyta a Chl*c* and fucoxanthin in Pheophyta (e.g., Han et al. 2008; dos Santos et al. 2014; Patel et al. 2018; Celekli and Bulut 2020; Dong et al. 2020). These pigments absorb the solar energy and transfer it to the reaction centers of PSII and PSI, where charge separation takes place resulting in the conversion of light energy into electrochemical potential. The $P680^{\bullet+}$ species generated in the PSII reaction center drives the splitting of water at the water oxidizing center (Whitmarsh and Govindjee 1999; Barber and Tran 2013). In photosynthesizing organisms following redox components of PSII are known to be involved in transferring electrons from H_2O to the plastoquinone pool, namely, the water oxidizing manganese cluster (Mn_4CaO_5), the amino acid tyrosine (Y_Z), the reaction center Chl (P680), pheophytin and two plastoquinone molecules, Q_A and Q_B (Whitmarsh 1998). Figure 3.2 shows the structure of PSII of vascular plants corresponding also to the PSII of green algae.

The PSII extrinsic proteins shield the catalytic Mn_4CaO_5 cluster from the outside bulk solution and enhance the binding of inorganic cofactors, such as Ca^{2+} and Cl^-, in the oxygen-evolving complex (OEC) (Barber 1999, 2016), whereby PSII extrinsic proteins specific to higher plants and green algae are PsbP and PsbQ (Allen et al. 2011; Ifuku and Noguchi 2016) (Fig. 3.2), while red algae and diatoms have unique PSII extrinsic proteins, such as PsbQ′ and Psb31 (Ifuku and Noguchi 2016), and PsbO (required for stabilization of the water-splitting complex of PSII), which is commonly found in all oxygenic organisms, is most strongly bound to PSII and stabilizes the Mn cluster (Ifuku and Noguchi 2016). The control of the energy

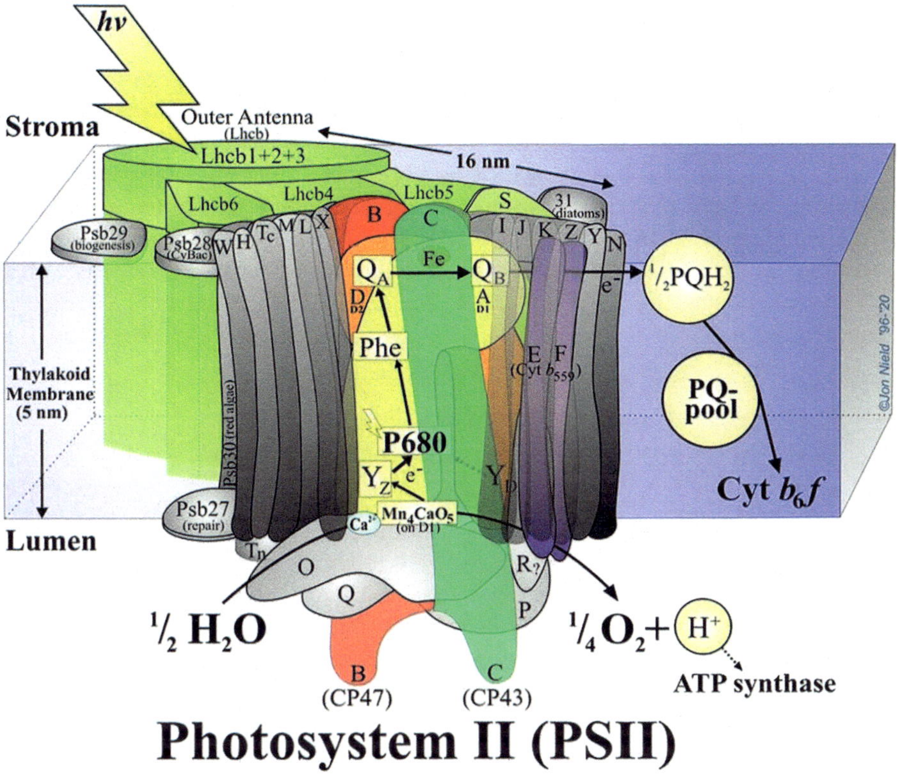

Fig. 3.2 Structure of photosystem II

distributions of PSI and PSII in red algae occurs by changing the energy transfer among PBS, PSI and PSII, and it was concluded that spillover occurs in vivo in PBS–PSII–PSI mega complexes of both cyanobacteria and red algae (Ueno et al. 2016).

As the possible sites of action of toxic metal ions (e.g. Cu^{2+}, Cd^{2+}, Zn^{2+}, Ni^{2+}) causing at higher concentrations inhibition of PET in PSII (i) the OEC (releasing Mn^{2+} ions from manganese cluster); (ii) the tyrosine intermediates Yz and Y_D situated on the donor side of PSII in D_1 and D_2 proteins or their close vicinity; (iii) the core of PSII (P680); (iv) redox-active quinones Q_A and Q_B on the acceptor side of PSII; (v) displacement of the non heme Fe^{2+} on the acceptor side of PSII by toxic metal ions were reported (Fig. 3.2; in detail see in Masarovičová et al. 2010; Kráľová et al. 2019).

Fluorescence spectroscopy is a suitable method to investigate the degree of PET inhibition caused by metal ions both in vitro and in vivo. The inhibition of PET is reflected in reduced maximum quantum yield (F_v/F_m) of PSII, where F_v is variable fluorescence and F_m maximum fluorescence (Holzinger et al. 2015; Zhang et al. 2017; Wang et al. 2020). Using fluorescence measurements also interaction of metal ions with Chl*a* and aromatic amino acids contained in pigment-protein complexes can be estimated (Masarovičová et al. 2010; Kráľová et al. 2019). The sites of action of metal ions in the photosynthetic apparatus can be determined, for example, using EPR spectroscopy (Masarovičová et al. 2010; Šeršeň and Kráľová 2013; Kráľová et al. 2019).

Inhibition of the PET in algal PSII by toxic metal ions was observed in many fresh and marine microalgal species (Oukarroum et al. 2012; Ou-Yang et al. 2013; Huang et al. 2016; Cabrita et al. 2018; Ji et al. 2018; Gan et al. 2019; Kráľová and Jampílek 2021; Chen et al. 2016; Jiang et al. 2016; Yong et al. 2018) as well as in seaweeds (Unal et al. 2010; Zakeri and Abu Bakar 2013; Wang et al. 2020).

4 Responses of Seaweeds to Toxic Metals Used as Biomarkers for Monitoring Metal Pollution

Excess concentrations of toxic metals generating harmful ROS and causing oxidative stress adversely affect growth and development of freshwater and marine micro- and macroalgae reflected in the inhibition of photosynthetic processes, reduced growth, altered algal cell ultrastructure, up- or down-regulation of gene expression and increased accumulation of toxic metals in algal biomass (e.g. Cabrita et al. 2018; Ji et al. 2018; Gan et al. 2019; Kráľová and Jampílek 2021). Sensitive responses of seaweeds to the presence of toxic metals enable to use their morphological and physiological characteristics (Fig. 3.3) for monitoring of metal pollution in the marine environment originating predominantly from anthropogenic activities in coastal regions.

Endemic seaweed *Halimeda jolyana* (Bryopsidales, Chlorophyta) from tropical southwestern Atlantic reefs responded to metal exposure by a pronounced decrease of biomass yield, and its photosynthetic performance declined. Partial recovery of

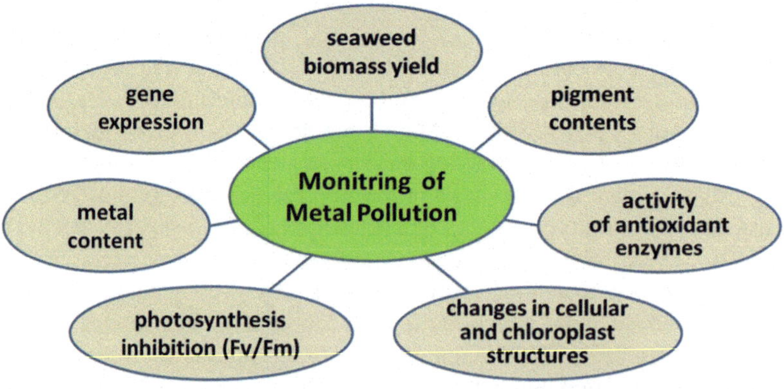

Fig. 3.3 Morphological and physiological characteristics suitable for monitoring of metal pollution using seaweeds

photosynthesis was observed only at a higher temperature when metal enrichment was ceased suggesting higher vulnerability of this seaweed at lower temperatures. Considerable ultrastructural alterations in chloroplasts as a result of metal enrichment were observed regardless of temperatures (Scherner et al. 2018).

In an in situ experiment the thalli of *Ulva australis* were transplanted offset locations on various levels of contamination with metal. After 12 days, the accumulation of As, Cu, Pb and Zn was detected, while the accumulation of Zn was significantly the highest. An increase in electron-dense bodies in the algal cell walls and vacuoles reflected metal accumulation, although the growth rate, photosynthesis and content of photosynthetic pigments in transplanted seaweed did not differ from the control. The fact that the rate of metal absorption in *U. australis* has been steadily increasing over time confirms that this macroalga can bioaccumulate metals. On the other hand, the degradation of thalli over time indicated that the deployment time (20 days) could be limited (Farias et al. 2017a). Concentrations of toxic metals in algal tissues of green seaweeds *Cladophora glomerata* and *Ulva compressa* collected from El-Mex and Sidi Kirayr locations correlated with their soluble concentrations in seawater. In seaweeds from El-Mex site the bioconcentration factors (BCFs) decreased in the following order, Fe > Cu > Cr > Co > Cd > Ni> Pb > Zn >Mn in *C. glomerata* and Fe > Cu > Cd >Pb >Ni >Co >Mn >Cr >Zn in *U. compressa*, while in seaweeds from Sidi Kirayr location they decreased as follows, Cd > Ni > Co > Pb > Mn > Fe > Cu >Zn >Cr (*C. glomerata*) and Cd > Pb > Zn > Cu > Mn > Ni > Co > Fe > Cr (*U. compressa*), and metal-induced stress resulted in apparent alternation in algal thalli morphology. *C. glomerata* producing non-enzymatic and enzymatic compounds for ROS scavenging was capable to adapt excellently to metal excess, much better than *U. compressa*, by producing non-enzymatic and enzymatic compounds for scavenging of the produced ROS. However, both seaweed species were found to be suitable bioindicators of metal pollution and could be used for monitoring marine environment quality. They are also excellent biomarkers for oxidative damage assessment (Ismail and Ismail 2017).

Exposure of *Gelidium floridanum* (Rhodophyta) to 50 μM and 100 μM Cd, Cu and Pb for 7 days resulted in discoloration of thallus pigmentation, chloroplast alteration, degeneration of thylakoids and decreased contents of Chl*a* and phycobiliproteins. Exposure to metals resulted in increased cell wall thickness and the volume of plastoglobuli and absorption of tested metals in the cell wall was detected. From three tested metals, Cu showed the greatest toxicity to *G. floridanum* (dos Santos et al. 2014). Mendes et al. (2014) studied toxic effects of binary mixtures of metal cations to the seaweed *Gracilaria domingensis* (Rhodophyta) after exposure for 48 h. Mixtures of Cd^{2+}/Cu^{2+} and Zn^{2+}/Ca^{2+} showed additive effect, while those of Cu^{2+}/Zn^{2+}, Cu^{2+}/Mg^{2+}, Cu^{2+}/Ca^{2+}, Zn^{2+}/Mg^{2+}, and Ca^{2+}/Mg^{2+} exhibited synergistic effect. On the other hand, all interactions studied with Cd^{2+} were found to be antagonistic. Red marine macroalga *Acanthophora spicifera* and two green seaweeds *Chaetomorpha antennina* and *Ulva reticulata* accumulated Cu to a considerable higher extent than Cd and Cu was also found to be much more toxic to macroalgae than Cd and caused greater DNA damage. Oxidative stress caused by both tested metals accompanied by ROS induction resulted in reduced levels of antioxidants and induced antioxidant defense systems. The toxic impact of metals on tested macroalgae, which could be used as bioindicators of marine pollution, decreased as follows: *U. reticulata* > *A. spicifera* > *C. antennina* (Babu et al. 2014).

The evaluation of the effects of nutrients and toxic metals (Cr, Pb, Cu and Cd) on the physiological characteristics of brown seaweed *Cystoseira tamariscifolia* grown along the Atlantic coast of Morocco showed that the toxic metal contents of macroalga (especially Cd) and the concentration of P correlated with stress physiological parameters and inversely correlated with pigment contents. In the less polluted areas, the physiology of *C. tamariscifolia* was pronouncedly affected, whereas in the highly polluted areas, this brown seaweed disappeared suggesting that this species is suitable to monitor the pollution degree in coastal areas (Boundir et al. 2019). At the exposure of *Sargassum thunbergii* to 0.1 mg/L Zn, slight enhancement of the specific growth rates at the first 5 d (followed with a gradual decrease), a pronounced rise in Chl*a* levels, and increased mRNA expression of rbcL gene were found. At treatment of macroalgae with higher Zn concentrations (0.5, 1.0 and 5.0 mg/L) or with Cd in the concentration range 0.1–5.0 mg/L the specific growth rates, Chl*a* contents and mRNA expression of rbcL gene were reduced. Moreover, at Cd treatment, the oxygen evolution rate and respiration rate declined with the exposure time. Consequently, it can be stated that effects of the tested metals on physiological and gene transcription levels of *S. thunbergii* differed from each other (Lu et al. 2018).

In brown seaweed, *Fucus ceranoides*, exposed to Ag at different salinity regimes (10 and 28 psu), inhibition of algal growth and increased ROS production, was observed with increasing Ag concentration, toxic effect being greater at lower salinity, suggesting better bioavailability of Ag^+ and AgCl species (Ramesh et al. 2015).

Ulva ohnoi, a green macroalga capable to form macroalgal blooms (or "green tides"), is a fast-growing species and can absorb large amounts of nitrogen and phosphorus. Using environmentally relevant Cd concentrations (0.625–15 μg/L) the effects of salinity (S15 and S 35) and temperature on the Cd toxicity and absorption efficiency by *U. ohnoi* were evaluated, and it was found that Cd affected photosynthetic parameters and reduced growth rate of algae. However, when algae were cultivated at 18 °C and S15 or in the temperature range 18–25 °C and S35 and Cd concentrations varied between 0.625 and 2.5 μg Cd/L, positive growth rate was sustained. Maximum estimated accumulation value was 4.20 μg Cd/g d.w. at 15 μg/L of Cd, 18 °C and S35; maximum value of the bioconcentration factor was 81.3±1.1% of Cd applied at the concentration of 0.625 μg/L at S15 and 18 °C, which allows the use of *U. ohnoi* in the phytoremediation of Cd in salt or brackish water (Bastos et al. 2019).

In red seaweed *Sarcodia suiae* exposed to Cd for 24 h, the bioabsorption was considerably higher than the bioaccumulation, and using treatment with 5 mg Cd^{2+}/L, the ratios of bioabsorption/bioaccumulation in light and dark achieved 2.17 and 1.74, respectively. Increasing Cd bioaccumulation was accompanied by a decrease of Chl*a* concentration, oxygen evolution rate and oxygen consumption rate (respiratory efficiency). The levels of bioaccumulation and bioabsorption in the light were pronouncedly higher than those in the dark, and similarly, also the ratios of phycoerythrin/Chl*a*, phycocyanin/Chl*a* and allophycocyanin/Chl*a* reached higher values at light conditions suggesting that photosynthesis increased both Cd bioaccumulation and bioabsorption (Han et al. 2020). In the red seaweed, *Gracilaria* exposed to toxic Cd concentration of 0.4 mM causing oxidative stress accompanied by a generation of ROS such as $O_2^{\bullet -}$ and H_2O_2, inhibition of antioxidant system and enhancement of the lipoxygenase (LOX) activity, and MDA level and DNA demethylation was observed. Addition of 50 μM Se showed a beneficial impact on the red seaweed resulting in suppression of ROS accumulation and MDA contents, improving the level of enzymatic and nonenzymatic antioxidants and their redox ratio, and enhancing phycobiliproteins and phytochelatins over the controls. A similar beneficial effect was obtained also with the application of 1 mM spermine, and consequently, it could be concluded that Se and spermine can regulate the stabilization of DNA methylation by reducing the events of cytosine demethylation resulting in mitigation of Cd-induced stress in seaweeds (Kumar et al. 2012).

Zhang et al. (2015) performed a comparative proteomic analysis of *Sargassum fusiforme* exposed to Cd stress (1 day and 5 days) and found that the metabolic activity of macroalga was adversely affected via the down-regulation of key metabolic enzymes involved in carbohydrate metabolism and energy metabolism. However, *S. fusiforme* was able to adapt to 5 days of Cd stress by stimulation of consumption of photoassimilates via the up-regulation of glycolysis and the citrate cycle to supply energy for survival. Changes in metabolism and cellular organization of *S. cymosum* after exposure to Cd for 1 and 2 weeks under laboratory-controlled conditions (0–0.8 mg/L) were investigated by Costa et al. (2017a). Retention capacity of macroalga was >90% but no effect on photosynthetic characteristics was observed. At exposure to Cd higher concentrations of Chl compared to control were found, and at 14 days considerable changes in total composition and proportion of fucoxanthin and β-carotene were observed but any lutein was not detected; major accumulation of phenolics and flavonoids was observed as well. Cd treatment resulted in particular disorganization of cell wall fibrils. Hence, *S. cymosum* responded to Cd-induced stress with physiological and structural alterations connected with the defense mechanisms against oxidative stress; however, at low exposure concentration acclimation mechanisms to Cd stress were proved.

In *Porphyra yezoensis* collected monthly from January to April in 2011 and *Laminaria japonica* collected monthly from March to July in 2010 inorganic As (iAs) to total As decreased in both macroalgae with the time due to the capability of both algae to metabolize toxic iAs form and transform it to organic forms resulting in As resistance. Moreover, the transformation of AsO_4^{3-} to organic arsenic increased with the growth and metabolic rate of macroalgae showing a rise with an increase of environmental temperature. On the other hand, the content of inorganic Cd in algae was minor, and Cd was predominantly associated with pectates and protein (Zhao et al. 2012).

The impact of Mn (10–80 mg/L) on *S. cymosum* during 1 and 2 weeks under laboratory-controlled conditions was investigated by Costa et al. (2017b). *S. cymosum* showed high Mn biosorption capacity and the presence of Mn stimulated the growth of seaweeds; higher Chl*a* and Chl*c* levels and modification in the composition of carotenoid profile were estimated. The contents of phenolic and flavonoid compounds were found to decrease with increasing exposure period and applied Mn concentration. Moreover, after 14 days of treatment with Mn changes on the thallus surface with disruption of cortical cell walls and disorganization of cell wall fibrils were observed. Medium and high concentrations of Mn acted as a metabolic stressor for *S. cymosum* with a detrimental impact on the organelle structure and cellular organization and mitigated acclimation mechanisms.

Based on the high negative correlations observed between Chl*a* and Chl*c* contents and the ratio of Chl*c*/Chl*a* in *S. angustifolium* and Ni concentration in this macroalgae, it could be concluded that these characteristics authentically reflected adverse impact of high concentrations of the toxic metal on *S. angustifolium* (Alahverdi and Savabieasfahani 2012).

Exposure of *Ulva australis* to 25 μg/L and 50 μg/L of Zn for 7 days resulted in the retraction of cytoplasm and a reduction of the number of starch granules without adverse impact on the photosynthetic performance and growth (Farias et al. 2017b). Investigation of the photosynthetic performance of the red alga *Pyropia yezoensis* exposed to 25 and 100 μg Zn/L and pCO_2 of 400 or 1000 μatm showed that the adverse impact of higher Zn concentrations on *P. yezoensis* was mitigated by ocean acidification, which improved the relative growth rate, the net photosynthetic rate and respiratory rate of macroalga thalli cultured under tested Zn concentrations. On the other hand, in the presence of Zn the MDA levels were reduced under ocean acidification compared to ambient CO_2 conditions, and SOD activity increased as well. Moreover, ocean acidification enhanced D_1 removal, with increasing expression levels of the PSII reaction center proteins D_2, CP47 and RbcL, suggesting that it could alleviate the toxic impact of Zn on the seaweed (Ma et al. 2020).

In *Ulva compressa* marine macroalga cultivated in the presence of Cu (2.5–10 μM) for up to 12 d, a linear correlation between intracellular Cu and the Cu concentration in the culture medium was observed. Moreover, increasing intracellular Cu concentrations were accompanied with increases in glutathione (GSH) and phytochelatins (PCs), followed by higher levels of metallothioneins expression indicating that thiol-containing peptides and proteins can participate in Cu accumulation. Cu release to culture medium from algal cells, which were cultivated with 10 μM Cu and then transferred to synthetic seawater without Cu and cultivated for further 3 d, was accompanied by a similar release of nanomolar amount of GSH, while PCs or small proteins were not detected. Consequently, it can be assumed that the release of Cu and GSH to the extracellular medium is involved in the detoxification mechanism (Navarrete et al. 2019). *Ulva lactuca* tested in laboratory experiments accumulated >90% of the total Cu intracellularly, without intra-specific differences. Its use for biomonitoring reflected the metal pollution, which was confirmed also with metal contents estimated in sediments. This species was characterized with low levels of metal exclusion suggesting that *U. lactuca* responded to metal stress with the syntheses of metal chelators and the antioxidant metabolism (Valdes et al. 2018). The exudates that bind metals released by spores of *U. lactuca* (Chlorophyta) and *Lessonia spicata* (Phaeophyta) were found to increase the 48 h EC_{50} values of the germination of spores from 8 to 23 μg Cu/L for *U. lactuca* and from 119 to 213 μg Cu/L for *L. spicata* (Fellous et al. 2017).

Gelidium floridanum (Rhodophyta) tetraspores cultivated on seawater enriched with 3.0 μM $CuCl_2$ and incubated under 30 °C for 24 h had lower viability and seriously altered cellular and chloroplast structures as well as cells with degenerated cytoplasm and cell walls suggesting an adverse combined effect of Cu and higher temperature on early stages of seaweed (Kreusch et al. 2019). Treatment of *G. floridanum* with Cu at 30 °C for 7 days resulted in a lower concentration of Chl*a*, smaller phycobiliprotein rods and lower concentration of soluble sugars, while after 14 d of cultivation, higher concentrations of Chl*a* and soluble sugars were observed. The accumulation of carotenoids and the release of phenolic compounds can be considered as specific protective mechanisms against temperature and Cu stress, respectively, although reduced growth of *G. floridanum* was observed at exposure to 3 μM Cu and increased seawater temperature of 30 °C (Kreusch et al. 2018). At exposure of *Macrocystis pyrifera* and *Undaria pinnatifida* meiospores to 100–400 μg/L Cu for 9 days 6–15% of the dissolved Cu was adsorbed by the macroalgal cells. Meiospores germination at all tested concentration was observed in both species; however it showed a decrease with increasing Cu concentration; gametophyte growth and sexual differentiation were arrested under all Cu treatments. The EC_{50} values related to the arrest of germination were estimated as 157 and 231 μg/L Cu for *M. pyrifera* and *U. pinnatifida*, respectively. Even though the EC_{50} value of the invasive species *U. pinnatifida* was higher, due to consecutive inhibition of gametogenesis in the presence of Cu, there was no difference in Cu tolerance between both tested seaweeds' early life stages (Leal et al. 2016).

In the meristem of juvenile sporophytes of *Saccharina japonica* cultured in seawater in the presence of 100 and 200 μg/L of Cu^{2+}, bleaching occurred on the third day indicating serious damage, and transcriptome profiling of the meristem exposed to 200 μg/L of Cu^{2+} showed 11,350 differentially expressed genes (4944 up- and 6406 down-regulated), whereby genes coding glutathione-*S*-transferase and vanadium-dependent bromoperoxidase and iodoperoxidase were found to be remarkably regulated; the down-regulated genes were associated, for example, with photosynthesis, protein synthesis, redox activity, and metabolism and biosynthesis of functional biomolecule (Zhang et al. 2019). Treatment of *Sargassum fusiforme* with 47 μM Cu for 1 day resulted in impaired photosynthesis accompanied with reduced levels of carbohydrates (e.g. mannitol), and ascorbate was assumed to play a substantial role in the antioxidant system because its level declined much more compared to treatment with 8 μM Cu lasting 7 days (considered as chronic stress) (Zou et al. 2014). In *S. fusiforme* cultured in fresh seawater in the presence of Cu (4–47 μM), the macroalga tolerated Cu stress at low concentrations, while with increasing Cu concentration the algal growth showed a decrease. At the exposure of *S. fusiforme* to high Cu concentration for 1 day, considerable reduction of proteins related to energy metabolism and photosynthesis was observed. On the other hand, induction of proteins related to carbohydrate metabolism, protein destination, RNA degradation and signalling regulation was observed. At exposure of *S. fusiforme* to lower sub-lethal Cu concentrations for 1 week, pronounced induction of energy metabolism-related proteins and only small quantities of proteins related to membranes and transport were estimated suggesting sensitivity of this brown seaweed to the acute or chronic Cu stress (Zou et al. 2015). A novel field transplantation technique using strains of brown alga *Ectocarpus siliculosus* incorporated into dialysis tubing was applied by Saez et al. (2015a) to investigate intra-specific responses of the macroalga to toxic metals. Higher metal accumulation in algal material from the metal-contaminated site was accompanied with increased levels of non-enzymatic antioxidants and activities of antioxidant enzymes (SOD, ascorbate peroxidase (APX), CAT, and GR) suggesting metal tolerance of this strain in contrast to the strain originating from the low-polluted site, which was adversely affected by oxidative stress. Similar results were obtained at investigation of Cu-induced intra-specific oxidative damage and responses of the antioxidant defense system in strains of *E. siliculosus* with different pollution histories (Saez et al. 2015b). The contamination histories of the locations from where the strains of *E. siliculosus* were isolated affected the response of these strains to Cu exposure. This was reflected also in Cu exclusion and production of PCs, whereby the strain originating from the area with lower metal contamination exhibited lower production of PCs and lower expression of enzymes involved in phytochelatin biosynthetic pathway, which supported the induction of oxidative stress resulting in its higher sensitivity to Cu (Roncarati et al. 2015). Juveniles of the canopy-forming brown alga *Fucus serratus* were found to be more susceptible to environmental stressors such as Cu and temperature than adult specimens. The negative effect of high Cu concentration on *F. serratus* juveniles was amplified by high temperature resulting in reduced survival suggesting that they could be used for assessment of the risk of exposure to toxic metals (Nielsen et al. 2014). The brown alga *Dictyota kunthii* exposed to 100 μg Cu/L for 4 days responded to metal excess by various mechanisms, including metal accumulation, activation of CAT, APX, dehydroascorbate reductase, glutathione peroxidase and peroxiredoxins and an induced release of Cu binding compounds suggesting that *Dictyota kunthii* is a Cu-tolerant species (Sordet et al. 2014).

5 Biosorption of Toxic Metals by Living and Dead Seaweed Biomass

Gracilaria sp. is red seaweeds able to accumulate toxic metals via adsorption on the surface of talus in the form of cations, anions or organic compounds. *G. lemaneiformis* showing powerful adsorption capacities for heavy metals from seawater, which culminated in May (highest bioconcentration factors), affected the levels of heavy metals in coastal sediment from a typical mariculture zone and was found to be suitable for bioremediation of heavy metals in the coastal sediments (Luo et al. 2020). Saldarriaga-Hernandez et al. (2020) in their review work emphasized the multifunctional potential of invasive seaweed biomass *Sargassum* sp. as a cheap but highly effective natural material for decontamination of the environment, especially of coastal ecosystems. The researchers analysed bioremediation mechanisms as well as absorption of pollutants by seaweed and presented prospective possibilities for the integrated use of the algal raw material. Using dead biomass from the seaweed *Sargassum swartzii*, 60% of the total biosorption capacity was achieved in 30 min, and considering regeneration feasibility of such biomass, it could be applied for the treatment of wastewater containing heavy metals (Gabruk et al. 2015).

Sargassum, *Ulva* and *Turbinaria* macroalgae were tested for removal of 8 toxic metals from complex synthetic laboratory wastewaters at 0.016 M total concentration. As the best biosorbent *Turbinaria conoides* showed total uptake capacities of 3.20 and 3.01 mmol/g in the batch, and continuous-mode of operation was estimated (Vijayaraghavan et al. 2012b). Seaweeds *Scytosiphon lomentaria* and *Ulva rigida* from coastal areas with mining activities in Chile were studied. Seaweeds originating from Quintay showed the greatest bioaccumulation factors for Fe and As. Bioaccumulation factor *in S. lomentaria* decreased as follows, Fe > Cu > Zn > Cd > Cr > As > Mo, while in *U. rigida* the descending order was Fe > Cu > Cd > Zn > Cr > Mo > As. Antioxidant activity levels of these algal species in areas with high mining activities achieved higher levels and the concentration of metals in waters were related with oxidative stress biomarkers in both species suggesting that they can be used for monitoring of heavy metal pollution in coastal waters (Olivares et al. 2016). Santos-Silva et al. (2018) estimated the background levels of trace elements in four seaweeds, *Dictyopteris delicatula* and *Canistrocarpus cervicornis* (Pheophyta) as well as *Ceratodictyon variabile* and *Palisada perforata* (Rhodophyta) from Trindade, an isolated island in the South Atlantic. The highest concentrations of Cd, Hg and As were observed in brown seaweeds; spatial differences for these metals in samples were detected. On the other hand, the highest concentrations of Zn, Pb and Cu were observed in seaweeds from the only populated beach, but the levels of trace elements in seaweeds were generally low. Ryan et al. (2012) performed a comprehensive study focused on Pb, Zn, As, Cd, Co, Cr, Cu, Mn and Ni distribution in *Polysiphonia lanosa* (L) Tandy, *Ascophyllum nodosum* (L) Jolis, *Fucus vesiculosus* (L) and *Ulva* sp. and found that intracellular and surface-bound metal levels depended both on the metal and seaweed species but the majority of metals was localized intracellularly. The highest concentration factor showed *P. lanosa* characterized with enhanced bioaccumulation ability, whereby in this macroalga the uptake of Mn was found to be inhibited by Zn. Biosorption of Cr^{6+}, Cr^{3+}, Pb^{2+} and Cd^{2+} by algal biomass of *Sargassum wightii* (brown) and *Caulerpa racemosa* (green) depended on biomass, concentration and pH. The maximal biosorption was observed at pH 5.0 and using the metal concentration of 100/mg, *S. wightii* showed 78% biosorption of Cr^{6+}, Cr^{3+}, Pb^{2+} and Cd^{2+} ions, while biosorption of Cd^{2+} and Cr^{6+} by *C. racemosa* achieved 85% and that of Cr^{3+} and Pb^{2+} only 50% (Tamilselvan et al. 2012).

The concentrations of metals in ten green seaweed species collected along the Palk Bay decreased in the following order of Pb > Cu > Zn > Cd, and it was found that the exposure to metals of these seaweeds represented a moderate hazard risk to humans. It could be mentioned that the trace element accumulation in four seasons decreased in the order summer > monsoon > pre-monsoon > post-monsoon (Rajaram et al. 2020). *Ulva armoricana* (Chlorophyta) can reduce the environmental impact of aquaculture effluent in coastal ecosystems. When the seaweed was cultured in various enriched media at a stocking density of 500 kg wet weight/pond, it bioaccumulated following amounts of toxic metals: 9.908–32.942 mg Zn/kg, 1.893–5.927 mg Cu/kg and 0.254–1.500 mg Cd/kg suggesting that this green macroalga can be successfully used as a biological filter capable of absorbing dissolved nutrients from aquaculture wastewater (Amosu et al. 2016). In green macroalgae *Ulva lactuca* from six areas receiving effluents from industries located in a coastal region, the concentration of Pb (8.32 μg/g) was higher than in water and sediment samples. On the other hand, high Cd and Pb concentrations were determined in sediment. The abundance of toxic metals in sediment decreased in order Cd > Cr > Pb, while in water it decreased as follows: Cr > Pb > Cd (Kamala-Kannan et al. 2008).

Gracilaria caudata and *Gracilaria cervicornis* (Rhodophyta) showed the ability to capture Cu, Ni and Zn from the petrochemical wastewater on the principle of ion exchange (cation exchange resin). Both red algae were found to have an ion-exchange capacity of 1.2 mEq/g. The equilibrium affinity constants for the functional groups decreased as follows: Cu >> Zn> Ni. Ca-loaded *G. cervicornis* had a lower affinity for Zn and Ni compared to Cu. Using a column full of crude *G. cervicornis*, a useful capacity of 0.25 and 0.24 mEq/g was found for the first and second operation cycles, and the scientists proposed connecting the column system in series for purification operations (Cechinel et al. 2018). Investigation of biosorption of Cu^{2+}, Ni^{2+}, Cd^{2+} and Pb^{2+} from aqueous solution by *Kappaphycus alvarezii* (Rhodophyta), an edible seaweed, performed at pH 4.5 indicated that the uptake of individual metals by macroalga decreased in the order: Pb (0.51 mmol/g) > Cd (0.48 mmol/g) ≈ Cu (0.47 mmol/g) > Ni (0.38 mmol/g). The most suitable experimental isotherm data were obtained using Toth's model; within 45 min, 90% of the metal was removed, and the use of 0.01 M HCl resulted in maximal desorption of the tested metal ions from the seaweed biosorbent (Praveen and Vijayaraghavan 2015). A waste material of seaweed *Ascophyllum* sp. applied as biosorbent for removal of toxic metals from aqueous solution using fixed-bed sorption column showed high removal efficiencies (RE) for Ni^{2+}, Zn^{2+} and Al^{3+} of 90, 90 and 74% when initial metal solutions of 10 mg/L were applied. However, in multicomponent metal solutions containing Sb^{3+} the RE of these metals was reduced to 28, 17 and 24%, respectively, whereby in Sb^{3+} binding only -COOH and $-SO_3^-$ groups were involved. The RE for Sb^{3+}, both alone and in combination with other metals estimated using red macroalga *Polysiphonia lanosa* as a biosorbent, was 67% and -COOH, -OH, $-SO_3^-$ and ether groups were involved in Sb^{3+} binding. While a larger amount of weakly acidic groups (mainly -COOH) was included in the binding of Sb^{3+} to *P. lanosa*, a larger proportion of strongly acidic (mainly $-SO_3^-$) groups was observed in the binding of Sb^{3+} to the waste material from the seaweed *Ascophyllum* (Bakir et al. 2009).

Shchukin et al. (2018) performed a comparative analysis of trace element contents in brown algae of the families Laminariales and Fucales of different origins and found that As accumulated better in Laminariales, while Al, Fe, Cu, and Mn in the brown macroalgae of the family Fucales; differences in accumulation of Cd, Hg, Pb, Ni, Zn, Cr and Sr in members of these two families were comparable with differences between species within a single family. While Al, Cd, Co, Cr, Cu, Fe, Hg, Mn, Ni, Pb and Zn concentrations in algal biomass depended predominantly on the technogenic factor, Sr concentrations were affected by the geographic latitude of the region in which the seaweeds grew. On the other hand, no relationship was found between As accumulation in algae and the level of pollution or the geographic region in which the algae were collected. Evaluation of the multi-elemental composition of brown seaweed *Macrocystis pyrifera* from San Jorge Gulf (Patagonia, Argentina) showed seasonal variations in the content for some metals: the highest As content was observed in summer and autumn, Fe concentration increased to the winter and Zn concentration was maximum in autumn. The sum of essential micronutrients (Fe + Zn + Mn + Cu) ranged from 114 to 171 mg/kg dry weight (d.w.), while As levels were in the range 36–66 mg/kg and Pb, Ni and Cu were not detected (Salomone et al. 2017). Using NaOH-pretreated biomass of brown seaweed *Hizikia fusiformis* contributing to the increase of the functional forms of carboxylate ester units, the maximum sorption for Pb^{2+}, Cd^{2+}, Ni^{2+} and Zn^{2+} was observed at pH ranging from 4 to 6; biosorption process was rapid, and the maximum metal adsorption capacities evaluated using Langmuir model decreased as follows: $Pb^{2+} > Cd^{2+} > Ni^{2+} > Zn^{2+}$ (Shin and Kim 2014). Silva et al. (2009) investigated uranium biosorption by *Sargassum filipendula* brown macroalga under dynamic conditions from real effluent contaminated beside uranium also with Ba, Cr, Fe, Mn, Pb, Ca and Mg. Whereas 64% of the uranium was absorbed, together with Cr, Pb and Ba, the absorption of Ca, Mg, Fe and Mn did not occur, and their concentrations in the solution showed an increase connected with mechanisms of ion-exchange with components of biomass polysaccharides. The 85–87% reduction of mass observed after drying and calcination of the biomass suggested that such algal biomass loaded with radionuclides and other toxic metals could be stored for long period.

Investigation of Cd biosorption performance by dried biomass of the *Sargassum thunbergii*, *Laminaria japonica* (Phaeophyta), *Ulva pertusa*, *Enteromorpha linza* (Chlorophyta) and *Chondrus ocellatus* (Rhodophyta) showed that the highest removal efficiency at different initial Cd^{2+} concentrations exhibited *S. thunbergii*, whereby 90% of adsorption occurred within 15 min and for the biosorption interaction of Cd^{2+} ions with -COOH, $-NH_2$, $-SO_3$ and -OH occurring on the seaweed surface was crucial (Jadeja and Zhou 2018).

Ulva fasciata and *Ulva lactuca* showed maximum biosorption capacities for Cd^{2+} of 8.353 and 8.804 mg/g d.w., respectively. Different algal weights were immobilized using 10 mL of 4% Ca alginate and in such pretreated algal biomass as well as in fresh algal biomass the crucial role in biosorption played -OH groups and amide groups via hydrogen bond and carbonyl extension in -COOH groups. Immobilization of different weights of algae using Ca alginate resulted in improved biosorption capacity in case of low weight; morphological changes were observed among immobilized algal beads before and after Cd^{2+} biosorption, whereby ion-exchange mechanism was involved in Cd^{2+} biosorption on both green seaweed species (El-Sheekh et al. 2020). Exposure of *Ulva lactuca* for 48 h to 100 μg/L of either Cd or Cu or to 100 and 1000 μg/L metal mixtures resulted in pronounced impairment of photosynthetic parameters (Jarvis and Bielmyer-Fraser 2015).

At investigating the subcellular distribution and chemical forms of Cd in *Porphyra yezoensis* (edible red seaweed) exposed to Cd (0.01–5.0 mg/L) for up to 96 h, the Cd portion in the cell wall increased with exposure time and increasing Cd concentrations, whereby the amount of Cd localized in the cell wall achieved 41.2–79.2% and as a crucial strategy to suppress Cd toxicity in this red seaweed cell wall deposition and forming of precipitates with phosphate were assumed (Zhao et al. 2015).

Cd biosorption by alginate extraction residue from *Sargassum filipendula*, an industrial waste which is often discharged into the sea, was found to be spontaneous and exothermic with maximum biosorption capacity of 0.394 and 0.429 mmoL/g, respectively, at 293 and 303 K. Based on the simplified life cycle assessment, it could be stated that by the use of dealginated residue the environmental impacts related to acidification, climate change, eutrophication, human toxicity and photochemical oxidation could be reduced (Nishikawa, et al. 2018). Continuous Cd removal from water solutions by *Sargassum angustifolium* in a packed-bed column (bed heights, 2.6–7.5 cm; feed flow rates: 15–30 mL/min) under consecutive sorption-desorption cycles was studied by Jafari and Jamali (2016), whereas a Cd concentration of 5 μg/L (standard limit for drinking water) was considered to be the key concentration. Using a bed length of 7.5 cm and flow rate of 15 mL/min resulted in 81% column performance. The bed height changes practically did not affect the maximum sorption capacity; nevertheless, it slightly decreased with increasing feed flow rate. At the end of the fourth consecutive sorption-desorption cycle, the column adsorption efficiency was reduced by 20%; desorption efficiencies >99% were observed in each cycle. Cd^{2+} biosorption on *S. angustifolium* seaweed included an initial rapid stage lasting first 30 min, in which ion exchange was the dominated mechanism, and after 40–50 min of contact, equilibrium was reached characterized with very low Cd^{2+} adsorption due to the intraparticle diffusion. The biosorption process was spontaneous and endothermic; the experimental data in the whole range of contact time could be perfectly described by the pseudo-second-order kinetic model, while the Langmuir isotherm model was found to be most suitable to fit the equilibrium data (Jafari et al. 2015). In batch experiments using dead biomass of the brown seaweed *Sargassum sinicola* as biosorbent of Cd and Cu under saline conditions, it was shown that Cd removal was pronouncedly reduced from 81.8% to 5.8% when salinity increased from 0 to 40 psu. On the other hand, salinity increase did not affect Cu removal. At 35 psu maximum capacity of biosorption was 3.44 mg/g for Cd and 116 mg/g for Cu. Saturation of ca. 90% for both metals was obtained in 60 min and was not pronouncedly affected by salinity (Patron-Prado et al. 2010). As good biosorbents of Cd also nonliving biomass of brown algae *Sargassum sinicola* and *Sargassum lapazeanum* was reported with maximum biosorption capacity (q_{max}) of Cd estimated according to the Langmuir adsorption isotherm

62.42±0.44 mg/g for *S. sinicola* and 71.20±0.80 mg/g for *S. lapazeanum*, respectively (Patron-Prado et al. 2011).

Investigation of Al^{3+} and Cd^{2+} biosorption on dead brown seaweed, *Turbinaria conoides*, in both single and binary systems showed that the macroalga achieved maximum Al^{3+} biosorption at pH 4 with a capacity of 2.37 mmol/g, while for Cd^{2+} maximum biosorption was observed at pH 5 with a capacity of 0.96 mmol/g. The -COOH groups on the surface of seaweed were found to be responsible for excellent uptake capacity against both tested ions via the ion-exchange mechanism. In binary systems, Al^{3+} uptake was reduced to 56% and to 27% for Cd^{2+} compared to single-solute systems. Using 0.1 M HCl as eluant *T. conoides* biomass was successfully reused in three repeated cycles (Vijayaraghavan et al. 2012a).

Pennesi et al. (2012) tested nonliving seaweeds *Cystoseira*, *Dictyopteris* and *Eisenia* sp. (Pheophyta), *Caulerpa* and *Ulva* sp. (Chlorophyta) and *Ceramium*, *Gracilaria* and *Porphyra* sp. (Rodophyta) as As^{5+} biosorbents and found that these species were characterized by significant sorption ability, which achieved a maximum value of ca. 1.3±0.1 mg/g for the red alga *Ceramium*, comparable with those of activated carbon and other low-cost adsorbents. As biosorption was found to depend on the composition and structure of the outer layer of the seaweeds, As speciation and functional group availability at different pH, and eventual counter-ion interactions with arsenate. The sorption of As was not pronouncedly affected by biosorbent acid washing; however, the basic washing was found to improve As sorption of green seaweeds. *Hizikia fusiforme* macroalga can contain large inorganic As concentrations, while *Fucus spiralis* contains mainly arsenosugars. When *F. spiralis* was exposed to 0, 1 and 10 mg/L arsenate solutions for 24 h, and *H. fusiforme* was analysed fresh, in all samples As^{3+}, glutathione and reduced PC_2 were detected. However, while in arsenate exposed *F. spiralis* traces of various As compounds were found, probably As^{3+}-GS or As^{3+}-PC_2 complexes, in *H. fusiforme* no As^{3+}-PC complexes were found. The researchers supposed that As^{3+}-PC complexes are not the foremost principal storage form for long-term As storage within marine macroalgae. It could be mentioned that glutathione concentrations in *H. fusiforme* were 40-fold higher than those in *F. spiralis* (Wood et al. 2011). The accumulation of As by the brown seaweed *Sargassum elegans* Suhr 1840 collected from seven sites along the coast of KwaZulu-Natal, South Africa, containing high amount of β-sitosterol, fucosterol and phaeophytin ranged from 42 to 105 mg/kg, however with a high portion of inorganic As (21 to 53 mg/kg), which could represent increased risk at consumption of this species (Magura et al. 2019). Inorganic arsenic from seawater accumulated in seaweeds in the form of $HAsO_4^{2-}$ instead of HPO_4^{2-}, and although it was rapidly metabolized to organoarsenic species, residual iAs in seaweed biomass could be harmful for consumers of seaweed products. The total As (As_{tot}) in thallus parts of *Laminaria digitata* ranged from 36 to 131 mg/kg d.w., while in *Ascophyllum nodosum* from 38 to 111 mg/kg d.w., with no statistically significant differences between different thallus parts. On the other hand, iAs represented <1% from As_{tot} in *A. nodosum*, while in *L. digitata* it achieved 2.2–87 mg/kg, increasing through the thallus from the stipe to the decaying distal blades; in the middle to decaying distal blades it represented >50% of As_{tot} (Ronan et al. 2017).

Hg levels accumulated in *Ulva lactuca* (green), *Gracilaria gracilis* (red) and *Fucus vesiculosus* (brown) macroalgae from contaminated waters with high salinity and Hg concentrations of 10–100 μg/L (comparable with those found in the environment) were 20.8 and 208 μg/g, corresponding to bioconcentration factor of ca. 2000. Application of living biomass was capable to reduce Hg levels by ca. 90% resulting in drinking water fulfilling the European criteria. The biosorption process was assumed to be essentially of chemical nature, and volatilization of Hg or its conversion to organomercuric compounds was negligible (Henriques et al. 2015). Activated carbons based on biomass of *Sargassum* and *Enteromorpha* seaweeds fabricated using activation temperature of 800 °C were found to show the best efficiency in gaseous Hg^0 removal. In general, Hg^0 removal efficiency of both samples raised also with increasing reaction temperature, and it was observed that the Hg^0 removal was controlled by the external mass transfer at 80 °C and controlled by the chemisorption at 120 °C and 160 °C, whereby the physisorption and chemisorption were mediated by the surface active sites (C-Cl groups and oxygen species) of the sorbent and its great specific surface area and pore structure (Liu et al. 2019). *Enteromorpha* chars prepared by pyrolysis and modified by NH_4Cl or NH_4Br using 5 wt% loading and reaction temperature of 130 °C were studied for the removal of gaseous Hg^0 by Xu et al. (2019). Optimal pyrolysis temperature was found to be 800 °C, and modification of chars resulted in pronouncedly improved Hg removal, especially when NH_4Br was used. Hg adsorption over modified seaweed chars was endothermic process, the experimental data fitted the pseudo-second-order kinetic model, and in the removal of Hg chemisorption played a crucial role. Whereas presence of O_2, NO and lower concentration of H_2O stimulated Hg removal, presence of SO_2 and higher H_2O concentrations showed inhibitory effect. As the predominant chemisorption sites for Hg^0 removal in NH_4Br-modified *Sargassum* chars, the C-Br and C=O covalent groups were considered (Liu et al. 2018b). Similar results related to removal Hg^0 were obtained using biochars derived from seaweed, which were impregnated with potassium iodine. As the dominant chemical adsorption sites on the surface of the KI modified seaweed biochars participating on Hg^0 removal, the covalent groups (C-I) and the chemisorbed O_2 and/or weakly bonded oxygen species were estimated (Yang et al. 2018). The brown seaweeds *T. conoides* and *Sargassum* sp. tested as biosorbents of Hg^{2+} removal from aqueous solution achieved at pH 5 maximum biosorption capacities of 170.3 and 145.8 mg/g, respectively, compared to 138.4 mg/g observed with the green seaweed *Ulva* sp. Biosorption was observed to be rapid, equilibrium was reached within 90 min, and the experimental data fitted well the Toth model and could be described by the pseudo-first-order model. Using 0.05 M HCl for desorption enabled reuse of *T. conoides* biomass; three consecutive sorption-desorption cycles resulted only in 8.8% reduction in Hg^{2+} biosorption capacity. On the other hand, pronounced biomass weight loss was observed at treatment of *Sargassum* sp. and *Ulva* sp. with 0.05 M HCl (Vijayaraghavan and Joshi 2012).

Osmium uptake measured in *Fucus vesiculosus* non-fertile tips cultured in the presence of different concentrations of Os applied in the form of isotopic composition with $^{187}Os/^{188}Os$ positively correlated with the concentration of the Os doped seawater, and $^{187}Os/^{188}Os$ composition of the seaweed was the same as that of the culture medium suggesting the potential of *F. vesiculosus* to be used for estimation of Os isotopic composition of the seawater (Racionero-Gomez et al. 2017).

In *Porphyra yezoensis* algae exposed to three different Pb concentrations (0.01, 0.1 and 1.0 mg/L) for up to 144 h, Pb was deposited in the cell wall, and formation of the precipitation of less active 2% acetic acid extractable form could be considered as one of the mechanisms for accumulation, transportation and detoxification of Pb in this red macroalga (Zhao et al. 2019). Investigation of Pb removal efficiency using marine macroalga biosorbents as alternative materials for permeable reactive barriers showed that in the batch experiments higher Pb removal efficiency exhibited *Undaria pinnatifida* compared to *Phragmites australis*, while in the column experiments the Pb removal efficiency showed a decrease for both tested materials at approximately the same time (Soto-Rios et al. 2014). The optimum biosorption conditions for Pb^{2+} ions biosorption on the brown seaweed *Sargassum ilicifolium* were found to be initial pH 3.7, biosorbent concentration 0.2 g/L and initial Pb^{2+} concentration 200 mg/L resulting in the maximum uptake capacity of 195±3.3 mg/g; the equilibrium biosorption data could be well described with both Langmuir and Freundlich isotherm models (Tabaraki et al. 2014). High-molecular alginates extracted from Djiboutian brown seaweeds, *Sargassum* sp. and *Turbinaria*, and isolated as sodium salts, which were deposited on the native Aerosil 200 SiO_2, NH_2-functionalized and COOH-functionalized SiO_2 particles, were tested as biosorbents of Pb^{2+} ions from aqueous solution. The highest uptake, 585 mg Pb^{2+}/g, was observed with the alginates originating from *Sargassum* sp. deposited on NH_2-functionalized SiO_2. On the other hand, using alginates originating from *Turbinaria* sp. deposited on SiO_2, several sorption-desorption cycles could be performed, and thus, they are suitable to be used as an inexpensive biosorbent for Pb^{2+} ions (Aden et al. 2019). Fucoidan, a sulfated polysaccharide extracted from brown seaweed, used in the form of a silica composite was found to be a prospective Pb^{2+}-imprinting matrix, and the fucoidan-based Pb^{2+} imprints were characterized with excellent sorption properties and showed higher capacity and higher binding strength for Pb^{2+} than other sulfated polysaccharides (Ferreira et al. 2017). The waste obtained after the sequential separation of polyphenols, fucoidan and alginate extracts from brown macroalgae *Fucus spiralis* seaweed, which was tested as biosorbent for Pb^{2+}, was able to remove Pb^{2+} ions almost completely, equilibrium being achieved in few hours, and such biosorbent showed maximum adsorption capacity of 132±14 mg/g (pH 4.5±0.5, 20 °C) predicted by Langmuir model (Filote et al. 2019).

Three brown algae *Sargassum vulgare*, *Cystosiera compressa* and *Turbinaria* were tested as biosorbents for purification of wastewater containing Cu^{2+} and Pb^{2+} ions under stirring (200 rpm), and it was found that the effectiveness of the individual biosorbents increased not only with increasing amount of biosorbents, but also with immersion time and low initial metal ions concentration. Changes of the pH of the medium significantly affected the effectiveness of the adsorption, maximum removal being observed at pH 8. The biosorption process could be described by Freundlich adsorption isotherm and second order kinetic model, whereas this process took place in three steps according to intraparticle diffusion process and the efficiency of the metal ions removal was affected by the biochemical structures of the tested biosorbents (Negm et al. 2018). *Chondracanthus chamissoi* (Rhodophyta) was found to be a suitable biosorbent for Pb^{2+} and Cd^{2+} showing optimum sorption efficiency at approximately pH 4 with maximum sorption capacity 1.37 mmol Pb/g and 0.76 mmol Cd/g, whereby the biosorbent exhibited considerable higher affinity for Pb^{2+} than for Cd^{2+}. Improvement of the sorption capacity and uptake kinetics with grinding the biomass was not observed. It could be assumed that during the sorption process, the metal ions interaction with carrageen belonging to the main constituents of this biosorbent occurred, whereby the affinity of the -SO_3^- groups on the sulfated polysaccharide was higher for Pb^{2+} than for Cd^{2+} (Yipmantin et al. 2011).

$CaCl_2$-pretreated biomass of brown seaweed, *Lobophora variegata* (Lamouroux), exhibited the maximum uptake capacities of 1.71 and 1.79 mmol/g for Cd^{2+} and Pb^{2+}, respectively; the sorption process could be described by the pseudo-second order kinetic model. Sorption mechanism was found to be complex and consisted of both surface adsorption and pore diffusion. Based on the fit of experimental data with Langmuir isotherm, it was supposed that Cd^{2+} and Pb^{2+} ions covered the surface of *L. variegata* by a monolayer. On the binding of metal ions by the seaweed participated predominantly -COOH groups accompanied by significant interactions with -NH_2 and amide groups of the biomass (Jha et al. 2009). *Sargassum horneri* (Pheophyta) biomass pretreated with $CaCl_2$ in order to increase Pb^{2+} adsorption, which was tested as biosorbent for Pb^{2+}, showed q_{max} of 0.696 mmol Pb/g evaluated using Langmuir adsorption isotherm equation, and ca. 1.5-fold higher amount of Ca^{2+} was released from the biosorbent compared to the adsorbed Pb^{2+} amount (Southichak et al. 2008). The total amount of light metals present on the surface of raw biomass of brown marine macroalgae *Ascophyllum nodosum* was estimated as 2.4 mEq/g, and comparable binding capacity was observed when the raw macroalgae were converted in different ionic forms, whereby as crucial functional groups responsible for cations binding -COOH (ca. to 1.3 mEq/g) and HSO_3^- (ca. to 1.1 mEq/g) groups were estimated. Using 0.1 M $CaCl_2$ at pH 8.0 the regeneration of biosorbent could be performed enabling its reuse in multiple cycles (Mazur et al. 2017). The advantage of the sequential application of macroalgal biosorbents, Fe-treated biochar and raw biochar fabricated from macroalgae, for the bioremediation of a complex industrial effluent was reported by Kidgell et al. (2014). Whereas Fe-biochar removed the metalloids from the wastewater, subsequent application of raw biochar removed the metals.

Based on temporal and spatial patterns of Zn and Pb content in *Ulva australis* collected from the Derwent Estuary, Tasmania, Australia, over 3 years (2013–2015), it was shown that both metals achieved high levels in the macroalga; however, Zn in seaweed varied seasonally (4.8–320.7 mg/g), while Pb levels were not affected. Highest Zn and Pb levels were observed in the middle-upper estuary, close to the zinc smelter, where seawater concentrations were higher suggesting that *U. australis* can be used for monitoring the effects of toxic metals in estuarine systems (Farias et al. 2019). Investigation of Zn bioaccumulation by *Gracilaria bursa-pastoris* (Rhodophyta) seaweed in the coastal section of the Evros River Delta, Aegean Sea, showed correlation with its sediment loads suggesting that accumulated Zn in macroalga reflected well the ambient Zn abundances, and the tested seaweed could be used as a bioindicator of Zn. On the other hand, observed negative correlation between accumulated Zn and water salinity was likely due to increasing Zn uptake with decreasing salinity. In contrast to Zn, bioaccumulated Cd in *G. bursa-pastoris* did not correlate with its content in the sediment and showed a negative correlation with Zn accumulated in the macroalga indicating that due to an antagonistic interaction between Cd and Zn, the use of *G. bursa-pastoris* as bioindicator of Cd would be not satisfactory (Boubonari et al. 2008). Formaldehyde-treated *Sargassum glaucescens* biomass used for biosorption of Zn^{2+} from aqueous solution achieved the maximum metal uptake capacity of 29.13 mg Zn/g in a batch experiment. Using *S. glaucescens* in a packed-bed column with 1.6 cm internal diameter, the increase in the height resulted in increased breakthrough and exhaustion time and a decrease of the breakthrough curves slope, whereby column with 18 cm height and 4 mL/min flow rate ensured the maximum dynamic capacity of 71.17 mg Zn/g (Dabbagh et al. 2016).

The residue of the alginate extraction from *Sargassum filipendula* seaweed as a biosorbent material in a fixed bed was reported to be suitable for removal of Ni^{2+} ions and can be used for treatment of effluents contaminated with toxic metals (Moino et al. 2017). Under optimum conditions *Enteromorpha* sp. was able to remove 87.16% Ni^{2+} at pH 4.79, biomass concentration of 1000 mg/L, contact time 70 min and temperature of 25 °C, while for maximum Cd removal (75.16%) pH of 4.88, biomass concentration of 1000 mg/L, contact time 50 min and temperature of 65 °C were necessary. Freundlich and Langmuir models described well Ni^{2+} and Cd^{2+} biosorption data, respectively, and using Langmuir model for *Enteromorpha* biomass, the maximum sorption capacities of 250 mg Ni/g and 167 mg Cd/g were predicted (Tolian et al. 2015). Using the alginate-based biosorbent produced from *Sargassum* sp. for biosorption of Ni^{2+} and Cu^{2+} ions from synthetic solutions and real electroplating effluents, the sorption equilibrium was observed within 180 min for Ni^{2+} ions and 360 min for Cu^{2+} ions; the adsorption kinetics data could be described by the pseudo-second-order and diffusion in spherical adsorbents, Langmuir model fitted well the experimental data, and the biosorbent showed the maximum sorption capacity 1.147 and 1.640 mmol/g for Ni^{2+} ions and Cu^{2+} ions, respectively. The biosorption of Ni^{2+} and Cu^{2+} ions was found to be mostly a chemical phenomenon, endothermic and spontaneous at temperatures ranging from 293 to 313 K. Using acidic and saline eluents, partial desorption of the Ni^{2+} and Cu^{2+} ions from the biosorbent was obtained allowing its reuse in new sorption/desorption cycles. Decreasing amounts of Ni^{2+} and Cu^{2+} ions biosorbed from real electroplating effluents containing high concentrations of light metals were observed (Barquilha et al. 2019).

In experiments with fixed bed column using *Sargassum glaucescens* brown alga, the highest biosorption capacity of *S. glaucescens* for Sb^{3+} was found to be 5.89 mg/g within a flow rate of 8 mL/min and a bed height of 15 cm, whereby with increasing bed height the breakthrough and exhaustion times increased, while showing a decrease with the enhancement of the flow rate (Dabbagh et al. 2019).

Murphy et al. (2008) compared Cr^{6+} and Cr^{3+} biosorption by red (*Palmaria palmata* and *Polysiphonia lanosa*), green (*Ulva compressa, Ulva linza, Ulva lactuca* and *Ulva intestinalis*) and brown seaweed (*Fucus vesiculosus* and *Fucus spiralis*) dried biomass. *P. palmata* was most effective in removing both Cr^{3+} and Cr^{6+} at low initial concentrations, while at high initial concentrations of metals, *F. vesiculosus* showed the highest efficiency in removing Cr^{3+} and *F. vesiculosus* together with *P. lanosa* had the highest efficiency of removing Cr^{6+}. For binding of Cr to *Ulva* spp. interactions of -NH_2, -COOH, -SO_3^- and -OH groups were responsible. In the rest tested seaweeds besides these function groups also the ether group participated in Cr^{6+} binding to the red seaweeds, and the ether group of brown seaweed was involved in Cr binding as well. Brown seaweeds, *Macrocystis pyrifera* and *Undaria pinnatifida*, tested for removal of Cr^{3+} from aqueous solutions exhibited the highest metal uptake at pH 4. The equilibrium data showed the best fitting with Langmuir model, and the maximum Cr^{3+} sorption capacities (q_{max}) of 0.77 mmol/g and 0.74 mmol/g were determined for *M. pyrifera* and *U. pinnatifida*, respectively (Cazon et al. 2012). Using dried biomass of the brown seaweed *Sargassum dentifolium* grinded to micro-size scale of 0.3868 μm, a 99.68% Cr^{6+} removal efficiency from 100 ppm Cr-contaminated wastewater using 1.5 g biosorbent per 100 mL, 1 h shaking continued to12 h static, pH 7.0 and 50 °C was reported and good fit of experimental data with the Langmuir isotherm suggested a monolayer adsorption on the biosorbent surface (Husien et al. 2019). Using *Laminaria digitata* seaweed biosorbent with -COOH and -OH groups as the major binding sites on the surface (qmax of 2.06 ±0.01 and 1.4±0.7 mmol/g; pKa of 3.28±0.01 and 11±1, respectively) tested for the removal of Cr^{3+} ions from aqueous solutions, it was found that depending on the pH the Cr species present in the solution, Cr^{3+} and $CrOH^{2+}$, exhibited different affinities for the -COOH groups occurring on the surface of the algal biomass. For a description of the kinetics at batch system, a mass transfer kinetics model was used enabling to obtain the distribution of $CrOH^{2+}$ and Cr^{3+} species in solution and at the binding sites (Dittert et al. 2013).

Deniz and Ersanli (2018) applied a coastal seaweed community composed of *Chaetomorpha* sp., *Polysiphonia* sp., *Ulva* sp. and *Cystoseira* sp. species to remove Cu^{2+} ions from synthetic aqueous medium by performing biosorption experiments in batch mode. The biosorption of Cu^{2+} ions was pronouncedly affected by pH, biosorbent amount, metal ion concentration and contact time, and the pore diffusion also played a role in the control of biosorption process. The Cu induced stress and ocean acidification was reflected also in physiological and morphological alterations of *Ulva prolifera*. Moderate enhancement of pCO_2 to 1000 μatm alleviated the toxic impact of Cu on seaweed, but at pCO_2 of 1400 μatm it was restored. At 1400 μatm pCO_2 the presence of 2 μM Cu resulted in branched thalli, likely due to defense mechanism against the abiotic stresses induced by high Cu concentration and high pCO_2 (Gao et al. 2017). The optimized combination of a composition consisting of seaweed (*Gracillaria* sp.), blood clamp (*Anadara granosa*) and zeolite, when used as a biofilter, was able to reduce the content of Cu in the aquatic environment from 1 mg/L to 0.119 mg/L (Achmadi et al. 2019). The optimum condition parameters for Cu^{2+} removal from wastewater by activated carbon fabricated using red seaweed *Gracilaria changii* as adsorbent in the batch experiment were pH 6.0, a contact time of 60 min and adsorbent dose 0.3 g resulting in maximum Cu^{2+} adsorption capacity of 0.07 mg/g. The experimental adsorption data could be fitted by Freundlich isotherm model and pseudo-second-order kinetic model (Lavania-Baloo et al. 2017). Using large brown algae seaweeds, *Saccharina japonica* and *Saccharina sculpera* for removal of Cu^{2+} ions from the aqueous solution, Cu uptake of ca. 0.3 mmol per gram of macroalga with particle sizes <250 μm was observed at the equilibrium Cu concentration of 19 mg/L, while at Cu concentration of 47 mg/L the Cu uptake reached even 1.9 mmol/g macroalga, and this rise could be connected with the diffusion of Cu^{2+} ions through the gel formed at the initial stage of the adsorption, into the inner alginates of the seaweed (Kuzuhara et al. 2018). Brown seaweed *Lessonia nigrescens* tested as biosorbent of Cu ions at pH 3.2±0.2 for 7 days of contact time achieved the maximum experimental uptake of 54.5 and 58.5 mg Cu/g by the blades and the stipes of the alga, respectively; however, the equilibrium sorption isotherms could be described by both Langmuir and Freundlich models only for stipes as biosorbent. On the other hand, the Ho and McKay pseudo-second-order model was suitable to describe the sorption kinetics for both stipes and blades (Hansen et al. 2017). Also, a study of the mechanisms of Cu^{2+} biosorption on *Lessonia nigrescens* dead biomass showed that the Cu^{2+} ions interacted with algal biomass predominantly via the abovementioned groups by ionic and coordinative bonds by ligand multidentism, and rearrangements of the cell wall stiffness (Cid et al. 2018). Maximum capacity for Cu^{2+} biosorption on the dead biomass of brown macroalgae *Durvillaea antarctica* at pH 5.0 was estimated as 1.44 mmol/g, and it was stated that the ion exchange mechanism of Cu^{2+} biomass adsorption can be specified as physisorption in a multilayer heterogeneous system (Cid et al. 2015). For dry algal biomass of brown seaweed *Cystoseira crinitophylla*, adsorption capacity of 160 mg Cu/g at 600 mg Cu/L equilibrium concentration and pH 4.5 was determined, whereby sorption isotherms could be well described with the Langmuir and Freundlich model equations. In the experiments using columns with *Cystoseira* protonated dry biomass, excellent regeneration of sorbent was achieved resulting in 100% efficiency even after 35 sorption/desorption cycles at pH 4.5 and 96% for pH 2.6 (Christoforidis et al. 2015).

The competitive biosorption of Cu^{2+} and Ag^{+} ions was studied in batch systems using acidified waste of *Sargassum filipendula* as biosorbent (2 g/L) at 25 °C for 12 h and pH ca. 5.0. Cu exhibited higher affinity for the biosorbent and a fast biosorption kinetic profile, while Ag equilibrium times depended on the Cu concentration. As the rate-limiting step in Cu^{2+} ion removal, external diffusion was considered, which can limit the kinetic rates of Ag^{+} ions as well. In simultaneous processes of chemisorption and physisorption processes, several mechanisms such as ion-exchange, complexation and electrostatic attraction played a role, whereby notable amounts of Ca^{2+} and Na^{+} were released by the ion-exchange mechanism. A homogenous covering of both metal ions on the surface of the particles was observed, and in the macroporous biosorbent a considerable amount of macropores was filled with tested metal cations (do Nascimento et al. 2019).

Table 3.1 Maximum metal biosorption capacity of selected seaweeds

Seaweed		Metal	pH	q_{max}	References
Green	*Ulva fasciata*	Cd^{2+}	4.0	18.87 mg/g	El-Sheekh et al. (2020)
	Ulva lactuca	Cd^{2+}	7.0	15.60 mg/g	El-Sheekh et al. (2020)
Red	*Kappaphycus alvarezii*	Cd^{2+}	4.5	0.48 mmol/g	Praveen and Vijayaraghavan (2015)
	Kappaphycus alvarezii	Cu^{2+}	4.5	0.47 mmol/g	Praveen and Vijayaraghavan (2015)
	Kappaphycus alvarezii	Ni^{2+}	4.5	0.38 mmol/g	Praveen and Vijayaraghavan (2015)
Brown	*Cystoseira trinodis*	Pb^{2+}	5.2	49.08 mg/g	Salehi et al. (2014)
	Cystoseira trinodis	Ni^{2+}	5.2	14.58 mg/g	Salehi et al. (2014)
	Sargassum ilicifolium	Pb^{2+}	3.7	195 mg/g	Tabaraki et al. (2014)
	Sargassum vulgare	Fe^{3+}	3.0	63.67 mg/g	Benaisa et al. (2016)
	Turbinaria conoides	Al^{3+}	4.0	2.37 mmol/g	Vijayaraghavan et al. (2012a)
	Turbinaria conoides	Cd^{2+}	5.0	0.96 mmol/g	Vijayaraghavan et al. (2012a)

A brief overview of the maximum metal biosorption capacity of selected seaweeds is given in Table 3.1.

6 Metal Levels in Edible Seaweed Species Intended for Consumption

Seaweeds contain pronounced amounts of pigments, minerals and some water-and fat-soluble vitamins, which after extraction could be utilized for enrichment of functional foods. The most important compounds in terms of human nutrition found in 11 European edible seaweed species and their benefits to human health were analyzed by Kraan (2013). Also, the potential use of prebiotics from marine macroalgae, which are rich in polysaccharides, for both humans and animals was discussed by O'Sullivan et al. (2010). On the other hand Chiocchetti et al. (2017) overviewed the occurrence of toxic metal(loid)s in seafood products and discussed the risk resulting from their consumption. The brown algae *Saccharina latissima* and *Alaria esculenta* and the red alga *Palmaria palmata* collected from natural populations and aquaculture in the NE Atlantic were found to be good sources of antioxidants, and the accumulated concentrations of toxic metals were below the upper limits set by the French recommendation and the EU Commission suggesting a low risk of their consumption for humans. Moreover, these seaweeds were good sources of antioxidants achieving the highest mean polyphenol content in winter (for *Alaria* and *Saccharina*) and spring (for *Palmaria*) (Roleda et al. 2019). Circuncisao et al. (2018) summarized the findings related to the mineral content of prevalent edible European macroalgae and focused attention also on the main factors interfering in their accumulation. Accumulation of Mg, and especially Fe, was found to be prevalent in Chlorophyta, in contrast to Rhodophyta and Phaeophyta, which accumulate higher concentrations of Mn and I, respectively. Moreover, the researchers stated that seaweeds can be used to replace NaCl in common foods resulting in increased content in elements that are frequently deficient in the European population. Taxonomic genus and to a lesser extent also the geographic origin strongly affected the trace element profile, and a remarkable risk related to seaweed consumption connected with increased levels of Al and Cd was estimated (Miedico et al. 2017).

Cherry et al. (2019) in their review paper analysed the nutritional composition of edible seaweeds; evaluated health benefits connected with consumption of whole seaweeds extracted bioactive components, and seaweed-based food products for humans; and drew attention also to the potential of negative impact of the consumption of edible seaweeds containing excess iodine and As. Determination of Cd, Cu, Hg, Pb, Zn, total As and inorganic As in 52 samples from 11 algae-based products commercialized in Spain for direct human consumption (*Gelidium* spp., *Porphyra umbilicales*, *Chondrus crispus*, *Undaria pinnatifida*, *Ulva rigida*, *Laminaria* spp., *Hizikia fusiforme*, *Eisenia bicyclis* and *Himanthalia elongata*) showed that the highest values of total and inorganic As contained samples of *Hizikia fusiforme* and that most Cd concentrations exceeded the French Legislation (Besada et al. 2009). Paz et al. (2019a) determined the content of toxic metals (Al, Cd, Pb and Hg) in edible seaweed samples consumed in Europa to assess the toxicological risk from the intake of these metals. The highest level of Al was recorded in seaweed salad (57.5 mg Al/kg d.w.), while the highest concentrations of Al (38.9 mg/kg d.w.), Cd (0.59 mg/kg d.w.) and Pb (0.40 mg/kg d.w.) were shown in Asian algae and the highest concentration of Hg (0.017 mg/kg d.w.) was found in European algae. The researchers stated that the consumption of 5 g a day of dehydrated seaweed would not pose a risk to the health of adults. Investigation of metal abundance in seaweeds from locations along the Firth of Forth and Forth Estuary in Scotland showed that within the same macroalgal species, abundances of As, I, Pb, Cu, Cd and U were affected by mixing between freshwater riverine and North Sea marine saltwater, whereby additional mixing of natural and anthropogenic inputs from the surrounding geology and industry had an impact on Zn, Ni, Co, Re and Os accumulation in macroalgae (Ownsworth et al. 2019).

Determination of the chemical composition of three edible seaweeds, *Hypnea spicifera*, *Codium capitatum* and *Sargassum elegans*, from the Indian Ocean, KwaZulu-Natal coast, South Africa, showed that the moisture level ranged from 85.4 to 89.5%, protein levels from 6.1 to 11.8%, lipids from 7.5 to 13.1% and carbohydrates from 37.8 to 71.9%. The concentrations of individual metals in seaweed decreased as follows: Ca > Mg > Fe > Cu > As > Zn > Ni > C r> Pb > Co ≈ Se. Whereas *C. capitatum* and *H. spicifera* containing low concentrations of toxic metals were found to be suitable as potential sources of most essential nutrients, high levels of As in *S. elegans* ranging from 94.70±6.6 μg/g in winter to 65.10±2.3 μg/g in summer indicate that the consumption of those species could represent certain risk (Magura et al. 2016). Ganesan et al. (2020) evaluated the nutritional value of underexploited edible seaweeds *Gracilaria edulis*, *Acanthophora spicifera*, *Padina gymnospora*, *Enteromorpha flexuosa* and *Ulva fasciata* with recommended dietary allowances and found high levels of micronutrients such as Fe (14.8–72 mg/100 g), iodine (38.8–72.2 mg/100 g), and Ca (410–870 mg/100 g). In *Ulva fasiciata* the levels of essential aminoacids ranged from 189.2 to 306 mg/g, and the seaweed contained 0.80% arachidonic acid, 0.59% of linoleic acid and monounsaturated fatty acids ranging from 3.05 to 14.08%. The observed contents of toxic metals in tested species were 0.012–0.076 ppm for As and 0.012–0.081 ppm for Cd, while 0.030 ppm for Hg (concentration within the tolerable limit) was found only in *A. spicifera*. The researches stated that the abovementioned nutrients correspond to > 70% of macro- and micronutrients in precise combination of recommended dietary allowances suggested for pregnant women and could be used as an alternative food source. Rubio et al. (2017) analysed concentrations of 20 metals in edible seaweeds (*Gelidium*, *Palmaria*, *Chondrus*, *Porphyra*, *Himanthalia*, *Eisenia*, *Laminaria*, *Undaria*) originating from Asia and EU cultivated using conventional or organic practices. Higher concentrations of trace and toxic elements were found in red seaweed, suggesting the potential use of *Porphyra* as a potential bioindicator for metals. Considerable differences were observed between the average metallic content of Asian and European seaweed. The average Cd concentration in seaweeds from the conventional cultivation was more than twofold compared with that from the organic cultivation (0.28 vs. 0.13 mg/kg). By consumption of seaweed at 4 g/day the dietary intake of Mg and Cr could be supported, and the mean intakes of Al, Cd and Pb of 0.064, 0.001 and 0.0003 mg/day, respectively, do not cause health problems when other toxic metals in seaweed are missing.

Paul et al. (2014) compared biomass productivities and biochemical properties of *Caulerpa lentillifera* and *C. racemosa*, edible varieties of the green seaweed genus *Caulerpa* known under the collective term "sea grapes". Productivity of tested species in 6 weeks yielded 2 and < 0.5 kg per week for *C. lentillifera* and *C. racemosa*, respectively. The nutritional value of *C. racemosa* was higher than that of *C. lentillifera* for both polyunsaturated fatty acids (10.6 vs. 5.3 mg/g d.w.) and pigments (9.4 vs. 4.2 mg/g d.w.), while the increasing frond size resulted in decreasing content of eicosapentaenoic acid in both species. Higher levels of Zn, Mg and Sr were estimated in *C. lentillifera*, while *C. racemosa* had higher levels of Se. As and Cd contents (1 vs. 0.1 ppm) were higher in *C. lentillifera* in comparison with *C. racemosa* showing higher Pb, Cu and V levels. Based on high biomass production potential in monoculture and nutritional properties the researchers recommended increased commercialization of *C. lentillifera* as aquaculture product in tropical Australia and in Southeast Asia.

Etemadian et al. (2018) evaluated nutritional properties of two dried brown seaweeds *Sirophysalis trinodis* and *Polycladia myrica* and found that they contained all of the essential amino acids, especially methionine, isoleucine, leucine, lysine, threonine, valine, phenylalanine and arginine as well as high amounts of important fatty acids (linolenic acid, palmitic acid and oleic acid), proline, Ca, Mg, Fe Mn, Cu an Zn, and thus, these seaweeds can be used as a valuable source of nutrient supplements for humans and animals. From the most consumed edible seaweeds in Europe, highest concentrations of K (57,480 mg/kg d.w.) were observed in *Himanthalia elongata*, while the highest Fe content (58.8 mg/kg d.w.) was determined in Asian *Undaria pinnatifida* species, whereby a dose of 5 g/day of dehydrated seaweed of this species can contribute to the admissible daily intake of Mg (9.32% adults) and Na (7.05% adults) (Paz et al. (2019b).

7 Conclusions

Seaweeds, which together with corals create marvelous underwater formations playing with countless colors, are photosynthesizing organisms producing a significant amount of oxygen in the atmosphere. However, excess concentrations of toxic metals generating harmful reactive oxygen species and causing oxidative stress adversely affect their growth and development reflected in the inhibition of photosynthetic processes, reduced growth, altered algal cell ultrastructure, up- or down-regulation of gene expression and increased accumulation of toxic metals in seaweed biomass. Sensitive responses of seaweeds to the presence of toxic metals enable to use them for monitoring of metal pollution in marine environment originating predominantly from anthropogenic activities in coastal regions by investigating their morphological and physiological characteristics. The surface of these marine macroalgae contains various functional groups able to interact with toxic metal ions and bulk or nanoscale metal particles resulting in biosorption, whereby the metal concentrations bioaccumulated by seaweeds often correlate not only with the metal concentrations in surrounding aqueous environment but also with those in the sediments. Some seaweed species belonging mainly to brown algae (*Phaeophyceae*), particularly Sargassum sp., which show a great rate of tolerance against toxic metals, can accumulate in the biomass considerable metal amounts and consequently can be used for the removal of metals from metal-contaminated environment, whether using living or non-living seaweed biomass. Removal of toxic metals using living seaweed biomass can be used in situ in coastal areas close to industrial effluent discharges. On the other hand, the use of cheap metal sorbents prepared from dead seaweed biomass, which are effective also after several sorption-desorption cycles, is advantageous even before metal-contaminated wastewaters are discharged into the sea or rivers. Considering the high nutritional value of edible seaweeds containing high amounts of essential minerals, several vitamins and other health-promoting compounds it is inevitable to ensure that the accumulated levels of toxic metals do not exceed the permissible levels and did not pose a health risk to the human population. In general, seaweed biomass originating from renewable resources could be considered as a very prospective inexpensive material for effective removal of metals from the aqueous environment, and widespread consummation of edible seaweed species in the form of food supplements can exhibit beneficial impact on the health of increasingly growing human population.

Acknowledgements The authors thank Doctor Jon Nield, Principal Research Fellow in the School of Biological and Chemical Sciences, The Queen Mary University of London, for courtesy to use the scheme of PSII (http://macromol.sbcs.qmul.ac.uk/oldsite/psIIimages/PSII.html). This study was supported by the Slovak Research and Development Agency (project APVV-17-0318), and by and the Slovak Grant Agency for Science (VEGA 1/0787/18).

References

Achmadi I, Setya B, Ahadi AA (2019) The effectiveness of combination of seaweed (*Gracillaria* sp.), blood clamp (*Anadara granosa*), and zeolite as biofilter in the reduction of heavy metal copper (Cu). 1st International Conference on Fisheries and Marine Science. In: Mukti AT, Sulmartiwi L, Mubarak AS (eds) IOP Conference Series-Earth and Environmental Science 236:UNSP 012058

Aden M, Husson J, Monney S, Franchi M, Knorr M, Euvrard M (2019) Biosorption of Pb(II) ions from aqueous solution using alginates extracted from Djiboutian seaweeds and deposited on silica particles. Pure Appl Chem 91(3):459–475

Alahverdi M, Savabieasfahani M (2012) Seaweed and chlorophyll as biomarkers of metals in the Persian Gulf, Iran. Bull Environ Contam Toxicol 89(3):501–506

Ali AYA, Idris AM, Ebrahim AM, Eltayeb MAH (2017) Brown algae (Phaeophyta) for monitoring heavy metals at the Sudanese Red Sea coast. Appl Water Sci 7(7):3817–3824

Allen JF, Wilson BM, de Paula WBM, Puthiyaveetil S, Nield J (2011) A structural phylogenetic map for chloroplast photosynthesis. Trends Plant Sci 16(12):645–655

Amosu AO, Robertson-Andersson DV, Kean E, Maneveldt GW, Cyster L (2016) Biofiltering and uptake of dissolved nutrients by *Ulva armoricana* (Chlorophyta) in a land-based aquaculture system. Int J Agric Biol 18(2):298–304

Arumugam N, Chelliapan S, Kamyab H, Thirugnana S, Othman N, Nasri NS (2018) Treatment of wastewater using seaweed: a review. Int J Environ Res Public Health 15(12):2851

Babu MY, Palanikumar L, Nagarani N, Devi VJ, Kumar SR, Ramakritinan CM, Kumaraguru AK (2014) Cadmium and copper toxicity in three marine macroalgae: evaluation of the biochemical responses and DNA damage. Environ Sci Pollut Res 21(16):9604–9616

Bakir A, McLoughlin P, Tofail SAM, Fitzgerald E (2009) Competitive sorption of antimony with zinc, nickel, and aluminum in a seaweed based fixed-bed sorption column. Clean 37(9):712–719

Barber J (1999) Organisation of the photosystem two light harvesting system. In: Argyroudi-Akoyunoglou H, Senger H (eds) The chloroplast: from molecular biology to biotechnology. Kluwer Academic, Dordrecht, pp 11–18

Barber J (2016) Photosystem II: the water splitting enzyme of photosynthesis and the origin of oxygen in our atmosphere. Q Rev Biophys 49:e14

Barber J, Tran PD (2013) From natural to artificial photosynthesis. J R Soc Interface 10(81):20120984

Barett J, Anderson JM (1980) The P-700-chlorophyl a-protein complex and two major light-harvesting complexes of *Acrocarpia paniculata* and other brown seaweeds. Biochim Biophys Acta 590:309–323

Barquilha CER, Cossich ES, Tavares CRG, da Silva EA (2019) Biosorption of nickel(II) and copper(II) ions from synthetic and real effluents by alginate-based biosorbent produced from seaweed *Sargassum* sp. Environ Sci Pollut Res 26(11):11100–11112

Bastos E, Schneider M, de Quadros DPC, Welz B, Batista MB, Horta PA, Rorig LR, Barufi JB (2019) Phytoremediation potential of *Ulva ohnoi* (Chlorophyta): Influence of temperature and salinity on the uptake efficiency and toxicity of cadmium. Ecotoxicol Environ Saf 174:334–343

Benaisa S, El Mail R, Jbari N (2016) Biosorption of Fe(III) from aqueous solution using brown algae *Sargassum vulgare*. J Mater Environ Sci 7(5):1461–1468

Berkaloff C, Duval JC (1980) Correlated influence of cation concentration and excitation intensity on PS II activity-II. Comparative study between green plant and brown-alga chloroplasts. Photosynth Res 1:127–135

Besada V, Andrade JM, Schultze F, Gonzalez JJ (2009) Heavy metals in edible seaweeds commercialised for human consumption. J Mar Syst 75(1–2):305–313

Betancor S, Dominguez B, Tuya F, Figueroa FL, Haroun R (2015) Photosynthetic performance and photoprotection of *Cystoseira humilis* (*Phaeophyceae*) and *Digenea simplex* (*Rhodophyceae*) in an intertidal rock pool. Aquat Bot 121:16–25

Bonanno G, Veneziano V, Orlando-Bonaca M (2020) Comparative assessment of trace element accumulation and biomonitoring in seaweed *Ulva lactuca* and seagrass *Posidonia oceanica*. Sci Total Environ 718:137413

Boubonari T, Malea P, Koyro HW, Kevrekidis T (2008) The red macroalga *Gracilaria bursa-pastoris* as a bioindicator of metals (Fe, Zn, Cu, Pb, Cd) in oligohaline coastal environments. Fresen Environ Bull 17(12B):2207–2216

Boundir Y, Hasni M, Rafik F, Sabri H, Bahammou N, Cheggour M, Achtak H, Cherifi O (2019) First study of the ecological status in the Atlantic coast of Morocco using the brown seaweed *Cystoseira tamariscifolia*. Appl Ecol Environ Res 17(6):14315–14331

Cabrita MT, Duarte B, Gameiro C, Godinho RM, Cacador I (2018) Photochemical features and trace element substituted chlorophylls as early detection biomarkers of metal exposure in the model diatom *Phaeodactylum tricornutum*. Ecol Indic 95(Part 2):1038–1052

Capdevila P, Hereu B, Salguero-Gomez R, la Rovira G, Medrano A, Cebrian E, Garrabou J, Kerstine DK, Linares C (2019) Warming impacts on early life stages increase the vulnerability and delay the population recovery of a long-lived habitat-forming macroalga. J Ecol 107(3):1129–1140

Cazon JPH, Benitez L, Donati E, Viera M (2012) Biosorption of chromium(III) by two brown algae *Macrocystis pyrifera* and *Undaria pinnatifida*: equilibrium and kinetic study. Eng Life Sci 12(1):95–103

Cechinel MAP, Mayer DA, Mazur LP, Silva LGM, Girardi A, Vilar VJP, de Souza AAU, de Souza SMAGU (2018) Application of ecofriendly cation exchangers (*Gracilaria caudata* and *Gracilaria cervicornis*) for metal ions separation and recovery from a synthetic petrochemical wastewater: batch and fixed bed studies. J Clean Prod 172:1928–1945

Celekli A, Bulut H (2020) Biochemical and morphological responses to cadmium-induced oxidative stress in *Cladophora glomerata*. Turk J Bot 44(3):222–231

Chalkley R, Child F, Al-Thaqafi K, Dean AP, White KN, Pittman JK (2019) Macroalgae as spatial and temporal bioindicators of coastal metal pollution following remediation and diversion of acid mine drainage. Ecotoxicol Environ Saf 182:UNSP 109458

Chan ECS (2003) Microbial nutrition and basic netabolism. In: Mara D, Horan N (eds) Handbook of water and wastewater microbiology. Academic, London, pp 3–36

Chen ZW, Song SF, Wen Y, Zou Y, Liu HJ (2016) Toxicity of Cu (II) to the green alga *Chlorella vulgaris*: a perspective of photosynthesis and oxidant stress. Environ Sci Pollut Res 23(18):17910–17918

Cherry P, O'Hara C, Magee PJ, McSorley EM, Allsopp PJ (2019) Risks and benefits of consuming edible seaweeds. Nutr Rev 77(5):307–329

Chiocchetti G, Jadan-Piedra C, Velez D, Devesa V (2017) Metal(loid) contamination in seafood products. Crit Rev Food Sci Nutr 57(17):3715–3728

Christoforidis AK, Orfanidis S, Papageorgiou SK, Lazaridou AN, Favvas EP, Ach M (2015) Study of Cu(II) removal by *Cystoseira crinitophylla* biomass in batch and continuous flow biosorption. Chem Eng J 277:334–340

Cid H, Ortiz C, Pizarro J, Barros D, Castillo X, Giraldo L, Moreno-Pirajan JC (2015) Characterization of copper (II) biosorption by brown algae *Durvillaea antarctica* dead biomass. Adsorption 21(8):645–658

Cid HA, Flores MI, Pizarro JF, Castillo XA, Barros DE, Moreno-Pirajan JC, Ortiz CA (2018) Mechanisms of Cu^{2+} biosorption on *Lessonia nigrescens* dead biomass: functional groups interactions and morphological characterization. J Environ Chem Eng 6(2):2696–2704

Circuncisao AR, Catarino MD, Cardoso SM, Silva AMS (2018) Minerals from macroalgae origin: health benefits and risks for consumers. Mar Drugs 16(11):400

Costa GB, Simioni C, Pereira DT, Ramlov F, Maraschin M, Chow F, Horta PA, Bouzon ZL, Schmidt EC (2017a) The brown seaweed *Sargassum cymosum*: changes in metabolism and cellular organization after long-term exposure to cadmium. Protoplasma 254(2):817–837

Costa GB, Simioni C, Ramlov F, Maraschin M, Chow F, Bouzon ZL, Schmidt EC (2017b) Effects of manganese on the physiology and ultrastructure of *Sargassum cymosum*. Environ Exp Bot 133:24–34

Costa GB, Ramlov F, Koerich G, Navarro BB, Cabral D, Rodrigues ERO, Ramos B, Fadigas SD, Rorig LR, Maraschin M, Horta PA (2019) The effects of mining tailings in the physiology of benthic algae: understanding the relation between mud's inductive acidification and the heavy metal's toxicity. Environ Exp Bot 167:103818

Dabbagh R, Sharifipoor S, Keshtkar A, Vafajoo L (2016) Removal of zinc (II) from synthetic effluent using seaweeds: a review of modeling of fixed-bed columns. Desalin Water Treat 57(1):24509–24518

Dabbagh R, Mirkamali MS, Vafajoo L (2019) Removal of antimony metalloid from synthetic effluent using seaweed as a low-cost natural sorbent: adsorption on a fixed-bed column. J Water Chem Technol 41(1):21–28

Davis TA, Volesky B, Mucci A (2003) A review of the biochemistry of heavy metal biosorption by brown algae. Water Res 37(18):4311–4330

Deniz F, Ersanli ET (2018) An ecofriendly approach for bioremediation of contaminated water environment: potential contribution of a coastal seaweed community to environmental improvement. Int J Phytoremediation 20(3):256–263

Dittert IM, Vilar VJP, da Silva EAB, de Souza SMAGU, de Souza AAU, Botelho CMS, Boaventura RAR (2013) Turning *Laminaria digitata* seaweed into a resource for sustainable and ecological removal of trivalent chromium ions from aqueous solutions. Clean Techn Environ Policy 15(6):955–965

do Nascimento WJ, da Silva MGC, Vieira MGA (2019) Competitive biosorption of Cu^{2+} and Ag^{+} ions on brown macro-algae waste: kinetic and ion-exchange studies. Environ Sci Pollut Res 26(23):23416–23428

Dong LL, Zhang GQ, Li W, Ding T, Wang HX, Zhang G (2020) Effects of Cu^{2+} and Hg^{2+} on growth and photosynthesis of two *Scenedesmus* species. Pol J Environ Stud 29(2):1129–1135

dos Santos RW, Schmidt EC, Felix MRD, Polo LK, Kreusch M, Pereira DT, Costa GB, Simioni C, Chow F, Ramlov F, Maraschin M, Bouzon ZL (2014) Bioabsorption of cadmium, copper and lead by the red macroalga *Gelidium floridanum*: physiological responses and ultrastructure features. Ecotoxicol Environ Saf 105:80–89

Einav R (2020) Chlorophyta—green seaweed. Blue ecosystems. https://www.blue-ecosystems.com/racheliSeaWeed/English/Chlorophyta%2D%2D-green-seaweed

El-Sheekh M, El-Sabagh S, Abou Elsoud G, Elbeltagy A (2020) Efficacy of immobilized biomass of the seaweeds *Ulva lactuca* and *Ulva fasciata* for cadmium biosorption. Iran J Sci Technol A 44:37–49

Etemadian Y, Shabanpour B, Ramzanpour Z, Shaviklo AR, Kordjazi M (2018) Nutritional and functional properties of two dried brown seaweeds *Sirophysalis trinodis* and *Polycladia myrica*. J Aquat Food Prod Technol 27(2):219–235

Farias DR, Hurd CL, Eriksen RS, Simioni C, Schmidt E, Bouzon ZL, Macleod CK (2017a) In situ assessment of *Ulva australis* as a monitoring and management tool for metal pollution. J Appl Phycol 29(5):2489–2502

Farias DR, Schmidt E, Simioni C, Bouzon ZL, Hurd CL, Eriksen RS, Macleod CK (2017b) Photosynthetic and ultrastructural responses of *Ulva australis* to Zn stress. Micron 103:45–52

Farias DR, Hurd CL, Eriksen RS, Macleod CK (2019) *Ulva australis* as a tool for monitoring metal-polluted estuarine system; spatial and temporal considerations. Glob Nest J 21(4):449–454

Fellous A, Andrade S, Vidal-Ramirez F, Calderon R, Beltran J, Correa JA (2017) Modulatory effect of the exudates released by the brown kelp *Lessonia spicata* on the toxicity of copper in early developmental stages of ecologically related organisms. Environ Sci Pollut Res 24(4):3900–3911

Ferreira VRA, Azenha MA, Pereira CM, Silva AF (2017) Preparation and evaluation of Pb(II)-imprinted fucoidan-based sorbents. React Funct Polym 115:53–62

Figueroa FL, Celis-Pla PSM, Martinez B, Korbee N, Trilla A, Arenas F (2019) Yield losses and electron transport rate as indicators of thermal stress in *Fucus serratus* (Ochrophyta). Algal Res 41:101560

Filote C, Volf I, Santos SCR, Botelho CMS (2019) Bioadsorptive removal of Pb(II) from aqueous solution by the biorefinery waste of *Fucus spiralis*. Sci Total Environ 648:1201–1209

Gabruk NG, Lien VT, Duong BT (2015) Sorption of copper ions by brown algae *Sargassum swartzii*. Tomsk State Univ J 391:227–231

Gan TT, Zhao NJ, Yin GF, Chen M, Wang X, Liu JG, Liu WQ (2019) Optimal chlorophyll fluorescence parameter selection for rapid and sensitive detection of lead toxicity to marine microalgae *Nitzschia closterium* based on chlorophyll fluorescence technology. J Photochem Photobiol B 197:111551

Ganesan AR, Subramani K, Shanmugam M, Seedevi P, Park S, Alfarhan AH, Rajagopal R, Balasubramanian B (2020) A comparison of nutritional value of underexploited edible seaweeds with recommended dietary allowances. J King Saud Univ Sci 32(1):1206–1211

Gantt E, Grabowski B, Cunningham FX (2003) Antenna systems of red algae: phycobilisomes with photosystem II and chlorophyll complexes with photosystem I. In: Green BR, Parson WW (eds) Light-harvesting antennas in photosynthesis. Advances in photosynthesis and respiration, vol 13. Springer, Dordrecht, pp 307–322

Gao S, Niu JF, Chen WZ, Wang GC, Xie XJ, Pan GH, Gu WH, Zhu DL (2013) The physiological links of the increased photosystem II activity in moderately desiccated *Porphyra haitanensis* (Bangiales, Rhodophyta) to the cyclic electron flow during desiccation and re-hydration. Photosynth Res 116(1):45–54

Gao S, Huan L, Lu XP, Jin WH, Wang XL, Wu MJ, Wang GC (2016) Photosynthetic responses of the low intertidal macroalga *Sargassum fusiforme* (*Sargassaceae*) to saline stress. Photosynthetica 54(3):430–437

Gao G, Liu YM, Li XS, Feng ZH, Xu ZG, Wu HY, Xu JT (2017) Expected CO_2-induced ocean acidification modulates copper toxicity in the green tide alga *Ulva prolifera*. Environ Exp Bot 135:63–72

Gao S, Chi Z, Chen HL, Zheng ZB, Weng YX, Wang GC (2019) A supercomplex, of approximately 720 kDa and composed of both photosystem reaction centers, dissipates excess energy by PSI in green macroalgae under salt stress. Plant Cell Physiol 60(1):166–175

Han T, Kang SH, Park JS, Lee HK, Brown MT (2008) Physiological responses of *Ulva pertusa* and *U. armoricana* to copper exposure. Aquat Toxicol 86(2):176–184

Han TW, Tseng CC, Cai MG, Chen K, Cheng SY, Wang J (2020) Effects of cadmium on bioaccumulation, bioabsorption, and photosynthesis in *Sarcodia suiae*. Int J Environ Res Public Health 17(4):1294

Hansen HK, Gutierrez C, Madrid A, Jimenez R, Larach H (2017) Possible use of the algae *Lessonia nigrescens* as a biosorbent: Differences in copper sorption behavior using either blades or stipes. Waste Biomass Valori 8(4):1295–1302

Henriques B, Rocha LS, Lopes CB, Figueira P, Monteiro RJR, Duarte AC, Pardal MA, Pereira E (2015) Study on bioaccumulation and biosorption of mercury by living marine macroalgae: Prospecting for a new remediation biotechnology applied to saline waters. Chem Eng J 281:759–770

Heuzé V, Tran G, Giger-Reverdin S, Lessire M, Lebas F (2017) *Seaweeds (marine macroalgae)*. Feedipedia, a programme by INRA, CIRAD, AFZ and FAO. https://www.feedipedia.org/node/78

Holzinger A, Herburger K, Kaplan F, Lewis LA (2015) Desiccation tolerance in the chlorophyte green alga *Ulva compressa*: does cell wall architecture contribute to ecological success? Planta 242(2):477–492

Huan L, Wang C, Gao S, He LW, Lu XP, Wang XL, Liu XH, Wang GC (2018) Preliminary comparison of atmospheric CO_2 enhancement to photosynthesis of *Pyropia yezoensis* (Bangiales, Rhodophyta) leafy thalli and filamentous thalli. Phycol Res 66(2):117–126

Huang J, Cheng JP, Yi J (2016) Impact of silver nanoparticles on marine diatom *Skeletonema costatum*. J Appl Toxicol 36(10):1343–1354

Husien S, Labena A, El-Belely EF, Mahmoud HM, Hamouda AS (2019) Adsorption studies of hexavalent chromium [Cr(VI)] on micro-scale biomass of *Sargassum dentifolium*, seaweed. J Environ Chem Eng 7(6):UNSP 103444
Ifuku K, Noguchi T (2016) Structural coupling of extrinsic proteins with the oxygen-evolving center in photosystem II. Front Plant Sci 7:84
Ismail GA, Ismail MM (2017) Variation in oxidative stress indices of two green seaweeds growing under different heavy metal stresses. Environ Monit Assess 189(2):68
Jadeja RN, Zhou QX (2018) Comparative study of cadmium bio-sorption by red, green and brown seaweed biomass collected from Yellow sea, China. Indian J Geo Mar Sci 47(8):1561–1565
Jafari SA, Jamali A (2016) Continuous cadmium removal from aqueous solutions by seaweed in a packed-bed column under consecutive sorption-desorption cycles. Korean J Chem Eng 33(4):1296–1304
Jafari SA, Jamali A, Hosseini A (2015) Cadmium removal from aqueous solution by brown seaweed, *Sargassum angustifolium*. Korean J Chem Eng 32(10):2053–2066
Jarvis TA, Bielmyer-Fraser GK (2015) Accumulation and effects of metal mixtures in two seaweed species. Comp Biochem Physiol C Pharmacol 171:28–33
Jha B, Basha S, Jaiswar S, Mishra B, Thakur MC (2009) Biosorption of Cd(II) and Pb(II) onto brown seaweed, *Lobophora variegata* (Lamouroux): kinetic and equilibrium studies. Biodegradation 20(1):1–13
Ji Y, Xie XJ, Wang GC (2018) Effects of the heavy metal cadmium on photosynthetic activity and the xanthophyll cycle in *Phaeodactylum tricornutum*. J Oceanol Limnol 36(6):2194–2201
Jiang YG, Zhu YL, Hu ZL, Lei AP, Wang JX (2016) Towards elucidation of the toxic mechanism of copper on the model green alga *Chlamydomonas reinhardtii*. Ecotoxicology 25(7):1417–1425
Jiang JN, Yu YY, Zheng MS, Liu NN, Li YH, Xu NJ (2020) High light might alleviate inhibitory effects of high temperature on growth and physiological parameters of *Ulva prolifera*. Aquacult Res 51(5):2062–2070
Kamala-Kannan S, Batvari BPD, Lee KJ, Kannan N, Krishnamoorthy R, Shanthi K, Jayaprakash M (2008) Assessment of heavy metals (Cd, Cr and Pb) in water, sediment and seaweed (*Ulva lactuca*) in the Pulicat Lake, South East India. Chemosphere 71(7):1233–1240
Kanamarlapudi SLRK, Chintalpudi VK, Muddada S (2018) Application of biosorption for removal of heavy metals from wastewater. In: Derco J, Vrana B (eds) Biosorption. IntechOpen, London. https://www.intechopen.com/books/biosorption/application-of-biosorption-for-removal-of-heavy-metals-from-wastewater
Kaur R, Singh J, Khare R, Ali A (2012) Biosorption the possible alternative to existing conventional technologies for sequestering heavy metal ions from aqueous streams: a review. Int J Environ Public Health 15(12):2851
Kawai H, Henry EC (2017) Phaeophyta. In: Archibald J, Simpson A, Slamovits C (eds) Handbook of the protists. Springer, Cham, pp 267–304
Kidgell JT, de Nys R, Paul NA, Roberts DA (2014) The sequential application of macroalgal biosorbents for the bioremediation of a complex industrial effluent. PLoS One 9(7):e101309
Kim SK (2011) Handbook of marine macroalgae: biotechnology and applied phycology. Wiley-Blackwell, New York
Kraan S (2013) Pigments and minor compounds in algae. In: Dominguez H (ed) Functional ingredients from algae for foods and nutraceuticals. Woodhead Publishing, Cambridge, pp 205–251
Kráľová K, Jampílek J (2021) Impact of metal nanoparticles on marine and freshwater algae. In: Pessarakli M (ed) Handbook of plant and crop physiology, 4th edn. CRC, Boca Raton. in press
Kráľová K, Masarovičová E, Jampílek J (2019) Plant responses to stress induced by toxic metals and their nanoforms. In: Pessarakli M (ed) Handbook of plant and crop stress, 4th edn. CRC, Taylor and Francis, Boca Raton, pp 479–522
Kreusch M, Poltronieri E, Bouvie F, Batista D, Pereira DT, Ramlov F, Maraschin M, Bouzon ZL, Schmidt EC, Simioni C (2018) Ocean warming and copper pollution: implications for

metabolic compounds of the agarophyte *Gelidium floridanum* (Gelidiales, Rhodophyta). J Phycol 54(6):870–878

Kreusch M, Poltronieri E, Bouvie F, Pereira DT, Batista D, Ramlov F, Maraschin M, Bouzon ZL, Simioni C (2019) Cellular responses of *Gelidium floridanum* (Gelidiales, Rhodophyta) tetraspores under heat wave and copper pollution. J Phycol 55(6):1394–1400

Kumar M, Bijo AJ, Baghel RS, Reddy CRK, Jha B (2012) Selenium and spermine alleviate cadmium induced toxicity in the red seaweed *Gracilaria dura* by regulating antioxidants and DNA methylation. Plant Physiol Biochem 51:129–138

Kuzuhara S, Kudo K, Terakado O (2018) Biosorption of Cu(II) ions by kelps, large brown algae seaweeds, *Saccharina japonica* and *Saccharina sculpera*. J Sustain Metall 4(4):455–460

Lal MA (2018) Photosynthesis. In: Bhatla SC, Lal MA (eds) Plant physiology, development and metabolism. Springer, Singapore, pp 159–226

Lavania-Baloo, Idayu N, Salihi IU, Zainoddin J (2017) The use of macroalgae (*Gracilaria changii*) as bio-adsorbent for Copper (II) removal. In: 7th International Conference on Key Engineering Materials (ICKEM 2017). IOP Conference Series-Materials Science and Engineering 201:UNSP 012031

Leal PP, Hurd CL, Sander SG, Kortner B, Roleda MY (2016) Exposure to chronic and high dissolved copper concentrations impedes meiospore development of the kelps *Macrocystis pyrifera* and *Undaria pinnatifida* (Ochrophyta). Phycologia 55(1):12–20

Lin Z, Lin J, Luan Y, Dai W (2019) Application of algae for heavy metal adsorption: a 20-year meta-analysis. Ecotoxicol Environ Saf 190:110089

Liu L, Zou DH, Jiang H, Chen BB, Zeng XP (2018a) Effects of increased CO_2 and temperature on the growth and photosynthesis in the marine macroalga *Gracilaria lemaneiformis* from the coastal waters of South China. J Appl Phycol 30(2):1271–1280

Liu ZY, Yang W, Xu W, Liu YX (2018b) Removal of elemental mercury by bio-chars derived from seaweed impregnated with potassium iodine. Chem Eng J 339:468–478

Liu ZY, Adewuyi YG, Shi S, Chen H, Li Y, Liu DJ, Liu YX (2019) Removal of gaseous Hg^0 using novel seaweed biomass-based activated carbon. Chem Eng J 366:41–49

Lu F, Dind G, Liu W, Zhan DM, Wu HY, Guo W (2018) Comparative study of responses in the brown algae *Sargassum thunbergii* to zinc and cadmium stress. J Oceanol Limnol 36(3):933–941

Luo HT, Wang Q, Liu ZW, Wang SY, Long AM, Yang YF (2020) Potential bioremediation effects of seaweed *Gracilaria lemaneiformis* on heavy metals in coastal sediment from a typical mariculture zone. Chemosphere 245:125636

Ma J, Wang W, Liu XY, Wang ZQ, Gao G, Wu HL, Li XS, Xu JT (2020) Zinc toxicity alters the photosynthetic response of red alga *Pyropia yezoensis* to ocean acidification. Environ Sci Pollut Res 27(3):3202–3212

Magura J, Moodley R, Jonnalagadda SB (2016) Chemical composition of selected seaweeds from the Indian Ocean, KwaZulu-Natal coast, South Africa. J Environ Sci Health B 51(8):525–533

Magura J, Moodley R, Jonnalagadda SB (2019) Toxic metals (As and Pb) in *Sargassum elegans* Suhr (1840) and its bioactive compounds. Int J Environ Health 29(3):266–275

Malea P, Kevrekidis T (2014) Trace element patterns in marine macroalgae. Sci Total Environ 494:144–157

Masarovičová E, Kráľová K, Šeršeň F (2010) Plant responses to toxic metal stress. In: Pessarakli M (ed) Handbook of plant and crop stress, 3rd edn. CRC Press, Boca Raton, pp 595–634

Mazur LP, Pozdniakova TA, Mayer DA, de Souza SMAGU, Boaventura RAR, Vilar VJP (2017) Cation exchange prediction model for copper binding onto raw brown marine macro-algae *Ascophyllum nodosum*: batch and fixed-bed studies. Chem Eng J 316:255–276

Mendes LF, Stevani CV, Zambotti-Villela L, Yokoya NS, Colepicolo P (2014) Toxic effect of metal cation binary mixtures to the seaweed *Gracilaria domingensis* (Gracilariales, Rhodophyta). Environ Sci Pollut Res 21(13):8216–8223

Miedico O, Pompa C, Tancredi C, Cera A, Pellegrino E, Tarallo M, Chiaravalle AE (2017) Characterisation and chemometric evaluation of 21 trace elements in three edible seaweed species imported from south-east Asia. J Food Compos Anal 64(Part 2):188–197

Millar A (2011) Macroalgae: The Facts. The Fish Site. https://thefishsite.com/articles/macroalgae-the-facts

Moenne A, Gonzalez A, Saez CA (2016) Mechanisms of metal tolerance in marine macroalgae, with emphasis on copper tolerance in Chlorophyta and Rhodophyta. Aquat Toxicol 176:30–37

Moino BP, Costa CSD, da Silva MGC, Vieira MGA (2017) Removal of nickel ions on residue of alginate extraction from *Sargassum filipendula* seaweed in packed bed. Can J Chem Eng 95(11):2120–2128

Mouritsen OG (2013) The science of seaweeds. Marine macroalgae benefit people culturally, industrially, nutritionally, and ecologically. Am Sci 101:458–465

Murphy V, Hughes H, McLoughlin P (2008) Comparative study of chromium biosorption by red, green and brown seaweed biomass. Chemosphere 70(6):1128–1134

Navarrete A, Gonzalez A, Gomez M, Contreras RA, Diaz P, Lobos G, Brown MT, Saez CA, Moenne A (2019) Copper excess detoxification is mediated by a coordinated and complementary induction of glutathione, phytochelatins and metallothioneins in the green seaweed *Ulva compressa*. Plant Physiol Biochem 135:423–431

Negm NA, Abd El Wahed MG, Hassan ARA, Abou Kana MTH (2018) Feasibility of metal adsorption using brown algae and fungi: Effect of biosorbents structure on adsorption isotherm and kinetics. J Mol Liq 264:292–305

Nielsen HD, Nielsen SL (2010) Adaptation to high light irradiances enhances the photosynthetic Cu^{2+} resistance in Cu^{2+} tolerant and non-tolerant populations of the brown macroalgae *Fucus serratus*. Mar Pollut Bull 60(5):710–717

Nielsen SL, Nielsen HD, Pedersen MF (2014) Juvenile life stages of the brown alga *Fucus serratus* L. are more sensitive to combined stress from high copper concentration and temperature than adults. Mar Biol 161(8):1895–1904

Nishikawa E, da Silva MGC, Vieira MGA (2018) Cadmium biosorption by alginate extraction waste and process overview in life cycle assessment context. J Clean Prod 178:166–175

O'Sullivan L, Murphy B, McLoughlin P, Duggan P, Lawlor PG, Hughes H, Gardiner GE (2010) Prebiotics from marine macroalgae for human and animal health applications. Mar Drugs 8(7):2038–2064

Okuku EO, Peter HK (2012) Choose of heavy metals pollution biomonitors: a critic of the method that uses sediments total metals concentration as the benchmark. Int J Environ Res 6(1):313–322

Olivares HG, Lagos NM, Gutierrez CJ, Kittelsen RC, Valenzuela GL, Lillo MEH (2016) Assessment oxidative stress biomarkers and metal bioaccumulation in macroalgae from coastal areas with mining activities in Chile. Environ Monit Assess 188(1):25

Ortiz-Calderon C, Silva HC, Vásquez DB (2017) Metal removal by seaweed biomass. In: Tumuluru JS (ed) Biomass volume estimation and valorization for energy. IntechOpen, London. https://www.intechopen.com/books/biomass-volume-estimation-and-valorization-for-energy/metal-removal-by-seaweed-biomass

Oukarroum A, Polchtchikov S, Perreault F, Popovic R (2012) Temperature influence on silver nanoparticles inhibitory effect on photosystem II photochemistry in two green algae, *Chlorella vulgaris* and *Dunaliella tertiolecta*. Environ Sci Pollut Res 19(5):1755–1762

Ou-Yang HL, Kong XZ, Lavoie M, He W, Qin N, He QS, Yang B, Wang R, Xu FL (2013) Photosynthetic and cellular toxicity of cadmium in *Chlorella vulgaris*. Environ Toxicol Chem 32(12):2762–2770

Ownsworth E, Selby D, Ottley CJ, Unsworth E, Raab A, Feldmann J, Sproson AD, Kuroda J, Faidutti C, Bucker P (2019) Tracing the natural and anthropogenic influence on the trace elemental chemistry of estuarine macroalgae and the implications for human consumption. Sci Total Environ 685:259–272

Parrott SF (2017) Characteristics of Seaweed. Sciencing. https://sciencing.com/do-volvox-eat-8735995.html

Patel A, Tiwari S, Prasad SM (2018) Toxicity assessment of arsenate and arsenite on growth, chlorophyll a fluorescence and antioxidant machinery in *Nostoc muscorum*. Ecotoxicol Environ Saf 157:369–379

Patron-Prado M, Acosta-Vargas B, Serviere-Zaragoza E, Mendez-Rodriguez LC (2010) Copper and cadmium biosorption by dried seaweed *Sargassum sinicola* in saline wastewater. Water Air Soil Pollut 210(1–4):197–202

Patron-Prado M, Casas-Valdez M, Serviere-Zaragoza E, Zenteno-Savin T, Lluch-Cota DB, Mendez-Rodriguez L (2011) Biosorption capacity for cadmium of brown seaweed *Sargassum sinicola* and *Sargassum lapazeanum* in the gulf of California. Water Air Soil Pollut 221(1–4):137–144

Paul NA, Neveux N, Magnusson M, de Nys R (2014) Comparative production and nutritional value of "sea grapes"—the tropical green seaweeds *Caulerpa lentillifera* and *C. racemosa*. J Appl Phycol 26(4):1833–1844

Paz S, Rubio C, Frias I, Gutierrez AJ, Gonzalez-Weller D, Martin V, Revert C, Hardisson A (2019a) Toxic metals (Al, Cd, Pb and Hg) in the most consumed edible seaweeds in Europe. Chemosphere 218:879–884

Paz S, Rubio C, Frias I, Luis-Gonzalez G, Gutierrez AJ, Gonzalez-Weller D, Hardisson A (2019b) Human exposure assessment to macro- and trace elements in the most consumed edible seaweeds in Europe. Environ Sci Pollut Res 26(36):36478–36485

Pennesi C, Veglio F, Totti C, Romagnoli T, Beolchini F (2012) Nonliving biomass of marine macrophytes as arsenic(V) biosorbents. J Appl Phycol 24(6):1495–1502

Pinto E, Sigaud-Kutner TCS, Leitao MAS, Okamoto OK, Morse D, Colepicolo P (2003) Heavy metal-induced oxidative stress in algae. J Phycol 39(6):1008–1018

Praveen RS, Vijayaraghavan K (2015) Optimization of Cu(II), Ni(II), Cd(II) and Pb(II) biosorption by red marine alga *Kappaphycus alvarezii*. Desalin Water Treat 55(7):1816–1824

PRNSA (2013) Point Reyes National Seashore Association. Seaweeds: plants or algae? https://www.ptreyes.org/activities/seaweeds-plants-or-algae

Quintano E, Celis-Pla PSM, Martinez B, Diez I, Muguerza N, Figueroa FL, Gorostiaga JM (2019) Ecophysiological responses of a threatened red alga to increased irradiance in an in situ transplant experiment. Mar Environ Res 144:166–177

Racionero-Gomez B, Sproson AD, Selby D, Gannoun A, Grocke DR, Greenwell HC, Burton KW (2017) Osmium uptake, distribution, and $^{187}Os/^{188}Os$ and $^{187}Re/^{188}Os$ compositions in *Phaeophyceae* macroalgae, *Fucus vesiculosus*: implications for determining the $^{187}Os/^{188}Os$ composition of seawater. Geochim Cosmochim Acta 199:48–57

Rajaram R, Rameshkumar S, Anandkumar A (2020) Health risk assessment and potentiality of green seaweeds on bioaccumulation of trace elements along the Palk Bay coast, Southeastern India. Mar Pollut Bull 154:111069

Ramesh K, Berry S, Brown MT (2015) Accumulation of silver by *Fucus* spp. (*Phaeophyceae*) and its toxicity to *Fucus ceranoides* under different salinity regimes. Ecotoxicology 24(6):1250–1258

Reis PA, Cassiano J, Veiga P, Rubal M, Sousa-Pinto I (2014) *Fucus spiralis* as monitoring tool of metal contamination in the northwest coast of Portugal under the European Water Framework Directives. Environ Monit Assess 186(9):5447–5460

Rodriguez-Rojas F, Celis-Pla PSM, Mendez L, Moenne F, Munoz PT, Gabriela LM, Diaz P, Sanchez-Lizaso JL, Brown MT, Moenne A, Saez CA (2019) MAPK pathway under chronic copper excess in green macroalgae (Chlorophyta): involvement in the regulation of detoxification mechanisms. Int J Mol Sci 20(18):4546

Roleda MY, Marfaing H, Desnica N, Jonsdottir R, Skjermo J, Rebours C, Nitschke U (2019) Variations in polyphenol and heavy metal contents of wild-harvested and cultivated seaweed bulk biomass: health risk assessment and implication for food applications. Food Control 95:121–134

Ronan JM, Stengel DB, Raab A, Feldmann J, O'Hea L, Bralatei E, McGovern E (2017) High proportions of inorganic arsenic in *Laminaria digitata* but not in *Ascophyllum nodosum* samples from Ireland. Chemosphere 186:17–23

Roncarati F, Saez CA, Greco M, Gledhill M, Bitonti MB, Brown MT (2015) Response differences between *Ectocarpus siliculosus* populations to copper stress involve cellular exclusion and induction of the phytochelatin biosynthetic pathway. Aquat Toxicol 159:167–175
Rubio C, Napoleone G, Luis-Gonzalez G, Gutierrez AJ, Gonzalez-Weller D, Hardisson A, Revert C (2017) Metals in edible seaweed. Chemosphere 173:572–579
Ryan S, McLoughlin P, O'Donovan O (2012) A comprehensive study of metal distribution in three main classes of seaweed. Environ Pollut 167:171–177
Rybak A, Messyasz B, Leska B (2013) The accumulation of metal (Co, Cr, Cu, Mn and Zn) in freshwater *Ulva* (Chlorophyta) and its habitat. Ecotoxicology 22(3):558–573
Saez CA, Gonzalez A, Contreras RA, Moody AJ, Moenne A, Brown MT (2015a) A novel field transplantation technique reveals intra-specific metal-induced oxidative responses in strains of *Ectocarpus siliculosus* with different pollution histories. Environ Pollut 199:130–138
Saez CA, Roncarati F, Moenne A, Moody AJ, Brown MT (2015b) Copper-induced intra-specific oxidative damage and antioxidant responses in strains of the brown alga *Ectocarpus siliculosus* with different pollution histories. Aquat Toxicol 159:81–89
Saldarriaga-Hernandez S, Hernandez-Vargas G, Iqbal HMN, Barcelo D, Parra-Saldivar R (2020) Bioremediation potential of *Sargassum* sp. biomass to tackle pollution in coastal ecosystems: circular economy approach. Sci Total Environ 715:136978
Salehi P, Tajabadi FM, Younesi H, Dashti Y (2014) Optimization of lead and nickel biosorption by *Cystoseira trinodis* (brown algae) using response surface methodology. Clean 42:243–250
Salomone VN, Riera M, Cerchietti L, Custo G, Muniain C (2017) Seasonal determination of trace and ultra-trace content in *Macrocystis pyrifera* from San Jorge Gulf (Patagonia) by total reflection X-ray fluorescence. Spectrochim Acta B 131:74–78
Santos-Silva MC, Machado EC, Wallner-Kersanach M, Camargo MG, Andrade C, Sa F, Pellizzari F (2018) Background levels of trace elements in brown and red seaweeds from Trindade, a remote island in South Atlantic Ocean. Mar Pollut Bull 135:923–931
Saravanakumar K, Senthilkumar R, Prasad DMR, Prasad BSN, Manickam S, Gajendiran V (2020) Batch and column arsenate sorption using *Turbinaria ornata* seaweed derived biochar: experimental studies and mathematical modeling. ChemistrySelect 5(12):3661–3668
Scherner F, Bastos E, Rover T, Oliveira ED, Almeida R, Itokazu AG, Bouzon ZL, Roerig LR, Pereira SMB, Horta PA (2018) *Halimeda jolyana* (Bryopsidales, Chlorophyta) presents higher vulnerability to metal pollution at its lower temperature limits of distribution. Environ Sci Pollut Res 25(12):11775–11786
Schmidt EC, dos Santos RW, de Faveri C, Horta PA, Martins RD, Latini A, Ramlov F, Maraschin M, Bouzon ZL (2012) Response of the agarophyte *Gelidium floridanum* after in vitro exposure to ultraviolet radiation B: changes in ultrastructure, pigments, and antioxidant systems. J Appl Phycol 24(6):1341–1352
Schofield O, Evens TJ, Millie DF (1998) Photosystem II quantum yields and xanthophyll-cycle pigments of the macroalga *Sargassum natans* (*Phaeophyceae*): Responses under natural sunlight. J Phycol 34(1):104–112
Seaweeds (2020) The Seaweed Site: information on marine algae. http://www.seaweed.ie
Šeršeň F, Kráľová K (2013) EPR spectroscopy—a valuable tool to study photosynthesizing organisms exposed to abiotic stresses. In: Dubinsky Z (ed) Photosynthesis. Rijeka, IntechOpen, pp 248–283. https://www.intechopen.com/books/photosynthesis/epr-spectroscopy-a-valuable-tool-to-study-photosynthesizing-organisms-exposed-to-abiotic-stresses
Shamim S (2018) Biosorption of heavy metals. In: Derco J, Vrana B (eds) Biosorption. IntechOpen, London. https://www.intechopen.com/books/biosorption/biosorption-of-heavy-metals
Shchukin VM, Kuz'mina NE, Erina AA, Yashkir VA, Merkulov VA (2018) Comparative analysis of the heavy metal, aluminum, and arsenic contents in brown algae of various origins. Pharm Chem J 52(7):627–634
Shin WS, Kim YK (2014) Biosorption characteristics of heavy metals (Ni^{2+}, Zn^{2+}, Cd^{2+}, Pb^{2+}) from aqueous solution by *Hizikia fusiformis*. Environ Earth Sci 71(9):4107–4114

Silva JIR, Ferreira ACD, da Costa ACA (2009) Uranium biosorption under dynamic conditions: Preliminary tests with *Sargassum filipendula* in real radioactive wastewater containing Ba, Cr, Fe, Mn, Pb, Ca and Mg. J Radioanal Nucl Chem 279(3):909–914

Sordet C, Contreras-Porcia L, Lovazzano C, Goulitquer S, Andrade S, Potin P, Correa JA (2014) Physiological plasticity of *Dictyota kunthii* (*Phaeophyceae*) to copper excess. Aquat Toxicol 150:220–228

Soto-Rios PC, Nakano K, Fujibayashi M, Leon-Romero M, Nishimura O (2014) Lead removal efficiency using biosorbents as alternative materials for permeable reactive barriers. Water Sci Technol 70(2):307–314

Southichak B, Nakano K, Nomura M, Chiba N, Nishimura O (2008) Marine macroalga *Sargassum horneri* as biosorbent for heavy metal removal: roles of calcium in ion exchange mechanism. Water Sci Technol 58(3):697–704

Sun X, Liu ZW, Jiang QC, Yang YF (2019) Concentrations of various elements in seaweed and seawater from Shen'ao Bay, Nan'ao Island, Guangdong coast, China: environmental monitoring and the bioremediation potential of the seaweed. Sci Total Environ 659:632–639

Tabaraki R, Nateghi A, Ahmady-Asbchin S (2014) Biosorption of lead (II) ions on *Sargassum ilicifolium*: application of response surface methodology. Int Biodeter Biodegr 93:145–152

Tala F, Penna-Diaz MA, Luna-Jorquera G, Rothausler E, Thiel M (2017) Daily and seasonal changes of photobiological responses in floating bull kelp *Durvillaea antarctica* (Chamisso) Hariot (Fucales: *Phaeophyceae*). Phycologia 56(3):271–283

Tamilselvan N, Saurav K, Kannabiran K (2012) Biosorption of Cr (VI), Cr (III), Pb (II) and Cd (II) from aqueous solutions by *Sargassum wightii* and *Caulerpa racemosa* algal biomass. J Ocean Univ China 11(1):52–58

Terada R, Yuge T, Watanabe Y, Mine T, Morikawa T, Nishihara GN (2020) Chronic effects of three different stressors, irradiance, temperature, and desiccation, on the PSII photochemical efficiency in the heteromorphic life-history stages of cultivated *Pyropia yezoensis* f. *narawaensis* (Bangiales) from Japan J Appl Phycol 32:3273–3284

Thomsen MS, Mondardini L, Alestra T, Gerrity S, Tait L, South PM, Lilley SA, Schiel DR (2019) Local extinction of bull kelp (*Durvillaea* spp.) due to a marine heatwave. Front Mar Sci 6:UNSP 84

Tolian G, Jafari SA, Zarei S (2015) Optimization of biosorption of nickel(II) and cadmium(II) by indigenous seaweed *Enteromorpha* using response surface methodology. Water Qual Res J Can 50(2):109–122

Tonon AP, Zaini PA, Falcao VR, Oliveira MC, Collen J, Boyen C, Colepicolo P (2018) *Gracilaria tenuistipitata* (Rhodophyta) tolerance to cadmium and copper exposure observed through gene expression and photosynthesis analyses. J Appl Phycol 30(3):2129–2141

Tsuchiya Y, Nishihara GN, Terada R (2012) Photosynthetic and temperature characteristics of five *Sargassum* species (Fucales), *S. piluliferum*, *S. patens*, *S. fusiforme*, *S. crispifolium*, and *S. alternato-pinnatum* from Kagoshima, Japan, using dissolved oxygen sensor and pulse-amplitude-modulated (PAM) fluorometer. Nippon Suisan Gakkaishi 78(2):189–197

Ueno Y, Aikawa S, Kondo A, Akimoto S (2016) Energy transfer in cyanobacteria and red algae: confirmation of spillover in intact megacomplexes of phycobilisome and both photosystems. Phys Chem Lett 7(18):3567–3571

Unal D, Isik NO, Sukatar A (2010) Effects of chromium VI stress on green alga *Ulva lactuca* (L.). Turk J Biol 34(2):119–124

Valdes FA, Lobos MG, Diaz P, Saez CA (2018) Metal assessment and cellular accumulation dynamics in the green macroalga *Ulva lactuca*. J Appl Phycol 30(1):663–671

Vijayaraghavan K, Joshi UM (2012) Interaction of mercuric ions with different marine algal species. Biorem J 16(4):225–234

Vijayaraghavan K, Gupta S, Joshi UM (2012a) Comparative assessment of Al(III) and Cd(II) biosorption onto *Turbinaria conoides* in single and binary systems. Water Air Soil Pollut 223(6):2923–2931

Vijayaraghavan K, Joshi UM, Kamala-Kannan S (2012b) An attempt to develop seaweed-based treatment technology for the remediation of complex metal-bearing laboratory wastewaters. Ecol Eng 47:278–283

Volesky B (2007) Biosorption and me. Water Res 41(18):4017–4029
Wang XM, Shan TF, Pang SJ (2020) Effects of cobalt on spore germination, gametophyte growth and development, and juvenile sporophyte growth of *Saccharina japonica* (*Phaeophyceae*). J Appl Phycol 32(1):511–518
Whitmarsh J (1998) Electron transport and energy transduction. In: Raghavendra AS (ed) Photosynthesis: a comprehensive treatise. Cambridge University Press, Cambridge, pp 87–110
Whitmarsh J, Govindjee (1999) The photosynthetic process. In: Singhal GS, Renger G, Sopory SK, Irrgang K-D, Govindjee (eds) Concepts in photobiology: photosynthesis and photomorphogenesis. India, Narosa Publishing House, New Delhi, pp 11–51
Wood BA, Miyashita S, Kaise T, Raab A, Meharg AA, Feldmann J (2011) Arsenic is not stored as arsenite-phytochelatin complexes in the seaweeds *Fucus spiralis* and *Hizikia fusiforme*. Environ Chem 8(1):30–43
Xu JT, Gao KS (2016) Photosynthetic contribution of UV-A to carbon fixation by macroalgae. Phycologia 55(3):318–322
Xu W, Pan JF, Fan BW, Liu YX (2019) Removal of gaseous elemental mercury using seaweed chars impregnated by NH_4Cl and NH_4Br. J Clean Prod 216:277–287
Yang W, Liu ZY, Xu W, Liu YX (2018) Removal of elemental mercury from flue gas using *Sargassum* chars modified by NH_4Br reagent. Fuel 214:196–206
Yipmantin A, Maldonado HJ, Ly M, Taulemesse JM, Guibal E (2011) Pb(II) and Cd(II) biosorption on *Chondracanthus chamissoi* (a red alga). J Hazard Mater 185(2–3):922–9291
Yong WK, Sim KS, Poong SW, Wei D, Phang SM, Lim PE (2018) Interactive effects of temperature and copper toxicity on photosynthetic efficiency and metabolic plasticity in *Scenedesmus quadricauda* (*Chlorophyceae*). J Appl Phycol 30(6):3029–3041
Zakeri HA, Abu Bakar L (2013) Copper-, lead- and mercury-induced changes in maximum quantum yield, chlorophyll a content and relative growth of three Malaysian green macroalgae. Mal J Fund Appl Sci 9(1):16–21
Zhang AQ, Xu T, Zou HX, Pang QY (2015) Comparative proteomic analysis provides insight into cadmium stress responses in brown algae *Sargassum fusiforme*. Aquat Toxicol 163:1–15
Zhang D, Zhang QS, Yang XQ (2017) Seasonal dynamics of photosynthetic activity in the representive brown macroalgae *Sagrassum thunbergii* (*Sargassaceae* Phaeophyta). Plant Physiol Biochem 120:88–94
Zhang YR, Wang XM, Shan TF, Pang SJ, Xu NJ (2019) Transcriptome profiling of the meristem tissue of *Saccharina japonica* (*Phaeophyceae*, Laminariales) under severe stress of copper. Mar. Genomics 47:UNSP 100671
Zhao YF, Shang DR, Ning JS, Zhai YX (2012) Arsenic and cadmium in the marine macroalgae (*Porphyra yezoensis* and *Laminaria japonica*)—forms and concentrations. Chem Spec Bioavailab 24(3):197–203
Zhao YF, Wu JF, Shang DR, Ning JS, Zhai YX, Sheng XF, Ding HY (2015) Subcellular distribution and chemical forms of cadmium in the edible seaweed, *Porphyra yezoensis*. Food Chem 168:48–54
Zhao YF, Shang DR, Ning JS, Zhai YX, Sheng XF, Ding HY (2019) Subcellular distribution and chemical forms of lead in the red algae, *Porphyra yezoensis*. Chemosphere 227:172–178
Zhu L, Xiao H, Wang Y, Jian XY, Zhang ZP, Zhang HX, Tang XX (2015) Physiological responses of macroalga *Gracilaria lemaneiformis* (Rhodophyta) to UV-B radiation exposure. Chinese J Oceanol Limnol 33(2):389–399
Zhu XF, Zou DH, Huang YH, Cao JM, Sun YP, Chen B, Chen XY (2017) Physiological responses of *Porphyra haitanensis* (Rhodophyta) to copper and cadmium exposure. Bot Mar 60(1):27–37
Zou HX, Pang QY, Lin LD, Zhang AQ, Li N, Lin YQ, Li LM, Wu QQ, Yan XF (2014) Behavior of the edible seaweed *Sargassum fusiforme* to copper pollution: short-term acclimation and long-term adaptation. PLoS One 9(7):e101960
Zou HX, Pang QY, Zhang AQ, Lin LD, Li N, Yan XF (2015) Excess copper induced proteomic changes in the marine brown algae *Sargassum fusiforme*. Ecotoxicol Environ Saf 111:271–280

Chapter 4
Role of Microorganisms in Managing Soil Fertility and Plant Nutrition in Sustainable Agriculture

Heba I. Mohamed, Mahmoud R. Sofy, Abdulwareth Abdulkader Almoneafy, Magdi T. Abdelhamid, Abdul Basit, Ahmed R. Sofy, Rafiq Lone, and Maged M. Abou-El-Enain

Contents

H. I. Mohamed (✉) · M. M. Abou-El-Enain
Department of Biological and Geological Sciences, Faculty of Education, Ain Shams University, Cairo, Egypt

M. R. Sofy · A. R. Sofy
Botany and Microbiology Department, Faculty of Science, Al-Azhar University, Nasr City, Cairo, Egypt

A. A. Almoneafy
Department of Biology Sciences, College of Education and Science at Rada'a, Albaydaa University, Rada'a, Yemen

M. T. Abdelhamid
Botany Department, National Research Centre, Dokki, Giza, Egypt

A. Basit
Department of Horticulture, Faculty of Crop Production Sciences, The University of Agriculture, Peshawar, Pakistan

R. Lone
Centre of Research for Development, University of Kashmir, Srinagar, Jammu and Kashmir, India

© The Author(s), under exclusive license to Springer Nature Switzerland AG 2021

H. I. Mohamed et al. (eds.), *Plant Growth-Promoting Microbes for Sustainable Biotic and Abiotic Stress Management*,
https://doi.org/10.1007/978-3-030-66587-6_4

1 Introduction

As of now assessed at 7.4 billion, the total populace is dependent on agriculture results of which field crops have an enormous offer. The human populace is relied upon to arrive at nine billion in the following 30 years, which is required to raise an extended 60% food request (Muller et al. 2017). Cereals, potatoes, and legumes have incredibly added to the world's tremendous populace's food prerequisites. In any case, these harvests' yields are undermined by helpless soil quality, warmth and cold pressure, salinity and dry season, and different pathogen infections (Mohamed et al. 2018, 2019; Ghonaim et al. 2020). Thorough utilization of pesticides and composts is done to fulfill the foreseen interest for food creation, yield, and gainful limit of significant harvests; however, their utilization frequently brings about different contamination and natural surroundings' tainting. To safeguard the sustainability of our natural ecosystem and environment, increase in yields of agricultural crops for addressing future food issues needs some novel and eco-friendly aspects of agricultural practices (Majeed et al. 2017).

Greatest arable land revelation and extreme utilization of chemical fertilizer have demonstrated yield stagnation and a decline in soil quality, a significant worry in developing and industrialized nations (Singh 2015). Plant nutrients are drained from the soil by an intensive crop method, and the signs of nutrient deficiency have been seen in plant parts and crop yield and quality have also been reduced. Current farming practices in food grain production have reduced the recycling of organic and other waste generated by the farm, resulting in a decline in soil organic matter (Shukla et al. 2013). Thus, age-old practices, the use of natural resources, agricultural inhabitants, microbes, and organic improvements are remembered by the global science community (Singh and Strong 2016). The use of microorganisms in this process increases at a tremendous rate, and agricultural production increases due to stagnation. The dynamics of nutrient emanates from any source in the field. The soil microorganisms are primarily responsible for the pattern of crop supplements, soil attributes, organic deteriorations and soil degradation.

Soil microorganisms change the hurtful impacts of substantial metals, pesticides, herbicides, and other concoction of substances. They debase the mixes into littler, less unstable portions. They improve soil ripeness levels, for example, obsession of environmental nitrogen, in situ assemblies of fixed phosphorus (P) and potassium (K), and production of siderophores, notwithstanding their capacity as a natural scavenger (Rao 2014). In root nodules of legume crops, more than half were needed to fasten atmospheric nitrogen by soil microorganism (Singh et al. 2016). Microbial biomass in soil plays an important role in soil health improvements and crop production, thus maintaining sustainable agricultural and environmental efficiency (Singh et al. 2010; Singh and Singh 2012). The present chapter is focused on different microbes in the soil, agricultural applications as bioinoculants for plant growth promotion, enhancing production for agro-environmental sustainability, and roles in soil fertility and health.

2 The Rhizosphere

Microorganisms represent an assortment of life in the soil and assume a significant part in natural cycles by influencing the biogeochemical cycles, supplement procurement, and pedogenesis of the earth (Schulz et al. 2013; Smith et al. 2015). They become essential elements of the arrangement of cultivation when a seed is germinated in the soil for the beginning of its life cycle as essential live components of the soils (Meena et al. 2017a). The soil close to the plant roots (rhizosphere) is a vital living and climate for these microorganisms (Ullah et al. 2015; Bender et al. 2016).

Soil microorganisms play a critical role in sustaining and improving soil basics by joining natural and low soil particles to form aggregates (Bach et al. 2010). Earth's microorganisms are responsible for generating gene, glycoproteins, and polysaccharides, discharging the soil to create the foundation for soil structure (Hassink 1994). The provision of natural acids will affect pH for the microbes of the rhizosphere, thereby raising the availability of the additives for plants (Dimkpa et al. 2009). Also, the level of carbon and nitrogen in soil increases microorganisms to contribute towards soil saving surface and welfare (Bhatia 2008). Soil microbes are presently likewise utilized in the bioremediation of natural contaminants in soils through the mineralization cycle (Garbisu et al. 2017).

Some endophytic (bacteria and fungi) are microorganisms that assume a fundamental function in the control of abiotic and biotic stress with a specific spotlight on plant growth and development advancing rhizobacteria (PGPR) (e.g., *Azospirillum*, *Enterobacter*, *Pseudomonas*, *Rhizobium*, *Bacillus*, *Cyanobacteria*, and so forth) and arbuscular mycorrhizal organisms (AM) (Hassen et al. 2016). PGPR upgrades the openness of iron (Fe) in soil by delivering Fe-chelating specialists "siderophores" and delivering organic acids "malate, citrate, and so forth" that decline pH, to energize the development of plants by controlling nitrogen fixation and plant hormones and production enzymes, improving the activity of antioxidants and editing volatile organic compounds (Singh and Jha 2016; Cohen et al. 2017).

In unfavorable circumstances, hardly any microbe is fitted with sigma variables that can determine impairment of gene expression upregulation (Taniguchi and Wendischen 2015). Nonetheless microbes, AM, adjust the rhizosphere by aggregating glomalin-related soil proteins (GRSP) and extending the area of absorption through arbuscules and cortical root vesicle production and water-holding capability (Chowdhury et al. 2015; Kuan et al. 2016). Fungal hyphae also increase the supply of nutrients by upregulating P transporters and modifying gene transcription levels and activity related to stress-related antioxidants (Talaat and Shawky 2014; Maya and Matsubara 2013). Plant roots send the microorganisms signals by generating a wide range of exudates, such as amino acids, sugars, and various dicarboxyl acids "fumarate and succinate, malate," which fill microorganisms as chemoattractors and influence their rhizosphere selection (Audrain et al. 2015; Schmidt et al. 2015; Rosier et al. 2016).

3 Microorganisms in Soil

Soil is a dynamic medium and supports different microbial communities such as bacteria, fungi, actinomycetes, algae, viroids, viruses, protozoans, nematodes, etc., which play a vital role in maintaining soil fertility, cycling of nutrient elements in the biosphere, humus formation, biological conversions, geochemical cycling, ecosystem sustenance, etc., besides supporting plant life and plant productivity (Singh et al. 2011a, b, c; Davinic et al. 2012; Meliani et al. 2012). In general, microorganisms are known as fauna and flora that are smaller than 0.1 mm (Ellouze et al. 2014; Qiao et al. 2017). The population and variety of soil microorganisms differ widely and are influenced by the degree of fertility, organic matter, hazardous material presence, and climate influences. They are in size for a minute and cannot be seen with the naked eyes. In a soil having approximately 109 soil bacteria, with the help of a microscope, can see only 1 % of the total population (Bhatti et al. 2017; Bhat et al. 2017; Buscardo et al. 2018).

PGPRs are involved in various useful soil activities, such as destruction of crop residues, mineralization and soil organic matter synthesis, immobilization of mineral supplements, nitrification and nitrogen fixation, phosphate solubilization, and plant hormone formation that helps in crop productivity and protection. PGPRs also produce auxins that contribute to the growth of mycorrhizal mycelium (Fernández-Bidondo et al. 2011). This mycorrhizal mycelium colonizes plant roots which leads to changes in the exudates of the roots which have a selective effect on the rhizosphere populations of other microbes (negative or positive) (Ye et al. 2015; Cornejo et al. 2017). In plant growth and development, microorganisms are involved through direct mechanisms, including hormone signalling modulation and phytopathogenic protection. Root exudate metabolites are responsible for contact between plants and microbes (Fig. 4.1).

3.1 Arbuscular Mycorrhiza (AM)

Because of their capacity to build up an interface for the trading of nutrients inside the living cells of the plant, mycorrhizal organisms contrast from other plant-fungus affiliations (Coats and Rumpho 2014). AM fungi have a broader absorbent surface than root hairs, along these lines assisting with retaining stable ions in the soil external the depletion zones (Pichardo et al. 2012). In light of the presence of different extraradical or intraradical hyphal structures, seven unique sorts of mycorrhizal affiliations have been recognized such as "ectomycorrhiza, ecto-endomycorrhiza, monotropoid mycorrhiza, ericoid mycorrhiza, arbutoid mycorrhiza, orchid mycorrhiza and vesicular-arbuscular mycorrhiza" (Smith and Read 2008). The development of arbuscular and vesicles inside cortex cells recognizes AM fungi. Their presence has led to the former standard name of the vesicular-arbuscular

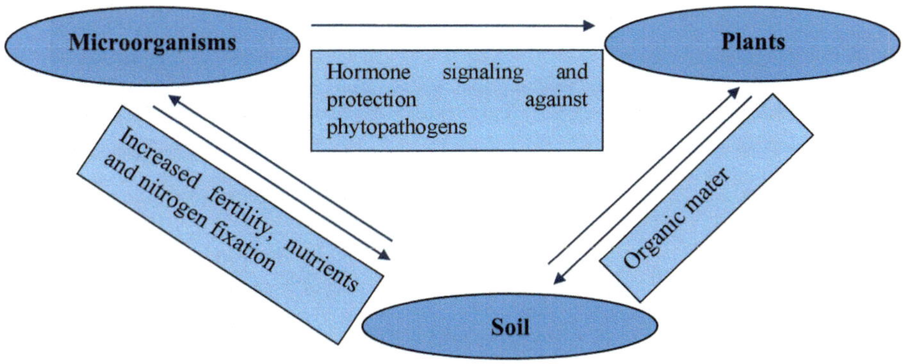

Fig. 4.1 Soil, microorganisms, and plant relationships

mycorrhizal organisms, but currently, not all fungi produce vesicles arbuscular mycorrhizal (Friberg 2001).

3.1.1 Role of Arbuscular Mycorrhiza in Soil Fertility

3.1.1.1 Increasing Phosphorus Availability

Inorganic phosphate (Pi) soil is important for the correct development and working of all living things; normally plant development and yield are influenced when supplies of Pi are decreased (Niu et al. 2013; López-Arredondo et al. 2014; Rawat et al. 2016) as appeared in Table 4.1. The approach is to build the root-soil interface to enhance access and retention of the usable Pi (Yasin et al. 2016), and to solubilize Pi found in organic mineral complexes (Nouri et al. 2014; Bahadur et al. 2017; Verma et al. 2017b; Kumar et al. 2017). The monitoring of AM symbiosis according to phosphorus (P) availability is probably the most recognized example of such regulations (Balzergue et al. 2011). Pi can be consumed by the AM symbioses using root epidermal cells and root hairs, and through AM organisms (Nouri et al. 2014; Watts-Williams et al. 2015; Meena et al. 2015b, 2016; Bahadur et al. 2016). Along these lines, deciphering the synthetic flagging code in AM beneficial interaction stays an overwhelming yet unmistakable objective for the not so distant future (Nath et al. 2017; Verma et al. 2017a).

The different components proposed to represent improved nutrients take-up incorporate (1) expanded investigation of soil; (2) expanded phosphorus movement by arbuscular into plants; (3) root condition adjustment; (4) productive utilization of P inside plants; (5) dynamic move of P to the foundations of plants; and (6) expanded capacity of consumed P. The root take-up of phosphate is a lot speedier than the dissemination of particles through the root assimilation surfaces (Bhat et al. 2017). This prompts an area of phosphate consumption around the roots.

Table 4.1 List of different phosphorus-solubilizing fungal strains used for stimulation of plant growth

Fungus	References
Aspergillus terreus	Abdel-Ghany and Alawlaqi (2018)
Aspergillus, *Fusarium*, *Penicillium*	Elias et al. (2016)
Aspergillus ustus *Aspergillus tamarii*	Pany et al. (2018)
Byssochlamys nivea	Dolatabad et al. (2017)
Aspergillus niger	Prajapati and Modi (2012)
Aspergillus terreus	Prajapati and Modi (2012)
Penicillium sp.	Sangeeth et al. (2012)
Glomus intraradices, *Glomus mosseae*	Wu et al. (2005)
Ectomycorrhizal fungi	Alves et al. (2010)
Talaromyces funiculosus	Kanes et al. (2015)
Trichosporon beigelii, *Pichia norvegensis*, *Cryptococcus albidus* var. aerius, *Candida etchellsii*, *Rhodotorula aurantiaca*, *Cryptococcus luteolus*, *Neosartorya fischeri* var. fischeri, *Cryptococcus terreus A*, *Candida montana*, *Penicillium purpurogenum* var. rubrisclerotium and yeast isolate	Gizaw et al. (2017)
Rhizopus stolonifera, *R. oryzae*	Patel et al. (2015)
Trichoderma, *Paecilomyces*, *Beauveria*, *Metarhizium*	Shukla and Vyas (2014)
Mortierella sp.	Xueming et al. (2014)
Trichoderma sp. and *Papulaspora* sp.	Vitorino et al. (2012)
Piriformospora indica	Yadav et al. (2010)
Sporotrichum thermophile	Singh and Satyanarayana (2010)

3.1.1.2 Fixing Atmospheric Nitrogen

Nitrogen (N) is essential for amino acid manufacture and enters into protein and nucleic acid synthesis indirectly. In shoots, AM-related plants have an expanded N content. There are various systems proposed for this impact: (1) nitrogen fixation enhancement; (2) the immediate absorption of nitrogen; (3) facilitation of the transfer of nitrogen, a mechanism by which non-nodulated plants gain from a portion of nitrogen fixed by nodulated plants; and (4) increased enzymatic processes such as "pectinase, xyloglucanase and cellulose" that are capable of decomposing soil organic matter involved in nitrogen metabolism (Barea 1997). AM hyphae tend to strip nitrogen and bring it to soil plants. They contain organic nitrogen breakdown chemicals and contain an enzyme that also modifies nitrogen structures of the soil. AM increases growth, nodulation, and fixation of nitrogen in the legume-rhizobial symbiosis. Mycorrhizal associations provide more than 50% of the plant N requirement (McFarland et al. 2010; Hindumathi and Reddy 2012; Hindumathi et al. 2016).

3.1.1.3 Improving Soil Structure

Ecosystem disruptions impact the soil physical, chemical, and biological processes. AM helps to bind soil particles and enhance soil aggregation and conservation of the soil (Bhat et al. 2017). AM is also known to boost soil fertility, as they generate glomalin that forms micro aggregates and finally macro aggregates after accumulation in soil, along with the AM hyphae, and therefore, it explicitly serves as a backbone to aggregate the soil and stabilize the soil. It also releases exudates into the land and thereby promotes full stability so that other microorganisms grow better (Khanday et al. 2016).

3.1.2 Potential Benefits of AM Fungi in Plant Nutrition

In all plants, the root is known as the essential channel for water assimilation and mineral components. The utilization of harmonious microorganisms, for example, mycorrhizal fungi, alongside appropriate concoction and natural contributions to the region of the plant root system is one of the logical arrangements recommended to expand the development and efficiency of the root arrangement of plants (Redecker et al. 2013). Nutrient absorption, for example, P, performed by the diffusion method and heading toward the root, depends on how rapidly and at what intervals to cover the root-absorbing surfaces, the diffusion of nutrients in the soil (Datta et al. 2017; Meena et al. 2017b; Gogoi et al. 2018). AM fungal hyphae also produce an exogenous enzyme, such as "phosphatases, phytases and nitrate reductase" which are critical for nutrient absorption and metabolism in addition to being structurally effective in extracting nutrients from soil exchange sites (Antibus et al. 1992). The absorbed "P" can be transformed into polyphosphate granules in the external hyphae and transported to the shrubs for transfer to the host (Smith and Smith 2011).

Exogenous enzymes such as phosphatic enzymes hydrolyze P from unavailable sources and release P from organic P complexes and promote P absorption in humid tropical conditions generated from AM fungal extraradical hypha (Carlile et al. 2001). Extraradical amino acids, peptides, and ions (NO^{3-} or NH^{4+}) obtained nitrogen of AM in different shapes (Hawkins et al. 2000; Giri and Mukerji 2004). AM fungal hyphae have been reported to absorb inorganic N and convert it into intraradical hyphae with amino acids (arginine). Intraradical hyphae have also shown amino acids are decomposed to join the C and transfer the remainder N as ammonium to the host plant. These fungi, instead, release P from insolvent metal compounds via the segment of organic acids such as oxalic acid/oxalates which have a higher affinity with P to combine Ca, Fe, and Al ions, absorbing P (Miyasaka and Habte 2001).

AM fungi release in the soils glomalin-related proteins (GRSP) that are important for the activity of the ecosystem and for ecological restore (Wu and Zou 2017) to increase soil texture (by regulating soil water-plant relationships). Glomalin serves as a hydrophobic glue to help the soil fungus cope and mitigate macro-disruption during the aerial growth of tissue in the plant (Singh et al. 2010). It can have a direct impact on the environment by improving soil aggregation by constructing macroaggregate structures by physically linking soil particles and organic materials (Leifheit et al. 2014, 2015). These aggregates improve the storage of carbon and nutrients and create a conducive environment for soil microorganisms to survive and grow. They are important for improving the root development, root growth, and microbial activity which promote water aeration and movement, and thus stimulate the structures and productivity of plants in a community (Rillig et al. 2015). Another big benefit of AM is enhanced dry season and saltiness impacts, high soil temperatures, weak pH, and heavy-metal harm relief (Augé et al. 2015).

3.2 *Plant Growth Promoting Bacteria (PGPB)*

Among rhizosphere microorganisms, plant growth promoting bacteria (PGPB) hold a key position in natural ecosystem as well as in agriculture because they can stimulate plant growth by fixing nitrogen, improving soil nutrients, suppressing pathogenic microbes, and modifying plants' responses to biotic and abiotic stresses. Endophytic or free-living helpful microscopic bacteria stimulate plant growth and development by direct and indirect methods (Shameer and Prasad 2018). PGPB plays a key role in both the natural environment and agriculture among rhizosphere microorganisms because they can boot up plant growth by fixing nitrogen, improvement of nitrate uptakes by plant roots (Stokstad 2016), enhancing soil nutrients (Jacoby et al. 2017), decreasing pathogenic microbes, altering the responses of plants to different stresses (Shameer and Prasad 2018; Gouda et al. 2018; Etesami 2018; Berendsen et al. 2018), upgrading phosphorus solubilization (Otieno et al. 2015), and release of hormones, vitamins, volatile compounds, and other growth regulatory substances (Glick 2014) (Fig. 4.2).

Moreover, improving soil quality by changing over phosphates and nitrogen and remediating heavy metals with the utilization of some PGPBs is notable in farming fields (Rahman et al. 2018). Some PGPB strains have been demonstrated to be positively correlated with the suppression of different plant microbes by delivering antagonistic metabolites (Pineda et al. 2017; Berendsen et al. 2018) and inducing disease tolerance and improving insusceptibility reactions of hosts (Saraf et al. 2014; Chung et al. 2016; Rahman et al. 2018). Through PGPB, it was also argued that the conditions of drought, salinity, and photosynthetic activity can be managed effectively, and the biomass of some crops can be increased by responding properly to the stress requirements imposed (East 2013; Shrivastava and Kumar 2015; Ma et al. 2016).

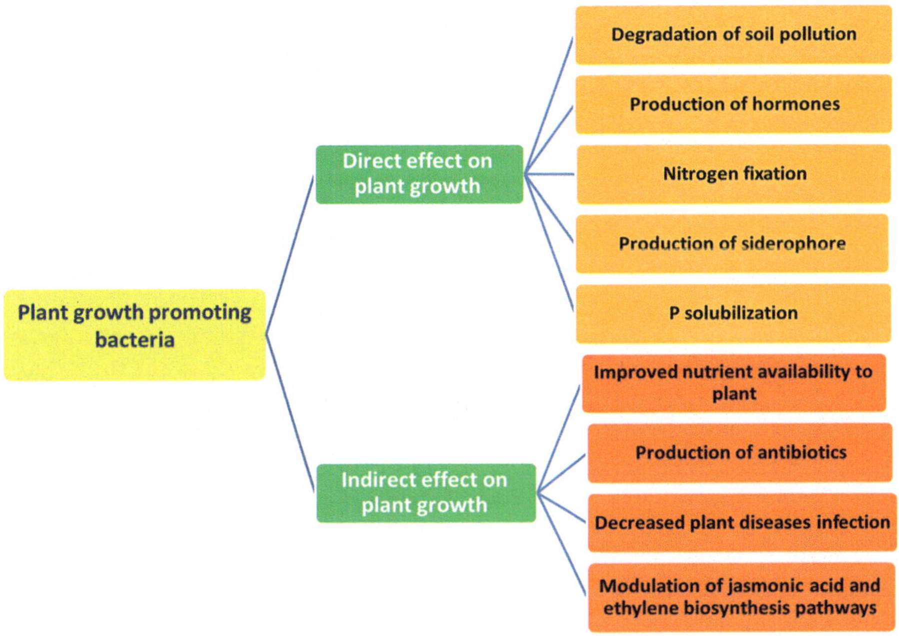

Fig. 4.2 An illustration of the plant growth promoting activities by PGPB

PGPB was also argued that the conditions of drought, salinity and photosynthetic activity can be managed effectively, and the biomass of some crops can be increased by responding properly to the stress requirements imposed (East 2013; Shrivastava and Kumar 2015; Ma et al. 2016).

3.2.1 Role of PGPB in Enhancing Soil Fertility

3.2.1.1 Phosphorus-Solubilizing Microorganisms

Soil microbes assume a key function in plant soil P collection and their capacity to solubilize and mineralize inorganic and natural soil P parts (Adhya et al. 2015). For some reasons, the utilization of phosphate-solubilizing species in agronomic practices is upheld. For instance, through their consistent activities in the soil, they support soil fertility, increment plant development, and harvest yield through accessibility and absorbed minerals, don't cause natural contamination, improve soil warmth and cooling, ensure plants against some soil-borne microbes, and provide low-cost technologies with a strong cost-benefit ratio for their production (Parani and Saha 2012; Paul and Sinha 2017). For agronomic purposes, phosphorus is second just to nitrogen as the most restricting component for plant development. Phosphorus empowers the fixation of nitrogen in legume crops and is significant for the creation of photosynthesis and sugar (Sridevi and Mallaiah 2009; Saeid et al. 2018).

Furthermore, the third plant macronutrient to help or even elevate the seed, grain efficiency, protein fusion, enzyme production, and starch creation, control root development, and control the movement of stomata in plant cells and photosynthesis is potassium (K) (Britzke et al. 2012; Zorb et al. 2014). A soil K deficiency prevents plants from consuming enough K, which in turn contributes to K deficiency syndromes. K deficiency in plants causes the leaf edges to turn yellow with a burnt appearance. It may also be responsible for slow growth and incomplete root growth. The K function is a series of tasks including activation of enzymes, adenosine triphosphate and sugar, photosynthesis speeds, and photosynthesis-based sugar transport to other parts of the plant through the phloem (Kour et al. 2020). K stimulates the starch synthase enzyme and thus plays a crucial role in the supply and transport of water, nutrients, and nutrients (Marschner 1995; Pettigrew 2008). In the ordinary P and K cycles, microorganisms have a significant part as P or K solubilizing microorganisms in the plant soil (Diep and Hieu 2013). These microorganisms discharge organic acids that fill in as chelating specialists and break down rock and silicon particles quickly and then release the K particles into the soil (Bennett et al. 1998).

Some strains of bacteria such as *Bacillus* spp. can release polysaccharides and carboxylic acids to solubilize K compounds that can be accessed by plants, such as "tartaric acid and citric acid" (Yi et al. 2008). The use of potassium-solubilizing microorganisms (KSM) as biofertilizers can increase the level of available K^+ in the soil, thereby mitigating K deficiency in a more environmentally friendly manner (Barker et al. 1998).

KSBs are heterotrophic and use organic materials to extract cellular and energy. They perform meaningful soil functions, such as the formation of humus, the cycling of other organic rocks, and the prevention of the build-up of dead organic materials. KSBs are furthermore aerobic bacteria that play an important role in preserving the soil structure by contributing to the formation and stabilization of soil water-stable aggregates. In addition to this, these bacteria stimulate the growth of plants and/or antagonize soil-borne root pathogens (Archana et al. 2013). The component for mineral K solubilization relies upon the creation of organic acids that assume an essential part in soil natural K mineralization (Prajapati and Modi 2012). Microorganisms framed natural corrosive with a low sub-atomic weight, which chiefly serves to deliver K from minerals during physiological movement and incites an ascent in the fixation slope during mineral hydrolysis employing its retention of dissolvable K (Bahadur et al. 2014). After decomposition, organic matter contains acids such as "citric acid, formic acid, malic acid, and oxalic acid." The dissolution of K compounds is increased by these organic acids, made up of the protons and Ca^{2+} ion complexing (Shanware et al. 2014). The aggregation of complexes between organic acids and metal ions, including Fe^{2+}, Al^{3+}, and Ca^{2+}, is the solution for K solubilization (Styriakova et al. 2003).

Another study showed that K was solubilized by the release of inorganic and organic acids and by the development of *Bacillus*, *Clostridium*, and *Thiobacillus* mucilaginous capsules containing exopolysaccharides (Diep and Hieu 2013). The microbes, for example, *Cladosporium cladosporioides*, enduring capacity incorporates the age of protons, organic acids, siderophores, and organic ligands, which fill in as chelating specialists and play out the chelation cycle by liberating the K mineral for crop take-up (Meena et al. 2014a, b; Kumar et al. 2015). In three distinctive however related stages known as a triple activity, organic acid molecules influence mineral enduring: (1) acids tie to the mineral surface and eliminate minerals by electron move response from the mineral particles; (2) they break the oxygen connections, and (3) via its carboxyl and hydroxyl gatherings, the chelate particles present in an arrangement (Welch et al. 2002). Moreover, the discharge of amino acids, vitamins, and pytohormone substances, for example, indole-3-acitic acid (IAA) and gibberellic acid (GA3), is additionally known to assist increment with plant growth and development (Mohamed and Gomaa 2012). The addition of the bacterial strain *Enterobacter hormaechei* expanded the okra root and shoot length and adequately assembled K in the plant when feldspar was applied to the soil (Prajapati et al. 2013; Etesami 2020) (Table 4.2).

Table 4.2 List of different phosphorus-solubilizing bacteria strains used for stimulation of plant growth

Bacteria	References
Bacillus subtilis RJ46	Saikia et al. (2018)
Acinetobacter spp.	Rokhbakhsh-Zamin et al. (2011)
Pseudomonas fluorescens	Ghorchiani et al. (2018)
Coccus sp. *Streptococcus* sp. *Bacillus* sp.	Wahid et al. (2016)
Bacillus mucilaginosus	Zarjani et al. (2013)
Bacillus edaphicus	Zarjani et al. (2013)
Enterobacter hormaechei (KSB-8)	Prajapati et al. (2013)
Paenibacillus mucilaginosus	Liu et al. (2012)
Paenibacillus glucanolyticus	Sangeeth et al. (2012)
Agrobacterium tumefaciens	Meena et al. (2015a)
Bacillus spp. PSB9 and PSB16	Panhwar et al. (2011)
Ewingella americana and *Enterobacter aerogenes*	Ribeiro and Cardoso (2012)
Pseudomonas cedrina, Rahnella aquatilis, Rhizobium nepotum, and *Rhizobium tibeticum*	Rfaki et al. (2015)
Alcaligens, Klebsiella, and *Azotobacter*	Behera et al. (2017)
Bacillus megaterium, B. methylotrophicus, B. subtilis, B. aryabhattai, B. simplex, *B. pumilus, Arthrobacter chlorophenolicus, Providencia rettgeri, Enterobacter cloacae,* *E. ludwigii, Proteus penneri, P. vulgaris, Psychrobacter alimentarius, Acinetobacter* *johnsonii, Serratia marcescens, Rhodococcus erythropolis, Pseudomonas geniculata,* and *Ochrobactrum haematophilum*	Gao et al. (2016)
Aneurinibacillus aneurinilyticus	Chauhan et al. (2017)
Azospirillum lipoferum and *Azospirillum brasilense*	Mohamed et al. (2017)
Arthrobacter sp.	Zarjani et al. (2013)
Paenibacillus glucanolyticus	Sangeeth et al. (2012)

3.2.2 Role of PGPB in Nitrogen Fixation

As it is a basic piece of nucleic acids, enzymes, and proteins, nitrogen (N) is one of the most significant components for plant development. 78% of N is in vaporous form. Nonetheless, N isn't accessible to plants and is hence known to be one of the most growth restricting nutrients (Dalton and Kramer 2006). Atmospheric nitrogen (N_2) must be altered or fixed to ammonia (NH_3) by nitrogen-fixing microorganisms in order to become accessible to plants (Kim and Rees 1994). Biological nitrogen fixation microorganisms are classified as being (a) N_2-fixing symbiotic bacteria (Ahemad and Khan 2012); (b) non-leguminous (e.g., *Frankia*); and (c) nitrogen-fixing types, such as cyanobacteria, and are non-symbiotic (free-living and endophytes) (Bhattacharyya and Jha 2012).

In the root nodule, the bacteria do not have a cell wall. By an enzyme called the enzyme nitrogenase, they regulate the nitrogen of the atmosphere and contain ammonia (Olanrewaju et al. 2017). The complex of nitrogenase is described as a metalloenzyme consisting of two elements: (1) the metal cofactor dinitrogenase and (2) the iron protein reductase, the iron protein. The reductase in dinitrogenase provides electrons with a high strength reduction while the reduction in N2 to NH3 is accomplished by the dinitrogenase. This method requires a large amount of energy, requiring 16 ATP moles to reduce the nitrogen content by 1 mole. The microbial carbon is allocated oxidative phosphorylation for more ATP production instead of storing energy in the form of glycogen by synthesizing the glycogen synthesis. A nitrogenase gene (nif) is required to be an oxygen-sensitive gene for this process (Kundan et al. 2015).

3.3 Actinomycetes

Actinomycetes, gram-positive filamentous bacteria, grow in the rhizosphere and colonize plant tissues in agriculture for a longer period after sleeping sporulation. Actinomycetes for their PGP and biocontrol activities have been studied (Shaharokhi et al. 2005). Few actinomycetes are also reported as endophytes, colonizing plant tissue from various plants, including legumes of grain, wheat, corn, and several medicinal plants. Actinomycetes are well recognized as one of the most important groups and play a key role in maintaining the rhizosphere ecosystem and recycling of soil nutrients (Benizri et al. 2005). Actinomycetes were recorded in abundance using both cultivable and molecular plant rhizosphere techniques (Khamna et al. 2009). Actinomycetes are an important group of soil bacteria, mainly *Streptomyces*, due to their ability to produce PGP substances and secondary metabolites, such as antibiotics and enzymes.

3.3.1 Role of Actinomycetes in Phosphate Solubilization

Several scientists have reported rare but very high P-solubility actinomycetes with buffered tricalcium phosphate (TCP) and rock phosphate medium (RP). Some are related to highly cited strains of *Bacillus* or *Pseudomonas*, such as the *Arthrobacter* and *Streptomyces* sp. under similar conditions (Hamdali et al. 2012; Jog et al. 2014). The secretion of organic acid (Table 4.3) results in the acidification of the microbial cells and their surroundings, resulting in the proton replacement of Ca^{+2} and the solubilization of mineral phosphate (Rodríguez and Fraga 1999).

3.3.2 Role of Actinomycetes in Nitrogen Fixation

Nitrogen is a critical nutrient for virtually all lifeforms. While nitrogen makes up about 79% of our atmosphere, plants cannot use nitrogen in its gaseous state. It first must be fixed or combined into either ammonia, NH_3 or Nitrate, NO_3. The natural nitrogen cycle relies on nitrogen-fixing bacteria like those found in the Frankia family of actinobacteria, to supply the fixed nitrogen. Fixed nitrogen is often the limiting factor for growth, both in your garden and in the general environment.Fixed nitrogen is regularly the restricting element for development, both in the garden and in the overall condition. About 15% of the world's nitrogen fixed normally is from advantageous connections between different types of the *Frankia* group of actinobacteria and their host plants. The plants that structure harmonious associations with *Frankia* are called actinorhizal plants. *Frankia* can give most or the entirety of the host plant's nitrogen needs (Sayed 2011). *Frankia* can repair N_2 by forming a vesicle, a spherical cell dedicated to N_2 attachment, even during free-living circumstances. The vesicle is enclosed in several dozen hopanoid lipid envelopes that serve as an oxygen barrier that disables nitrogenase (Berry et al. 1993).

Table 4.3 Some organic acids created by potassium-solubilizing microbes

Microbes	Organic acids
Bacillus edaphicus	Citric, tartaric, oxalic acids
Enterobacter hormaechei	Oxalic acid, citric acid
Paenibacillus mucilaginosus	Oxalic, tartaric, citric acids
Cladosporium sp.	Oxalic, citric, gluconic acids
Aspergillus niger	Citric acid
B. megaterium, *Pseudomonas* sp.	Malic, lactic, lactic, oxalic acids

4 Conclusion

The soil fertility and health are very important for plant growth and development; the primary supplement needs of the yields production are satisfied by mineral manures. The unjustifiable utilization of mineral manure alongside supplement mining in serious farming has brought about the development of multi-nutrient inadequacy alongside undermined soil wellbeing, eventually deteriorating crop efficiency. In this regard, soil microorganisms are the main retreat liable for various soil measures influencing the change of supplements and consequently affecting the resulting accessibility to plant foundations of these nutrients. The capacity for microorganisms to solubilize and mineralize nutrients from inorganic and organic pools is now very much seen, and their utilization could open another skyline for better harvest creation and profitability with improved soil fertility. Improving soil productivity by adding beneficial microbes and enzymes without disrupting the ecological structure of the soil is also one of the main challenges in the current scenario, as different anthropogenic activities contributing to environmental problems increased. In addition, microbes play an essential role in plant growth's nutrient availability as different forms of rhizobacteria live in or around soil rhizosphere, are associated with plant tissues, and play an incredible role in crop development. In this manner, steady and continuous exploration endeavors are needed to investigate and characterize more organisms with higher supplement transformation productivity and to plan improved inoculums with higher time span of usability and feasibility for their last application under field conditions.

References

Abdel-Ghany MT, Alawlaqi MM (2018) Molecular identification of thermo-halotolerant *Aspergillus terreus* and its correlation in sustainable agriculture. Bioresources 13:8012–8023

Adhya TK, Kumar N, Reddy G, Podile RA, Bee H, Samantaray B (2015) Microbial mobilization of soil phosphorus and sustainable P management in agricultural soils. Special section: sustainable phosphorus management. Curr Sci 108(7):1280–1287

Ahemad M, Khan MS (2012) Effects of pesticides on plant growth-promoting traits of Mesorhizobium strain MRC4. J Saudi Soc Agric Sci 11:63–71

Alves CJ, Figueiredo SM, Azevedo SS, Clementino IJ, Keid LB, Vasconcellos SA, Batista CSA, Rocha VCM, Higino SS (2010) Detection of Brucella ovis in ovine from Paraib State, in the Northeast region of Brazil. Braz J Microbiol 41:65–367

Antibus RK, Sinsabaugh RL, Linkins AE (1992) Phosphatase activities and phosphorus uptake from inositol phosphate by ectomycorrhizal fungi. Can J Bot 70:794–801

Archana DS, Nandish MS, Savalagi VP, Alagawadi AR (2013) Characterization of potassium solubilizing bacteria (KSB) from rhizosphere. Soil Bioinfolet 10(1B):248–257

Audrain B, Farag MA, Ryu CM, Ghigo JM (2015) Role of bacterial volatile compounds in bacterial biology. FEMS Microbiol Rev 39:222–233

Augé RM, Toler HD, Saxton AM (2015) Arbuscular mycorrhizal symbiosis alters stomatal conductance of host plants more under drought than under amply watered conditions: a meta-analysis. Mycorrhiza 25:13–24. https://doi.org/10.1007/s00572-014-0585-4

Bach EM, Baer SG, Meyer CK, Six J (2010) Soil texture affects soil microbial and structural recovery during grassland restoration. Soil Biol Biochem 42:2182–2191

Bahadur I, Meena VS, Kumar S (2014) Importance and application of potassic biofertilizer in Indian agriculture. Int Res J Biol Sci 3:80–85

Bahadur I, Maurya BR, Meena VS, Saha M, Kumar A, Aeron A (2016) Mineral release dynamics of tricalcium phosphate and waste muscovite by mineral-solubilizing rhizobacteria isolated from Indo-Gangetic Plain of India. Geomicrobiol J 34(5):2017. https://doi.org/10.1080/01490451.2016.1219431

Bahadur I, Maurya BR, Meena VS, Saha M, Kumar A, Aeron A (2017) Mineral release dynamics of tricalcium phosphate and waste muscovite by mineral-solubilizing rhizobacteria isolated from Indo-Gangetic Plain of India. Geomicrobiol J. https://doi.org/10.1080/01490451.2016.1219431

Balzergue C, Puech-Pagès V, Bécard G, Rochange SF (2011) The regulation of arbuscular mycorrhizal symbiosis by phosphate in pea involves early and systemic signalling events. J Exp Bot 62:1049–1060

Barea JM (1997) Mycorrhiza/bacteria interactions on plant growth promotion. In: Ogoshi A, Kobayashi L, Homma Y, Kodama F, Kondon N, Akino S (eds) Plant growth-promoting rhizobacteria, present status and future prospects. OECD, Paris, pp 150–158

Barker WW, Welch SA, Chu S, Banfield JF (1998) Experimental observations of the effects of bacteria on aluminosilicate weathering. Am Mineral 83:1551–1563

Behera BC, Yadav H, Singh SK, Mishra RR, Sethi BK, Dutta SK, Thatoi HN (2017) Phosphate solubilization and acid phosphatase activity of Serratia sp. isolated from mangrove soil of Mahanadi river delta, Odisha, India. J Genet Eng Biotechnol 15:169–178

Bender SF, Wagg C, van der Heijden MG (2016) An underground revolution: biodiversity and soil ecological engineering for agricultural sustainability. Trends Ecol Evol 31:440–452

Benizri E, Piutti S, Verger S, Pages L, Vercambre G, Poessel JL, Michelot P (2005) Replant diseases: bacterial community structure and diversity in peach rhizosphere as determined by metabolic and genetic fingerprinting. Soil Biol Biochem 37:1738–1746

Bennett PC, Choi WJ, Rogera JR (1998) Microbial destruction of feldspars. Mineral Manag 8:149–150

Berendsen RL, Vismans G, Yu K, Song Y, Jonge R, Burgman WP, Pieterse CM (2018) The disease-induced assemblage of a plant beneficial bacterial consortium. ISME J 12:1496

Berry AM, Harriott OT, Moreau RA, Osman SF, Benson DR, Jones AD (1993) Hopanoid lipids compose the *Frankia* vesicle envelope, presumptive barrier of oxygen diffusion to nitrogenase. Proc Natl Acad Sci U S A 90:6091–6094

Bhat RA, Dervash MA, Mehmood MA, Bhat MS, Rashid A, Bhat JIA, Singh DV, Lone R (2017) Mycorrhizae: a sustainable industry for plant and soil environment. In: Varma A et al (eds) Mycorrhiza-nutrient uptake, biocontrol, ecorestoration. Springer International, Berlin, pp 473–502

Bhatia CR (2008) Role of microbial diversity for soil, health and plant nutrition. In: Nautiyal CS, Dion P (eds) Molecular mechanisms of plant and microbe coexistence. Springer, Berlin, pp 53–74

Bhattacharyya PN, Jha DK (2012) Plant growth-promotig rhizobacteria (PGPR): emergence in agriculture. World J Microbiol Biotechnol 28:1327–1250

Bhatti AA, Haq S, Bhat RA (2017) Actinomycetes benefaction role in soil and plant health. Microb Pathog 111:458–467

Britzke D, Da Silva LS, Moterle DF, Rheinheimer DDS, Bortoluzzi EC (2012) A study of potassium dynamics and mineralogy in soils from subtropical Brazilian lowlands. J Soil Sediment 12:185–197

Buscardo E, Geml J, Schmidt SK, Silva AL, Ramos RT, Barbosa SM, Andrade SS, Dalla Costa R, Souza AP, Freitas H, Cunha HB (2018) Of mammals and bacteria in a rainforest: temporal dynamics of soil bacteria in response to simulated N pulse from mammalian urine. Funct Ecol 32:773–784

Carlile MJ, Watkinson SC, Gooday GW (2001) The fungi, 2nd edn. Academic Press, San Diego

Chauhan A, Guleria S, Balgir PP, Walia A, Mahajan R, Mehta P, Shirkot CK (2017) Tricalcium phosphate solubilization and nitrogen fixation by newly isolated Aneurinibacillus aneurinilyticus CKMV1 from rhizosphere of *Valeriana jatamansi* and its growth promotional effect. Braz J Microbiol 48:294–304

Chowdhury SP, Hartmann A, Gao X, Borriss R (2015) Biocontrol mechanism by root-associated *Bacillus amyloliquefaciens* FZB42—a review. Front Microbiol 6:780. https://doi.org/10.3389/fmicb.2015.00780

Chung JH, Song GC, Ryu CM (2016) Sweet scents from good bacteria: case studies on bacterial volatile compounds for plant growth and immunity. Plant Mol Biol 90:677–687

Coats VC, Rumpho ME (2014) The rhizosphere microbiota of plant invaders: an overview of recent advances in the microbiomics of invasive plants. Front Microbiol 5:368. https://doi.org/10.3389/fmicb.2014.00368

Cohen AJ, Brauer M, Burnett R, Anderson HR, Frostad J, Estep K, Balakrishnan K, Brunekreef B, Dandona L, Dandona R, Feigin V (2017) Estimates and 25-year trends of the global burden of disease attributable to ambient air pollution: an analysis of data from the Global Burden of Diseases Study 2015. Lancet 389:1907–1918

Cornejo P, Seguel A, Aguilera P, Meier S, Larsen J, Borie F (2017) Arbuscular mycorrhizal fungi improve tolerance of agricultural plants to cope abiotic stress conditions. In: Singh DP, Singh HB, Prabha R (eds) Plant-microbe interactions in agroecological perspectives. Springer, Singapore, pp 55–80

Dalton DA, Kramer S (2006) Nitrogen-fixing bacteria in non-legumes. Springer, Dordrecht, pp 105–113

Datta R, Kelkar A, Baraniya D, Molaei A, Moulick A, Meena RS, Formanek P (2017) Enzymatic degradation of lignin in soil: a review. Sustain MDPI 1163(9):1–18

Davinic M, Fultz LM, Acosta-Martinez V, Calderon FJ, Cox SB, Dowd SE, Allen VG, Zak JC, Moore-Kucera J (2012) Pyrosequencing and mid-infrared spectroscopy reveal distinct aggregate stratification of soil bacterial communities and organic matter composition. Soil Biol Biochem 46:63–72

Diep CN, Hieu TN (2013) Phosphate and potassium solubilizing bacteria from weathered materials of denatured rock mountain, Ha Tien, Kiên Giang province, Vietnam. Am J Life Sci 1(3):88–92

Dimkpa C, Weinand T, Asch F (2009) Plant–rhizobacteria interactions alleviate abiotic stress conditions. Plant Cell Environ 32:1682–1694

Dolatabad HK, Javan-Nikkhah M, Shier WT (2017) Evaluation of antifungal, phosphate solubilisation, and siderophore and chitinase release activities of endophytic fungi from *Pistacia vera*. Mycol Prog 16:777–790

East R (2013) Soil science comes to life. Nature 501:S18

Elias F, Woyessa D, Muleta D (2016) Phosphate solubilization potential of rhizosphere fungi isolated from plants in Jimma zone, Southwest Ethiopia. Int J Microbiol 2016:1–11

Ellouze W, Esmaeili Taheri A, Bainard LD, Yang C, Bazghaleh N, Navarro-Borrell A, Hanson K, Hamel C (2014) Soil fungal resources in annual cropping systems and their potential for management. Biomed Res Int 2014:531824. https://doi.org/10.1155/2014/531824

Etesami H (2018) Can interaction between silicon and plant growth-promoting rhizobacteria benefit in alleviating abiotic and biotic stresses in crop plants? Agric Ecosyst Environ 253:98–112

Etesami H (2020) Enhanced phosphorus fertilizer use efficiency with microorganisms. In: Meena RS (ed) Nutrient dynamics for sustainable crop production. https://doi.org/10.1007/978-981-13-8660-2_8

Fernández-Bidondo L, Silvani V, Colombo R, Pérgola M, Bompadre J, Godeas A (2011) Pre-symbiotic and symbiotic interactions between *Glomus intraradices* and two Paenibacillus species isolated from AM propagules. In vitro and in vivo assays with soybean (AG043RG) as plant host. Soil Biol Biochem 43:1866–1872

Friberg S (2001) Distribution and diversity of arbuscular mycorrhizal fungi in traditiona agriculture on the Niger inland delta, Mali, West Africa. CBM's Skriftserie 3:53–80

Gao L, Kong F, Feng C, Wang J, Gao J, Shen G, Zhang C (2016) Isolation, characterization, and growth promotion of phosphatesolubilizing bacteria associated with Nicotiana tabacum (tobacco). Pol J Environ Stud 25:993–1003

Garbisu C, Garaiyurrebaso O, Epelde L, Grohmann E, Alkorta I (2017) Plasmid-mediated bioaugmentation for the bioremediation of contaminated soils. Front Microbiol 8:1966. https://doi.org/10.3389/fmicb.2017.01966

Ghonaim MM, Mohamed HI, Omran AA (2020) Evaluation of wheat salt stress tolerance using physiological parameters and retrotransposon-based markers. Genet Resour Crop Evol 68(1). https://doi.org/10.1007/s10722-020-00981-w

Ghorchiani M, Etesami H, Alikhani HA (2018) Improvement of growth and yield of maize under water stress by co-inoculating an arbuscular mycorrhizal fungus and a plant growth promoting rhizobacterium together with phosphate fertilizers. Agric Ecosyst Environ 258:59–70

Gizaw B, Tsegay Z, Tefera G, Aynalem E, Wassie M, et al. (2017) Phosphate Solubilizing Fungi Isolated and Characterized from Teff Rhizosphere Soil Collected from North Showa and Gojam, Ethiopia. J Fertil Pestic 8: 180. https://doi.org/10.4172/2471-2728.1000180

Giri B, Mukerji KG (2004) Mycorrhizal inoculant alleviates salt stress in *Sesbania aegyptiaca* and *Sesbania grandiflora* under field conditions: evidence for reduced sodium and improved magnesium uptake. Mycorrhiza 14:307–312

Glick BR (2014) Bacteria with ACC deaminase can promote plant growth and help to feed the world. Microbiol Res 169:30–39

Gogoi N, Baruah KK, Meena RS (2018) Grain legumes: impact on soil health and agroecosystem. In: Meena RS et al (eds) Legumes for soil health and sustainable management. Springer, Singapore. https://doi.org/10.1007/978-981-13-0253-4_16

Gouda S, Kerry RG, Das G, Paramithiotis S, Shin HS, Patra JK (2018) Revitalization of plant growth-promoting rhizobacteria for sustainable development in agriculture. Microbiol Res 206:131–140

Hamdali H, Moursalou K, Tchangbedji G, Ouhdouch Y, Hafidi M (2012) Isolation and characterization of rock phosphate solubilizing actinobacteria from a Togolese phosphate mine. Afr J Biotechnol 11:312–320

Hassen AI, Bopape FL, Sanger LK (2016) Microbial inoculants as agents of growth promotion and abiotic stress tolerance in plants. In: Singh DP, Singh HB, Prabha R (eds) Microbial inoculants in sustainable agricultural productivity. Springer, New Delhi, pp 23–36

Hassink J (1994) Effect of soil texture on the size of the microbial biomass and on the amount of C and N mineralized per unit of microbial biomass in Dutch grassland soils. Soil Biol Biochem 26:1573–1581

Hawkins HJ, Johansen A, George E (2000) Uptake and transport of organic and inorganic nitrogen by arbuscular mycorrhizal fungi. Plant and Soil 226:275–285

Hindumathi A, Reddy BN (2012) Synergistic effect of arbuscular mycorrhizal fungi and Rhizobium on the growth and charcoal rot of soybean [Glycine max (L.) Merr.]. World J Sci Technol 2:63–70

Hindumathi A, Reddy BN, Sabitha Rani A, Reddy AN (2016) Associative effect of arbuscular mycorrhizal fungi and Rhizobium on plant growth and biological control of charcoal rot in green gram [*Vigna radiata* L. (Wilczek)]. In: Bhima B, Anjana Devi T (eds) Microbial biotechnology: technological challenges and developmental trends. Apple Academic Press, Milton, pp 155–170

Jacoby R, Peukert M, Succurro A, Koprivova A, Kopriva S (2017) The role of soil microorganisms in plant mineral nutrition—current knowledge and future directions. Front Plant Sci 8:1617

Jog R, Pandya M, Nareshkumar G, Rajkumar S (2014) Mechanism of phosphate solubilisation and antifungal activity of Streptomyces spp. isolated from wheat roots and rhizosphere and their application in improving plant growth. Microbiology 160:778–788

Kanes OS, Weckert MW, Kadam TA, Bhosale HJ (2015) Phosphate solubilization by stress-tolerant soil fungus *Talaromyces funiculosus* SLS8 isolated from the neem rhizosphere. Ann Microbiol 65:85–93

Khamna S, Yokota A, Lumyong S (2009) Actinomycetes isolated from medicinal plant rhizosphere soil: diversity and screening of antifungal compound, indole-3-acetic acid and siderophore production. World J Microbiol Biotechnol 25:649–655

Khanday M, Bhat RA, Haq S, Dervash MA, Bhatti AA, Nissa M, Mir MR (2016) Arbuscular mycorrhizal fungi boon for plant nutrition and soil health. In: Hakeem KR et al (eds) Soil science: agricultural and environmental prospective. Springer International, Cham, pp 317–332

Kim J, Rees DC (1994) Nitrogenase and biological nitrogen fixation. Biochem 33:389–397

Kour D, Rana KL, Kaur T, Sheikh I, Yadav AN, Kumar V et al (2020) Microbe-mediated alleviation of drought stress and acquisition of phosphorus in great millet (*Sorghum bicolour* L.) by drought-adaptive and phosphorus-solubilizing microbes. Biocatal Agric Biotechnol 23:101501. https://doi.org/10.1016/j.bcab.2020.101501

Kuan KB, Othman R, Rahim KA, Shamsuddin ZH (2016) Plant growth-promoting rhizobacteria inoculation to enhance vegetative growth, nitrogen fixation and nitrogen remobilisation of maize under greenhouse conditions. PLoS One 11:e0152478. https://doi.org/10.1371/journal.pone.0152478

Kumar A, Bahadur I, Maurya BR, Raghuwanshi R, Meena VS, Singh DK, Dixit J (2015) Does a plant growth-promoting rhizobacteria enhance agricultural sustainability? J Pure Appl Microbiol 9:715–724

Kumar A, Maurya BR, Raghuwanshi R, Meena VS, Islam MT (2017) Co-inoculation with Enterobacter and rhizobacteria on yield and nutrient uptake by wheat (*Triticum aestivum* L.) in the alluvial soil under Indo-Gangetic Plain of India. J Plant Growth Regul 36(3). https://doi.org/10.1007/s00344-016-9663-5

Kundan R, Pant G, Jadon N, Agrawal PK (2015) Plant growth-promoting rhizobacteria: mechanism and current perspective. J Fertil Pest 6:1–9

Leifheit EF, Veresoglou SD, Lehmann A, Morris EK, Rillig MC (2014) Multiple factors influence the role of arbuscular mycorrhizal fungi in soil aggregation—a meta-analysis. Plant and Soil 374:523–537

Leifheit EF, Verbruggen E, Rillig MC (2015) Arbuscular mycorrhizal fungi reduce decomposition of woody plant litter while increasing soil aggregation. Soil Biol Biochem 81:323–328

Liu D, Lian B, Dong H (2012) Isolation of Paenibacillus sp. and assessment of its potential for enhancing mineral weathering. Geomicrobiol J 29:413–421

López-Arredondo DL, Leyva-González MA, González-Morales SI, López-Bucio J, Herrera-Estrella L (2014) Phosphate nutrition: improving low-phosphate tolerance in crops. Annu Rev Plant Biol 65:95–123

Ma Y, Rajkumar M, Zhang C, Freitas H (2016) Inoculation of *Brassica oxyrrhina* with plant growth promoting bacteria for the improvement of heavy metal phytoremediation under drought conditions. J Hazard Mater 320:36–44

Majeed A, Muhammad Z, Islam S, Ullah Z, Ullah R (2017) Cyanobacterial application as biofertilizers in rice fields: role in growth promotion and crop productivity. PSM Microbiol 2:47–50

Marschner H (1995) Mineral nutrition of higher plants. Academic Press, London

Maya MA, Matsubara Y (2013) Influence of arbuscular mycorrhiza on the growth and antioxidative activity in cyclamen under heat stress. Mycorrhiza 23:381–390

McFarland J, Ruess R, Keilland K, Pregitzer K, Hendrick R, Allen M (2010) Cross-ecosystem comparisons of in situ plant uptake of amino acid-N and NH4+. Ecosystems 13:177–193

Meena VS, Maurya BR, Bahadur I (2014a) Potassium solubilization by bacterial strain in waste mica. Bangladesh J Bot 43:235–237

Meena VS, Maurya BR, Verma JP (2014b) Does a rhizospheric microorganism enhance K+ availability in agricultural soils? Microbiol Res 169:337–347

Meena RS, Meena VS, Meena SK, Verma JP (2015a) The needs of healthy soils for a healthy world. J Clean Prod 102:560–561

Meena VS, Meena SK, Verma JP, Meena RS, Ghosh BN (2015b) The needs of nutrient use efficiency for sustainable agriculture. J Clean Prod 102:562–563

Meena SK, Rakshit A, Meena VS (2016) Effect of seed bio-priming and N doses under varied soil type on nitrogen use efficiency (NUE) of wheat (*Triticum aestivum* L.) under greenhouse conditions. Biocatal Agric Biotechnol 6:68–75

Meena KK, Sorty AM, Bitla UM, Choudhary K, Gupta P, Pareek A, Singh DP, Prabha R, Sahu PK, Gupta VK, Singh HB (2017a) Abiotic stress responses and microbe-mediated mitigation in plants: the omics strategies. Front Plant Sci 8:172. https://doi.org/10.3389/fpls.2017.00172

Meena RS, Kumar S, Pandey A (2017b) Response of sulfur and lime levels on productivity, nutrient content and uptake of sesame under guava (*Psidium guajava* L.) based agri-horti system in an acidic soil of eastern Uttar Pradesh, India. J Crop Weed 13(2):222–227

Meliani A, Bensoltane A, Mederbel K (2012) Microbial diversity and abundance in soil: related to plant and soil type. Am J Plant Nutr Fertil Technol 2:10–18

Miyasaka SC, Habte M (2001) Plant mechanisms and mycorrhizal symbioses to increase phosphorus uptake efficiency. Commun Soil Sci Plant Anal 32:1101–1147

Mohamed HI, Gomaa EZ (2012) Effect of plant growth-promoting Bacillus subtilis and *Pseudomonas fluorescens* on growth and pigment composition of radish plants (Raphanus sativus) under NaCl stress. Photosynthetica 50(2):263–272

Mohamed HME, Shaieb FMA, El-Komy HMA (2017) Solubilization of inorganic phosphates by isolated *Azospirillum lipoferum* (H3) as free or alginate immobilized inoculation. ContROL 1:1–7

Mohamed HI, Akladious SA, El-Beltagi HS (2018) Mitigation the harmful effect of salt stress on physiological, biochemical and anatomical traits by foliar spray with trehalose on wheat cultivars. Fresen Environ Bull 27(10):7054–7065

Mohamed HI, Ashry NA, Ghonaim MM (2019) Physiological analysis for heat shock-induced biochemical (responsive) compounds and molecular characterizations of ESTs expressed for heat tolerance in some Egyptian maize hybrids. Gesunde Pflanzen 71:213–222

Muller A, Schader C, Scialabba NEH, Brüggemann J, Isensee A, Erb KH, Niggli U (2017) Strategies for feeding the world more sustainably with organic agriculture. Nat Commun 8:1290

Nath D, Maurya BR, Meena VS (2017) Documentation of five potassium- and phosphorus-solubilizing bacteria for their K and P-solubilization ability from various minerals. Biocatal Agric Biotechnol 10:174–181

Niu YF, Chai RS, Jin GL, Wang H, Tang CX, Zhang YS (2013) Responses of root architecture development to low phosphorous availability: a review. Ann Bot 112:391–408

Nouri E, Breuillin-Sessoms F, Feller U, Reinhardt D (2014) Phosphorus and nitrogen regulate arbuscular mycorrhizal symbiosis in Petunia hybrida. PLoS One 9:e90841. https://doi.org/10.1371/journal.pone.0090841

Olanrewaju OS, Glick BR, Babalola OO (2017) Mechanisms of action of plant growth-promoting bacteria. World J Microbiol Biotechnol 33:197

Otieno N, Lally RD, Kiwanuka S, Lloyd A, Ryan D, Germaine KJ, Dowling DN (2015) Plant growth promotion induced by phosphate solubilizing endophytic Pseudomonas isolates. Front Microbiol 6:745

Panhwar QA, Radziah O, Rahman AZ, Sariah M, Razi IM, Naher UA (2011) Contribution of phosphate-solubilizing bacteria in phosphorus bioavailability and growth enhancement of aerobic rice. Span J Agric Res 9:810–820

Pany S, Mishra S, Gupta N (2018) Evaluation of native rhizospheric and phosphate solubilizing microbes for growth and development of Pongamia *pinnata* under nursery condition. Adv Biores 9:92–101

Parani K, Saha BK (2012) Prospects of using phosphate-solubilizing Pseudomonas as biofertilizer. EJBS 4:40–44

Patel S, Panchal B, Karmakar N, Rajkumar JS (2015) Solubilization of rock phosphate by two Rhizopus species isolated from coastal areas of south Gujar at and its effect on chickpea. Ecol Environ Conserv 21:229–237

Paul D, Sinha SN (2017) Isolation and characterization of phosphate solubilizing bacterium Pseudomonas aeruginosa KUPSB12 with antibacterial potential from river Ganga, India. Ann Agrar Sci 15:130–136

Pettigrew WT (2008) Potassium influences on yield and quality production for maize, wheat, soybean and cotton. Physiol Plant 133:670–681

Pichardo ST, Su Y, Han FX (2012) The potential effects of arbuscular mycorrhizae (AM) on the uptake of heavy metals by plants from contaminated soils. J Bioremed Biodegr 3:e124. https://doi.org/10.4172/2155-6199.1000e124

Pineda A, Kaplan I, Bezemer TM (2017) Steering soil microbiomes to suppress aboveground insect pests. Trends Plant Sci 22:770–778

Prajapati KB, Modi HA (2012) Isolation and characterization of potassium solubilizing bacteria from ceramic industry soil. CIB Tech J Microbiol 1:8–14

Prajapati K, Sharma MC, Modi HA (2013) Growth promoting effect of potassium solubilizing microorganisms on Abelmoschus esculentus. Int J Agric Sci 3:181–188

Qiao Q, Wang F, Zhang J, Chen Y, Zhang C, Liu G, Zhang J (2017) The variation in the rhizosphere microbiome of cotton with soil type, genotype and developmental stage. Sci Rep 7:3940

Rahman M, Sabir AA, Mukta JA, Khan MMA, Mohi-Ud-Din M, Miah MG, Islam MT (2018) Plant probiotic bacteria Bacillus and Paraburkholderia improve growth, yield and content of antioxidants in strawberry fruit. Sci Rep 8:2504

Rao DLN (2014) Recent advances in biological nitrogen fixation in agricultural systems. Proc Ind Natl Sci Acad 80(2):359–378

Rawat J, Sanwal P, Saxena J (2016) Potassium and its role in sustainable agriculture. In: Meena VS, Maurya BR, Verma JP, Meena RS (eds) Potassium solubilizing microorganisms for sustainable agriculture. Springer, New Delhi, pp 235–253. https://doi.org/10.1007/978-81-322-2776-2_17

Redecker D, Schüßler A, Stockinger H, Stürmer SL, Morton JB, Walker C (2013) An evidence-based consensus for the classification of arbuscular mycorrhizal fungi (*Glomeromycota*). Mycorrhiza 23:515–531

Rfaki A, Nassiri L, Ibijbijen J (2015) Isolation and characterization of phosphate solubilizing bacteria from the rhizosphere of faba bean (*Vicia faba* L.) in Meknes region, Morocco. Br Microbiol Res J 6:247–254

Ribeiro CM, Cardoso EJBN (2012) Isolation, selection and characterization of root-associated growth-promoting bacteria in Brazil pine (*Araucaria angustifolia*). Microbiol Res 167:69–78

Rillig MC, Aguilar-Trigueros CA, Bergmann J, Verbruggen E, Veresoglou SD, Lehmann A (2015) Plant root and mycorrhizal fungal traits for understanding soil aggregation. New Phytol 205:1385–1388

Rodríguez H, Fraga R (1999) Phosphate solubilizing bacteria and their role in plant growth promotion. Biotechnol Adv 17:319–339

Rokhbakhsh-Zamin F, Sachdev D, Kazemi-Pour N, Engineer A, Pardesi KR, Zinjarde S, Dhakephalkar PK, Chopade BA (2011) Characterization of plant-growth-promoting traits of Acinetobacter species isolated from rhizosphere of Pennisetum glaucum. J Microbiol Biotechnol 21:556–566

Rosier A, Bishnoi U, Lakshmanan V, Sherrier DJ, Bais HP (2016) A perspective on inter-kingdom signaling in plant-beneficial microbe interactions. Plant Mol Biol 90:537–548

Saeid A, Prochownik E, Dobrowolska-Iwanek J (2018) Phosphorus solubilization by Bacillus species. Molecules 23:2897

Saikia J, Sarma RK, Dhandia R, Yadav A, Gupta VK, Bharali R, Saikia R (2018) Alleviation of drought stress in pulse crops with ACC deaminase producing rhizobacteria isolated from acidic soil of Northeast India. Sci Rep 8:3560

Sangeeth KP, Bhai RS, Srinivasan V (2012) *Paenibacillus glucanolyticus*, a promising potassium solubilizing bacterium isolated from black pepper (Piper nigrum L.) rhizosphere. J Spic Aromat Crops 21:118–124

Saraf M, Pandya U, Thakkar A (2014) Role of allelochemicals in plant growth-promoting rhizobacteria for biocontrol of phytopathogens. Microbiol Res 169:18–29

Sayed WF (2011) Improving *Casuarina* growth and symbiosis with *Frankia* under different soil and environmental conditions-review. Folia Microbiol 56(1):1–9

Schmidt R, Cordovez V, De Boer W, Raaijmakers J, Garbeva P (2015) Volatile affairs in microbial interactions. ISME J 9:2329. https://doi.org/10.1038/ismej.2015.42
Schulz S, Brankatschk R, Dümig A, Kögel-Knabner I, Schloter M, Zeyer J (2013) The role of microorganisms at different stages of ecosystem development for soil formation. Biogeosciences 10:3983–3996
Shaharokhi S, Bonjar S, Saadoun GHI (2005) Biological control of potato isolates of *Rhizoctonia solani* by *Streptomyces olivaceus* strain 115. Australas Biotechnol 4:132–138
Shameer S, Prasad T (2018) Plant growth-promoting rhizobacteria for sustainable agricultural practices with special reference to biotic and abiotic stresses. Plant Growth Regul:1–13
Shanware AS, Kalkar SA, Trivedi MM (2014) Potassium solublisers: occurrence, mechanism and their role as competent biofertilizers. Int J Curr Microbiol App Sci 3(9):622–629
Shrivastava P, Kumar R (2015) Soil salinity: a serious environmental issue and plant growth-promoting bacteria as one of the tools for its alleviation. Saudi J Biol Sci 22(2):123–131
Shukla RM, Vyas RV (2014) Phosphate solubilizing efficiency of mycopesticides. IJAEB 7:705–710
Shukla M, Patel RH, Verma R, Deewan P, Dotaniya ML (2013) Effect of bio-organics and chemical fertilizers on growth and yield of chickpea (*Cicer arietinum* L.) under middle Gujarat conditions. Vegetos 26(1):183–187
Singh RP, Jha PN (2016) Mitigation of salt stress in wheat plant (*Triticum aestivum*) by ACC deaminase bacterium Enterobacter sp. SBP-6 isolated from *Sorghum bicolor*. Acta Physiol Plant 38:110. https://doi.org/10.1007/s11738-016-2123-9
Singh B, Satyanarayana T (2010) Plant growth promotion by an extracellular HAP-phytase of a thermophilic mold Sporotrichum thermophile. Appl Biochem Biotechnol 160:1267–1276
Singh JS, Singh DP (2012) Reforestation: a potential approach to mitigate the excess CH4 build-up. Ecol Manage Restor 13(3):245–248
Singh JS, Strong PJ (2016) Biologically derived fertilizer: a multifaceted bio-tool in methane mitigation. Ecotoxicol Environ Saf 124:267–276
Singh JS, Pandey VC, Singh DP, Singh RP (2010) Influence of pyrite and farmyard manure on population dynamics of soil methanotroph and rice yield in saline rain-fed paddy field. Agric Ecosyst Environ 139:74–79
Singh JS, Abhilash PC, Singh HB, Singh RP, Singh DP (2011a) Genetically engineered bacteria: an emerging tool for environmental remediation and future research perspectives. Gene 480:1–9
Singh JS, Pandey VC, Singh DP (2011b) Efficient soil microorganisms: a new dimension for sustainable agriculture and environmental development. Agric Ecosyst Environ 140:339–353
Singh JS, Singh DP, Dixit S (2011c) Cyanobacteria: an agent of heavy metal removal. In: Maheshwari DK, Dubey RC (eds) Bioremediation of pollutants. IK International, New Delhi, pp 223–243
Singh JS (2015) Plant-microbe interactions: a viable tool for agricultural sustainability. Appl Soil Ecol 92:45–46
Singh JS, Kumar A, Rai AN, Singh DP (2016) Cyanobacteria: a precious bio-resource in agriculture, ecosystem, and environmental sustainability. Front Microbiol 7(529):1–19
Smith SE, Read DJ (2008) Mycorrhizal symbiosis. Academic Press, London
Smith SE, Smith FA (2011) Roles of arbuscular mycorrhizas in plant nutrition and growth: new paradigms from cellular to ecosystem scales. Annu Rev Plant Biol 62:227–250
Smith P, Cotrufo MF, Rumpel C, Paustian K, Kuikman PJ, Elliott JA, McDowell R, Griffiths RI, Asakawa S, Bustamante M, House JI (2015) Biogeochemical cycles and biodiversity as key drivers of ecosystem services provided by soils. Soil 1:665–685
Sridevi M, Mallaiah KV (2009) Phosphate solubilizatin by Rhizobium sXG/dv'vstrains. Indian J Microbiol 49:98–102
Stokstad E (2016) The nitrogen fix. Science 353:1225–1227
Styriakova I, Styriak I, Hradil D, Bezdicka P (2003) The release of iron-bearing minerals and dissolution of feldspar by heterotrophic bacteria of Bacillus species. Ceram Silic 47(1):20–26

Talaat NB, Shawky BT (2014) Protective effects of arbuscular mycorrhizal fungi on wheat (*Triticum aestivum* L.) plants exposed to salinity. Environ Exp Bot 98:20–31

Taniguchi H, Wendisch VF (2015) Exploring the role of sigma factor gene expression on production by Corynebacterium glutamicum: sigma factor H and FMN as example. Front. Microbiol 6:740. https://doi.org/10.3389/fmicb.2015.00740

Ullah A, Heng S, Munis MFH, Fahad S, Yang X (2015) Phytoremediation of heavy metals assisted by plant growth-promoting (PGP) bacteria: a review. Environ Exp Bot 117:28–40

Verma R, Maurya BR, Meena VS, Dotaniya ML, Deewan P (2017a) Microbial dynamics as influenced by bio-organics and mineral fertilizer in alluvium soil of Varanasi. India Int J Curr Microbiol App Sci 6(2):1516–1524

Verma R, Maurya BR, Meena VS, Dotaniya ML, Deewan P, Jajoria M (2017b) Enhancing production potential of cabbage and improves soil fertility status of Indo-Gangetic Plain through the application of bio-organics and mineral fertilizer. Int J Curr Microbiol App Sci 6(3):301–309

Vitorino LC, Silva FG, Soares MA, Souchie EL, Costa AC (2012) Solubilization of calcium and iron phosphate and in vitro production of indoleacetic acid by endophytic isolates of Hyptis marrubioides Epling (Lamiaceae). Int Res J Biotechnol 3:47–54

Wahid F, Sharif M, Steinkellner S, Khan MA, Marwat K, Khan S (2016) Inoculation of arbuscular mycorrhizal fungi and phosphate solubilizing bacteria in the presence of rock phosphate improves phosphorus uptake and growth of maize. Pak J Bot 48:739–747

Watts-Williams SJ, Jakobsen I, Cavagnaro TR, Grønlund M (2015) Local and distal effects of arbuscular mycorrhizal colonization on direct pathway Pi uptake and root growth in *Medicago truncatula*. J Exp Bot 66:4061–4073

Welch SA, Taunton AE, Banfield JF (2002) Effect of microorganisms and microbial metabolites on apatite dissolution. Geophys J Roy Astron Soc 19:343–367

Wu QS, Zou YN (2017) Arbuscular mycorrhizal fungi and tolerance of drought stress in plants. In: Wu Q-S (ed) Arbuscular mycorrhizas and stress tolerance of plants. Springer, Singapore, pp 25–41

Wu SC, Cao ZH, Li ZG, Cheung KC, Wong MH (2005) Effects of biofertilizer containing N-fixer, P and K solubilizers and AM fungi on maize growth: a greenhouse trial. Geoderma 125:155–166

Xueming Z, Zhenping H, Yu Z, Huanshi Z, Pei Q (2014) Arbuscular mycorrhizal fungi (AMF) and phosphate-solubilizing fungus (PSF) on tolerance of beach plum (Prunus maritima) under salt stress. Aust J Crop Sci 8:945–950

Yadav V, KumarM DDK, Kumar H, Sharma R, Tripathi T, Tuteja N, Saxena AK, Johri AK (2010) A phosphate transporter from the root endophytic fungus *Piriformospora indica* plays a role in phosphate transport to host plant. J Biol Chem 285:26532–26544

Yasin M, Munir I, Faisal M (2016) Can Bacillus spp. enhance K+ uptake in crop species. In: Meena VS, Maurya BR, Verma JP, Meena RS (eds) Potassium solubilizing microorganisms for sustainable agriculture. Springer, New Delhi, pp 163–170. https://doi.org/10.1007/978-81-322-2776-2_12

Ye S, Yanga Y, Xin G, Wang Y, Ruan L, Ye G (2015) Studies of the Italian ryegrass–rice rotation system in southern China: arbuscular mycorrhizal symbiosis affects soil microorganisms and enzyme activities in the *Lolium multiflorum* L. rhizosphere. Appl Soil Ecol 90:26–34

Yi Y, Huang W, Ge Y (2008) Exopolysaccharide: a novel important factor in the microbial dissolution of tricalcium phosphate. World J Microbiol Biotechnol 24:1059–1065

Zarjani JK, Aliasgharzad N, Oustan S, Emadi M, Ahmadi A (2013) Isolation and characterization of potassium solubilizing bacteria in some Iranian soils. Arch Agro Soil Sci 77:7569

Zorb C, Senbayram M, Peiter E (2014) Potassium in agriculture–status and perspectives. J Plant Physiol 171:656–669

Chapter 5
Role of Endophytic Bacteria in the Alleviation of Heavy Metals from an Ecosystem

Parteek Prasher and Mousmee Sharma

Contents

1 Introduction

Rapid urbanization and industrialization yielded unanticipated disadvantages to the environment, which serves both as a source of commercial raw materials and sink for the industrial waste products (Commoner et al. 1971). The waste materials such as heavy metals and organic dyes present hazardous effects due to their non-degradation to the harmless products by natural processes, further increasing their environmental persistence (Srivastava et al. 2017; Prasher et al. 2019). The heavy metals readily accumulate in the various trophic levels of an ecosystem, and while integrating into

P. Prasher (✉)
Department of Chemistry, UGC Sponsored Centre for Advanced Studies, Guru Nanak Dev University, Amritsar, India

Department of Chemistry, University of Petroleum and Energy Studies, Energy Acres, Dehradun, India
e-mail: pprasher@ddn.upes.ac.in

M. Sharma
Department of Chemistry, UGC Sponsored Centre for Advanced Studies, Guru Nanak Dev University, Amritsar, India

Department of Chemistry, Uttaranchal University, Arcadia Grant, Dehradun, India

© The Author(s), under exclusive license to Springer Nature Switzerland AG 2021

H. I. Mohamed et al. (eds.), *Plant Growth-Promoting Microbes for Sustainable Biotic and Abiotic Stress Management*,
https://doi.org/10.1007/978-3-030-66587-6_5

the consumers, they exert significant mutagenic and carcinogenic effects (Kloke et al. 1984). The strategies adopted to alleviate the metal-based pollutants from the contaminated sites include the application of metal chelators, leaching with acids, or chemical treatment of the polluted site with lime and fly ash (Mishra et al. 2017). However, these approaches apply for the sites having a smaller area requiring rapid and complete decontamination and adversely affect the soil biota, its structure, and fertility (Castaldi et al. 2005). The identification of hyper-accumulating plants that accrue high amounts of heavy metals boosted phytoremediation as an emergent doctrine for highly effective, eco-friendly metal sequestration from the contaminated environment (Shrivastava et al. 2019; Rascio and Navari-Izzo 2011). The endophytic bacteria residing in these plants benefit the plant development and their optimal thriving by nitrogen fixation, biosynthesis of auxins, and plant hormones such as indole-3-acetic acid, generation of siderophore, and improved phosphate solubilization, in addition to affording heavy metal resistance (Ma et al. 2016). Nevertheless, the endophytes improve mineral nutrition in the host plant and provide resistance against the abiotic and biotic stresses (Santoyo et al. 2016). Besides, the endophytic siderophores mediate phytoremediation of the heavy metal stressed soils by complexation with the bivalent heavy metal ions that further assimilate within the host plant parts (Khare et al. 2018). Hence, the endophytic bacteria hold a high prominence in the phytoremediation of soils contaminated with heavy metals without producing any adverse impact on the ecosystem (Zhang et al. 2019a). The contemporary phytoremediation efforts obligate the utility of endophytic bacteria in improving soil health due to these beneficial manifestations. The current chapter epigrammatically discusses the manifestations of heavy metal toxicity and beneficial symbiotic association of endophytic bacteria with plants that prompts the uptake and accumulation of essential metal ions, and nutrients. Also, the chapter deals with the role of endophytic bacteria for alleviating the heavy metal toxicity from the ecosystem.

2 Host Plant Colonization by Endophytic Bacteria

Only certain strains of endophytic bacteria afford beneficial effects to the host plant when inoculated in the fields. The main effects include maintaining the plant health in a metal stressed environment. However, due to the underprivileged survival rate and weaker ability to colonize, several endophytic strains fail to produce the desired advantages (Rajkumar et al. 2009). The culture-based methods appraise the potency of endophyte to endure and inhabit in the host plant cells and tissues; however, the approach suffers limitations for not being able to differentiate between the inoculated strains and native endophytic population (Stecher and Johnson 2004). Further, the antibiotic-resistant mutant strains of parent endophyte customarily used to differentiate from native endophytic population require a careful predetermination of resistance in native endophyte population to produce validated results (Ma et al. 2015). Besides, the biochemical methods incorporating the introduction of marker genes, and molecular methods including FISH, molecular fingerprinting, and real-time PCR, provide the application for targeting, detection, and visualization of endophytic bacteria inoculated in the plant tissues or rhizosphere (Ruppel et al. 2006). These methods also enable the differentiation of inoculated endophyte over the native microflora. The marker genes including chromogenic markers and immunofluorescence techniques including confocal laser scanning microscopy enable the visualization of localized endophytic bacteria on plant surface (Pickup 1991). The labelling by green fluorescent proteins further enables appraisal of in vivo localization and in situ analysis and identification of endophytic cells in biofilms (Elbeltagy et al. 2001). Techniques such as quantitative real-time PCR efficiently quantify and detect the endophytic bacteria colonies in plants. Notably, the coupling of advanced fingerprinting techniques such as PCR-DGGE, BOX-PCR, ADSRRS, and RAPD, in addition to multivariate analysis enables the determination of colonization behavior of inoculated endophytic bacteria and the configuration of microbial communities associated with the host plant.

To recruit the favourable endophytic bacterial species from the rhizosphere, the host plants release marked volumes of photosynthates and exudates to establish the primary communication, which further accelerates the colonization process (Hartmann et al. 2014). Some plant species release specific chemicals such as oxalate for the same purpose. Similarly, the compounds released by the host plant are involved in the quorum sensing of endophytic bacteria protected against the pathogenic strains in the rhizosphere, in addition to stimulating the growth in the host plant (Zinniel et al. 2002). The endophytic bacteria successfully intrude different parts of the plant including seeds and embryo, hence thriving in the seedlings during their initial growth and germination. The communication between the microbiota present in rhizosphere and plant roots emerging out of the seedlings begins lately, which prompts the association of soil microbiota with plant roots (Liu et al. 2017). Alternately, a few endophytic bacterial strains capture the tissues of the plant shoot system eventually colonizing the entire plant endosphere. The persistent biotic and abiotic stress countered by plants in the form of salinity, heat, and diseases presents trivial effect on the thriving of endophytes as the latter readily manipulate their functional behaviour to avert the adverse conditions in the host (Abisado et al. 2018). The genomic analysis confirmed that the genes participating in the production of biofilm, its adhesion, and subsequent motility contribute significantly to decide the colonization of the plant tissues by endophytic bacteria and their lifestyle within the host plant. In addition, the genes involved in bacterial chemotaxis and secretion express themselves during the colonization process, which mainly takes place in the root system of the host plant (Table 5.1).

3 Heavy Metal Stress in Plants

The farmlands worldwide face slightly to moderate contamination by toxic heavy metals including As, Pb, Zn, Co, Cu, Cd, and Hg owing to the application of phosphate fertilizers, smelters, industrial waste, and sludge from sewage, metallurgical processes, and unauthorized watering practices (Li et al. 2019; Prasher et al. 2020).

Table 5.1 Recently reported diversity of endophytic bacteria

Plant	Endophytic genera	Ref.
Distichlis spicata, Pluchea absinthioides	*Bacillaceae* and *Enterobacteriacea*	Zhang et al. (2019a)
Panax notoginseng	*Holophaga, Conexibacter, Luteolibacter, Gemmatimonas, Methylophilus, Solirubrobacter*, and *Prosthecobacter*	Dong et al. (2018)
Pennisetum sinese	*Pantoea, Pseudomonas, Burkholderia, Arthrobacter, Psychrobacter*, and *Neokomagataea*	Deng et al. (2019)
Dendrobium stems	*Ralstonia, Comamonas*, and *Lelliottia*	Wang et al. (2019)
Tinospora cordifolia	*Bacillus, Aneurinibacillus*, and *Pseudomonas*	Duhan et al. (2020)
Pseudowintera colorata	*Pseuomonas, Bacillus* sp.	Purushotham et al. (2020)
Aerobic rice (variety ARB6)	*Gammaproteobacteria, Betaproteobacteria, Alphaproteobacteria, Firmicutes*, and *Bacteroidetes*	Vishwakarma and Dubey (2020)
Pseudobrickellia brasiliensis	*Pseudomonas aeruginosa*, and *Bacillus subtilis*	Cardoso et al. (2020)
Vernonia anthelmintica	*Micrococcus endophyticus, Bacillus megaterium, Pseudomonas chlororaphis, Stenotrophomonas pavanii, Pantoea ananatis*	Rustamova et al. (2020)
Messerschmidia sibirica	*Phytophthora nicotianae, Alternaria alternata, Ralstonia solanacearum, Acidovorax citrulli*	Zhang et al. (2019b; 2019c)
Oryzia sativa	*Bacillus subtilis* subsp. *inaquosorum*	Nasrollahi et al. (2020)
Ephedra foliata	*Pseudomonas aeruginosa, Staphylococcus aureus, Escherichia coli, Bacillus subtilis*	Ghiasvand et al. (2019)
Lycium ruthenicum Murr	*Actinobacteria, Streptomyces, Micrococcales*	Liu et al. (2019)
Polygonum hydropiper	*Arthrobacter koreensis, Bacillus subtilis*	Ye et al. (2020)
Cucumis sativus	*Methylobacterium, Frigoribacterium, Lysinimonas*	Mahmood et al. (2019)
Lupinus luteus, Lupinus angustifolius	*Paenibacillus glycanilyticus, Pseudomonas brenneri*	Ferchichi et al. (2019)
Coffea canephora	*Enterobacter* sp., *Streptomyces* sp., *Bacillus* sp., *Serratia* sp., *Paenibacillus* sp.	Hoang et al. (2020)
Paeonia ludlowii	*Alphaproteobacteria, Acidobacteria-6, Betaproteobacteria, Chloracidobacteria, Gammaproteobacteria, Spartobacteria*	Lu et al. (2020)
Zea mays L.	*Bacillus cereus* and *Enterobacter cloacae*	Abedinzadeh et al. (2019)
Piper nigrum L.	*Bacillus siamensis, Bacillus amyloliquefaciens, Bacillus velezenis, Bacillus methylotrophiycus*	Ngo et al. (2020)
Aloe vera	*Paraburkholderia, Enterobacter, Proteobacteria, Firmicutes, Bacteroidetes, Actinobacteria*	Silva et al. (2019)

(continued)

Table 5.1 (continued)

Plant	Endophytic genera	Ref.
Zea mays L.	*Enterobacter*, *Shigella*, *Pseudomonas*, *Achromobacter*	Lu et al. (2020)
Hordeum vulgare L.	*Paraburkholderia tropica*	Garcia et al. (2019)
Panax ginseng Meyer	*Bacillus amyloliquefaciens*	Sun et al. (2019b)
Oryzia sativa L.	*Bacillus altitudinis*, *Bacillus subtilis*	Nasrollahi et al. (2020)
Manihot esculenta	*Bacillus amyloliquefaciens*, and *Microbacterium imperiale*	Freitas et al. (2019)
Camellia oleifera	*Bacillus subtilis*	Xu et al. (2020)
Hybrid pennisetum	*Bacillus megaterium*	Shah et al. (2020)
Rice sprout	*Enterobacter* sp.	Liu et al. (2020a, b)
Sinapis alba	*Pseudomonas* sp.	Plociniczak et al. (2020)
Ocimum sanctum	*Enterobacter cloacae*	Panigrahi et al. (2019)

Nevertheless, the various natural sources such as landslides, volcanic eruptions, soil erosion, and weathering of rocks contribute a substantial amount of heavy metals in the environment (Mohammed et al. 2011). The metals and metal ions serve as micronutrients essential for the plant development and as cofactors for catalyzing several critical biological pathways (Sessitsch et al. 2013). Conversely, the heavy metal strain adversely affects the plant growth and development by systemic impairment of critical morphological, metabolic, and biomolecular processes (Mithofer et al. 2004; Prasher et al. 2018). A higher amount of heavy metals destroys plant proteins and macromolecules by generating physiologically hazardous reactive oxygen species, superoxide anion radical, and hydroxyl radical, which manifests redox imbalance, thereby resulting in functional loss of proteins and vital organelles (Maksymiec 2007). The redox balance in the plants sustains processes such as the formation of tracheary elements, lignification, and related cross-linking processes in the plant cell wall (Barcelo and Poschenrieder 1990). Therefore, the loss in this balance results in plant wilting and necrosis (Dutta et al. 2018). The disruption in redox homeostasis serves as the main reason for heavy metal toxicity in plants. The co-contamination by heavy metals interferes with the beneficial mutualistic relation between the plant and rhizobia, thereby restraining an optimal plant growth (Ghori et al. 2019). Notably, the heavy metal toxicity inhibits the metabolic and enzymatic activity of soil microbiota, eventually discouraging the efficiency to degrade hazardous organic pollutants (Tiwari and Lata 2018). Other potentially damaging effects caused by heavy metal stress include lipid peroxidation that eventually leads to the disruption of biomembranes to yield malondialdehyde as a decomposition product (Yadav 2010). Hence, the plants need to maintain an optimum cellular level of heavy metals to endure the stressed soils (Figs. 5.1 and 5.2; Table 5.2).

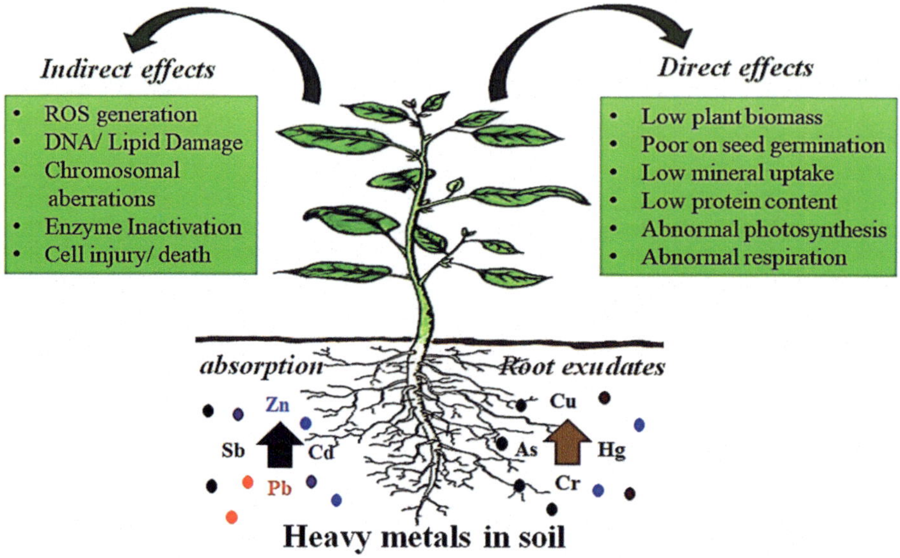

Fig. 5.1 Effect of heavy metal hyperaccumulation in plants

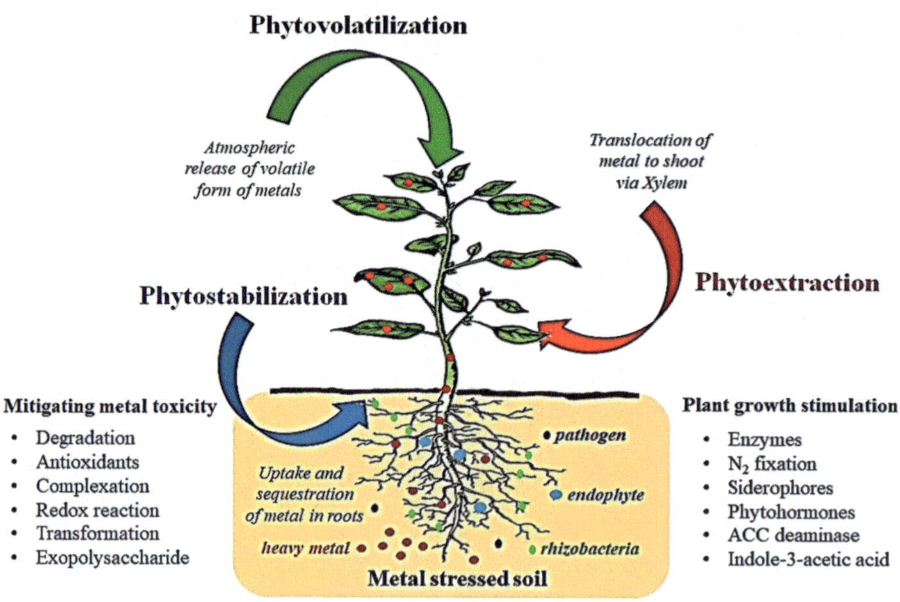

Fig. 5.2 The fate of heavy metal after internalization in the plant

Table 5.2 Toxic effects of heavy metals in plants

Heavy metals	Source	Effect on plant	Ref.
Cu	Smelting of ores, mining	Retarded plant growth, leaf chlorosis, triggers redox imbalance	Chen et al. (2015)
Hg	Hydroelectric discharge, paper and pulp industry, mining of cinnabar (ore)	Binding to water channel proteins, obstructing water flow in plants	Carvalho et al. (2019)
Cr	Tanning industry	Inhibits chlorophyll biosynthesis, plant wilting, root injury, metabolic alterations	Shanker et al. (2005)
Pb	Sewage sludge, paper and paint industry, mining and smelting activities	Reacts with sulfhydryl group of enzymes and inhibits their activity, alters membrane permeability	Zhou et al. (2018)
As	Natural geochemical processes, phosphate fertilizers, glass manufacturing	Disturbance in cellular redox homeostasis, irreversible binding with phytochelatins, and metallothioneins	Kalita et al. (2018)
Co	Fossil fuels, incineration, alloys wearing, sewage sludge	Disturbs catalase activity, affects translocation of essential elements, decrease in water potential and rate of transpiration	Lwalaba et al. (2020)
Ni	Mining and smelting, pesticides, phosphate fertilizers	Affects H-ATPase activity of the plasma membrane, nutrient imbalance, impaired cell membrane function	Parlak (2016)

4 Phytoremediation by Endophytic Bacteria

The endophytic bacteria afford a mutualistic association with the host plant as obligate and facultative dependence on the latter. Obligate endophytes rigorously depend on the host plant for their development and persistence, whereas the facultative endophytes spend some life cycle stages outside the host plant (Afzal et al. 2019). After entry in the host plant, the endophytic bacteria remain localized at the point of entry in the host or spread to the vascular system and intercellular spaces (Lodewyckx et al. 2002). The endophytes customarily demonstrate a modest specificity towards plant tissues and readily acclimatize to the variable conditions in plant tissues and the rhizosphere. To facilitate phytoremediation, the endophytes utilize both direct and indirect strategies (Walia et al. 2017). The direct strategies involve the production of functional biomolecules that directly manage pollutant degradation, hyper-accumulation, and co-metabolization. The biomolecules include enzymes such as cytochrome P450 monooxygenases, polyphenol peroxidases, glutathione S-transferases, and nitroreductases that regulate the heavy metal stabilization, and their deposition, eventually followed by detoxification (Singh et al. 2017). The mechanism of direct phytoremediation by endophytes involves covalent modification of the organic pollutants by supplementing polar, hydrophilic groups such as sulfides, amino acids, malonic acid, carbohydrates, glutathione, and malonic acid that improves the polarity and aqueous solubility of the pollutant, hence rendering it degradable and assimilable to the endophytes in the host plant (He et al. 2020). Co-metabolization of complex pollutants with plants and another microorganism present in the soil serves as another direct phytoremediation strategy adopted by endophytes. Notably, the extracellular chelating compounds produced by endophytes including siderophores, biosurfactants, and acidic organic matter facilitate the soil demineralization and accelerate metal solubilization (Rajkumar et al. 2012). Nevertheless, the indirect heavy metal phytoremediation approaches by endophytic bacteria constitute an enhanced supply of plant nutrients, plant growth regulation, alleviation of toxic stress, and communication with the microbial communities (Ma et al. 2011).

5 Role of Endophytic Bacteria for the Accumulation of Heavy Metals in Plants

The hyperaccumulator plant species aggregate an abundant amount of heavy metals in the shoot system compared to the surrounding soils, thereby providing suitable thriving conditions for the persistence of endophytic bacteria adapted to endure the heavy metal stress (Kramer 2010). In addition to promoting the growth of the host plant, the endophytes attenuate heavy metal toxicity by acidification mechanism, the formation of siderophores, metal chelation, and mobilization of metal ions as phosphates (Grobelak and Hiller 2017). Several endophytic bacterial strains produce 5-ketogluconic acid that promoted zinc solubilization (Mumtaz et al. 2017). Similarly, the lead resistant endophytic bacteria enhance the uptake of the heavy metal in the host plants to 75–130% by metal release from the non-soluble phases in soil (Sheng et al. 2008). Interestingly, the seed or rhizosphere inoculation with metal-resistant endophytic bacteria considerably improves the efficiency of metal extraction by endophytes (Manara 2012). Contrarily, in some cases the metal-resistant endophytes downregulated the heavy metal uptake and increase the plant biomass (Mesa et al. 2015). As such, the endophytic bacteria isolated from rice tissues lower the uptake of nickel and cadmium in roots and shoot system mainly due to the heavy metal immobilization in rhizosphere by the bacteria (Ahemad 2019). The endophytic bacteria display resistance to heavy metals or their combination in the natural environment, however, the resistance occasionally diminishes for specific coexisting metals such as nickel and cobalt (Khare et al. 2018). The endophytic bacteria contain characteristic features such as cation influx systems, efflux pumps and detoxification systems acquired through the horizontal gene transfer that provides resistant against the heavy metals (Blanco and Lugtenberg 2014; Mudila et al. 2019). Due to these features, the hyperaccumulator plants select the heavy metal resistant endophytic bacteria while directly retorting to the soil contamination caused by metal stress, which promotes optimal thriving of the host plant in a specific niche (Kandel et al. 2017). Nevertheless, the correct mechanism of heavy metal resistance offered by endophytic bacteria to a host plant is still under investigation (Table 5.3).

Table 5.3 Recent reports on heavy metal alleviation by endophytic bacteria

Hyperaccumulator plant	Endophytic bacteria	Metal accumulated	Ref.
Pteris vittata	*Proteobacteria*, *Actinobacteria*, *Firmicutes*, *Agrobacterium*, *Stenotrophomonas*, *Pseudomonas*, *Rhodococcus*, and *Bacillus* sp.	As	Gu et al. (2018)
Pteris vittata	*Proteobacteria*, *Actinobacteria*	As	Xu et al. (2016)
Pteris vittata	*Exiguobacterium* sp. and *Aeromonas* sp.	As	Han et al. (2016)
Oryza sativa japonica	*Ochrobactrum tritici*	As	Moens et al. (2020)
Brachiaria mutica	*Pantoea stewartii*, *Microbacterium arborescens*, and *Enterobacter* sp.	Cd	Ahsan et al. (2019)
Sedum alfredii	*Endophyte SaMR12 (ES)*	Cd	Wu et al. (2018a)
Sedum alfredii	*Buttiauxella* sp. *SaSR13*	Cd	Wu et al. (2018b)
Chromolaena odorata	*Exiguobacterium*, *Bacillus* sp., *Paenibacillus* sp., *Alcaligenes* sp.	Cd	Siripan et al. (2018)
Rinorea aff. Bengalensis, *Ficus trachypison*, and *Trichospermum morotaiense*	*Proteobacteria*, *Acidobacteria*, *Actinobacteria*	Ni	Lopez et al. (2019)
Arabis alpina and *Dysphania ambrosioides*	*Pseudomonas*, *Microbacterium* sp.	Pb-Zn	Sun et al. (2019a)
Aeschynomene fluminensis and *Polygonum acuminatum*	*Lysobacter soli*, *Pantoea* sp., *Burkholderia* sp., *Bacillus* sp., *Enterobacter* sp., *Klebsiella pneumoniae*	Hg	Mello et al. (2019)
Solanum lycopersicum	*Klebsiella* sp. and *Enterobacter* sp.	Cr	Gupta et al. (2020)
Adiantum capillus-veneris	*Paracoccus versutus* and *Aeromonas caviae*	As	Marwa et al. (2020)
Chromolaena odorata	*Micrococcus luteus*	Pb	Jampasri et al. (2020)
Cucumis sativus	*Aureobasidium pullulans* BSS6	Cd-Pb	Ali et al. (2019)
Sedum alfredii	*Sphingomonadaceae* and *Streptomycetaceae*	Cd-Zn	Luo et al. (2017)

(continued)

Table 5.3 (continued)

Hyperaccumulator plant	Endophytic bacteria	Metal accumulated	Ref.
Pteris vittata	*Proteobacteria*, *Firmicutes*, and *Bacteroidetes*	As	Tiwari et al. (2016)
Prosopis laevigata	*Bacillus* sp.	Cr	Ramirez et al. (2019)
Helianthus annuus	*Klebsiella* sp. and *Enterobacter* sp.	Cr	Gupta et al. (2019)
Leptochloa fusca and *Brachiaria mutica*	*Enterobacter* sp. *Microbacterium arborescens*, and *Pantoea stewartii*	Cr	Ahsan et al. (2018)
Cicer arietinum	*Bacillus thuringiensis*, *Stenotrophomonas maltophilia*, *B. subtilis*, and *B. cereus*	Cr	Shreya et al. (2020)
Glycine max L.	*Sphingomonas* sp. LK11	Cr	Bilal et al. (2018)
Sphaeralcea angustifolia	*Kocuria palustris*	As	Vital et al. (2019)
Pteris vittata	*Pseudomonas* sp.	As-Pb	Manzoor et al. (2019)
Oryza sativa	*Bacillus* sp., *Acinetobacter* sp.	As	Rahman et al. (2020)
Oryza sativa	*Enterobacter ludwigii*, *Exiguobacterium indicum*	Cd-Ni	Jan et al. (2019)
Salix matsudana Koidz	*Sphingobium yanoikuyae*	Cd	Zeng et al. (2020)
Sedum alfredii	*Sphingomonas SaMR12*	Cd	Wang et al. (2020a)
Oryza sativa	*Burkholderia* sp.	Cd	Wang et al. (2020b)
Solanum lycopersicum, *Lolium perenne*, *Glycine max*	*Burkholderia* sp.	Cd	Guo et al. (2020)
Linum usitatissimum	*Serratia* sp.	Cd	Shahid et al. (2019)

6 Conclusion

The endophytic bacteria provide a symbiotic association with the host plant by promoting the development of the latter in unfavourable soils containing heavy metals. The endophytic bacteria, which possess extraordinary tolerance towards the hazardous heavy metals, afford several mechanisms to adsorb the metals, without producing any harmful effects on the host plants. Besides, the endophytic bacteria improve the uptake and storage of essential trace elements in the host plant, which further benefit the latter. However, the appraisal of in-field application of endophytic bacteria must consider the genetic diversity and the molecular mechanisms contributing towards the metal hyperaccumulation. Importantly, the rationalization of the mechanistic basis of interaction between the endophytic bacteria and root exudates in heavy metal polluted soils further validates the application of endophytic bacteria for the high-performance phytoremediation on a large scale.

Acknowledgement The authors thank the Department of Chemistry, University of Petroleum and Energy Studies; Department of Chemistry, Uttaranchal University; and Department of Chemistry, Guru Nanak Dev University for providing the necessary resources. SEED grant by UPES is duly acknowledged.

References

Abedinzadeh M, Etesami H, Alikhani HA (2019) Characterization of rhizosphere and endophytic bacteria from roots of maize (*Zea mays* L.) plant irrigated with wastewater with biotechnological potential in agriculture. Biotech Rep 21:e00305

Abisado RG, Benomar S, Klaus JR, Dandekar AA, Chandler JR (2018) Bacterial quorum sensing and microbial community interactions. MBio 9:e02331–e02317

Afzal I, Shinwari ZK, Sikandar S, Shahzad S (2019) Plant beneficial endophytic bacteria: mechanisms, diversity, host range and genetic determinants. Microbiol Res 221:36–49

Ahemad M (2019) Remediation of metalliferous soils through the heavy metal resistant plant growth promoting bacteria: paradigms and prospects. Arab J Chem 12:1365–1377

Ahsan MT, Najam Ul-Haq M, Saeed A, Mustafa T, Afzal M (2018) Augmentation with potential endophytes enhances phytostabilization of Cr in contaminated soil. Environ Sci Pollut Res 25:7021–7032

Ahsan MT, Tahseen R, Ashraf A, Mahmood A, Najam-Ul-Haq MN, Arslan M, Afzal M (2019) Effective plant-endophyte interplay can improve the cadmium hyperaccumulation in Brachiaria mutica. World J Microbiol Biotechnol 35:188

Ali A, Bilal S, Khan AL, Mabood F, Al-Harrasi A, Le I-J (2019) Endophytic *Aureobasidium pullulans* BSS6 assisted developments in phytoremediation potentials of *Cucumis sativus* under cd and Pb stress. J Plant Interact 14:303–313

Barcelo J, Poschenrieder C (1990) Plant water relations as affected by heavy metal stress: a review. J Plant Nutr 13:1–37

Bilal S, Khan AL, Shahzad R, Kim Y-H, Imran M, Khan MJ, Al-Harrasi A, Kim TH, Lee I-J (2018) Mechanisms of Cr(VI) resistance by endophytic *Sphingomonas* sp. LK11 and its Cr(VI) phytotoxic mitigating effects in soybean (*Glycine max* L.). Ecotoxicol Environ Saf 164:648–658

Blanco J-M, Lugtenberg BJJ (2014) Biotechnological applications of bacterial endophytes. Curr Biotechnol 3:60–75

Cardoso VM, Campos FF, Santos ARO, Ottoni MHF, Rosa CA, Almeida VG, Grael CFF (2020) Biotechnological applications of the medicinal plant Pseudobrickellia brasiliensis and its isolated endophytic bacteria. J Appl Microbiol 129:926. https://doi.org/10.1111/jam.14666

Carvalho LVB, Hacon SS, Vega CM, Vieira JA et al (2019) Oxidative stress levels induced by mercury exposure in Amazon juvenile populations in Brazil. Int J Environ Res Public Health 16:2682

Castaldi P, Santona L, Melis P (2005) Heavy metal immobilization by chemical amendments in a polluted soil and influence on white lupin growth. Chemosphere 60:365–371

Chen J, Shafi M, Li S, Wang Y, Wu J, Ye Z, Peng D, Yan W, Liu D (2015) Copper induced oxidative stresses, antioxidant responses and phytoremediation potential of Moso bamboo (Phyllostachys pubescens). Sci Rep 5:13554

Commoner B, Corr M, Stamler PJ (1971) The causes of pollution. Environ Sci Policy Sustain Dev 13:2–19

Deng Z-S, Zhang B-C, Qi X-Y, Sun Z-H, He X-L, Liu Y-Z, Li J, Chen K-K, Lin Z-X (2019) Root-associated endophytic bacterial community composition of Pennisetum sinese from four representative provinces in China. Microorganisms (MDPI Basel) 7:47

Dong L, Cheng R, Xiao L, Wei F, Wei G, Xu J, Wang Y, Guo X, Chen Z, Chen S (2018) Diversity and composition of bacterial endophytes among plant parts of *Panax notoginseng*. Chin Med 13:41

Duhan P, Bansal P, Rani S (2020) Isolation, identification and characterization of endophytic bacteria from medicinal plant Tinospora cordifolia. S Afr J Bot 134:43. https://doi.org/10.1016/j.sajb.2020.01.047

Dutta S, Mitra M, Agarwal P, Mahapatra K, De S, Sett U, Roy S (2018) Oxidative and genotoxic damages in plants in response to heavy metal stress and maintenance of genome stability. Plant Signal Behav 13:e1460048

Elbeltagy A, Nishioka K, Sato T, Suzuki H, Ye B, Hamada T, Isawa T, Mitsui H, Minamisawa K (2001) Endophytic colonization and in planta nitrogen fixation by a *Herbaspirillum sp.* isolated from wild rice species. Appl Environ Microbiol 67:5285–5293

Ferchichi N, Toukabri W, Boularess M, Smaoui A, Mhamdi R, Trabelsi D (2019) Isolation, identification and plant growth promotion ability of endophytic bacteria associated with lupine root nodule grown in Tunisian soil. Arch Microbiol 201:1333–1349

Freitas MA, Medeiros FHV, Melo IS, Pereira PF, Penaflor MFGV, Bento JMS, Pare PW (2019) Stem inoculation with bacterial strains Bacillus amyloliquefaciens (GB03) and microbacterium imperiale (MAIIF2a) mitigates Fusarium root rot in cassava. Phytoparasitica 47:135–142

Garcia SS, Bernabeu PR, Vio SA, Cattelan N, Garcia JE, Puente ML, Galar ML, Prieto CI, Luna MF (2019) *Paraburkholderia tropica* as a plant-growth–promoting bacterium in barley: characterization of tissues colonization by culture-dependent and -independent techniques for use as an agronomic bioinput. Plant and Soil 451:89. https://doi.org/10.1007/s11104-019-04174-y

Ghiasvand M, Makhdoumi A, Matin MM, Vaezi J (2019) Exploring the bioactive compounds from endophytic bacteria of a medicinal plant: *Ephedra foliata* (Ephedrales: Ephedraceae). Adv Trad Med 20:61–70

Ghori N-H, Ghori T, Hayat MQ, Imadi SR, Gul A, Altay V, Ozturk M (2019) Heavy metal stress and response in plants. Int J Environ Sci Technol 16:1807–1828

Grobelak A, Hiller J (2017) Bacterial Siderophores promote plant growth: screening of catechol and Hydroxamate Siderophores. Int J Phytoremediation 19:825–833

Gu Y, Wang Y, Sun Y, Zhao K, Xiang Q, Yu X, Zhang X, Chen Q (2018) Genetic diversity and characterization of arsenic-resistant endophytic bacteria isolated from *Pteris vittata*, an arsenic hyperaccumulator. BMC Microbiol 18:42

Guo JK, Zhao J, Ren XH, Jia HL, Muhammad H, Lv X, Wei T, Hua L (2020) Effects of *Burkholderia sp.* D54 on growth and cadmium uptake of tomato, ryegrass and soybean plants. Int J Environ Sci Technol 17:1149–1158

Gupta P, Kumar V, Usmani Z, Rani R, Chandra A, Gupta VK (2019) A comparative evaluation towards the potential of *Klebsiella sp.* and *Enterobacter sp.* in plant growth promotion, oxidative stress tolerance and chromium uptake in *Helianthus annuus* (L.). J Hazard Mater 377:391–398

Gupta P, Kumar V, Usmani Z, Rani R, Chandra A, Gupta VK (2020) Implications of plant growth promoting *Klebsiella sp.* CPSB4 and *Enterobacter sp.* CPSB49 in luxuriant growth of tomato plants under chromium stress. Chemosphere 240:124944

Han Y-H, Fu J-W, Chen Y, Rathinasbapathi B, Ma LQ (2016) Arsenic uptake, arsenite efflux and plant growth in hyperaccumulator *Pteris vittata*: role of arsenic-resistant bacteria. Chemosphere 144:1937–1942

Hartmann A, Rothballer M, Hense BA, Schroder P (2014) Bacterial quorum sensing compounds are important modulators of microbe-plant interactions. Front Plant Sci 5:131

He W, Megharaj M, Wu C-Y, Subhashchandrabose SR, Dai C-C (2020) Endophyte-assisted phytoremediation: mechanisms and current application strategies for soil mixed pollutants. Crit Rev Biotechnol 40:31–45

Hoang H, Tran LH, Nguyen TH et al (2020) Occurrence of endophytic bacteria in Vietnamese Robusta coffee roots and their effects on plant parasitic nematodes. Symbiosis 80:75–84

Jampasri K, Pokethitiyook P, Poolpak T, Kruatrachue M, Ounjai P, Kumsopa A (2020) Bacteria-assisted phytoremediation of fuel oil and lead co-contaminated soil in the salt-stressed condition by *chromolaena odorata* and *Micrococcus luteus*. Int J Phytoremediation 22:322–333

Jan R, Khan MA, Asaf S, Lee L, Kim I-J, K.M. (2019) Metal resistant endophytic bacteria reduces cadmium, nickel toxicity, and enhances expression of metal stress related genes with improved growth of *Oryza Sativa*, via regulating its antioxidant machinery and endogenous hormones. Plan Theory 8:363

Kalita J, Pradhan AK, Shandilya ZM, Tanti B (2018) Arsenic stress responses and tolerance in rice: physiological cellular and molecular approaches. Ric Sci 25:235–249

Kandel SL, Joubert PM, Doty SL (2017) Bacterial endophyte colonization and distribution within plants. Microorganisms (MDPI Basel) 5:77

Khare E, Mishra J, Arora NK (2018) Multifaceted interactions between endophytes and plant: developments and prospects. Front Microbiol 9:2732

Kloke A, Sauerbeck DR, Vetter H (1984) The contamination of plants and soils with heavy metals and the transport of metals in terrestrial food chains. In: Nriagu JO (ed) Changing metal cycles and human health. Dahlem workshop reports, life sciences research report, vol 28. Springer, Berlin, Heidelberg

Kramer U (2010) Metal hyperaccumultion in plants. Annu Rev Plant Biol 61:517–534

Li C, Zhou K, Qin W, Tian C, Qi M, Yan X, Han W (2019) A review on heavy metals contamination in soil: effects, sources, and remediation techniques. Soil Sedim Contam 28:380–394

Liu H, Carvalhais LC, Crawford M, Singh E, Dennis PG, Pieterse CMJ, Schenk PM (2017) Inner plant values: diversity, colonization and benefits from endophytic bacteria. Front Microbiol 8:2552

Liu Y-H, Wei Y-Y, Mohamad OAA, Salam N, Zhang Y-G, Guo J-W, Li L, Egamberdieva D, Li W-J (2019) Diversity, community distribution and growth promotion activities of endophytes associated with halophyte Lycium ruthenicum Murr. 3 Biotech 9:144

Liu Y, Tan H, Cao L, Zhang R (2020a) Rice sprout endophytic *Enterobacter sp.* SE-5 could improve tolerance of mature rice plants to salt or Cd^{2+} in soils. Arch Agronomy Soil Sci 66:873–883

Liu Y, Yan H, Zhang X, Zhang R, Li M, Xu T, Yang F, Zheng H, Zhao J (2020b) Investigating the endophytic bacterial diversity and community structures in seeds of genetically related maize (*Zea mays* L.) genotypes. 3 Biotech 10:27

Lodewyckx C, Vangronsveld J, Porteous F, Moore ERB, Taghavi S, Mezgeay M, Lelie D (2002) Endophytic bacteria and their potential applications. Crit Rev Plant Sci 21:583–606

Lopez S, Goux X, Ent A, Erskine PD, Echevarria G, Calusinska M, Morel JL, Benizri E (2019) Bacterial community diversity in the rhizosphere of nickel hyperaccumulator species of Halmahera Island (Indonesia). Appl Soil Ecol 133:70–80

Lu Y, Zhang E, Hong M, Yin X, Cai H, Yuan L, Yuan F, Li L, Zhao K, Lan X (2020) Analysis of endophytic and rhizosphere bacterial diversity and function in the endangered plant Paeonia ludlowii. Arch Microbiol 202:1717. https://doi.org/10.1007/s00203-020-01882-3

Luo J, Tao Q, Wu K, Li J, Qian J, Liang Y, Yang X, Li T (2017) Structural and functional variability in root-associated bacterial microbiomes of cd/Zn hyperaccumulator *Sedum alfredii*. Appl Microbiol Biotechnol 2017:7961–7976

Lwalaba JLW, Louis LT et al (2020) Physiological and molecular mechanisms of cobalt and copper interaction in causing phyto-toxicity to two barley genotypes differing in co tolerance. Ecotoxicol Environ Saf 187:109866

Ma Y, Prasad MNV, Rajkumar M, Freitas H (2011) Plant growth promoting rhizobacteria and endophytes accelerate phytoremediation of metalliferous soils. Biotechnol Adv 29:248–258

Ma Y, Oliveira RS, Nai FJ, Rajkumar M, Luo YM, Rocha I, Freitas H (2015) The hyperaccumulator Sedum plumbizincicola harbors metal-resistant endophytic bacteria that improve its phytoextraction capacity in multi-metal contaminated soil. J Environ Manage 156:62–69

Ma Y, Rajkumar M, Zhang C, Freitas H (2016) Beneficial role of bacterial endophytes in heavy metal phytoremediation. J Environ Manage 174:14–25

Mahmood A, Takagi K, Ito K, Kataoka R (2019) Changes in endophytic bacterial communities during different growth stages of cucumber (*Cucumis sativus* L.). World J Microbiol Biotechnol 35:104

Maksymiec W (2007) Signaling responses in plants to heavy metal stress. Acta Physiol Plant 29:177

Manara A (2012) Plant responses to heavy metal toxicity. In: Furini A (ed) Plants and heavy metals. SpringerBriefs in molecular science. Springer, Dordrecht

Manzoor M, Abid R et al (2019) Metal tolerance of arsenic-resistant bacteria and their ability to promote plant growth of *Pteris vittata* in Pb-contaminated soil. Sci Total Environ 660:18–24

Marwa N, Mishra N, Singh N, Mishra A, Saxena G, Pandey V, Singh S (2020) Effect of rhizospheric inoculation of isolated arsenic (as) tolerant strains on growth, as-uptake and bacterial communities in association with Adiantum capillus-veneris. Ecotoxicol Environ Saf 196:110498

Mello IS, Souza WP, Barros BM, Silva GF, Campos ML, Soares MA (2019) Endophytic bacteria mitigate mercury toxicity to host plants. Symbiosis 79:251–262

Mesa J, Naranjo EM, Caviedes MA, Gomez SR, Pajuelo E, Llorente IDR (2015) Endophytic cultivable Bacteria of the metal bioaccumulator Spartina maritima improve plant growth but not metal uptake in polluted marshes soils. Front Microbiol 6:1450

Mishra J, Singh R, Arora NK (2017) Alleviation of heavy metal stress in plants and remediation of soil by rhizosphere microorganisms. Front Microbiol 8:1706

Mithofer A, Schulze B, Boland W (2004) Biotic and heavy metal stress response in plants: evidence for common signals. FEBS Lett 566:1–5

Moens M, Branco R, Morais PV (2020) Arsenic accumulation by a rhizosphere bacterial strain *Ochrobactrum tritici* reduces rice plant arsenic levels. World J Microbiol Biotechnol 36:23

Mohammed AS, Kapri A, Goel R (2011) Heavy metal pollution: source, impact, and remedies. In: Khan M, Zaidi A, Goel R, Musarrat J (eds) Biomanagement of metal-contaminated soils. Environmental pollution, vol 20. Springer, Dordrecht

Mudila H, Prasher P, Kumar M, Kapoor H, Kumar A, Zaidi MGH, Verma A (2019) An insight into cadmium poisoning and its removal from aqueous sources by graphene adsorbents. Int J Environ Health Res 29:1–21

Mumtaz MZ, Ahmad M, Jamil M, Hussain T (2017) Zinc solubilizing Bacillus spp. potential candidates for biofortification in maize. Microbiol Res 202:51–60

Nasrollahi M, Pourbabaei AA, Etesami H, Talebi K (2020) Diazinon degradation by bacterial endophytes in rice plant (Oryzia sativa L.): a possible reason for reducing the efficiency of diazinon in the control of the rice stem–borer. Chemosphere 246:125759

Ngo VA, Wang S-L et al (2020) Phytophthora antagonism of endophytic Bacteria isolated from roots of black pepper (Piper nigrum L.). Agronomy (MDPI Basel) 10:286

Panigrahi S, Mohanty S, Rath CC (2019) Characterization of endophytic bacteria *Enterobacter cloacae* MG00145 isolated from *Ocimum sanctum* with indole acetic acid (IAA) production and plant growth promoting capabilities against selected crops. S Afr J Bot 134:17. https://doi.org/10.1016/j.sajb.2019.09.017

Parlak KU (2016) Effect of nickel on growth and biochemical characteristics of wheat (Triticum aestivum L.) seedlings. NJAS 76:1–5

Pickup RW (1991) Development of molecular methods for the detection of specific bacteria in the environment. J Gen Microbiol 137:1009–1019

Plociniczak T, Plociniczak M-P, Kwasniewski M et al (2020) Response of rhizospheric and endophytic bacterial communities of white mustard (Sinapis alba) to bioaugmentation of soil with the *Pseudomonas* sp. H15 strain. Ecotoxicol Environ Saf 194:110434

Prasher P, Singh M, Mudila H (2018) Silver nanoparticles as antimicrobial therapeutics: current perspectives and future challenges. 3 Biotech 8:411

Prasher P, Sharma M, Mudila H, Khati B (2019) Uptake, accumulation, and toxicity of metal nanoparticles in autotrophs. In: Panpatte D, Jhala Y (eds) Nanotechnology for agriculture. Springer, Singapore

Prasher P, Sharma M et al (2020) Emerging trends in clinical implications of bio-conjugated silver nanoparticles in drug delivery. Colloid Interface Sci Commun 35:100244

Purushotham N, Jones E, Monk J, Ridgway H (2020) Community structure, diversity and potential of endophytic bacteria in the primitive New Zealand medicinal plant Pseudowintera colorata. Plants (MDPI Basel) 9:156

Rahman MS, Jamal MAHM, Biswas PK, Rahman SM, Sharma SP, Saha SK (2020) Arsenic remediation in Bangladeshi rice varieties with enhance plant growth by unique arsenic-resistant bacterial isolates. Geomicrobiol J 37:130–142

Rajkumar M, Ae N, Freitas H (2009) Endophytic bacteria and their potential to enhance heavy metal phytoextraction. Chemosphere 77:153–160

Rajkumar M, Sandhya S, Prasad MNV, Freitas H (2012) Perspectives of plant-associated microbes in heavy metal phytoremediation. Biotechnol Adv 30:1562–1574

Ramirez V, Baez A, Lopez P (2019) Chromium hyper-tolerant Bacillus sp. MH778713 assists phytoremediation of heavy metals by Mesquite trees (Prosopis laevigata). Front Microbiol 10:1833

Rascio N, Navari-Izzo F (2011) Heavy metal hyperaccumulating plants: how and why do they do it? And what makes them so interesting? Plant Sci 180:169–181

Ruppel S, Rühlmann J, Merbach W (2006) Quantification and localization of bacteria in plant tissues using quantitative real-time PCR and online emission fingerprinting. Plant and Soil 286:21–35

Rustamova N, Wubulikasimu A et al (2020) Endophytic bacteria associated with medicinal plant Vernonia anthelmintica: diversity and characterization. Curr Microbiol 77:1457. https://doi.org/10.1007/s00284-020-01924-5

Santoyo G, Moreno-Hagelsieb G et al (2016) Plant growth promoting bacterial endophytes. Microbiol Res 183:92–99

Sessitsch A, Kuffner M, Kidd P, Vangronsveld J, Wenzel WW, Fallman K, Puschenreiter M (2013) The role of plant-associated bacteria in the mobilization and phytoextraction of trace elements in contaminated soils. Soil Biol Biochem 60:182–194

Shah AA, Liu Z, Qian C, Wu J, Zhong X, Kalsoom U (2020) Effect of endophytic Bacillus megaterium colonization on structure strengthening, microbial community, chemical composition and stabilization properties of hybrid Pennisetum. J Sci Food Agric 100:1164–1173

Shahid M, Javed MT, Masood S, Akram MS, Azeem M, Ali Q, Gilani R, Basit F, Abid A, Lindberg S (2019) Serratia sp. CP-13 augments the growth of cadmium (cd)-stressed Linum usitatissimum L. by limited cd uptake, enhanced nutrient acquisition and antioxidative potential. Appl Microbiol 126:1708–1721

Shanker AK, Cervantes C, Loza-Tavera H, Avudainayagam S (2005) Chromium toxicity in plants. Environ Int 31:739–753

Sheng X-F, Xia J-J, Jiang C-Y, He L-Y, Qian M (2008) Characterization of heavy metal-resistant endophytic bacteria from rape (Brassica napus) roots and their potential in promoting the growth and lead accumulation of rape. Environ Pollut 156:1164–1170

Shreya D, Jinal HN, Kartik VP, Amaresan N (2020) Amelioration effect of chromium-tolerant bacteria on growth, physiological properties and chromium mobilization in chickpea (*Cicer arietinum*) under chromium stress. Arch Microbiol 202:887–894

Shrivastava M, Khandelwal A, Srivastava S (2019) Heavy metal Hyperaccumulator plants: the resource to understand the extreme adaptations of plants towards heavy metals. In: Srivastava S, Srivastava A, Suprasanna P (eds) Plant-metal interactions. Springer, Cham

Silva CF, Vitorino LC et al (2019) Screening of plant growth-promoting endophytic bacteria from the roots of the medicinal plant Aloe vera. S Afr J Bot 134:3. https://doi.org/10.1016/j.sajb.2019.09.019

Singh M, Kumar A, Singh R, Pandey KD (2017) Endophytic bacteria: a new source of bioactive compounds. 3 Biotech 7:315

Siripan O, Thamchaipenet A, Surat W (2018) Enhancement of the efficiency of cd phytoextraction using bacterial endophytes isolated from *Chromolaena odorata*, a cd hyperaccumulator. Int J Phytoremediation 20:1096–1105

Srivastava V, Sarkar A, Singh S, Singh P, Araujo ASF, Singh RP (2017) Agroecological responses of heavy metal pollution with special emphasis on soil health and plant performances. Front Environ Sci 5:64

Stecher BFB, Johnson EA (2004) Single-cell microbiology: tools, technologies, and applications. Microbiol Mol Biol Rev 68:538–559

Sun W, Xiong Z, Chu L, Li W, Soares MA, White JF, Li H (2019a) Bacterial communities of three plant species from Pb-Zn contaminated sites and plant-growth promotional benefits of endophytic *Microbacterium* sp. (strain BXGe71). J Hazard Mater 370:225–231

Sun Z, Yang L-M, Han M, Han Z-M, Yang L, Cheng L, Yang X, Lv Z-L (2019b) Biological control ginseng grey mold and plant colonization by antagonistic bacteria isolated from rhizospheric soil of *Panax ginseng* Meyer. Biol Cont 138:104048

Tiwari S, Lata C (2018) Heavy metal stress, signaling, and tolerance due to plant-associated microbes: an overview. Front Plant Sci 9:452

Tiwari S, Sarangi BK, Thul ST (2016) Identification of arsenic resistant endophytic bacteria from *Pteris vittata* roots and characterization for arsenic remediation application. J Environ Manage 180:359–365

Vishwakarma P, Dubey SK (2020) Diversity of endophytic bacterial community inhabiting in tropical aerobic rice under aerobic and flooded condition. Arch Microbiol 202:17–29

Vital TZ, Roman-Ponce B, Orduna FNR, Santos PE, Murrieta MSV, Deng Y, Yuan HL, Wang ET (2019) An endophytic *Kocuria palustris* strain harboring multiple arsenate reductase genes. Arch Microbiol 201:1285–1293

Walia A, Guleria S, Chauhan A, Mehta P (2017) Endophytic bacteria: role in phosphate solubilization. In: Maheshwari D, Annapurna K (eds) Endophytes: crop productivity and protection. Sustainable development and biodiversity, vol 16. Springer, Cham

Wang S-S, Liu J-M, Sun J, Sun Y-F, Liu J-N, Jia N, Fan B, Dai X-F (2019) Diversity of culture-independent bacteria and antimicrobial activity of culturable endophytic bacteria isolated from different Dendrobium stems. Sci Rep 9:10389

Wang Q, Ge C, Xu S, Wu Y, Sahito ZA, Ma L, Pan F, Zhou Q, Huang L, Feng Y, Yang X (2020a) The endophytic bacterium *Sphingomonas* SaMR12 alleviates cd stress in oilseed rape through regulation of the GSH-AsA cycle and antioxidative enzymes. BMC Plant Biol 20:63

Wang C, Huang Y et al (2020b) *Burkholderia* sp. Y4 inhibits cadmium accumulation in rice by increasing essential nutrient uptake and preferentially absorbing cadmium. Chemosphere 252:126603

Wu K, Li J, Luo J, Liu Y, Song Y, Liu N, Rafiq MT, Li T (2018a) Effects of elevated CO2 and endophytic bacterium on photosynthetic characteristics and cadmium accumulation in *Sedum alfredii*. Sci Total Environ 643:357–366

Wu K, Luo J, Li J, An Q, Yang X, Liang Y, Li T (2018b) Endophytic bacterium *Buttiauxella sp. SaSR13* improves plant growth and cadmium accumulation of hyperaccumulator *Sedum alfredii*. Environ Sci Pollut Res 25:21844–21854

Xu J-Y, Han Y-H, Chen Y, Zhu L-J, Ma LQ (2016) Arsenic transformation and plant growth promotion characteristics of as-resistant endophytic bacteria from as-hyperaccumulator *Pteris vittata*. Chemosphere 144:1233–1240

Xu J-X, Li Z-Y, Lv X, Yan H, Zhou G-Y, Cao L-X et al (2020) Isolation and characterization of *Bacillus subtilis* strain 1-L-29, an endophytic bacteria from *Camellia oleifera* with antimicrobial activity and efficient plant-root colonization. PLoS One 15:e0232096

Yadav SK (2010) Heavy metals toxicity in plants: An overview on the role of glutathione and phytochelatins in heavy metal stress tolerance of plants. S Afr J Bot 76:167–179

Ye D, Li T, Liu J, Yi Y, Zou L, Zhang X, Xie M (2020) Characteristics of endophytic bacteria from *Polygonum hydropiper* and their use in enhancing P-phytoextraction. Plant and Soil 448:647–663

Zeng X, Pang L, Chen Y, Kong X, Chen J, Tian X (2020) Bacteria *Sphingobium yanoikuyae Sy310* enhances accumulation capacity and tolerance of cadmium in *Salix matsudana* Koidz roots. Environ Sci Pollut Res 27:19764. https://doi.org/10.1007/s11356-020-08474-0

Zhang Q, Acuna JJ, Inostroza NG, Mora ML, Radic S, Sadowsky MJ, Jorquera MA (2019a) Endophytic bacterial communities associated with roots and leaves of plants growing in Chilean extreme environments. Sci Rep 9:4950

Zhang C, Tian XY, Zhang C-S (2019b) Diversity and probiotic activities of endophytic bacteria associated with the coastal halophyte *Messerschmidia sibirica*. Appl Soil Ecol 143:35–44

Zhang Y, Yu X, Zhang W, Lang D, Zhang X, Cui G, Zhang X (2019c) Interactions between endophytes and plants: beneficial effect of endophytes to ameliorate biotic and abiotic stresses in plants. J Plant Biol 62:1–13

Zhou J, Zhang Z, Zhang Y, Wei Y, Jiang Z (2018) Effects of lead stress on the growth, physiology, and cellular structure of privet seedlings. PLoS One 13:e0191139

Zinniel DK, Lambrecht P, Harris NB, Feng Z, Kuczmarski D, Higley P, Ishimaru CA, Arunakumari A, Barletta RG, Vidaver AK (2002) Isolation and characterization of endophytic colonizing Bacteria from agronomic crops and prairie plants. Appl Environ Microbiol 68:2198–2208

Chapter 6
Microbial Enzymes and Soil Health

Glacy Jaqueline da Silva, Wesley Ribeiro Rivadavea, Juliane Destro de Lima, Pedro Henrique Riboldi Monteiro, and Franquiéle Bonilha da Silva

Contents

1 Introduction

Soil represents the main support and regulator of energy and nutrient flows, and it has seven main functions, three of which are ecological (production of biomass; filtration, buffering and transformation of matter with respect to environmental protection against groundwater pollution; habitat biological and genetic reserve of plants, animals and organisms); three functions linked to human activity (basis for

G. J. da Silva (✉) · W. R. Rivadavea · J. D. de Lima
Department of Biotechnology Applied to Agriculture, Universidade Paranaense—UNIPAR, Umuarama, Paraná, Brazil
e-mail: glacyjaqueline@prof.unipar.br

P. H. R. Monteiro
Department of Agronomy, Federal University of Vales do Jequitinhonha and Mucuri, UFVJM, Diamantina, Minas Gerais, Brazil

F. B. da Silva
Department of Soils, Faculty of Agronomy, Federal University of Rio Grande do Sul, Porto Alegre, Rio Grande do Sul, Brazil

© The Author(s), under exclusive license to Springer Nature Switzerland AG 2021

H. I. Mohamed et al. (eds.), *Plant Growth-Promoting Microbes for Sustainable Biotic and Abiotic Stress Management*,
https://doi.org/10.1007/978-3-030-66587-6_6

industrial structures and socioeconomic activities; source of particulate material; cultural heritage of humanity); and a function related to agriculture and the environment (providing the medium for plant growth, habitat for animals and microorganisms, regulation of water flow in the environment, environmental buffer in the attenuation and degradation of chemical compounds harmful to the environment) (Araujo and Monteiro 2007).

In view of the numerous properties and functions performed by the soil, actions for its conservation should routinely be undertaken, which is only possible with adequate knowledge and management, since the changes in land use associated with the management system, the excessive use of pesticides and fertilizers the application of urban and industrial residues have led to changes in the biological properties of the soil, and the biological indicators are great auxiliary for verifying the effect of these agricultural practices on the quality of the soil (Araujo and Monteiro 2007).

Soil health is intrinsically related to the synergy of all these functions. Just as evaluating a patient's clinical exams is essential to diagnose their problems, understanding and knowing the health of the soil is an important strategy for the adoption of conservation and recovery strategies for degraded and agricultural soils. Enzymes play a key role in this assessment, as they provide very sensitive data on the biological quality of soils.

Thus, the objective of this literature review is to survey the most recent and updated information on studies involving soil health, with an emphasis on enzymatic activity. In addition, we intend to seek the latest updates on the practical application of these studies.

2 Origin of Soil Enzyme Studies

Soil microbiology aims to answer questions related to the types of microorganisms present in the soil, activities, metabolic capacity and frequency of nutrient input oscillations such as C, N, P, K, etc. (Dubeux et al. 2006). Factors such as temperature, humidity, hydrogen ion concentrations, content, composition and age of organic matter influence the local microbiota (Foster et al. 2018). However, the infinity of microorganisms present in the soil allows a functional redundancy in the establishment of nutrient cycling, since they belong to different biological groups, acting in different biological fractions.

The microorganisms present in the soil are abundant, and when in a balanced system, they can act synergistically to assist the growth and development of plants, providing a healthy rhizosphere that can assist in the prevention of pathogen attacks. However, when an imbalance occurs, configured mainly by inadequate soil management, these microorganisms can make room for pathogenic microorganisms and cause an imbalance in soil health and quality, which can lead to low crop productivity (Van Elsas et al. 2012) (Fig. 6.1).

The first studies on soil quality and health started in the middle of the 1980s, and it was from the turn of the century that guidelines such as soil safety and carbon

Fig. 6.1 The figure on the left shows a healthy rhizosphere, composed mostly of beneficial microorganisms. In this same figure, we can see a healthy plant. In the figure on the right, we can see a rhizosphere represented by a large majority of pathogenic microorganisms. In this figure, we identify a plant that is practically lifeless and sick

sequestration became relevant in the scientific community, i.e. the interest in the subject is recent (FAO-ISRIC-ISSS 2015).

Scientific knowledge on soil microbiota has provided more advanced studies on biological nitrogen fixation, organic matter production, microflora, rhizosphere, microorganisms acting on soil formation and structuring, pesticide degradation (soil remediation or bioremediation), recalcitrant products and enzymatic activity (Abiraami et al. 2020).

At the beginning of these studies, assessing the biological quality of the soil was a very difficult task, as the amount of organic matter varied very slowly, which did not reflect the current health of the system, and this was not necessarily correlated with microbial activity. It was then suggested to quantify the enzymes present in the soil, due to their ease, rapid response to changes in the environment and mainly because they are highly related to soil biodiversity (Bandick and Dick 1999).

The first enzymatic studies aimed to locate the fraction of the soil with the highest enzyme activity and the highest activity of specific enzymes per fraction of the soil (Hoffmann and Seegerer 1950). Also, responses to specific enzymes and their correlation with biological activity and cellular soil respiration began to be sought (Frankenberger Jr and Dick 1983). Little by little, studies have intensified, and today we can find patterns of recommended enzyme activity (Mendes et al. 2018; Abiraami et al. 2020) for agricultural soils.

In this regard, recently Embrapa Cerrados, in Brazil, launched a new technology based on the physical and chemical evaluation of the soil with the inclusion of the evaluation of the enzymes arylsulfatase and beta-glucosidase, in addition to several other associated parameters and linked in an information network. This new technology is called "Soil Bioanalysis", or just BioAs (Embrapa 2020). This is an answer of about 30 years of studies of biological quality assessment in the Brazilian savannahs and promises to become a world reference in parameters of soil evaluation.

This evolution is an important step towards the implementation of increasingly sustainable and productive agriculture since the world still has about six billion hectares of degraded soil (Ayub et al. 2020). Once the soil studies at the enzymatic level are intensified, it will be possible to monitor these degraded areas more closely and, consequently, recover more significant portions of these soils, and thus produce more food for the population.

3 Enzymes as Bioindicators

Soil is an indispensable part of terrestrial ecosystems because it houses biological processes and reactions, performing numerous key functions (Chaer et al. 2009; Welc et al. 2012; Wahsha et al. 2017). In recent years the study of soil biology has become an important ally in confronting agriculture in the face of constant change and climate adaptation. With the increasing environmental degradation, due to the advances in searches for agricultural land, several studies have provided advances for innovations by sensitive indicators of biological soil quality. These provide an accurate and sensitive assessment of an ecosystem, in addition to providing a basis for indicating sustainable practices that favor soil recovery (Sardans and Penuelas 2005; Hungria et al. 2009; Porto et al. 2009; Araújo et al. 2013; Singh and Sharma 2020; Huera-Lucero et al. 2020).

Precisely, the study of the enzymatic activity of microorganisms in the face of climate change has become a common activity among the scientific community, since these studies have shown significant and important data that assist in several decision-makers (Bell et al. 2010; Kardol et al. 2010; Weedon et al. 2011; Wahsha et al. 2017; Huera-Lucero et al. 2020). Given these circumstances, maintaining soil quality, or even recovering it, is fundamental to sustainability, aiming at agricultural production and environmental conservation.

The accumulation of organic matter particulate, animal waste and rhizospheric depositions lead to the existence of microhabitats with high biological activity and a diversity hot spot (Kuzyakov and Blagodatskaya 2015). Enzymes are the proximate agents of organic matter decomposition, i.e. when enzyme activity is measured, quantitatively and qualitatively together, it is possible to infer changes in microbial demand for both carbon, nitrogen and phosphorus (Henry 2013), since these activities are directly related to the acquisition, fixation and retention of alternative sources of nutrients in the soil, contributing significantly to greater and better acquisition of these for the plant (Nguyen et al. 2017). For this reason, the highest enzyme concentration, for most soil enzymes, is found in the rhizosphere of plants when compared to soil mass (Singh and Sharma 2020).

The enzymes present in the soil can also be essential for the control of biotic and abiotic stresses. In a study conducted by Jetiyanon (2007) it was observed that plants under climatic or pathogenic stress generated different enzyme responses, as well as resistance. Therefore, the presence of certain enzymes in high concentrations in plants indicates not only the presence of pathogenic stress but also an active

Table 6.1 Likely direct impact of agricultural management practices on the microbial biomass and key C and N chemical transformation processes in an arable cropping system

	Microbial biomass	Mineralisation	Nitrification	CO_2 emissions	Methane production	Denitrification
Crop rotation including grassland or green manures instead of continuous arable cropping.	+	+		+		
Retention of crop residues instead of burning.	+	+		+		
Minimum tillage practices instead of full soil cultivation techniques.	+	-		-	-	+
Irrigation of crops in rain-limited cropping environments.	+	+	+	+		+
Drainage of agricultural land in high rainfall/waterlogged environments.		+	+	+	-	-
Application of fungicides to soil.	-					
Application of inorganic N fertilisers to soil.			+			+
Application of organic amendments to soil.	+	+		+		+
Liming of soil to raise pH on acidic soils.	+	+	+	+		+

(+) = increase in pool/process, (−) = decrease in pool/process (adapted by Murphy et al. 2007)

defense and resistance mechanisms. In this line of reasoning, Prasannath (2017) found that the induction of enzymatic synthesis by microorganisms, as well as the application of certain enzymes (peroxidase, β-1,3-glucanase, chitinase, phenylalanine ammonia-lyase and polyphenol oxidase), can promote resistance of the plant against pathogens.

Soil enzymes can be synthesized, accumulated, inactivated and/or decomposed, playing a very important role for both natural and agricultural systems. Most of the biochemical transformations of the soil are dependent or related to the presence of enzymes and, consequently, of the microorganisms that produce them (Table 6.1).

In this way, the evaluation of enzymatic activities can be of great value to indicate whether a soil is adequately performing processes that are closely linked to its quality (da Silva et al. 2009; Badiane et al. 2001); the maintenance of soil quality also should be evaluated when exposed to xenobionts that aim to combat the biotic stress (Kumari et al. 2017). This is due to the sensitivity, coherence, cost and precision of the enzymatic activity; recently, some studies with specific biomes started to be carried out, aiming at the determination of parameters of the main enzymes related to soil health. These studies aim to facilitate a more practical approach to enzymatic assessments as a routine for farmers (Mendes et al. 2018) (Table 6.2).

Until very recently, these assessments at practical levels were not taken into account when making decisions on a farm. But little by little these activities are being implemented, since the quality of the soil is closely related to productivity, efficiency and consequently greater profit for farmers. Maintaining and/or improving soil health is essential for greater sustainability in agriculture, and the enzymes present in the soil are fundamental parts of this process.

Table 6.2 Systematization of the main enzymes present in the soil, their importance in agriculture and their mode of catalysis

Enzyme	Importance for agriculture	Method of catalysis	References
Arylsulfatase	Detect management problems and natural regeneration of degraded soils, supply of microbial diseases, detection of heavy metals, detection of herbicides	Hydrolysis of arylsulfate by fission of the O-S bond	Xian et al. (2015), Wyszkowska et al. (2016), Medeiros et al. (2017) and Aponte et al. (2020)
Glucanases	Phytopathogen control, thermostability	Hydrolyzes 1,4-β-D-glucosidic bonds adjacent to 1,3-β-linkages in mix-linked β-glucans	Chen et al. (2009, 2015, 2019) and Gonçalves et al. (2020)
Proteases	Nitrogen mineralization, phytopathogen control, nematicidal action, detection of management problems and natural regeneration of degraded soils	Hydrolyzes the terminal amino acids of polypeptide chains	Majumdar and Chakraborty (2017), Lori et al. (2017), Ding et al. (2018), Sulaiman et al. (2020) and Hu et al. (2020)
Cellulases	Recycling nature, studies with irrigation soil and cooper stress, CH4 emission evaluation, supression of pathogens	Hydrolysis of β-1-4 glycosidic linkages within cellulose	Cao et al. (2019), Wang et al. (2020), Inayati et al. (2020) and Saravanakumar and Wang (2020)
B-glucosidase	Soil quality indicator; nutrient cycling; soil pH indicator	Hydrolysis of β-1-4 glycosidic linkages within short dextran chains, such as cellobiose, cellotriose and cellotetraose	Ferreira et al. (2017, 2018), Muzangwa et al. (2020), Acosta-Martinez and Tabatabai (2000), Godoy (2020) and Ndiaye et al. (2000)
Phosphatases	Indicator of the availability of inorganic phosphorus; soil pH indicator; soil quality indicator	Hydrolysis of ester bonds involving a phosphate group in organic material	Piotrowska-Długosz and Charzyński (2015) and Margalef et al. (2017)

3.1 *Proteases*

Proteases are a class of enzymes that hydrolyze proteins. For this reason, they play a fundamental role in the nitrogen cycle, in the most varied types of soil (Vranova et al. 2013). They are sensitive indicators of the mineralization capacity of nitrogen present in the soil. The largest producers of proteases are fungi and bacteria, such as *Bacillus* spp., and every day, mainly due to advanced molecular biology techniques, more protease-producing microorganisms are discovered and purified for large-scale use in various industrial and agricultural sectors (Sevinc and Demirkan 2011; Langklotz et al. 2012; Woźniak 2019).

Proteases are enzymes of great importance for the control of phytopathogenic fungi, since they use chitin and β-glucan fibers, essential components of the fungal cell wall, to compose their matrix (Elad and Kapat 1999). Studies with genetic transformation into beet have shown that protease production in the soil is closely related to the biocontrol of the pathogen *Pythium ultimum* (Dunne et al. 1997). *Trichoderma* spp. fungi are excellent sources of proteases that act significantly against pathogenic fungi (Schuster and Schmoll 2010; Sulaiman et al. 2020). Another study that indicates bioprotection against pathogens by proteases was demonstrated through secreted proteases from *Bacillus amyloliquefaciens*, which was effective against *Macrophomina phaseolina*, *F. oxysporum*, *Fusarium semitectum* and *Alternaria alternata* (Majumdar and Chakraborty 2017). Several other proteases have been researched for having nematicidal action. One of the most recent studies shows the action of the secreted protease of *Bacillus cereus* with a very effective nematicidal action against *Meloidogyne incognita* (Hu et al. 2020).

Proteases are also efficient in assessing the biological quality of the soil. In a study developed to assess soil quality through crop rotation with garlic (*Allium sativum* L.) and pumpkin (*Cucumis sativus* L.), it was found that this rotation provided a significant increase in some soil enzymes, including proteases (Ding et al. 2018). In a study comparing organic and conventional cultivation, proteases had an 84% increase in activity in organic soil, and, consequently, greater nitrogen mineralization (Lori et al. 2017).

3.2 Arylsulphatase

The enzyme arylsulfatase has as its main source of substrate the sulfate esters, found mainly in organic matter. For this reason, it is common to observe a high value of this enzyme in soils rich in organic matter, especially in rainy seasons (Miguel et al. 2020).

The arylsulphatase enzyme has an excellent relationship with the labile carbon of the soil and is very sensitive to detect natural regeneration of degraded soils and management from organic nutrients (Medeiros et al. 2017; Ghosh et al. 2020). Also, high levels of C and arylsulfatase can be considered, together, excellent bioindicators microbial diseases suppression. In a study developed by Leon et al. (2006), an excellent negative correlation was found with the incidence of the snap bean common root rot disease (*Aphanomyes euteiches*), and the enzyme activity + C, with high enzyme and C activity found in treatments with a low incidence of the disease.

When comparing arylsulfatase activity in different soil types and different depths, it is possible to verify that in forest soils, the activity is greater than in removable soils, mainly due to the deposition of organic matter (Balota et al. 2014). Due to the positive correlation of organic matter, carbon and activity of the enzyme arylsulfatase, it is expected that soils that have little turnover, crop rotation and no-tillage present higher productivity when compared to removable soils (Borase et al. 2020). However, this high productivity is not always correlated with the enzyme arylsulfatase, as is the case found in the study by Da Silva et al. (2020), which did not obtain a high correlation of the activity of this specific enzyme, with high productivity in coffee, in the Brazilian cerrado region.

The enzyme arylsulphatase, among several evaluated enzymes, was the one that was most sensitive to evaluate heavy metals in the soil such as copper, arsenic, cadmium, zinc and nickel, and for this reason, it can be considered an excellent bioindicator for contaminated soils (Xian et al. 2015; Wyszkowska et al. 2016; Aponte et al. 2020). It was also shown to be sensitive when evaluated under herbicide overdose (Baćmaga et al. 2014).

We verified from these results, the great versatility of the enzyme arylsulfatase for use as a bioindicator in several situations that affect soils around the world.

3.3 Glucanases

The β-1,3-glucanases enzymes are important controllers of phytopathogens, as they break down the cell wall of fungi and bacteria, through the hydrolysis of glucose residues in the polysaccharide chains (Zhongcun et al. 2004; Vancov and Keen 2009).

Chen et al. (2015) suggest that the application of β-1,3-glucanases increases the concentration of *Bacillus* in the soil and decreases the concentration of the fungi *Fusarium*. This result makes a lot of sense since this class of enzymes is released by several microorganisms in the soil, including *Bacillus subtilis*, which explains the increase in the concentration of *Bacillus* in the soil, in the addition of the enzyme. It has been shown to induce resistance against various pathogenic fungi, such as *Phytophthora fragariae* var. *rubi*, *Aspergillus flavus*, *Fusarium culmorum*, *Fusarium oxysporum* f. sp. *cucumerinum*, *Fusarium udum*, *Macrophomina phaseolina* and *Treptomyces sioyaensis* (Valois et al. 1996; Rezzonico et al. 1998; Chun-Ta and Bradford 2003; Hong and Meng 2003; Wróbel-Kwiatkowska et al. 2004, Liang et al. 2005; Roy-Barman et al. 2006; Chen et al. 2009). Recently, several studies on soil proteomes have discovered new microorganisms that produce new glucanases. However, most of these studies are still restricted to the chemical and molecular identification of enzymes (Zhou et al. 2016; Angelov et al. 2017; Borshchevskaya et al. 2019; Edison and Pradeep 2020). However, dos Gonçalves et al. (2020) identified a β-glucanase in mangrove soils, which is thermostable and active at basic pH. Another study recently developed found a new endo-1,4-β-glucanase extracted from *Burkholderia pyrrocinia* JK, very active at a temperature of 10 °C (Chen et al. 2019).

All the results listed give a comprehensive dimension of the activity of glucanases in several biocontrol mechanisms, both of pathogens in the soil and abiotic stresses.

3.4 Cellulases

Cellulases are enzymes responsible for the hydrolysis of 1,4-β-D-glucosidic bonds of cellulose, present mainly in the cell wall of vegetables and thus play a significant role in the recycling nature (Khoshnevisan et al. 2017). This enzyme does not act in isolation. For the complete breakdown of cellulose molecules, it acts in synergy with other enzymes (endoglucanases, exo-cellobiohydrolase, exo-glucanases and β-glucosidases).

Changes in microbial biomass in irrigated rice paddies, treated with different irrigation and N fertilization systems, showed that cellulase had a high negative correlation with CH4 emissions from these sites (Wang et al. 2020). Another study involving irrigated soils was carried out by Cao et al. (2020), which verified the influence of copper accumulated in irrigated soils and its high negative correlation with cellulase quantification. These studies show the importance of this enzyme to assist in studies of irrigation, CH4 emission and copper stress.

Several studies have correlated the action of the cellulase enzyme as an adjuvant in the suppression of pathogens in agriculture. The use of *Micromonospora carbonacea*, a cellulase producer, proved to be efficient in the biocontrol of *Phytophthora cinnamomi*, a fungus that causes root rot disease in some cultures (El-Tarabily et al. 1996). The fungus *Trichoderma virens*, a major producer of cellulase and chitinase, is very efficient in the biocontrol of pathogens. Studies have already proved its efficiency against *Rhizoctonia solani*, an extremely harmful and difficult to control fungus in mungbean (Inayati et al. 2020): *Macrophomina phaseolina*, *Fusarium graminearum*, and B*otrytis cinerea* (Saravanakumar and Wang 2020) and *Meloidogyne incognita* (Zhang et al. 2015).

Currently, due to greater access to renewable technologies, several studies have been conducted in the search for cellulase-producing microorganisms that can be used on a large scale for the commercial production of this enzyme. Bhadrecha et al. (2020) found 17 strains with the same capacity to degrade cellulose and pectinase. Aslam et al. (2017) characterized a cellulase-producing *Bacillus amyloliquefaciens*-ASK11 in the presence of high concentrations of cadmium. Fasiku et al. (2020) found 8 potential bacteria that produce a large amount of cellulase, amylase and protease.

These and hundreds of other scientific works available show the importance of microorganisms as natural recyclers of the environment, suppression of pathogens and consequently increased productivity.

3.5 β-Glucosidase

The activity of microorganisms and the enzyme β-glycosidase are constantly associated, which makes it one of the most common enzymes found in the soil, acting in the final stage of the cellulose decomposition process (Stieven et al. 2020).

β-Glycosidase is one of the most important soil glycosidase enzymes, occurring in prokaryotic and eukaryotic organisms (Waldrop et al. 2000). This is used as an indicator of quality (Ndiaye et al. 2000; Ferreira et al. 2017) and nutrient cycling (Muzangwa et al. 2020), directly influenced by the carbon content in the soil, acting on the production of glucose (final product of degradation), which is an essential component for energy sources for other microorganisms. Its involvement with the carbon cycle makes it essential to assess the biological quality of the soil (Prosser et al. 2011; Godoy 2020). β-Glycosidases act as quality assessors under different soil management practices (Doni et al. 2012) and determine microbial activities present in it (Godoy 2020).

The pH variations greatly influence the enzymatic activity of β-glycosidase, and in this way, it has become a reliable indicator to measure acidity and basicity in the soil (Acosta-Martinez and Tabatabai 2000). For this reason, several studies that evaluate this soil enzyme used different pH rates (Foster et al. 2018).

The use of agro-industrial residues to replace chemical fertilizers for the production of organic matter is efficient in increasing the production of this enzyme in several studies (Chang et al. 2007; Godoy 2020; Martins et al. 2020). Likewise, in a monoculture system, the production of B-glucosidase was shown to be inferior when compared to crop rotation systems (Sarto et al. 2020; Muzangwa et al. 2020), corroborating the premise of a biological indicator of soil quality of this enzyme, even in soils cultivated with fruit trees (Mondal et al. 2020), since in these, the soil turnover is very low. No-till systems positively influence the production of B-glucosidase, when compared mainly with conventional cultivation systems (Miralles et al. 2012).

In a study developed by Stieven et al. (2020), where he compared the biological activity of a soil integrated with crop farming and forest, with pasture area and native forest area, for 2 years, he observed that the greatest activity of β-glycosidase came from the presence of fungi. Also, it identified that the period of soil collection has a great influence on its activity. A similar result was found in a study carried out in the Caatinga biome, in Brazil, where the enzyme activity was monitored in soils originating from environmental preservation forests and in an area with constant anthropic activity, for 3 years, in the dry and rainy periods. Again, a significant reduction in the activity of the enzyme β-glycosidase (and the other enzymes involved in the study) can be observed in soils with high anthropic activity (Cavalcante et al. 2020). Already in work carried out in the Pantanal biome, in Brazil, on native pasture soils classified with conservation status as excellent, regular and marginal, it can be seen that, as in other biomes, the enzyme β-glycosidase showed reduced activity, when correlated with conservation status. In other words, soils classified as excellent had higher enzymatic activity, while soils classified as marginal had the lowest activity averages (Pelissaro et al. 2020).

We know that only greater or lesser enzyme activity depends on several factors. What is considered greater for sandy soil is not considered greater for clay soil. When we talk about different biomes, these differences are much more pronounced. In this way, we were able to verify that there is a need for a global soil mapping, taking into account mainly the soil texture. We can see that the research is intense, but there are still many gaps to be filled. In Brazil, due to the high correlation of the activity of β-glycosidases with biological quality of the soil, many research centers have already routinely adopted this enzyme for different soil types and regions (Mendes et al. 2003, 2012, 2018). We will talk more about this in Sect. 4.

3.6 Phosphatases

Phosphatases are a group of specific enzymes that interfere with phosphoric acid hydrolysis (Condron et al. 2005). The phosphatase reference in the soil is used as an indicator of the availability of inorganic phosphorus, essential for plants since they use only this form of phosphorus (Piotrowska-Długosz and Charzyński 2015). Acid phosphatase is found in acidic soils and alkaline phosphatase in alkaline soils (Dodor and Tabatabai 2003). The activity of both (acidic and alkaline) can be used to check the optimum pH for implanting cultures, and the acid/alkaline ratio is proven to be more efficient than detection by chemical method (Acosta-Martinez et al. 2003).

The study developed by Margalef et al. (2017) showed that the higher the weathering of the soil, the lower the activity of phosphatases. The same study showed that forest soils have high enzymatic activity, whereas, in savanna soils and conventional planting, the enzyme activity has dropped. In angiosperm forests, enzyme activity was higher than in gymnosperm forests.

The study conducted by Mndzebele et al. (2020) quantified the activity of acid and alkaline phosphatase as an indicator of availability of P for the soil and plant, from the supplementation of different concentrations of chemical fertilizers (NPK) in different cultivation systems (a consortium of cowpea and amarantus). This study demonstrated that the less concentrated the fertilizer was (25% of what is required for the crop), the more was the activity of phosphatases. This indicates that the low concentration of fertilizer induces the production of this enzyme, to possibly supply the required needs of the crops. On the other hand, we also see that the high concentration of fertilizers inhibits the production of this enzyme. This can be a problem if we think of the enzyme phosphatase as an indicator of soil health.

Another very interesting study conducted in the long term (13 years) with legumes and other crops proposed the monitoring of the soil from four different crop rotations and three different levels of integrated nutrient management. This study evaluated several biological aspects, such as soil organic carbon, soil microbial biomass carbon, microbial biomass nitrogen and some enzymes, including acid and alkaline phosphatases. This study demonstrated that the use of fertilizers in the long term did not increase the enzyme indexes, mainly of phosphatases, demonstrating the need to add organic matter in the soil, since the production of these enzymes happens from microorganisms. In addition, it has shown that the addition of organic matter is closely related to increased crop productivity (Borase et al. 2020).

Several studies report the extraction of new phosphatase enzymes in several microorganisms, such as *Serratia* sp. (Behera et al. 2017), *Leclercia adecarboxylata* and *Pseudomonas putida*, extracted from soils contaminated with heavy metals (Teng et al. 2019) and *Aspergillus niger* (Nelofer et al. 2016). These studies show the diversity of microorganisms producing phosphatases and envision infinite possibilities for their use.

3.7 Chitinases

Chitinases are enzymes produced by a wide spectrum of bacteria and eukaryotes, including plants and animals (Adrangi and Faramarzi 2013). These enzymes are the second biomaterial in greater quantity in nature, behind only cellulose (Gasmi et al. 2019), which is widely distributed, particularly as a structural polysaccharide in the skeleton insects and crustaceans and on the fungal cell wall (Shinya and Fukamizo 2017).

Chitinases are a group of enzymes related to the pathogenicity of plants, as they, as the name suggests, degrade chitin, which is a fundamental component of the fungal cell wall (Jalil et al. 2015).

As this enzyme, together with 1–3 glucanase, degrades chitin in the fungi cell wall, they become osmotically sensitive and more prone to cell lysis. Chitinases are produced by different microorganisms, such as *Trichoderma* spp., *Bacillus cereus* and *Pantoea agglomerans*, and are proven to be very efficient in the biocontrol of pathogens. Studies from the 1970s proved its efficiency in the control of *Rhizoctonia solani*, *Sclerotium rolfsii*, *Pythium ultimum*, *Alternaria* spp., *Bipolaris oryzae*, *Botrytis cinerea*, *Curvularia lunata*, *Fusarium oxysporum*, *Fusarium udum*, Mycosphaerella arachidicola and Pestalotiopsis theae (Abeles et al. 1970, Chernin et al. 1995; Pleban et al. 1997, Chu and Ng 2005;Kirubakaran and Sakthivel 2007). Studies such as Lacombe-Harvey et al. (2018) show that GH19 chitinases were found in plants, nematodes and some members of *Streptomycetaceae*.

Currently, one of the most commercialized microorganisms and seen as a potential in agriculture against phytopathogenic fungi is fungi of the type *Trichoderma* sp. These are famous for producing an amount and variety of chitinases that are very effective against other types of fungi. A study conducted by Loc et al. (2020) showed that extracellular chitinases extracted from *T. asperellum* PQ34 completely inhibited the in vitro growth of *Colletotrichum* sp. and *Sclerotium rolfsii*. In vivo it already had a high inhibitory power against *Sclerotium rolfsii* and acted as an excellent controller of fungal infections of seeds. Another important chitinase-producing microorganism is *Bacillus cereus*. Madkour et al. (2019) isolated the strain *B. cereus* S3C and from it managed to obtain cell lysis of several phytopathogenic fungi such as *Fusarium oxysporum* and *Rhizoctonia solani*.

Fungal diseases are difficult to control, and for large crops, such as soybeans, they are true pests. That is why there are many studies today that demonstrate the role of chitinase enzymes in the degradation of the cell wall of these phytopathogens. The use of isolated enzymes or chitinase-producing microorganisms is an excellent strategy to combat these diseases since they do not generate residues and contribute to a higher biological quality of the soil.

4 Brazil: Pioneering in Bioanalysis of Soil

Brazil is an essentially agricultural country. In 2020, Brazil estimates that it will reach 247 million tons of grains, that is, 2.2% above the result of the previous year (IBGE 2020). These results are essential to keep the Brazilian surplus positive. Much of this positive result is due to the good agricultural practices adopted, such as correct soil management, use of seeds recommended for planting among the most diverse types of climate and topography, use of suitable cultivation methods for each specific crop, control of invasive plants, use of registered fertilizers and inoculants, handling and use of recommended and registered pesticides, giving priority to integrated pest management, fertilization and adequate liming, taking into account the chemical and physical analysis of the soil, among others.

The physical attributes of the soil considered as indicators of quality are the parameters of texture, infiltration rate and water holding capacity, depth of rooting and density of the soil. Chemical attributes include pH, total carbon, electrical conductivity and nutrients (Micros and Macronutrients).

However, even with so much care and minimally calculated handling practices, the producer ends up finding low productivity in areas with optimal concentrations of chemical fertilizers (Cherubin et al. 2016). In a study conducted by Santi et al. (2016), it was found places with optimal amounts of phosphorus in the soil, with regular monitoring, with low productivity in corn, proposing a low correlation between these characteristics. Conversely, places evaluated as having low biological diversity, presented in three different crops (corn, oats and soybeans) a high correlation with low productivity. These characteristics are increasingly common, especially in soils produced with monocultures, showing that the current practices adopted, although efficient, are becoming insufficient, as we know them.

Biological indicators are extremely sensitive and are excellent indicators of soil quality, as mentioned several times in this study. The quantification of microorganisms using NGS techniques has enabled greater knowledge of functional groups in the soil. However, routine molecular analyses of microbiome or metataxonomy are still restricted due to the high cost and demand for highly qualified labor. The biological parameters evaluated most frequently, mainly for their reliability, low cost and simplicity, are microbial biomass, dehydrogenase, N fixation, phosphatase activity, β-glycosidase and urease, in that respective order (Gil-Sotres et al. 2005). Interpreting all these indicators, individually, has always been a complex job. Unlike chemical and physical compounds, where parameters can be measured as little, medium or high, the values of biological indicators are very variable, as they can be influenced by climate, soil type, management, etc. (Lopes et al. 2013).

Due to all the complexity of the soil environment, there is no fixed quantification that accurately determines what a top-quality soil is. For this purpose, those that are in environmental balance are established as parameters of maximum quality, that is, soils of native forests are usually used, as these are of the maximum standard. Another approach used is to establish soils of maximum quality with those that reach high productivity with less ecological damage (Gil-Sotres et al. 2005). Both strategies have been used and defended by several authors, while criticized by others since these "ideal standards" will not always match the maximum quality standards. In addition, when considering the soils with the highest productivity as the best (in the second strategy), interest in research on recovering soils and contaminated soils is reduced (Gil-Sotres et al. 2005).

In this regard, for more than 20 years Brazilian researchers have been hard at studying the biological components of the soil, to understand this complex system and thus be able to provide Brazilian producers with an optimized and precise way of analyzing soil quality taking into account chemical, physical and biological parameters.

Throughout this period, several advances have been made both from a scientific and a practical point of view. From a scientific point of view, we can cite numerous scientific articles that report the efforts to evaluate and achieve levels of assessments that would generate qualitative responses concerning individual assessments of biological parameters. The first major result came from the study published by Lopes et al. (2013), who developed three long-term experiments (17, 12 and 12 years, respectively), with 24 treatments, conducted in the Brazilian cerrado biome. This work aimed to study the management of P-based fertilizers, using various sources and application rates, and for that, several biological indicators were evaluated, such as P-melich, organic carbon, microbial biomass, basal respiration and enzymes, cellulase, β-glycosidase, arylsulfatase and phosphatase, in addition to average productivity. This was the first study to provide tables of interpretation of reference values for soil microbial indicators based on crop productivity. This work was of fundamental importance because it showed a high correlation, positively, between the enzymes β-glycosidase, arylsulfatase and acid phosphatase with productivity and microbial biomass.

Since then, several Brazilian cooperatives and foundations have started to adopt the use of these enzymes in their analyses and have proven, crop after crop, that the practices adopted after these analyses contributed to the increase of soil resilience and increase of productivity in crops. Also, the production values were not correlated with chemical and physical analysis and always had a high correlation with biological analysis and organic matter index. In other words, the increase in organic matter in the soil increases enzyme activity, and it is extremely sensitive to climatic variations, and management (Acosta-Martinez and Tabatabai 2000), in addition to being highly correlated with respiration and with alkaline phosphatase enzymes, acid phosphatase enzyme, dehydrogenase enzyme and cellulase enzyme (Lopes et al. 2013).

However, adding these enzymes together with chemical and physical analyses in routine evaluations was still not a simple task, since it was necessary to evaluate different treatments, which often became impracticable due to costs. Most studies report loss of enzyme activity if they are not stored correctly, or processed quickly, which often makes analysis in commercial laboratories unfeasible, since the time between collection and analysis needs to be fast, often increasing the costs of hiring a larger number of employees, buying coolers and refrigerators, etc. The use of refrigerators and freezers is also not a very reliable practice. Several studies show a loss or increase in enzyme activity after cooling or freezing (Lopes et al. 2015). In addition, information on standards for tropical soils is lacking.

To try to solve this problem, Lopes et al. (2015) evaluated the activity of the enzymes arylsulfatase, β-glycosidase and acid phosphatase in samples of cultivated soil and soil from a native forest of the Cerrado biome (clayey oxisol), air-dried and stored for a long time. The results of this study showed that there was an average reduction of 26% in β-glycosidase, 53% in arylsulfatase and 72% in acid phosphatase in the activity of these enzymes. Besides, the results proved that for the enzymes arylsulfatase and β-glycosidase, there was no reduction in the ability to detect variations in treatments in the soil. From this work, it was possible to establish and recommend air-drying for analysis of the enzymes β-glycosidase and arylsulfatase, for the soil in question. In this way, Embrapa Cerrados developed the fertBio concept (fertility + biology). This concept is based on the premise of evaluating both biological and chemical aspects in the same soil collection (Lopes et al. 2015). These samples are collected in the post-harvest phase and air-dried before the evaluations are carried out in the laboratory and sieved through a 2 mm sieve.

From the aforementioned studies, through the observation made that the enzymes β-glycosidase and arylsulfatase were not affected by liming and fertilization (like the enzyme phosphatase acid), they had a high correlation with organic matter and productivity, they were super sensitive to climatic variations, and thus, they functioned as a "soil diary", Embrapa Cerrados developed a technology called Bioanalysis (BioAs). Bioanalysis consists of integrating the evaluation of the enzymes arylsulfatase and β-glycosidase, together with physical and chemical analyses of the soil. More than just evaluating these enzymes and making them standard, accredited laboratories to carry out this analysis have access to an Embrapa database, where there is a compilation of validated algorithms to assist and interpret this data.

From the parameters evaluated by BioAs (activity of β-glycosidase enzymes, arylsulfatase, organic matter, potential CTC, acidity, supplies of K, Ca, Mg, V, S and P), it is possible to assign a soil quality index (SQI), based on these chemicals, physical and biological parameters. This SQI is made available to the producer through values (high, medium and low) of nutrient cycling, nutrient storage, nutrient supply, biological SQI and chemical SQI. In this way, the producer has a more complete tool and at the same time with a more simplified interpretation.

Bioanalysis is an important advance in the interpretation of validation of biological analyses for soil. It put Brazil at the forefront of technology. It still has many limitations, as we must take into account that Brazil has six different biomes and it is validated only for the Cerrado biome, and all the annual crops present in it. However, with the adherence and accreditation of laboratories around Brazil, it is intended to extend BioAs to all biomes and locations, even the most remote of Brazil.

5 Conclusions and Future Perspectives

Understanding the mechanisms of action of enzymes in agriculture is an important step towards their systematized use, which allows the insertion of new technologies based on enzymes in agriculture. Given the knowledge we already have about enzymes in agriculture, we know that together these are extremely important tools for maintaining soil health and fertility and, consequently, increasing productivity and sustainably decreasing pests. The enzymes act together, like true gears that, if moved correctly, are true allies of good soil health (Fig. 6.2).

The use of enzymes in the soil has different aspects and is of great importance from a scientific, environmental and social point of view. The application of enzymes in contaminated soils, that is, the use of enzymes for bioremediation is a strategy with proven effectiveness for the most diverse problems (contamination with heavy metals, petroleum, oil, etc.). However, its use on a large scale, faces great challenges, such as high production costs. Studies with agricultural by-products, mainly in countries with high agricultural production, and thus, a great generator of by-products, has been greatly expanded, with promising results, which promise to serve as a basis for enzyme production, increasing efficiency and lowering these costs.

Another aspect that hinders the production of new enzymes is the lack of discovery of new microorganisms. Based on the availability of new technologies and biotechnological advances, such as new generation sequencing (NGS) and metagenomics studies, this problem has been gradually overcome. Today the available databases have a significant collection of microorganisms. However, when we talk about soil, this collection is still limited. In a basic survey, we can see that many

Fig. 6.2 Showing that the set of enzymes available in the soil, and not just one enzyme, helps in the proper functioning and health of the soil

times, 50% of the microorganisms present in a soil sample have not yet been identified. Another problem is that the vast majority of these microorganisms, even when they are identified, are extremely difficult to grow in the laboratory. The development of new strategies for the in vitro cultivation of these microorganisms will treat incalculable gains for the sector since they will open thousands of new possibilities. This information confirms that the challenges are still very great, but that we have a vast diversity of microorganisms to be explored and this generates countless possibilities. Thus, it is expected soon, we can obtain new commercial products based on these new microorganisms.

The use of enzymes to measure the biological quality of the soil is another important factor that is gaining greater notoriety every day. Every day more agricultural producers from around the world are becoming aware that the use of this type of analysis adds significant gains in terms of productivity. Today we can say that the use of analyses for enzymatic evaluation of the soil is a sure strategy and that it tends to follow a worldwide progression, from the creation of technologies for the joint evaluation of these analyses, together with chemical and physical analyses. The launch of BioAs in Brazil is a proof of this, and we hope to soon adopt this type of technology in other parts of the world. This type of technology is essential to raise awareness of the use of sustainable tools for agriculture.

References

Abeles FB, Bosshart RP, Forrence LE, Habig WH (1970) Preparation and purification of glucanase and chitinase from bean leaves. Plant Physiol 47:129–134

Abiraami TV, Singh S, Nain L (2020) Soil metaproteomics as a tool for monitoring functional microbial communities: promises and challenges. Rev Environ Sci Biotechnol 19:1–30

Acosta-Martinez V, Tabatabai M (2000) Enzyme activities in a limed agricultural soil. Biol Fertil Soils 31:85–91

Acosta-Martinez V, Zobeck T, Gill T, Kennedy A (2003) Enzyme activities and microbial community structure in semiarid agricultural soils. Biol Fertil Soils 38:216–227

Adrangi S, Faramarzi MA (2013) From bacteria to human: a journey into the world of chitinases. Biotechnol Adv 31:1786–1795

Angelov A, Pham VTT, Übelacker M, Brady S, Leis B, Pill N, Liebl W (2017) A metagenome-derived thermostable β-glucanase with an unusual module architecture which defines the new glycoside hydrolase family GH148. Sci Rep 7:1–13

Aponte H, Herrera W, Cameron C, Black H, Meier S, Paolini J, Cornejo P (2020) Alteration of enzyme activities and functional diversity of a soil contaminated with copper and arsenic. Ecotoxicol Environ Safe 192:110264

Araujo ASF, Monteiro RTR (2007) Indicadores biológicos de qualidade do solo. Biosci J 23(3):66–75

Araújo ASF, Cesarz S, Leite LFC, Borges CD, Tsai SM, Eisenhauer N (2013) Soil microbial properties and temporal stability in degraded and restored lands of Northeast Brazil. Soil Biol Biochem 66:175–181

Aslam S, Hussain A, Qazi JI (2017) Production of cellulase by *Bacillus amyloliquefaciens*-ASK11 under high chromium stress. Waste Biomass Valori 10:53–61

Ayub MA, Usman M, Faiz T, Umair M, ul Haq MA, Rizwan M, ur Rehman MZ (2020) Restoration of degraded soil for sustainable agriculture. In: Soil health restoration and management. Springer, Berlin, pp 31–81

Baćmaga M, Wyszkowska J, Borowik A, Tomkiel M, Kucharski J (2014) Response of Fungi, β-glucosidase, and arylsulfatase to soil contamination by Alister Grande 190 OD, Fuego 500 SC, and Lumax 537.5 SE herbicides. Pol J Environ Stud 23:19–25

Badiane NNY, Chotte JL, Pate E, Masse D, Rouland C (2001) Use of soil enzyme activities to monitor soil quality in natural and improved fallows in semi-arid tropical regions. Appl Soil Ecol 18:229–238

Balota EL, Yada IF, Amaral H, Nakatani AS, Dick RP, Coyne MS (2014) Long-term land use influences soil microbial biomass P and S, phosphatase and arylsulfatase activities, and S mineralization in a Brazilian Oxisol. Land Degrad Dev 25:397–406

Bandick AK, Dick RP (1999) Field management effects on soil enzyme activities. Soil Biol Biochem 31:1471–1479

Behera BC, Yadav H, Singh SK, Mishra RR, Sethi BK, Dutta SK, Thatoi HN (2017) Phosphate solubilization and acid phosphatase activity of Serratia sp. isolated from mangrove soil of Mahanadi river delta, Odisha, India. Genet Eng Biotechn N 15:169–178

Bell TH, Klironomos JN, Henry HAL (2010) Seasonal responses of extracellular enzyme activity and microbial biomass to warming and N addition. Soil Sci Soc Am J 74:828–838

Bhadrecha P, Bala M, Khasa YP, Arshi A, Singh J, Kumar M (2020) Hippophae rhamnoides L. rhizobacteria exhibit diversified cellulase and pectinase activities. Physiol Mol Biol Plants 26:1075–1085

Borase DN, Nath CP, Hazra KK, Senthilkumar M, Singh SS, Praharaj CS, Kumar N (2020) Long-term impact of diversified crop rotations and nutrient management practices on soil microbial functions and soil enzymes activity. Ecol Indic 114:106322

Borshchevskaya LN, Gordeeva TL, Kalinina AN, Bulushova NV, Sineoky SP (2019) Cloning and expression of Bacillus pumilis Bg57 β-Glucanase gene in Pichia pastoris: purification and characteristics of recombinant enzyme. Biochem 55:771–779

Cavalcante WF, Silva LRCD, Silva EGD, Oliveira JTC, Moreira KA (2020) Enzymatic activity of Caatinga biome with and without anthropic action. Revista Caatinga 33(1): 142–150.

Cao Y, Zheng F, Zhang W, Meng X, Liu W (2019) Trichoderma reesei XYR1 recruits SWI/SNF to facilitate cellulase gene expression. Molecular microbiology, 112(4), 1145–1162.

Cao Y, Ma C, Chen H, Chen G, White JC, Xing B (2020) Copper stress in flooded soil: impact on enzyme activities, microbial community composition and diversity in the rhizosphere of Salix integra. Sci Total Environ 704:135350

Chaer G, Fernandes M, Myrold D, Bottomley P (2009) Comparative resistance and resilience of soil microbial communities and enzyme activities in adjacent native forest and agricultural soils. Microb Ecol 58:414–424

Chang EH, Chung RS, Tsai YH (2007) Effect of different application rates of organic fertilizer on soil enzyme activity and microbial population. J Soil Sci 53:132–140

Chen F, Wang M, Zheng Y, Luo J, Yang X, Wang X (2009) Quantitative changes of plant defense enzymes and phytohormone in biocontrol of cucumber Fusarium wilt by Bacillus subtilis B579. World J Microbiol Biotechnol 25:675–684

Chen Y, Xu H, Zhou M, Wang Y, Wang S, Zhang J (2015) Salecan enhances the activities of β-1, 3-glucanase and decreases the biomass of soil-borne fungi. PLoS One 10

Chen F, Ye J, Kameshwar AKS, Wu X, Ren J, Qin W, Li DW (2019) A novel cold-adaptive endo-1, 4-β-glucanase from Burkholderia pyrrocinia JK-SH007: gene expression and characterization of the enzyme and mode of action. Front Microbiol 10:3137

Chernin L, Ismailov Z, Haran S, Chet I (1995) Chitinolytic Enterobacter agglomerans antagonistic to fungal plant pathogens. Appl Environ Microbiol 61:1720–1726

Cherubin MR, Santi AL, Pias OHC (2016) Amostragem de solo na agricultura de precisão. In: Santi AL, Giotto E, Sebem E, Amado TJC (eds) Agricultura de Precisão no Rio Grande do Sul, 1st edn. CESPOL, Santa Maria, RS, pp 79–98

Chu KT, Ng TB (2005) Purification and characterization of a chitinase-like antifungal protein from black turtle bean with stimulatory effect on nitric oxide production by macrophages. Biol Chem 386:19–24

Chun-Ta W, Bradford KJ (2003) Class I chitinase and [beta]-1, 3-glucanase are differentially regulated by wounding, methyl jasmonate, ethylene, and gibberellin tomato seeds and leaves1. Plant Physiol 133:263

Condron LM, Turner BL, Cade-Menun BJ (2005) Chemistry and dynamics of soil organic phosphorus. In: Sims T, Sharpley AN (eds) Phosphorus: agriculture and the environment. ASA, Madison

da Silva LG, Mendes IDC, Reis Junior FB, Fernandes MF, de Melo JT, Kato E (2009) Atributo físicos, químicos e biológicos de um Latossolo de Cerrado sob plantio de espécies florestais. Embrapa Tabuleiros Costeiros-Boletim de Pesquisa e Desenvolvimento (INFOTECA-E).

Da Silva OOA, De Oliveira-Longatti SM, De Castro Caputo PS, Rufini M, Carvalho GR, De Carvalho TS, De Souza Moreira FM (2020) Microbiological indicators of soil quality are related to greater coffee yield in the Brazilian Cerrado region. Ecol Indic 113:106205

Ding H, Ali A, Cheng Z (2018) Dynamics of a soil fungal community in a three-year green garlic/cucumber crop rotation system in Northwest China. Sustainability 10:1391

Dodor DE, Tabatabai MA (2003) Effect of cropping systems on phosphatases in soils. J Plant Nutr Soil Sci 166:7–13

Doni S, Macci C, Chen H, Masciandaro G, Ceccanti B (2012) Isoelectric focusing of β-glucosidase humic-bound activity in semi-arid Mediterranean soils under management practices. Biol Fertil Soils 48:183–190

Dubeux J, Lira MDA, Santos MD, Cunha MD (2006) Fluxo de nutrientes em ecossistemas de pastagens: impactos no ambiente e na produtividade. Simpósio sobre o Manejo da Pastagem 23:439–506

Dunne C, Crowley JJ, Moënne-Loccoz Y, Dowling DN, de Bruijn FJ, O'Gara F (1997) Biological control of Pythium ultimum by Stenotrophomonas maltophilia W81 is mediated by an extracellular proteolytic activity. Microbiology 143:3921–3931

Edison LK, Pradeep NS (2020) Functional screening of β-Glucanase producing Actinomycetes strains from Western Ghats ecosystems of Kerala, India bioRxiv

Elad Y, Kapat A (1999) The role of Trichoderma harzianum protease in the biocontrol of Botrytis cinerea. Eur J Plant Pathol 105:177–189

El-Tarabily KA, Sykes ML, Kurtböke ID, Hardy Gest J, Barbosa AM, Dekker RFH (1996) Synergistic effects of a cellulase-producing Micromonospora carbonacea and an antibiotic-producing Streptomyces violascens on the suppression of Phytophthora cinnamomi root rot of Banksia grandis. Can J Bot 74:618–624

Embrapa Cerrados (2020). Available at https://youtu.be/IBJYc30aFas

FAO-ISRIC-ISSS (2015) Status of the world's soil resources: main report. Food and Agriculture Organization of the United Nations, Rome

Fasiku SA, Ogunsola OF, Fakunle A, Olanbiwoninu AA (2020) Isolation of Bacteria with potential of producing extracellular enzymes (amylase, cellulase and protease) from soil samples. Adv Appl Microbiol:21–26

Ferreira EPB, Pedra LF, Martin-didonet CCG (2017) População e atividade microbiana do solo em sistema de produção agroecológica. Rev Ciênc Agron 48:22–31

Ferreira ME, Caramori SS, Zago LM, Moreira AK, Silva-Neto CM, Nabout JC (2018) Biochemical activity in Brazilian Cerrado soils is differentially affected by perennial and annual crops. Australian J Crop Sci 12(2), 235–242

Foster EJ, Fogle EJ, Cotrufo MF (2018) Sorption to biochar impacts β-glucosidase and phosphatase enzyme activities. Agriculture 8(10):158

Frankenberger WT Jr, Dick WA (1983) Relationships between enzyme activities and microbial growth and activity indices in soil. Soil Sci Soc Am J 47:945–951

Gasmi M, Kitouni M, Carro L (2019) Chitinolytic actinobacteria isolated from an Algerian semi-arid soil: development of an antifungal chitinase-dependent assay and GH18 chitinase gene identification. Ann Microbiol 69:395–405

Ghosh A, Singh AB, Kumar RV, Manna MC, Bhattacharyya R, Rahman MM, Misra S (2020) Soil enzymes and microbial elemental stoichiometry as bio-indicators of soil quality in diverse cropping systems and nutrient management practices of Indian Vertisols. Appl Soil Ecol 145:103304

Gil-Sotres F, Trasar-Cepeda C, Leiros MC, Seoane S (2005) Different approaches to evaluating soil quality using biochemical properties. Soil Biol Biochem 37:877–887

Godoy FA (2020) Identificação de bactérias promotoras de crescimento de trigo e seu impacto na comunidade bacteriana da rizosfera. ESALQ

Gonçalves ACS, Rezende RP, Marques EDLS, Soares MR, Dias JCT, Romano CC, Pirovani CP (2020) Biotechnological potential of mangrove sediments: identification and functional attributes of thermostable and salinity-tolerant β-glucanase. Int J Biol Macromol 147:521–526

Henry HA (2013) Reprint of "soil extracellular enzyme dynamics in a changing climate". Soil Biol Biochem 56:53–59

Hoffmann E, Seegerer A (1950) Biochem Z 321, 97

Hong TY, Meng M (2003) Biochemical characterization and antifungal activity of an endo-1, 3-β-glucanase of Paenibacillus sp. isolated from garden soil. Appl Microbiol Biotechnol 61:472–478

Hu H, Gao Y, Li X, Chen S, Yan S, Tian X (2020) Identification and Nematicidal characterization of proteases secreted by endophytic Bacteria Bacillus cereus BCM2. Phytopathology 110:336–344

Huera-Lucero T, Labrador-Moreno J, Blanco-Salas J, Ruiz-Téllez T (2020) A framework to incorporate biological soil quality indicators into assessing the sustainability of territories in the Ecuadorian Amazon. Sustainability 12(7):3007

Hungria M, Franchini JC, Brandão O Jr, Kaschuk G, Souza RA (2009) Soil microbial activity and crop sustainability in a long-term experiment with three soiltillage and two crop-rotation systems. Appl Soil Ecol 42:288–296

IBGE. Instituto Brasileiro de Geografia e Estatistica (2020) Levantamento Sistemático da produção agrícola. Available at https://www.ibge.gov.br/estatisticas/economicas/agricultura-e-pecuaria/9201-levantamento-sistematico-da-producao-agricola.html?edicao=27643&t=resultados

Inayati A, Sulistyowati L, Aini LQ, Yusnawan E (2020) Mycoparasitic potential of Trichoderma virens against mung bean soil borne pathogen Rhizoctonia solani: hyper parasite and hydrolytic enzymes production. J Agric Sci

Jalil SU, Mishra M, Ansari MI (2015) Current view on chitinase for plant defence. Trends Biosci 8:6733–6743

Jetiyanon K (2007) Defensive-related enzyme response in plants treated with a mixture of Bacillus strains (IN937a and IN937b) against different pathogens. Biol Control 42:178–185

Kardol P, Cregger MA, Campany CE, Classen AT (2010) Soil ecosystem functioning under climate change: plant species and community effects. Ecology 91:767–781

Khoshnevisan K, Vakhshiteh F, Barkhi M, Baharifar H, Poor-Akbar E, Zari N, Bordbar AK (2017) Immobilization of cellulase enzyme onto magnetic nanoparticles: applications and recent advances. Mol Catal 442:66–73

Kirubakaran SI, Sakthivel N (2007) Cloning and overexpression of antifungal barley chitinase gene in Escherichia coli. Protein Expr Purif 52:159–166

Kumari M, Pandey S, Bhattacharya A, Mishra A, Nautiya CS (2017) Protective role of biosynthesized silver nanoparticles against early blight disease in Solanum lycopersicum. Plant Physiol Biochem 121:216–225

Kuzyakov Y, Blagodatskaya E (2015) Microbial hotspots and hot moments in soil: concept & review. Soil Biol Biochem 83:184–199

Lacombe-Harvey M, Brzezinski R, Beaulieu C (2018) Chitinolytic functions in actinobacteria: ecology, enzymes, and evolution. Appl Microbiol Biotechnol 102:7219–7230
Langklotz S, Baumann U, Narberhaus F (2012) Structure and function of the bacterial AAA protease FtsH. Biochim Biophys 1823:40–48
Leon MCC, Stone A, Dick RP (2006) Organic soil amendments: impacts on snap bean common root rot (Aphanomyes euteiches) and soil quality. Appl Soil Ecol 31:199–210
Liang XQ, Holbrook CC, Lynch RE, Guo BZ (2005) GENETICS AND RESISTANCE-b-1, 3-Glucanase activity in Peanut seed (Arachis hypogaea) is induced by inoculation with Aspergillus flavus and Copurifies with a Conglutin-like protein. Phytopathology 95:506–511
Loc NH, Huy ND, Quang HT, Lan TT, Thu Ha TT (2020) Characterisation and antifungal activity of extracellular chitinase from a biocontrol fungus, Trichoderma asperellum PQ34. Mycology 11(1):38–48
Lopes ACA, Souza GDM, Chaer GM et al (2013) Interpretation of microbial soil indicators as a function of crop yield and organic carbon. Soil Sci Soc Am J 77:461–472
Lopes CAA, de Sousa DMG, dos Reis Junior FB, Mendes IC (2015) Air-drying and long-term storage effects on β-glucosidase, acid phosphatase and arylsulfatase activities in a tropical savannah oxisol. App Soil Ecol 93; 68–77
Lori M, Symnaczik S, Mäder P, De Deyn G, Gattinger A (2017) Organic farming enhances soil microbial abundance and activity—a meta-analysis and meta-regression. PLoS One 12(7):e0180442
Madkour MA, Afifi MA, Metry EA, Ismail IM (2019) Biological control of soil-borne fungal pathogens. Arab Univ J Agric Sci 27(1):749–760
Majumdar S, Chakraborty U (2017) Optimization of protease production from plant growth promoting Bacillus amyloliquefaciens showing antagonistic activity against phytopathogens. Int J Pharm Biol Sci 8:635–642
Margalef O, Sardans J, Fernández-Martínez M, Molowny-Horas R, Janssens IA, Ciais P, Peñuelas J (2017) Global patterns of phosphatase activity in natural soils. Sci Rep 7:1–13
Martins ME Da M, Martins E Da S, Martins HL (2020) Production and characterization of a thermostable β-glucosidase from Myceliophthora heterothallica. Bioscience 36
Medeiros EV, Duda GP, Santos LAR, Sousa Lima JR, Almeida Cortêz JS, Hammecker C, Cournac L (2017) Respostas do carbono orgânico do solo, biomassa microbiana e enzimas à regeneração natural em uma região seca tropical do Nordeste do Brasil. Catena 151:137–146
Mendes IC, Souza LV, Resck DVS, Gomes AC (2003) Biological properties of aggregates from a Cerrado oxisol under conventional and no-till management systems. Rev Bras Ciênc Solo 27:435–443
Mendes IDC, Fernandes MF, Chaer GM, Reis FB Jr (2012) Biological functioning of Brazilian Cerrado soils under different vegetation types. Plant and Soil 359:183–195
Mendes IDC, De Sousa DMG, Dos Reis Junior FB, Lopes ADC (2018) Bioanálise de solo: como acessar e interpretar a saúde do solo. Embrapa Cerrados
Miguel DL, da Silva EMR, da Silva CF, Pereira MG, Leite LFC (2020) Soil microbiological properties and enzyme activity in agroforestry systems compared with monoculture, natural regeneration, and native Caatinga. Embrapa Agrobiol 36
Miralles I, Ortega R, Almendros G, Gil-Sotres F, Trasar-Cepeda C, Leirós M, Soriano M (2012) Modifications of organic matter and enzymatic activities in response to change in soil use in semi-arid mountain ecosystems (southern Spain). Eur J Soil Sci 63:272–283
Mndzebele B, Ncube B, Fessehazion M, Mabhaudhi T, Amoo S, du Plooy C, Modi A (2020) Effects of cowpea-Amaranth intercropping and Fertiliser application on soil phosphatase activities, available soil phosphorus, and crop growth response. Agronomy 10(1):79
Mondal T, Bisht JK, Mishra PK, Pandey BM, Mahanta D, Meena VS, Pattanayak A (2020) Soil enzymes: indicator for soil health under fruit based agri-horti system In: Roy MM, Malaviya VK, Yadav TS, Sah D Radhakrishna A. (ed). International Grassland Congress Proceedings. University of Kentuchy

Murphy DV, Stockdale EA, Brookes PC, Goulding KW (2007) Impact of microorganisms on chemical transformations in soil. In: Soil biological fertility. Springer, Berlin, pp 37–59

Muzangwa L, Mnkeni PNS, Chiduza C (2020) The use of residue retention and inclusion of legumes to improve soil biological activity in maize-based no-till Systems of the eastern Cape Province, South Africa. Agric Res 9:66–76

Ndiaye EL et al (2000) Integrative biological indicators for detecting change in soil quality. Am J Alternat Agric 15:26–36

Nelofer R, Syed Q, Nadeem M, Bashir F, Mazhar S, Hassan A (2016) Isolation of phosphorus-solubilizing fungus from soil to supplement biofertilizer. Arab J Sci Eng 41:2131–2138

Nguyen TTN, Xu CY, Tahmasbian I et al (2017) Efects of biochar on soil available inorganic nitrogen: a review and meta-analysis. Geoderma 288:79–96

Pelissaro H, Zanella M, Santos S, Cardoso E, Brasil M (2020) Atributos microbiológicos do solo como indicadores de conservação das pastagens nativas do Pantanal. Embrapa Pantanal-Capítulo em livro científico (ALICE)

Piotrowska-Długosz A, Charzyński P (2015) The impact of the soil sealing degree on microbial biomass, enzymatic activity, and physicochemical properties in the Ekranic Technosols of Toruń (Poland). J Soil Sediment 15:47–59

Pleban S, Chernin L, Chet I (1997) Chitinolytic activity of an endophytic strain of Bacillus cereus. Lett Appl Microbiol 25:284–288

Porto ML, Alves JC, Diniz AA, Souza AP, Santos D (2009) Indicadores biológicos de qualidade do solo em diferentes sistemas de uso no brejo paraibano. Cienc Agrotec 33:1011–1017

Prasannath K (2017) Plant defense-related enzymes against pathogens: a review. J Agric Sci 11:38

Prosser JA, Speir TW, Stott DE (2011) Oxidoredutases do solo e hidrólise do FDA. In: Dick RP (ed) Métodos de enzimologia do solo, série no 9. Sociedade de Ciências do Solo da América, pp 103–124

Rezzonico E, Flury N, Meins F, Beffa R (1998) Transcriptional down-regulation by abscisic acid of pathogenesis-related β-1, 3-glucanase genes in tobacco cell cultures. Plant Physiol 117:585–592

Roy-Barman S, Sautter C, Chattoo BB (2006) Expression of the lipid transfer protein ace-AMP1 in transgenic wheat enhances antifungal activity and defense responses. Transgenic Res 15:435

Santi AL, Silva DAA, Corassa GM, Basso CJ (2016) Manejo de precisão: Planos de manejo inteligentes e estratégias multi-plantas. In: Santi AL, Giotto E, Sebem E (eds) Agricultura de Precisão no Rio Grande do Sul, vol 1, 1st edn. Triunfal Gráfica e Editora, Santa Maria, pp 251–283

Saravanakumar K, Wang MH (2020) Isolation and molecular identification of Trichoderma species from wetland soil and their antagonistic activity against phytopathogens. Physiol Mol Plant Pathol 109:101458

Sardans J, Penuelas J (2005) Drought decreases soil enzyme activity in a Mediterranean Quercus ilex L. forest. Soil Biol Biochem 37:455–461

Sarto MV, Borges WL, Sarto JR, Pires CA, Rice CW, Rosolem CA (2020) Soil microbial community and activity in a tropical integrated crop-livestock system. Appl Soil Ecol 145:103350

Schuster A, Schmoll M (2010) Biology and biotechnology of Trichoderma. Appl Microbiol Biotechnol 87:787–799

Sevinc N, Demirkan E (2011) Production of protease by Bacillus sp. N-40 isolated from soil and its enzymatic properties. J Biol Environ Sci 5:95–103

Shinya S, Fukamizo T (2017) Interaction between chitosan and its related enzymes: a review. Int J Biol Macromol 104:1422–1435

Singh S, Sharma S (2020) Temporal changes in rhizosphere biological soil quality indicators of wheat in response to nitrogen and straw incorporation. Trop Ecol:1–17

Stieven AC, Mesquita Mendes W, Wruck F, Guimarães Couto E, Tiago Da Silva Campos D (2020) Atributos do Solo em Sistemas Diferenciados de Uso e Manejo do Solo. Colloquium Agrariae 16:1–15

Sulaiman MM, Yass STA, Aish AA, Basheer L, Yasir SJA, Youssef SA (2020) Activity of Trichoderma Spp. against Erwinia Carotovora causal agent of potato tuber soft rot. Plant Arch 20:115–118

Teng Z, Shao W, Zhang K, Huo Y, Li M (2019) Characterization of phosphate solubilizing bacteria isolated from heavy metal contaminated soils and their potential for lead immobilization. J Environ Manage 231:189–197

Valois D, Fayad K, Barbasubiye T, Garon M, De'ry C, Brzezinski R, Beaulieu C (1996) Glucanolytic actinomycetes antagonistic to Phytophthora fragariae var. rubi, the causal agent of raspberry root rot. Appl Environ Microbiol 62:1630–1635

van Elsas JD, Chiurazzi M, Mallon CA, Elhottovā D, Krištůfek V, Salles JF (2012) Microbial diversity determines the invasion of soil by a bacterial pathogen. PNAS 109:1159–1164

Vancov T, Keen B (2009) Rapid isolation and high-throughput determination of cellulase and laminarinase activity in soils. J Microbiol Methods 79(2):174–177

Vranova V, Rejsek K, Formanek P (2013) Proteolytic activity in soil: a review. Appl Soil Ecol 70:23–32

Wahsha M, Nadimi-Goki M, Fornasier F, Al-Jawasreh R, Hussein EI, Bini C (2017) Microbial enzymes as an early warning management tool for monitoring mining site soils. Catena 148:40–45

Waldrop MP, Balser TC, Firestone MK (2000) Linking microbial community composition to function in a tropical soil. Soil Biol Biochem 32:18371846

Wang K, Li F, Dong Y (2020) Methane emission related to enzyme activities and organic carbon fractions in Paddy soil of South China under different irrigation and nitrogen management. J Soil Sci Plant Nutr:1–14

Weedon JT, Aerts R, Kowalchuk GA, van Bodegom PM (2011) Enzymology under global change: organic nitrogen turnover in alpine and sub-Arctic soils. Biochem Soc Trans 39:309–314

Welc M, Bünemann EK, Fließbach A, Frossard E, Jansa J (2012) Soil bacterial and fungal communities along a soil chronosequence assessed by fatty acid profiling. Soil Biol Biochem 49:184–192

Woźniak A (2019) Chemical properties and enzyme activity of soil as affected by tillage system and previous crop. Agriculture 9(12):262

Wróbel-Kwiatkowska M, Lorenc-Kukula K, Starzycki M, Oszmiański J, Kepczyńska E, Szopa J (2004) Expression of β-1, 3-glucanase in flax causes increased resistance to fungi. Physiol Mol Plant Pathol 65:245–256

Wyszkowska J, Wieczorek K, Kucharski J (2016) Resistance of arylsulfatase to contamination of soil by heavy metals. Pol J Environ Stud 25:365

Xian Y, Wang M, Chen W (2015) Quantitative assessment on soil enzyme activities of heavy metal contaminated soils with various soil properties. Chemosphere 139:604–608

Zhang S, Gan Y, Xu B (2015) Biocontrol potential of a native species of Trichoderma longibrachiatum against Meloidogyne incognita. Appl Soil Ecol 94:21–29

Zhongcun P, Kodo O, Yasunori S et al (2004) Purification and characterization of an endo–1,3–β–glucanase from Arthrobacter sp. J Biol Macromol 4:57–66

Zhou Y, Wang X, Wei W, Xu J, Wan W, Xie Z, Wei C (2016) A novel efficient β-glucanase from a paddy soil microbial metagenome with versatile activities. Biotechnol Biofuels 9

Chapter 7
Pseudomonas as Plant Growth-Promoting Bacteria and Its Role in Alleviation of Abiotic Stress

Tahira Yasmeen, Abdul Aziz, Mohsin Tariq, Muhammad Saleem Arif, Sher Muhammad Shahzad, Muhammad Riaz, Aqsa Javed, Shafaqat Ali, and Muhammad Rizwan

Contents

T. Yasmeen (✉) · A. Aziz · M. S. Arif · M. Riaz · S. Ali · M. Rizwan
Department of Environmental Sciences and Engineering, Government College University Faisalabad, Faisalabad, Pakistan
e-mail: tahirayasmeen@gcuf.edu.pk

M. Tariq · A. Javed
Department of Bioinformatics and Biotechnology, Government College University Faisalabad, Faisalabad, Pakistan

S. M. Shahzad
Department of Soil and Environmental Sciences, University College of Agriculture, University of Sargodha, Sargodha, Pakistan

© The Author(s), under exclusive license to Springer Nature Switzerland AG 2021

H. I. Mohamed et al. (eds.), *Plant Growth-Promoting Microbes for Sustainable Biotic and Abiotic Stress Management*,
https://doi.org/10.1007/978-3-030-66587-6_7

1 Introduction

Plant growth-promoting rhizobacteria are the soil bacteria that reside in rhizosphere and rhizoplane and carry the intrinsic potential to promote the plant growth by utilizing different mechanisms, i.e., nitrogen fixation, different phytohormones (Abbass and Okon 1993; Egamberdiyeva 2005; Gravel et al. 2007; Gutiérrez-Mañero et al. 2001; García de Salamone et al. 2001) and enzyme production, mineral solubilization (Basak and Biswas 2009; Panhwar et al. 2012), siderophore productivity, and biocontrol activity, and induce systemic resistance (Adesemoye et al. 2009; Gopalakrishnan et al. 2011a, b, c). Various operative and a taxonomical cluster of PGPR encompass *Pseudomonas, Erwinia, Bacillus, Caulobacter, Serratia, Arthrobacter, Micrococcus, Flavobacterium, Chromobacterium, Agrobacterium, Hyphomicrobium, Azorhizobium, Rhizobium, Bradyrhizobium, Sinorhizobium, Azorhizobium, Mesorhizobium, Allorhizobium,* etc. As the demand for food has increased with expansion in population, the practices to boost agriculture production have also been upgraded in several parts of the world (Clair and Lynch 2010). The application of PGPR as bioinoculant due to their substantial influences on crop growth and yield has attained considerable attention; furthermore, the emergence of PGPR used as an alternative source of chemical fertilizers and pesticides has increased its acceptance in farmer community (Vessey 2003).

Environmental stresses like climate change, water scarcity, increased salinity, and induction of hazardous chemicals have decreased the soil fertility (He et al. 2018); these are the major factors to become the reason for deficient growth and productivity of plants (Clair and Lynch 2010). Although PGPR is frequent and their spreading is principally controlled by specificities of environment nevertheless, *Pseudomonas,* in particular, are getting substantial attention due to multifarious features that support plant growth under abiotic stresses (Sitaraman 2015; Saravanakumar and Samiyappan 2007). So, in this chapter, we discussed the potential of *Pseudomonas* as PGPR to alleviate abiotic stresses in detail.

2 Plant Growth-Promoting Traits of *Pseudomonas*

Pseudomonas spp. have been studied largely for their involvement in plant production, biological control, antibiotic aptitudes, siderophore, indoleacetic acid production, toxic compound degradation, and phosphate solubilization (Bensidhoum et al. 2016; Jain and Pandey 2016). To express the potential of plant development promotion, the PGPR should colonize in the host plant roots. Root colonization is important because without fixed attachment the bacteria can easily be detached via water flow and the valuable extracellular secretions of a bacterial cell could diffuse in the rhizosphere being available to other microorganisms rather than plant roots (Mia et al. 2010). *Pseudomonas* is complicated in different mechanisms of plant development, which are summarized in the next section.

2.1 Nitrogen Fixation

Like other PGPB, several *Pseudomonas* spp. possess the capability of binding the biological N_2 to compensate for the soil nitrogen loss. In chickpea, *Pseudomonas* was reported to contribute as plant development promoter, nodulation, and nitrogen fixation (Gopalakrishnan et al. 2015; Parmar and Dadarwal 1999). *Pseudomonas* carrying potential of phosphate solubilization have also been found promoting nutrient uptake in host plants through fixing nitrogen and dissolving phosphorus (Karimi et al. 2012). Coincubation of nitrogen-fixing *Pseudomonas* with other bacterial species has also been successful on different crops including common bean where *Pseudomonas* inoculation in combination with *Rhizobium* increased the plant growth yield by improving nitrogen fixation. This study exposed that *Pseudomonas and Rhizobium* co-inoculation increased the nitrogen ration derived from the nitrogen fixation mechanism up to 50% of the plant nitrogen demand and reduced the implementation of inorganic N fertilizer (Yadegari et al. 2010). The inoculation of nitrogen-fixing *P. stutzeri A1501* in maize crop also augmented plant growth resulting from reduced application of inorganic N-fertilizer beside promoting N_2-fixing and ammonium-oxidizing bacterial communities in the rhizosphere (Ke et al. 2019).

2.2 Nutrient Solubilization

Numerous studies describe that plant growth-promoting *Pseudomonas* have significant capabilities to solubilize different important nutrients in soil. Nutrient-solubilizing microorganisms perform a significant role to convert insoluble nutrient into soluble by releasing extracellular enzymes, which increase the availability of soluble nutrients in the soil (Krishnakumar et al. 2014). These nutrients are solubilized by *Pseudomonas* which help the plants to increase their growth and productivity. Inoculation of *Pseudomonas Pf-5* and *CHA0* strains has been characterized, and these strains imposed virtuous influence on the growth of tomato plant as a result of nutrient solubilization and siderophore production. Also, to promote plant growth, these growth-promoting *Pseudomonas* enhanced the dry biomass, yield and nutrient content in tomato plant tissues (Hu et al. 2017).

Bacterial strains of *Pseudomonas* are considered among the most powerful nutrient solubilizers. Mostly, they play an important role in mineral phosphorus (P) solubilization. PGPRs are going to assimilate numerous insoluble as well as a soluble form of phosphate. *Pseudomonas* are prominent among these P-solubilizing PGPRs (Singh et al. 2018). The assessment of *Pseudomonas putida* strain Rs-198 inoculation on pepper plants proved effective that increased the plant biomass and enhanced nutrient uptake (P and Fe) by solubilizing nutrients and producing phytohormones (He et al. 2019). Zarei et al. (2019) described that inoculation of *P. fluorescens* strains promoted plant development and productivity, but also play an important role to compensate for the adverse factor of the ecosystem. The inoculated strains of

Pseudomonas fluorescens could solubilize nutrient especially phosphates besides the abilities to produce ACC deaminase in the configuration of 1-aminocyclopropane-1-carboxylic acid, auxin, and siderophores, and these helped to enhance the growth of crops by involving different mechanisms.

2.3 ACC Deaminase and IAA Production

Pseudomonas carrying PGP characteristics also express ACC deaminase activity which reduces (but not prevents entirely) ethylene level in plants that rises during exposure to environmental stresses. This enzyme supports plants against environment stress, promotes the productivity of plant (enhances the root dimension and aerial structure of plants), and facilitates the plants for adaptation and survival. The PGP *Pseudomonas* produce the ACC deaminase which behaves like a biological promoter for ACC and considered as one of the chief mechanisms that are involved in the plant development under stress conditions. In addition to ACC deaminase activity, PGP *Pseudomonas* produce IAA which enhances root tissue uptake, resultantly boosting the growth of plants and initiating the enzyme ACC synthase transcription, which eventually increases the ethylene production. The high level of ethylene reduces IAA signal transduction thus reducing IAA-catalyzed growth of plant. Therefore, the presence of a PGP *Pseudomonas* which encompass the enzyme ACC deaminase reduce the plant ethylene level thus reducing the above-mention feedback inhibition (del Carmen Orozco-Mosqueda et al. 2020).

Auxin controlled the numerous biological developments, i.e., cell development, cell division, tissue distinction, and light response in plants. Normally, root length increases by the excretion of IAA, augmentation of root surface aptitude and capability of plant to access soil nutrients occurs (Santoro et al. 2015). Bacteria that produce IAA are acknowledged to uplift the seed germination, pledge adventitious and lateral root formation thus increasing the surface area to facilitate the host plant for better absorption of nutrients and water from soil (Ahemad and Kibret 2014).

2.4 Siderophore Production

Siderophore is a low-weight-molecule and chelating representative which bounds with the iron and is carried to a cell by cooperating with membranous receptors (Johnstone and Nolan 2015; Saha et al. 2016). The PGPRs that produce siderophore are considered as an auspicious substitute to the traditional farming predominantly in the perspective of increasing the development and productivity of plant by dropping the use of chemical fertilizer. However, there is still a need for the scientific demonstration of a direct advantageous consequence of siderophore-producing PGPR on iron acquirement and plant growth in abiotic stress. Indirect indications of siderophores enhance the development as well as a biological control system in nutrient-lacking soil.

The growth-promoting character of *Pseudomonas* has been related to the viable impounding of iron by siderophore production. Siderophore plays a vital role to upsurge the bacterial fitness in diverse environmental conditions. Several cellular processes including repairing of DNA, redox reactions, electron transport, metabolic activities, and regulation of gene expression require iron for normal functioning, and several bacteria can sense and uptake the iron from surrounding environment. Bacteria scavenge the iron by producing siderophores (Green and Paget 2004; Puig et al. 2017; Braun and Hantke 2011; Frawley and Fang 2014). Parmar and Chakraborty (2016) demonstrated formation of extracellular water resolvable yellow-green siderophore by *Pseudomonas fluorescens* that proved beneficial for plant growth development due to enhancement in root length, shoot height, and leaf number in an inoculated plant. Siderophore-forming *Pseudomonas* also influence physicochemical properties of soil (Sayyed et al. 2019). The physicochemical factor influences the production of PGPR's siderophore. Consequently, the studies carried out to discover the potential of siderophore-producing *Pseudomonas* appraise their plant growth promotion efficiency as observed in *Arachis hypogaea L.* (Subramanium and Sundaram 2020). Barrientos-Moreno et al. (2019) illustrated a connection between siderophore production, arginine metabolism, and oxidative stress tolerance in *Pseudomonas putida*. Some other investigations are described in Table 7.1 showing the effect of plant growing promoter *Pseudomonas* species on different plants.

3 *Pseudomonas* and Alleviation of Abiotic Stress

Some studies have identified a broad picture of special effects of abiotic stresses for the productivity of numerous plants (Almansouri et al. 2001; Li et al. 2011). However, the PGPR having extremely multifaceted and fascinating mechanisms related to stress alleviation improve the plants to better survive against stress environments (Gopalakrishnan et al. 2015). Numerous PGP *Pseudomonas* strains can increase stress-bearing capacity of plants under various abiotic stress conditions (Fig. 7.1). Growth hormone production by PGPR improves plant root morphology which is a key biological mechanism that can increase the water and nutrient absorption in plants under severe edaphic circumstances. These groups of bacteria have specific plant growth-promoting traits like cellulase and protease glucanases which cause cell lysis and fungal cell wall degradation (Mabood et al. 2014). Moreover, few PGPR strains have sigma factor which supports the plant to diminish the adversative effect of abiotic strain by using the specific amendments in gene appearance. Indoleacetic acid is a key signalling molecule, produced by PGP *Pseudomonas,* thus establishing a synergistic relationship with plant roots and working for phytostimulation (Ashwitha et al. 2013; Duca et al. 2014). Domination of ACC deaminase movement and IAA productivity contribution can exert positive effects on the

Table 7.1 Effect of stress-tolerant PGP *Pseudomonas* inoculation on different plants

Bacterial strains	Inoculated plants	PGP traits	Effect on plants	References
P. putida; *P. aeruginosa*; *P. fluorescens* *P. vulgaris*	Bean, Soybean, Mung bean, Chickpea, Common bean	IAA production, ACC deaminase production, siderophore productivity	Increase plant productivity Increased efficiency of biological nitrogen fixation Significant solicitation of N and P in grains	Yadegari et al. (2010), Kang et al. (2014), Sarma and Saikia (2014), Singh et al. (2018), Verma et al. (2010) and Younesi and Moradi (2014)
P. stutzeri; *P. protegens*; *P. putida*; *P. plecoglossicida*	Maize	IAA production, siderophore production, Chitinase, b-l,3-glucanase, ACC deaminase activity, P solubilization	Increased plant growth; Positive effect on the population of the N-fixing and NH_4-oxidizing microbial communities in the rhizosphere; Increased the biomass, nitrogen content, and nitrogen fixation rate; Increased nutrient solubilization; Significant effect on grain yield and soil fertility	Mahajan et al. (2020), Fox et al. (2016), Pandey and Maheshwari (2007), Kaur and Reddy (2014) and Singh et al. (2018)
P. protegens; *Pseudomonas* sp.; *P. putida*; *P. plecoglossicida*; *P. aeruginosa*	Wheat	IAA production, ACC deaminase activity, siderophore, gibberellin synthesis, and P solubilization	Increased the plant growth, biomass, nitrogen content, and nitrogen availability; expressively improved the grain yield and soil fertility Inhibited ethylene levels; increased nutrient solubilization and root colonization	Fox et al. (2016), Govindasamy et al. (2009), Aloni et al. (2006), Kaur and Reddy (2014), Nadeem et al. (2010) and del Carmen Orozco-Mosqueda et al. (2020)
Pseudomonas sp.; *P. stutzeri*	Sorghum, rice	IAA production, N_2 complex	Increased plant growth and yield by phytohormone production	Ashraf et al. (2011), Pham et al. (2017) and Lu et al. (2020)
P. aeruginosa; *P. fluorescens*	Peanut, groundnut	IAA assembly, ACC deaminase productivity	Stimulated the plant growth; Induced tolerance against abiotic stress	Gupta et al. (2020), Saravanakumar and Ramasamy (2007), Gupta and Pandey (2019) and Estévez et al. (2009)

(continued)

Table 7.1 (continued)

Bacterial strains	Inoculated plants	PGP traits	Effect on plants	References
P. koreensis and P. entomophila	Sugarcane	ACC deaminase activity, nitrogen fixation	Enhanced the plant growth, development, and nitrogenase activity; Induced tolerance against abiotic stress	Estévez et al. (2009)
P. fluorescens; Pseudomonas sp.; *P. putida*	Canola	IAA production, ACC deaminase activity	Enhanced the plant growth by reducing ethylene production; Induced tolerance against abiotic stress	Akhgar et al. (2014), Ali and Kim (2018), del Carmen Orozco-Mosqueda et al. (2020) and Cheng et al. (2007)
P. fluorescens; P. migulae Pseudomonas sp.; *P. aeruginosa; P. stutzeri*	Tomato	ACC deaminase activity, siderophore productivity, gibberellin synthesis, P solubilization	Increased the biomass and assimilation of nutrients into the plant tissues; increased the plant growth by solubilization of nutrients; Increased the growth of plants under stress conditions	Ali et al. (2011), Gupta and Pandey (2019), del Carmen Orozco-Mosqueda et al. (2020), Ali et al. (2012, 2014), Orozco-Mosqueda et al. (2019), Hernández-León et al. (2015), Hu et al. (2017), Tank and Saraf (2010) and Ferreira et al. (2019)
P. putida GAP-P45	Sunflower	IAA production, ACC deaminase activity	Protected sunflower seedlings from drought stress through exopolysaccharide secretion and improving soil structure	Sandhya et al. (2009a, b)
P. fluorescens	Sweet corn	Siderophore production, ACC deaminase activity	Increased the plant growth by reducing the ethylene level; Increased the iron and phosphate uptake	Zarei et al. (2019)
Pseudomonas sp. A3R3	Wild plants	IAA production, siderophore production, heavy metal mobilization	Increased the plant growth by solubilization of nutrients	Ahn et al. (2004)

(continued)

Table 7.1 (continued)

Bacterial strains	Inoculated plants	PGP traits	Effect on plants	References
P. putida, P. fluorescens	Tea	Siderophore production, P solubilization, IAA production	Enhanced the plant development; Improved the quality and quantity of crops; significant effect on grain yield and soil fertility	Singh et al. (2018) and Esitken et al. (2010)
Pseudomonas BA-8	Strawberry	IAA production, P solubilization	Enrichment of soluble solids and sugar	Singh et al. (2018)
P. synxantha	Aloe vera	IAA production, P solubilization	Enhanced the growth level and biosynthesis level	Guo et al. (2010)
P. migulae 8R6 *and Pseudomonas* sp. UW4	Sesame	IAA production, ACC deaminase activity	Induction of genes preventing cell damage	del Carmen Orozco-Mosqueda et al. (2020)
P. fluorescens UM270	Medicago	IAA and Siderophore productivity	Increased plant productivity for stress conditions	Hernández-León et al. (2015)
P. aeruginosa FP6	Chilli	Siderophore production	Increased growth yield	Sasirekha and Srividya (2016)

abiotically stressed plants by improving their tolerance against stress factors (Saleem et al. 2007; Yun-Xiu and Xiao-dong 2007).

Survival of *Pseudomonas* under abiotic stresses may be attributed to the construction of exopolysaccharides (EPS) that help to defend the bacteria after water stress and water potential fluxes. Exopolysaccharides support biofilm formation which improves water retention and has binding potential to form soil aggregates that regulate water and nutrient flow from soil to plant roots (Sandhya et al. 2009a, b; Roberson and Firestone 1992; Tisdall and Oades 1982). The PGP *Pseudomonas* bacteria are also acknowledged for the generation of various growth promoters of plants like rhizobitoxine, exopolysaccharides (Vardharajula et al. 2011), and specific signal molecules like lumichrome (Dakora et al. 2015) and lipochitooligosaccharides (Tanaka et al. 2015). By reducing ethylene production, rhizobitoxine can increase the development of plant under stress circumstances. Correspondingly, lipochitooligosaccharides and lumichrome support different plants in detecting ecosystem abiotic stresses. Furthermore, they behave like plant growth promoter which improves biomass production, root and shoot growth, and embryogenesis. In specific, these signal molecules help to develop a synergistic relationship between the plants and rhizospheric microorganisms, therefore shielding the plant from the confrontational things of abiotic stress (Tanaka et al. 2015). Colonization of *P. chlororaphis* (O6) in *Arabidopsis thaliana* roots could also induce abiotic stress tolerance by producing 2R, 3R-butanediol (volatile metabolite) (Cho et al. 2008). Mitigation

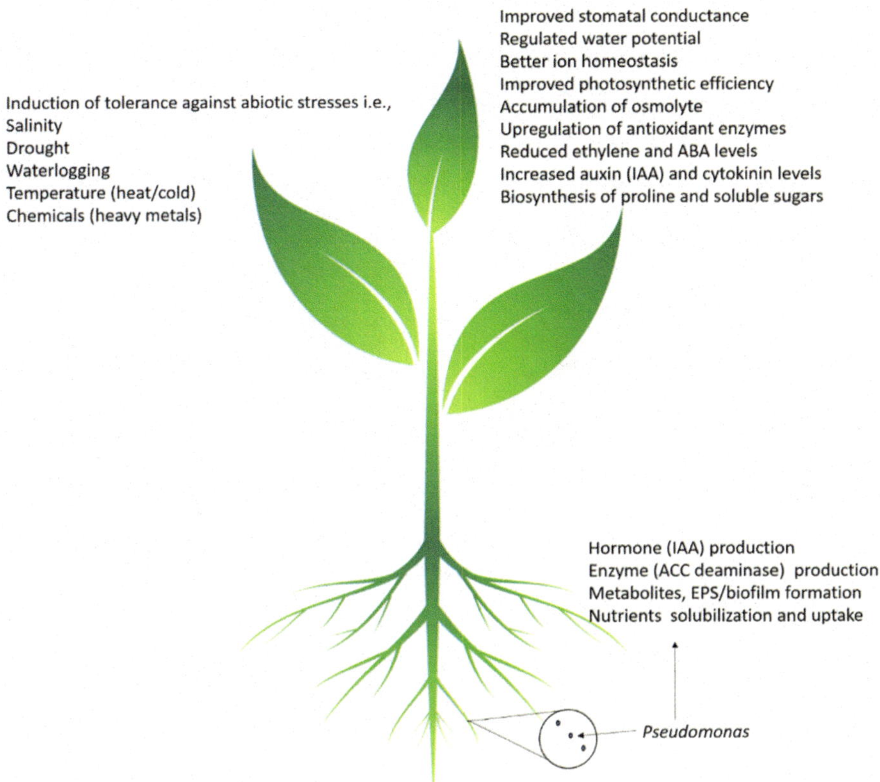

Fig. 7.1 Demonstration of abiotic stress tolerance induced by abiotic stress-tolerant plant growth-promoting *Pseudomonas*

of abiotic stress by inoculation of plant growth-promoting stress-tolerant *Pseudomonas* is summarized in Table 7.2.

3.1 Pseudomonas Under Water Stress

Stress associated with water or moisture in the larger sense can be acknowledged as flooding and drought which causes considerable damage to crops over the world. Water deficiency and overflow both cause different morphological and physiological changes in the plants which finally reduce the productivity and quality of plant yield. Under waterlogging, plant roots and soil submerge in excess water, thus creating hypoxic and anoxic environments which results in reducing the accessibility of O_2 to the plants (Salazar et al. 2015). Under waterlogged conditions, different physiological responses of plants alter to cope with the insufficient supply of oxygen.

Table 7.2 Effect of salt-tolerant PGP *Pseudomonas* inoculation on different plants under abiotic stresses

Bacterial inoculation	Plant	Stress factor	Effects on plants	References
Pseudomonas sp.; *P. fluorescens*	*Pisum sativum* L.	Drought	Increased the development of plant, yield, and maturing; Increased the germination of a seed, shoot, and root length, production of antioxidant enzymes, and dry biomass; Decreased ethylene production	Arshad et al. (2007), Gupta and Pandey (2019), Zahir et al. (2008) and Saikia et al. (2018)
Pseudomonas 6-P	*Cicer arietinum*	Drought	Increased plant tolerance against stress factors	Sharma et al. (2013)
Pseudomonas sp. RJ15	*Vigna mungo* L.	Drought	Increased the seed germination and root and shoot length; Upregulation of antioxidant enzymes	Saikia et al. (2018)
Pseudomonas sp.	Finger millet (*Eleusine coracana* L.)	Drought	Significantly increased antioxidant activity; improved plant fitness against oxidative stress	Chandra et al. (2018)
P. fluorescens	Foxtail millet	Drought	Efficiently colonized in rhizosphere and enhanced plant development	Niu et al. (2018)
Pseudomonas	Groundnut and pigeon pea	Drought	Positive phosphatase activity induced drought tolerance; Increased relative water content; Increased osmotic regulation; modulated cell wall elasticity	Ashwitha et al. (2013) and Keyvan (2010)
Pseudomonas	Mung bean	Drought	An elevated level of ROS-scavenging enzymes and cellular osmolytes; upregulated drought stress-responsive genes; Augmented root, shoot length, and dry biomass	Sharma et al. (2013)

(continued)

Table 7.2 (continued)

Bacterial inoculation	Plant	Stress factor	Effects on plants	References
P. putida GAP-P45	Sunflower	Drought	Produced biofilm and improved soil aggregation to facilitate water and nutrient supply	Sandhya et al. (2009a, b)
Pseudomonas sp. UW4	*Lycopersicon esculentum*	Waterlogging	Significantly increased plant growth and tolerance to waterlogging	Ali and Kim (2018)
Pseudomonas sp. UW4	*Cucumis sativus*	Waterlogging	Increased the growth of plants	Li et al. (2012) and Ali and Kim (2018)
Pseudomonas	Sesame (*Sesamum indicum* L.)	Waterlogging	Mitigated waterlogging-related damage and enhanced plant progression	Ali et al. (2018)
Pseudomonas sp. UW4	*Brassica napus*	Waterlogging	Enhanced the plant growth under waterlogging by producing ACC deaminase	Ali and Kim (2018)
P. putida WT	*Rumex palustris*	Waterlogging	Altered plants' response to long-term and short-term submergence generating ACC deaminase	Ravanbakhsh et al. (2017) and Ali and Kim (2018)
P. putida	Tomato	Waterlogging	ACC deaminase synthesis ameliorated the flooding impacts	Grichko and Glick (2001)
Pseudomonas	Soybean	Salinity	Increased activities of ion transport and proline accumulation	Yasmin et al. (2020)
P. fluorescens	Maize	Salinity	Enhanced seedling development and plant growth	Kausar and Shahzad (2006) and Gupta and Pandey (2019)
P. fluorescens TDK1	Groundnut (*Arachis hypogaea*)	Salinity	Increased salt tolerance by retaining the ACC deaminase action and enhanced growth of plants	Saravanakumar and Ramasamy (2007) and Gupta and Pandey (2019)
P. fluorescens; P. migulae	Tomato	Salinity	Enhanced the health and growth of plant	Ali et al. (2011) and Gupta and Pandey (2019)

(continued)

Table 7.2 (continued)

Bacterial inoculation	Plant	Stress factor	Effects on plants	References
P. fluorescens	Canola (*Brassica napus* L.)	Salinity	Increased the growth of plants and resistance against salinity by producing ACC deaminase	Akhgar et al. (2014)
P. fluorescens	Lettuce	Temperature	Induced systemic tolerance; Promoted leaf development	Aponte et al. (2017)
Pseudomonas spp.	Gramineae and legume	Temperature	Improved plant growth and yield	Höflich et al. (1994) and Höflich and Kühn (1996)
Pseudomonas spp.	Wheat	Temperature	Stimulated growth along with low levels of ROS	Ali et al. (2011)
Pseudomonas sp. AMK-P6	Sorghum	Temperature	Created heat revelation proteins; Improved plant biological status	Ali et al. (2009)
Pseudomonas sp.; *P. fluorescens*	Wheat	Heavy metal (Cd, Pb)	Increased root length and plant growth; Reduced the ethylene production	Govindasamy et al. (2009), He et al. (2009), Singh et al. (2018) and Sheng et al. (2008)
Pseudomonas	Wheat	Heavy metal (Cd, Cr, Cu)	Declined the catalase (CAT), glutathione reductase (GR), superoxide dismutase SOD), proline, and malondialdehyde (MDA) levels	Rizvi and Khan (2017)
P. moraviensis	Wheat	Heavy metal (Co, Ni, Cr)	Decreased heavy metal uptake; decreased the biological accumulation coefficient (BAC) and translocation factor	Hassan et al. (2017)
P. aeruginosa	Chickpea	Heavy metal (Cr)	Improved symbiotic attributes, plant growth, and yield	Oves et al. (2013)

Overactivity of ACC synthase enzyme under hypoxic and anoxic conditions was reported in different plants (He et al. 1994; Sairam et al. 2008).

Inoculation of ACC enzymes producing *Pseudomonas* has been described to decrease the negative impact of produced ethylene and, consequently, improved the growth of the plant in excessive water circumstances. The water scarcity, on the other hand, is related to reduction in stomatal conductance and chlorophyll contents which create disturbance in the photosynthetic metabolism (Vurukonda et al. 2016;

Flexas et al. 2013). Different investigations have been reported at the worldwide level to exploit the importance and role of *Pseudomonas*, which improved the plant growth under water-stressed conditions (Gupta and Pandey 2019). Waterlogging severely affects physiological characteristics of terrestrial plants prominent to short photochemical productivity and stunted growth. ACC deaminase enzymes cleaved the ACC substrate into α-ketobutyrate, NH_4, and mitigate the hostile effects of persistent water stress. Ali et al. (2018) observed mitigation of waterlogging-related damage and improved plant growth in sesame (*Sesamum indicum* L.) plants that were inoculated with ACC deaminase forming *Pseudomonas.* Chandra et al. (2018) observed improvement in finger millet (*Eleusine coracana* (L.) Gaertn.) plant growth inoculated with ACC deaminase producing *Pseudomonas* spp. under drought-stressed conditions. Inoculated plants showed a significant increase in antioxidant activity that ultimately improved plant fitness by protecting it from oxidative stress induced by drought. Overexpression of ACC deaminase and exopolysaccharide (EPS) production potential in *Pseudomonas fluorescens* also helped to efficiently colonize in the rhizosphere of foxtail millet roots as reported by Niu et al. (2018). Effective bioinoculants (*Pseudomonas*) enhanced plant development in drought stress and proved as potential contributor to endure cultivation in arid precincts. Ashwitha et al. (2013) isolated indoleacetic acid (IAA) and 1-amino cyclopropane-1-carboxylic acid (ACC) deaminase-generating *Pseudomonas* bacteria with an additional character of positive phosphatase activity. Inoculation of these bacteria on groundnut and pigeon pea induced a certain degree of tolerance in drought-affected plants. High relative water content (RWC) as observed in drought-stressed *Pseudomonas*-inoculated plants could be the part of a resistance mechanism to drought stress that helps in osmotic regulation and/or modulation of cell wall elasticity in plant tissues (Keyvan 2010; Ritchie et al. 1990; Ashwitha et al. 2013). Under drought conditions, proline is accumulated in exposed plants as a physiological response to stress and helps to improve membrane stability (Delauney and Verma 1993). The elevated level of reactive oxygen species (ROS) scavenging enzymes and cellular osmolytes, upregulation of drought stress-responsive genes, augmented root and shoot length, and higher dry biomass are some other attributes that were observed in osmotic stress-tolerant *Pseudomonas* inoculated for mung bean in contrast with the uninoculated control plants under drought-stressed condition (Sharma et al. 2013). Plant development in water-stressed cucumber plants with *P. fluorescens* inoculation alone and/or in combination with compost/biochar was described by Nadeem et al. (2017). Fioreze et al. (2020) observed nonsignificant effect of *Pseudomonas* inoculation (alone or in combination with other bacterial species) on SPAD index under normal irrigation; however, under water deficit condition, notably increased values of SPAD (Soil Plant Analysis Development) catalogue were recorded with co-inoculation of *Pseudomonas* spp. and *Azospirillum brasilense* in wheat plants. Therefore it may be concluded that co-inoculation of *Pseudomonas* spp. with other beneficial bacterial species could effectively be utilized to alleviate the deleterious special effects of water stress in different plants.

3.2 *Pseudomonas* Under Salinity Stress

Soil salinity is also one of the severe issues related to the soil that has adversely influenced the yield and productivity of crops. It primarily increases the concentration of ions (Na^+ and Cl^-) in the soil that results in an osmotic and nutrient imbalance in plants leading to disturbing normal plant functions (Munns and Tester 2008; Tavakkoli et al. 2010; Yaish et al. 2016; Moradi et al. 2011; Yaish and Kumar 2015). Soil salinity induces necrosis on plant roots and shoots, delays leaf exterior, and reduces leaf amplitude (Rajendran et al. 2009; Rahneshan et al. 2018). Moreover, ionic imbalance in plants, damaged of stomatal appearance, reduction in the CO_2 accommodation and photosynthetic efficiency, and induction of reactive oxygen species (ROS) and ethylene levels have also been reported in salinity-deficient circumstances (Chaves et al. 2009; Sarabi et al. 2017; Chatterjee et al. 2017; Heydarian et al. 2016; Stearns and Glick 2003; Ali et al. 2014). Undesirable changes in plant growth may occur due to distressed transport of phytohormones and photosynthetic metabolites to the emerging plant tissues (Ashraf 2004). Inoculation of PGPR during the salinity-stressed condition for plants can promote the growth and development through a range of mechanisms including production of growth hormones and ACC deaminase enzyme (Siddikee et al. 2015), colonization in rhizosphere (Subramanian et al. 2015), and upregulation of ROS-scavenging enzymes (Nautiyal et al. 2013; Sarkar et al. 2018). thus ameliorating the undesirable effects of salinity on subjected plant. Salt-tolerant PGP *Pseudomonas* have been reported to increase nutrient uptake, photosynthetic pigments, and rate of photosynthesis besides decreasing extent of membrane damage that is investigated by non-enzymatic oxidation of fatty acids (lipid peroxidation) (Samaddar et al. 2019). Increased activities of ion transport, proline accumulation, and reduced concentration of K in *Pseudomonas*-inoculated soybean plants were also observed in hydroponically grown salinity-stressed plants (Yasmin et al. 2020). Likewise, plant growth promotion and instigation of salinity tolerance in maize, groundnut, tomato, and canola plants were observed with *Pseudomonas* inoculation in different salinity stress-related studies (Kausar and Shahzad 2006; Gupta and Pandey 2019; Saravanakumar and Ramasamy 2007; Ali et al. 2011; Akhgar et al. 2014). Inoculation of *P. putida* to cotton seeds before sowing also increased seedling biomass under the saline condition and prevented salinity-induced abscisic acid (ABA) accumulation in seedlings (Yao et al. 2010). Studies have also explored the synergistic capabilities of *Pseudomonas* species by inoculating as consortia or co-culture. Samaddar et al. (2019) testified that co-inoculation of *P. frederiksbergensis* (OB139) and *P. vancouverensis* (OB155) under salinity-stressed red pepper exhibited reduced emission of ethylene and improved contents of photosynthetic pigments compared to single inoculation.

3.3 *Pseudomonas* Under Temperature Stress

Analogous to other abiotic stress, temperature stress also adversely affects plant growth and developmental processes. In several crops it is an unusual constraint that happens during grain filling. Heat stress affects the flexibility of membrane lipids which leads toward loss of membrane integrity. Furthermore, it also stimulates the inactivation of chloroplast and mitochondrial enzymes besides inhibition of protein synthesis, reduction in ion flux, and production of ROS (Schöffl et al. 1999; Howarth 2005). Interseasonal climatic variation especially in temperature affects different sensitive crops. For example, the reduced yield of wheat plants even in well-watered conditions was observed due to increase in mean seasonal temperature (Wheeler et al. 1996; Batts et al. 1997).

Temperature disparities influence the enzymatic activities of cellular proteins after structural abnormalities in cell organelles (Ruelland and Zachowski 2010). Some thermotolerant plant growth-promoting species of *Pseudomonas* induce systemic tolerance in the plants that were exposed to heat stress as observed by Aponte et al. (2017) in lettuce plants. *P. fluorescens* along with *Azospirillum* sp. promoted leaf development in this experiment suggesting occurrence of synergist relationship among both PGPR strains. The rhizobacterial strains modulated the high-temperature stress and suggested application of this potential biotechnological tool to partially overcome the impacts of this abiotic stress. Höflich et al. (1994) and Höflich and Kühn (1996) isolated *Pseudomonas* spp., along with other PGPR, and inoculated to gramineae and legume plants under temperature stress circumstances. Inoculated plants were observed to possess high growth and yield. Stimulated growth along with low levels of ROS in *Pseudomonas*-inoculated heat-stressed wheat plants was also observed by Ali et al. (2011). However, Yarzábal et al. (2018) reported root and shoot elongation in *Pseudomonas*-inoculated wheat plants grown in cold temperature suggesting formulation of cold-active biofertilizers for colder regions. Ali et al. (2009) observed improved level of cellular metabolites (i.e., sugar, amino acids, proteins, prolines, and photosynthetic pigments) in *Pseudomonas* (AKM-P6)-inoculated pigeon pea under heat stress. Subramanian et al. (2015) also reported expression of certain proteins in *Solanum lycopersicum* Mill with *P. vancouverensis* and *P. frederiksbergensis* inoculation under cold stress. These proteins could protect the plant cells from cold stress. Moreover, reduction in membrane damage and oxidative stress was also observed in this investigation.

3.4 *Pseudomonas* Under Heavy Metal Toxicity

The metals and metalloids having relatively high solidity and considered toxic even at ppb levels are called heavy metals. The biosphere is getting polluted by heavy metals due to their consumption in industrial, agricultural, and domestic activities (Vijayaraghavan and Yun 2008). Heavy metal stress is of serious international concern because of their nondegradable nature. Once they enter into the ecosystem, they persist there and accumulate by passing through the food chain (Igwe et al. 2005). Higher meditations of heavy metals in rhizosphere/soil reduce microbial population and metabolic activities (Ahamed et al. 2004). Generally, noxious heavy metals damage cell membranes, cause enzyme inhibition in cytoplasm, and consequently reduce plant growth and development (Chibuike and Obiora 2014). The bacteria groups having PGPR traits can secrete different metabolites including antibiotics, proteins, and acids that help to alleviate the lethal effects of heavy metals (Denton 2007). Biofilms are collections of microbial cells that are attached to any surface (Flemming 1995) and can be used to remediate heavy metal pollution. Microbial biofilms either modify the heavy metals biochemically or accumulate them thus reducing their spread in the environment (Muñoz et al. 2006; Chang et al. 2006). Biofilms can minimize the expected threat of heavy metals by immobilizing or partitioning to different environmental compartments. Metal-resistant and immobilizing PGPRs have recently been reported with characteristic plant growth stimulation, heavy metal accumulation within plant tissues, and reduction in heavy metal bioavailability (Wang et al. 2018; Yuan et al. 2017; Mallick et al. 2018; Han et al. 2018). *Pseudomonas* either directly or indirectly play an imperative role to mitigate heavy metal stress by influencing metal accumulation or bioavailability. Plant growth-promoting heavy metal-resistant *Pseudomonas* have potential to tolerate the presence of heavy metals up to a varying extent and help to ameliorate the stress exerted by heavy metals. Meliani and Bensoltane (2016) examined the capabilities of biofilm and biosurfactant-producing *Pseudomonas* to minimize the toxic effects of zinc and lead. A decline in phytotoxic effect of Cd, Cr, and Cu was also observed in wheat plants by inoculating IAA and ACC deaminase-producing *Pseudomonas aeruginosa*. The plants in this investigation resulted from a decline in glutathione reductase (GR), superoxide dismutase (SOD), catalase (CAT), malondialdehyde (MDA), and proline levels with *Pseudomonas* inoculation in wheat plants (Rizvi and Khan 2017). Application of *Pseudomonas* in combination with arbuscular mycorrhizal fungi has also been evaluated for its role for bioremediation (Li et al. 2020) and resulted in a decrease in Zn concentration in Zn-stressed maize plants. *Pseudomonas moraviensis* in another study produced promising results for heavy metal (Co, Ni, Cr) reduction in rhizosphere of wheat plants. However, co-inoculation with other PGPR further augmented the decrease in Ni, Cr, and Mn over its single inoculation. *P. moraviensis* inoculation also contributed to decrease the BAC (biological accumulation coefficient) and TF (translocation factor) for Cd, Cr, Cu, Mn, and Ni (Hassan et al. 2017). Evidence of improved symbiotic attributes, growth, and yield of chickpea plants with *P. aeruginosa* inoculation in Cr-contaminated soil has also been articulated (Oves et al. 2013).

3.5 *Pseudomonas* Under Nutrient Deficiency

Water scarcity caused by drought and salinity stress leads to insufficient nutrient supply to the stress-affected plants. Nutrients are important factors besides water, air, light, and carbon that play an imperative role in the regulation of different plant growth stages (Lata et al. 2018). PGPRs develop a mutualistic relationship with plant roots by establishing bidirectional movements of nutrients, water, and metabolites. Root exudates are absorbed and nitrogen is delivered back to plant roots by nitrogen fixers; thus soil fertility is improved as a result of this synergistic relationship. Similarly, phosphate solubilizers provide phosphorus to plants and get root exudates having water and carbohydrates as metabolites from plants. Siderophores produced by PGPRs under iron-limiting conditions help in iron sequestration to make it available to the plants (Whipps 2001; Compant et al. 2005). The productivity of plant development hormones such as auxins, cytokinins, and gibberellins (Spaepen and Vanderleyden 2011; Glick 2012; Shilev 2013; Kang et al. 2010) by PGPR stimulates root proliferation that results in extended absorption area for nutrient uptake (Sharma et al. 2013; Ahemad and Kibret 2014). Nordstedt et al. (2020) in an investigation grew ornamental plants in low nutrient regimes and observed that the plants inoculated with PGP *Pseudomonas* sp. accumulated higher nutrient contents (N, P, K, Ca, Mg, and S) in shoot compared to uninoculated plants. Plant growth promotion by inoculating *Pseudomonas* sp. in *Pelargonium peltatum, Dahlia variabilis*, and *Chrysanthemum* has been reported by Göre and Altin (2006). Srivastava and Srivastava (2020) observed a strong correlation among different combinations of *Pseudomonas* and salt stress and morpho-physiological attributes of *Arabidopsis thaliana.* The adversative effect of salinity in P-limiting conditions was well administered with the application of *P. putida*, thus suggesting its application in nutrient-deficient conditions to improve crop productivity.

4 Use of *Pseudomonas* for Bioremediation and Phytoremediation

Remediation of polluted sites following bioremediation has gained attention for being an environmentally friendly and cost-effective nature of operation. Microbial potential to tolerate, sequester, immobilize, mobilize, and transform the contaminants is monitored to detoxify the contaminated sites (Bruins et al. 2000; Gibbons et al. 2011). It is observed by various researchers that microorganism especially bacterial species can effectively absorb different heavy metals from the contaminated source such as soil or water (Fig. 7.2). *Pseudomonas aeruginosa*, a Gram-negative pathogenic bacterium, can precipitate cadmium when anaerobic conditions are prevailing. *Pseudomonas putida* has been worked upon by different researchers, with observations that it has the highest levels of passive biosorption for heavy metals. *Pseudomonas syringae* can bind with copper ions as it produces

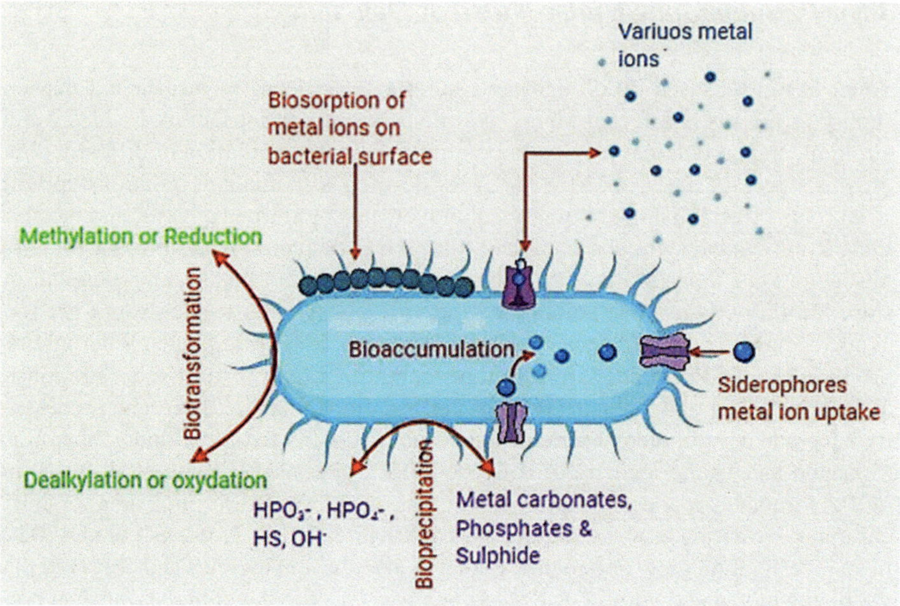

Fig. 7.2 Various mechanisms of bioremediation of heavy metals by bacteria (i.e., *Pseudomonas* spp.)

copper-inducible proteins such as CopA, CopB, and CopC (Igiri et al. 2018). Microbial biofilms are more crucial for the bioremediation of the environment as these have more potential to counter heavy metals compared to individual bacterial cells. *Pseudomonas aeruginosa* biofilm cells have more resistance toward a variety of heavy metal ions such as Cu^{2+}, Cd^{2+}, and Pb^{2+}. The cells around the biofilm that encompasses these heavy metals die and permanently capture these ions and as a result help to bioremediate these toxic ions from the environment. It has been noted during the research by different scientists that methylation of heavy metals increases their permeability across the bacterial cell. Microbial methylation is one of the most important aspects of bioremediation of metals and metal ions. *Bacillus* spp., *Clostridium* spp., and *Pseudomonas* spp. are responsible for the bio-methylation of heavy metals such as arsenic (As), mercury (Hg), and selenium (Se). Some researchers have used mercury-resistant bacterial strains for the bioremediation of heavy metals from different contaminated environments. De Jaysankar et al. (2008) showed that *P. aeruginosa* removed 17.4 mg/L of cadmium (Cd) from the contaminated source. Six genes are found in *Pseudomonas* for resistance of cadmium that was recognized from three genes assembled as cadA2R, czcCBA1, and colRS. Metal efflux systems were predicted from the first two homologous genes (Nowicki et al. 2015). Their study confirms that microbial strains like that of *Pseudomonas* spp. are potential candidates for the bioremediation of heavy metals from the environment to build a more sustainable ecosystem. Al-Dhabi et al. (2019) confirmed that *Pseudomonas* spp. have a crucial role in the remediation of cadmium in the contaminated soil and can, therefore, be used as candidates for the bioremediation of various metal ions from the environment. Among six different strains that were isolated, *Pseudomonas* strain Al-Dhabi-126 had the maximum tolerance for the cadmium, which was 2100 μg/mL. They concluded that *Pseudomonas* sp. can act as bioremediation agents for industrially generated effluents. Apart from acting as the direct source of bioremediation by absorption of heavy metals, bacterial strains particularly *Pseudomonas* sp. aid plants to bear environmental stress and thus enhance their ability to absorb different minerals and salt from the soil and in this way increase the quality of the soil. *Pseudomonas putida* as bio-stimulant support *T. sativum* to induce tolerance, and in this way, it indirectly participates in bioremediation through plants more commonly known as phytoremediation (Oosten et al. 2017). Laccase enzyme produced by *P. putida* MTCC 7525 has shown maximum activity against industrial effluents and dyes with 16–84% decolorization of these effluents and synthetic dyes within 24 h of incubation (Kuddus et al. 2013).

Gong et al. (2018) showed that the engineered strain of *P. putida* KT2440 effectively degrades pesticides in the soil, which also simultaneously degrades organophosphate, carbamates, and pyrethroids. Their study further confirms that these engineered strains can be used for in situ bioremediation of highly effective agricultural land due to excessive use of toxic fertilizers. Butachlor is an active ingredient in the preparation of herbicides and is considered as an environmental contaminant. High levels of butachlor in the soil decrease its fertility and retard plant growth that is why soils contaminated with this chemical compound must be treated. Mohanty and Jena (2019) showed that *P. putida* G3 strain is highly tolerant of butachlor soils. They showed that this bacterial strain can help in the effective bioremediation of contaminated soils at a rate of 700 mg/L within 360 h. Apart from bioremediation of butachlor, this strain has shown that it can remove up to 500–700 mg/L of alachlor and glyphosate. Studies have proved that *P. aeruginosa* TPHK-4 can be effectively employed for the removal of weathered petroleum hydrocarbons by biostimulation and bioaugmentation mechanisms (Ramadass et al. 2018). Quinclorac (QNC) is an environmental persistent herbicide which is used in rice fields. Less degradability and high persistency in the environment make QNC an environmental constraint that must be removed for a sustainable environment. *P. putida* II-2 can mineralize QNC into different metabolic energy sources in the form of different carbon compounds that are used as an energy source by various soil deviling bacterial species, and in this way the concentration of quinclorac is maintained within the normal limits (Yang et al. 2020). Combined application of chelating agents (i.e., citric, oxalic, and amino acids) and *P. fluorescens* to remediate contaminated soil (with metals) also produces promising results (Gómez-Garrido et al. 2018). Gupta et al. (2018) observed ameliorative effect of PGP *Pseudomonas* sp. CPSB21 that mobilized Cr from contaminated soil to sunflower plants and revealed its potential contribution in the practice of microbe-assisted phytoremediation. Inoculation of CPSB21 strain was observed with an increased ability of Cr^{6+} uptake. Many other *Pseudomonas* sp. have more excellent tolerance capabilities to survive in highly toxic environments. Appanna et al. (1996) described that *P. fluorescens* strain could persist in the manifestation of either Mn, CO, or Cs with multiple metal stress tolerance mechanism. *P. fluorescens* strain can instantaneously stimulate plant development and metal uptake in the plants having high potential to absorb metals (Wu et al. 2020). From the above discussion, it is evident that *Pseudomonas* spp. can be effectively used, employed, and engineered for the bioremediation and phytoremediation of various environmental constraints and pollutants in a sustainable and eco-friendly manner.

5 Conclusion

It is concluded that different potent strains of *Pseudomonas* help plants to better survive under various abiotic stresses like salinity, drought, flooding, temperature, nutrient deficiency, and heavy metal exposure. *Pseudomonas*, after successful colonization in plant roots, modify different plant physiological and biochemical pathways that are linked with induction of tolerance in plants and support for their better survival under abiotic stresses. *Pseudomonas* appeared to activate defense mechanisms under stressful conditions besides regulating osmotic and redox potential. Its evident contribution in bioremediation and assistance in phytoremediation provides the basis for exploitation of this valuable bioresource as a sustainable approach to administer abiotic stresses and to upsurge crop productivity in an eco-friendly way.

References

Abbass Z, Okon Y (1993) Plant growth promotion by *Azotobacter paspali* in the rhizosphere. Soil Biol Biochem ozco25(8):1075–1083

Adesemoye AO, Torbert HA, Kloepper JW (2009) Plant growth-promoting rhizobacteria allow reduced application rates of chemical fertilizers. Microb Ecol 58(4):921–929

Ahemad M, Kibret M (2014) Mechanisms and applications of plant growth promoting rhizobacteria: current perspective. J King Saud Univ Sci 26(1):1–20

Ahamed I, Hayat S, Ahmad A, Samiullah A (2004) Effect of heavy metal on survival of certain group of indigenous soil microbial population. J Appl Sci Environ Engag 9:115–121

Ahn KS, Ha U, Jia J, Wu D, Jin S (2004) The truA gene of *Pseudomonas aeruginosa* is required for the expression of type III secretory genes. Microbiology 150(3):539–547

Akhgar AR, Arzanlou M, Bakker PAHM, Hamidpour M (2014) Characterization of 1-aminocyclo propane-1-carboxylate (ACC) deaminase-containing *Pseudomonas* spp. in the rhizosphere of salt-stressed canola. Pedosphere 24(4):461–468

Al-Dhabi NA, Esmail GA, Ghilan AKM, Arasu MV (2019) Optimizing the management of cadmium bioremediation capacity of metal-resistant *Pseudomonas* sp. strain Al-Dhabi-126 isolated from the industrial city of Saudi Arabian environment. Int J Environ Res Public Health 16:4788. https://doi.org/10.3390/ijerph16234788

Ali S, Kim WC (2018) Plant growth promotion under water: decrease of waterlogging-induced ACC and ethylene levels by ACC deaminase-producing bacteria. Front Microbiol 9:1096

Ali SZ, Sandhya V, Grover M, Kishore N, Rao LV, Venkateswarlu B (2009) *Pseudomonas* sp. strain AKM-P6 enhances tolerance of sorghum seedlings to elevated temperatures. Biol Fertil Soils 46(1):45–55

Ali SZ, Sandhya V, Grover M, Linga VR, Bandi V (2011) Effect of inoculation with a thermotolerant plant growth promoting *Pseudomonas putida* strain AKMP7 on growth of wheat (Triticum spp.) under heat stress. J Plant Interact 6(4):239–246
Ali S, Charles TC, Glick BR (2012) Delay of flower senescence by bacterial endophytes expressing 1-aminocyclopropane-1-carboxylate deaminase. J Appl Microbiol 113(5):1139–1144
Ali S, Charles TC, Glick BR (2014) Amelioration of high salinity stress damage by plant growth-promoting bacterial endophytes that contain ACC deaminase. Plant Physiol Biochem 80:160–167
Ali S, Khan MA, Kim WC (2018) *Pseudomonas veronii* KJ mitigates flood stress-associated damage in Sesamum indicum L. Appl Biol Chem 61(5):575–585
Almansouri M, Kinet JM, Lutts S (2001) Effect of salt and osmotic stresses on germination in durum wheat (*Triticum durum* Desf.). Plant and Soil 231(2):243–254
Aloni R, Aloni E, Langhans M, Ullrich CI (2006) Role of cytokinin and auxin in shaping root architecture: regulating vascular differentiation, lateral root initiation, root apical dominance and root gravitropism. Ann Bot 97(5):883–893
Aponte A, Castillo O, Cabrera G, Pernia M, Hernandez Y (2017) Rhizobacteria *Pseudomonas fluorescens* and *Azospirillum* sp. association enhances growth of Lactuca sativa L. under tropical conditions. J Cent Eur Agric 18(2):424–440
Appanna VD, Gazso LG, Pierre MS (1996) Multiple-metal tolerance in *Pseudomonas fluorescens* and its biotechnological significance. J Biotechnol 52(2):75–80
Arshad M, Saleem M, Hussain S (2007) Perspectives of bacterial ACC deaminase in phytoremediation. Trends Biotechnol 25(8):356–362
Ashraf M (2004) Some important physiological selection criteria for salt tolerance in plants. Flora Morphol Distrib Funct Ecol Plants 199(5):361–376
Ashraf MA, Rasool M, Mirza MS (2011) Nitrogen fixation and indole acetic acid production potential of bacteria isolated from rhizosphere of sugarcane (*Saccharum officinarum* L.). Adv Biol Res 5(6):348–355
Ashwitha K, Rangeshwaran R, Vajid NV, Sivakumar G, Jalali SK, Rajalaksmi K, Manjunath H (2013) Characterization of abiotic stress tolerant *Pseudomonas* sp. occurring in Indian soils. J Biol Control 27(4):45–48
Barrientos-Moreno L, Molina-Henares MA, Pastor-García M, Ramos-González MI, Espinosa-Urgel M (2019) Arginine biosynthesis modulates pyoverdine production and release in *Pseudomonas putida* as part of the mechanism of adaptation to oxidative stress. J Bacteriol 201:e00454–e00419. https://doi.org/10.1128/JB.00454-19
Basak BB, Biswas DR (2009) Influence of potassium solubilizing microorganism (*Bacillus mucilaginosus*) and waste mica on potassium uptake dynamics by Sudan grass (*Sorghum vulgare* Pers.) grown under two Alfisols. Plant and Soil 317(1–2):235–255
Batts GR, Morison JIL, Ellis RH, Hadley P, Wheeler TR (1997) Effects of CO_2 and temperature on growth and yield of crops of winter wheat over four seasons. Eur J Agron 7(1–3):43–52
Bensidhoum L, Nabti E, Tabli N, Kupferschmied P, Weiss A, Rothballer M, Schmid M, Keel C, Hartmann A (2016) Heavy metal tolerant *Pseudomonas protegens* isolates from agricultural well water in northeastern Algeria with plant growth promoting, insecticidal and antifungal activities. Eur J Soil Biol 75:38–46
Braun V, Hantke K (2011) Recent insights into iron import by bacteria. Curr Opin Chem Biol 15:328–334
Bruins M, Kapil S, Oehme F (2000) Microbial resistance to metals in the environment. Ecotoxicol Environ Saf 45:198–207
Chandra D, Srivastava R, Glick BR, Sharma AK (2018) Drought-tolerant *Pseudomonas* spp. improve the growth performance of finger millet (*Eleusine coracana* (L.) Gaertn.) under non-stressed and drought-stressed conditions. Pedosphere 28(2):227–240
Chang WC, Hsu GS, Chiang SM, Su MC (2006) Heavy metal removal from aqueous solution by wasted biomass from a combined AS-biofilm process. Bioresour Technol 97(13):1503–1508

Chatterjee P, Samaddar S, Anandham R, Kang Y, Kim K, Selvakumar G, Sa T (2017) Beneficial soil bacterium *Pseudomonas frederiksbergensis* OS261 augments salt tolerance and promotes red pepper plant growth. Front Plant Sci 8:705

Chaves MM, Flexas J, Pinheiro C (2009) Photosynthesis under drought and salt stress: regulation mechanisms from whole plant to cell. Ann Bot 103(4):551–560

Cheng Z, Park E, Glick BR (2007) 1-Aminocyclopropane-1-carboxylate deaminase from Pseudomonas putida UW4 facilitates the growth of canola in the presence of salt. Can J Microbiol 53(7):912–918

Chibuike GU, Obiora SC (2014) Heavy metal polluted soils: effect on plants and bioremediation methods. Appl Environ Soil Sci 2014:752708. https://doi.org/10.1155/2014/752708

Cho SM, Kang BR, Han SH, Anderson AJ, Park JY, Lee YH, Cho BH, Yang KY, Ryu CM, Kim YC (2008) 2R, 3R-butanediol, a bacterial volatile produced by Pseudomonas chlororaphis O6, is involved in induction of systemic tolerance to drought in Arabidopsis thaliana. Mol Plant Microbe Interact 21(8):1067–1075

Clair SBS, Lynch JP (2010) The opening of Pandora's box: climate change impacts on soil fertility and crop nutrition in developing countries. Plant and Soil 335(1–2):101–115

Compant S, Duffy B, Nowak J, Clement C, Barka EA (2005) Use of plant growth promoting bacteria for biocontrol of plant diseases: principles, mechanisms of action, and future prospects. Appl Environ Microbiol 71(9):4951–4959

Dakora FD, Matiru V, Kanu AS (2015) Rhizosphere ecology of lumichrome and riboflavin, two bacterial signal molecules eliciting developmental changes in plants. Front Plant Sci 6:700

De Jaysankar, Ramaiah N, Vardanyan L (2008) Detoxification of toxic heavy metals by marine bacteria highly resistant to mercury. Marine Biotechnol 10(4):471–477

del Carmen Orozco-Mosqueda M, Glick BR, Santoyo G (2020) ACC deaminase in plant growth-promoting bacteria (PGPB): an efficient mechanism to counter salt stress in crops. Microbiol Res 235:126439

Delauney AJ, Verma DPS (1993) Proline biosynthesis and osmoregulation in plants. Plant J 4(2):215–223

Denton B (2007) Advances in phytoremediation of heavy metals using plant growth promoting bacteria and fungi. MMG 445 Basic Biotechnol 3: 1–5

Duca D, Lorv J, Patten CL, Rose D, Glick BR (2014) Indole-3-acetic acid in plant-microbe interactions. Antonie Van Leeuwenhoek 106(1):85–125

Egamberdiyeva D (2005) Plant-growth-promoting rhizobacteria isolated from a Calcisol in a semi-arid region of Uzbekistan: biochemical characterization and effectiveness. J Plant Nutr Soil Sci 168(1):94–99

Esitken A, Yildiz HE, Ercisli S, Donmez MF, Turan M, Gunes A (2010) Effects of plant growth promoting bacteria (PGPB) on yield, growth and nutrient contents of organically grown strawberry. Sci Hortic 124(1):62–66

Estévez J, Dardanelli MS, Megías M, Rodríguez-Navarro DN (2009) Symbiotic performance of common bean and soybean co-inoculated with rhizobia and Chryseobacterium balustinum Aur9 under moderate saline conditions. Symbiosis 49(1):29–36

Ferreira MJ, Silva H, Cunha A (2019) Siderophore-producing rhizobacteria as a promising tool for empowering plants to cope with iron limitation in saline soils: a review. Pedosphere 29(4):409–420

Fioreze SL, Pinheiro MG, Pereira YD, da Cruz SP (2020) Inoculation of wheat plants with *Pseudomonas* spp. and *Azospirillum brasilense* under drought stress. J Exp Agric Int 42(2):1–7

Flemming HC (1995) Sorption sites in biofilms. Water Sci Technol 32(8):27

Flexas J, Niinemets Ü, Gallé A, Barbour MM, Centritto M, Diaz-Espejo A, Douthe C, Galmés J, Ribas-Carbo M, Rodriguez PL, Rosselló F (2013) Diffusional conductance to CO_2 as a target for increasing photosynthesis and photosynthetic water-use efficiency. Photosynth Res 117(1–3):45–59

Fox AR, Soto G, Valverde C, Russo D, Lagares A Jr, Zorreguieta Á, Alleva K, Pascuan C, Frare R, Mercado-Blanco J, Dixon R (2016) Major cereal crops benefit from biological nitrogen fixation when inoculated with the nitrogen-fixing bacterium *Pseudomonas protegens* Pf-5 X940. Environ Microbiol 18(10):3522–3534

Frawley ER, Fang FC (2014) The ins and outs of bacterial iron metabolism. Mol Microbiol 93:609–616

García de Salamone IE, Hynes RK, Nelson LM (2001) Cytokinin production by plant growth-promoting rhizobacteria and selected mutants. Can J Microbiol 47(5): 404–411

Gibbons SM, Feris K, McGuirl MA, Morales SE, Hynninen A, Ramsey PW, Gannon JE (2011) Use of microcalorimetry to determine the costs and benefits to *Pseudomonas putida* strain KT2440 of harboring cadmium efflux genes. Appl Environ Microbiol 77:108–113

Glick BR (2012) Plant growth-promoting bacteria: mechanisms and applications. Scientifica 2012:963401

Gómez-Garrido M, Navarro JM, Navarro FJM, Cano AF (2018) The chelating effect of citric acid, oxalic acid, amino acids and *Pseudomonas fluorescens* bacteria on phytoremediation of cu, Zn, and Cr from soil using *Suaeda vera*. Int J Phytoremediation 20(10):1033–1042

Gong T, Xu X, Dang Y, Kong A, Wu Y, Liang P, Wang S, Yu H, Xu P, Yang C (2018) An engineered *Pseudomonas putida* can simultaneously degrade organophosphates, pyrethroids and carbamates. Sci Total Environ 628–629:1258–1265

Gopalakrishnan S, Humayun P, Kiran BK, Kannan IGK, Vidya MS, Deepthi K, Rupela O (2011a) Evaluation of bacteria isolated from rice rhizosphere for biological control of charcoal rot of sorghum caused by *Macrophomina phaseolina* (Tassi) Goid. World J Microbiol Biotechnol 27(6):1313–1321

Gopalakrishnan S, Kiran BK, Humayun P, Vidya MS, Deepthi K, Jacob S, Vadlamudi S, Alekhya G, Rupela O (2011b) Biocontrol of charcoal-rot of sorghum by actinomycetes isolated from herbal vermicompost. Afr J Biotechnol 10(79):18142–18152

Gopalakrishnan S, Pande S, Sharma M, Humayun P, Kiran BK, Sandeep D, Vidya MS, Deepthi K, Rupela O (2011c) Evaluation of actinomycete isolates obtained from herbal vermicompost for the biological control of *Fusarium* wilt of chickpea. Crop Prot 30(8):1070–1078

Gopalakrishnan S, Srinivas V, Prakash B, Sathya A, Vijayabharathi R (2015) Plant growth-promoting traits of *Pseudomonas geniculata* isolated from chickpea nodules. 3 Biotech 5(5):653–661

Göre ME, Altin N (2006) Growth promoting of some ornamental plants by root treatment with specific fluorescent pseudomonads. Aust J Biol Sci 6:610–615

Govindasamy V, Senthilkumar M, Mageshwaran V, Annapurna K (2009) Detection and characterization of ACC deaminase in plant growth promoting rhizobacteria. J Plant Biochem Biotech 18(1):71–76

Gravel V, Antoun H, Tweddell RJ (2007) Growth stimulation and fruit yield improvement of greenhouse tomato plants by inoculation with *Pseudomonas putida* or *Trichoderma atroviride*: possible role of indole acetic acid (IAA). Soil Biol Biochem 39(8):1968–1977

Green J, Paget MS (2004) Bacterial redox sensors. Nat Rev Microbiol 2:954–966

Grichko VP, Glick BR (2001) Amelioration of flooding stress by ACC deaminase-containing plant growth-promoting bacteria. Plant Physiol Biochem 39(1):11–17

Guo Y, Ni Y, Huang J (2010) Effects of rhizobium, arbuscular mycorrhiza and lime on nodulation, growth and nutrient uptake of lucerne in acid purplish soil in China. Trop Grassl 44:109–114

Gupta S, Pandey S (2019) Unravelling the biochemistry and genetics of ACC deaminase-an enzyme alleviating the biotic and abiotic stress in plants. Plant Gene 18:100175

Gupta P, Rani R, Chandra A, Kumar V (2018) Potential applications of *Pseudomonas* sp. (strain CPSB21) to ameliorate Cr^{6+} stress and phytoremediation of tannery effluent contaminated agricultural soils. Sci Rep 8:4860. https://doi.org/10.1038/s41598-018-23322-5

Gupta V, Kumar GN, Buch A (2020) Colonization by multi-potential *Pseudomonas aeruginosa* P4 stimulates peanut (*Arachis hypogaea* L.) growth, defense physiology and root system functioning to benefit the root-rhizobacterial interface. J Plant Physiol 248:153144

Gutiérrez-Mañero FJ, Ramos-Solano B, Probanza AN, Mehouachi JR, Tadeo F, Talon M (2001) The plant-growth-promoting rhizobacteria *Bacillus pumilus* and *Bacillus licheniformis* produce high amounts of physiologically active gibberellins. Physiol Plant 111(2):206–211

Han H, Sheng X, Hu J, He L, Wang Q (2018) Metal-immobilizing *Serratia liquefaciens* CL-1 and *Bacillus thuringiensis* X30 increase biomass and reduce heavy metal accumulation of radish under field conditions. Ecotoxicol Environ Saf 161:526–533

Hassan TU, Bano A, Naz I (2017) Alleviation of heavy metals toxicity by the application of plant growth promoting rhizobacteria and effects on wheat grown in saline sodic field. Int J Phytoremediation 19(6):522–529

He C-J, Drew MC, Morgan PW (1994) Plant induction of enzymes associated with lysigenous aerenchyma formation in roots of *Zea mays* L. during hypoxia and nitrogen starvation. Plant Physiol 105:861–865

He M, He CQ, Ding NZ (2018) Abiotic stresses: general defenses of land plants and chances for engineering multi stress tolerance. Front Plant Sci 9:1771

He Y, Wu Z, Wang W, Liu X, Ye BC (2019) Bacterial community and phosphorus species changes in pepper rhizosphere soils after *Pseudomonas putida* Rs-198 inoculation. Rhizosphere 11:100164

Hernández-León R, Rojas-Solís D, Contreras-Pérez M, del Carmen Orozco-Mosqueda M, Macías-Rodríguez LI, Reyes-de la Cruz H, Valencia-Cantero E, Santoyo G (2015) Characterization of the antifungal and plant growth-promoting effects of diffusible and volatile organic compounds produced by *Pseudomonas fluorescens* strains. Biol Control 81:83–92

Heydarian Z, Yu M, Gruber M, Glick BR, Zhou R, Hegedus DD (2016) Inoculation of soil with plant growth promoting bacteria producing 1-aminocyclopropane-1-carboxylate deaminase or expression of the corresponding acdS gene in transgenic plants increases salinity tolerance in *Camelina sativa*. Front Microbiol 7:1966

Höflich G, Kühn G (1996) Förderung das Wachstums und der Nährstoffaufnahme bei kurziferen Öl-und Zwischenfruhten durch inokulierte Rhizospherenmikroorganismen. Zeischrift für Pflanzenernährung und Bodenkunde 159:575–578

Höflich G, Wiehe W, Kühn G (1994) Plant growth stimulation with symbiotic and associative rhizosphere microorganisms. Experientia 50:897–905

Howarth CJ (2005) Genetic improvements of tolerance to high temperature. In: Ashraf M, Harris PJC (eds) Abiotic stresses: plant resistance through breeding and molecular approaches. Howarth Press, New York, pp 277–300

Hu J, Wei Z, Weidner S, Friman VP, Xu YC, Shen QR, Jousset A (2017) Probiotic *Pseudomonas* communities enhance plant growth and nutrient assimilation via diversity-mediated ecosystem functioning. Soil Biol Biochem 113:122–129

Igiri BE, Okoduwa SIR, Idoko GO, Akabuogu EP, Adeyi AO, Ejiogu IK (2018) Toxicity and bioremediation of heavy metals contaminated ecosystem from tannery wastewater: a review. Hindawi J Toxicol 2018:2568038. https://doi.org/10.1155/2018/2568038

Igwe JC, Nnororm IC, Gbaruko BC (2005) Kinetics of radionuclides and heavy metals behavior in soils: implications for plant growth. Afr J Biotechnol 4(13)

Jain R, Pandey A (2016) A phenazine-1-carboxylic acid producing polyextremophilic *Pseudomonas chlororaphis* (MCC2693) strain, isolated from mountain ecosystem, possesses biocontrol and plant growth promotion abilities. Microbiol Res 190:63–71

Johnstone TC, Nolan EM (2015) Beyond iron: non-classical biological functions of bacterial siderophores. Dalton Trans 44(14):6320–6339

Kang BG, Kim WT, Yun HS, Chang SC (2010) Use of plant growth-promoting rhizobacteria to control stress responses of plant roots. Plant Biotechnol Rep 4(3):179–183

Kang SM, Khan AL, Waqas M, You YH, Kim JH, Kim JG, Hamayun M, Lee IJ (2014) Plant growth-promoting rhizobacteria reduce adverse effects of salinity and osmotic stress by regulating phytohormones and antioxidants in *Cucumis sativus*. J Plant Interact 9(1):673–682

Karimi K, Amini J, Harighi B, Bahramnejad B (2012) Evaluation of biocontrol potential of *Pseudomonas* and *Bacillus* spp. against Fusarium wilt of chickpea. Aust J Crop Sci 6(4):695

Kaur G, Reddy MS (2014) Influence of P-solubilizing bacteria on crop yield and soil fertility at multilocational sites. Eur J Soil Biol 61:35–40

Kausar R, Shahzad SM (2006) Effect of ACC-deaminase containing rhizobacteria on growth promotion of maize under salinity stress. J Agric Soc Sci 2(4):216–218

Ke X, Feng S, Wang J, Lu W, Zhang W, Chen M, Lin M (2019) Effect of inoculation with nitrogen-fixing bacterium *Pseudomonas stutzeri* A1501 on maize plant growth and the microbiome indigenous to the rhizosphere. Syst Appl Microbiol 42(2):248–260

Keyvan S (2010) The effects of drought stress on yield, relative water content, proline, soluble carbohydrates and chlorophyll of bread wheat cultivars. J Anim Plant Sci 8(3):1051–1060

Krishnakumar S, Bai VDM, Rajan RA (2014) Evaluation of phosphate solubilizing microorganisms (PSMs) from rhizosphere soil of different crop plants and its antagonistic activity. J Microbiol Biotech Food Sci 3(5): 412–415

Kuddus M, Joseph B, Ramteke PW (2013) Production of laccase from newly isolated *Pseudomonas putida* and its application in bioremediation of synthetic dyes and industrial effluents. Biocatal Agric Biotechnol 2:333–338

Lata R, Chowdhury S, Gond SK, White JF Jr (2018) Induction of abiotic stress tolerance in plants by endophytic microbes. Lett Appl Microbiol 66(4):268276

Li J, Yin LY, Jongsma MA, Wang CY (2011) Effects of light, hydropriming and abiotic stress on seed germination, and shoot and root growth of pyrethrum (*Tanacetum cinerariifolium*). Ind Crop Prod 34(3):1543–1549

Li J, Sun J, Yang Y, Guo S, Glick BR (2012) Identification of hypoxic-responsive proteins in cucumber roots using a proteomic approach. Plant Physiol Biochecm, 51; 74–80

Li Y, Zeng J, Wang S, Lin Q, Ruan D, Chi H, Zheng M, Chao Y, Qiu R, Yang Y (2020) Effects of cadmium-resistant plant growth-promoting rhizobacteria and *Funneliformis mosseae* on the cadmium tolerance of tomato (*Lycopersicon esculentum* L.). Int J Phytoremediation 22(5):451–458

Lu C, Yang Z, Liu J, Liao Q, Ling W, Waigi MG, Odinga ES (2020) Chlorpyrifos inhibits nitrogen fixation in rice-vegetated soil containing *Pseudomonas stutzeri* A1501. Chemosphere 2020:127098

Mabood F, Zhou X, Smith DL (2014) Microbial signaling and plant growth promotion. Can J Plant Sci 94(6):1051–1063

Mahajan SG, Nandre VS, Salunkhe RC, Shouche YS, Kulkarni MV (2020) Chemotaxis and physiological adaptation of an indigenous abiotic stress tolerant plant growth promoting Pseudomonas stutzeri: Amelioration of salt stress to *Cicer arietinum*. Biocatalysis Agric Biotechnol 101652

Mallick I, Bhattacharyya C, Mukherji S, Dey D, Sarkar SC, Mukhopadhyay UK, Ghosh A (2018) Effective rhizoinoculation and biofilm formation by arsenic immobilizing halophilic plant growth promoting bacteria (PGPB) isolated from mangrove rhizosphere: a step towards arsenic rhizoremediation. Sci Total Environ 610:1239–1250

Meliani A, Bensoltane A (2016) Biofilm-mediated heavy metals bioremediation in PGPR pseudomonas. J Bioremed Biodegr 7(370):2

Mia MB, Shamsuddin Z, Mahmood M (2010) Use of plant growth promoting bacteria in banana: a new insight for sustainable banana production. Int J Agric Biol 12(3):459–467

Mohanty SS, Jena HM (2019) Degradation kinetics and mechanistic study on herbicide bioremediation using hyper butachlor-tolerant *Pseudomonas putida* G3. Proc Saf Environ Protect 125:172–181

Moradi A, Tahmourespour A, Hoodaji M, Khors F (2011) Effect of salinity on free living-diazotroph and total bacterial populations of two saline soils. Afr J Microbiol Res 5(2):144–148

Munns R, Tester M (2008) Mechanisms of salinity tolerance. Annu Rev Plant Biol 59:651–681

Muñoz R, Alvarez MT, Muñoz A, Terrazas E, Guieysse B, Mattiasson B (2006) Sequential removal of heavy metals ions and organic pollutants using an algal-bacterial consortium. Chemosphere 63(6):903–911

Nadeem SM, Zahir ZA, Naveed M, Asghar HN, Arshad M (2010) Rhizobacteria capable of producing ACC-deaminase may mitigate salt stress in wheat. Soil Sci Soc Am J 74(2):533–542

Nadeem S, Imran M, Naveed M, Khan MY, Ahmad M, Zahir Z, Crowley D (2017) Synergistic use of biochar, compost and plant growth promoting rhizobacteria for enhancing cucumber growth under water deficit conditions. J Sci Food Agric 97. 10.1002/jsfa.8393.

Nautiyal CS, Srivastava S, Chauhan PS, Seem K, Mishra A, Sopory SK (2013) Plant growth-promoting bacteria *Bacillus amyloliquefaciens* NBRISN13 modulate gene expression profile of leaf and rhizosphere community in rice during salt stress. Plant Physiol Biochem 66:1–9

Niu X, Song L, Xiao Y, Ge W (2018) Drought-tolerant plant growth-promoting rhizobacteria associated with foxtail millet in a semi-arid agroecosystem and their potential in alleviating drought stress. Front Microbiol 8:2580

Nordstedt NP, Chapin LJ, Taylor CG, Jones ML (2020) Identification of *Pseudomonas* Spp. that increase ornamental crop quality during abiotic stress. Front Plant Sci 10:1754. https://doi.org/10.3389/fpls.2019.01754

Nowicki EM, O'Brien JP, Brodbelt JS, Trent MS (2015) Extracellular zinc induces phosphoethanolamine addition to *Pseudomonas aeruginosa* lipid a via the ColRS two-component system. Mol Microbiol 97(1):166–178

Oosten MJV, Pepe O, De Pascale S, Silletti S, Maggio A (2017) The role of biostimulants and bioeffectors as alleviators of abiotic stress in crop plants. Chem Biol Technol Agric 4:5. https://doi.org/10.1186/s40538-017-0089-5

Orozco-Mosqueda M, Duan J, DiBernardo M, Zetter E, Campos-García J, Glick BR, Santoyo G (2019) The production of ACC deaminase and trehalose by the plant growth promoting bacterium *Pseudomonas* sp. UW4 synergistically protect tomato plants against salt stress. Front Microbiol 10:1392

Oves M, Khan MS, Zaidi A (2013) Chromium reducing and plant growth promoting novel strain *Pseudomonas aeruginosa* OSG41 enhance chickpea growth in chromium amended soils. Eur J Soil Biol 56:72–83

Pandey P, Maheshwari DK (2007) Two-species microbial consortium for growth promotion of *Cajanus cajan*. Curr Sci:1137–1142

Panhwar QA, Othman R, Rahman ZA, Meon S, Ismail MR (2012) Isolation and characterization of phosphate-solubilizing bacteria from aerobic rice. Afr J Biotechnol 11(11):2711–2719

Parmar HY, Chakraborty H (2016) Effect of siderophore on plant growth promotion. Int J Appl Pure Sci Agric 2(3):60–68

Parmar N, Dadarwal KR (1999) Stimulation of nitrogen fixation and induction of flavonoid-like compounds by rhizobacteria. J Appl Microbiol 86(1):36–44

Pham V, Rediers H, Ghequire M, Nguyen H, Mot R, Vanderleyden J, Spaepen S (2017) The plant growth-promoting effect of the nitrogen-fixing endophyte Pseudomonas stutzeri A15. Arch Microbiol 199(3):513–517

Puig S, Ramos-Alonso L, Romero AM, Martínez-Pastor MT (2017) The elemental role of iron in DNA synthesis and repair. Metallomics 9:1483–1500

Rahneshan Z, Nasibi F, Moghadam AA (2018) Effects of salinity stress on some growth, physiological, biochemical parameters and nutrients in two pistachio (*Pistacia vera* L.) rootstocks. J Plant Interact 13(1):73–82

Rajendran K, Tester M, Roy SJ (2009) Quantifying the three main components of salinity tolerance in cereals. Plant Cell Environ 32(3):237–249

Ramadass K, Megharaj M, Venkateswarlu K, Naidu R (2018) Bioavailability of weathered hydrocarbons in engine oil-contaminated soil: impact of bioaugmentation mediated by pseudomonas spp. on bioremediation. Sci Total Environ 636:968–974

Ravanbakhsh M, Sasidharan R, Voesenek LA, Kowalchuk GA, Jousset A (2017) ACC deaminase-producing rhizosphere bacteria modulate plant responses to flooding. J Ecol 105(4):979–986

Ritchie SW, Nguyen HT, Holaday AS (1990) Leaf water content and gas-exchange parameters of two wheat genotypes differing in drought resistance. Crop Sci 30(1):105–111

Rizvi A, Khan MS (2017) Biotoxic impact of heavy metals on growth, oxidative stress and morphological changes in root structure of wheat (*Triticum aestivum* L.) and stress alleviation by *Pseudomonas aeruginosa* strain CPSB1. Chemosphere 185:942–952

Roberson EB, Firestone MK (1992) Relationship between desiccation and exopolysaccharide production in a soil pseudomonas sp. Appl Environ Microbiol 58(4):1284–1291

Ruelland E, Zachowski A (2010) How plants sense temperature. Environ Exp Bot 69(3):225–232

Saha M, Sarkar S, Sarkar B, Sharma BK, Bhattacharjee S, Tribedi P (2016) Microbial siderophores and their potential applications: a review. Environ Sci Pollut Res 23(5):3984–3999

Saikia J, Sarma RK, Dhandia R, Yadav A, Bharali R, Gupta VK, Saikia R (2018) Alleviation of drought stress in pulse crops with ACC deaminase producing rhizobacteria isolated from acidic soil of Northeast India. Sci Rep 8(1):1–16

Sairam RK, Kumutha D, Ezhilmathi K, Deshmukh PS, Srivastava GC (2008) Physiology and biochemistry of waterlogging tolerance in plants. Biol Planta 52:401–412

Salazar C, Hernández C, Pino MT (2015) Plant water stress: associations between ethylene and abscisic acid response. Chil J Agric Res 75:71–79

Saleem M, Arshad M, Hussain S, Bhatti AS (2007) Perspective of plant growth promoting rhizobacteria (PGPR) containing ACC deaminase in stress agriculture. J Ind Microbiol Biotechnol 34(10):635–648

Samaddar S, Chatterjee P, Choudhury AR, Ahmed S, Sa T (2019) Interactions between *Pseudomonas* spp. and their role in improving the red pepper plant growth under salinity stress. Microbiol Res 219:66–73

Sandhya V, Ali S, Grover M, Kishore N, Venkateswarlu B (2009a) *Pseudomonas* sp. strain P45 protects sunflowers seedlings from drought stress through improved soil structure. J Oilseed Res 26:600–601

Sandhya VZ, Ali S, Grover M, Reddy G, Venkateswarlu B (2009b) Alleviation of drought stress effects in sunflower seedlings by the exopolysaccharides producing *Pseudomonas putida* strain GAP-P45. Biol Fertil Soils 46(1):17–26

Santoro MV, Cappellari LDR, Giordano W, Banchio E (2015) Plant growth-promoting effects of native pseudomonas strains on Mentha piperita (*peppermint*): an in vitro study. Plant Biol 17(6):1218–1226

Sarabi B, Bolandnazar S, Ghaderi N, Ghashghaie J (2017) Genotypic differences in physiological and biochemical responses to salinity stress in melon (*Cucumis melo* L.) plants: prospects for selection of salt tolerant landraces. Plant Physiol Biochem 119:294–311

Saravanakumar D, Samiyappan R (2007) ACC deaminase from *Pseudomonas fluorescens* mediated saline resistance in groundnut (*Arachis hypogea*) plants. J Appl Microbiol 102(5):1283–1292

Sarkar A, Kumar P, Pramanik K, Mitra S, Soren T (2018) A halotolerant Enterobacter sp. displaying ACC deaminase activity promotes rice seedling growth under salt stress. Res Microbiol 169:20–32. https://doi.org/10.1016/j.resmic.2017.08.005

Sarma RK, Saikia R (2014) Alleviation of drought stress in mung bean by strain *Pseudomonas aeruginosa* GGRJ21. Plant and Soil 377(1–2):111–126

Sasirekha B, Srividya S (2016) Siderophore production by *Pseudomonas aeruginosa* FP6, a biocontrol strain for *Rhizoctonia solani* and *Colletotrichum gloeosporioides* causing diseases in chilli. Agric Nat Resour 50(4):250–256

Sayyed RZ, Ilyas N, Tabassum B, Hashem A, Abd-Allah EF, Jadhav HP (2019) Plausible role of plant growth-promoting rhizobacteria in future climatic scenario. Environmental biotechnology: for sustainable future. Springer, Singapore, pp 175–197

Schöffl F, Prandl R, Reindl A (1999) Molecular responses to heat stress. In: Shinozaki K, Yamaguchi-Shinozaki K (eds) Molecular responses to cold, drought, heat and salt stress in higher plants. RG Landes, Austin, TX, pp 81–98

Sharma P, Khanna P, Kumar PI (2013) Efficacy of aminocyclopropane-1-carboxylic acid (ACC)-deaminase-producing rhizobacteria in ameliorating water stress in chickpea under axenic conditions. Afr J Microbiol Res 7: 5749-5757

Sharma SB, Sayyed RZ, Trivedi MH, Gobi TA (2013) Phosphate solubilizing microbes: sustainable approach for managing phosphorus deficiency in agricultural soils. Springer Plus 2(1):587
Sheng XF, Xia JJ, Jiang CY, He LY, Qian M (2008) Characterization of heavy metal-resistant endophytic bacteria from rape (*Brassica napus*) roots and their potential in promoting the growth and lead accumulation of rape. Environ Pollut 156(3):1164–1170
Shilev S (2013) Soil rhizobacteria regulating the uptake of nutrients and undesirable elements by plants. In: Plant microbe symbiosis: fundamentals and advances. Springer, Berlin, pp 147–167
Siddikee MA, Sundaram S, Chandrasekaran M, Kim K, Selvakumar G, Sa T (2015) Halotolerant bacteria with ACC deaminase activity alleviate salt stress effect in canola seed germination. J Korean Soc Appl Biol Chem 58(2):237–241
Singh VK, Singh AK, Singh PP, Kumar A (2018) Interaction of plant growth promoting bacteria with tomato under abiotic stress: a review. Agric Ecosyst Environ 267:129–140
Sitaraman R (2015) *Pseudomonas* spp. as models for plant-microbe interactions. Front Plant Sci 6:787
Spaepen S, Vanderleyden J (2011) Auxin and plant-microbe interactions. Cold Spring Harb Perspect Biol 3(4):a001438
Srivastava S, Srivastava S (2020) Prescience of endogenous regulation in *Arabidopsis thaliana* by *Pseudomonas putida* MTCC 5279 under phosphate starved salinity stress condition. Sci Rep 10:5855. https://doi.org/10.1038/s41598-020-62725-1
Stearns JC, Glick BR (2003) Transgenic plants with altered ethylene biosynthesis or perception. Biotechnol Adv 21(3):193–210
Subramanian P, Mageswari A, Kim K, Lee Y, Sa T (2015) Psychrotolerant endophytic *Pseudomonas* sp. strains OB155 and OS261 induced chilling resistance in tomato plants (*Solanum lycopersicum* mill.) by activation of their antioxidant capacity. Mol Plant Microbe Interact 28(10):1073–1081
Subramanium N, Sundaram L (2020) Siderophore producing *Pseudomonas* spp. isolated from rhizospheric soil and enhancing iron content in *Arachis hypogaea* L. plant. Int J Agric Technol 16(2):429–442
Tanaka K, Cho SH, Lee H, Pham AQ, Batek JM, Cui S, Qiu J, Khan SM, Joshi T, Zhang ZJ, Xu D (2015) Effect of lipo-chito oligosaccharide on early growth of C4 grass seedlings. J Exp Bot 66(19):5727–5738
Tank N, Saraf M (2010) Salinity-resistant plant growth promoting rhizobacteria ameliorates sodium chloride stress on tomato plants. J Plant Interact 5(1):51–58
Tavakkoli E, Rengasamy P, McDonald GK (2010) High concentrations of Na^+ and Cl^- ions in soil solution have simultaneous detrimental effects on growth of faba bean under salinity stress. J Exp Bot 61(15):4449–4459
Tisdall JM, Oades J (1982) Organic matter and water-stable aggregates in soils. J Soil Sci 33(2):141–163
Vardharajula S, Ali SZ, Grover M, Reddy G, Bandi V (2011) Drought-tolerant plant growth promoting *Bacillus* spp.: effect on growth, osmolytes, and antioxidant status of maize under drought stress. J Plant Interact 6(1):1–14
Verma JP, Yadav J, Tiwari KN, Lavakush S, Singh V (2010) Impact of plant growth promoting rhizobacteria on crop production. Int J Agric Res 5(11):954–983
Vessey JK (2003) Plant growth promoting rhizobacteria as biofertilizers. Plant and Soil 255(2):571–586
Vijayaraghavan K, Yun YS (2008) Bacterial biosorbents and biosorption. Biotechnol Adv 26(3):266–291
Vurukonda SSKP, Vardharajula S, Shrivastava M, Sk ZA (2016) Enhancement of drought stress tolerance in crops by plant growth promoting rhizobacteria. Microbiol Res 184:13–24
Wang Q, Zhang WJ, He LY, Sheng XF (2018) Increased biomass and quality and reduced heavy metal accumulation of edible tissues of vegetables in the presence of cd-tolerant and immobilizing *Bacillus megaterium* H3. Ecotoxicol Environ Saf 148:269–274

Wheeler TR, Batts GR, Ellis RH, Hadley P, Morison JIL (1996) Growth and yield of winter wheat (*Triticum aestivum*) crops in response to CO_2 and temperature. J Agric Sci 127(1):37–48

Whipps JM (2001) Microbial interactions and biocontrol in the rhizosphere. J Exp Bot 52(1):487511

Wu Y, Ma L, Liu Q, Sikder MM, Vestergård M, Zhou K, Wang Q, Yang X, Feng Y (2020) *Pseudomonas fluorescens* promote photosynthesis, carbon fixation and cadmium phytoremediation of hyperaccumulator *Sedum alfredii.* Sci Total Environ 726:138554

Yadegari M, Rahmani HA, Noormohammadi G, Ayneband A (2010) Plant growth promoting rhizobacteria increase growth, yield and nitrogen fixation in *Phaseolus vulgaris.* J Plant Nutr 33(12):1733–1743

Yaish MW, Kumar PP (2015) Salt tolerance research in date palm tree (*Phoenix dactylifera* L.), past, present, and future perspectives. Front Plant Sci 6:348

Yaish MW, Al-Lawati A, Jana GA, Vishwas Patankar H, Glick BR (2016) Impact of soil salinity on the structure of the bacterial endophytic community identified from the roots of caliph medic (Medicago truncatula). PLoS One 11(7):e0159007

Yang Y, Singh RP, Song D, Chen Q, Zheng X, Zhang C, Zhang M, Li Y (2020) Synergistic effect of *Pseudomonas putida* II-2 and *Achromobacter* sp. QC36 for the effective biodegradation of the herbicide quinclorac. Ecotoxicol Environ Saf 188:109826

Yao L, Wu Z, Zheng Y, Kaleem I, Li C (2010) Growth promotion and protection against salt stress by *Pseudomonas putida* Rs-198 on cotton. Eur J Soil Biol 46(1):49–54

Yarzábal LA, Monserrate L, Buela L, Chica E (2018) Antarctic *Pseudomonas* spp. promote wheat germination and growth at low temperatures. Polar Biol 41(11):2343–2354

Yasmin H, Naeem S, Bakhtawar M, Jabeen Z, Nosheen A, Naz R, Keyani R, Mumtaz S, Hassan MN (2020) Halotolerant rhizobacteria *Pseudomonas pseudoalcaligenes* and *Bacillus subtilis* mediate systemic tolerance in hydroponically grown soybean (*Glycine max* L.) against salinity stress. PLoS One 15(4):e0231348

Younesi O, Moradi A (2014) Effects of plant growth-promoting rhizobacterium (PGPR) and arbuscular mycorrhizal fungus (AMF) on antioxidant enzyme activities in salt-stressed bean (*Phaseolus vulgaris* L.). Agric (Pol'nohospodárstvo) 60(1):10–21

Yuan Z, Yi H, Wang T, Zhang Y, Zhu X, Yao J (2017) Application of phosphate solubilizing bacteria in immobilization of Pb and Cd in soil. Environ Sci Pollut Res 24(27):21877–21884

Yun-xiu JI, Xiao-dong H (2007) Effects of plant growth-promoting rhizobacteria on the seedling growth of oat and annual ryegrass under salt stress. Int Conf Agric Eng:661–665

Zahir ZA, Munir A, Asghar HN, Shaharoona B, Arshad M (2008) Effectiveness of rhizobacteria containing ACC deaminase for growth promotion of peas (*Pisum sativum*) under drought conditions. J Microbiol Biotechnol 18(5):958–963

Zarei T, Moradi A, Kazemeini SA, Farajee H, Yadavi A (2019) Improving sweet corn (*Zea mays* L. var saccharata) growth and yield using *Pseudomonas fluorescens* inoculation under varied watering regimes. Agric Water Manag 226:105757

Chapter 8
Plant Growth-Promoting Rhizobacteria (PGPR) as Biocontrol Agents for Viral Protection

Abdul Basit, Syed Tanveer Shah, Sidra Tul Muntha, and Heba I. Mohamed

Contents

A. Basit (✉) · S. T. Shah
Department of Horticulture, Faculty of Crop Production Sciences, The University of Agriculture, Peshawar, Pakistan

S. T. Muntha
Laboratory of Germplasm Innovation and Molecular Breeding, Institute of Vegetable Sciences, Zhejiang University, Hangzhou, China

H. I. Mohamed
Department of Biological and Geological Sciences, Faculty of Education, Ain Shams University, Cairo, Egypt

© The Author(s), under exclusive license to Springer Nature Switzerland AG 2021
H. I. Mohamed et al. (eds.), *Plant Growth-Promoting Microbes for Sustainable Biotic and Abiotic Stress Management*,
https://doi.org/10.1007/978-3-030-66587-6_8

1 Introduction

Sustainable agriculture particularly ecological sustainability can be achieved without the use of chemical fertilizers that spoil fertility of the soil and biological diversity. There is a need for alternative measures to cope up with these problems using natural agents. Plants and microorganism interact with each other in the ecosystem. This interaction with a different degree is due to various attractions toward secretions of different metabolites which are generally referred to as root exudates. Bacteria usually live in a larger quantity in root influencing soil compared with the contiguous bulk soil, generally called the rhizosphere. Keeping in view the above information, the role of rhizosphere microbiology has great importance in sustainable agriculture as well as the world economy. In this context, there is a need to search beneficial microorganism (in which fungi, bacteria, *Actinobacteria*, and algae are very common). Due to the rapid growth and availability, there are up to one billion bacterial counts in each gram of soil. The plant-bacteria interaction may be neutral, injurious, or sometimes beneficial. These beneficial bacteria (usually associative, free-living, or symbiotic) usually make a symbiotic relationship with plant roots and also can bound themselves with the soil particles and soil aggregates and hence are rhizospheric (Maheshwari and Annapurna 2017) and make them fit. The metabolic processes of plant are interrupted due to any abnormality in form of either biotic or abiotic stress making the plant unhealthy. This disruption can be minimized by the use of crop rotation or organic composts (biological ways). PGPR (plant growth-promoting rhizobacteria) is another method to compete with stresses influencing plant growth and development (Maheshwari et al. 2019). PGPRs are generally defined as those bacteria living inside plant roots or their locale that promote the growth of plant (Kloepper et al. 2004b). The major source of microbial biocontrol agents is usually soil bacteria (mostly from *Bacillus*, *Pseudomonas*,

Agrobacterium, *Streptomyces* origin) (Hofte and Altier 2010). PGPR strains and their products not only promote growth but also hinder the attack of pathogens to plants (Beneduzi et al. 2012) and nowadays are excessively used in agricultural formulations (Mishra and Arora 2018).

Plant growth is affected by PGPR in two ways, i.e., indirect and direct. PGPRs directly provide various compounds to a bacterium that is synthesized by PGPR, thus enhancing directly plant growth, e.g., improvement in uptake of nutrients or provision of phytohormones, whereas during an indirect process, PGPR reduces the harmful effect of phytopathogens by producing an antagonistic substance that resists pathogen growth, thereby promoting plant growth (Glick 1995). Interaction of PGPR and plants is commercially used in sustainable agriculture, and their applications are studied in many horticultural and agronomic crops (Gray and Smith 2005).

There are different management strategies to control plant virus disease that include an amalgamation of selected cultural practices, using genetically resistant varieties, reducing insect vectors by applying insecticidal spray, and a combination of all these (Hull 1994). Cross-protection and developing genetically engineered plant are two additional methods to manage viruses (Denholm et al. 1996). The most environmentally sound and economical option to minimize viral diseases is by far the use of genetically resistant varieties though not always available. However, cross-protection is a successful method against virus-host systems but not possible in some crops (due to threat related to infectious agent inoculation). In case of non-availability of resistant varieties, genetically cultured crops can help against targeted viruses (Tricoli et al. 1995). The effective control of the infectious virus and its vector using insecticidal application depends on transmission mode and requires knowledge about the ecology of an area but has major environmental concerns. The effective and economic measure to manage viral diseases is systemic acquired resistance (SAR) (Ryals et al. 1994), a kind of plant natural defense mechanism. Biological and chemical agents are used against viral infections (Kessman et al. 1994). Nonpathogenic microorganisms are used to prompt plant defenses, generally referred to as induced systemic resistance (ISR) (Ryals et al. 1994). For example, attempts to induce ISR have been done by Mann (1965) and found a significant reduction of tobacco mosaic virus using *Bacillus uniflagellatus*. Tobacco necrosis virus was significantly controlled by a bacterium living in the root colonization, i.e., *Pseudomonas fluorescens* (Maurhofer et al. 1994). Furthermore, cucumber mosaic virus (CMV) was significantly reduced by using PGPR (Raupach et al. 1996).

The use of biocontrol agents (BCAs) and PGPR is well-thought-out to be the most effective method to safeguard plants against plant pathogens. Many PGPRs and BCA are extensively studied against different plant pathogens and showed promising results to promote plant growth (Glick et al. 2007) particularly against different viral diseases in various crops (Srinivasan and Mathivanan 2009).

There is an increasing demand of microorganism's mixture against plant viruses in recent times where PGPR strains (individually or in combination) against numerous phytopathogens for biological control (Srinivasan and Mathivanan 2009). PGPR and BCAs control diseases caused by viruses using ISR (induced systemic resistance) mechanism in plants that are investigated in the field as well as under

Table 8.1 Effects of PGPR on viral diseases of plants

PGPR	Viruses	Effects	References
Bacillus	Tobacco mosaic virus	Cultures and extracts from cultures reduced numbers of lesions from TMV	Mann (1965)
P. fluorescens CHAO	Tobacco necrosis virus	Reduction in TNV leaf necrosis in *P. fluorescens*-treated tobacco plants	Maurhofer et al. (1994)
P. fluorescens, *Serratia marcescens*	Cucumber mosaic virus	Treatment of cucumber or tomato plants with PGPR-induced systemic resistance against CMV	Raupach et al. (1996)
Bacillus amyloliquefaciens, *B. subtilis*, *B. pumilus*	Tomato mottle virus	Disease severity ratings were significantly less in all PGPR powder-based treatments	Murphy and Zehnder (2000)
Bacillus Amyloliquefaciens, *B. subtilis*, *B. pumilus*	Cucumber mosaic cucumovirus (CCMV)	PGPR-mediated ISR occurred against CCMV following mechanical inoculation on tomato	Zehnder et al. (2000)
Bacillus amyloliquefaciens	Pepper mild mottle virus (PMMoV)	*Bacillus*-induced systemic resistance against PMMoV in tobacco via salicylic acid- and jasmonic acid-dependent pathways	Ahn et al. (2002)

greenhouse (Murphy et al. 2003) especially bunchy top virus in banana (Harish et al. 2009), mottle and spotted viruses in tomato (Kandan et al. 2005), and mosaic virus in cucumber (Kloepper et al. 2004a) with certain exceptions (Ton et al. 2002). Table 8.1 represents the effect of PGPR on viral diseases of plants. This chapter aimed to understand the basic knowledge about plant growth-promoting rhizobacteria and its role as a biocontrol agent, to understand the role of PGPR in sustainable development especially in agriculture sector, to comprehend the interaction of PGPR with plants in inducing resistance against various plant pathogens, to study the role of PGPR in enhancing plant resistance against various viral diseases, and lastly, to elaborate the role of PGPR in disease management in various horticultural crops.

2 Plant Growth-Promoting Rhizobacteria (PGPR)

Bhattacharya and Jha (2012) defined PGPR as plant rhizoshperic colonizing bacteria that promote growth through either phosphate solubilization, fixation of nitrogen, sensing of quorums, or various other mechanisms. The application of PGPR started way back when Theophrastus (372–287 BC) added life to soil and defects were removed by mixing of different soils (Tisdale and Nelson 1975) which was technically proven when microscopy was introduced. The atmospheric nitrogen was converted to usable form by plants using soil bacteria and was confirmed during an investigation of root colonization in grasses (Hellriegel and Wilfarth 1888). Kloepper and Schroth (1978), while performing experiments on radishes, coined the term "rhizobacteria" for the first time, thus defined as bacterial community that competitively colonizes around plant roots that reduce plant diseases and promote growth. Some of the properties related to PGPR show their ability as biocontrol agent and stimulate plant growth (Vessey 2003). Rhizobacteria with plants have a negative, neutral, and positive type of relation and are further categorized as intracellular (iPGPR) or extracellular (ePGPR) PGPR depending upon the type of interaction (Martinez-Viveros et al. 2010). The ePGPR consists of different bacterial genera like *Azorhizobium*, *Mesorhizobium*, *Bradyrhizobium*, *Allorhizobium*, etc. and *Frankia* species mostly living in rhizoplane, cortex cells in root or rhizosphere (Verma et al. 2010). PGPRs promote yield and yield-related attributes of crop both directly and indirectly where colonization in the rhizosphere helps in hydrogen cyanide (Stutz et al. 1986), antibiotic production (Weller et al. 2002), and siderophore (Schippers et al. 1988). Figure 8.1 represents a bacterial community in rhizosphere showing their sole importance.

2.1 *Taxonomic Classification of PGPR*

Organisms are compared accurately through a reliable taxonomic system. However, during the last 30 years, many new characterization techniques are developed but the identification principle remains the same. The existing identification of bacterial strains is broadly classified into three categories which include (1) customary biochemical, physiological, and morphological attributes; (2) conventional biochemical tests in contracted form; and (3) genomic and chemotaxonomic characters. It has become clear that none of the phenotypic methods are appropriate to categorize all strains of bacteria. Hence, the need to study chemotaxonomic and nucleic acid analyses has emerged; however, it is difficult to study all species of bacterial strains in standard condition. Therefore, nowadays, polyphasic technique is becoming vital to classify bacterial strains. Polyphasic technique is generally defined as the reliable characterization of organisms by an amalgamation of phenotypic, genotypic, and chemotypic characters of microorganisms (Colwell 1970).

2.2 *Phenotypic Characters*

Phenotypic characters are comprised of physiological, biochemical, and morphological characteristics of microbes (de Vos et al. 2009). The investigations for conventional phenotypes are characterized by microscopic cell appearance, the

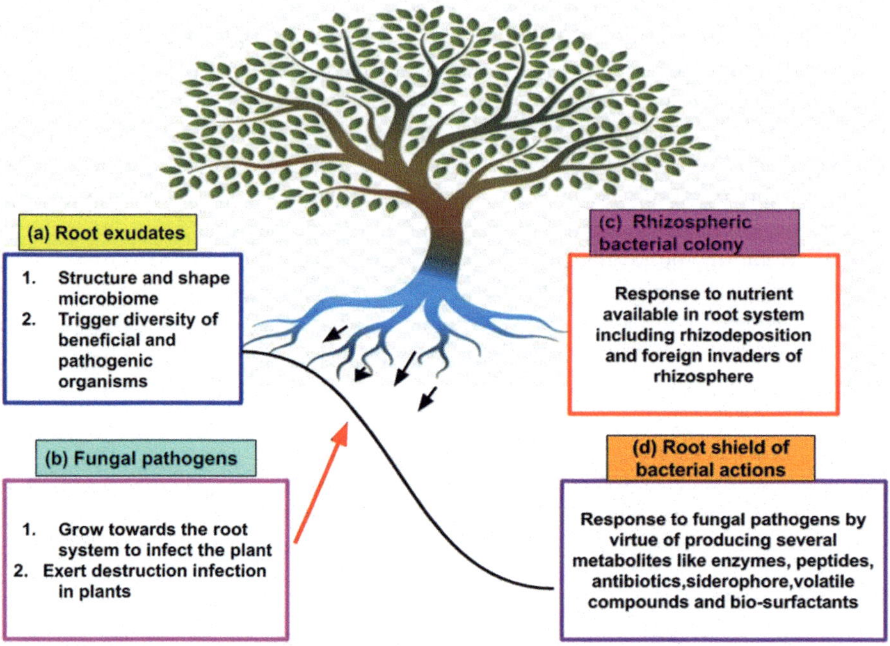

Fig. 8.1 Bacterial community in rhizosphere showing their sole importance from **a** to **d**

morphology of colony, growth substrate and salt conditions for microorganism growth, temperature, pH, vulnerability to various antimicrobial agents, etc. One of the vital diagnostic features is gram reaction if the analysis of cell wall composition is performed. Biochemical tests comprised of nitrogen metabolism, oxygen relationship, and fermentation reactions to identify different bacterial strains (Rodriguez-Diaz et al. 2008). However, the authenticity of the outcomes varies from laboratory to laboratory which is a big issue, and standardized procedure should be followed while experimenting. Restrictive nature of expressing genes (similar individuals express various phenotypic features under different environment) is another drawback of phenotypic techniques and hence should be compared with data set of closely rated organisms. A battery of dehydrated reagents is used in contracted form of conventional biochemical tests (for taxonomical studies) where the reaction is started by adding standardized inoculum. For identification of *Paenibacillus* (Seldin and Penido 1986) and *Bacillus* (Logan and Berkeley 1984), API 50CH (a phenotypic fingerprinting system) is used which contains 1 control and 49 carbohydrates, while API 20 NE are used to identify species of *Pseudomonas* (Barr et al. 1989). Later Miller and Rhoden (1991) used Biolog assay (very much less laborious) to identify bacterial strains which is very useful to identify PGPR strains that belong to *P. azotofixans* (Pires and Seldin 1997).

2.3 Chemotaxonomic Properties

FAME profiling, Fourier converted infrared spectroscopy, matrix-assisted laser desorption/ionization time-of-flight (MALDI-TOF) mass spectroscopy, PAGE analysis of protein of the whole cell, diamino acid component of a cell wall, polar lipid analysis, Raman spectroscopy, and pyrolysis mass spectroscopy are some of the chemotaxonomic fingerprinting methods to determine PGPR. Fatty acids (major constituents of lipopolysaccharides and lipids) are widely used for taxonomic purposes. Different chemotaxonomic fingerprinting techniques used for PGPR detection include FAME profiling, PAGE analysis of whole-cell proteins, polar lipid analysis, quinone content, cell wall diamino acid content, pyrolysis mass spectrometry, Fourier transform infrared spectroscopy, Raman spectroscopy, and matrixassisted laser desorption/ionization time-of-flight (MALDI-TOF) mass spectrometry. Fatty acids are the major constituents of lipids and lipopolysaccharides and have been used extensively for taxonomic purposes. FAME analysis is presently the only chemotaxonomic technique that is linked to a commercial database for identification purposes. Fatty acid profiles showing variability in chain length, double-bond position, and substituent groups are perfectly suitable for taxon description and also for comparative analyses of profiles that have been obtained under identical growth conditions (Suzuki et al. 1993). Sodium dodecyl sulfate-PAGE for proteins of the whole cell along with normalizations for comparison of data for computers and standard procedure analysis requires standardized growth condition making important inputs to polyphase taxonomic studies among the formers of aerobic endospore (Logan et al. 2009). Furthermore, polar lipids and quinones are now widely used to find unique genera of *Bacillaceae*. In the end, the whole bacterial cell chemical composition is being examined through UV resonance Raman spectroscopy, pyrolysis mass spectrometry, and Fourier transform infrared spectroscopy (Logan et al. 2009).

2.4 Genetic Techniques

Genetic techniques are determined by DNA or RNA molecules and made revolutionary contributions to identify the taxonomy of bacterial strains. Plasmid profiling, fragment length polymorphism (RFLP), pulsed-field gel electrophoresis (PFGE), randomly amplified polymorphic DNA (RAPD), and ribotyping amplified ribosomal DNA restriction analysis (ARDRA) are some of the methods to determine bacterial strains and their subspecies (Depret and Laguerre 2008). One or more of these methods are used to determine PGPRs (Monteiro et al. 2009). The technique in which DNA can be denatured at a very high temperature but by lowering the temperature molecules can be brought to its original state is known as reassociation of DNA-DNA technique, which is effective to compare two strains of bacteria with complete genome (Stackebrandt and Liesack 1993). Another method of taxonomic classification of bacterial species is DNA microarray. It uses fragments of DNA instead of the whole genomic DNA. A single microarray is used to hybridize different DNA fragments but is a very expansive method. On a revolutionary point of view, when gene sequences of rRNA molecules were discovered, it becomes very easy to compare evolutionary similarities among strains which are known as phylogenetic comparisons (Stackebrandt and Goebel 1994). Related bacterial species are then analyzed by constructing dendrograms or phylogenetic tree. The phylogenetic tree ascertains the genus to which the strain belongs, and its closest neighbors, i.e., those sharing the clade or showing >97 % 16S rRNA gene sequence similarity, are obtained from various culture collections to perform further genotypic, chemotaxonomic, and phenotypic analysis. At present, by correlation with experimental data obtained in the comparison of total genomic DNA (DNA–DNA hybridization), it is proposed that a similarity below 98.7–99 % on the 16S rRNA gene sequences of two bacterial strains is sufficient to consider them as belonging to different species. On the other hand, two strains showing similarities above the 98.7 % threshold may represent two different species. In these cases, total genome DNA–DNA hybridization must be performed, and those strains for which similarities are below 70 % are considered to belong to different species (Stackebrandt and Goebel 1994). In the end, housekeeping or other protein-coding genes (recA, ropB, gyrB) are of great importance for the taxonomic analysis of species. For instance, the basis for clustering arrangement of *Paenibacillus* is a sequential comparison of rpoB (Mota et al. 2005). Furthermore, Wang et al. (2007) and Cerritos et al. (2008) reported gyrB and recA sequential comparison for *B. Subtilis* and *Bacillus* species, respectively. Figure 8.1 represents bacterial community in rhizosphere showing their sole importance.

3 PGPR Interactions with Plants

3.1 Induced Resistance

A condition where plant develops a defensive strategy when needed against harmful organisms is called induced resistance (Van Loon et al. 1998). PGPR can reduce the activity of pathogenic microorganisms not only through microbial antagonism, but also by activating the plant to better defend itself. This phenomenon, termed 'induced systemic resistance' (ISR) was first described by Van Peer et al. (1991) and effective in many agronomic (tobacco and beans) as well as Horticultural crops (tomato, radish, cucumber and carnations etc.) and *Arabidopsis thaliana* (Van Loon et al. 1998). Phenotypically, rhizobacteria-induced ISR shows greater similarity with pathogen-mediated resistance. In pathogen-induced resistance, the infected plant part shows greater resistance, if the pathogen again attacked it which is generally called systemic acquired resistance (SAR) (Ross 1961). ISR is different from SAR from the fact that SAR is systemically prompted with necrotizing pathogen after inoculation while nonpathogenic rhizobacteria induce ISR. Moreover, signals of a salicylic acid molecule are required in SAR, while signals of salicylic acid are not necessary but instead require ethylene and jasmonic acid signals. Regarding differences in gene expression between the two resistance methods, SAR induces pathogenesis-related proteins, while ISR expresses PR set of genes, but both SAR and ISR are effective against broad-spectrum plant pathogens (Van Loon et al. 1998). *Arabidopsis thaliana* L. is a model plant that favors its use in PGPR studies, thus showing importance of SAR and ISR to test against fungal, viral, and bacterial pathogens (O'Callaghan et al. 2000). Using this model system, ISR is triggered in different plant species using *P. fluorescens* WCS417r as an inducing agent (a nonpathogenic rhizobacterial strains) (Bigirimana and Hofte 2002) and also its use against different plant pathogens in *Arabidopsis* root colonization (Ton et al. 2002).

Rhizobacteria suppress diseases in plants using ISR which is spatially separated from pathogens. ISR is dependent on combinations of rhizobacteria (Pieterse et al. 2002) or host suggesting that plant- and ISR-inducing rhizobacterium recognition is important in ISR induction. Iron-regulated siderophores, lipopolysaccharides, and outer membrane are some bacterial constituents used as ISR inducers (Van Loon et al. 1998). Various changes in plant roots to induce ISR are increased enzyme levels, i.e., polyphenol oxidase, peroxidase, and chitinase (Chen et al. 2000); expression of genes during stress condition (Timmusk and Wagner 1999); increased phytoalexin production (Ongena et al. 1999); strengthened cell wall of cortex and epidermis; and depositions of recently produced obstruction at infection sites (Benhamou et al. 2000), but not all these changes are induced in plant-bacteria interaction (Steijl et al. 1999). The consistency of biocontrol agents for disease protection is not enough to contest with disease control using traditional methods. However, efficiency of biocontrol agents is improved by using antagonistic organism with several mechanisms of action (De Boer et al. 1999). Furthermore, combined use of ISR and SAR offers great importance in agricultural practices and is helpful in an increased level of protection against specific bacterial pathogens (Van Wees et al. 2000). Finally, induced resistance is very useful against viral diseases.

3.2 Colonization in Roots

Colonization in roots is not only vital to apply microorganism for useful purpose but also the utmost step to manage the pathogenicity of soilborne microbes (Lugtenberg et al. 2001). PGPRs generally colonize in the root system and suppress lethal microbes in plant rhizosphere which ultimately improved plant growth (Schroth and Hancock 1982). Colonization is largely considered important for biocontrol (Parke 1991). The inability of PGPR to colonize them around plant roots limits their effectiveness (Benizri et al. 2001). Suppression of diseases and colonization or even primary population size of biocontrol agents are generally correlated with each other (Bull et al. 1991) where colonization ability is correlated with osmotolerance (Loper et al. 1985). Percolating water under field condition is an essential role to distribute bacteria passively on plant roots (Liddell and Parke 1989). Attachment of bacteria to roots is greatly affected by features of cell surface and secondary metabolite accumulation by mutants (Carroll et al. 1995) which include rapid growth rate, lipopolysaccharide properties, and phototrophy for amino acid and vitamin B1 (Lugtenberg et al. 1996), which is vital for colonization. When roots are under disease controlling conditions in a gnotobiotic system, both biocontrol agent and pathogen can be simultaneously visualized using autofluorescent proteins and confocal laser scanning microscopy which is helpful to understand biocontrol processes and root colonization (Benizri et al. 2001).

3.3 Genetic Variations in Host

The response and support of plant to beneficial microbes and biocontrol agents changes with plant species and cultivars (Handelsman and Stabb 1996) where some support and attract biocontrol agents that have an antagonistic relationship with pathogens. For example, isolates of *Bacillus* in wheat roots enhanced its growth (Chanway et al. 1988), while the response of *B. polymyxa* is better to legumes (Chanway et al. 1988). Induction of genes for biosynthesis of pyoluteorin in *P. fluorescens* varies with plant species (Kraus and Loper 1995) due to variation in root exudates. Furthermore, survival rate and disease incidence in the existence of biocontrol agent and pathogen also vary from specie to specie (King and Parke 1996). *P. fluorescens* can suppress disease more as compared with parent host-pathogen combinations and not others due to overproduction of 2, 4-diacetyl-phloroglucinol and pyoluteorin (Maurhofer et al. 1995).

4 Interaction of PGPR in the Rhizosphere

4.1 Interactions with the Microbial Community

There are variations for disease suppression by biocontrol agent in laboratory and field where these are effective to suppress disease in the laboratory but ineffective to control in the field. Indigenous soil microbial communities affect biocontrol agents. Biocontrol agents are further affected by the communities in which they were introduced. The harmful populations of microorganism are decreased due to the introduction of fluorescent pseudomonads which in some cases displace tenant microflora (Yuen and Schroth 1986). Soilborne pathogens are suppressed by the introduction and activity of PGPR population which can also be achieved by cultural practices (Kloepper et al. 1999), i.e., plant straw and organic manures (Siddiqui 2004), other integrated techniques of pest management, cropping system with antagonistic plants, and shills in microbial community structure (Kloepper et al. 1999).

4.2 Interactions of PGPR Strains

Biocontrol agents are generally used against one pathogen to control diseases in a plant (Wilson and Backman 1999) but due to the ineffectiveness of one agent against all pathogens that attack a host plant in soil environments show erratic performance. Therefore, a combination of different biocontrol agents may be effective to suppress diseases in plant. Similarly, two or more biocontrol agents in different taxonomical organism (need different conditions, i.e., moisture, temperature, pH, etc.) colonize more aggressively which enhances the efficiency of biocontrol agent and growth of plant. Biocontrol agents are not from a single organism having more populations but are from naturally arose mixture of biocontrol agents. The use of different strains of PGPR mixtures has a greater tendency against multiple cucumber pathogens (Raupach and Kloepper 1998). The targeted pathogens as well as PGPRs are sometimes suppressed due to the incompatibility of co-inoculants (Leeman et al. 1996). Therefore, for compatibility of co-inoculated microbes, successful strain mixture development is essential (De Boer et al. 1997). The cost required to make a product of a single strain of biocontrol agent is much less than the product produced by a mixture of biocontrol agents due to production and registration issues. But one has to keep in mind that a mixture of biocontrol agents is well needed due to well adaptation to environmental changes throughout the growing season and hence protects plant against a variety of pathogens. Improvement in genetic diversity, stability, and level of biocontrol agent over a variety of environmental conditions is one of the other advantages of using microorganism mixtures (Pierson and Weller 1994), i.e., bacteria and fungi in particular.

5 Role of Plant Growth-Promoting Rhizobacteria in Biological Control

Plant growth-promoting rhizobacteria are a group of rhizobacteria including rod-shaped Gram-negative bacteria but also a very lower percentage of cocci, rods, and pleomorphic and Gram-positive bacteria. *Bradyrhizobium japonicum* (Guerinot and Chelm 1984), *Rhizobium ciceri* (Nour et al. 1994), *Rhizobium galegae* (Lindstrom 1989), *Sinorhizobium fredii* (Chen et al. 1988), *Azorhizobium caulinodans* (Dreyfus et al. 1988), *Mesorhisobium chacoense* (Velazquez et al. 2001), *Rhizobium fredii* (Scholla and Elkan 1984), *Rhizobium gallicum*, *Rhizobium giardinii* (Amarger et al. 1997), *Sinorhizobium medicae* (Rome et al. 1996), *Mesorhizobium pluriforium* (de Lajudie et al. 1998b), *Rhizobium etli* (Segovia et al. 1993), *Allorhizobium undicola* (de Lajudie et al. 1998a), and *Sinorhizobium arboris* (Nick et al. 1999) are some of the examples included in this group. PGPR living in the rhizosphere has an instant effect on the root system of plant and serves to form a variety of active microbial community. Evolution of relationship between rhizospheric microbes and plants can be predicted due to the allocation of carbon (deposition of 20%) to roots

(Handelsman and Stabb 1996), thus leading to control the diseases in the rhizosphere. *Streptomyces* spp. control fungal root diseases and produce plant-promoting hormones and siderophores under in vitro condition hence considered as an antifungal biocontrol agent and colonizing bacteria in the rhizosphere (Miller et al. 1990). PGPRs especially *Pseudomonas*, *Bacillus*, *Azospirillum*, and *Azotobacter* promote root colonization, root development and emergence of seedling, seed germination, water utilization, mineral nutrition, and disease suppression and hence improve overall growth of plant (Nelson 2004). PGPR strains use different mechanisms, i.e., lowering ethylene levels; inducing pathogen resistance and nitrogen fixation; promoting mycorrhizal functioning and siderophore and phytohormone production; decreasing the toxicity of pollutants, etc., to promote growth of plants (Glick et al. 1999) directly or indirectly (Castro et al. 2009). The group of PGPR strains including *Bacillus subtilis* and *P. fluorescens* is usually considered the best for an indirect effect to promote growth (Damayanti et al. 2007). Therefore, PGPR based on the mode of action is divided into biofertilizers, biopesticide, and phytostimulator. Experimental verifications have shown that multiple mechanisms are used to improve plant growth which can be instantaneously triggered (Martinez-Viveros et al. 2010). PGPRs, with regular collaboration with microbial population (residing in the rhizosphere), affect the expression of each trait given above (Lugtenberg and Kamilova 2009). New insights for importance and biological control and genetic basis of the pathways are provided by biochemical and molecular approaches (Joshi and Bhatt 2011). A critical density of population of PGPR must be retained to be effective in the rhizosphere, although PGPR inoculation in plants may be improved for the time being.

Regarding the role of *Pseudomonas*, inoculations of canola seeds *P. putida* (GR122) significantly increase seedling's root length grown under sterile conditions (Lifshitz et al. 1987) which was confirmed by El-Khawas et al. (2000) using spectrophotometer. There are several bacterial strains (*Azospirillum*, *Klebsiella*, *Pseudomonas*, *Bacillus*, *Sarcina*, and *Azobacter*) which are known as auxin-producing strains, isolated from the rhizosphere of various crops (Forlani et al. 1995). Another biocontrol agent living in the rhizosphere microbiota is filamentous *Actinobacteria*. *Actinobacteria* are reported to produce cytokinin dihydrozeatin riboside which is important to synthesize cytokinin in pure culture (Garcia de Salamone et al. 2001). Due to the formation of endospores, resistance of environmental stresses is possible by a Gram-positive aerobic organism, *Bacillus*, which is also reported to promote plant growth (Kumar et al. 2011). In contrast to *Pseudomonas*, *Bacillus* is not rhizospheric competent, but some strains are reported to be rhizosphere competent (Kumar et al. 2011). *Bacillus cereus* and *P. fluorescens* are reported to produce the highest amount (8.3 and 4.4 mg L^{-1}) of auxin grown as a batch culture on a shaker. It is also reported that tryptophan and zinc is also considered to increase the biosynthesis of auxin using *P. fluorescens* (Nasr 2002). Probanza et al. (2002) reported promoting seedling growth of *P. pinea* using the genus *Bacillus* due to the production of gibberellin. Interaction of plants with beneficial soilborne microbes is essential to acquire nutrients and water in the plant rhizosphere (Ryan et al. 2009). Rhizobacteria are interdependent on other microbes as one microbe can convert plant exudates from one form to another which can be utilized by another microorganism. Therefore, the interaction of plant and microbes is dealt to a flexible and vibrant biological environment of the rhizosphere (Mayak et al. 2004), thereby extracting macro- and micronutrients that affect plant growth. PGPR can also produce different phytohormones like cytokinins, auxins, and gibberellins that can change the architecture of root and promote plant growth (Kloepper et al. 2007). Furthermore, root biomass, root surface area, and root hair number of tomato-inoculated PGPR are increased by IAA-mediated ethylene production (Ribaudo et al. 2006).

6 PGPR as Biocontrol Agents

Plant growth and development is badly affected by harmful soilborne pathogens. It is essential to find alternative ways to protect the plant from different diseases. PGPR is used to induce systemic resistance in plants against different pathogens (Wei et al. 1996) especially against bacterial, viral, fungal diseases,nematodes and insects (Maurhofer et al. 1998). Biocontrol using PGPRs produces bacterial metabolites (siderophores binding Fe) that minimize the density and activities of harmful pathogens in the rhizosphere of plants (Kloepper 1996). For example, the deleterious activities of *F. oxysporum* (causative agent of tomato wilt disease) can be minimized using *B. subtilis* (Ghonim 1999).

A resistance mechanism where nonpathogenic rhizobacteria suppress disease in plants is known as induced systemic resistance (ISR) (Van Loon et al. 1998) where plant excites an improved state of ability against pathogens. It is previously reported in carnation that was protected against *F. oxysporum* by *P. fluorescens* strain WCS417r (Van Peer et al. 1991) and rhizobacteria strain in cucumber plants against anthracnose (Wei et al. 1991). The resemblance of rhizobacteria-facilitated ISR is the same as that of pathogen-induced systemic acquired resistance (SAR) wherein in both the cases the uninfected plant part shows more resistance to the pathogen (Van Loon et al. 1998) including viruses, bacteria, fungi, nematodes, and insects (Pozo and Azcon-Aguilar 2007). The most studied rhizobacteria that induce ISR are *Bacillus* and *Pseudomonas* (Van Wees et al. 2008). The term ISR to describe nonpathogenic PGPR involvement to promote resistance regardless of the pathway used in the process was used by Vleesschauwer and Hofte (2009), while salicylic acid-induced resistance against localized infection was termed as SAR (Vleesschauwer and Hofte 2009). The signaling pathways for ISR and SAR are different, where SAR-induced pathway involves salicylic acid while ISR needs ethylene and jasmonic acid (Van Loon et al. 1998). Stronger protection is mediated by SAR as compared to ISR (Van Loon 2007), but the protection mediated by SAR and ISR when used together is more than those provided alone (Van Wees et al. 2000).

SAR gets activated when a tissue develops the symptoms of necrosis (Vleesschauwer and Hofte 2009), but it was reported that without the development of necrotic symptom it also triggered SAR as in case of *Arabidopsis thaliana*

(Mishina and Zeier 2007). Resistance to further attacks is prompted by SAR after the initial infection occurs. Pathogenesis-related proteins (PRs) which are a specific set of defense-related genes are then activated, and their accumulation is generally associated with SAR (Van Loon 2007). Some PRs are from 1,3-glucanases and chitinases which can hydrolyze cell wall of fungus (Van Loon et al. 1998). Those plants which are not capable to hoard SA can activate ISR. Hence, one can say that PRs are generally induced alongside SAR while ISR pathway does not use SA or activated PR genes (Pieterse et al. 1996).

An activator in the form of a regulatory protein NPR1 is required in SA signal transduction which is an essential part of SAR signaling pathway (Van Loon et al. 1998). NPR1 existed as a multimer in non-induced plants and is converted to monomeric form by SA induction (Verhagen et al. 2006). In this connection, *npr1* (*Arabidopsis* mutant) does not show SAR due to non-expression of PR genes by *npr1*. It is worth mentioning that ISR is not associated with SA and hence independent of PRs (Van Loon et al. 1998). NPR1 overexpression in A. thaliana was reported to result from an increase in the transcript levels of PR genes, hence proves that NPR1-dependent PR gene-mediated disease resistance (Friedrich et al., 2001). Most recently, overexpression of NPR1 was also revealed to confer disease resistance against a broad range of pathogens in different crops (Sundaresha et al., 2016). These results revealed that NPR1 is a potential candidategene for developing disease-resistant transgenic crops against multiple pathogens. An active sign of defense is the increased amount of jasmonic acid (JA) and ethylene (ET) by the infected plants (Mauch et al. 1994). ISR does not depend on the increasing production rate of JA and ET but the increased sensitivity of these hormones (Pieterse et al. 2001). During ISR development, *Arabidopsis* ET responds to mutant *etr1* and JA to mutant *jar*1. Resistance in tomato DC 3000 is promoted against *P. syringae* using methyl jasmonate (MeJA) and 1-aminocyclopropane 1-carboxylate (ACC) (an ethylene precursor) in SA non-accumulating NahG plants (Pieterse et al. 1998).

7 Role of Allelochemicals in PGPR and Their Mode of Action for Suppression of Phytopathogens

Some rhizobia act as biocontrol agents that refers to suppress one or more phytopathogens, while others directly affect plant growth by activating plant hormone production, nitrogen fixation, phosphorus solubilization, and siderophores. PGPRs improve the growth of plant indirectly through phytopathogen suppression while producing some secondary substances known as allelochemicals which prevent phytopathogenic growth. Antibiotics, siderophores, detoxification enzymes, lytic enzymes, and biocidal volatiles are the known allelochemicals which are produced by microbes in the soil. Siderophoreschelate available iron from the soil; antibiotics discourage bacte-rial colonization; lytic enzymes degrade many organic compoundsincluding chitin (fungal cell walls); detoxification enzymes pre-vent damage from pathogenic toxins. Production of volatiles suchas hydrogen cyanide, suppress the

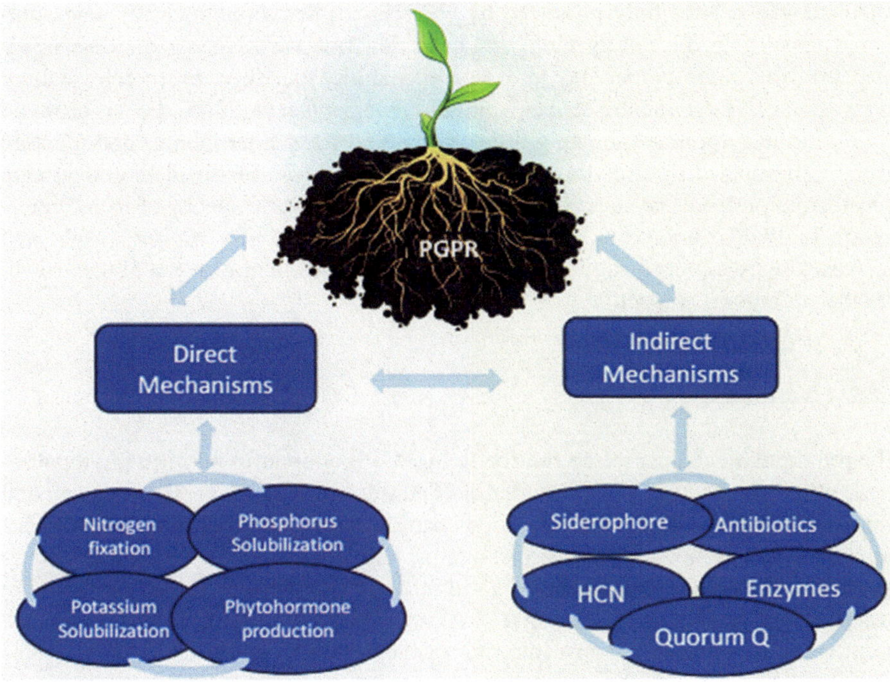

Fig. 8.2 Mechanisms of interaction between PGPR and phytopathogens for plant growth and development

growth of fungal pathogens the ability to successfully compete with pathogens for nutrientsor specific niches on the root; and the ability to induce systemicresistance (ISR) (Compant et al. 2005). Mechanisms are briefly explained in Fig. 8.2.

7.1 An Approach for Plant Pathogen Suppression by PGPRs

PGPRs may utilize indirect mechanisms to suppress deleterious effects of pathogens by production and effect of various allelochemicals, i.e., antibiotics, siderophores, degrading enzymes of a cell wall, hydrogen cyanide/volatile compounds, and multiple other indirect mechanisms like quorum quenching and competition between PGPRs and pathogens.

7.1.1 Siderophores

Fungi and bacteria produce compounds with low molecular weight, i.e., 500–1000 Da, which chelate with ions (Fe^{3+}) to facilitate its transport inside the cell (Ramadan et al. 2016). Based on their structure, siderophores are named as catecholate; the ones only produced by bacteria are known as hydroxymate compound which is produced by fungi and bacteria. Biocontrol agents synthesize siderophores in a quantity that may limit iron availability to pathogens, thereby leading to resistance of host counter to the pathogens (Meziane et al. 2005). Fe^{3+} is changed to Fe^{2+} in the cytoplasm, showing little attraction toward siderophores and released in the cell supporting growth of microbes. Such release involved the enzymatic transformation from the state of ferric ion to ferrous. The efficiency of iron sequestration in PGPRs may enhance plant growth by availability of iron inside and decrease the iron concentration in rhizosphere that leads to prevention of the growth of other microbes, respectively.

7.1.2 Antibiotics

The principal mechanism used indirectly by PGPB counter to harmful phytopathogenic effects leads to antibiotic synthesis (Olanrewaju et al. 2017; Raaijmakers and Mazzola 2012). However, an antibiotic is pathogen-specific; therefore, one particular antibiotic may not show any effect against any other phytopathogen of the same species, whereas PGPR-synthesized antibiotics may contribute altered responses at different field conditions (Glick 2007). Among all the PGPR strains, Bacillus and Pseudomonas are the two most important genera well studiedextensively for antibiosis mechanisms in the disease management practices (Jayaprakashvel and Mathivanan 2011). Antibiotics gene regulation studied according to three lev-els for synthesis: environmental sensing, global regulation thatties antibiotic production and regulatory loci linked to genes forpathway enzymes. Gene tends to be cluster and at least some reg-ulatory genes are linked. With discovery of phenazine derivatives (the first antibiotics) from *Pseudomonas* in 1983, a wave of studies started to the direction of bacteria as biocontrol agents. Since three decades, a variety of derivatives are extracted and being experimented in laboratories and in fields which serve as antibacterial, antihelminthic, antiviral, and antifungal agents. From *Bacillus* only, there are multiple derivatives, namely, sublancin, Tas A, bacilysin, subtilin, subtilosin, chlorotetain, surfactin, bacillaene, fengycin, and iturin, whereas the derivatives from pseudomonas are 2,4-diacetylphloroglucinol (DAPG), aerugine, butyrolactones, cepaciamide A, ecomycins, zwittermycin A, kanosamine, rhamnolipids, pyoluteorins, viscosinamide, and many others (Goswami et al. 2016).

7.1.3 Enzymatic Degradation by Cell Wall/Antifungal Mechanism

Several plants possess antifungal mechanisms by synthesizing enzymes to degrade fungal cell wall. These enzymes included protease, β-1,3-glucanase, lipase and chitinase degrading chitin, an essential part of the fungus cell wall (Husson et al. 2017), carbohydrate cell wall (Vaddepalli et al. 2017), cell wall proteins, and cell wall-associated lipid, respectively. Few PGPRs mimic the mechanism of plants against fungi to produce these antifungal enzymes of the cell wall (Chernin et al. 1995).

These biocontrol agents proved to be more effective when genetically engineered for genes encoding degrading enzymes in laboratories (Koby et al. 1994). The overexpressed chitinase gene mutant stains and co-transformed with *amds* gene showed an improved response against fungal protection (Kowsari et al. 2016). JS, a *Bacillus* sp., proved to affect the overexpression of PR-3 and PR-2 genes encoding chitinase and β-1,3-glucanase consecutively and persuaded plant ISR (Kim et al. 2015).

7.1.4 Competition Between PGPR and Pathogens

There is a general mechanism of competition between phytopathogens and PGPRs either for binding sites to roots of plants or nutrient uptake (Innerebner et al. 2011; Barahona et al. 2011). This kind of competition may delay the phytopathogenic proliferation by limiting the phytopathogenic binding to the plants. Since it is generally not possible to create PGPR mutants for binding to plant surface with comparable efficiency than phytopathogen, therefore the absolute manifestation of ability and function of biocontrol PGPR is rather missing. It is predominantly believed that other biocontrol mechanisms work parallel to competitiveness to inhibit the functioning of phytopathogens. The first-ever research study was carried out for nutriens competition in biological control of damping-off (*Pythium aphanidermatum*) during 1987 (Elad and Chet 1987).

7.1.5 Quorum Sustaining

Quorum sensing is a pathway of sensing the existence of similar bacteria. During proliferation, once bacterial cells have attained an optimum density, a particular set of genes turn on to alter cell metabolism (Cornforth et al. 2014; Zhang et al. 2019), and bacteria begin to synthesize and secrete out chemical molecules called autoinducers. As the cell density increases, the level of autoinducers also increases to threshold outside cell environment. The molecules bind to receptors of cells in the vicinity and trigger a signaling pathway that leads to population-wise altered gene expression and eventually unified growth of cells high virulence (Huang et al. 2016). Stimulating the quorum sensing at a specific point can inhibit the virulence of phytopathogen and prevent plant growth inhibition (Pei and Lamas-Samanamud 2014; Chan et al. 2011). An enzyme known as lactonase, produced by PGPR, lowers autoinducer which may be used as a quenching tool and pretreat seedling with such PGPR to activate inhibitory pathway against phytopathogens (Glick 2015a). However, it has been a successful strategy observed in lab and tested on eggplant, Chinese cabbage, and potatoes with mutant *aiiA* gene that hindered the soft rot symptoms in crops (Dong et al. 2000).

8 Spectrum of Protection by PGPR Against Virus Through ISR

Rhizosphere bacteria confer defense counter to several phytopathogens (Compant et al. 2010). Different studies reported the antagonistic of PGPR against many pathogens besides having the ability of growth promotion in plants. A wide-ranging action of PGPR can be more encouraging in the application as compared to those with antagonistic activity against individual or couple of pathogens (Zhang et al. 2017; Sun et al. 2017).

P. fluorescens exhibited resistance against a broad spectrum of viruses such as cucumber mosaic virus (CMV), bean yellow mosaic virus (BYMV), tobacco necrosis virus (TNV), tomato spotted wilt virus (TSWV) and tomato mosaic tobamovirus (ToMV), and Urdbean leaf crinkle virus (ULCV). Similarly *B. amyloliquefaciens* strain showed the effectiveness against a range of virus infections like potato virus Y (PVY), pepper mild mottle virus (PMMoV), cucumber mosaic virus (CMV), tobacco necrosis virus (TNV), tobacco mosaic virus (TMV), tomato chlorotic spot virus, (TCSV), tomato spotted wilt virus (TSWV), and tomato mottle virus (TMV). This shows that similar PGPR strain induces resistance counter to various pathogens of the same crop. On the other hand, biological control is obtained from the group of PGPRs in natural rhizosphere rather than from a population group of an individual strain (Zhang et al. 2017; Sun et al. 2017). The abovementioned PGPRs have been applied in a combination to fight against viral diseases and showed resistance against virulence in plants (Table 8.1).

9 Biochemical and Molecular Networks of PGPR in Useful Plant-Microbe Interactions

9.1 Enhancement of Plant Growth and Nutritional Acquisition by Bacteria

Bacterial-derived chemicals of rhizosphere, i.e., NOD factors of rhizobial (LCOs or lipochito oligosaccharides), were successfully merged into the products of crop protection, enhancing the growth of plant both in leguminous and nonleguminous crops, and also stimulate defense of plant (Subramanian et al. 2016). PGPRs is an uncherished resource for elicitors of new plant defense but could be highly effective as compared to the products derived synthetically for the protection of crops (Wiesel et al. 2014; Bektas and Eulgem 2014). Plants developed in a symbiotic relationship with microorganisms colonizing terrestrial ecologies successfully (Werner et al. 2014). Considering PGPRs occurred naturally, its interaction and useful features could substitute the agro-systematic design with reduced inputs of fertilizers and enhanced or sustained yields of plant. Broadly explored useful trait of plant is the biologically mutual symbiosis of nitrogen fixation by rhizobia (Udvardi and Poole 2013).

9.2 Nitrogen (N_2) Fixation

Nitrogen is considered as the essential nutrient for better growth and yield of plant (Imriz et al. 2014; Gupta et al. 2015). It is also known as the basic constituent of amino acids, membrane lipids, and nitrogen nucleotides (Marschner 1995). Fixed nitrogen use is called biological nitrogen fixation (BNF) count for two-thirds of the atmosphere (Shridhar 2012). Mostly studied PGPRs for nitrogen fixation are *Azoarcus* sp., *Rhizobium* sp., *Klebsiella pneumoniae*, *Beijerinckia* sp., and *Pantoea agglomerans* where these are applied in the form of mixtures (Damam et al. 2016). Biological nitrogen fixation generally occurs during the unreactive atmospheric nitrogen through nitrogenase enzymatic action by bacteria to NH_3 form utilized further by the plants (Bhattacharjee et al. 2008).

9.3 Solubilization of Phosphorus

Phosphorous plays a key role in the development of plant in all the metabolic processes, i.e., plant receptor signaling, energy transfer, photosynthesis, macromolecular biosynthesis, and respiration (Anand et al. 2016). Insoluble phosphorus solubilization is the key feature of PGPRs although they synthesize organic acids with low molecular weight (Sharma et al. 2013; Bahadur et al. 2018). Utmost usual genera of PGPR bacteria, i.e., *Bacillus megaterium*, *B. polymyxa*, *B. subtilis*, *B. circulans*, *B. firmus*, *Pseudomonas rathonia*, *P. striata*, *Rhizobium meliloti*, and *R. leguminosarum*, could have phosphorylated by organic acid production (Imriz et al. 2014).

9.4 Phytohormone Production

Plant hormones mainly consist of cytokinin (CK), gibberellic acid (GA), auxin (IAA), ethylene, and abscisic acid. These are considered as the significant factors in alleviation of plant stress (Bücker-Neto et al. 2017). About 80% of the PGPRs can synthesize IAA (Patten and Glick 2002) that plays a key role in formation of nodules and has a primary effect on cell differentiation and growth of plant roots (Gopalakrishnan et al. 2015). A research study on using IAA inoculation was carried out, in which the produced fungi *Massariosphaeria* sp. and Aspergillus sp. reported significant effects to the host plants *Aeschynomene fluminensis* and *Polygonum acuminatum* against antagonistic mercury effects (Pietro-Souza et al.

2017). Nowadays, strains of bacteria, i.e., *S. rochei* and *Streptomyces sundarbansensis*, have been reported playing a key role in the plant development and growth through biosynthesis of IAA (Han et al. 2018). The derivative compound of purine, cytokinin, also plays key role in cell differentiation and cell division. *Bacillus megaterium*, a strain of bacteria, is concerned with plant growth by cytokinin production (Numan et al. 2018).

10 Possible Influence of PGPR on Enhancing Resistance Counter to Viral Diseases

Biotic and abiotic stresses badly affect the quantity, quality, and parameters of plant growth grounded on plant stages, where average plant productivity may be minimized between 65 and 87% (Gursoy et al. 2012). Different viruses of plants have been reported globally and considered to be very important plant pathogens which are responsible for high economic losses of both quality and productivity of various crops (Balconi et al. 2012). On the other hand, fungi, bacteria, and pests, where not even a single control method functions directly, have developed until now to resist against viruses. Management of viral disease in plants specifically depends on host plant genetic resistance, their environment, and pesticide which has been produced synthetically for controlling vectors, which can be an essential viral management strategy. Pesticides which are used effectively against viral diseases of plants are available in abundance, but due to persevering of poisoning residues, they cannot be deliberated useful in sense of long-term solution as they risk the health of the environment. In contrast, some other diseases caused by viroids and viruses are having very few inexistent and ineffective solutions chemically (El-Dougdoug et al. 2012), hence the investigating for viral disease management of plants by natural defense induction of plants, for example, systematic acquired resistance (SAR). Currently, PGPR systematic resistance to plant viruses has been developed for viral disease management, despite the point that various ISR-based PGPR studies were carried out on several fungal and bacterial pathogens (Kloepper et al. 2004a, b), but the available studied reported very limited information on the management of viral diseases while using different beneficial microorganisms and PGPR specifically. Moreover, plant growth is stimulated by some PGPRs, which in turn results in a good hand economic revenue for the growers (Babalola 2010).

10.1 Induced Systematic Resistance of PGPR Against Viruses in Plant

Two different types of induced resistance of pathogen [induced systematic resistance (ISR) and systematic acquired resistance (SAR)] are well defined as "the physical and chemical stimulation of defense tool of host plants by the inducer which leads to various pathogens that have been controlled." The appearance of ISR in response to inoculation challenge of a pathogen is having similarity to that of SAR, because of reduced number of infected plants and infection severity. Often, this reduction is linked to invaded tissue reduction and pathogen growth reduction in tissues which are induced, reporting that plants can better resist to pathogens (Van Loon and Bakker 2005). PGPRs, called the microorganisms of bacteria which are saprophytic, live freely in the rhizosphere and aggressively inhabit the root system (Ramjegathesh et al. 2013). Beneficial effects to plants can be provided through supplementary secretions, i.e., hormones, vitamins, and other growth factors enhancing plant productivity and growth (Babalola 2010). Many bacteria species, i.e., *Agrobacterium, Aeromonas, Alcaligenes, Arthrobacter, Azoarcus, Azospirillum, Azotobacter, Burkholderia, Bacillus, Bradyrhizobium, Cyanobacteria* (mainly *Nostoc* and *Anabaena*), *Comamonas, Gluconacetobacter, Enterobacter, Herbaspirillum, Klebsiella, Pseudomonas, Paenibacillus, Rhizobium, Variovorax, Serratia, Xanthomonas*, and *Streptomyces*, are reported as PGPRs. Few of them like *Burkholderia, Azoarcus, Herbaspirillum*, and *Gluconacetobacter* are found in various tissues and cells of higher plants known as endophytes (Vessey 2003).

Salicylic acid (SA) is recognized as plant growth regulator which functions as a signaling molecule both in abiotic and biotic stresses (Basit et al. 2018), nearly in signal transduction intracellularly. SA helps in enhancing release of H_2O_2 and derived active oxygen of H_2O_2, and defense-related genes induced activities. Two different mechanisms after signalling pathways of SA i.e. a pathway of initiating bacterial and fungal resistance through genes expression of NPR1 and PR, while the other one triggers resistance against viral infection through substitutive oxidase. However, hormones mainly regulate defense in plants against viruses, depending on salicylic acid (SA) (Alazem and Lin 2015). In contrast, as there is a link between defense mediated by SA and antiviral mechanism of siRNA, SA could be considered as significant for local and systematic resistance; however, it contributes in elementary responses of immune system and resistance of R-gene (Beris et al. 2018).

Conversely, *Arabidopsis thaliana* endangered from cucumber mosaic virus (CMV) through a strain *Serratia marcescens* (90-166) by virus protection signaling passageway where JA could be found dependent and SA and NPR1 as independent (Ryu et al. 2004). Phaseolus vulgaris leaves treated with *Bacillus globisporus* and *Pseudomonas fluorescens* reported an enhanced activity of peroxidase and B-1,3-glucanase, individually in tobacco necrosis virus (TNV) and leaf sheaths inoculated from pathogens (Shoman et al. 2003).

PAL plays a key role in synthesis of various secondary metabolites, i.e., phenols and phenylpropanoids, and also lignin and derivatives of salicylic acid, which stimulates the triggered resistance of PGPR and imparts plant immunity (Li et al. 2016a). Since then, the accumulation of secondary metabolites is proposed to control the viruses' invasion. Thus, defense enzyme stimulation (PPO, PAL, and POD), as well as PR proteins by *P. fluorescens* strain application, can enhance plant resistance to a bunchy top virus of banana (BBTV) (Harish et al. 2008) and spotted wilt virus of tomto (TSWV). Similarly, in tobacco plants treated with *Bacillus* tested with spotted wilt virus of tomato, the amount of defense enzymes (PPO, PAL, and POD) and

Table 8.2 Plant protection for various horticultural crops against virus pathogens using PGPRs

Virus	Plants	Recent potential PGPR	References
Banana bunchy top virus	Banana	*P. fluorescens* (Pf1+CHA0)	Kavino et al. (2009)
		P. fluorescens (Pf1) + *Bacillus* (EPB22)	Harish et al. (2009)
Bitter gourd yellow mosaic virus	Bitter gourd	*P. chlororaphis*	Rajinimala et al. (2009)
Cucumber mosaic virus (CMV)	Cucumber	*Streptomyces griseorebens* Streptomyces cavourensis	Shafie et al. (2016)
	Tomato (*Solanum lycopersicon* L.)	*Stenotrophomonas rhizophila* + *P. aeruginosa*	Dashti et al. (2012)
	Pepper (*Capsicum annuum*)	*B. amyloliquefaciens* (5B6)	Lee and Ryu (2016)
	Tobacco	*Paenibacillus lentimorbus* (B-30488)	Kumar et al. (2016)
	Arabidopsis thaliana	*B. amyloliquefaciens* (IN937a) + *B. subtilis* (GB03)	Ryu et al. (2007b)
Tomato mosaic tobamovirus (ToMV)	Tobacco	*Streptomyces afghanensis*	Hussein (1992)
	Datura metel	*P. fluorescens* 2	Megahed et al. (2013)
Watermelon mosaic virus (WMV)	Pumpkin	*B. pumilus 293* (B2)	El-beshehy et al. (2015)
Cucumber green mottle mosaic virus (CGMMV)	Cucumber	*Stenotrophomonas maltophilia* (HW2)	Li et al. (2016b)
Tomato mottle virus (ToMoV)	Tomato	*B. amyloliquefaciens* (IN 937a)	Zehnder et al. (2001)
Pepper mild mottle virus	Pepper	*P. oleovorans* (KBPF-004)	Kim et al. (2017)
Tomato chlorotic spot virus	Tomato	*B. amyloliquefaciens* (IN937a) + *B .pumilus* (SE34) *B. pumilus* (SE34) + *B. amyloliquefaciens* (IN937a) + *B. sphaericus* (SE56)	Abdalla et al. (2017)
Potato virus Y (PVY)	Tomato	*B. amyloliquefaciens* (*MBI600*)	Beris et al. (2018)

proteins PR was observed to significantly increase as compared to untreated/control plants (Lian et al. 2011). Strains of *Bacillus cereus* (I-35), *Brevibacterium sanguinis* (I-16), and *Bacillus* sp. (I-6) enhanced the activity of POD in the plants of hot pepper after inoculation of TMV and showed that such rhizobacteria may improve the defense response of plants through the activity of POD (Shafie et al. 2016).

Cv. viceae of *Rhizobium leguminosarum* (composition of ARC-202 and ICARDA-441) carried systematic resistance counter to yellow mosaic virus of bean (BYMV) through treatment of seeds, where the improved levels of polyphenol oxidase, peroxidase, free proline, and total phenols were reported in bean faba.

11 Disease Management by PGPR in Horticultural Crops

PGPRs are known to be particular strains of root inhibiting bacteria which provoke the improved rate of growth in plants; induced systematic resistance counter to insects, pests, and diseases; and suppressed soil pathogens (Zebelo et al. 2016). The utmost common bacteria which are entomopathogenic are *Xenorhabdus*/*Photorhabdus* and *Bacillus thuringiensis* species. These have been developed alternate to chemical pesticides and are effective in insect pest control and are host-specific (Kupferschmied et al. 2013). Thus, the ability of PGPR to protect plants against pathogens and pests makes them a significant resource for the development of biocontrol agents against plant viruses by induced the ISR intermediated studies of the significance of the PGPR against viruses in various plant species have acquired importance across the defense spectrum of PGPR viruses. Various PGPR species have been used as microbial inoculants for plant protection against viral infections and to enhance the yield of crops as mentioned in Table 8.2.

Strategies for disease management of plant viruses usually include particular cultural practice incorporation and insecticide application for controlling insects and varieties which are genetically resistant. The efficiency of reduction in viral infection through vector control by application of insecticides depends on the transmission method. Moreover, viral disease management can be carried out through initiation of natural defenses of plants, i.e., SAR. In the majority of the cases, the agents used biologically include fungi, phytopathogenic bacteria, and viruses.

11.1 Bunchy Top Virus of Banana (BBTV)

BBTV, *Babuvirus* genus, is a member of family *Nanoviridae* which causes the disease of bunchy top, a serious viral disease of banana (*Musa* spp.; family, Musaceae). It is usually transferred in a non-replicated and determined circulative manner by an aphid vector called *Pentalonia nigronervosa* Coq. This disease is difficult to be reduced easily; however, a way to disease management of bunchy top is the use of ISR in in vitro micro-propagated banana virus-free plantlets with endophytic bacteria (PGPE) and rhizobacteria (PGPR) strengthening the banana plants against infection of viruses (Harish et al. 2008). Effectively used mixture of useful microbes (CHA0+PF1+EPB22) reduced the occurrence of disease, BBTV, in the plantlets of banana produced through tissue culture method (Kavino et al. 2007a). However, Harish et al. (2009) reported that in field conditions and inside a greenhouse, application of (EPB22+Pf1) combination has increased crop yield while reducing the incidence of BBTV.

11.2 Tomato Chlorotic Spot Virus (TCSV)

TCSV (genus: *Tospovirus*), belonging to family *Bunyaviridae*, is usually transferred through Thysanoptera thrips in a persistent and propagative way; however, the high effective TSCV vectors are *Frankliniella schultzei* in its dark form tailed by *Frankliniella occidentalis* form (Martínez et al. 2018). TCSV causes different infections interlinked to heavy losses of quality and yield in crops of both agricultural and ornamental (Polston et al. 2013). Severe effect of TCSV disease can be reduced by strain IN937a, *B. amyloliquefaciens*, to nearly 50% in comparison with control and untreated infected crop (Abdalla et al. 2017).

11.3 Tomato Mosaic Tobamovirus (ToMV)

ToMV belongs to family Virgaviridae and genus *Tobamovirus*, which could be found in tomato crops globally. It may be either seed-borne or transmitted mechanically through contaminated tools and by workers while propagating plants (Adams et al. 2012). *Streptomyces afghanensis*, an Egyptian isolate, is having a strong antiviral influence against ToMV inhibiting an indigenous wound on *N. tabacum* L (cultivar: White Burley). Culture filtrate, cell and liquid culture of *B. circulans* and *P. fluorescens* 2, which individually communicated ISR on tomato crops reduces the symptoms of ToMV and local wounds of ToMV formed on *Datura metel* as a host indicator of ToMV (Megahed et al. 2013).

11.4 Tomato Mottle Virus (ToMoV)

ToMoV, genus *Begomovirus*, belongs to family *Geminiviridae* and is usually transmitted by *Bemisia tabaci* (*Gennadius*), B biotype whitefly, and was noted to be a primarily limiting agent to yield of tomato crop. Management of ToMoV was restricted because of its vector's ability to acquire unavailability of genetically resistant tomatoes and resistance of insecticide. As compared to control plants, the plants treated with strains of *Bacillus subtilis* (IN937b) and *B. amyloliquefaciens* (IN937a) have lower amount of severe ToMoV disease ratings (Zehnder et al. 2001). Consequently, in all treatments which are powder-based, the analysis of Southern blot reported tomato plants infected with lower ToMoV in percentage as compared to plants in control treatment or going through seed treatment singly (Murphy et al. 2000).

11.5 Tomato Spotted Wilt Virus (TSWV)

TSWV, a *Tospovirus* genus, is known to be a member of family *Bunyaviridae*, which ranks among the most economically significant top 10 viruses of plants globally (Naidu et al. 2008). Commonly, it is transmitted by *Frankliniella occidentalis*

(Pergande), a western flower thrip, and also through various other thrip species causing damages to crops grown either in an open field or greenhouses in all kind of climatic zones. Handling of seedlings, seeds, and foliar and soil application treatment with different strains of *P. fluorescens* (CoT-1, CoP-1, and CHA0) either singly or in the form of mixtures have instigated ISR against infection of TSWW in tomato crop both in the field and in greenhouse. It was also reported that mixture of CoP-1+CoT-1+CHA0 has decreased infection by a maximum of 84% while CHA0 strain alone or in combination with C0T-1 strain by 80% as compared to control plants of tomato (untreated). *P. fluorescens* strains which were treated with the plants of tomato reported an improved promotion of growth as compared to the untreated or control plants, in both of greenhouse and field conditions (Kandan et al. 2002). Beris et al. (2018) studied as well that soil amended, drench, or foliar application of strain of *Bacillus amyloliquefaciens* (MB1600) has minimized the TSW virus occurrence at a maximum of 80%.

11.6 Potato Virus X (PVX)

PVX, genus *Potexvirus*, belongs to *Alphaflexiviridae* family and is a well-known potato crop pathogen globally. It causes an infection widely in potato plant host, specifically in family *Solanaceae* (Aboul-Ata et al. 2011; King et al. 2011). Both the concerted metabolites and their acetone source which is an Egyptian isolate (*Streptomyces afghanensis*) inhibited the local wound development triggered by PVX disease cv. whitefly, *N. tabacum*. However, seed treatment of potato crop with *Bacillus vallismortis* (strain: EXTN-1) reported the yield increase compared to control plants (untreated) at 45% of maximum (Park et al. 2006).

11.7 Potato Virus Y (PVY)

PVY, genus *Potyvirus*, fits in the family *Potyviridae*. Potato virus Y is a severe pathogen which causes infection to various significant crop species of *Solanaceae* family (nightshade), specifically in tomato, potato, pepper, and tobacco, and results in a high yield loss and degradation in quality (El-Dougdoug et al. 2014). Treatment of potato plants with a strain of *Bacillus vallismortis* (EXTN-1) improved yield at a maximum up to 45%, and its chlorophyll content also improved as compared to control plants (untreated) while guarding potato crop against the potato virus Y. Accretion of potato virus Y has decreased with the application of strain *B. amyloliquefaciens* (MB1600) at the time of very early infection and late PVY detection in apical leaves of the plants Beris et al. 2018).

11.8 Pepper Mild Mottle Virus (PMMoV)

PMMoV, genus *Tobamovirus*, a well-known member of the family *Virgaviridae*, is an important pepper crop pathogen with a 100% infection on the field. This virus may easily spread mechanically either through seed coats infected with PMMoV or by grafting methods (Svoboda et al. 2006). The activity of *Pseudomonas oleovorans*, strain KBPF-004, counter to pepper mild mottle virus was observed by Kim et al. (2017) parallel to control/untreated strain, ATCC-8062. The PMMoV-infected seeds of pepper harvested from the infected plants were left for a supernatant culture of free cells of every single strain before the plantation as compared to control PMMoV-infected seeds left untreated. It was observed that KBPF-004 strain has reduced the transmission rate of PMMoV seeds up to a maximum of 15.5% as compared to ATCC-8052 strain, which was found to be 61.9%.

11.9 Bitter Gourd Yellow Mosaic Virus (BGYMV)

Momordica charantia L., locally named as bitter gourd, is a well-known old species of Cucurbitaceae and is native to Africa and tropical Asia, far reached to Malaysia, tropical Africa, India, and China (Behera et al. 2010). BGYMV, being a *Geminiviridae* family member (*Begomovirus* genus), is a disease-causing virus causing an effect on the loss of yield due to vector whitefly, *Bemisia tabaci*. To control the disease, initiation of systematic infection resistance method in the bitter gourd crop counter to BGYMV is used (Rajinimala et al. 2009). The infection rate has reliably decreased at 45 days after planting DAS by seed treatment with *P. fluorescens* and *P. chlororaphis*, and further, it is essentially diminished at 75 DAS contrasted with immunized control plants (untreated) (Rajinimala et al. 2009).

11.10 Watermelon Mosaic Virus (WMV)

WMV, genus *Potyvirus*, belonging to family *Potyviridae*, is the virus that is distributed mostly in Mediterranean and temperate regions with a wide range of host than many potyviruses, causing heavy yield loss in all kind of cucurbits (Moradi 2011). For the management of mosaic virus disease of watermelon plant, a tactical approach relays on insecticidal use for controlling its vectors, i.e., aphids and whiteflies, and the parallel protection of hereditarily produced plants (El-beshehy et al. 2015). In addition, systemic resistance to *Watermelon virus* (WMV) in the cultivation of pumpkin is triggered by application in the soil of two distinct PGPR strains, i.e strain B1: *B. subtilis* and strain B2: *B. pumilus*, either in combination or alone. It was noted that strain B2 inhibited the disease up to 77.7% approximately which was found significantly higher as compared to the alone effect of strain B1 or the combination B1+B2 suppressing infection up to 33.3% and 66.6%, consecutively.

12 Challenges and Future Scenarios in the Selection and Classification of PGPR Strains

In 1958 the Soviet Union pioneered the method of applying rhizobacteria in soil and removing bacterial and fungal pathogens from plant parts although the determination of effective PGPR strains during that time was complicated. Choosing the proper strain is basic, so the most valuable microscopic organisms are screened for the experiment to be successful. Effective methodologies should be considered for this reason. For the number of roots colonizing microorganism and testing their effect on plant growth improvement, the procedure can be selected for PGPR strain. With the selected strain, the plant parts can be then treated to observe the impact. Recently, for the selection/determination of effective PGPR strains, mass screening method has been used (Compant et al. 2005). The specificity of adaptation of host plant in specific soil and climatic condition or microorganism is ought to be considered in choosing the segregation conditions and screening tests (Bowen and Rovira 1999).

Various methodologies can be chosen dependent on characteristics like antimicrobial siderophores and root colonization production. The determination of superior strains can be encouraged by the improvement of high-throughput examine frameworks and amazing bioassays (Spadden et al. 2002). For horticulture, PGPR is presently considered as a protected method because of expanding yield as it holds a promising arrangement in being good for nature. To protect plants from chemicals that are used to kill pests and effect, the biological system is most significant. By controlling the plant and pest diseases, PGPRs likewise affect yield due to which one-third of plant losses. PGPR appears to valuably influence on the laboratory similarly in the greenhouse experiment. A developing field to improve and investigate the PGPR strain is by a hereditary building which empowers to overexpress the characteristics so strains with required characters are acquired. Other than all the progress, there are ecological obstructions and unfavorable conditions that impact the action of PGPR. The issues of differing adequacy can be achieved by strain blending, improved immunization methods, or gene transfer of active genetic source of antagonists to the host plant. Different conditions can likewise influence PGPR as biocontrol because biocontrol agents need an indistinguishable environmental niche for existence and development. Subsequently, under different natural conditions, the effect of biocontrol agent could be enhanced through the use of compatible blended inoculum of biocontrol agent which could have a reliable achievement (Guetsky et al. 2001).

Other than being advantageous, there are a few difficulties faced by PGPR. The normal variety is a problem since it is hard to estimate how microscopic organisms will act in the lab and when put in the field. These varieties can be abrupt and influence the entire examination. Another test is that under field conditions, PGPR should be proliferated to recover their practicality and natural activity. This propagation can be as indicated by the plant type and could be occasional. The test could be as far as a working place that ought to be profoundly clean and fitting devices ought to be utilized because isolating and characterizing PGPR in vitro seems not to be easy.

13 Conclusions

PGPR is an effective growth-promoting bacterium of plants colonizing in roots and enhances the induced and systematic resistance and genetic diversity either directly or indirectly. PGPR is a diverse study; thus, rhizobacteria can be taken into account as a biocontrol for plant growth. Plant growth can be effective either directly while synthesizing a compound by bacterium, i.e., aiding various nutrient uptake from the surrounding environment or plant hormones, or indirectly, reducing the harmful effect of plant pathogens by PGPR. Two different types of pathogens induced resistance (induced and acquired systematic resistance (ISR, SAR)), defined as "physical or chemical stimulation of host plant defense mechanism by the inducer before various pathogens which have been controlled." Two different mechanisms after signaling pathways of SA i.e. a pathway of initiating bacterial and fungal resistance through genes expression of NPR1 and PR, while the other one triggers resistance against viral infection through substitutive oxidase. Various PGPR species have been used as microbial inoculant for plant protection against viral infections and to enhance the yield of various crops. The efficiency of reducing viral infection through vector control by application of insecticides depends on the transmission method. Moreover, viral disease management can be carried out through the initiation of natural defenses of plants, i.e., SAR. In the majority of the cases, the agents used biologically include fungi, phytopathogenic bacteria, and viruses.

References

Abdalla OA, Bibi S, Zhang S (2017) Application of plant growth-promoting rhizobacteria to control Papaya ringspot virus and Tomato chlorotic spot virus. Arch Phytopathol Plant Prot 50:584–597

Aboul-Ata A-AE et al (2011) Diagnosis and control of cereal viruses in the Middle East. Adv Virus Res 81:33–61

Adams MJ, Heinze C, Jackson AO, Kreuze JF, Macfarlane SA, Torrance L (2012) Tobamovirus. In: King AMQ, Adams MJ, Carstens EB, Lefkowitz EJ (eds) Virus taxonomy: ninth report of the international committee on taxonomy of viruses. Elsevier/Academic Press, London, pp 1153–1156

Ahn IP, Park K, Kim CH (2002) Rhizobacteria-induced resistance perturbs viral disease progress and triggers defense related gene expression. Mol Cells 13:302–308

Alazem M, Lin N-S (2015) Roles of plant hormones in the regulation of host-virus interactions. Mol Plant Pathol 16:529–540

Amarger N, Macheret V, Laguerre G (1997) Rhizobium gallicum sp. nov. and Rhizobium giardinii sp. nov., from Phaseolus vulgaris nodules. Int J Syst Bacteriol 47(4):996–1006

Anand K, Kumari B, Mallick MA (2016) Phosphate solubilizing microbes: an effective and alternative approach as bio-fertilizers. Int J Pharm Sci 8:37–40

Babalola OO (2010) Beneficial bacteria of agricultural importance. Biotechnol Lett 32:1559–1570

Bahadur PS, Liaqat F, Eltem R (2018) Plant growth pFromoting properties of phosphate solubilizing Bacillus species isolated from the Aegean Region of Turkey. Turk J Bot 42:183–196

Balconi C, Stevanato P, Motto M, Biancardi E (2012) Breeding for biotic stress resistance/ tolerance in plants. In: Ashraf M, Öztürk M, Ahmad M, Aksoy A (eds) Crop production for agricultural improvement. Springer, Dordrecht, pp 57–114. https://doi.org/10.1007/978-94-007-4116-4_4

Barahona E, Navazo A, Martínez-Granero F, Zea-Bonilla T, PérezJiménez RM, Martín M, Rivilla R (2011) *Pseudomonas fluorescens* F113 mutant with enhanced competitive colonization ability and improved biocontrol activity against fungal root pathogens. Appl Environ Microbiol 77:5412–5419

Barr JG, Emmerson AM, Hogg GM, Smyth E (1989) API-20NE and sensititre autoidentification systems for identifying Pseudomonas spp. J Clin Pathol 42:1113–1114

Basit A, Shah K, Rahman MU, Xing L, Zuo X, Han M, Alam N, Khan F, Ahmed I, Khalid MA (2018) Salicylic acid an emerging growth and flower inducing hormone in marigold (Tagetes sp. L.). Pure Appl Biol 7(4):1301–1308. https://doi.org/10.19045/bspab.2018.700151

Behera TK, Behera S, Bharathi LK, John KJ, Simon PW, Staub JE (2010) Bitter gourd: botany, horticulture, breeding. In: Janick J (ed) Horticultural reviews, vol 37. https://doi.org/10.1002/9780470543672.ch2

Bektas Y, Eulgem T (2014) Synthetic plant defense elicitors. Front Plant Sci 5:804

Beneduzi A, Ambrosini A, Passaglia LMP (2012) Plant growth-promoting rhizobacteria (PGPR): their potential as antagonists and biocontrol agents. Genet Mol Biol 35:1044–1051

Benhamou N, Gagne S, Quere DL, Dehbi L (2000) Bacterial-mediated induced resistance in cucumber: Beneficial effect of the endophytic bacterium *Serratia plymuthica* on the protection against infection by *Pythium ultimum*. Phytopathology 90:45–56

Benizri E, Baudoin E, Guckert A (2001) Root colonization by inoculated plant growth promoting rhizobacteria. Biocontr Sci Technol 11:557–574

Beris D, Theologidis I, Skandalis N, Vassilakos N (2018) Bacillus amyloliquefaciens strain MBI600 induces salicylic acid dependent resistance in tomato plants against Tomato spotted wilt virus and Potato virus Y. Sci Rep 8:10320. https://doi.org/10.1038/s41598-018-28677-3

Bhattacharjee RB, Singh A, Mukhopadhyay SN (2008) Use of nitrogen-fixing bacteria as biofertiliser for non-legumes: prospects and challenges. Appl Microbiol Biotechnol 80:199–209

Bhattacharya PN, Jha DK (2012) Plant growth-promoting rhizobacteria (PGPR): emergence in agriculture. World J Microbiol Biotechnol 28:1327–1350

Bigirimana J, Hofte M (2002) Induction of systemic resistance to *Colletotrichum lindemuthianum* in bean by a benzothiadiazole derivative and rhizobacteria. Phytoparasitica 30:159–168

Bowen GD, Rovira AD (1999) The rhizosphere and its management to improve plant growth. Adv Agron 66:1–102

Bücker-Neto L, Paiva AL, Machado RD, Arenhart RA, Margis-Pinheiro M (2017) Interactions between plant hormones and heavy metals responses. Genet Mol Biol 40:373–386

Bull CT, Weller DM, Thomashow LS (1991) Relationship between root colonization and suppression of *Gaeumannomyces graminis* var. tritici by *Pseudomonas fuorescens* strain 2-79. Phytopathology 81:954–959

Carroll H, Moenne-Loccoz Y, Dawling DN, Gara FO (1995) Mutational disruption of the biosynthesis genes coding for the antifungal metabolite2,4 diacetylphloroglucinol does not influence the ecological fitness of *Pseudomonas fluorescens* F113 in the rhizosphere of sugar beets. Appl Environ Microbiol 61:3002–3007

Castro RO, Cornejo HAC, Rodriguez LM, Bucio JL (2009) The role of microbial signals in plant growth and development. Plant Signal Behav 4(8):701–712

Cerritos R, Vinuesa P, Eguiarte LE, Herrera-Estrella L, Alcaraz-Peraza LD, Arvizu-Gomez JL, Olmedo G, Ramirez E, Siefert JL, Souza V (2008) *Bacillus coahuilensis* sp. nov., a moderately halophilic species from a desiccation lagoon in the Cuatro Cienegas Valley in Coahuila, Mexico. Int J Syst Evol Microbiol 58:919–923

Chan K-G, Atkinson S, Mathee K, Sam C-K, Chhabra SR, Camara M, Koh C-L, Williams P (2011) Characterization of N acylhomoserine lactone-degrading bacteria associated with the Zingiber officinale (ginger) rhizosphere: co-existence of quorum quenching and quorum sensing in Acinetobacter and Burkholderia. BMC Microbiol 11:51

Chanway CP, Nelson LM, Holl FB (1988) Cultivar-specific growth promotion of spring wheat (*Triticum aestivum* L.) by co-existent Bacillus species. Can J Microbiol 34:925–929

Chen WX, Yan GH, Li JL (1988) Numerical taxonomic study of fast growing soybean rhizobia and a proposal that *Rhizobium fredii* be assigned to *Sinorhizobium* gen. nov. Int J Syst Bacteriol 38(4):392–397

Chen C, Belanger RR, Benhamou N, Paulitz TC (2000) Defense enzymes induced in cucumber roots by treatment with plant growth-promoting rhizobacteria (PGPR) and *Pythium aphanidermatum*. Physiol Mol Plant Pathol 56:13–23

Chernin L, Ismailov Z, Haran S, Chet I (1995) Chitinolytic *Enterobacter agglomerans* antagonistic to fungal plant pathogens. Appl Environ Microbiol 61:1720–1726

Colwell RR (1970) Polyphasic taxonomy of the genus Vibrio: numerical taxonomy of Vibrio cholerae, Vibrio parahaemolyticus and related Vibrio species. J Bacteriol 104:410–433

Compant S, Duffy B, Nowak J, Clement C, Barka EA (2005) Use of plant growth-promoting bacteria for biocontrol of plant diseases: principles, mechanisms of action and future prospects. Appl Environ Microbiol 71:4951–4959

Compant S, Clément C, Sessitsch A (2010) Plant growth-promoting bacteria in the rhizo- and endosphere of plants: their role, colonization, mechanisms involved and prospects for utilization. Soil Biol Biochem 42:669–678. https://doi.org/10.1016/j.soilbio.2009.11.024

Cornforth DM, Popat R, McNally L, Gurney J, Scott-Phillips TC, Ivens A, Diggle SP, Brown SP (2014) Combinatorial quorum sensing allows bacteria to resolve their social and physical environment. Proc Natl Acad Sci 111(11):4280–4284

Damam M, Kaloori K, Gaddam B, Kausar R (2016) Plant growth promoting substances (phytohormones) produced by rhizobacterial strains isolated from the rhizosphere of medicinal plants. Int J Pharm Sci Rev 37:130–136

Damayanti TA, Pardede H, Mubarik NR (2007) Utilization of root colonizing bacteria to protect hot-pepper against Tobacco Mosaic Tobamovirus. Hayati J Biosci 14(3):105–109

Dashti NH, Ali NY, Cherian VM, Montasser MS (2012) Application of plant growth-promoting rhizobacteria (PGPR) in combination with a mild strain of Cucumber mosaic virus (CMV) associated with viral satellite RNAs to enhance growth and protection against a virulent strain of CMV in tomato. Can J Plant Pathol 34:177–186. https://doi.org/10.1080/07060661.2012.685495

De Boer M, van der Sluis I, van Loon LC, Bakker PAHM (1997) In vitro compatibility between fluorescent Pseudomonas spp. strains can increase effectivity of Fusarium wilt control by combinations of these strains. In: Ogoshi A, Kobayashi K, Homma Y, Kodama F, Kondo N, Akino S (eds) Plant growth-promoting rhizobacteria—present status and future prospects. Proceedings of the International workshop on plant growthpromoting rhizobacteria, 4th edn. Nakanishi Printing, Sappora, pp 380–382

De Boer M, Van der Sluis I, van Loon LC, Bakker PAHM (1999) Combining fluorescent Pseudomonas spp. strains to enhance suppression of Fusarium wilt of radish. Eur J Plant Pathol 105:201–210

de Lajudie P, Laurent-Fulele E, Willems A (1998a) Allorhizobium undicola gen. nov., sp. nov., nitrogen-fixing bacteria that efficiently nodulate *Neptunia natans* in Senegal. Int J Syst Bacteriol 48:1277–1290

de Lajudie P, Willems A, Nick G, Moreira F (1998b) Characterization of tropical tree rhizobia and description of *Mesorhizobium plurifarium* sp. Nov. Int J Syst Bacteriol 48:369–382

de Vos P, Garrity G, Jones D, Krieg NR, Ludwig W, Rainey FA, Schleifer K-H, Whitman WB (2009) Bergey's manual of systematic bacteriology. The Firmicutes, vol 3, 2nd edn XXVI, 1450 p. 393 illus., Hardcover. Originally published by Williams & Wilkins, 1984

Denholm I, Cahil M, Byrne FJ, Devonshire AL (1996) Progress with documenting and combating insecticide resistance in Bemisia. In: Gerling D, Mayer RT (eds) Bemisia 1995: taxonomy, biology, damage, control and management. Intercept Ltd, Andover, Hants, pp 577–603

Depret G, Laguerre G (2008) Plant phenology and genetic variability in root and nodule development strongly influence genetic structuring of Rhizobium leguminosarum biovar viciae populations nodulating pea. New Phytol 179:224–235

Dong YH, Xu JL, Li XZ, Zhang LH (2000) AiiA, an enzyme that inactivates the acylhomoserine lactone quorum-sensing signal and attenuates the virulence of Erwinia carotovora. Proc Natl Acad Sci U S A 97:3526–3531

Dreyfus B, Garcia JL, Gillis M (1988) Characterization of Azorhizobium caulinodans gen. nov., sp. nov., a stem-nodulating nitrogen-fixing bacterium isolated from Sesbania rostrata. Int J Syst Bacteriol 38:89–98

Elad Y, Chet I (1987) Possible role of competition for nutrients in biocontrol of Pythium damping-off by bacteria. Phytopathology 77:190–195
El-beshehy EKF, Youssef SA, Elazzazy AM (2015) Resistance induction in pumpkin Cucurbita maxima L. against Watermelon mosaic potyvirus by plant growth-promoting rhizobacteria. Biocontrol Sci Technol 25:525–542
El-Dougdoug KA, Dawoud RA, Rezk AA, Sofy AR (2012) Incidence of fruit trees viroid diseases by tissue pint hbridization in Egypt. Int J Virol 8:114–120. https://doi.org/10.3923/ijv.2012.114.120
El-Dougdoug KA, Sofy AR, Mousa AA, Refaey EE (2014) Monitoring variability responses of cultivated potato varieties infected with Potato virus Y pepper isolate. Egypt J Virol 11:82–101
El-Khawas HM, Ibrahim IA, Anwar HM, Hegazi NA (2000) Isolation and characterization of plant growth promoting rhizobacteria producing indole-3-acetic acid from plants growing in Egypt. J Agric Sci Mansoura Univ 25:6493–6502
Forlani G, Mantelli M, Branzoni M, Nielsen E, Favilli F (1995) Root colonization efficiency, plant growth promoting activity and potentially related properties associated bacteria. J Gene Breed 49:343–351
Friedrich L, Lawton K, Dietrich R, Willits M, Cade R, Ryals J (2001) NIM1 over expression in Arabidopsis potentiates plant disease resistance and results in enhanced effectiveness of fungicides. Mol Plant Microbe Interact 14:1114–1124
Garcia de Salamone IE, Hynes RK, Nelson LM (2001) Cytokinin production by plant growth promoting rhizobacteria and selected mutants. Can J Microbol 47:404–411
Ghonim MI (1999) Induction of systemtic resistance against Fusarium wilt in tomato by seed treatment with the biocontrol agent *Bacillus subtilis*. Bull Fac Agric Cairo Univ 50:313–328
Glick BR (1995) The enhancement of plant growth by free-living bacteria. Can J Microbiol 41:109–117
Glick BR (2015a) Biocontrol mechanisms beneficial plant-bacterial interactions. Springer, Berlin, pp 123–157
Glick BR, Cheng Z, Czarny J, Duan J (2007) Promotion of plant growth by ACC deaminase-producing soil bacteria. Eur J Plant Pathol 119:329–39
Glick BR, Patten CL, Holguin G, Penrose DM (1999) Biochemical and genetic mechanisms used by plant growth-promoting bacteria. Imperial College Press, London
Gopalakrishnan S, Sathya A, Vijayabharathi R, Varshney RK, Gowda CL, Krishnamurthy L (2015) Plant growth promoting rhizobia: challenges and opportunities. 3. Australas Biotechnol 5(355):377
Goswami D, Thakker JN, Dhandhukia PC, Tejada Moral M (2016) Portraying mechanics of plant growth promoting rhizobacteria (PGPR): a review. Cogent Food Agric 2:1127500
Gray EJ, Smith DL (2005) Intracellular and extracellular PGPR: Commonalities and distinctions in the plant-bacterium signaling processes. Soil Biol Biochem 37:395–412
Guerinot ML, Chelm BK (1984) Isolation and expression of the Bradyrhizobium japonicum adenylate cyclase gene (cya) in *Escherichia coli*. J Bacteriol 159:1068–1071
Guetsky R, Shtienberg D, Elad Y, Dinoor A (2001) Combining biocontrol agents to reduce the variability of biological control. Phytopathology 91:621–627
Gupta G, Parihar SS, Ahirwar NK, Snehi SK, Singh V (2015) Plant growth promoting rhizobacteria (PGPR): current and future prospects for development of sustainable agriculture. J Microb Biochem Technol 7(2):096–102
Gursoy M, Balkan A, Ulukan H (2012) Ecophysiological responses to stresses in plants: a general approach. Pak J Biol Sci 15:506–516. https://doi.org/10.3923/pjbs.2012.506.516
Han D, Wang L, Luo Y (2018) Isolation, identification, and the growth promoting effects of two antagonistic actinomycete strains from the rhizosphere of Mikania micrantha Kunth. Microbiol Res 208:1–11
Handelsman J, Stabb EV (1996) Biocontrol of soil-borne plant pathogens. Plant Cell 8:1855–1869
Harish S, Kavino M, Kumar N, Saravanakumar D, Soorianathasundaram K, Samiyappan R (2008) Biohardening with plant growth promoting rhizosphere and endophytic bacteria induces sys-

temic resistance against Banana bunchy top virus. Appl Soil Ecol 39:187–200. https://doi.org/10.1016/j.apsoil.2007.12.006

Harish S, Kavino M, Kumar N, Balasubramanian P, Samiyappan R (2009) Induction of defense-related proteins by mixtures of plant growth promoting endophytic bacteria against Banana bunchy top virus. Biol Control 51:16–25. https://doi.org/10.1016/j.biocontrol.2009.06.002

Hellriegel H, Wilfarth H (1888) Untersuchungen uber die Stickstoffnahrung der Gramineen und Leguminosen. In: Beilageheft zu der Zeitschrift des Vereins fur Rubenzucker-Industrie Deutschen Reichs, p 234

Hofte M, Altier N (2010) Fluorescent pseudomonads as biocontrol agents for sustainable agricultural systems. Res Microbiol 161:464–471

Huang J, Shi Y, Zeng G, Gu Y, Chen G, Shi L, Hu Y, Tang B, Zhou J (2016) Acyl-homoserine lactone-based quorum sensing and quorum quenching hold promise to determine the performance of biological wastewater treatments: an overview. Chemosphere 157:137–151

Hull R (1994) The movement of plant viruses. Annu Rev Phytopathol 27:213–240

Hussein ME (1992) The effect of an Egyptian isolate of Streptomyces afghanensis on some plant viruses. Acta Virol 36:479–482

Husson E, Hadad C, Huet G, Laclef S, Lesur D, Lambertyn V, Jamali A, Gottis S, Sarazina C, Nguyen Van Nhien A.(2017) The effect of room temperature ionic liquids on the selective biocatalytic hydrolysis of chitin via sequential or simultaneous strategies. Green Chem 19: 4122–4131.

Imriz G, Özdemir F, Topal BE, Taş MN, Yakışır E, Okur O (2014) Bitkisel üretimde bitki gelişimini teşvik eden rizobakteri (pgpr)'ler ve etki mekanizmaları. Elektronik Mikrobiyoloji Dergisi 12:1–19

Innerebner G, Knief C, Vorholt JA (2011) Protection of *Arabidopsis thaliana* against leaf-pathogenic *Pseudomonas syringae* by Sphingomonas strains in a controlled model system. Appl Environ Microbiol 77:3202–3210

Jayaprakashvel M, Mathivanan N (2011) Management of plant diseases by microbial metabolites. D.K. Maheshwari (Ed.), Bacteria in agrobiology: plant nutrient management, Springer-Verlag, Berlin/Heidelberg, pp. 237–265, 10.1007/978-3-642-21061-7_10

Joshi P, Bhatt AB (2011) Diversity and function of plant growth promoting rhizobacteria associated with wheat rhizosphere in North Himalayan region. Int J Environ Sci 1(6):1135–1143

Kandan A, Commare RR, Nandakumar R, Ramiah M, Raguchander T, Samiyappan R (2002) Induction of phenylpropanoid metabolism by *Pseudomonas fluorescens* against tomato spotted wilt virus in tomato. Folia Microbiol 47:121–129. https://doi.org/10.1007/bf02817669

Kandan A, Ramaiah M, Vasanthi VJ, Radjacommare R, Nandakumar R, Ramanathan A, Samiyappan R (2005) Use of *Pseudomonas fluorescens* based formulations for management of tomato spotted wilt virus and enhanced yield in tomato. Biocontrol Sci Technol 15:553–569

Kavino M, Harish S, Kumar N, Saravanakumar D, Damodaran T, Samiyappan R (2007a) Potential implications of biopriming in banana (Musa spp) plantlets against Banana bunchy top virus (BBTV). J Plant Interact 2:149–158. https://doi.org/10.1080/17429140701586365

Kavino M, Harish S, Kumar N, Samiyappan R (2009) Rhizobacteria-mediated growth promotion of banana Leads to protection against Banana bunchy top virus under field conditions. Acta Hortic:69–76. https://doi.org/10.17660/actahortic.2009.828.5

Kessman H, Staub T, Hofmann C, Ward E, Uknes S, Ryals J (1994) Induction of systemic acquired resistance in plants by chemicals. Annu Rev Phytopathol 32:439–459

Kim J-S, Lee J, Lee C-h, Woo SY, Kang H, Seo S-G, Kim S-H (2015) Activation of pathogenesis-related genes by the hizobacterium, *Bacillus* sp. JS, which induces systemic resistance in tobacco plants. Plant Pathol J 31:195–201

Kim N-G et al (2017) Pseudomonas oleovorans strain KBPF-004 culture supernatants reduced seed transmission of Cucumber green mottle mosaic virus and Pepper mild mottle virus, and remodeled aggregation of 126 kDa and subcellular localization of movement protein of Pepper mild mottle virus. Plant Pathol J 33:393–401

King EB, Parke JL (1996) Population density of the biocontrol agent *Burkholderia cepacia* AMMDRl on four pea cultivars. Soil Biol Biochem 28:307–312

King AMQ, Adams MJ, Carstens EB, Lefkowitz EJ (2011) Virus taxonomy: classification and nomenclature of viruses. In: Ninth report of the international committee on taxonomy of viruses. Elsevier/Academic Press, London

Kloepper JW (1996) Biological control agents vary in specificity for host, pathogen control, ecological habitat and environmental conditions. Biol Sci 46:406–409

Kloepper JW, Schroth MN (1978) Plant growth-promoting rhizobacteria on radishes. In: Proceedings of the 4th international conference on plant pathogenic bacteria. Gilbert Clarey Tours, pp 879–882

Kloepper JW, Rodriguez-Ubana R, Zehnder GW, Murphy JF, Sikora E, Fernadez C (1999) Plant root-bacterial interactions in biological control of soil borne diseases and potential extension to systemic and foliar diseases. Aust Plant Pathol 28:21–26

Kloepper JW, Reddy MS, Rodríguez-Kabana R, Kenney DS, Kokalis-Burelle N, Martinez-Ochoa N (2004a) Application for rhizobacteria in transplant production and yield enhancement. Acta Hortic 631:219–229. https://doi.org/10.17660/actahortic.2004.631.28

Kloepper JW, Ryu CM, Zhang S (2004b) Induced systemic resistance and promotion of plant growth by Bacillus spp. Phytopathology 94:1259–1266. https://doi.org/10.1094/PHYTO.2004.94.11.1259

Kloepper JW, Gutierrez-Estrada A, Mclnroy JA (2007) Photoperiod regulates elicitation of growth promotion but not induced resistance by plant growth-promoting rhizobacteria. Can J Microbiol 53(2):159–167

Koby S, Schickler H, Ilan C, Oppenheim AB (1994) The chitinase encoding Tn7-based chiA gene endows *Pseudomonas fluorescens* with the capacity to control plant pathogens in soil. Gene 147:81–83

Kowsari M, Zamani M, Motallebi M (2016) Overexpression of chimeric chitinase 42 enhanced antifungal activity of *Trichoderma harzianum* against *Fusarium graminearum*. Mycol Iran 3:15–23

Kraus J, Loper JE (1995) Characterization of a genomic region required for production of the antibiotic pyoluteorin by the biological control agent *Pseudomonas fluorescens* Pf-5. Appl Environ Microbiol 61:849–854

Kumar A, Prakash A, Johri BN (2011) Bacillus as PGPR in crop ecosystem. In: Maheshwari DK (ed) Bacteria in agrobiology: crop ecosystems. Springer, Berlin

Kumar S et al (2016) Paenibacillus lentimorbus inoculation enhances tobacco growth and extenuates the virulence of Cucumber mosaic virus. PLoS One 11:e0149980. https://doi.org/10.1371/journal.pone.0149980

Kupferschmied P, Maurhofer M, Keel C (2013) Promise for plant pest control: root-associated pseudomonads with insecticidal activities. Front Plant Sci 4:287. https://doi.org/10.3389/fpls.2013.00287

Lee GH, Ryu C-M (2016) Spraying of leaf-colonizing Bacillus amyloliquefaciens protects pepper from Cucumber mosaic virus. Plant Dis 100:2099–2105. https://doi.org/10.1094/pdis-03-16-0314-re

Leeman M, Den Ouden FM, Van Pelt JA, Dirkx FPM, Steijl H, Bakker PAHM, Schippers B (1996) Iron availability affects induction of systemic resistance to Fusarium wilt of radish by *Pseudomonas fluorescens*. Phytopathology 86:149–155

Li H et al (2016a) Control of Tomato yellow leaf curl virus disease by Enterobacter asburiae BQ9 as a result of priming plant resistance in tomatoes. Turk J Biol 40:150–159. https://doi.org/10.3906/biy-1502-12

Li H, Huang W, Xu L, Zhou X, Liu H, Cheng Z (2016b) Stenotrophomonas maltophilia HW2 enhanced cucumber resistance against Cucumber green mottle mosaic virus. J Plant Biol 59:488–495. https://doi.org/10.1007/s12374-016-0246-6

Lian L, Xie L, Zheng L, Lin Q (2011) Induction of systemic resistance in tobacco against Tobacco mosaic virus by Bacillus spp. Biocontrol Sci Technol 21:281–292. https://doi.org/10.1080/09583157.2010.543667

Liddell CM, Parke JL (1989) Enhanced colonization of pea taproots by a fluorescent pseudomonad biocontrol agent by water infiltration into soil. Phytopathology 79:1327–1332

Lifshitz R, Kloepper JW, Mozlowski M, Simonson C, Carlson J, Tipping EM, Zaleska I (1987) Growth promotion of canola (rapeseed) seedlings by a strain of Pseudomonas putida under gnotobiotic conditions. Can J Microbiol 33:390–395

Lindstrom K (1989) Rhizobiumgalegae, a new species of legume root nodule bacteria. Int J Syst Bacteriol 39:365–367

Logan NA, Berkeley RCW (1984) Identification of Bacillus strains using the API system. J Gen Microbiol 130:1871–1882

Logan NA, Berge O, Bishop AH, Busse HJ, De Vos P, Fritze D, Heyndrickx M, Kampfer P, Rabinovitch L, Salkinoja-Salonen MS, Seldin L, Ventosa A (2009) Proposed minimal standards for describing new taxa of aerobic, endospore-forming bacteria. Int J Syst Evol Microbiol 59(8):2114–2121

Loper JE, Haack C, Schroth MN (1985) Population dynamics of soil pseudomonads in the rhizosphere of potato (*Solanum tuberosum* L.). Appl Environ Microbiol 49:416–422

Lugtenberg B, Kamilova F (2009) Plant-growth-promoting rhizobacteria. Annu Rev Microbiol 63:541–556

Lugtenberg BJJ, van der Bij A, Bloemberg G, Chin-A-Woeng T, Dekkers L, Kravchenko L, Mulders I, Phoelich C, Simons M, Spaink H, Tikhonovich I, de Weger L, Wijffelman C (1996) Molecular basis of rhizosphere colonization by Pseudomonas bacteria. In: Stacey G, Mullin B, Gresshoff PM (eds) Biology of plant-microbe interactions. ISPMB, St. Paul, MN, pp 433–440

Lugtenberg BJJ, Dekkers L, Bloemberg GV (2001) Molecular determinants of rhizosphere colonization by Pseudomonas. Annu Rev Phytopathol 39:461–490

Maheshwari DK, Annapurna K (2017) Endophytes: crop productivity and protection, vol 2. Springer International, Cham

Maheshwari DK, Saraf M, Dheeman S (2019) Plant growth-promoting rhizobacteria (PGPR) as protagonists of ever-sustained agriculture: an introduction. In: Maheshwari DK, Dheeman S (eds) Field crops: sustainable management by PGPR, vol 3. Springer Nature, Cham, pp 1–10

Mann EW (1965) Inhibition of tobacco mosaic virus by a bacterial extract. Phytopathology 59:658–662

Marschner H (1995) Mineral nutrition of higher plants. Academic Press, San Diego, CA

Martínez RT, de Almeida MMS, Rodriguez R, de Oliveira AS, Melo FL, Resende RO (2018) Identification and genome analysis of Tomato chlorotic spot virus and dsRNA viruses from coinfected vegetables in the Dominican Republic by high-throughput sequencing. Virol J 15:24. https://doi.org/10.1186/s12985-018-0931-9

Martinez-Viveros O, Jorquera MA, Crowley DE, Gajardo G, Mora ML (2010) Mechanisms and practical considerations involved in plant growth promotion by rhizobacteria. J Soil Sci Plant Nutr 10:293–319

Mauch F, Hadwiger LA, Boller T (1994) Ethylene: Symptom, not signal for the induction of chitinase and-1,3-glucanase in pea pods by pathogens and elicitors. Plant Physiol 76(607):611

Maurhofer M, Hase C, Meuwly P, Métraux JP, Defago G (1994) Induction of systemic resistance of tobacco to tobacco necrosis virus by the root-colonizing *Pseudomonas fluorescens* strain CHA0: influence of the gacA gene and of pyoverdine production. Phytopathology 84:139–146

Maurhofer M, Keel C, Haas D, Defago G (1995) Influence of plant species on disease suppression by *Pseudomonas fluorescens* strain CHAO with enhanced antibiotic production. Plant Pathol 44:40–50

Maurhofer M, Reimann C, Sacherer SP, Heebs S, Haas D, Defago G (1998) Salicylic acid biosynthetic genes expressed in Pseudomonas fluorescens strain P3 improve the induction of systemic resistance in tobacco against necrosis virus. Phytopathology 88:678–684

Mayak S, Tirosh T, Glick BR (2004) Plant growth-promoting bacteria confer resistance in tomato plants to salt stress. Plant Physiol Biochem 42(6):565–572

Megahed AA, El-Dougdoug KA, Othman BA, Lashin SM, Ibrahim MA, Sofy AR (2013) Induction of resistance in tomato plants against Tomato mosaic tobamovirus using beneficial microbial isolates. Pak J Biol Sci 16:385–390

Meziane H, Van Der Sluis I, Van Loon LC, Höfte M, Bakker PAHM (2005) Determinants of Pseudomonas putida WCS358 involved in inducing systemic resistance in plants. Mol Plant Pathol 6:177–185

Miller JM, Rhoden DL (1991) Preliminary evaluation of Biolog, a carbon source utilization method for bacterial identification. J Clin Microbiol 29:1143–1147

Miller JR, Hare EW, Wu J (1990) Quantitative characterization of the vegetation red edge reflectance. I. An inverted-Gaussian reflectance model. Int J Rem Sens 11:1755–1773

Mishina TE, Zeier J (2007) Pathogen-associated molecular pattern recognition rather than development of tissue necrosis contributes to bacterial induction of systemic acquired resistance in Arabidopsis. Plant J 50:500–513

Mishra J, Arora NK (2018) Secondary metabolites of fluorescent pseudomonads in biocontrol of phytopathogens for sustainable agriculture. Appl Soil Ecol 125:35–45

Monteiro JM, Vollu RE, Coelho MRR, Alviano CS, Blank AF, Seldin L (2009) Culture-dependent and -independent approaches to analyze the bacterial community of different genotypes of *Chrysopogon zizanioides* (L.) Roberty (vetiver) rhizospheres. J Microbiol 47:363–370

Moradi Z (2011) Diagnosis and molecular variability of Watermelon mosaic virus isolates from North, East, North-east and North-west regions of Iran. Asian J Plant Pathol 5:115–125. https://doi.org/10.3923/ajppaj.2011.115.125

Mota FF, Gomes EA, Paiva E, Seldin L (2005) Assessment of the diversity of Paenibacillus species in environmental samples by a novel rpoB-based PCR-DGGE method. FEMS Microbiol Ecol 53:317–328

Murphy JF, Zehnder GW, Schuster DJ, Sikora EJ, Polston JE, Kloepper JW (2000) Plant growth-promoting rhizobacterial mediated protection in tomato against Tomato mottle virus. Plant Dis 84:779–784. https://doi.org/10.1094/pdis.2000.84.7.779

Murphy JF, Reddy MS, Ryu CM, Kloepper JW, Li R (2003) Rhizobacteria mediated growth promotion of tomato leads to protection against cucumber mosaic virus. Phytopathology 93:1301–1307

Murphy JF, Zehnder GW (2000) Plant growth-promoting rhizobacterial mediated protection in tomato against tomato mottle virus. Plant Dis 84:779–784

Naidu RA, Sherwood JL, Deom CM (2008) Characterization of a vector-non-transmissible isolate of Tomato spotted wilt virus. Plant Pathol 57:190–200. https://doi.org/10.1111/j.1365-3059.2007.01707.x

Nasr SA (2002) Stimulation of auxin biosynthesis by some bacterial, and yeast strains. Arab Univ J Agric Sci Ain Shams Univ Cairo 10:89–107

Nelson LM (2004) Plant growth promoting rhizobacteria (PGPR): prospects for new inoculants. Online. Crop Management https://doi.org/10.1094/CM-2004-0301-05-RV

Nick G, de Lajudie P, Eardly BD (1999) Sinorhizobium arboris sp. nov. and Sinorhizobium kostiense sp. nov., isolated from leguminous trees in Sudan and Kenya. Int J Syst Bacteriol 49:1359–1368

Nour SM, Fernandez MP, Normand P, Cleyet-Marel JC (1994) Rhizobium ciceri sp. nov., consisting of strains that nodulate chickpeas (*Cicer arietinum* L.). Int J Syst Bacteriol 44:511–522

Numan M, Bashir S, Khan Y, Mumtaz R, Shinwari ZK, Khan AL, Ahmed AH (2018) Plant growth promoting bacteria as an alternative strategy for salt tolerance in plants: a review. Microbiol Res 209:21–32

O'Callaghan KJ, Dixon RA, Cocking EC (2000) Arabidopsis thaliana: a model for studies of colonization by non-pathogenic and plant growth promoting bacteria. Aust J Plant Physiol 28:975–982

Olanrewaju OS, Glick BR, Babalola OO (2017) Mechanisms of action of plant growth promoting bacteria. World J Microbiol Biotechnol 33(11):197

Ongena M, Daayf F, Jacques P, Thonart P, Benhamou N, Paulitz TC, Cornelis P, Koedam N, Belanger RR (1999) Protection of cucumber against Pythium root rot by fluorescent pseudomonads: predominant role of induced resistance over siderophores and antibiosis. Plant Pathol 48:66–76

Park KS, Paul D, Ryu KR, Kim EY, Kim YK (2006) Bacillus vallismortis strain EXTN-1 mediated systemic resistance against Potato virus Y and X in the field. Plant Pathol J 22:360–363. https://doi.org/10.5423/ppj.2006.22.4.360

Parke JL (1991) Root colonization by indigenous and introduced microorganisms. In: Keister DL, Cregan PB (eds) The rhizosphere and plant growth. Kluwer Academic, Boston, MA, pp 33–42

Patten CL, Glick BR (2002) Role of Pseudomonas putida indole acetic acid in development of host plant root system. Appl Environ Microbiol 48:3795–3801

Pei R, Lamas-Samanamud GR (2014) Inhibition of biofilm formation by T7 bacteriophages producing quorum-quenching enzymes. Appl Environ Microbiol 80:5340–5348

Pierson EA, Weller DM (1994) Use of mixtures of fluorescent pseudomonads to suppress take-all and improve the growth of wheat. Phytopathology 84:940–947

Pieterse CMJ, Van Wees SCM, Hoffland E, Van Pelt JA, Van Loon LC (1996) Systemic resistance in Arabidopsis induced by biocontrol bacteria is independent of salicylic acid accumulation and pathogenesis-related gene expression. Plant Cell 8:1225–1237

Pieterse CMJ, Van Wees SCM, Van Pelt JA, Knoester M, Laan R, Gerrits H, Weisbeek PJ, Van Loon LC (1998) A novel signaling pathway controlling induced systemic resistance in Arabidopsis. Plant Cell 10:1571–1580

Pieterse CMJ, Ton J, Van Loon LC (2001) Cross-talk between plant defence signalling pathways: boost or burden? AgBiotechNet 3:ABN068

Pieterse CMJ, Van Wees SCM, Ton J, Van Pelt JA, van Loon LC (2002) Signalling in rhizobacteria-induced systemic resistance in Arabidopsis thaliana. Plant Boil 4:535–544

Pietro-Souza W, Mello IS, Vendruscullo SJ, da Silva GF, da Cunha CN, White JF, Soares MA (2017) Endophytic fungal communities of Polygonum acuminatum and Aeschynomene fluminensis are influenced by soil mercury contamination. PLoS One 12:e0182017

Pires MN, Seldin L (1997) Evaluation of Biolog system for identification of strains of *Paenibacillus azotofixans*. Antonie Leeuwenhoek 71:195–200

Polston JE, Wood E, Palmateer AJ, Zhang S (2013) Tomato chlorotic spot virus. University of Florida Institute of Food and Agricultural Sciences. http://edis.ifas.ufl.edu/pp306

Pozo MJ, Azcon-Aguilar C (2007) Unraveling mycorrhiza-induced resistance. Curr Opin Plant Biol 10:393–398

Probanza A, Garcia JAI, Palomino MR, Ramos B, Manero FJG (2002) *Pinus pinea* L. seedling growth and bacterial rhizosphere structure after inoculation with PGPR Bacillus (*B. Licheniformis* CECT 5106 and B. pumilus CECT 5105). Appl Soil Ecol 20:75–84

Raaijmakers JM, Mazzola M (2012) Diversity and natural functions of antibiotics produced by benefcial and plant pathogenic bacteria. Annu Rev Phytopathol 50:403–424

Rajinimala N, Rabindran R, Ramaiah M (2009) Management of Bittergourd yellow mosaic virus (BGYMV) by using virus inhibiting chemical, biocontrol agents, antiviral principles (AVP) and insecticide. Arch Phytopathol Plant Prot 42:738–750. https://doi.org/10.1080/03235400701390729

Ramadan EM, AbdelHafez AA, Hassan EA, Saber FM (2016) Plant growth promoting rhizobacteria and their potential for biocontrol of phytopathogens. Afr J Microbiol Res 10(15):486–504

Ramjegathesh R, Samiyappan R, Raguchander T, Prabakar K, Saravanakumar D (2013) Plant–PGPR interactions for pest and disease resistance in sustainable agriculture. In: Maheshwari D (ed) Bacteria in agrobiology: disease management. Springer, Berlin, pp 293–320. https://doi.org/10.1007/978-3-642-33639-3_11

Raupach GS, Kloepper JW (1998) Mixture of plant growth-promoting rhizobacteria enhance biological control of multiple cucumber pathogens. Phytopathology 88:1158–1164

Raupach GS, Liu L, Murphy JF, Tuzun S, Kloepper JW (1996) Induced systemic resistance of cucumber and tomato against cucumber mosaic cucumovirus using plant growth-promoting rhizobacteria (PGPR). Plant Dis 80:891–894

Ribaudo C, Krumpholz E, Cassan F, Bottini R, Cantore M, Cura A (2006) *Azospirillum sp.* promotes root hair development in tomato plants through a mechanism that involves ethylene. J Plant Growth Regul 24:175–185

Rodriguez-Diaz M, Rodelas B, Pozo C, Martinez-Toledo MV, Gonzalez-Lopez J (2008) A review on the taxonomy and possible screening traits of plant growth promoting rhizobacteria. In: Ahmad I, Pichtel J, Hayat S (eds) Plant-bacteria interactions: strategies and techniques to promote plant growth. Wiley, London

Rome S, Fernandez MP, Brunel B, Normand P, Cleyet-Marel JC (1996) Sinorhizobium medicae sp. nov., isolated from annual *Medicago spp.* Int J Syst Bacteriol 46:972–980

Ross AF (1961) Systemic acquired resistance induced by localized virus infections in plants. Virology 14:340–358

Ryals J, Uknes S, Ward E (1994) Systemic acquired resistance. Plant Physiol 104:1109–1112

Ryan PR, Dessaux Y, Thomashow LS, Weller DM (2009) Rhizosphere engineering and management for sustainable agriculture. Plant and Soil 321:363–383

Ryu C-M, Murphy JF, Mysore KS, Kloepper JW (2004) Plant growth-promoting rhizobacteria systemically protect Arabidopsis thaliana against Cucumber mosaic virus by a salicylic acid and NPR1-independent and jasmonic acid-dependent signaling pathway. Plant J 39:381–392

Ryu C-M, Murphy JF, Reddy MS, Kloepper JW (2007b) A two-strain mixture of rhizobacteria elicits induction of systemic resistance against Pseudomonas syringae and Cucumber mosaic virus coupled to promotion of plant growth on Arabidopsis thaliana. J Microbiol Biotechnol 17:280–286

Schippers B et al (1988) Biological control of pathogens with rhizobacteria. Philos Trans R Soc B Biol Sci 318:283–293. https://doi.org/10.1098/rstb.1988.0010

Scholla MH, Elkan GH (1984) Rhizobium fredii sp. nov., a fastgrowing species that effectively nodulates soybeans. Int J Syst Bacteriol 34(484):486

Schroth MN, Hancock JG (1982) Disease suppressive soil and root colonizing bacteria. Science 216:1376–1381

Segovia L, Young JPW, Matinez-Romero E (1993) Reclassification of American Rhizobium *leguminosarum Biovar* Phaseoli type I strains as Rhizobium etli sp. nov. Int J Syst Bacteriol 43:374–377

Seldin L, Penido EGC (1986) Identification of Bacillus azotofixans using API tests. Antonie Leeuwenhoek 52:403–409

Shafie RM, Hamed AH, El-Sharkawy HHA (2016) Inducing systemic resistance against Cucumber mosaic cucumovirus using Streptomyces spp. Egypt J Phytopathol 44:127–142

Sharma A, Johri BN, Sharma AK, Glick BR (2013) Plant growth-promoting bacterium Pseudomonas sp. strain GRP3 influences iron acquisition in mung bean (Vignaradiata L. Wilzeck). Soil Biol Biochem 35:887–894

Shoman SA, Abd-Allah NA, El-Baz AF (2003) Induction of resistance to Tobacco necrosis virus in bean plants by certain microbial isolates. Egypt J Biol 5:10–18

Shridhar BS (2012) Review: nitrogen fixing microorganisms. Int J Microb Res 3:46–52

Siddiqui ZA (2004) Effects of plant growth promoting bacteria and composted organic fertilizers on the reproduction of *Meloidogyne incognita* and tomato growth. Bioresour Technol 95:223–227

Sikora RA (1992) Management of the antagonistic potential in agricultural ecosystems for the biological control of plant parasitic nematodes. Annu Rev Phytopathol 30:245–270

Spadden MC, Gardener BB, Fravel DR (2002) Biological control of plant pathogens: research, commercialization, and application in the USA. Online Plant Health Prog

Srinivasan K, Mathivanan N (2009) Biological control of Sunflower necrosis virus disease with powder and liquid formulations of plant growth promoting microbial consortia under field conditions. Biol Control 51:395–402. https://doi.org/10.1016/j.biocontrol.2009.07.013

Stackebrandt E, Goebel BM (1994) Taxonomic note: a place for DNA-DNA re-association and 16S rRNA sequence analysis in the present species definition in bacteriology. Int J Syst Bacteriol 44:846–849

Stackebrandt E, Liesack W (1993) Nucleic acids and classification. In: Goodfellow M, O'Donnell AG (eds) Handbook of new bacterial systematics. Academic, London, pp 151–194

Steijl H, Niemann GJ, Boon JJ (1999) Changes in chemical composition related to fungal infection and induced resistance in carnation and radish investigated by pyrolysis mass spectrometry. Physiol Mol Plant Pathol 55:297–311

Stutz EW, Defago G, Kern H (1986) Naturally occurring fluorescent pseudomonads involved in suppression of black root rot of tobacco. Phytopathology 76:181–185

Subramanian S, Souleimanov A, Smith DL (2016) Proteomic studies on the effects of lipo-chitooligosaccharide and thuricin 17 under unstressed and salt stressed conditions in Arabidopsis thaliana. Front Plant Sci 7:1314

Sun GZ, Yao T, Feng CJ, Chen L, Li JH, Wang LD (2017) Identification and biocontrol potential of antagonistic bacteria strains against *Sclerotiniasclerotiorum* and their growth-promoting effects on *Brassica napus*. Biol Control 104:35–43

Suzuki K, Goodfellow M, O'Donnell AG (1993) Cell envelopes and classification. In: Goodfellow M, O'Donnell AG (eds) Handbook of new bacterial systematics. Academic, London, pp 195–250

Svoboda J, Červená G, Rodová J, Jokeš M (2006) First report of Pepper mild mottle virus in pepper seeds produced in the Czech Republic. Plant Prot Sci 42:34–37. https://doi.org/10.17221/2694-pps

Timmusk S, Wagner EGH (1999) The plant-growth-promoting rhizobacterium *Paenibacillus polymyxa* induces changes in *Arabidopsis thaliana* gene expression: a possible connection between biotic and abiotic stress responses. Mol Plant Microbe Interact 12:951–959

Tisdale SL, Nelson WL (1975) Soil fertility and fertilizers, 3rd edn. Macmillan, New York, p 694

Ton J, Van Pelt JA, Van Loon LC, Pieterse CMJ (2002) The Arabidopsis ISR1 locus is required for rhizobacteria mediated induced systemic resistance against different pathogens. Plant Biol 4:224–227

Tricoli DM, Careny KJ, Russell PF, McMaster JR, Groff DW, Hadden KC, Himmel PT, Hubbard JP, Boeshore ML, Quemada HD (1995) Field evaluation of transgenic squash containing single or multiple virus coat protein gene constructs for resistance to cucumber mosaic virus, watermelon mosaic virus 2, and zucchini yellow mosaic virus. BioTechnology 13:1458–1464

Udvardi M, Poole PS (2013) Transport and metabolism in legume-rhizobia symbioses. Annu Rev Plant Biol 64:781–805

Vaddepalli P, Fulton L, Wieland J, Wassmer K, Schaeffer M, Ranf S, Schneitz K (2017) The cell wall-localized atypical β-1, 3 glucanase ZERZAUST controls tissue morphogenesis in *Arabidopsis thaliana*. Development 144:2259–2269

Van Loon LC (2007) Plant responses to plant growth-promoting rhizobacteria. Eur J Plant Pathol 119:243–254

Van Loon LC, Bakker PAHM (2005) Induced systemic resistance as a mechanism of disease suppression by rhizobacteria. In: Siddiqui ZA (ed) PGPR: Biocontrol and biofertilization. Springer, Dordrecht, pp 39–66. https://doi.org/10.1007/1-4020-4152-7_2

Van Loon LC, Bakker PAHM, Pieterse CMJ (1998) Systemic resistance induced by rhizosphere bacteria. Annu Rev Phytopathol 36:453–483

Van Peer R, Niemann GJ, Schippers B (1991) Induced resistance and phytoalexin accumulation in biological control of Fusarium wilt of carnation by Pseudomonas sp. strain WCS417r. Phytopathology 81:728–733

Van Wees SCM, De Swart EAM, Van Pelt JA, Van Loon LC, Pieterse CMJ (2000) Enhancement of induced disease resistance by simultaneous activation of salicylate and jasmonate dependent defense pathways in *Arabidopsis thaliana*. Proc Natl Acad Sci U S A 97:8711–8716

Van Wees SCM, Van der Ent S, Pieterse CMJ (2008) Plant immune responses triggered by beneficial microbes. Curr Opin Plant Biol 11:443–448

Velazquez E, Igual JM, Willems A (2001) Mesorhizobium chacoense sp. nov.: a novel species that nodulates Prosopis albain the Chaco Arido region (Argentina). Int J Syst Evol Microbiol 51:1011–1021

Verma JP, Yadav J, Tiwari KN, Lavakush, Singh V (2010) Impact of plant growth promoting rhizobacteria on crop production. Int J Agric Res 5(11):954–983

Verhagen BWM, Van Loon LC, Pieterse CMJ (2006) Induced disease resistance signaling in plants. In: Silva JAT (ed) Floriculture, ornamental and plant biotechnology, vol 3. Global Science Books, Gainesville, pp 334–343

Vessey JK (2003) Plant growth promoting rhizobacteria as biofertilizers. Plant and Soil 255:571–586. https://doi.org/10.1023/a:1026037216893

Vleesschauwer D, Hofte M (2009) Rhizobacteria-induced systemic resistance. Adv Bot Res 51:223–281

Wang LT, Lee FL, Tai CJ, Kasai H (2007) Comparison of gyrB gene sequences, 16S rRNA gene sequences and DNA-DNA hybridization in the *Bacillus subtilis* group. Int J Syst Evol Microbiol 57:1846–1850

Wei G, Kloepper JW, Tuzun S (1991) Induction of systemic resistance of cucumber to Colletotrichum orbiculare by select strains of plant growth promoting rhizobacteria. Phytopathology 81: 1508–1512

Wei G, Kloepper JW, Tuzun S (1996) Induced systemic resistance to cucumber diseases and increased plant growth by Plant growth promoting rhizobacteria under field conditions. Phytopathology 86:221–224

Weller DM, Raaijmakers JM, Gardener BB, Thomashow LS (2002) Microbial populations responsible for specific soil suppressiveness to plant pathogens. Annu Rev Phytopathol 40:309–348

Werner GDA, Cornwell WK, Sprent JI, Kattge J, Kiers ET (2014) A single evolutionary innovation drives the deep evolution of symbiotic N2-fixation in angiosperms. Nat Commun 5:4087

Wiesel L, Newton AC, Elliott I, Booty D, Gilroy EM, Birch PRJ, Hein I (2014) Molecular effects of resistance elicitors from biological origin and their potential for crop protection. Front Plant Sci 5:655

Wilson M, Backman PA (1999) Biological control of plant pathogens. In: Ruberson JR (ed) Handbook of pest management. Marcel Dekker, New York, pp 309–335

Yuen GY, Schroth MN (1986) Interactions of Pseudomonas flourescens strain E6 with ornamental plants and its effect on the composition of root-colonizing microflora. Phytopathology 76:176–180

Zebelo S, Song Y, Kloepper JW, Fadamiro H (2016) Rhizobacteria activates (+)-delta-cadinene synthase genes and induces systemic resistance in cotton against beet armyworm (Spodoptera exigua). Plant Cell Environ 39:935–943. https://doi.org/10.1111/pce.12704

Zehnder GW, Yao C, Murphy JF, Sikora ER, Kloepper JW (2000) Induction of resistance to tomato against cucumber mosaic cucumo virus by plant growth promoting rhizobacteria. Biocontrol 45:127–137

Zehnder GW, Murphy JF, Sikora EJ, Kloepper JW (2001) Application of rhizobacteria for induced resistance. Eur J Plant Pathol 107:39–50

Zhang X, Zhou YY, Li Y, Fu XC, Wang Q (2017) Screening and characterization of endophytic *Bacillus* for biocontrol of grapevine downy mildew. Crop Prot 96:173–175

Zhang Z, Cao R, Jin L, Zhu W, Ji Y, Xu X, Zhu L (2019) The regulation of N-acyl-homoserine lactones (AHLs)-based quorum sensing on EPS secretion via ATP synthetic for the stability of aerobic granular sludge. Sci Total Environ 673:83–91

Chapter 9
Potential Role of Endophytes in Weeds and Herbicide Tolerance in Plants

Krutika Lonkar and Ragini Bodade

Contents

1 Introduction

In the agricultural system, plant-microbial interactions have been studied extensively for the benefits towards the host plants. The plant-associated microbial community (phytobiome) contributes mainly towards plant growth, resistance to the biotic and abiotic stress and disease resistance by different undefined mechanisms. Endophytes are microbes that inhabit the interior of plant tissues without producing any harmful effects to them. Reported studies on endophytic microbial (bacteria/fungi) and mycorrhizal community revealed their involvement in soil structure stabilization for efficient water and nutrient uptake towards host plant. Rhizobacteria and endophytic bacteria/fungi promote the plant growth by nitrogen fixation, phosphate solubilization, induction of biotic and abiotic stress tolerance and by secondary metabolite production, viz. siderophores for ferrous chelation, phytohormones and antibiotic/insecticide/alkaloids against pathogens/insect/herbivore pests (Fig. 9.1). Therefore, these microorganisms in agricultural system could be used to regulate the application of fertilizers, pesticides and herbicides (Yu et al. 2019; Andrews et al. 2010).

K. Lonkar · R. Bodade (✉)
Department of Microbiology, Savitribai Phule Pune University, Pune, Maharashtra, India
e-mail: ragini.bodade@unipune.ac.in

© The Author(s), under exclusive license to Springer Nature Switzerland AG 2021

H. I. Mohamed et al. (eds.), *Plant Growth-Promoting Microbes for Sustainable Biotic and Abiotic Stress Management*,
https://doi.org/10.1007/978-3-030-66587-6_9

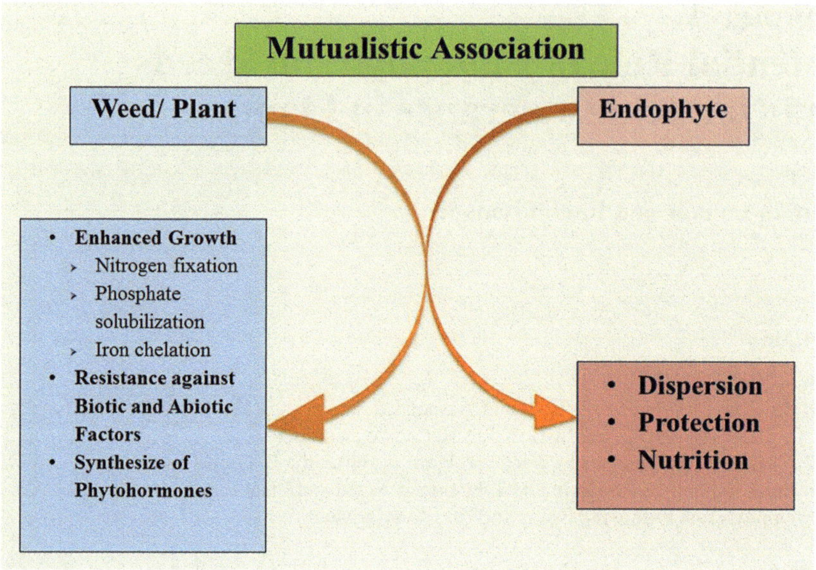

Fig. 9.1 Mutualistic interaction between endophyte and host plant

A weed may be defined as any plant that interferes with the objective of farming and accounts not more than 1% of total plant species on the earth. More broadly, 'weed' signifies the species that are distributed evenly, can tolerate and grow in unfavourable conditions and reproduce quickly. Moreover, they compete with the crop plants for nutrients, sunlight, water and space in a farm and carry diseases inducing insect-pest, which affects the growth of the crops. They are characterized with vegetative reproduction, ease of dissemination of their reproductive organs, ample seed production, seed modification and long seed dormancy irrespective of human hazard activity and environmental conditions. More than 120 common weeds are reported, viz. ryegrass (*Lolium perenne* L.), black-grass (*Alopecurus myosuroides*), cheatgrass (*Bromus tectorum*), wild carrot (*Daucus carota*), ground-ivy (*Glechoma hederacea*), goosegrass (*Eleusine indica*), congress grass (*Parthenium hysterophorus*), timothy grass (*Phleum pratense*), creeping thistle (*Cirsium arvense*), couch grass (*Elymus repens*) and dandelion (*Taraxacum officinale*), *Amaranthus* spp., kochia (*Kochia scoparia*), wall barley (*Hordeum glaucum*), bluegrass (*Poa annua*), green foxtail (*Setaria viridis*) and *Chenopodium album* (Betts et al. 1992; Malinowski and Belesky 2000; Qasem and Foy 2001; Harding and Raizada 2015). Among these, some weeds have the potential to induce allergic reactions to humans/animals and are responsible for a decrease in crop quality/quantity or animal products with a 20–37% loss in the crop yield. In the USA the loss of the annual crop due to weeds is more than 26 billion USD. While in India, it contributes to the total economic loss of around 11 billion USD in ten major crops of agriculture system. To establish in the field over the crops, weed acquired various strategies like production of allelochemicals, viz. carbohydrate and aldehyde molecules, quinines, complex conjugated flavonoids, tannins, terpenes, sesquiterpene, coumarin, organic acids, phenolics, steroids, alkaloids, simple unsaturated lactones, long-chain fatty acids, cinnamic acid derivatives, cyanogenic glucosides and sulphide that impede the germination, nutrient uptake, growth and development of crop plants. Moreover, agricultural practices like tillage and monoculture help the weed to persist in the field for a long time (Qasem and Foy 2001; Suryanarayanan 2019; Sindhu and Sehrawat 2017; Gharde et al. 2018). Therefore, many measures have been taken to limit the weed infestation worldwide, such as crop competition, crop rotation, tillage, mechanical uprooting of weeds, herbicide usage and the burial of growing points.

Herbicides are a class of chemical compounds or pesticides that alone or in the mixture used to destroy, repel, control or mitigate unwanted weeds growing in or around the crop. The usage of herbicides for weed management has rapidly developed since 1944. Under the umbrella of the Weed Science Society of America, they have been classified into 29 different classes based on mechanism of action. Clodinafop propargyl (CF), atrazine, glyphosate (GP), 2,4-dichlorophenoxyacetic acid (2,4-D), 2-methyl-4-chlorophenoxyacetic acid (MCPA), metolachlor, tribenuron-methyl, tralkoxydim, sulphonylureas, fenoxaprop, molinate, diuron, pendimethalin, imazapyr and paraquat (PQ) are examples of the commonly used herbicides throughout the world. During the 1960s–2005, there is a consistent increase in herbicide consumption globally form 1960 (20%) till 2005 (48%). The global pesticide including herbicide, insecticide and fungicides/bactericides revealed sales of 850 million US dollars to 31,191 million US dollars, where the herbicides stand first place after the 1990s (44%). The prolonged applications of herbicides and alteration in weed management might have further emerged the herbicide tolerance in the weeds as well as desired plants. An excessive amount of herbicide further leads to soil and water pollution as well (Sindhu and Sehrawat 2017; Singh and Singh 2016; Marin-Morales et al. 2013; Zhang et al. 2011). The possible ways through which the desired plant or weed have acquired herbicide resistance could be by the plants' ability to metabolize and detoxify the herbicide alone. Excessive herbicide usage could induce biotic and abiotic stress in plants that leads to the gene expression required for herbicide degradation. However new evidence proposes that the microorganisms, both free-living and the microorganisms residing inside the host plants' living tissues (endophytes), can accord to the herbicide tolerance either by its detoxification or by inducing plant chemical stress via latent stress signalling pathway. This revealed a biotrophic relationship in between microorganisms and plants (Kremer 2005; Tétard-Jones and Edwards 2016).

In the current review, we highlight the endophyte's role in herbicide degradation and tolerance in plants/weeds.

2 Role of Free-Living and Weed/Plant-Associated Microbes in Herbicide Degradation

Xenobiotics are the organic contaminants released into the environments due to human activities. Its amount accelerated further due to industrialization, agricultural practices, population growth and military activities. They are mainly pesticides, fuels and polycyclic aromatic hydrocarbons. Introduction of the earliest selective herbicides MCPA and 2,4-D in the late 1940s significantly changed the weed management without harming the crop yield. However, the introduction of herbicide-resistant crops (HRCs) against a specific chemical herbicide leads to the wide applications of nonselective herbicides and reason for the development of diverse herbicide-resistant (HR) weed biotypes. Currently, more than 25 herbicide target sites are investigated at molecular level, viz. photosystem II (PS II) inhibitors (triazine, simazine), auxin pathway (2,4-D) inhibitors, acetolactate synthase (ALS) inhibitors (sulphonylureas, imidazolinone, triazolopyrimidine, pyrimidinylthiobenzoates), EPSPS inhibitors (GP) and ACCase inhibitors (aryloxyphenoxypropionate, phenylpyrazolin and cyclohexanedione) of a metabolic pathway. So far, more than 478 HR weed biotypes belonging to the 252 weed species from 67 countries have been reported. The resistance pattern of weeds is ALS inhibitor (133) > triazine (71) > ACC inhibitor (43) > GP (24). More than 220 identified weed species show resistance to one or more herbicides. The HR weed species are predominantly revealed from the families, viz. Alismataceae (seven species), Asteraceae (39 species), Amaranthaceae (11 species), Brassicaceae (22 species), Caryophyllaceae (six species), Chenopodiaceae (eight species), Cyperaceae (12 species), Poaceae (80 species), Polygonaceae (seven species) and Scrophulariaceae (nine species) (Heap 2014; Harding and Raizada 2015; Hussain et al. 2018; Vrbnicanin et al. 2017).

In 1937, Hans Molisch first introduced the term 'allelopathy' meaning *allelon* 'of each other' and *pathos* 'to suffer'. In natural and anthropogenic ecosystems, any biochemical interaction occurring between the plants/weed and microorganisms via chemical compounds (allelochemicals, allelopathins or allelopathic compounds) is termed as allelopathy. The plant/weed allopathy comprises the inhibitory action against cultivated weed/plants, but also other plant/weed species and plant pathogens (Mishra et al. 2013; Soltys et al. 2013). Therefore, in the ecosystem highly diverse microbial communities are harboured in vicinity of both weeds and crops. Different interactions observed between the microbes with their host range from mutualism, symbiosis, commensalism or pathogenic forms thereby contributing in soil structure and quality, plant health and productivity, induction of plant defence mechanism and control of phytopathogens (Table 9.1). Moreover, the plant variety and soil are both determinants that define the soil microbial community structure (Fatema et al. 2019). Although pesticides like herbicides are beneficial for agricultural productivity, their excessive usage creates many environmental problems and lethal effects on human health. Thus, knowing the greater applications of microbial strains, we could retard the usage of herbicides and fertilizers without affecting yield. Here, bioherbicides allow restoration of the fertility and efficiency of the

Table 9.1 Some examples of plant/weeds and its mutually associated microbes

Host plant/ weed	Endophyte	Beneficial effects	References
Brassica napus	*Microbacterium* sp., *Bacillus endophyticus, Bacillus amyloliquefaciens, Paenibacillus polymyxa, Azorhizobium caulinodans, Alcaligenes* sp.	Heavy metal and antibiotic resistance, plant growth, promotion and antagonism	Card et al. (2015)
Solanum tuberosum	*Bacillus* sp., *Streptomyces* spp., *Pseudomonas putida, Serratia plymuthica, Burkholderia phytofirmans*	ACC deaminase activity, phosphate solubilization, siderophore production, indoleacetic acid (IAA) production	Degrassi and Carpentieri-Pipolo (2020)
Oryza sativa	*Bacillus* sp., *Azospirillum* sp., *Pseudomonas stutzeri, Burkholderia* sp., *Rhizobium* sp., *Rhodococcus* sp., *Ralstonia*	IAA, N-fixing, P solubilization, ACC deaminase, antifungal activity, growth promotion	
Glycine max	*Bacillus subtilis, Bacillus thuringiensis, Serratia, Pseudomonas, Kosakonia, Stenotrophomonas, Acinetobacter, Ralstonia, Enterobacter, Agrobacterium, Rhizobium*	Antifungal activity; phytases; N fixation; phosphate solubilization, production of siderophores, IAA synthesis and ACC deaminase	
Triticum aestivum	*B. subtilis, Bacillus cereus, B. thuringiensis, Azospirillum* sp., *Arthrobacter* sp., *Burkholderia cepacia*	Plant growth promotion, siderophore production and Zn solubilization, biocontrol, phytohormone synthesis: IAA, GA, ABA; phosphate solubilization	
Solanum lycopersicum	*B. subtilis, B. phytofirmans, Sphingomonas* sp.	Production of gibberellins and IAA, IAA synthesis, ACC deaminase, IAA synthesis, ACC deaminase	
Zea mays	*Bacillus* spp., *Pseudomonas* spp., *Azospirillum brasilense, Enterobacter asburiae, Enterobacter* sp., *P. polymyxa, Sinorhizobium meliloti*	Antifungal activity, N fixation and growth promotion, biocontrol agent	
Vitis vinifera	*Bacillus pumilus, Paenibacillus* sp., *B. subtilis, Curtobacterium* sp.	Biocontrol agent	

(continued)

Table 9.1 (continued)

Host plant/ weed	Endophyte	Beneficial effects	References
Psoralea corylifolia	*Bacillus* sp., *Marinorhizobium* sp., *Sinorhizobium* sp.	Plant growth promotion and salinity stress tolerance	Fatema et al. (2019)
Leucaena leucocephala	*Microbacterium proteolyticum, Sphingomonas paucimobilis, Rhodococcus kroppenstedtii, Sphingomonas pseudosanguinis, Pseudomonas oryzihabitans*	Degrades mimosine and N fixation	
Lepidium draba	*Pseudomonas viridiflava, Bacillus* sp., *Pseudomonas* sp., *Arthrobacter* sp.	Biocontrol agent, hydrogen cyanide production, phosphate solubilization	
Urtica dioica	*Bacillus methylotrophicus, B. pumilus, B. cereus, B. amyloliquefaciens*	Plant growth promotion and biocontrol agent	
Plantago lanceolata	*Bacillus* sp.		
Calendula arvensis	*Pseudomonas brassicacearum, B. amyloliquefaciens*		
P. hysterophorus	*Bacillus* sp.	Biocontrol agent	
Eupatorium adenophorum	*Stenotrophomonas maltophilia, Stenotrophomonas rhizophila*	Secondary metabolite production, plant growth promotion, bioremediation	
Lactuca dissecta	*Pseudomonas mendocina*	Plant growth promotion	
Solanum surattense	*Pseudomonas stutzeri*		
Sonchus arvensis	*P. putida*		
Nicotiana glauca	*B. cereus, Alcaligenes faecalis*	Biocontrol agent	
Spartina pectinata	*Herbaspirillum frisingense*	N fixation	
Cyperus conglomeratus	*Micrococcus luteus*	Salinity and stress tolerance	

(continued)

Table 9.1 (continued)

Host plant/ weed	Endophyte	Beneficial effects	References
L. perenne L.	*Pseudomonas* spp.	Production of ACC deaminase, IAA, siderophores, HCN, phosphate solubilization, biocontrol agent	Feng et al. (2017)
Cannabis sativa L.	*Paecilomyces lilacinus*	Biocontrol agent	
Helianthus annuus L.	*Aspergillus terreus* LWL5 and *Penicillium citrinum* LWL4	Production of GAs, siderophores, oxidative stress responses, biocontrol	
Tinospora cordifolia	*Cladosporium velox*	Biocontrol agent	
Dendrobium candidum	*Pseudomonas saponiphila*	Production of IAA, siderophores, HCN, phosphate solubilization	
Convolvulus arvensis	*Yersinia ruckeri, Aspergillus flavus, Aspergillus niger, Drechslera biseptata, A. terreus*	Plant growth promotion	Mukhtar et al. (2010)
Euphorbia helioscopia	*Azospirillum lipoferum, Acinetobacter lwoffii, Cladosporium cladosporioides, Aspergillus sydowii, Alternaria alternata*		
Chenopodium album	*Curtobacterium albidum, Acinetobacter lwoffii, Aspergillus phoenicis, A. flavus, Cuvularia clavata,*		
Elymus dauricus	*Pseudomonas aeruginosa* and *Pseudomonas savastanoi*	Phytoremediation	Siciliano et al. (1998)

affected ecosystems and further aid in alleviating the herbicide-resistant and invasive weeds in diversified cropping systems. Moreover, they can make a wide impact for bioremediation of water and land system in the future as well.

2.1 *Role of Free-Living Microbes in Herbicide Degradation*

Plant rhizosphere hosts complex microbial communities, which has affected plant root exudate and agricultural practices. It varies with plant species, genotype, plant age, stress exposure and nutritional status. Plant root exudates contain flavonoids, strigolactones, benzoxazinoid and malic acids that could attract the beneficial microbes involved in plant growth promotion by achieving nitrogen fixation, biofilm formation and pathogen retardation. Moreover, selective enrichment of microbes in rhizosphere can be attained by alteration in root exudate compositions by the plants.

Secretion of phenazine-1-carboxylic acid and 2,4-diacetylphloroglucinol (2,4-DAPG) by *Pseudomonas* spp. suppresses the growth of pathogenic *Rhizoctonia solani*, while secretion of lipoproteins by *Bacillus* spp. and *Pseudomonas* spp. acts against many soil pathogens. Moreover, production of pyrrolnitrin, oomycin A, hydrogen cyanide, phenazine and 2,4-diacetylphloroglucinol from genus *Pseudomonas* protects plants from many diseases. Secondary metabolites like IAA found to show an antagonistic effect on rhizospheric microbes and plant growth. Secondary metabolites, viz. antibiotics, siderophores, lytic enzymes and toxins, from soil microbes assist them to establish in the plant roots and rhizosphere. Majority of the free-living rhizospheric microorganism has the potential to use the herbicide as a source of carbon, nitrogen, sulphur, phosphorus and energy. Biotransformation of herbicides has been achieved through the reduction, oxidation, hydrolysis and lyase reactions. The rates of degradation reactions are accelerating under environmental conditions (temperature, soil organic matter) and agrochemical practices (history of herbicide-treated soil) (Tétard-Jones and Edwards 2016; Sindhu and Sehrawat 2017). Anderson et al. (1994) proved the importance of rhizosphere soil in the degradation of herbicides (Anderson et al. 1994). A group of microbes use sulfentrazone herbicide as a carbon source and metabolize it to 3-hydroxymethyl-sulfentrazone (HMS). The bacterial strains *Ralstonia pickettii*, *Rhizobium radiobacter* and *Methylobacterium radiotolerans* and fungi, viz. *Cladosporium* sp., *Eupenicillium* sp. *Paecilomyces* sp., are the potential sulfentrazone degraders and had a tolerance level up to 7.0 μg/ml concentration (Martinez et al. 2008).

The phenylurea herbicides are used worldwide either pre- or post-emergence in fruit, cotton and cereal crops. Several soil fungi including *Mortierella isabellina, Rhizoctonia solani*, *Cunninghamella elegans*, *Talaromyces wortmanii*, *Rhizopus japonicus* and *A. niger* have potential to metabolize phenylurea herbicides, viz. chlorobromuron, isoproturon (IPU), linuron, metobromuron, fluometuron and diuron. Various bacterial strains, viz. *Pseudomonas fluorescens*, *Delftia acidovorans, Bacillus sphaericus*, *Sphingomonas* sp. and *Variovorax paradoxus*, contribute to the detoxification of the above-listed herbicide as well (Sørensen et al. 2003; EI Fantoussi et al., 1999). Aryloxyphenoxypropionate herbicide like clodinafop-propargyl (CF), a widely used herbicide for post-emergence control of cereal weeds, viz. *Setaria, Alopecurus* spp., *Lolium* sp., *Phalaris* sp. and *Avena* sp., is degrading by *Pseudomonas* sp. strain B2. Herbicide fenoxaprop-*P*-ethyl (FE) degrade by *Rhodococcus ruber* JPL2, *Sphingomonas, Chryseomonas, Actinobacteria, Stenotrophomonas, Aquamicrobium, Alcaligenes, Pseudomonas, Agromyces and Microbacterium* (Singh 2013; Hongming et al. 2015). Moreover, molinate a world-widely used thiocarbamate herbicide in rice crop protection serves as a nutrient for actinobacterium *Gulosibacter molinativorax* ON4T. Molinate (thiocarbamate) herbicide is degraded by bacterial and fungal species. Rhizospheric microbes degrade organophosphorus herbicide malathion and GP by multiple pathways. GP is non-selective systemic herbicide is degrading by *Arthrobacter atrocyaneus*, *Pseudomonas* sp., *Enterobacter cloacae K7.*and Flavobacterium sp. Fungi Trichoderma viride FRP3 use GP as source of phosphorus. The culture growth in Czapek broth containing GP reached maximum after 8 days with decease in GP and continue till 28 days (Arfarita et al. 2013; Nunes et al. 2013; Kryuchkova et al., 2014). Herbicide imazapyr is degrade by soil *Pseudomonas*, *Streptomyces* sp. strain PSI/5 and *Bacillus* sp. Fungi *Trichoderma viride* FRP3 use GP as source of phosphorus. The culture growth in Czapek broth containing GP reached maximum after 8 days with decease in GP and continued till 28 days (Arfarita et al. 2013; Nunes et al. 2013; Kryuchkova et al., 2014). Atrazine has been used as nitrogen source and degraded by *Agrobacterium radiobacter* J14a *Comamonas* sp. and *Klebsiella* sp. A1 soil bacteria. In another study, atrazine degradation by isolate *Burkholderia* sp. and *Enterobacter* sp. from sugarcane-cultivated soil showed 82.1% degradation after 62 days via intermediate desethyl atrazine (DEA) and deisopropylatrazine (DIA) (Ngigi et al. 2012). Phenoxyalkanoic acid (PAA) herbicides such as MCPA and 2,4-D are used worldwide. MCPA-degrading bacteria from the soil are mainly of alpha, beta and gammaproteobacteria. Alphaproteobacteria dominate the active degradation of MCPA (Liu et al. 2011). The enhanced mineralization of [U-^{14}C] 2,4-D is reported in rhizosphere soil of *Trifolium pratense* (Shaw and Burns 2004). Soil bacteria, viz. *Pseudomonas, Arthrobacter, Ralstonia eutropha* JMP134, *Delftia, Alcaligenes, Cupriavidus pinatubonensis* JMP134, *Burkholderia cepacia* and *Mortierella* metabolize 2,4-D as carbon and energy source and degrade it by oxidative reaction (Singh et al. 2014; Sviridov et al. 2015; Han et al. 2015; Ellegaard-Jensen et al. 2013; Kumar et al., 2016; Sandoval-Carrasco et al., 2013).

FE is an aryloxyphenoxy propionate (AOPP) herbicide used for the control of annual and perennial weeds of wheat and soybean crop. An efficient FE-degrading isolate *Rhodococcus sp.* strain T1 metabolizes 94% of 100 mg/L FE in 24 h into fenoxaprop acid (FA) by cleaving the ester bond. A similar type of conversion of FE to FA also occurs in *P. fluorescens* and *Alcaligenes* sp. Herbicide diclofop-methyl (DM) can be metabolized by *Sphingomonas paucimobilis* and *Chryseomonas luteola* (Hou et al. 2011; Serfling et al. 2007). Rhizospheric microbe of herbicide-resistant plant *Kochia* sp. was isolated for degradation of three herbicide mixture atrazine, metolachlor and trifluralin. The results revealed maximum microbial activity in rhizosphere soli as compared to edaphosphere soil (nonvegetated) (Anderson et al. 1994; Saxena et al., 1987). Importance of these free-living herbicide degradation microbes would help to attain the allocation of trait for herbicidal chemical production/degradation and to confer tolerance in transgenic crop plants. Recently the Liberty Link™ GM crops have achieved by transferring the bialaphos-resistant gene and phosphinothricin acetyltransferase gene from *Streptomyces hygroscopicus* and *Streptomyces viridochromogenes*, respectively. The transfer of glyphosate acetyltransferase gene from *Bacillus licheniformis* and acetolactate synthase gene in detoxifying bacteria revealed resistance to almost all the types of herbicide using GAT/HRA crop technology (Tétard-Jones and Edwards 2016).

2.2 *Role of Endophytes in Weed and Plant as Herbicide Degrader*

De Bary in 1866 first proposed the term 'endophyte'. It is simply a Greek-derived word in which 'endon' means within and 'phyton' referred to as plant. More broadly, endophytes are a group of microorganisms that include bacteria, fungi and viruses, which reside mostly within the plant tissues. The inhabitation of these microbial populations does not show any disease-like symptoms in plants. It is believed that genetic factors play a significant role in permitting a specific bacterium to become endophyte (Fatema et al. 2019). Endophytes, which make up the endobiome of plant, evolve with the plant and together constitute plants' holobiome. This type of symbiosis is a selective adaptation between the microorganism and the host, which is not easily reversible. Some studies revealed that the microbial species discovered within the host plant roots, shoots, leaves and other living parts are similar to the microbial population adjacent to the roots, hence, concluding the soil as a prime source of endophytes. The endophytes have expected to be motile for the successive transmission to other plant hosts. They could transmit by vascular and nonvascular plant tissues or through seeds contaminated with the external environment. Moreover, potential bacterial/fungal endophyte can migrate from soil to the host body parts by various modes like agricultural equipment, wind action and by vectors, viz. insects, mites and birds. The various entry points in host plant have been identified, such as stomata, hydathodes, germinating radicles, nectarthodes, lenticels, tissue wounds, broken trichomes, foliar damage from wind blown, rain or hail, soil particles or through undifferentiated meristematic root tissue and abrasion sustained during root growth. Being in a mutualistic association with plants, endophytes contribute to various benefits like host plant growth and reproduction, enhancement of nutrient uptake from soil, nitrogen fixation and induction of tolerance to biotic/abiotic stress. Microbial diversity of weed endophytes is mostly from families *Pseudomonadaceae, Bacillaceae, Micrococcaceae, Rhizobiaceae, Alcaligenaceae, Xanthomonadaceae, Nocardiaceae and* Microbacterium (Sturz et al. 2000; Frank et al. 2017; Fatema et al. 2019).

Endophyte *B. licheniformis* strain SDS12 isolated from *Parthenium* weed revealed 85.60 ± 1.36% diuron degradation via intermediate 3,4-dichloroaniline (3,4-DCA) formation (Singh and Singla 2019). Endophytic strain *Neurospora intermedia* DP8-1 from sugarcane plant metabolizes diuron, monuron, fenuron, isoproturon, linuron, metobromuron, chlortoluron and chlorobromuron. Two important intermediates *N*-(3,4-dichlorophenyl)-urea and *N*-(3,4-dichlorophenyl)-*N*-methylurea were primarily identified by liquid chromatography-mass spectrometry analysis (Wang et al. 2017). The endophytic bacterial isolates *Burkholderia gladioli* and *P. oryzihabitans* from soybean (*G. max*) showed different sensibility profiles to the GP (Kuklinsky-Sobral et al. 2005). Endophytic quinclorac-degrading *Bacillus megaterium* strain Q3 isolated from tobacco roots (*Nicotiana tabacum*) degrades 93% (initial concentration 20 mg/L) of quinclorac via 3-chlorin-8-quinoline-carboxylic,3,7-dichloro-8-methyl-quinoline and 8-quinoline-carboxylic acid (Liu et al. 2014). Rhizobium inoculation of pea plants (*Pisum sativum*) with POPHV6 showed clearance of the 2,4-D from the soil and thereby reduced herbicide translocation into aerial tissues (Germaine et al. 2006). Seed coating of the barley (*Hordeum vulgare*) with *B. cepacia* DBO1 (Pro101) allows effective seed germination and plant protection by degrading 2,4-D form contaminated soil (Jacobsen 1997). Ozawa et al. (2004) isolated simazine (2-chloro4,6-bis(ethylamino)-*s*-triazine) herbicide degrading bacteria *Agrobacterium radiobacter* and *Bradyrhizobium japonicum* from the corn plant roots and soybean plant nodules. Faster degradation has achieved in both the plants after inoculation with respective isolates in vermiculite soil (Ozawa et al. 2004). DM or 2-[4-(2,4-dichlorophenoxy)phenoxy] propanoate herbicide is applied for common food crops around the world, while 2,4-D is selective for the control of broadleaved weeds. *S*-triazine herbicide has been applied commonly in sugarcane-cultivated soil. Atrazine (2-chloro-4-ethylamino-6--isopropyl amino-*s*-triazine) is used as a nitrogen source by two sugarcane-associated bacterial endophytes *Burkholderia cepacia* and *Enterobacter cloacae* (Jordan et al. 2000; Wang et al.,1999).

Understanding the role of symbiotic endophytes in pesticide metabolism and herbicide tolerance in plants by direct and indirect route, genetically modified crops for pesticide degradation can retard the usage of herbicide in the field, but with some conventional limitations.

3 Role of Endophytes in Herbicide Tolerance in Weed and Plants

During the 1940s weed has been controlled extensively by herbicides. In the environment, microbial adaptation to chemicals stimulates its degradations or biotransformations by co-metabolism or growth-linked metabolism. Soon after due to detection of first herbicide resistance in triazine-resistant *Senecio vulgaris* weed in 1968, a steady increase in HR weed has occurred. The global worst HR weed species are from the *Amaranthus*, *Conyza*, *Lolium* and *Echinochloa*. The greatest number of weed-resistant species towards the herbicides is confirmed in the order by atrazine (66) > imazethapyr (44) > tribenuron methyl (43) > imazamox (37) > chlorsulfuron (36) > metsulfuron-methyl (35) > GP (34) > iodosulfuron-methyl-sodium (33) > simazine (31) > FE (31) > bensulfuron-methyl (29) > thifensulfuron-methyl (27) > pyrazosulfuron-ethyl (25) and fluazifop-*P*-butyl (25). The HR weed prevalence revealed more in the USA (144) followed by Australia (62), Canada (59), France (35), China (34), Spain (33), Germany (33), Brazil (31), Israel (29), Japan (18) and Chile (16). Crops have the most herbicide-resistant weed species than non-crop plants, viz. wheat (59) > corn (58) > soybean (46) > rice (39) > roadsides (31) > orchids (27) > barley (18) > cotton (14) (Vrbnicanin et al. 2017; Heap 2014).

HR or tolerance has often been used interchangeably by the researches. The resistance defines as any genetic change in response to selection by toxic chemicals that may impair control in the field (Moss and Rubin 1993). HR is considered as an adaptive response of weed population and conferred through the repeated usage of herbicide at a site of application. Several mechanisms for herbicide resistance are hypothesized including abundances and diversity of herbicide-degrading microbial community, mutations/horizontal gene transfer and recombination which are within the microbial populations, alteration of herbicide binding site of an enzyme, decreased translocation or absorption of herbicide at the action site and herbicide sequestering at the cell surface. In weed population, alteration of enzyme target site confers resistance to more than a single herbicide called as cross-resistance. When more than one HR mechanisms have adapted by the plant due to mutations, multiple resistance phenomenons occur (Poursat et al. 2019; Heap 2014). Therefore, HR in weeds/plants could be broadly categorized as target site resistance (TSR) and non-target site resistance (NTSR). TSR is conferred mostly by dominant, semi-dominant and recessive alleles mainly involving the microtubule polymerization and fatty acid biosynthesis pathway. The herbicide target protein structure can alter due to amino acid substitution at a target binding site or change in the herbicide molecule. TSR towards the fungicides has attributed to increased production of target proteins. However, NTSR has revealed as predominant type of the resistance mechanism adapted under abiotic stress conditions, mainly against GP and ACC inhibitors. Here the herbicide mode of action is either by a diversion of electrons from PS I ferredoxin, inhibition of fatty acid elongase, stimulation of transport inhibition response protein 1 (TIR1), inhibition of ACCase/ALS/EPSP synthase and by alteration in a hormonal signalling pathway. In HR weeds, several proteins have identified for herbicide degradation (glutathione-*s*-transferase, hydrolases, esterases,

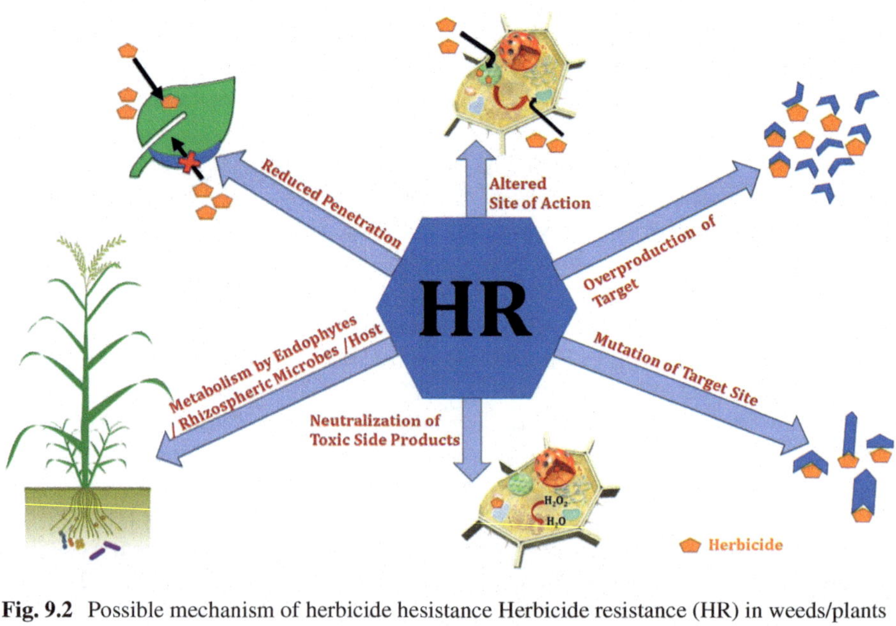

Fig. 9.2 Possible mechanism of herbicide hesistance Herbicide resistance (HR) in weeds/plants

cytochrome P450s) and supplementary oxidase, peroxidase and transport proteins. The TSR- or NTSR-induced resistance in the weed is likelihood depending on herbicide type and its mode of action, its usage rate, weed species and population size, plant fitness and environment (Busi et al. 2013; Delye et al. 2013). Figure 9.2 describes the possible mechanisms of herbicide actions, which after alterations can induce the HR in plants/weeds.

Moreover, endophytes that evolve with the host plant are associated in increasing the ecological fitness of plant by upregulating various defence genes including the herbicide degradation gene. Global Herbicide Resistance Action Committee (HRAC) reclassified herbicides into 18 groups depending on the mode of action. A plethora of research confirmed the role of weed-associated microbes in herbicide degradation (Prather et al. 2000; Powles and Yu 2010; Clay 2016) (Table 9.2).

Weed species, such as *E. indica, Echinochloa crus-galli, Amaranthus* sp., *P. annua, Echinochloa colona, A. myosuroides* and *L. rigidum*, developed herbicide resistance especially due to their congenital genetic variability. Additionally, herbicides (e.g. synthetic auxins and sulphonylurea) of different chemical groups and modes of action can significantly differ in their risk levels for resistance. However, herbicides (sulphonylurea, pyrimidinyl (thio)benzoate, sulfonyl-amino-carbonyl-triazolinone, imidazolinone) with different chemical groups but with the same mode of action can also be distinguished in their risk level for resistance. In recent studies, it has revealed that infection with a fungal endophyte gains the property of herbicide tolerance in the host plant. *L. rigidum* infection with *Neotyphodium* spp. resulting in increased tolerance to the graminicide DM in populations that were normally prone to the herbicide (Vila-Aiub et al. 2003). Relationships between the endophyte *Neotyphodium* spp. and tall fescue (*Festuca arundinacea*) have been observed as defensive mutualism. This interaction led to the alkaloid production by the endophyte or secondary metabolite by encouraging the host system. The endophytic metabolite protects the weed or host plant from the attack of insects, nematode or herbivore. However, there is uncertainty about the existence of the endophyte without its host, as it is completely dependent on the host for protection, a supply of the nutrition and water (Malinowski and Belesky 2000; Tozer et al. 2007; Perez et al. 2013). Alkaloid from fungal endophytes induced herbivore resistance in the host plants. The endophyte *Cryptosporiopsis* sp. confirm pathogen resistance in larch (*Larix decidua*) and barley (*H. vulgare*), while *Fusarium* sp. relief *Pyrenophora tritici-repentis* infection in wheat (*Triticum* sp.). Endophytic fungus increased crop yield in apples (*Malus domestica*) by increasing the amount of soluble sugars. In another study, endophytes *Penicillium minioluteum* and *Ampelomyces* sp. were verified for enhanced tomato (*S. lycopersicum*) and soybean (*G. max*) plant growth under the drought and salt tolerance condition. Isolate *Fusarium culmorum* from the coastal dune grass (*Leymus mollis*) confers salinity tolerance (300–500 mM NaCl) in soybean plant (Morsy et al. 2020). Plant growth has shunted when exposed to high or low temperatures as it disrupts the cell membrane and the photosynthetic apparatus. A mutualistic association between tropical weed panic grass/rosette grasses and a fungal endophyte allows both organisms to grow at high soil temperatures. *Dichanthelium lanuginosum* plants infected with fungal endophyte *Curvularia*

Table 9.2 Herbicide, site of action and resistant weed as per a report of International Survey of Herbicide-Resistant Weeds (www.weedscience.org)

Sr. no.	Herbicide type	Site of action	Herbicide-resistant weeds/ plants	Herbicide-degrading microbes
1	Clethodim, quizalofop, clodinafop, DM, FE	ACCase inhibitor	*Digitaria sanguinalis, Setaria faberi, S. viridis* var. *major, Avena fatua, A. myosuroides, Lolium multiflorum, Lolium rigidum, Avena sterilis, Echinochloa phyllopogon, Phalaris minor*	*Rhodococcus* sp. *T1*
2	Imazethapyr, cloransulam, bispyribac-sodium, chlorimuron, propoxycarbazone, chlorsulfuron	ALS/AHSH inhibitor	*Sinapis arvensis, Digitaria sanguinalis, Iva xanthifolia, Sorghum bicolor, Sinapis arvensis, Solanum ptycanthum, S. faberi, Setaria pumila, Echinochloa phyllopogon, S. viridis* var. *major, K. scoparia, H. annuus, Ambrosia artemisiifolia, Ambrosia trifida, Avena fatua, Amaranthus tuberculatus, Amaranthus retroflexus, Amaranthus hybridus, B. tectorum, A. myosuroides, L. rigidum, Phalaris minor, Rumex dentatus*	*Streptomyces* sp. strain PSI/5 *Pseudomonas and Bacillus*
3	Pendimethalin, trifluralin	Microtubule inhibitor	*L. rigidum, Eleusine indica, Poa annua*	*B. circulans, B. subtilis, Azotobacter chroococcum, B. megaterium, Fusarium oxysporum, Paecilomyces varioti, Rhizoctonia bataticola*
4	2,4-D, clopyralid, dicamba, mecoprop	Growth regulator (synthetic auxin)	*A. tuberculatus, K. scoparia, Stellaria media*	*Pseudomonas, Alcaligenes, Ralstonia, Delftia, Arthrobacter and Burkholderia*
5	Atrazine, metribuzin, chlorotoluron, isoproturon,	Photosynthesis inhibitor (PSII inhibitor) (triazine)	*Polygonum pensylvanicum, S. faberi, K. scoparia, A. tuberculatus, A. retroflexus, A. myosuroides, Phalaris minor, L. rigidum*	*Arthrobacter aurescens* TC-1, *Pseudomonas* sp. *strain* ADP

(continued)

Table 9.2 (continued)

Sr. no.	Herbicide type	Site of action	Herbicide-resistant weeds/ plants	Herbicide-degrading microbes
6	Bentazon, diuron	Photosynthesis inhibitor	*Sagittaria montevidensis, A. retroflexus, A. hybridus* (syn: *quitensis*)	*Beauveria bassiana, Caenorhabditis elegans, Phanerochaete chrysosporium, Mordellistena isabellina*
7	Glyphosate	EPSPS inhibitor	*K. scoparia, Ambrosia artemisiifolia, Ambrosia trifida, A. tuberculatus*	*Pseudomonas* sp., *Arthrobacter atrocyaneus, Flavobacterium* sp.
8	Glufosinate (phosphinothricin)	Glutamine synthetase inhibitor	*E. indica, L. perenne* ssp. *multiflorum, L. perenne, L. rigidum*	*Rhodococcus* sp.
9	Clomazone	HPPD inhibitor	*L. rigidum, Echinochloa crus-galli* var. *crus-galli*	*Aspergillus niger (UI-X172) and Cunninghamella echinulate (NRRL-3655)*
10	Carfentrazone, lactofen	Cell membrane disrupter (PPO inhibitor)	*A. tuberculatus, S. vernalis, Ambrosia artemisiifolia, Descurainia sophia, Amaranthus palmeri*	*Bacillus* sp.
11	Acetochlor, metolachlor, pendimethalin	Seedling shoot inhibitor (very long-chain fatty acid inhibitor) (VLCFA)	*Sorghum halepense, E. indica, P. annua, A. myosuroides, A. palmeri, Echinochloa crus-galli* var. *crus-galli, L. rigidum, A. tuberculatus*	*Achromobacter* sp. D 12, *Rhodococcus* sp. T3-1, *Bacillus* sp. ACD-9, *Delftia* sp.T3-6, *Sphingobium* sp.MEA31
12	Paraquat	Cell membrane disrupter (photosystem (PS)1 inhibitor)	*Arctotheca calendula* (L.) Levyns, *Epilobium adenocaulon* Hausskn, *Conyza bonariensis* (L.) Cronq., *Erigeron canadensis* L., *Erigeron philadelphicus* L., *Erigeron sumatrensis* Retz., *H. glaucum* Steud, *Hordeum leporinum* Link, *L. perenne*	*Lipomyces starkeyi, A. aerogenes, A. tumefaciens, P. fluorescens, B. cereus*
13	Bromoxynil	Pigment synthesis inhibitor	*C. album* L., *A. retroflexus, A. hybridus* (syn: *quitensis*), *S. vulgaris*	*Streptomyces felleus, Flexibacterium* sp., *Klebsiella pneumoniae, Flavobacterium* sp.

protuberata, exhibited tolerance to soil temperature at 50 °C for 3 days and as high as 65 °C for 10 days while non-infected plants died at 50 °C. Fungal endophytes have reported producing secondary metabolite pigment 'melanin' that may disperse in the hyphae or form a complex with oxygen radicals. Virus residing in a fungus can also associate to mediate the temperature tolerance (Redman et al. 2002; Marquez et al. 2007). Therefore, many endophytic bacteria and rhizobacteria increase the herbicide resistance of their host plants by degrading them, by producing toxic secondary metabolites or by inducing the inherent plant defence responses against pathogens also called induced systemic resistance (ISR) (Suryanarayanan 2019; Tétard-Jones and Edwards 2016). Systemic acquired resistance (SAR) allows ISR study in biotic or abiotic stress conditions. ISR induced by different endophyte factors are flagella, salicylic acid, siderophores, lipopolysaccharides, antimetabolites, jasmonic acid and *N*-acylhomoserine lactones (Fatema et al. 2019). Chen et al. (2017) demonstrated that at low concentration of DM, the rhizosphere microbes get affected and consequently allow cross talk between microbes and rice plant. This results in the stimulation of organic acid pathway (jasmonic acid or salicylic acid) and its release in root exudates. Most of the microbes utilize them as a carbon and nitrogen source and mark microbial abundance and diversity in the rhizosphere. Further induction of the stress response by the plant decreases its susceptibility to disease-associated pathogens. In another study, malic acid has been excreted by affected plant thereby inducing capsulated *B. subtilis* FB17. The *B. subtilis* FB17 further degrades the herbicide directly as well as stimulates plants' endogenous defence mechanisms (Rudrappa et al. 2008; Chen et al. 2017). *Piriformospora indica,* a fungal isolate from the Indian Thar desert, was investigated for growth-promoting activity in barley plant (*H. vulgare* L). The fungus induced plant salt stress tolerance and enhanced plant growth. Moreover, it conferred resistance to pathogenic fungus *Fusarium culmorum* (root rot) and the biotrophic fungus *Blumeria graminis* (Waller et al. 2005). Soil isolate *Klebsiella ozaenae* uses bromoxynil as sole nitrogen source and converts it to 3,5-dibromo-4-hydroxybenzoic acid by nitrilase enzyme (bxn). The plasmid-coded bxn gene is transferred to make transgenic herbicide-resistant tobacco and tomato plant. Phosphinothricin-degrading tripeptide (bar gene) from *S. hygroscopicus* was introduced into the tomato, potato and tobacco plant (Mazur and Falco 1989). Buss and Callaghan (2008) explained the possible role and interaction of pesticides with p-glycoprotein (p-gp) of ABC family in inducing herbicide, insecticide and fungicide resistance in plants. Upregulation of p-gp induces beneficial herbicide resistance in genetically modified (GM) plants. Applications of chemomodulators (natural and synthetic) and herbicide safeners further maximize the efficacy of herbicide actions for weed control (Buss and Callaghan 2008). Endophyte *Neotyphodium coenophialum* (ex *Acremonium*) from family Clavicipitaceae and their sexual antecedents in genus *Epichloe* are estimated to infect 20–30% of grass species asymptomatically. Saikkonen et al. (2013) demonstrated that endophytes *N. coenophialum* promote the competitive dominance of meadow fescue (*Scherodonus pratensis* and *Festuca pratensis*) and prevent weed invasion in the field (Saikkonen et al. 2013). The mutualistic association of fungi with the plant root system has been known as arbuscular mycorrhizal fungi (AMF). The association of AMF has several benefits to host weed (mycotrophic weeds). AMF aids in increasing surface area using the fungal mycelium network in the soil and allows better nutrient absorption for the host such as accumulation of ammonium nitrogen, potassium, calcium, manganese, iron, copper, zinc and nickel. Apart from the nutrient's absorption from the soil, AMF enhances seed germination, seed quality and seed production in mycotrophic weed. *Plantago lanceolata* L., an agricultural important weed and mycotrophic host for AMF, has strongly benefitted from the mycelium presence. Whereas, AMF infestation to non-mycotrophic weeds (non-host) like *C. album* L. and *Spergula arvensis* L. has shown an antagonistic effect. The presence of AMF mycelia to these non-host weeds has reduced survival rates and has resulted in the stunting effect of seedlings. AMF has benefitted host weed by reducing the biomass of the non-host (non-mycotrophic weeds like *Rumex obtusifolius* L., *Polygonum lapathifolium* L., *C. album, Brassica kaber, Portulaca oleracea* L.) by 50–90%. AMF has also proved to be beneficial for host plant to combat environmental stress like drought, high soil temperature, low light levels, adverse soil pH, saline soil and toxic metals. Mycorrhizal symbiosis with *Abutilon theophrasti* revealed greatly enhanced vegetative growth and flower, fruit and seed production (Stanley et al. 1993; Jordan et al. 2000; Begum et al. 2019).

Further, almost all plants are in the mutualistic association with microorganisms, and those who lack endophytes are more vulnerable to environmental stress and pathogenic attacks. Moreover, several findings on the mutualistic association revealed the importance of microorganisms in diverse conditions of the field with a net benefit to each participant.

4 Endophytes: A Promising Factor in Weed Management

Apart from producing metabolites supporting herbicide resistance, few bacterial as well as fungal endophytes have the potential of synthesizing toxic metabolites that show herbicidal activity. These metabolites induce chlorosis followed by necrosis in *L. minor*. Moreover, several rhizobacteria such as *Pseudomonas*, *Xanthomonas* sp., *Enterobacter* and *Serratia* had been developed as foliar and soil application bioherbicides. In another similar study, strains *P. viridiflava* CDRTc14 and *P. fluorescens* WH6 demonstrated similar phytotoxic effects on both *Lepidium draba* and *Lactuca sativa* seedlings. The results revealed germination arrest of grasses due to biosynthesis of 4-formylaminooxyvinylglycine (FVG) metabolite production using in vitro assay. Isolate *P. viridiflava,* associated with the weed *L. draba* L., revealed inhibitory effect towards its host by the lytic activity against pectin and also involved in heavy metal stress tolerance (Hoagland 1990; Harding and Raizada 2015; Samad et al. 2017; Sindhu et al. 2018; Suryanarayanan 2019). Previous studies reported that endophytic actinomycetes could be a source of herbicidal metabolites too (Singh et al. 2018). Thus, inoculation of such rhizosphere microorganisms and endophytes aids to minimize competition of weeds with crops and possibly reduces the usage of chemical herbicides that could benefit agriculture by contributing to

Fig. 9.3 Interaction between weed, pathogen and environment

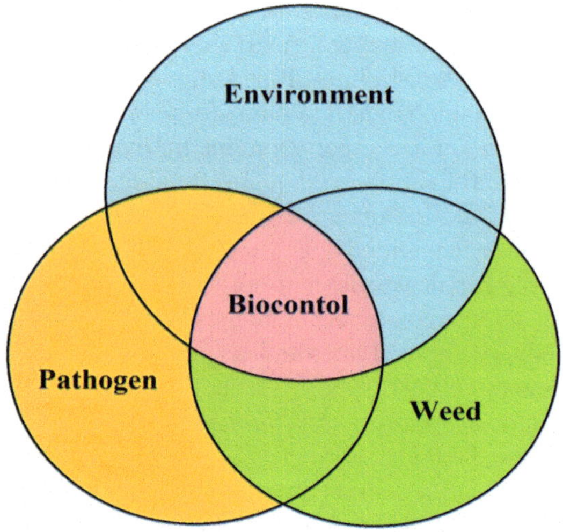

improve crop yields. Figure 9.3 represents the interactions of weed, pathogen and environment for successful biocontrol of HR plants/weeds (Sindhu et al. 2018).

The endophytic bacterial strains *B. pumilus, P. brassicacearum*, *B. methylotrophicus*, *B. amyloliquefaciens* and *B. cereus* were isolated from *P. lanceolata* and *U. dioica* plant. Isolate *Euphorbia helioscopia* was found effective against pathogenic *Agrobacterium* spp. and *Pectobacterium* spp. These strains enhanced tomato seed germination and plant growth (Krimi et al. 2016). The capability of *Stenotrophomona*s spp. as a biocontrol agent of plant pathogens has mentioned in various systems such as monocot and dicot crops as hosts. *S. maltophilia* strain C3 has a remarkable high hydrolytic potential. These species were found to be a promising source of various enzymes such as glucanases, DNases, chitinases, RNases, lipases, proteases and laccases (Berg et al. 1996; Galai et al. 2009; Islam 2011). The property of chitinolytic and proteolytic activity in *S. maltophilia* contributes to its biocontrol activity (Zhang and Yuen 1999; Zhang et al. 2001). *P. fluorescens* ISR 34 and *Bacillus* sp. ISR 37 isolated from *P. hysterophorus* inhibit downy mildew of pearl millet caused by *Sclerospora graminicola* by developing antimicrobial compounds or phytoalexins (Chandrashekhara et al. 2007). Thus, the endophyte screening for herbicidal (weedicide) properties should carry out to endorse the potentials in weed control. Screening of host-specific and non-host-specific phytotoxins as bioherbicides from pathogenic and non-pathogenic microbes (bacteria/fungi) may provide the new herbicides with novel characteristics. An array of these phytotoxins are identified form the weed microbes, viz. curvulins, eremophilanes, maculosin, ophiobolin, zinniol, AAL-toxins, AK-toxins and AM-toxins. Mycoherbicides are commercially developed for control of weed and for foliar application (curve surface of leaf/plant). A total of 14 fungi and one bacterium had been registered as bioherbicides in Canada, the Netherlands, China, Japan, South Africa and the USA, and mode of action of each biocontrol agent varies based on the type of microorganisms. These herbicides range from simple compounds (cyanide and organic acids) to complex molecule (plant growth regulators and secondary metabolites) such as auxins and ethylene (Sindhu et al. 2018). Several biocontrol agents have registered under the trade names of DeVine®, Collego™ and BioMal™. The mycoherbicide named DeVine® (*Phytophthora palmivora*) has been used against strangler vine (*Morrenia odorata*) that targets mainly the basal stems and the seedling roots and thereby promotes anthracnose (a plant disease involving lesions, necrosis and hypoplasia). AAL-toxin, a natural metabolite of *Alternaria alternata* f. sp. *lycopersici* pathogen, has been patented as herbicides against a range of crops and weeds species (Kennedy 1999; Kennedy and Kremer 1996). Furthermore, a biological control agent must have the following properties: (1) the microorganism must specifically target the weed and should not have any negative impact on the crop, environment and human life, (2) the microorganism should grow in larger mass, (3) the microorganism must be genetically stable and (4) it should kill a significant percentage of weed and should be stable at any environmental conditions. Several bacteria, fungi and some virus-based bioherbicides have been developed and registered till today (Trognitz et al. 2016).

5 Conclusion

Weed management plays a crucial role in preventing economic loss in agriculture. The review highlighted the potential role of endophytes in various kinds of biotic and abiotic stress tolerance conditions. Several findings support the statement that the plant along with their endophyte has a maximum survival rate than that which lacks such kind of mutual interaction. Certain endophytes have shown some antagonistic effects on the non-host plants as well, and such inhibitory effect of microbes can be used to control weeds for sustainable agriculture practices. Also, it has seen that excess use of herbicides resulted in the pollution of soil and water bodies. Soil pollution leads to the soil infertility that can hamper agricultural processes. In recent studies, it has shown that certain free-living rhizospheric microbes and endophytes possess the ability to degrade herbicide and thus are promising species in biomineralization and phytoremediation. Moreover, excess of herbicides can give rise to the herbicide-tolerant weed or plant species; however, the role of endophytes in herbicide tolerance in weeds and plants has been shadowed. Different factors can influence the HR evolution in weed and plants which are genetic (type/frequency/number of resistance gene), the biology of weed (seed production capacity and longevity in soil), type of herbicide (chemical structure, residual activity and action site), crop rotation, environmental conditions and herbicide dose. Regardless of all these reasons, biochemical and agroecological research has still desired to understand the causes and consequences of herbicide resistance in weeds and crops, as well as to design the strategies for weed management.

Acknowledgement The authors are thankful to the Department of Science and Technology-Science and Engineering Research Board (DST-SERB), New Delhi, India, for providing the financial assistance in the form of JRF fellowship to one of the author KL. The authors are thankful to the Head of Department of Microbiology, Savitribai Phule Pune University, Pune, MS, India, for providing the necessary facility to conduct the DST-SERB project SERB/F/8336/2019-2020.

References

Anderson TA, Kruger EL, Coats JR (1994) Enhanced degradation of a mixture of three herbicides in the rhizosphere of an herbicide-tolerant plant. Chemosphere 28(8):1551–1557

Andrews M, Hodge S, Raven JA (2010) Positive plant microbial interactions. Ann Appl Biol 157(3):317–320

Arfarita N, Imai T, Kanno A, Yarimizu T, Xiaofeng S, Jie W, Higuchi T, Akada R (2013) The potential use of *Trichoderma viride* strain FRP3 in biodegradation of the herbicide glyphosate. Biotechnol Biotechnol Equip 27(1):3518–3521

Begum N, Qin C, Ahanger MA, Raza S, Khan MI, Ahmed N, Ashraf M, Zhang L (2019) Role of arbuscular mycorrhizal fungi in plant growth regulation: implications in abiotic stress tolerance. Front Plant Sci 10(1068):1–15

Berg G, Marten P, Ballin G (1996) *Stenotrophomonas maltophilia* in the rhizosphere of oilseed rape occurrence, characterization and interaction with phytopathogenic fungi. Microbiol Res 151(1):19–27

Betts KJ, Ehlke NJ, Wyse DL, Gronwald JW, Somers DA (1992) Mechanism of inheritance of diclofop resistance in Italian ryegrass (*Lolium multiflorum*). Weed Sci 40:184–189

Busi R, Vila-Aiub MM, Beckie HJ, Gaines TA, Goggin DE, Kaundun SS, Lacoste M, Neve P, Nissen SJ, Norsworthy JK, Renton M (2013) Herbicide-resistant weeds: from research and knowledge to future needs. Evol Appl 6(8):1218–1221

Buss DS, Callaghan A (2008) Interaction of pesticides with p-glycoprotein and other ABC proteins: a survey of the possible importance to insecticide, herbicide and fungicide resistance. Pestic Biochem Physiol 90(3):141–153

Card SD, Hume DE, Roodi D, McGill CR, Millner JP, Johnson RD (2015) Beneficial endophytic microorganisms of Brassica–a review. Biol Control 90:102–112

Chandrashekhara SN, Deepak SA, Amruthesh KN, Shetty PN, Shetty SH (2007) Endophytic bacteria from different plant origin enhance growth and induces downy mildew resistance in pearl millet. Asian J Plant Pathol 1(1):1–11

Chen S, Li X, Lavoie M, Jin Y, Xu J, Fu Z, Qian H (2017) Diclofop-methyl affects microbial rhizosphere community and induces systemic acquired resistance in rice. J Environ Sci 51:352–360

Clay S (2016) Chapter 43: identification and control of herbicide-resistant weeds. In: Clay DE, Carlson CG, Clay SA, Byamukama E (eds) iGrow corn: best management practices. South Dakota State University, Brookings

Degrassi G, Carpentieri-Pipolo V (2020) Bacterial endophytes associated to crops novel practices for sustainable agriculture. Adv Biochem Biotechnol 5(1):1099

Delye C, Jasieniuk M, Le Corre V (2013) Deciphering the evolution of herbicide resistance in weeds. Trends Genet 29(11):649–658

El Fantroussi S, Verschuere L, Verstraete W, Top EM (1999) Effect of phenylurea herbicides on soil microbial communities estimated by analysis of 16S rRNA gene fingerprints and community-level physiological profiles. Appl Environ Microbiol 65(3):982–988

Ellegaard-Jensen L, Aamand J, Kragelund BB, Johnsen AH, Rosendahl S (2013) Strains of the soil fungus Mortierella show different degradation potentials for the phenylurea herbicide diuron. Biodegradation 24(6):765–774

Fatema K, Mahmud NU, Islam MT (2019) Beneficial effects of weed endophytic bacteria: diversity and potentials of their usage in sustainable agriculture. In: Agronomic crops. Springer, Singapore, pp 349–364

Feng NX, Yu J, Zhao HM, Cheng YT, Mo CH, Cai QY, Li YW, Li H, Wong MH (2017) Efficient phytoremediation of organic contaminants in soils using plant–endophyte partnerships. Sci Total Environ 583:352–368

Frank AC, Saldierna Guzmán JP, Shay JE (2017) Transmission of bacterial endophytes. Microorganisms 5(4):70

Galai S, Limam F, Marzouki MN (2009) A new *Stenotrophomonas maltophilia* strain producing laccase. Use in decolorization of synthetics dyes. Appl Biochem Biotechnol 158(2):416–431

Germaine KJ, Liu X, Cabellos GG, Hogan JP, Ryan D, Dowling DN (2006) Bacterial endophyte-enhanced phytoremediation of the organochlorine herbicide 2, 4-dichlorophenoxyacetic acid. FEMS Microbiol Ecol 57(2):302–310

Gharde Y, Singh PK, Dubey RP, Gupta PK (2018) Assessment of yield and economic losses in agriculture due to weeds in India. Crop Prot 107:12–18

Han L, Zhao D, Li C (2015) Isolation and 2, 4-D-degrading characteristics of *Cupriavidus campinensis* BJ71. Braz J Microbiol 46(2):433–441

Harding DP, Raizada MN (2015) Controlling weeds with fungi, bacteria and viruses: a review. Front Plant Sci 6:659

Heap I (2014) Global perspective of herbicide-resistant weeds. Pest Manag Sci 70(9):1306–1315

Hoagland RE (1990) Microbes and microbial products as herbicides: an overview. 439:2–52

Hoagland RE (1990) Microbes and microbial products as herbicides: an overview. ACS Symposium Series 439:2–52

Hongming L, Xu L, Zhaojian G, Fan Y, Dingbin C, Jianchun Z, Jianhong X, Shunpeng L, Qing H (2015) Isolation of an aryloxyphenoxy propanoate (AOPP) herbicide-degrading strain *Rhodococcus ruber* JPL-2 and the cloning of a novel carboxylesterase gene (feh). Braz J Microbiol 46(2):425–432

Hou Y, Tao J, Shen W, Liu J, Li J, Li Y, Cao H, Cui Z (2011) Isolation of the fenoxaprop-ethyl (FE)-degrading bacterium *Rhodococcus sp.* T1, and cloning of FE hydrolase gene feh. FEMS Microbiol Lett 323(2):196–203

Hussain I, Aleti G, Naidu R, Puschenreiter M, Mahmood Q, Rahman MM, Wang F, Shaheen S, Syed JH, Reichenauer TG (2018) Microbe and plant assisted-remediation of organic xenobiotics and its enhancement by genetically modified organisms and recombinant technology: a review. Sci Total Environ 628:1582–1599

Islam MT (2011) Potentials for biological control of plant diseases by *Lysobacter spp.*, with special reference to strain SB-K88. In: Bacteria in agrobiology: plant growth responses. Springer, Berlin, pp 335–363

Jacobsen CS (1997) Plant protection and rhizosphere colonization of barley by seed inoculated herbicide degrading *Burkholderia (Pseudomonas) cepacia* DBO1 (pRO101) in 2, 4-D contaminated soil. Plant Soil 189(1):139–144

Jordan NR, Zhang J, Huerd S (2000) Arbuscular-mycorrhizal fungi: potential roles in weed management. Weed Res 40(5):397–410

Kennedy AC (1999) Soil microorganisms for weed management. J Crop Prod 2(1):123–138

Kennedy AC, Kremer RJ (1996) Microorganisms in weed control strategies. J Prod Agric 9(4):480–485

Kremer RJ (2005) The role of bioherbicides in weed management. Biopestic Int 1(3):4

Krimi Z, Alim D, Djellout H, Tafifet L, Mohamed-Mahmoud F, Raio A (2016) Bacterial endophytes of weeds are effective biocontrol agents of *Agrobacterium spp., Pectobacterium spp.*, and promote growth of tomato plants. Phytopathol Mediterr 55(2):184–196

Kryuchkova YV, Burygin GL, Gogoleva NE, Gogolev YV, Chernyshova MP, Makarov OE, Fedorov EE, Turkovskaya OV (2014) Isolation and characterization of a glyphosate-degrading rhizosphere strain *Enterobacter cloacae* K7. Microbiol Res 69(1):99–105

Kuklinsky-Sobral J, Araujo WL, Mendes R, Pizzirani-Kleiner AA, Azevedo JL (2005) Isolation and characterization of endophytic bacteria from soybean (*Glycine max*) grown in soil treated with glyphosate herbicide. Plant Soil 273(1–2):91–99
Kumar A, Trefault N, Olaniran AO (2016) Microbial degradation of 2, 4-dichlorophenoxyacetic acid: insight into the enzymes and catabolic genes involved, their regulation and biotechnological implications. Crit Rev Microbiol 42(2):194–208
Liu Y, Liu SJ, Drake HL, Horn MA (2011) Alphaproteobacteria dominate active 2-methyl-4-chlorophenoxyacetic acid herbicide degraders in agricultural soil and drilosphere. Environ Microbiol 13(4):991–1009
Liu M, Luo K, Wang Y, Zeng A, Zhou X, Luo F, Bai L (2014) Isolation, identification and characteristics of an endophytic quinclorac degrading bacterium *Bacillus megaterium* Q3. PLoS One 9(9):e108012
Malinowski DP, Belesky DP (2000) Adaptations of endophyte-infected cool-season grasses to environmental stresses: mechanisms of drought and mineral stress tolerance. Crop Sci 40(4):923–940
Marin-Morales MA, De Campos Ventura-Camargo B, Hoshina MM (2013) Toxicity of herbicides: impact on aquatic and soil biota and human health. In: Herbicides–current research and case studies in use. IntechOpen, London, pp 399–443
Marquez LM, Redman RS, Rodriguez RJ, Roossinck MJ (2007) A virus in a fungus in a plant: three-way symbiosis required for thermal tolerance. Science 315(5811):513–515
Martinez CO, Silva CMM, Fay EF, Maia ADHN, Abakerli RB, Durrant LR (2008) Degradation of the herbicide sulfentrazone in a Brazilian Typic Hapludox soil. Soil Biol Biochem 40(4):879–886
Mazur BJ, Falco SC (1989) The development of herbicide resistant crops. Annu Rev Plant Biol 40(1):441–470
Mishra S, Upadhyay RS, Nautiyal CS (2013) Unravelling the beneficial role of microbial contributors in reducing the allelopathic effects of weeds. Appl Microbiol Biotechnol 97(13):5659–5668
Morsy M, Cleckler B, Armuelles-Millican H (2020) Fungal endophytes promote tomato growth and enhance drought and salt tolerance. Plan Theory 9(7):877
Moss SR, Rubin B (1993) Herbicide-resistant weeds: a worldwide perspective. J Agric Sci 120(2):141–148
Mukhtar I, Khokhar I, Mushtaq S, Ali A (2010) Diversity of epiphytic and endophytic microorganisms in some dominant weeds. Pak J Weed Sci Res 16(3):287–297
Ngigi AN, Getenga ZM, Boga HI, Ndalut PK (2012) Biodegradation of s-triazine herbicide atrazine by *Enterobacter cloacae* and *Burkholderia cepacia* sp. from long-term treated sugarcane-cultivated soils in Kenya. J Environ Sci Health B 47(8):769–778
Nunes OC, Lopes AR, Manaia CM (2013) Microbial degradation of the herbicide molinate by defined cultures and in the environment. Appl Microbiol Biotechnol 97(24):10275–10291
Ozawa T, Yoshida R, Wakashiro Y, Hase H (2004) Improvement of simazine degradation by inoculation of corn and soybean plants with rhizobacteria. Soil Sci Plant Nutr 50(8):1295–1299
Perez LI, Gundel PE, Ghersa CM, Omacini M (2013) Family issues: fungal endophyte protects host grass from the closely related pathogen *Claviceps purpurea*. Fungal Ecol 6(5):379–386
Poursat BA, van Spanning RJ, de Voogt P, Parsons JR (2019) Implications of microbial adaptation for the assessment of environmental persistence of chemicals. Criti Rev Environ Sci Techol 49(23):2220–2255
Powles SB, Yu Q (2010) Evolution in action: plants resistant to herbicides. Annu Rev Plant Biol 61:317–347
Prather TS, Ditomaso JM, Holt JS (2000) History, mechanisms, and strategies for prevention and management of herbicide resistant weeds. In: Proceedings of the California Weed Science Society (CWSS), vol 52, pp 155–163
Qasem JR, Foy CL (2001) Weed allelopathy, its ecological impacts and future prospects: a review. J Crop Prod 4(2):43–119

Redman RS, Sheehan KB, Stout RG, Rodriguez RJ, Henson JM (2002) Thermotolerance generated by plant/fungal symbiosis. Science 298(5598):1581–1581
Rudrappa T, Czymmek KJ, Pare PW, Bais HP (2008) Root-secreted malic acid recruits beneficial soil bacteria. Plant Physiol 148(3):1547–1556
Saikkonen K, Ruokolainen K, Huitu O, Gundel PE, Piltti T, Hamilton CE, Helander M (2013) Fungal endophytes help prevent weed invasions. Agric Ecosyst Environ 165:1–5
Samad A, Antonielli L, Sessitsch A, Compant S, Trognitz F (2017) Comparative genome analyses of the vineyard weed endophyte *Pseudomonas viridiflava* CDRTc14 showing selective herbicidal activity. Sci Rep 7(1):1–15
Sandoval-Carrasco CA, Ahuatzi-Chacón D, Galíndez-Mayer J, Ruiz-Ordaz N, Juárez-Ramírez C, Martínez-Jerónimo F (2013) Biodegradation of a mixture of the herbicides ametryn, and 2, 4-dichlorophenoxyacetic acid (2, 4-D) in a compartmentalized biofilm reactor. Bioresour Technol 145:33–36
Saxena AD, Zhang RW, Bollag JM (1987) Microorganisms capable of metabolizing the herbicide metolachlor. Appl Environ Microbiol 53(2):390–396
Serfling A, Wirsel SG, Lind V, Deising HB (2007) Performance of the biocontrol fungus *Piriformospora indica* on wheat under greenhouse and field conditions. Phytopathology 97(4):523–531
Shaw LJ, Burns RG (2004) Enhanced mineralization of [U-14C] 2, 4-dichlorophenoxyacetic acid in soil from the rhizosphere of *Trifolium pratense*. Appl Environ Microbiol 70(8):4766–4774
Siciliano SD, Goldie H, Germida JJ (1998) Enzymatic activity in root exudates of Dahurian wild rye (*Elymus dauricus*) that degrades 2-chlorobenzoic acid. J Agric Food Chem 46(1):5–7
Sindhu SS, Sehrawat A (2017) Rhizosphere microorganisms: application of plant beneficial microbes in biological control of weeds. In: Microorganisms for green revolution. Springer, Singapore, pp 391–430
Sindhu SS, Khandelwal A, Phour M, Sehrawat A (2018) Bioherbicidal potential of rhizosphere microorganisms for ecofriendly weed management. In: Role of rhizospheric microbes in soil. Springer, Singapore, pp 331–376
Singh B (2013) Degradion of clodinafop propargyl by *Pseudomonas sp.* strain B2. Bull Environ Contam Toxicol 91(6):730–733
Singh B, Singh K (2016) Microbial degradation of herbicides. Crit Rev Microbiol 42(2):245–261
Singh AK, Singla P (2019) Biodegradation of diuron by endophytic *Bacillus licheniformis* strain SDS12 and its application in reducing diuron toxicity for green algae. Environ Sci Pollut Res 26(26):26972–26981
Singh B, Kaur J, Singh K (2014) Microbial degradation of an organophosphate pesticide, malathion. Crit Rev Microbiol 40(2):146–154
Singh H, Naik B, Kumar V, Bisht GS (2018) Screening of endophytic actinomycetes for their herbicidal activity. Ann Agrar Sci 16(2):101–107
Soltys D, Krasuska U, Bogatek R, Gwiazdowski A (2013) Allelochemicals as bioherbicides—present and perspectives. In Herbicides-current research and case studies in use IntechOpen: London, UK. https://doi.org/10.5772/56185
Sørensen SR, Bending GD, Jacobsen CS, Walker A, Aamand J (2003) Microbial degradation of isoproturon and related phenylurea herbicides in and below agricultural fields. FEMS Microbiol Ecol 45(1):1–11
Stanley MR, Koide RT, Shumway DL (1993) Mycorrhizal symbiosis increases growth, reproduction and recruitment of *Abutilon theophrasti* medic. in the field. Oecologia 94(1):30–35
Sturz AV, Christie BR, Nowak J (2000) Bacterial endophytes: potential role in developing sustainable systems of crop production. Crit Rev Plant Sci 19(1):1–30
Suryanarayanan TS (2019) Endophytes and weed management: a commentary. Plant Physiol Rep 24:576–579
Sviridov AV, Shushkova TV, Ermakova IT, Ivanova EV, Epiktetov DO, Leontievsky AA (2015) Microbial degradation of glyphosate herbicides. Appl Biochem Microbiol 51(2):188–195

Tétard-Jones C, Edwards R (2016) Potential roles for microbial endophytes in herbicide tolerance in plants. Pest Manag Sci 72(2):203–209

Tozer KN, Lucas RJ, Edwards GR (2007) Suppression of annual grass weeds by AR542 endophyte infection in dryland tall fescue pastures. N Z Plant Prot 60:164–167

Trognitz F, Hackl E, Widhalm S, Sessitsch A (2016) The role of plant–microbiome interactions in weed establishment and control. FEMS Microbiol Ecol 92(10):fiw138

Vila-Aiub MM, Martinez-Ghersa A, Ghersa CM (2003) Evolution of herbicide resistance in weeds: vertically transmitted fungal endophytes as genetic entities. Evol Ecol 17(5–6):441–456

Vrbnicanin S, Pavlovi D, Bozic D (2017) Weed resistance to herbicides. Herbicide resistance in weeds and crops, p 7 IntechOpen: London, UK. https://doi.org/10.5772/67979

Waller F, Achatz B, Baltruschat H, Fodor J, Becker K, Fischer M, Heier T, Hückelhoven R, Neumann C, Von Wettstein D, Franken P (2005) The endophytic fungus *Piriformospora indica* reprograms barley to salt-stress tolerance, disease resistance, and higher yield. PNAS 102(38):13386–13391

Wang Q, Xie S, Hu R (2013) Bioaugmentation with Arthrobacter sp. strain DAT1 for remediation of heavily atrazine-contaminated soil. Int Biodeterior Biodegrad 77:63–67

Wang Y, Li H, Feng G, Du L, Zeng D (2017) Biodegradation of diuron by an endophytic fungus *Neurospora intermedia* DP8-1 isolated from sugarcane and its potential for remediating diuron-contaminated soils. PLoS One 12(8):e0182556

Yu X, Zhang W, Lang D, Zhang X, Cui G, Zhang X (2019) Interactions between endophytes and plants: beneficial effect of endophytes to ameliorate biotic and abiotic stresses in plants. J Plant Biol 62(1):1–13

Zhang Z, Yuen GY (1999) Biological control of *Bipolaris sorokiniana* on tall fescue by *Stenotrophomonas maltophilia* strain C3. Phytopathology 89(9):817–822

Zhang Z, Yuen GY, Sarath G, Penheiter AR (2001) Chitinases from the plant disease biocontrol agent, *Stenotrophomonas maltophilia* C3. Phytopathology 91(2):204–211

Zhang W, Jiang F, Ou J (2011) Global pesticide consumption and pollution: with China as a focus. Proc Int Acad Ecol Environ Sci 1(2):125

Chapter 10
The Auspicious Role of Plant Growth-Promoting Rhizobacteria in the Sustainable Management of Plant Diseases

Abdulwareth Abdulkader Almoneafy, Mohamed Moustafa-Farag, and Heba I. Mohamed

Contents

A. A. Almoneafy (✉)
Department of Biology Sciences, College of Education and Science at Rada'a, Albaydaa University, Rada'a, Yemen

M. Moustafa-Farag
Institute of Agricultural Resources and Environment, Guangdong Academy of Agricultural Sciences, Guangzhou, Guangdong, China

Horticulture Research Institute, Agriculture Research Center, Vegetable Seed Production and Technology, Giza, Egypt

H. I. Mohamed
Department of Biological and Geological Sciences, Faculty of Education, Ain Shams University, Cairo, Egypt

© The Author(s), under exclusive license to Springer Nature Switzerland AG 2021

H. I. Mohamed et al. (eds.), *Plant Growth-Promoting Microbes for Sustainable Biotic and Abiotic Stress Management*,
https://doi.org/10.1007/978-3-030-66587-6_10

1 Introduction

The plant can interact with surrounding microflora either positively or negatively (Glick 2020b). The positive interaction occurs when adjacent microorganisms can stimulate plant growth and fortify plant health; however, the negative interaction can intervene rarely due to man-made disturbance in natural vegetation (Schippers 1992). The negative interaction of some microorganisms with plants leads to the reduction of plant growth and induction of a relative fault in some/all biological and physiological activities in plants. This dramatically results in a classical phenomenon called plant disease (Agrios 2005). Disease development in plants is among several factors that cause a reduction in crop yield and quality of agricultural products and corruption of cereal crops (Singh et al. 2019). To deal with such serious threats against our economic crops, several approaches have been applied to eliminate or alleviate the damages caused by those diseases such as chemical approach by using pesticides as the prime tactic in this concern (He et al. 2016). In contrast, an indiscriminate use of these synthetic chemicals by agricultural producers resulted in rising several drawbacks like development of phytopathogens resistance toward such chemicals, the high costs of using these chemicals in the agricultural sector, and their residues' great risk on the environment and human being soundness (Compant et al. 2005; Fortunati et al. 2019; Lamichhane et al. 2018). Therefore necessity arises to looking for another management approach that is safe to non-target organisms, sustainable, and highly recyclable and has fewer costs than chemical pesticides (Fortunati et al. 2019; Wyckhuys et al. 2019). Interestingly some microorganisms that have positive interactions with plants have also the capability to suppress soil-borne pathogens through the exploitation of many mechanisms, and such competence can be used as an environmentally friendly means to manage those kinds of plant diseases. This type of management approach is termed as a biological control (Akhtar and Siddiqui 2011; Siddiqui 2006). Generally, bacterial species that can effectively colonize the plant roots or rhizosphere, stimulate plant growth, and strengthen plant health besides their ability to suppress plant disease are named plant growth-promoting rhizobacteria (PGPR) (Glick 2012; Olanrewaju et al. 2017). PGPR can provide the plants with those benefits in two ways, directly and indirectly (Köhl et al. 2019). Direct way represents the direct antagonistic effect of those bacteria against plant pathogens, and this can be attained whether by contrariness or nutrient uptake competition, infection site colonization, and physical niche occupation (Pal and Gardener 2006). While indirect interaction involves plant resistance induction toward phytopathogens and growth facilitation of plants (Prasad et al. 2015; Vos et al. 2015). This chapter highlights PGPR's role in the biological management of plant diseases.

2 Biological Management Approaches of Plant Diseases: An Overview

Biological control, in its broad definition by Raymaekers et al. (2020), is the use of living organisms (plants, animals, and microorganisms) or their secreted compounds or even their biological extracts for suppressing or mitigating phytopathogen growth and proliferation. This definition elaborated that we can employ three levels of biological management of plant diseases. The first one uses the whole organism whether it is an animal, plant, or microorganism; for instance, there are many investigations reported on application of nematodes as biological agents against phytopathogenic fungi due to their ability to feed on those fungal pathogens (Askary 2010; Chen and Ferris 2000; Ishibashi and Choi 1991; Okada 2001). Parallel to this, intercropping/cultivar mixtures approach (recently termed as *biocontrol plants*) is a useful means for reducing exposed plant area to phytopathogen inoculum in the field particularly in case of foliar diseases (Almoneafy 2006; Parolin et al. 2014; Vidal et al. 2020). Furthermore, microorganisms (fungi, bacteria, and viruses) were widely exploited in plant disease biological control, and they proved their proficiency to lessen the severe effect of diseases on plants in many studies (Almoneafy et al. 2012, 2014; Frampton et al. 2014; Horinouchi et al. 2007; Kering et al. 2019; Oskiera et al. 2015). The second direction in the biological management of phytopathogens is with the use of secreted compounds obtained from living organisms that revealed antimicrobial activities against plant pathogens; in this regard, antimicrobial peptides have been regarded as plant protectant products. These peptides have a short sequence with usually less than 50 amino acid residues (Montesinos 2007). Plants, insects, marine invertebrates, bacteria, amphibians, and mammals could secrete their derivatives as the first defensive line against deleterious microbial invasion (Toke 2005). Many kinds of research have demonstrated the antimicrobial effect of these bioactive compounds in bacteria (Ahsan et al. 2017; Ongena and Jacques 2008; Raaijmakers et al. 2006; Sarwar et al. 2018), plants (Lay and Anderson 2005), and animals (Zasloff 2002, 2019). The third way in this control approach includes the application of biological extracts gained from living organisms to diseased plants to reduce/mitigate disease incidence. Although most literature concerned with this approach focused on using and even formulating bioactive constituents extracted from plants to control phytopathogens (Borges et al. 2018; Osorio et al. 2010; Simonetti et al. 2020; Uppal et al. 2008). However, many studies have reported on the biocontrol activities of biological extracts of other organisms such as yeast (Zhang et al. 2020c), fungi (Tomas-Grau et al. 2020), and seaweed (El-Sheekh et al. 2020). Likewise, chitosan (a chemical treatment derivative of chitinous shell of crustaceans) was widely used to prevent plant diseases as a biocontrol agent (Almoneafy et al. 2014; Kakar et al. 2014; Lin et al. 2020; Prasad et al. 2020; Vanti et al. 2020; Zhang et al. 2020d).

3 The Most Documented Species of PGPR for the Biological Management of Plant Diseases

Investigators on PGPR research are constantly working either to unravel novel species or to verify new features found in pre-discovered species. However, several PGPR species are frequently mentioned in relevant literatures on this topic. In this section, we will discuss to some extent the role of the most documented PGPR species in the biological control of plant diseases.

3.1 Bacillus and Related Species

In agricultural systems, the *Bacillus* genus is aerobic, ubiquitous, and endospore-forming bacteria. Native populations of this genus occur abundantly in most soils of the rhizosphere and colonize various parts of the plants inconsistently (Kloepper et al. 2004; Mahaffee and Kloepper 1997). Recently, due to its advantages over other PGPR strains, *Bacillus* spp. have gained a great deal of attention in inoculant formulations, stable rhizosphere soil conservation, and improved sustainability potential for agriculture (Saxena et al. 2020).

The ability to generate a broad range of suppressive structural components contributes to host defense response and spore formation, making them one of the best candidates for efficient plant disease control, making it easier to formulate their commercial products (Andrić et al. 2020; Fan et al. 2017; Fira et al. 2018; Francis et al. 2010; Köhl et al. 2019). Antimicrobial activities of *Bacillus* species are well reported in a large number of related studies; some of the recent *Bacillus* biocontrol activity-related studies are summarized in Table 10.1. The distinctive characteristic of *Bacillus subtilis* strains is that approximately 4–5% of their entire genome is devoted to secondary metabolites synthesis with a wide range of antagonistic capabilities (Stein 2005). Similarly, a *Bacillus amyloliquefaciens* FZB42 genome sequence analysis has shown that approximately 8.5% were dedicated to the production of secondary antimicrobial metabolites (Chen et al. 2007). Latterly, some isolates such as FZB42, QST713, or SQR9, previously referred to as *Bacillus subtilis* or *Bacillus amyloliquefaciens*, have been reclassified as *Bacillus velezensis* illustrating the plant-associated model species *Bacillus* (Dunlap et al. 2016; Fan et al. 2017). Plant-associated bacillus species maintain their high potential as a strong competitor in the rhizosphere niche by either emitting a vast range of volatile compounds or synthesizing secondary soluble bioactive metabolites. Regarding volatile compounds, Kai (2020) reported that about 231 volatiles have been investigated to date; some are exclusive for certain *Bacillus* isolate with non-antimicrobial capabilities, while others are more generally involved with antimicrobial activities. Whereas bioactive secondary metabolites secreted by *Bacillus* may be either ribosomally synthesized and modified after translational processes, such as bacteriocins and lantibiotics, or produced enzymatically by multimodular mega-enzymes, as with polyketides (PK), dipeptides or cyclic lipopeptides (Fig. 10.2, Caulier et al. 2019; Fira et al. 2018; Kaspar et al. 2019; Rabbee et al. 2019). There are several functions of *Bacillus* volatiles and secondary metabolites. First is their contribution in direct antagonism with phytopathogens, second is their capability to induce host systemic resistance, and third is their association with *Bacillus* efficient root colonization and biofilm formation (Fig. 10.1) (Caulier et al. 2019; Fan et al. 2018; Kai 2020; Nayak et al. 2020; Rabbee et al. 2019).

Table 10.1 Biocontrol activities of *Bacillus*-associated plants against some phytopathogens

PGPR strain	Delivery method	Targeted pathosystem	Biocontrol activity	References
Bacillus megaterium AB4	Spraying of PGPR bio preparations	Broccoli *Alternaria japonica*	≈84% Reduction of disease index in broccoli leaves and ≈62% reduction in postharvest florets	Acurio Vásconez et al. (2020)
Bacillus subtilis ALBA01	Seed immersion/seedling treatment	Onion *Setophoma terrestris*	About 50% reduction of disease incidence	Sayago et al. (2020)
Bacillus methylotrophicus DR-08	Soil drenching	Tomato *Ralstonia solanacearum*	Significant reduction of disease incidence in both pot and field trials	Im et al. (2020)
Bacillus subtilis V26	In vitro/in vivo treatment with PGPR biosurfactant	Grapes *Botrytis cinerea*	Antifungal activity of biosurfactant under laboratory/in vivo conditions	Khedher et al. (2020)
Bacillus velezensis B-36	Plant injection/foliar spraying	Lotus *Fusarium oxysporum*	About 77.1% reduction of disease incidence when delivering PGPR as plant injection	Wang et al. (2020)
Paenibacillus polymyxa IMA5	Seed dipping/soil drenching	Eggplant *Ralstonia solanacearum*	Reduction by ≈70% of wilt incidence in PGPR treated plants	Alamer et al. (2020)
Bacillus pumilus HR10	Seedling inoculation	Pine, *Rhizoctonia solani*	Decrease of disease incidence equal to 76.88% compared with untreated PGPR plants	Zhu et al. (2020)
Bacillus sp. WB	Soil drenching	Watermelon *Fusarium oxysporum* f. sp. *niveum*	Reduction of *Fusarium* wilt incidence and increase in the activities of beneficial bacteria in the rhizosphere	Xu et al. (2020)
B. vallismortis (Ps) *B. amyloliquefaciens* (PsL) *B. thuringiensis* (IMC8)	Seed dipping	Sweet pepper *Phytophthora capsici*	Reducing disease severity particularly with (PsL) strain	Bhusal and Mmbaga (2020)
Bacillus velezensis AR1	Soil drenching	Sesame *Alternaria sesami*	Reduction in disease severity to less than 10% due to application with PGPR combined with fungicide	Bayisa (2020)

(continued)

Table 10.1 (continued)

PGPR strain	Delivery method	Targeted pathosystem	Biocontrol activity	References
B. paralicheniformis (EAL)	Seed immersion	*Enicostema axillare Fusarium oxysporum*	Increase of defense enzymes production and reduction of wilt incidence	Jinal et al. (2020)
B. amyloliquefaciens (VB7)	Bulb dipping/soil drenching/foliar application	Lilium *Botrytis cinerea* (SEL)	Significant decrease in the incidence of disease depending on the method of delivery	Nakkeeran et al. (2020)
Bacillus velezensis strain XT1	Foliar spraying + root irrigation	Strawberry and tomato, *Botrytis cinerea*	Reduction of incidence and severity of disease by 50% and 60%, respectively Stimulation of the immune responses through jasmonic acid and ethylene pathway	Toral et al. (2020)
Bacillus subtilis/Bacillus amyloliquefaciens	Stem inoculation	Tomato *Agrobacterium tumefaciens* C58	Reduction in gall weight on tomato infected plants	Frikha-Gargouri et al. (2017)
Paenibacillus elgii JCK-5075	Fermentation broth of PGPR was applied by soil drenching	Cabbage *Pectobacterium carotovorum*/tomato *Ralstonia solanacearum*/red pepper, *Xanthomonas euvesicatoria*	Suppression of three disease development, in a dose-depending manner	Le et al. (2020)
Bacillus velezensis strain HC6	Treating maize grains with PGPR cell culture or cell-free supernatant	Maize *Aspergillus* spp./*Fusarium* spp.	Growth inhibition of multi-pathogenic fungi and reduction of aflatoxin and ochratoxin production	Liu et al. (2020)
Bacillus cereus KTMA4	Seed dipping	Tomato *Fusarium oxysporum/Alternaria solani*	Reduction of *Fusarium* wilt by 66% and *Alternaria* blight by 54%	Karthika et al. (2020)
Bacillus velezensis FJAT-46737	Soil drenching with PGPR culture or its twofold diluted supernatants or seedling immersion with its crude lipopeptide	Tomato *Ralstonia solanacearum*	Biocontrol efficiency = 66% with PGPR culture, 82% with diluted supernatant, and 96% with crude lipopeptide	Chen et al. (2020b)

PGPR strain	Delivery method	Targeted pathosystem	Biocontrol activity	References
Bacillus subtilis (SSR2I) *Bacillus flexus* (AIKDL)	Seed immersion	Tomato *Ralstonia solanacearum/Fusarium oxysporum*	Reduction of wilt incidence + activation of plant resistance treated with SSR2I strain	Jinal and Amaresan (2020)
Bacillus velezensis Ba168	Seed treatment	Tobacco *Phytophthora nicotianae*	Reduction in disease index on PGPR tobacco treated plants	Guo et al. (2020)

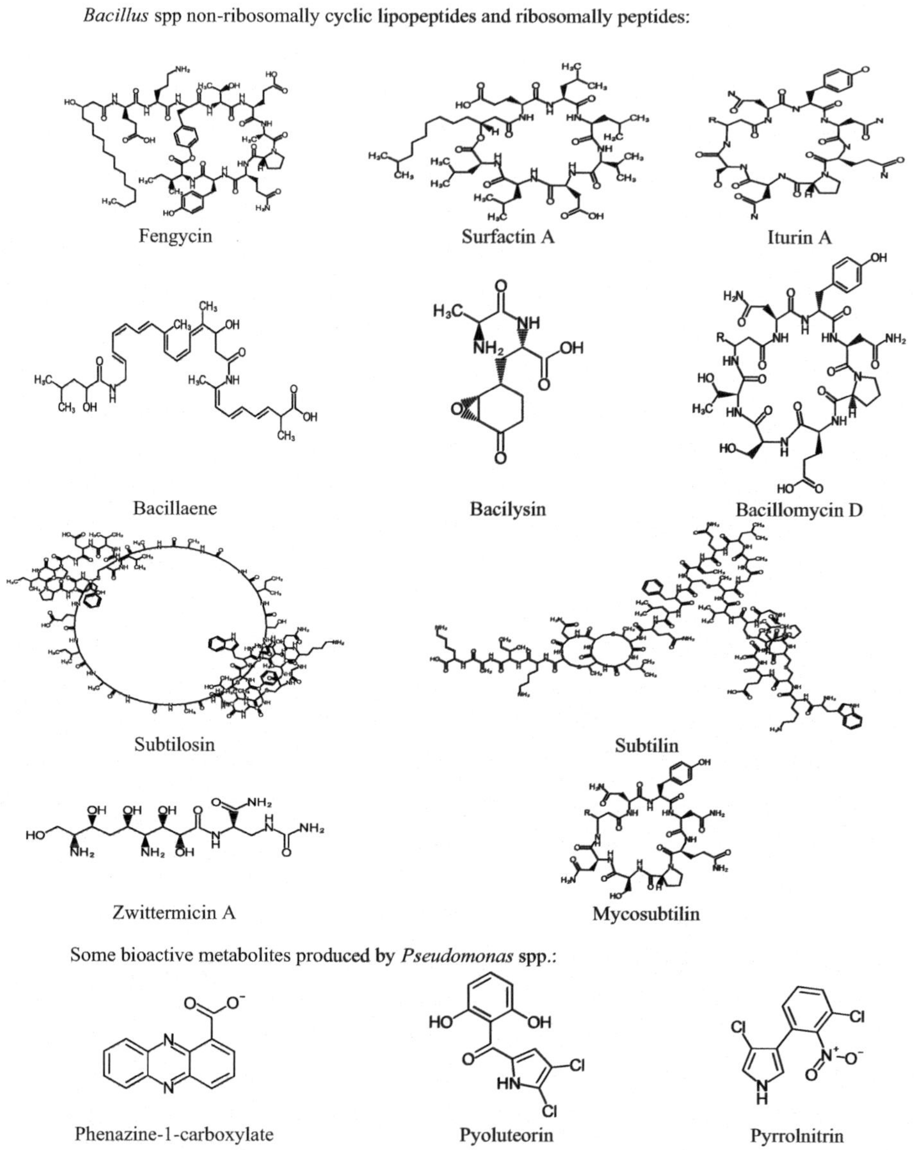

Fig. 10.2 The chemical structures of some antibiotics that are produced by different PGPR biocontrol agents. Some parts in this figure are created with Biorender.com

3.2 Pseudomonads

The *Pseudomonas* genus contains more than 100 species with many known plant-associated microbes that support plant growth and conquer plant disease (Nelkner et al. 2019). Plant-associated bacteria in this genus are aerobic, gram-negative, widespread in agricultural soil, and well suited for growth in the rhizosphere (Höfte

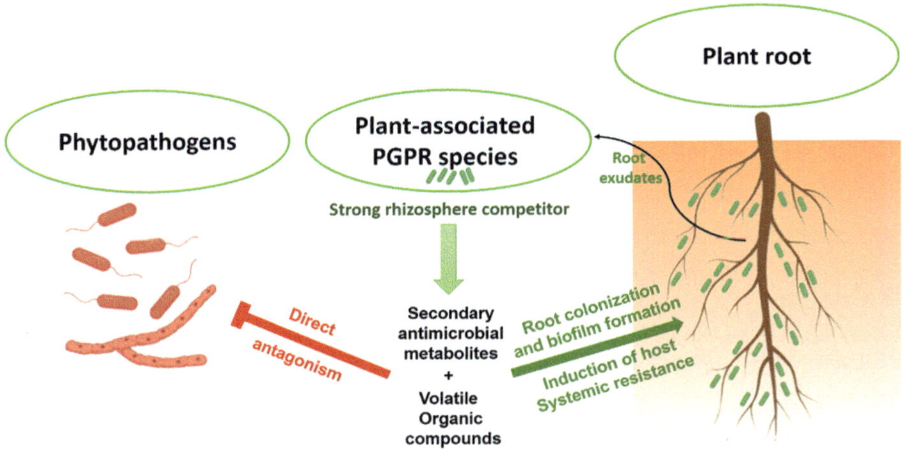

Fig. 10.1 Tripartite interaction between plant, PGPR, and phytopathogens in the rhizosphere. Some parts in this figure are created with Biorender.com

and Altier 2010; Mulet et al. 2010). The biocontrol capacities of *Pseudomonas* species associated with the rhizosphere have been widely investigated (Omoboye et al. 2019; Oni et al. 2015; Stringlis et al. 2018). *Pseudomonas* spp. pose many special properties which makes them suitable means for biological management of plant disease including their ability to (1) grow fast and their tendency for mass production; (2) nourish many kinds of root exudates and rapidly utilize them besides their capability to utilize a diverse range of available organic molecules in different niches as a source of energy; (3) colonize effectively spermosphere, rhizosphere, and phyllosphere; (4) compete effectively with other microorganisms for niche and nutrients due to their production of a wide range of antagonistic metabolites; and (5) acclimatize easily and broadly with different environments (Höfte and Altier 2010; Weller 2007). *Pseudomonas* spp. in general and *Pseudomonas fluorescens* in particular are extremely heterogeneous; for instance, sequencing of three *Pseudomonas fluorescens* strains revealed that they share about 60–64% of their genome, while a large fraction of their proteome (1146–1574 genes) is unique for each strain (Gross and Loper 2009).

Pseudomonas strains compete for ferric iron in an iron-deficient environment such as soil as most of them contain complex peptide siderophores (i.e., pyoverdines). Such siderophores are used for catching iron from the soil, and specific receptors in the bacterial outer membrane can recognize and take up the resulting siderophore-Fe complex. Subsequently, *Pseudomonas* siderophores enable them to colonize diverse ecological niches and to compete with pathogens for iron successfully (Haas and Défago 2005; Morrissey et al. 2004).

Moreover, plant-beneficial pseudomonads are extensively reported for their high capacity to produce several antibiotic compounds that have been comprehensively elaborated by Morrissey et al. (2004) and Raaijmakers et al. (2002). However, Haas and Défago (2005) demonstrated six classes of *Pseudomonas* antibiotic compounds as the most experimentally verified for their bioactivity against soil-borne pathogens including pyrrolnitrin, phloroglucinols, pyoluteorin, phenazines, cyclic lipopeptides (diffusible compounds; Fig. 10.2), and hydrogen cyanide (HCN; volatile compound). Another way for biocontrol activity of plant-beneficial *Pseudomonas* is represented with their capability to induce defense response in the plant (Alizadeh et al. 2013; Bakker et al. 2007; Couillerot et al. 2009; Höfte and Altier 2010; Morrissey et al. 2004; Oni et al. 2015; Ran et al. 2005). As biocontrol agents, the key disadvantage of *Pseudomonas* is their inability to develop endospores (as many *Bacillus* spp. do), which makes the formulation of these bacteria relatively difficult for commercial purposes (Weller 2007). Some of the recent studies related to *Pseudomonas* biocontrol efficacy against different plant pathogens were compiled in Table 10.2.

3.3 Streptomyces

Streptomyces is a group of thread-like (fungi-like), gram-positive, non-motile, non-acid-fast, aerobic, and positive catalase bacteria (Flärdh and Buttner 2009; Gopalakrishnan et al. 2020; Hasani et al. 2014). This genus with its peculiar terrestrial smell, high G+C (≈75%) in its DNA, and its large genomes is the most abundant in soil bacteria and actinomycetes that make up approximately 10% of the total soil microbial flora (Gopalakrishnan et al. 2020; Janssen 2006). Likewise, other soil microorganisms and their population are interacting negatively or positively with plants, and even part of them are free-living within the soil (Rey and Dumas 2017; Seipke et al. 2012). Since they can produce a wide variety of secondary bioactive metabolites, like antibiotics (more than two-thirds of the antibiotics discovered are totally or mainly produced by this bacteria), they have been used as commercial biocontrol products in agriculture, in addition to their involvement in the pharmaceutical industry (Luo et al. 2016; Peláez 2006). *Streptomyces* are reported to be able to degrade organic matter, solubilize phosphate, and produce organic acids, siderographs, and phytohormones beside their antimicrobial activities toward phytopathogens (Gonzalez-Franco and Hernandez 2009; Law et al. 2017; Passari et al. 2015). In the regard of biocontrol capability of *Streptomyces*, several *Streptomyces*-based biocontrol agents (secondary metabolites) are commonly used to manage plant diseases and insects, including blasticidin, kasugamycin, streptomycin, oxytetracycline, validamycin, polyoxins, natamycin, actinovate, mycostope, abamectin/avermectin, emamectin benzoate, polynactin, and milbemycin (Aggarwal et al. 2016; Flärdh and Buttner 2009). For their highly specific, readily degradable, and less environmentally toxic properties, these bioactive compounds are preferable, and these secondary metabolites are essential for effective competition with other microorganisms that come into contact (Aggarwal et al. 2016; de Lima Procópio et al. 2012; Gopalakrishnan et al. 2020). Moreover, *Streptomyces* can produce cell wall lytic enzymes such as chitinase, cellulase, protease, lipase, and β-1,3-glucanase

Table 10.2 Biocontrol determinants of plant-beneficial *Pseudomonas* against some plant disease-causing agents

PGPR strain	Delivery method	Targeted pathosystem	Biocontrol determinants	References
Pseudomonas chlororaphis ToZa7	Seed treatment	Tomato—*Fusarium oxysporum* f. sp. *radicis-lycopersici*	Pathogenesis-related (PR) genes *PR-1a* and *GLUA* upregulation + root colonization	Kamou et al. (2020)
Pseudomonas chlororaphis subsp. *aurantiaca* Pcho10	Spraying with PGPR cell suspension	Wheat—*Fusarium graminearum*	Phenazine-1-carboxamide (PCN) production	Hu et al. (2014)
Pseudomonas spp.	Soil and seedling treatment with crude CLPs PGPR extracts	Rice—*Magnaporthe oryzae*	Induction of host resistance (ISR) by cyclic lipopeptides (CLPs)	Omoboye et al. (2019)
Pseudomonas putida BP25	Seed treatment	Rice—*Magnaporthe oryzae*	Alter morphological change in roots + enhancing peroxidase and total phenol activities in rice	Ashajyothi et al. (2020)
Pseudomonas orientalis F9	Treatment with PGPR cell suspension on to the hypanthium of flowers	Apple—*Erwinia amylovora*	Competition potential for resources and space	Kron et al. (2020)
Pseudomonas fluorescens UM16, UM240, UM256, and UM270	Seedling treatments	*Medicago truncatula*—*Botrytis cinerea*	Diffusible volatile organic compounds	Hernández-León et al. (2015)
Pseudomonas aeruginosa PM12	Soil drenching with HMB	Tomato—*Fusarium oxysporum*	Production of ISR determinant (HMB)	Fatima and Anjum (2017)
Pseudomonas aeruginosa BRp3	Seed dipping	Rice—*Xanthomonas oryzae* pv. *oryzae*	Several secondary metabolites and defense-related enzymes	Yasmin et al. (2017)
P. fluorescens PICF7	Soil treatment	*Arabidopsis thaliana*—*Verticillium dahliae* Kleb	Induction of systemic resistance	Maldonado-González et al. (2015)

(continued)

Table 10.2 (continued)

PGPR strain	Delivery method	Targeted pathosystem	Biocontrol determinants	References
P. chlororaphis ssp. *aureofaciens* M71	Wound bacterization	Cypress—*Seiridium cardinale*	Phenazine-1-carboxylic acid (PCA) production	Raio et al. (2017)
Pseudomonas chlororaphis PCL1606	Root bacterization	Avocado—*Rosellinia necatrix*	2-hexyl, 5-propyl resorcinol (HPR) production	Calderón et al. (2014)
Pseudomonas sp. Ps14	Soil treatment	Cucumber—*Fusarium oxysporum* f. sp. *radicis cucumerinum* + *A. thaliana*, *Botrytis cinerea*	Activation of systemic resistance	Alizadeh et al. (2013)
P. aeruginosa 2apa	Seed immersion	Tomato foliar pathogens	Systemic resistance activation, accumulation of phenolics, elicitation of lipoxygenase activity	Hariprasad et al. (2014)
Pseudomonas spp.	Seed treatment	Bean—*Xanthomonas axonopodis* pv. *phaseoli*	Root colonization	Giorgio et al. (2016)
Pseudomonas strain Psd	Seedling treatment with pathogen spore suspension mixed with PGPR culture extract	Tomato—*Fusarium oxysporum* + *Verticillium dahliae*	Bioactivity of small RNAs (RsmX, RsmY, and RsmZ)	Upadhyay et al. (2017)
Pseudomonas fluorescens SBW25	Dual culture	Potato—*Streptomyces scabies* + *Phytophthora infestans*	Production of HCN and cyclic lipopeptides (CLPs)	Stefanato et al. (2019)
Pseudomonas putida RRF3	Root inoculation	Rice—PGPR interaction	Salicylic acid (SA)-mediated defense signaling mechanism in rice	Kandaswamy et al. (2019)
Pseudomonas fluorescens (P142)	Seed immersion	Tomato—*Ralstonia solanacearum* (B3B)	Priming plant defense response against bacterial wilt	Elsayed et al. (2020)
Pseudomonas putida MGP1	Fruit dipping with PGPR cell suspension	Papaya—*Phytophthora nicotianae*	ISR by activation of defense enzymes + PR1, NPR1	Shi et al. (2013)
Pseudomonas putida NH-50	Soil drenching	Sugarcane—*Glomerella tucumanensis*	Production of pyoluteorin antibiotic	Hassan et al. (2011)

in addition to the lignocellulolytic enzymes (Bhattacharya et al. 2009; Chamberlain 1997; Singh et al. 1999). Although *Streptomyces* spores are more sensitive to adverse conditions than *Bacillus* species endospores, they can survive and remain dormant for long periods under desiccated conditions (Filippova et al. 2005; Flärdh and Buttner 2009). This added advantage is also appropriate and necessary for the successful bioformulation of these microorganisms. *Streptomyces*-based biocontrol agents are usually formulated and applied as culture extract, spore suspension, emulsifiable concentrate, wettable powder, and wettable granules (Aggarwal et al. 2016; Arul Jose et al. 2013; Tamreihao et al. 2016). Several recent investigations concerned with the application of *Streptomyces*-based biocontrol agents in agriculture are given in Table 10.3.

4 Involved Mechanisms in the Biocontrol Activities of PGPR Species

A great deal of research has been accomplished to understand and elucidate the complex tripartite interactions between plants, pathogens, and PGPR. On the way to comprehend such interaction, many related mechanisms have verified and categorized into direct antagonistic mechanisms and indirect ones (Vos et al. 2015). Four biocontrol bioactivities are primarily based on PGPR direct antagonistic mechanisms including:

1. Parasitism of hyphae by certain *Streptomyces* strains, which is aided by secretion of lytic enzymes (e.g., chitinases and β-1,3-glucanases) (Singh and Gaur 2016)
2. Delivery of a significant number of non-volatile and/or volatile allelochemical inhibitors (antibiotics) such as iturin, butenolides, pyrones, siderophores, terpenoids, and peptaibols (Cao et al. 2018; Hou and Kolodkin-Gal 2020; Salwan and Sharma 2020)
3. Ecological niche or substrate competition which results in reduced pathogen infection pressure (Bloemberg and Lugtenberg 2001; Dobbelaere et al. 2003)
4. Interference with bacterial pathogen quorum sensing by secretion of specific enzymes which can degrade *N*-acyl homoserine lactones (AHLs), a bacterial quorum-sensing regulating molecule, such activity called as quorum quenching (Boyer et al. 2008; Dong et al. 2002; Kumar et al. 2011)

Inversely, indirect antagonisms related mechanisms arise from activities that do not include any physical interaction between PGPR and pathogen in inducing of host systemic immune (Bakker et al. 2007; Kloepper et al. 2004; Pieterse et al. 2014). Further discussion with some additional details on these biocontrol involved mechanisms is presented below.

Table 10.3 *Streptomyces*-based biocontrol agents and their biocontrol involved activities against some plant disease-causing agents

PGPR strain	Delivery method	Targeted Pathosystem	Biocontrol-involved activity	References
Streptomyces sp. MBFA-172	Spray with PGPR spore suspension	Strawberry—*Glomerella cingulata*	Effective colonization of aboveground plant parts	Marian et al. (2020)
Streptomyces palmae CMU-AB204^T	Mixing spore suspension or spores encapsulating alginate beads with soil	Palm oil—*Ganoderma boninense*	Production of bioactive compounds, i.e., actinopyrone A (1), anguinomycin A (2), and leptomycin A (3)	Sujarit et al. (2020)
Streptomyces jietaisiensis strain A034	Root treatment with a spore suspension	Chili—*Meloidogyne incognita*	Reduction of egg mass and decrease in the number of J2 of nematode	Ruanpanun and Nimnoi (2020)
Streptomyces pactum Act12	Soil and seedling treatment	Tomato—*Phelipanche aegyptiaca*	Inhibition of seed germination and germ tube elongation + reduction of parasitic plants emergence in greenhouse and field trials	Chen et al. (2020a)
Streptomyces badius gz-8	Incorporation of dinactin with culture media of pathogen + Leaves spraying with dinactin	Rubber—*Colletotrichum gloeosporioides*	Dinactin inhibited conidial formation and germination and decreased the spread of disease spots on the rubber leaves	Zhang et al. (2020a)
Streptomyces rubrogriseus HDZ-9-47	Soil treatment	Tomato—*Meloidogyne incognita*	Reduction of root galls + enrichment of beneficial microbes + reduction of certain soil-borne fungal phytopathogens	Jin et al. (2019)
Streptomyces sp. CNS-42	Soil treatment with mycelia suspension	Cucumber—*Fusarium oxysporum* f. sp. *cucumerinum*	Reduction of disease index due to production of bioactive compound staurosporine	Li et al. (2014)
Streptomyces sp. DHV3-2	Soil treatment with mycelia suspension	Tomato—*Verticillium dahliae*	Significant reduction of wilt incidence due to effective colonization of diseased roots by PGPR	Cao et al. (2016)
Streptomyces sp. MR14	Soil drenching with PGPR cells, supernatant and extract	Tomato—*Fusarium moniliforme*	Significant destruction of *Fusarium* wilt aided by PGPR cells, supernatant, and extract application	Kaur et al. (2019)

(continued)

Table 10.3 (continued)

PGPR strain	Delivery method	Targeted Pathosystem	Biocontrol-involved activity	References
Streptomyces griseorubens E44G	Incorporation of culture filtrate with pathogen PDA medium	Tomato—*Fusarium oxysporum* f. sp. *lycopersici*	Antifungal activity against *Fusarium* by secretion of chitinase	Al-Askar et al. (2015)
Streptomyces enissocaesilis		Sunflower—*Orobanche cumana*	Reduction of seed germination + number of weed tubercles + increase of host polyphenol oxidase activity and improve beneficial microflora	Chen et al. (2016a)
Streptomyces anulatus S07	Soil treatment with S07 wheat/sand inoculum	Wheat—cereal cyst nematode	Significant drop in female nematode population densities	Zhang et al. (2016)
Streptomyces blastmyceticus JZB130180	Fruit injection with PGPR fermentation broth	Peach—*Monilinia fructicola*	Antifungal activity due to secretion of bioactive compounds, i.e., chitinase, protease, cellulase, siderophores	Ni et al. (2019)
Streptomyces alboflavus TD-1	Pathogen culture treated with TD-1 wheat bran culture by two inverse face-to-face Petri dishes assay	*Aspergillus flavus*	Causing antifungal activity and downregulation of aflatoxin B1 biosynthesis genes	Yang et al. (2019)
Streptomyces palmae PC 12	Spraying rice seedlings with PGPR spore suspension	Rice—*Pyricularia* sp.	Significant reduction in disease severity due to effective foliar PGPR colonization and secretion of cell wall degrading enzymes	Chaiharn et al. (2020)
Streptomyces sp. NEAU-HV9	Seedling root dipping with PGPR suspension	Tomato—*Ralstonia solanacearum*	Effective disease suppression attributed to actinomycin D production	Ling et al. (2020)
Streptomyces sp. HAAG3-15	Soil drenching with PGPR spore suspension	Cucumber—*F. oxysporum* f. sp. *cucumerinum*	Decrease of disease severity and incidence + production of azalomycin B	Cao et al. (2020)
Streptomyces sp. PBSH9	Seed treatment was more effective than other delivery methods	Potato—*Streptomyces galilaeus*	Disease control up to 81% in greenhouse and between 47.64% and 73.97% in 3-year field trials	Zhang et al. (2020b)

(continued)

Table 10.3 (continued)

PGPR strain	Delivery method	Targeted Pathosystem	Biocontrol-involved activity	References
Streptomyces sp. LH4	Treatment plants with cultured PGPR	Cucumber—*Sclerotinia sclerotiorum*	Induction of disease resistance due to modulation of jasmonic acid and salicylic acid defense-related hormones	Mun et al. (2020)
Streptomyces sp. AgN23	Plants spraying with PGPR culture broth	*Arabidopsis thaliana*—*Alternaria brassicicola*	Colonization of host phyllosphere + induction of resistance by biosynthesis of salicylic acid	Vergnes et al. (2019)
Streptomyces katrae NB20	Fruits soaking with PGPR cultural suspension	Banana—*Colletotrichum musae*	Reduction of disease severity and pathogen conidial germination	Shu et al. (2017)
Streptomyces JD211	Soil mixing with PGPR powder formulation	Rice—*Magnaporthe oryzae*	Increased defensive enzyme activities and the expression of PR1 genes	Shao et al. (2018)

4.1 Direct Antagonistic Mechanisms

4.1.1 Parasitism

Parasitism is the connection between two species of organisms in which one profits to the detriment of another, often without harming the host organism (Lewin 1982). Our concern with parasitism is the ability of biocontrol agents to parasitize phytopathogens, which is called hyperparasitism. The classical example in this regard is an obligate parasitic relationship between bacterial species *Pasteuria penetrans* and root-knot nematodes (Chen and Dickson 1998; Davies et al. 1988; Lewin 1982). However, there is also a facultative parasitism in the case of *Streptomyces* bacteria on some fungal phytopathogens (Chen et al. 2016b; Tapio and Pohto-Lahdenperä 1991; Ziedan et al. 2010). *Streptomyces* isolate CC53 showed mycoparasitic activity against *Sclerotium rolfsii*, represented in coiling of CC53 filaments around pathogen hypha and spore proliferation resulting finally in the destruction of a pathogen (Singh and Gaur 2016). Similarly, Ziedan et al. (2010) reported hyperparasitism of *Streptomyces alni* on *Fusarium oxysporum* by inhibition of fungal growth colonization over pathogen hypha and suppression of conidia formation. As mentioned before, *Streptomyces*'s mycoparasitic capability was attributed to their cell wall degrading enzyme (e.g., chitinases and β-1,3-glucanases) production, since these enzymes initiate the physical destruction of the fungal cell wall (González-Franco and Robles-Hernandez 2009; Shrivastava et al. 2017).

4.1.2 Nutrient and Spatial Competition

Limited and specific nutrients and spaces in the rhizosphere-associated surfaces lead to severe competition and even exclusive selection of the resident population of soil microbes for these essential requirements (Ab Rahman et al. 2018; Ram et al. 2018). The competence of biocontrol agents in the rhizosphere requires a high potential of effective root colonization by native microflora (Whipps 1997, 2001). Effective root colonization can be achieved through three essential factors that must include in PGPR biocontrol agent; firstly Efficient competitors should be able to thrive and proliferate in the presence of a great diversity of microorganisms and phytopathogens via the metabolism of seed and root exudates, along with the plant roots, for a considerable time. Secondly, they should produce bioactive compounds that can inhibit the growth of a broad spectrum of neighboring microorganisms (Fukui et al. 1994). Thirdly, they should have the capability for biofilm formation along with their colonized niche; this ability enables them to establish an operative physical barrier that prevents any chemical or physical communications between colonized surfaces and other competitors (Davey and O'Toole 2000). Many studies reported that there are genetic and environmental criteria controlling the colonization process between plant roots and PGPR biocontrol agents (Bishnoi 2015; Bruto et al. 2014; Barriuso et al. 2008). For instance, root exudates could become a selective substrate for the stimulation of certain groups of microorganisms but not others. This can be attained by the fact that the exudates are more favorable to certain microorganisms than others are or that the exudates are toxic to other groups of microflora so that the enzyme of the colonizers is appropriate for the detoxification of these phytotoxic materials (Bais et al. 2004). Moreover, it is found that rice root exudates induce a stronger chemical response of endophytic bacteria than other rhizobacterial organisms (Bacilio-Jiménez et al. 2003). On the other side, rhizobacteria use several metabolic approaches such as glucose extracellular conversion to gluconic acid and 2-ketogluconic acid that effectively allow certain bacteria, like many *Pseudomonas* species, to conceal glucose and therefore provide a competitive advantage over other microorganisms without their ability to take advantage of those compounds (Gottschalk 1986).

4.1.3 Antimicrobial Metabolites (Siderophores, Antibiotics, Lytic Enzymes, and Volatile Organic Compounds)

The antimicrobial activity of PGPR biocontrol agents is the most significant mechanism used to suppress plant pathogens to harm the plant (Handelsman and Stabb 1996). This activity is attributed to the action of three types of rhizobacterial bioactive compounds, i.e., siderophores, antibiotics, and volatile organic compounds (Glick 2020a).

Siderophores are ferric-ion high-affinity, low-molecular mass compounds synthesized by siderophores producing PGPR (Chaiharn et al. 2009; Glick 2020a). These compounds can effectively chelate with Fe^{3+} ion found in the soil and take it back to the PGPR bacterial cell which can attach with bacterial membrane receptor and used for bacterial growth (Glick 2020a). Consequently, PGPR siderophores will bind with a large part of the limited concentration of soil Fe^{3+} ion and prevent surrounded phytopathogens (their siderophores have lower iron affinity than PGPR siderophores) from acquiring this ion sufficiently for their growth and proliferation. While the host plant is less effected with the iron-limited situation in soil than phytopathogens, also it can uptake the ferric-siderophore complexes and release their reductive iron for use in plant growth (Fig. 10.3; Ghosh et al. 2020; Glick 2020a).

PGPR antibiotics are secondary bioactive metabolites produced by PGPR biocontrol agents to suppress the growth of other competitor microorganisms as well as phytopathogens; they are synthesized either ribosomally or non-ribosomally in PGPR bacterial cells and secreted in their adjacent niche (Hou and Kolodkin-Gal 2020; Mavrodi et al. 2006; Tarkka and Hampp 2008).

A considerable number of PGPR-mediated antibiotics are well reported and characterized for their antimicrobial effect, chemical structure, and mode of action in many related studies such as lipopeptides and bacteriocins of *Bacillus*, i.e., fengycin, zwittermicin A, and bacillomycin D (Gu et al. 2017; Ongena and Jacques 2008; Qin et al. 2019; Villegas-Escobar et al. 2018); *Pseudomonas*, phenazine-1-carboxylate, 2,4-diacetylphloroglucinol, pyocyanine, and viscosin (Mavrodi et al. 2006; Mishra and Arora 2018; Omoboye et al. 2019); and *Streptomyces*, dioctatin, streptomycin, and kasugamycin (Fig. 10.2; Salwan and Sharma 2020; Tarkka and

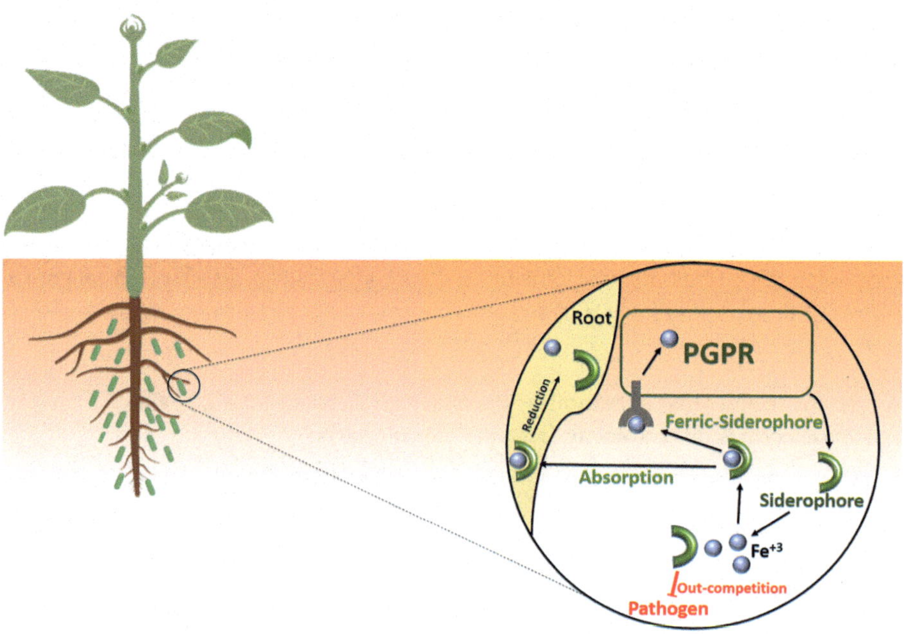

Fig. 10.3 Suggested pathways of exertion siderophores, soil iron-chelating, and iron acquisition by the plant and how these pathways lead to suppressing soil-borne pathogens by siderophores producing PGPR. Some parts in this figure are created with Biorender.com

Hampp 2008). Such antibiotics have antagonistic interaction with the plant pathogenic microbes causing fungosis and sporogenic inhibition, disturb the permeability of the cellular membrane, damage macromolecules, impede electron transport, and lyse fungal hyphae (Haas and Défago 2005; Kannojia et al. 2019; Kenawy et al. 2019; Niranjana and Hariprasad 2014). It is demonstrated that cyclic lipopeptides perform a crucial position in competition with other soil microbes, seeds/root colonization, and formation of biofilm (Nielsen et al. 2005; Raaijmakers et al. 2010; Tran et al. 2007). Another biocontrol activity of PGPR antibiotics, particularly under in vivo conditions, represented the activation of plant systemic resistance since low concentrations of these metabolites do not have enough suppressive activity against plant pathogens in the rhizosphere compared to their high antagonistic activity under laboratory conditions (Fernando et al. 2005).

Cellular lytic enzymes (e.g., chitinases and β-1,3-glucanases) are also considered as other important antimicrobial substances implicated in the plant pathogen suppression (Frikha-Gargouri et al. 2017; Singh and Gaur 2016). The mode of action of these enzymes ranges from the breakdown of fungal cell wall carbohydrates like chitin (the chief component in the phytopathogenic fungal cell wall) to the induction of some abnormal structures in fungal hyphae like bulges and/or rupture of hyphae tip and hyphal curling (Budi et al. 2000). These enzymes are secreted from both gram-positive PGPR like *Streptomyces* spp., *B. circulans*, *B. cereus*, and *B. licheniformis* (Sadfi et al. 2001; Singh and Gaur 2016) and gram-negative ones such as *P. fluorescens*, *Enterobacter agglomerans*, *Pseudomonas aeruginosa*, and *Serratia marcescens* (Kamensky et al. 2003; Neiendam Nielsen and Sørensen 1999).

Microbial volatiles are signal molecules characterized as having low molecular weight with a lipophilic nature, low boiling point, and high vapor pressure in normal conditions that are produced by catabolic pathways such as glycolysis and can travel through different ecological niches (Fincheira and Quiroz 2018). To date, about 231 volatile secondary metabolites have been characterized from *Bacillus* consisting mainly of nitrogen-containing compounds, ketones, hydrocarbons, aldehydes, alcohols, acids, ester, sulfur, and aromatics (Kai 2020). Similarly, several volatile inhibitory metabolites from *Pseudomonas* have been reported, like hydrogen cyanide (HCN), ketones, alcohols, sulfides, and aldehydes; however, HCN is the major volatile among them (Raza et al. 2016). Due to their eco-friendly properties and long-term protective capabilities on crops against plant pathogens, further attention was given to PGPR-producing volatiles by researchers (Hernández-León et al. 2015; Yang et al. 2019). These molecules have been shown to have antimicrobial and nematicidal activity and may induce systemic resistance in plants (Audrain et al. 2015; Fincheira and Quiroz 2018; Naznin et al. 2014).

4.1.4 Quorum Quenching

Several bacterial populations can sense their population density and control gene expression by a biological phenomenon called quorum sensing. Quorum sensing chemical signals can organize and synchronize many bacterial behaviors, such as microbial interactions and even interactions with their hosts in different environments (Cornforth et al. 2014). Quorum-sensing main signal molecules in gram-positive bacteria are oligopeptides and substituted gamma-butyrolactones, whereas in most gram-negative bacteria, the signal molecules for this phenomenon are *N*-acyl homoserine lactones (AHLs) (Faure et al. 2009). According to Whitehead et al. (2001), *N*-acyl homoserine lactones (AHLs) have been widely investigated as the most quorum-sensing regulating molecule. Several biological tools could quench the quorum-sensing phenomenon. For instance, many PGPR can produce an enzyme named as acyl homoserine lactonase (AiiA) able to degrade AHL molecules by hydrolyzing the lactone ring of these molecules; this enzyme was first reported in soil *Bacillus* isolates by Dong et al. (2000). They confirmed that this enzyme was able to inhibit AHL activity and alleviate the virulence of *Erwinia carotovora*. In other investigations, researchers identified two enzyme types which inactivate AHLs in several bacterial species/genera, i.e., AHL acylases and AHL lactonases, both of which have been shown to inhibit the biological activity of AHLs (Zhang and Dong 2004). PGPR quorum-quenching ability was shown to be the main biocontrol mechanism against phytopathogens (Molina et al. 2003).

4.2 Indirect Antagonistic Mechanisms

4.2.1 Induced Systemic Resistance

Van Loon et al. (1998) defined induced systemic resistance (ISR) as a state of improved defenses capabilities established by a plant that is exposed to specified biological or chemical inducements. In the opposite of the above-pointed mechanisms, ISR need not any physical contact between PGPR biocontrol agent and phytopathogen. Two researcher groups (Van Peer et al. 1991; Wei et al. 1991) discovered PGPR biocontrol agents as resistance inducers against phytopathogens. The advantages of ISR are its non-specificity for plant disease suppression, while other biological control activities are used for selected antagonists against only one or a few pathogens. For example, Hariprasad et al. (2014) report widespread suppression of Pseudomonas aeruginosa strain 2apa against many root and foliar pathogens of tomato by inducing systemic resistance in infected plants Several PGPR structural or secreted bioactive metabolites were shown to be determinants of ISR such as lipopolysaccharides (Van Peer and Schippers 1992), siderophores (Meziane et al. 2005), salicylic acid (De Meyer et al. 1999), pyochelin and pyocyanin antibiotics (Audenaert et al. 2002), 2,4 DAPG (Weller et al. 2004), volatiles (Song and Ryu 2013), *N*-acyl homoserine lactone (AHL) (Han et al. 2016), and flagella (Sumayo et al. 2013). Moreover, PGPR-related ISR bring about several structural and physiological alterations as defense responses in host plant like reinforcement of cell wall, suberization, formation of papillae, increase of defensive enzyme production (i.e., polyphenol oxidase and peroxidase), and production of pathogenesis-related proteins, thaumatin-like proteins, PR1, PR-2, chitinase, and β-1,3-glucanase (Raj et al. 2012; Van Loon et al. 1998; Zdor and Anderson 1992).

Finally, Glick (2020a) discussed another indirect biocontrol mechanism that alleviates the resulting damage of disease in plants by lowering stress ethylene in infected tissues at length; he reported that PGPR producing 1-aminocyclopropane-1-carboxylate (ACC) deaminase could reduce ethylene levels in both developed and stressed plant tissues.

5 Conclusion and Future Insights

PGPR is considered a promised means for sustaining ecological resources in different ecosystems particularly those ecosystems found in arid and semi-arid areas. Furthermore, these groups of bacteria can be applied as an alternative eco-friendly means of synthetic pesticides or at least can involve in integrated pest management programs against targeted plant diseases. Those PGPR species with spore formation capabilities are more favorable and applicable to low-cost bioformulation with high potential for the long shelf life of their products. More studies are still necessary to explore many of the ambiguous aspects of PGPR-pathogens-plant interactions. For instance, recently, a great interest was directed to the role of melatonin (one of the plant hormones) in the mitigation of biotic stress on plants (Moustafa-Farag et al. 2020). In this regard, it is demonstrated that some PGPR strains can produce melatonin in plants or stimulate the plant to produce endogenous melatonin (Jiao et al. 2016). So the key question here is to what extent will such melatonin-producing PGPR enhance plant resistance to phytopathogens? Based on the fact that root exudates can determine bacterial species in the rhizosphere, it is important to investigate which appropriate means can direct plant ability to exert selective root exudates to attract beneficial microorganisms. Few recent research projects have been allocated with the ecological impact of the delivery of PGPR biological agents on rhizosphere microflora, although this concern still has a considerable number of obscure facets to uncover. To reduce the performance variability of PGPR biocontrol agents, several endeavors could be helpful, such as applying PGPR consortiums with different biocontrol determinants or following certain delivery methods or even manipulating PGPR strains genetically to improve their biocontrol efficacies. As it has non-specific biocontrol activities against plant pathogens, PGPR-mediated ISR can be used to protect crops against a wide range of phytopathogens. Eventually, despite a few studies carried out with PGPR-plant-pathogen tripartite interactions using proteomics and genomic approaches, more research work is needed in this regard to clarify the exact role of each component in these interactions, which metabolites are involved, and how we can manage these interactions to obtain desirable findings.

References

Ab Rahman SFS, Singh E, Pieterse CMJ, Schenk PM (2018) Emerging microbial biocontrol strategies for plant pathogens. Plant Sci 267:102–111

Acurio Vásconez RD, Tenorio Moya EM, Collaguazo Yépez LA, Chiluisa-Utreras VP, Vaca Suquillo IDLÁ (2020) Evaluation of *Bacillus megaterium* strain AB4 as a potential biocontrol agent of Alternaria japonica, a mycopathogen of Brassica oleracea var. italica. Biotechnol Rep (Amst) 26:e00454

Aggarwal N, Thind SK, Sharma S (2016) Role of secondary metabolites of Actinomycetes in crop protection. In: Plant growth promoting actinobacteria. Springer, Berlin, pp 99–121

Agrios GN (2005) Plant pathology. Academic Press, San Diego

Ahsan T, Chen J, Zhao X, Irfan M, Wu Y (2017) Extraction and identification of bioactive compounds (eicosane and dibutyl phthalate) produced by *Streptomyces* strain KX852460 for the biological control of Rhizoctonia solani AG-3 strain KX852461 to control target spot disease in tobacco leaf. AMB Exp 7(1):54

Akhtar MS, Siddiqui ZA (2011) Role of plant growth promoting rhizobacteria in biocontrol of plant diseases and sustainable agriculture. In: Maheshwari DK (ed) Plant growth and health promoting bacteria. Springer, Berlin, pp 157–195

Alamer A, Sabah I, Tomah AA, Li B, Zhang J-Z (2020) Isolation, identification and characterization of rhizobacteria strains for biological control of bacterial wilt (Ralstonia solanacearum) of eggplant in China. Agriculture 10(2):37

Al-Askar AA, Baka ZA, Rashad YM, Ghoneem KM, Abdulkhair WM, Hafez EE, Shabana YM (2015) Evaluation of *Streptomyces griseorubens* E44G for the biocontrol of Fusarium oxysporum f. sp. lycopersici: ultrastructural and cytochemical investigations. Ann Microbiol 65(4):1815–1824

Alizadeh H, Behboudi K, Ahmadzadeh M, Javan-Nikkhah M, Zamioudis C, Pieterse CMJ, Bakker PAHM (2013) Induced systemic resistance in cucumber and Arabidopsis thaliana by the combination of Trichoderma harzianum Tr6 and *Pseudomonas* sp. Ps14. Biol Control 65(1):14–23

Almoneafy AA (2006) Pathological studies on Wheat Yellow Rust (Puccinia striiformis f. sp, tritici) in Yemen. Master thesis, Sana'a University, Sana'a, Yemen

Almoneafy AA, Xie GL, Tian WX, Xu LH, Zhang GQ, Ibrahim M (2012) Characterization and evaluation of *Bacillus* isolates for their potential plant growth and biocontrol activities against tomato bacterial wilt. Afr J Biotechnol 11(28):7193–7201

Almoneafy AA, Kakar KU, Nawaz Z, Li B, Saand MA, Chun-lan Y, Xie G-L (2014) Tomato plant growth promotion and antibacterial related-mechanisms of four rhizobacterial *Bacillus* strains against Ralstonia solanacearum. Symbiosis 63(2):59–70

Andrić S, Meyer T, Ongena M (2020) *Bacillus* responses to plant-associated fungal and bacterial communities. Front Microbiol 11:1350

Arul Jose P, Sivakala KK, Jebakumar SRD (2013) Formulation and statistical optimization of culture medium for improved production of antimicrobial compound by *Streptomyces* sp. JAJ06. Int J Microbiol

Ashajyothi M, Kumar A, Sheoran N, Ganesan P, Gogoi R, Subbaiyan GK, Bhattacharya R (2020) Black pepper (Piper nigrum L.) associated endophytic *Pseudomonas putida* BP25 alters root phenotype and induces defense in rice (Oryza sativa L.) against blast disease incited by Magnaporthe oryzae. Biol Control 143:104181

Askary TH (2010) Nematodes as biocontrol agents. In: Sociology, organic farming, climate change and soil science. Springer, Berlin, pp 347–378

Audenaert K, Pattery T, Cornelis P, Höfte M (2002) Induction of systemic resistance to Botrytis cinerea in tomato by *Pseudomonas aeruginosa* 7NSK2: role of salicylic acid, pyochelin, and pyocyanin. Mol Plant Microbe Int 15(11):1147–1156

Audrain B, Farag MA, Ryu C-M, Ghigo J-M (2015) Role of bacterial volatile compounds in bacterial biology. FEMS Microbiol Rev 39(2):222–233

Bacilio-Jiménez M, Aguilar-Flores S, Ventura-Zapata E, Pérez-Campos E, Bouquelet S, Zenteno E (2003) Chemical characterization of root exudates from rice (Oryza sativa) and their effects on the chemotactic response of endophytic bacteria. Plant Soil 249(2):271–277

Bais HP, Park S-W, Weir TL, Callaway RM, Vivanco JM (2004) How plants communicate using the underground information superhighway. Trends Plant Sci 9(1):26–32

Bakker PAHM, Pieterse CMJ, Van Loon LC (2007) Induced systemic resistance by fluorescent *Pseudomonas* spp. Phytopathology 97(2):239–243

Barriuso J, Solano BR, Lucas JA, Lobo AP, García-Villaraco A, Mañero FJG (2008) Ecology, genetic diversity and screening strategies of plant growth promoting rhizobacteria (PGPR). J Plant Nutr 4:1–17

Bayisa RA (2020) Enhancing resistance of Sesamum indicum against Alternaria sesami through *Bacillus velezensis* AR1. Pest Manage Sci

Bhattacharya A, Chanda S, Barik S (2009) Lipase and protease producing microbes from the environment of sugar beet field. Ind J Agric Biochem 22(1):26–30

Bhusal B, Mmbaga MT (2020) Biological control of Phytophthora blight and growth promotion in sweet pepper by *Bacillus* species. Biol Control 150:104373

Bishnoi U (2015) PGPR interaction: an ecofriendly approach promoting the sustainable agriculture system. Adv Bot Res 75:81–113. Elsevier

Bloemberg GV, Lugtenberg BJJ (2001) Molecular basis of plant growth promotion and biocontrol by rhizobacteria. Curr Opin Plant Biol 4(4):343–350

Borges DF, Lopes EA, Fialho Moraes AR, Soares MS, Visôtto LE, Oliveira CR, Moreira Valente VM (2018) Formulation of botanicals for the control of plant-pathogens: a review. Crop Prot 110:135–140

Boyer M, Bally R, Perrotto S, Chaintreuil C, Wisniewski-Dyé F (2008) A quorum-quenching approach to identify quorum-sensing-regulated functions in Azospirillum lipoferum. Res Microbiol 159(9):699–708

Bruto M, Prigent-Combaret C, Muller D, Moënne-Loccoz Y (2014) Analysis of genes contributing to plant-beneficial functions in plant growth-promoting rhizobacteria and related Proteobacteria. Sci Rep 4:6261

Budi SW, van Tuinen D, Arnould C, Dumas-Gaudot E, Gianinazzi-Pearson V, Gianinazzi S (2000) Hydrolytic enzyme activity of *Paenibacillus* sp. strain B2 and effects of the antagonistic bacterium on cell integrity of two soil-borne pathogenic fungi. Appl Soil Ecol 15(2):191–199

Calderón CE, de Vicente A, Cazorla FM (2014) Role of 2-hexyl, 5-propyl resorcinol production by *Pseudomonas chlororaphis* PCL1606 in the multitrophic interactions in the avocado rhizosphere during the biocontrol process. FEMS Microbiol Ecol 89(1):20–31

Cao P, Liu C, Sun P, Fu X, Wang S, Wu F, Wang X (2016) An endophytic *Streptomyces* sp. strain DHV3-2 from diseased root as a potential biocontrol agent against Verticillium dahliae and growth elicitor in tomato (Solanum lycopersicum). Anton Leeuw Int J G 109(12):1573–1582

Cao Y, Pi H, Chandrangsu P, Li Y, Wang Y, Zhou H, Xiong H, Helmann JD, Cai Y (2018) Antagonism of two plant-growth promoting *Bacillus velezensis* isolates against Ralstonia solanacearum and Fusarium oxysporum. Sci Rep 8(1):1–14

Cao P, Li C, Wang H, Yu Z, Xu X, Wang X, Zhao J, Xiang W (2020) Community structures and antifungal activity of root-associated endophytic actinobacteria in healthy and diseased cucumber plants and *Streptomyces* sp. HAAG3-15 as a promising biocontrol agent. Microorganisms 8(2)

Caulier S, Nannan C, Gillis A, Licciardi F, Bragard C, Mahillon J (2019) Overview of the antimicrobial compounds produced by members of the *Bacillus subtilis* group. Front Microbiol 10:302

Chaiharn M, Chunhaleuchanon S, Lumyong S (2009) Screening siderophore producing bacteria as potential biological control agent for fungal rice pathogens in Thailand. World J Microbiol Biotechnol 25(11):1919–1928

Chaiharn M, Theantana T, Pathom-aree W (2020) Evaluation of biocontrol activities of *Streptomyces* spp. against rice blast disease fungi. Pathogens 9(2):126

Chamberlain K (1997) Development of lignocellulolytic *Streptomyces* species as a biological agent to control thatch accumulation in turf. Master of Sciences Thesis, University of Idaho

Chen ZX, Dickson DW (1998) Review of Pasteuria penetrans: biology, ecology, and biological control potential. J Nematol 30(3):313

Chen J, Ferris H (2000) Growth and nitrogen mineralization of selected fungi and fungal-feeding nematodes on sand amended with organic matter. Plant Soil 218(1):91–101

Chen XH, Koumoutsi A, Scholz R, Eisenreich A, Schneider K, Heinemeyer I, Morgenstern B, Voss B, Hess WR, Reva O, Junge H, Voigt B, Jungblut PR, Vater J, Süssmuth R, Liesegang H, Strittmatter A, Gottschalk G, Borriss R (2007) Comparative analysis of the complete genome sequence of the plant growth–promoting bacterium *Bacillus amyloliquefaciens* FZB42. Nat Biotechnol 25(9):1007–1014

Chen J, Xue QH, McErlean CSP, Zhi JH, Ma YQ, Jia XT, Zhang M, Ye XX (2016a) Biocontrol potential of the antagonistic microorganism *Streptomyces enissocaesilis* against Orobanche cumana. Biocontrol 61(6):781–791

Chen Y-Y, Chen P-C, Tsay T-T (2016b) The biocontrol efficacy and antibiotic activity of *Streptomyces plicatus* on the oomycete *Phytophthora capsici*. Biol Control 98:34–42

Chen J, Xue Q, Ma Y, Chen L, Tan X (2020a) *Streptomyces pactum* may control Phelipanche aegyptiaca in tomato. Appl Soil Ecol 146:103369

Chen M, Wang J, Liu B, Zhu Y, Xiao R, Yang W, Ge C, Chen Z (2020b) Biocontrol of tomato bacterial wilt by the new strain *Bacillus velezensis* FJAT-46737 and its lipopeptides. BMC Microbiol 20(1):160

Compant S, Reiter B, Sessitsch A, Nowak J, Clément C, Ait Barka E (2005) Endophytic colonization of Vitis vinifera L. by plant growth-promoting bacterium Burkholderia sp. strain PsJN. Appl Environ Microbiol 71(4):1685–1693

Cornforth DM, Popat R, McNally L, Gurney J, Scott-Phillips TC, Ivens A, Diggle SP, Brown SP (2014) Combinatorial quorum sensing allows bacteria to resolve their social and physical environment. Proc Natl Acad Sci 111(11):4280–4284

Couillerot O, Prigent-Combaret C, Caballero-Mellado J, Moënne-Loccoz Y (2009) *Pseudomonas fluorescens* and closely-related fluorescent pseudomonads as biocontrol agents of soil-borne phytopathogens. Lett Appl Microbiol 48(5):505–512

Davey ME, O'Toole GA (2000) Microbial biofilms: from ecology to molecular genetics. Microbiol Mol Biol Rev 64(4):847–867

Davies KG, Kerry BR, Flynn CA (1988) Observations on the pathogenicity of Pasteuria penetrans, a parasite of root-knot nematodes. Ann Appl Biol 112(3):491–501

De Meyer G, Capieau K, Audenaert K, Buchala A, Métraux J-P, Höfte M (1999) Nanogram amounts of salicylic acid produced by the rhizobacterium *Pseudomonas aeruginosa* 7NSK2 activate the systemic acquired resistance pathway in bean. Mol Plant Microbe Interact 12(5):450–458

Dobbelaere S, Vanderleyden J, Okon Y (2003) Plant growth-promoting effects of diazotrophs in the rhizosphere. Crit Rev Plant Sci 22(2):107–149

Dong Y-H, Xu J-L, Li X-Z, Zhang L-H (2000) AiiA, an enzyme that inactivates the acylhomoserine lactone quorum-sensing signal and attenuates the virulence of Erwinia carotovora. Proc Natl Acad Sci 97(7):3526–3531

Dong Y-H, Gusti AR, Zhang Q, Xu J-L, Zhang L-H (2002) Identification of quorum-quenching N-acyl homoserine lactonases from *Bacillus* species. Appl Environ Microbiol 68(4):1754–1759

Dunlap CA, Kim S-J, Kwon S-W, Rooney AP (2016) *Bacillus velezensis* is not a later heterotypic synonym of *Bacillus amyloliquefaciens*; *Bacillus methylotrophicus*, *Bacillus amyloliquefaciens* subsp. *plantarum* and '*Bacillus oryzicola*' are later heterotypic synonyms of *Bacillus velezensis* based on phylogenomics. Int J Syst Evol Microbiol 66(3):1212–1217

Elsayed TR, Jacquiod S, Nour EH, Sørensen SJ, Smalla K (2020) Biocontrol of bacterial wilt disease through complex interaction between tomato plant, antagonists, the indigenous rhizosphere microbiota, and Ralstonia solanacearum. Front Microbiol 10:2835

El-Sheekh MM, Mousa ASH, Farghl AAM (2020) Biological control of fusarium wilt disease of tomato plants using seaweed extracts. Arab J Sci Eng 45(6):4557–4570

Fan B, Blom J, Klenk H-P, Borriss R (2017) *Bacillus amyloliquefaciens, Bacillus velezensis*, and *Bacillus siamensis* form an "Operational Group B. amyloliquefaciens" within the B. subtilis species complex. Front Microbiol 8:22

Fan B, Wang C, Song X, Ding X, Wu L, Wu H, Gao X, Borriss R (2018) *Bacillus velezensis* FZB42 in 2018: the Gram-positive model strain for plant growth promotion and biocontrol. Front Microbiol 9:2491

Fatima S, Anjum T (2017) Identification of a potential ISR determinant from *Pseudomonas aeruginosa* PM12 against Fusarium wilt in tomato. Front Plant Sci 8:848

Faure D, Vereecke D, Leveau JHJ (2009) Molecular communication in the rhizosphere. Plant Soil 321(1–2):279–303

Fernando WGD, Nakkeeran S, Zhang Y (2005) Biosynthesis of antibiotics by PGPR and its relation in biocontrol of plant diseases. In: PGPR: biocontrol and biofertilization. Springer, Berlin, pp 67–109

Filippova SN, Gorbatyuk EV, Poglazova MN, Soina VS, Kuznetsov VD (2005) Endospore formation by *Streptomyces avermitilis* in submerged culture. Microbiology 74(2):169–178

Fincheira P, Quiroz A (2018) Microbial volatiles as plant growth inducers. Microbiol Res 208:63–75

Fira D, Dimkić I, Berić T, Lozo J, Stanković S (2018) Biological control of plant pathogens by *Bacillus* species. J Biotechnol 285:44–55

Flärdh K, Buttner MJ (2009) *Streptomyces* morphogenetics: dissecting differentiation in a filamentous bacterium. Nat Rev Microbiol 7(1):36–49

Fortunati E, Mazzaglia A, Balestra GM (2019) Sustainable control strategies for plant protection and food packaging sectors by natural substances and novel nanotechnological approaches. J Sci Food Agric 99(3):986–1000

Frampton RA, Taylor C, Moreno AVH, Visnovsky SB, Petty NK, Pitman AR, Fineran PC (2014) Identification of bacteriophages for biocontrol of the kiwifruit canker phytopathogen *Pseudomonas syringae* pv. *actinidiae*. Appl Environ Microbiol 80(7):2216–2228

Francis I, Holsters M, Vereecke D (2010) The Gram-positive side of plant–microbe interactions. Environ Microbiol 12(1):1–12

Frikha-Gargouri O, Ben Abdallah D, Bhar I, Tounsi S (2017) Antibiosis and bmyB gene presence as prevalent traits for the selection of efficient *Bacillus* biocontrol agents against crown gall disease. Front Plant Sci 8:1363

Fukui R, Schroth MN, Hendson M, Hancock JG (1994) Interaction between strains of pseudomonads in sugar beet spermospheres and their relationship to pericarp colonization by Pythium ultimum in soil. Phytopathology 84(11):1322–1330

Ghosh SK, Bera T, Chakrabarty AM (2020) Microbial siderophore–a boon to agricultural sciences. Biol Control 144:104214

Giorgio A, Lo Cantore P, Shanmugaiah V, Lamorte D, Iacobellis NS (2016) Rhizobacteria isolated from common bean in southern Italy as potential biocontrol agents against common bacterial blight. Eur J Plant Pathol 144(2):297–309

Glick BR (2012) Plant growth-promoting bacteria: mechanisms and applications. Scientifica (Cairo) 2012:963401

Glick BR (2020a) Biocontrol of bacteria and fungi. In: Beneficial plant-bacterial interactions. Springer, Berlin, pp 181–230

Glick BR (2020b) Introduction to plant growth-promoting bacteria. In: Glick BR (ed) Beneficial plant-bacterial interactions. Springer, Cham, pp 1–37

Gonzalez-Franco AC, Hernandez LR (2009) Actinomycetes as biological control agents of phytopathogenic fungi. Techno Chih III 2:64–73

González-Franco AC, Robles-Hernandez L (2009) Actinomycetes as biological control agents of phytopathogenic fungi. Tecnociencia Chihuahua 3(2):64–73

Gopalakrishnan S, Srinivas V, Prasanna SL (2020) *Streptomyces*. In: Amaresan N, Kumar MS, Annapurna K, Kumar K, Sankaranarayanan A (eds) Beneficial microbes in agro-ecology. Academic Press, Boston, pp 55–71

Gottschalk G (1986) Regulation of bacterial metabolism. In: Bacterial metabolism. Springer, Berlin, pp 178–207

Gross H, Loper JE (2009) Genomics of secondary metabolite production by *Pseudomonas* spp. Nat Prod Rep 26(11):1408–1446

Gu Q, Yang Y, Yuan Q, Shi G, Wu L, Lou Z, Huo R, Wu H, Borriss R, Gao X (2017) Bacillomycin D produced by *Bacillus amyloliquefaciens* is involved in the antagonistic interaction with the plant-pathogenic fungus Fusarium graminearum. Appl Environ Microbiol 83(19):e01075

Guo D, Yuan C, Luo Y, Chen Y, Lu M, Chen G, Ren G, Cui C, Zhang J, An D (2020) Biocontrol of tobacco black shank disease (Phytophthora nicotianae) by *Bacillus velezensis* Ba168. Pest Biochem Phys 165:104523

Haas D, Défago G (2005) Biological control of soil-borne pathogens by fluorescent pseudomonads. Nat Rev Microbiol 3(4):307–319

Han S, Li D, Trost E, Mayer KF, Vlot AC, Heller W, Schmid M, Hartmann A, Rothballer M (2016) Systemic responses of barley to the 3-hydroxy-decanoyl-homoserine lactone producing plant beneficial endophyte Acidovorax radicis N35. Front Plant Sci 7:1868

Handelsman J, Stabb EV (1996) Biocontrol of soilborne plant pathogens. Plant Cell 8(10):1855

Hariprasad P, Chandrashekar S, Singh SB, Niranjana SR (2014) Mechanisms of plant growth promotion and disease suppression by *Pseudomonas aeruginosa* strain 2apa. J Basic Microbiol 54(8):792–801

Hasani A, Kariminik A, Issazadeh K (2014) Streptomycetes: characteristics and their antimicrobial activities. Int J Adv Biol Biomed Res 2(1):63–75

Hassan MN, Afghan S, Hafeez FY (2011) Biological control of red rot in sugarcane by native pyoluteorin-producing *Pseudomonas putida* strain NH-50 under field conditions and its potential modes of action. Pest Manage Sci 67(9):1147–1154

He D-C, Zhan J, Lian-Hui X (2016) Problems, challenges and future of plant disease management: from an ecological point of view. J Integr Agric 2016:60345–60352

Hernández-León R, Rojas-Solís D, Contreras-Pérez M, Orozco-Mosqueda MDC, Macías-Rodríguez LI, Reyes-de la Cruz H, Valencia-Cantero E, Santoyo G (2015) Characterization of the antifungal and plant growth-promoting effects of diffusible and volatile organic compounds produced by *Pseudomonas fluorescens* strains. Biol Control 81:83–92

Höfte M, Altier N (2010) Fluorescent pseudomonads as biocontrol agents for sustainable agricultural systems. Res Microbiol 161(6):464–471

Horinouchi H, Muslim A, Suzuki T, Hyakumachi M (2007) Fusarium equiseti GF191 as an effective biocontrol agent against Fusarium crown and root rot of tomato in rock wool systems. Crop Prot 26(10):1514–1523

Hou Q, Kolodkin-Gal I (2020) Harvesting the complex pathways of antibiotic production and resistance of soil bacilli for optimizing plant microbiome. FEMS Microbiol Ecol.

Hu W, Gao Q, Hamada MS, Dawood DH, Zheng J, Chen Y, Ma Z (2014) Potential of *Pseudomonas chlororaphis* subsp. *aurantiaca* strain Pcho10 as a biocontrol agent against Fusarium graminearum. Phytopathology 104(12):1289–1297

Im SM, Yu NH, Joen HW, Kim SO, Park HW, Park AR, Kim J-C (2020) Biological control of tomato bacterial wilt by oxydifficidin and difficidin-producing *Bacillus methylotrophicus* DR-08. Pestic Biochem Phys 163:130–137

Ishibashi N, Choi DR (1991) Biological control of soil pests by mixed application of entomopathogenic and fungivorous nematodes. J Nematol 23(2):175–181

Janssen PH (2006) Identifying the dominant soil bacterial taxa in libraries of 16S rRNA and 16S rRNA genes. Appl Environ Microb 72(3):1719–1728

Jiao J, Ma Y, Chen S, Liu C, Song Y, Qin Y, Yuan C, Liu Y (2016) Melatonin-producing endophytic bacteria from grapevine roots promote the abiotic stress-induced production of endogenous melatonin in their hosts. Front Plant Sci 7:1387

Jin N, Lu X, Wang X, Liu Q, Peng D, Jian H (2019) The effect of combined application of *Streptomyces rubrogriseus* HDZ-9-47 with soil biofumigation on soil microbial and nematode communities. Sci Rep 9(1):16886

Jinal NH, Amaresan N (2020) Evaluation of biocontrol *Bacillus* species on plant growth promotion and systemic-induced resistant potential against bacterial and fungal wilt-causing pathogens. Arch Microbiol.

Jinal HN, Sakthivel K, Amaresan N (2020) Characterisation of antagonistic *Bacillus paralicheniformis* (strain EAL) by LC–MS, antimicrobial peptide genes, and ISR determinants. Anton Leeuw Int J G 113(8):1167–1177

Kai M (2020) Diversity and distribution of volatile secondary metabolites throughout *Bacillus subtilis* isolates. Front Microbiol 11:559

Kakar KU, Duan YP, Nawaz Z, Sun G, Almoneafy AA, Hassan MA, Elshakh A, Li B, Xie G-L (2014) A novel rhizobacterium Bk7 for biological control of brown sheath rot of rice caused by *Pseudomonas fuscovaginae* and its mode of action. Eur J Plant Pathol 138(4):819–834

Kamensky M, Ovadis M, Chet I, Chernin L (2003) Soil-borne strain IC14 of Serratia plymuthica with multiple mechanisms of antifungal activity provides biocontrol of Botrytis cinerea and Sclerotinia sclerotiorum diseases. Soil Biol Biochem 35(2):323–331

Kamou NN, Cazorla F, Kandylas G, Lagopodi AL (2020) Induction of defense-related genes in tomato plants after treatments with the biocontrol agents *Pseudomonas chlororaphis* ToZa7 and Clonostachys rosea IK726. Arch Microbiol 202(2):257–267

Kandaswamy R, Ramasamy MK, Palanivel R, Balasundaram U (2019) Impact of *Pseudomonas putida* RRF3 on the root transcriptome of rice plants: insights into defense response, secondary metabolism and root exudation. J Biosci 44(4):98

Kannojia P, Choudhary KK, Srivastava AK, Singh AK (2019) PGPR bioelicitors: induced systemic resistance (ISR) and proteomic perspective on biocontrol. In: PGPR amelioration in sustainable agriculture. Elsevier, Amsterdam, pp 67–84

Karthika S, Midhun SJ, Jisha MS (2020) A potential antifungal and growth-promoting bacterium *Bacillus* sp. KTMA4 from tomato rhizosphere. Microb Pathogenesis 142:104049

Kaspar F, Neubauer P, Gimpel M (2019) Bioactive secondary metabolites from *Bacillus subtilis*: a comprehensive review. J Nat Prod 82(7):2038–2053

Kaur T, Rani R, Manhas RK (2019) Biocontrol and plant growth promoting potential of phylogenetically new *Streptomyces* sp. MR14 of rhizospheric origin. AMB Expr 9(1):125

Kenawy A, Dailin DJ, Abo-Zaid GA, Abd Malek R, Ambehabati KK, Zakaria KHN, Sayyed RZ, El Enshasy HA (2019) Biosynthesis of antibiotics by PGPR and their roles in biocontrol of plant diseases. In: Plant growth promoting rhizobacteria for sustainable stress management. Springer, Berlin, pp 1–35

Kering KK, Kibii BJ, Wei H (2019) Biocontrol of phytobacteria with bacteriophage cocktails. Pest Manage Sci 75(7):1775–1781

Khedher SB, Boukedi H, Laarif A, Tounsi S (2020) Biosurfactant produced by *Bacillus subtilis* V26: a potential biological control approach for sustainable agriculture development. Org Agric 1–8

Kloepper JW, Ryu C-M, Zhang S (2004) Induced systemic resistance and promotion of plant growth by *Bacillus* spp. Phytopathology 94(11):1259–1266

Köhl J, Kolnaar R, Ravensberg WJ (2019) Mode of action of microbial biological control agents against plant diseases: relevance beyond efficacy. Front Plant Sci 10:845

Kron AS, Zengerer V, Bieri M, Dreyfuss V, Sostizzo T, Schmid M, Lutz M, Remus-Emsermann MNP, Pelludat C (2020) *Pseudomonas orientalis* F9 pyoverdine, safracin, and phenazine mutants remain effective antagonists against Erwinia amylovora in apple flowers. Appl Environ Microbiol 86(8)

Kumar A, Prakash A, Johri BN (2011) *Bacillus* as PGPR in crop ecosystem. In: Bacteria in agrobiology: crop ecosystems. Springer, Berlin, pp 37–59

Lamichhane JR, Osdaghi E, Behlau F, Köhl J, Jones JB, Aubertot J-N (2018) Thirteen decades of antimicrobial copper compounds applied in agriculture. A review. Agron Sustain Dev 38(3):28

Law JW-F, Ser H-L, Khan TM, Chuah L-H, Pusparajah P, Chan K-G, Goh B-H, Lee L-H (2017) The potential of *Streptomyces* as biocontrol agents against the rice blast fungus, Magnaporthe oryzae (Pyricularia oryzae). Front Microbiol 8:3

Lay FT, Anderson MA (2005) Defensins-components of the innate immune system in plants. Curr Protein Pept Sci 6(1):85–101

Le KD, Kim J, Yu NH, Kim B, Lee CW, Kim J-C (2020) Biological control of tomato bacterial wilt, Kimchi cabbage soft rot, and red pepper bacterial leaf spot using *Paenibacillus elgii* JCK-5075. Front Plant Sci 11:775

Lewin RA (1982) Symbiosis and parasitism—definitions and evaluations. Bioscience 32(4):254–260

Li X, Huang P, Wang Q, Xiao L, Liu M, Bolla K, Zhang B, Zheng L, Gan B, Liu X, Zhang L, Zhang X (2014) Staurosporine from the endophytic *Streptomyces* sp. strain CNS-42 acts as a potential biocontrol agent and growth elicitor in cucumber. Anton Leeuw Int J G 106(3):515–525

de Lima Procópio RE, da Silva IR, Martins MK, de Azevedo JL, de Araújo JM (2012) Antibiotics produced by *Streptomyces*. Braz J Infect Dis 16(5):466–471

Lin Y-C, Chung K-R, Huang J-W (2020) A synergistic effect of chitosan and lactic acid bacteria on the control of cruciferous vegetable diseases. Plant Pathol J 36(2):157–169

Ling L, Han X, Li X, Zhang X, Wang H, Zhang L, Cao P, Wu Y, Wang X, Zhao J, Xiang W (2020) A *Streptomyces* sp. NEAU-HV9: isolation, identification, and potential as a biocontrol agent against Ralstonia solanacearum of tomato plants. Microorganisms 8(3):351

Liu Y, Teng K, Wang T, Dong E, Zhang M, Tao Y, Zhong J (2020) Antimicrobial *Bacillus velezensis* HC6: production of three kinds of lipopeptides and biocontrol potential in maize. J Appl Microbiol 128(1):242–254

Luo L, Cai J, Wang C, Lin J, Du X, Zhou A, Xiang M (2016) Purification and characterization of an alkaliphilic endo-xylanase from *Streptomyces althioticus* LMZM and utilization in the pulp paper industry. J Chem Technol Biotechnol 91(4):1093–1098

Mahaffee WF, Kloepper JW (1997) Temporal changes in the bacterial communities of soil, rhizosphere, and endorhiza associated with field-grown cucumber (Cucumis sativus L.). Microbiol Ecol 34(3):210–223

Maldonado-González MM, Bakker PAHM, Prieto P, Mercado-Blanco J (2015) Arabidopsis thaliana as a tool to identify traits involved in Verticillium dahliae biocontrol by the olive root endophyte *Pseudomonas fluorescens* PICF7. Front Microbiol 6:266

Marian M, Ohno T, Suzuki H, Kitamura H, Kuroda K, Shimizu M (2020) A novel strain of endophytic *Streptomyces* for the biocontrol of strawberry anthracnose caused by Glomerella cingulata. Microbiol Res 234:126428

Mavrodi DV, Blankenfeldt W, Thomashow LS (2006) Phenazine compounds in fluorescent *Pseudomonas* spp. biosynthesis and regulation. Annu Rev Phytopathol 44:417–445

Meziane H, Van Der Sluis I, Van Loon LC, Höfte M, Bakker PAHM (2005) Determinants of *Pseudomonas putida* WCS358 involved in inducing systemic resistance in plants. Mol Plant Pathol 6(2):177–185

Mishra J, Arora NK (2018) Secondary metabolites of fluorescent pseudomonads in biocontrol of phytopathogens for sustainable agriculture. Appl Soil Ecol 125:35–45

Molina L, Constantinescu F, Michel L, Reimmann C, Duffy B, Défago G (2003) Degradation of pathogen quorum-sensing molecules by soil bacteria: a preventive and curative biological control mechanism. FEMS Microbiol Ecol 45(1):71–81

Montesinos E (2007) Antimicrobial peptides and plant disease control. FEMS Microbiol Lett 270(1):1–11

Morrissey JP, Cullinane M, Abbas A, Mark GL, O'Gara F (2004) Biosynthesis and regulation of anti-fungal metabolites by Pseudomonads. In: *Pseudomonas*. Springer, Berlin, pp 637–670

Moustafa-Farag M, Almoneafy A, Mahmoud A, Elkelish A, Arnao MB, Li L, Ai S (2020) Melatonin and its protective role against biotic stress impacts on plants. Biomolecules 10(1)

Mulet M, Lalucat J, García-Valdés E (2010) DNA sequence-based analysis of the *Pseudomonas* species. Environ Microbiol 12(6):1513–1530

Mun B-G, Lee W-H, Kang S-M, Lee S-U, Lee S-M, Lee DY, Shahid M, Yun B-W, Lee I-J (2020) *Streptomyces* sp. LH 4 promotes plant growth and resistance against Sclerotinia sclerotiorum in cucumber via modulation of enzymatic and defense pathways. Plant Soil 448(1):87–103

Nakkeeran S, Priyanka R, Rajamanickam S, Sivakumar U (2020) *Bacillus amyloliquefaciens* alters the diversity of volatile and non-volatile metabolites and induces the expression of defence genes for the management of Botrytis leaf blight of Lilium under protected conditions. J Plant Pathol.

Nayak SK, Nayak S, Patra JK (2020) Rhizobacteria and its biofilm for sustainable agriculture: a concise review. In: Yadav MK, Singh BP (eds) New and future developments in microbial biotechnology and bioengineering: microbial biofilms. Elsevier, Amsterdam, pp 165–175

Naznin HA, Kiyohara D, Kimura M, Miyazawa M, Shimizu M, Hyakumachi M (2014) Systemic resistance induced by volatile organic compounds emitted by plant growth-promoting fungi in Arabidopsis thaliana. Plos One 9(1):e86882

Neiendam Nielsen M, Sørensen J (1999) Chitinolytic activity of *Pseudomonas fluorescens* isolates from barley and sugar beet rhizosphere. FEMS Microbiol Ecol 30(3):217–227

Nelkner J, Tejerizo GT, Hassa J, Lin TW, Witte J, Verwaaijen B, Winkler A, Bunk B, Spröer C, Overmann J, Grosch R, Pühler A, Andreas S (2019) Genetic potential of the biocontrol agent *Pseudomonas brassicacearum* (formerly *P. trivialis*) 3Re2-7 unraveled by genome sequencing and mining, comparative genomics and transcriptomics. Genes (Basel) 10(8):601

Ni M, Wu Q, Wang H-l, Liu W-C, Hu B, Zhang D-P, Zhao J, Liu D-W, Lu C-G (2019) Identification of a novel strain, *Streptomyces blastmyceticus* JZB130180, and evaluation of its biocontrol efficacy against Monilinia fructicola. J Zhejiang Univ Sci B 20(1):84–94

Nielsen TH, Nybroe O, Koch B, Hansen M, Sørensen J (2005) Genes involved in cyclic lipopeptide production are important for seed and straw colonization by *Pseudomonas* sp. strain DSS73. Appl Environ Microbiol 71(7):4112–4116

Niranjana SR, Hariprasad P (2014) Understanding the mechanism involved in PGPR-mediated growth promotion and suppression of biotic and abiotic stress in plants. In: Future challenges in crop protection against fungal pathogens. Springer, Berlin, pp 59–108

Okada H (2001) A fungal-feeding nematode of the family Tylenchidae from decomposing rice straw. Phytopathology.

Olanrewaju OS, Glick BR, Babalola OO (2017) Mechanisms of action of plant growth promoting bacteria. World J Microbiol Biotechnol 33(11):197

Omoboye OO, Oni FE, Batool H, Yimer HZ, De Mot R, Höfte M (2019) *Pseudomonas* cyclic lipopeptides suppress the rice blast fungus Magnaporthe oryzae by induced resistance and direct antagonism. Front Plant Sci 10:901

Ongena M, Jacques P (2008) *Bacillus* lipopeptides: versatile weapons for plant disease biocontrol. Trends Microbiol 16(3):115–125

Oni FE, Kieu Phuong N, Höfte M (2015) Recent advances in *Pseudomonas* biocontrol. In: Bacteria-plant interactions: advanced research and future trends. Caister Academic Press, Norfolk, pp 167–198

Oskiera M, Szczech M, Bartoszewski G (2015) Molecular identification of Trichoderma strains collected to develop plant growth-promoting and biocontrol agents. J Hortic Res 23(1):75–86

Osorio E, Flores M, Hernández D, Ventura J, Rodríguez R, Aguilar CN (2010) Biological efficiency of polyphenolic extracts from pecan nuts shell (Carya Illinoensis), pomegranate husk (Punica granatum) and creosote bush leaves (Larrea tridentata Cov.) against plant pathogenic fungi. Ind Crop Prod 31(1):153–157

Pal KK, Gardener BM (2006) Biological control of plant pathogens. Plant Health Instruct. https://doi.org/10.1094/PHI-A-2006-1117-02

Parolin P, Bresch C, Poncet C, Desneux N (2014) Introducing the term 'Biocontrol Plants' for integrated pest management. Sci Agric 71:77–80

Passari AK, Mishra VK, Gupta VK, Yadav MK, Saikia R, Singh BP (2015) In vitro and in vivo plant growth promoting activities and DNA fingerprinting of antagonistic endophytic actinomycetes associates with medicinal plants. PLoS One 10(9):e0139468

Peláez F (2006) The historical delivery of antibiotics from microbial natural products—can history repeat? Biochem Pharmacol 71(7):981–990

Pieterse CMJ, Zamioudis C, Berendsen RL, Weller DM, Van Wees SCM, Bakker PAHM (2014) Induced systemic resistance by beneficial microbes. Annu Rev Phytopathol 52:347–375

Prasad R, Kumar M, Varma A (2015) Role of PGPR in Soil Fertility and Plant Health. In: Egamberdieva D, Shrivastava S, Varma A (eds) Plant-growth-promoting rhizobacteria (PGPR) and medicinal plants. Springer, Cham, pp 247–260

Prasad RD, Chandrika KSVP, Godbole V (2020) A novel chitosan biopolymer based Trichoderma delivery system: storage stability, persistence and bio efficacy against seed and soil borne diseases of oilseed crops. Microbiol Res 237:126487

Qin Y, Wang Y, He Y, Zhang Y, She Q, Chai Y, Li P, Shang Q (2019) Characterization of subtilin L-Q11, a novel class I bacteriocin synthesized by *Bacillus subtilis* L-Q11 isolated from orchard soil. Front Microbiol 10:484

Raaijmakers JM, Vlami M, De Souza JT (2002) Antibiotic production by bacterial biocontrol agents. Anton Leeuw Int J G 81(1–4):537

Raaijmakers JM, De Bruijn I, de Kock MJD (2006) Cyclic lipopeptide production by plant-associated *Pseudomonas* spp.: diversity, activity, biosynthesis, and regulation. Mol Plant Microbe Int 19(7):699–710

Raaijmakers JM, De Bruijn I, Nybroe O, Ongena M (2010) Natural functions of lipopeptides from *Bacillus* and *Pseudomonas*: more than surfactants and antibiotics. FEMS Microbiol Rev 34(6):1037–1062

Rabbee MF, Ali MS, Choi J, Hwang BS, Jeong SC, K-h B (2019) *Bacillus velezensis*: a valuable member of bioactive molecules within plant microbiomes. Molecules 24(6)

Raio A, Reveglia P, Puopolo G, Cimmino A, Danti R, Evidente A (2017) Involvement of phenazine-1-carboxylic acid in the interaction between *Pseudomonas chlororaphis* subsp. aureofaciens strain M71 and Seiridium cardinale in vivo. Microbiol Res 199:49–56

Raj SN, Lavanya SN, Amruthesh KN, Niranjana SR, Reddy MS, Shetty HS (2012) Histo-chemical changes induced by PGPR during induction of resistance in pearl millet against downy mildew disease. Biol Control 60(2):90–102

Ram RM, Keswani C, Bisen K, Tripathi R, Singh SP, Singh HB (2018) Biocontrol technology: eco-friendly approaches for sustainable agriculture. In: Omics technologies and bio-engineering. Elsevier, Amsterdam, pp 177–190

Ran LX, Li ZN, Wu GJ, Van Loon LC, Bakker PH (2005) Induction of systemic resistance against bacterial wilt in Eucalyptus urophylla by fluorescent *Pseudomonas* spp. Eur J Plant Pathol 113(1):59–70

Raymaekers K, Ponet L, Holtappels D, Berckmans B, Cammue BPA (2020) Screening for novel biocontrol agents applicable in plant disease management – a review. Biol Control 144:104240

Raza W, Ling N, Liu D, Wei Z, Huang Q, Shen Q (2016) Volatile organic compounds produced by *Pseudomonas fluorescens* WR-1 restrict the growth and virulence traits of Ralstonia solanacearum. Microbiol Res 192:103–113

Rey T, Dumas B (2017) Plenty is no plague: *Streptomyces* symbiosis with crops. Trends Plant Sci 22(1):30–37

Ruanpanun P, Nimnoi P (2020) Evaluation on the efficiency and persistence of *Streptomyces jietaisiensis* strain A034 in controlling root knot disease and promoting plant growth in the plant-parasitic nematode infested soils. Biol Control 144:104221

Sadfi N, Chérif M, Fliss I, Boudabbous A, Antoun H (2001) Evaluation of bacterial isolates from salty soils and *Bacillus thuringiensis* strains for the biocontrol of fusarium dry rot of potato tubers. J Plant Pathol 83(2):101–118

Salwan R, Sharma V (2020) Bioactive compounds of *Streptomyces*: biosynthesis to applications. In: Studies in natural products chemistry, vol 64. Elsevier, Amsterdam, pp 467–491

Sarwar A, Hassan MN, Imran M, Iqbal M, Majeed S, Brader G, Sessitsch A, Hafeez FY (2018) Biocontrol activity of surfactin A purified from *Bacillus* NH-100 and NH-217 against rice bakanae disease. Microbiol Res 209:1–13

Saxena AK, Kumar M, Chakdar H, Anuroopa N, Bagyaraj DJ (2020) *Bacillus* species in soil as a natural resource for plant health and nutrition. J Appl Microbiol 128(6):1583–1594

Sayago P, Juncosa F, Albarracín Orio AG, Luna DF, Molina G, Lafi J, Ducasse DA (2020) *Bacillus subtilis* ALBA01 alleviates onion pink root by antagonizing the pathogen Setophoma terrestris and allowing physiological status maintenance. Eur J Plant Pathol 157(3):509–519

Schippers B (1992) Prospects for management of natural suppresiveness to control soilborne pathogens. In: Tjamos EC, Papavizas GC, Cook RJ (eds) Biological control of plant diseases: progress and challenges for the future. Springer, Boston, pp 21–34

Seipke RF, Kaltenpoth M, Hutchings MI (2012) *Streptomyces* as symbionts: an emerging and widespread theme? FEMS Microbiol Rev 36(4):862–876

Shao Z, Li Z, Fu Y, Wen Y, Wei S (2018) Induction of defense responses against Magnaporthe oryzae in rice seedling by a new potential biocontrol agent *Streptomyces* JD211. J Basic Microbiol 58(8):686–697

Shi J, Liu A, Li X, Chen W (2013) Control of Phytophthora nicotianae disease, induction of defense responses and genes expression of papaya fruits treated with *Pseudomonas putida* MGP1. J Sci Food Agric 93(3):568–574

Shrivastava P, Kumar R, Yandigeri MS (2017) In vitro biocontrol activity of halotolerant *Streptomyces aureofaciens* K20: a potent antagonist against Macrophomina phaseolina (Tassi) Goid. Saudi J Biol Sci 24(1):192–199

Shu C, Chen Q, Pi L, Zhang D, Panhwar QA, Zhou E (2017) Identification and antifungal activity analysis of two biocontrol antagonists to Colletotrichum musae. J Phytopathol 165(7–8):554–561

Siddiqui ZA (2006) PGPR: prospective biocontrol agents of plant pathogens. In: Siddiqui ZA (ed) PGPR: biocontrol and biofertilization. Springer, Dordrecht, pp 111–142

Simonetti G, Pucci N, Brasili E, Valletta A, Sammarco I, Carnevale E, Pasqua G, Loreti S (2020) In vitro antimicrobial activity of plant extracts against *Pseudomonas syringae* pv. actinidiae causal agent of bacterial canker in kiwifruit. Plant Biosyst 154(1):100–106

Singh SP, Gaur R (2016) Evaluation of antagonistic and plant growth promoting activities of chitinolytic endophytic actinomycetes associated with medicinal plants against Sclerotium rolfsii in chickpea. J Appl Microbiol 121(2):506–518

Singh PP, Shin YC, Park CS, Chung YR (1999) Biological control of Fusarium wilt of cucumber by chitinolytic bacteria. Phytopathology 89(1):92–99

Singh M, Singh D, Gupta A, Pandey KD, Singh PK, Kumar A (2019) Plant growth promoting rhizobacteria: application in biofertilizers and biocontrol of phytopathogens. In: Singh AK, Kumar A, Singh PK (eds) PGPR amelioration in sustainable agriculture. Woodhead Publishing, Cambridge, pp 41–66

Song GC, Ryu C-M (2013) Two volatile organic compounds trigger plant self-defense against a bacterial pathogen and a sucking insect in cucumber under open field conditions. Int J Mol Sci 14(5):9803–9819

Stefanato F L, Trippel C, Uszkoreit S, Ferrafiat L, Grenga L, Dickens R, Kelly N, Kingdon A D H, Ambrosetti L, Findlay K C, Cheema J, Trick M, Chandra G, Tomalin G, Malone J G, Truman A W (2019) Pan-genome analysis identifies intersecting roles for *Pseudomonas* specialized metabolites in potato pathogen inhibition. BioRxiv: 783258.

Stein T (2005) *Bacillus subtilis* antibiotics: structures, syntheses and specific functions. Mol Microbiol 56(4):845–857

Stringlis IA, Zhang H, Pieterse CMJ, Bolton MD, de Jonge R (2018) Microbial small molecules–weapons of plant subversion. Nat Prod Rep 35(5):410–433

Sujarit K, Pathom-aree W, Mori M, Dobashi K, Shiomi K, Lumyong S (2020) *Streptomyces palmae* CMU-AB204T, an antifungal producing-actinomycete, as a potential biocontrol agent to protect palm oil producing trees from basal stem rot disease fungus, Ganoderma boninense. Biol Control 148:104307

Sumayo M, Hahm M-S, Ghim S-Y (2013) Determinants of plant growth-promoting Ochrobactrum lupini KUDC1013 involved in induction of systemic resistance against Pectobacterium carotovorum subsp. carotovorum in tobacco leaves. Plant Pathol J 29(2):174–181

Tamreihao K, Ningthoujam DS, Nimaichand S, Singh ES, Reena P, Singh SH, Nongthomba U (2016) Biocontrol and plant growth promoting activities of a *Streptomyces corchorusii* strain UCR3-16 and preparation of powder formulation for application as biofertilizer agents for rice plant. Microbiol Res 192:260–270

Tapio E, Pohto-Lahdenperä A (1991) Scanning electron microscopy of hyphal interaction between *Streptomyces griseoviridis* and some plant pathogenic fungi. Agr Food Sci 63(5):435–441

Tarkka M, Hampp R (2008) Secondary metabolites of soil streptomycetes in biotic interactions. In: Secondary metabolites in soil ecology. Springer, Berlin, pp 107–126

Toke O (2005) Antimicrobial peptides: New candidates in the fight against bacterial infections. Peptide Sci 80(6):717–735

Tomas-Grau RH, Hael-Conrad V, Requena-Serra FJ, Perato SM, Caro MDP, Salazar SM, Díaz-Ricci JC (2020) Biological control of strawberry grey mold disease caused by Botrytis cinerea mediated by Colletotrichum acutatum extracts. Biocontrol.

Toral L, Rodríguez M, Béjar V, Sampedro I (2020) Crop protection against Botrytis cinerea by rhizhosphere biological control agent *Bacillus velezensis* XT1. Microorganisms 8(7)

Tran H, Ficke A, Asiimwe T, Höfte M, Raaijmakers JM (2007) Role of the cyclic lipopeptide massetolide A in biological control of Phytophthora infestans and in colonization of tomato plants by *Pseudomonas fluorescens*. New Phytol 175(4):731–742

Upadhyay A, Kochar M, Upadhyay A, Tripathy S, Rajam MV, Srivastava S (2017) Small RNAs regulate the biocontrol property of fluorescent *Pseudomonas* strain Psd. Microbiol Res 196:80–88

Uppal AK, El Hadrami A, Adam LR, Tenuta M, Daayf F (2008) Biological control of potato Verticillium wilt under controlled and field conditions using selected bacterial antagonists and plant extracts. Biol Control 44(1):90–100

Van Loon LC, Bakker P, Pieterse CMJ (1998) Systemic resistance induced by rhizosphere bacteria. Annu Rev Phytopathol 36(1):453–483

Van Peer R, Schippers B (1992) Lipopolysaccharides of plant-growth promoting *Pseudomonas* sp. strain WCS417r induce resistance in carnation to Fusarium wilt. Neth J Plant Pathol 98(2):129

Van Peer R, Niemann GJ, Schippers B (1991) Induced resistance and phytoalexin accumulation in biological control of Fusarium wilt of carnation by *Pseudomonas* sp. strain WCS 417 r. Phytopathology 81(7):728–734

Vanti GL, Masaphy S, Kurjogi M, Chakrasali S, Nargund VB (2020) Synthesis and application of chitosan-copper nanoparticles on damping off causing plant pathogenic fungi. Int J Biol Macromol 156:1387–1395

Vergnes S, Gayrard D, Veyssière M, Toulotte J, Martinez Y, Dumont V, Bouchez O, Rey T, Dumas B (2019) Phyllosphere colonization by a soil *Streptomyces* sp. promotes plant defense responses against fungal infection. Mol Plant Microbe Int 33(2):223–234

Vidal T, Saint-Jean S, Lusley P, Leconte M, Ben Krima S, Boixel A-L, Wheatamix C, de Vallavieille-Pope C (2020) Cultivar mixture effects on disease and yield remain despite diversity in wheat height and earliness. Plant Pathol n/a(n/a).

Villegas-Escobar V, González-Jaramillo LM, Ramírez M, Moncada RN, Sierra-Zapata L, Orduz S, Romero-Tabarez M (2018) Lipopeptides from *Bacillus* sp. EA-CB0959: active metabolites responsible for in vitro and in vivo control of Ralstonia solanacearum. Biol Control 125:20–28

Vos CMF, De Cremer K, Cammue BPA, De Coninck B (2015) The toolbox of Trichoderma spp. in the biocontrol of Botrytis cinerea disease. Mol Plant Pathol 16(4):400–412

Wang GF, Meng JF, Tian T, Xiao XQ, Zhang B, Xiao YN (2020) Endophytic *Bacillus velezensis* strain B-36 is a potential biocontrol agent against lotus rot caused by Fusarium oxysporum. J Appl Microbiol 128(4):1153

Wei G, Kloepper JW, Tuzun S (1991) Induction of systemic resistance of cucumber to Colletotrichum orbiculare by select strains of plant growth-promoting rhizobacteria. Phytopathology 81(11):1508–1512

Weller DM (2007) *Pseudomonas* biocontrol agents of soilborne pathogens: looking back over 30 years. Phytopathology 97(2):250–256

Weller DM, van Pelt JA, Mavrodi DV, Pieterse CMJ, Bakker P, van Loon LC (2004) ISR in Arabidopsis against P. syringae pv. tomato by 2, 4 DAPG-producing P. fluorescens. Phytopathology 94:5108
Whipps JM (1997) Developments in the biological control of soil-borne plant pathogens. In: Advances in botanical research, vol 26. Elsevier, Amsterdam, pp 1–134
Whipps JM (2001) Microbial interactions and biocontrol in the rhizosphere. J Exp Bot 52(Suppl_1):487–511
Whitehead NA, Barnard AML, Slater H, Simpson NJL, Salmond GPC (2001) Quorum-sensing in Gram-negative bacteria. FEMS Microbiol Rev 25(4):365–404
Wyckhuys KAG, Hughes AC, Buamas C, Johnson AC, Vasseur L, Reymondin L, Deguine JP, Sheil D (2019) Biological control of an agricultural pest protects tropical forests. Commun Biol 2:10
Xu W, Wang K, Wang H, Liu Z, Shi Y, Gao Z, Wang Z (2020) Evaluation of the biocontrol potential of *Bacillus* sp. WB against Fusarium oxysporum f. sp. niveum. Biol Control 147:104288
Yang M, Lu L, Pang J, Hu Y, Guo Q, Li Z, Wu S, Liu H, Wang C (2019) Biocontrol activity of volatile organic compounds from *Streptomyces alboflavus* TD-1 against Aspergillus flavus growth and aflatoxin production. J Microbiol 57(5):396–404
Yasmin S, Hafeez FY, Mirza MS, Rasul M, Arshad HMI, Zubair M, Iqbal M (2017) Biocontrol of bacterial leaf blight of rice and profiling of secondary metabolites produced by rhizospheric *Pseudomonas aeruginosa* BRp3. Front Microbiol 8:1895
Zasloff M (2002) Antimicrobial peptides of multicellular organisms. Nature 415(6870):389–395
Zasloff M (2019) Antimicrobial peptides of multicellular organisms: my perspective. In: Matsuzaki K (ed) Antimicrobial peptides: basics for clinical application. Springer, Singapore, pp 3–6
Zdor RE, Anderson AJ (1992) Influence of root colonizing bacteria on the defense responses of bean. Plant Soil 140(1):99–107
Zhang LH, Dong YH (2004) Quorum sensing and signal interference: diverse implications. Mol Microbiol 53(6):1563–1571
Zhang J, Wang LM, Li YH, Ding SL, Yuan HX, Riley IT, Li HL (2016) Biocontrol of cereal cyst nematode by *Streptomyces anulatus* isolate S07. Australas Plant Pathol 45(1):57–64
Zhang K, Gu L, Zhang Y, Liu Z, Li X (2020a) Dinactin from a new producer, *Streptomyces badius* gz-8, and its antifungal activity against the rubber anthracnose fungus *Colletotrichum gloeosporioides*. Microbiol Res 240:126548
Zhang X-Y, Li C, Hao J-J, Li Y-C, Li D-Z, Zhang D-M, Xing X, Liang Y (2020b) A novel *Streptomyces* sp. strain PBSH9 for controlling potato common scab caused by *Streptomyces galilaeus*. Plant Dis 104(7):1986–1993
Zhang X, Wu F, Gu N, Yan X, Wang K, Dhanasekaran S, Gu X, Zhao L, Zhang H (2020c) Postharvest biological control of Rhizopus rot and the mechanisms involved in induced disease resistance of peaches by Pichia membranefaciens. Postharvest Biol Technol 163:111146
Zhang Z, Zhao P, Zhang P, Su L, Jia H, Wei X, Fang J, Jia H (2020d) Integrative transcriptomics and metabolomics data exploring the effect of chitosan on postharvest grape resistance to Botrytis cinerea. Postharvest Biol Technol 167:111248
Zhu M-L, Wu X-Q, Wang Y-H, Dai Y (2020) Role of biofilm formation by *Bacillus pumilus* HR10 in biocontrol against pine seedling damping-off disease caused by Rhizoctonia solani. Forests 11(6):652
Ziedan ESH, Farrag ES, El-Mohamedy RS, Abd Alla MA (2010) *Streptomyces alni* as a biocontrol agent to root-rot of grapevine and increasing their efficiency by biofertilisers inocula. Arch Phytopathol Pflanzenschutz 43(7):634–646

Chapter 11
Microbial Bioactive Compounds Produced by Endophytes (Bacteria and Fungi) and Their Uses in Plant Health

Abdul Basit, Syed Tanveer Shah, Izhar Ullah, Inayat Ullah, and Heba I. Mohamed

Contents

A. Basit (✉) · S. T. Shah · I. Ullah
Department of Horticulture, Faculty of Crop Production Sciences, The University of Agriculture, Peshawar, Pakistan

I. Ullah
Department of Agricultural Mechanization, The University of Agriculture, Peshawar, Peshawar, Pakistan

H. I. Mohamed
Department of Biological and Geological Sciences, Faculty of Education, Ain Shams University, Cairo, Egypt

© The Author(s), under exclusive license to Springer Nature Switzerland AG 2021

H. I. Mohamed et al. (eds.), *Plant Growth-Promoting Microbes for Sustainable Biotic and Abiotic Stress Management*,
https://doi.org/10.1007/978-3-030-66587-6_11

1 Introduction

The bioactive compounds, also known as "biologically active compounds", are extra nutritious components found in a minute quantity in various products of plants and foods rich in lipids (Cammack et al. 2006). Bioactive compounds are mostly formed by microbes and plants, having broad pharmaceutical characteristics including cardiovascular, anti-thrombotic, anticancer, antidiabetic, anti-glycaemic and antihypertensive (Villaescusa et al. 2015; Atanasov et al. 2015; El-Beltagi et al. 2018, 2019a, b, c; Hamed et al. 2019; Aminkhani et al. 2020), and are used as favoured medicines, made synthetically to cure different diseases with minimum side effects (Chang et al. 2013). Currently, these are demanded highly in the naturopathy and pharmaceuticals, because of their benefits to plants and human's health. These compounds are synthesized by microorganisms and nearly by few enzymes either with plant association or alone. Microbes living inside the tissues of plants are called endophytes, producing a wide range of these compounds (Singh et al. 2017).

Endophytes are known to be the class of endosymbiotic microbes which are distributed widely among the plants and make colonies in intra- or intercellular spaces of entire plant parts. They do not cause deceptive disease infection or substantial change in morphology (Fouda et al. 2019). Plant endophytes extending from a range of bacteria to fungi form a quality class of organic compounds, volatile in nature, and considered to be significant for the symbiotic relationship development under an environment with high competition with their host (Chowdhury et al. 2015). Endophytes spend most of their life inside the tissues of a plant with no visible injury or elicitation in any defence reaction in plant host (Liarzi et al. 2016), and they exhibit wide-ranging symbiotic relationship with the host plants and different lifestyles, also possessing beneficial long-term association to both the host and microbes (Card et al. 2016). Endophytes could be found in most of the ecosystems while reducing biotic and abiotic stressors of plant crops; stimulate immunity responses, excluding pathogens of plants by niche competition; and take part in the metabolism of phenylpropanoid and antioxidant activities, the activation of which produce defence in plants, molecules for survival and structural support (Ek-Ramos et al. 2019).

Endophytic bacteria have reported with many different useful benefits to their host plant including plant metabolism modulation, activity of promoting growth and signalling of plant hormones leading to ecological biotic and abiotic stress adaptation. Their use grants special concern for agricultural application development ensuring the improved performance of crops under drought, cold, contaminated soil and salinity stress conditions and enhanced resistance to disease (Lata et al. 2018).

The need to uplift human lifestyle using advance, new and beneficial compounds is ever growing. Despite the advancement in research and so much efforts to cope up with many medical problems (appearance of bacterial drug resistance, viruses threatening life, enhancement in fungal infections, problems with organ transplant, etc.), mankind is still unable to control these problems. Also, mankind faces different problems like land and water pollution, environmental degradation and biodiversity loss, but more importantly, there are hurdles to produce enough food for people in certain parts of the world which has become a threat to human life. The endophytic population is greatly affected by climatic conditions and the location where the host plant grows. They produce a wide range of compounds useful for plants for their growth, protection to environmental conditions, and sustainability, in favor of a good dwelling place within the hosts. They protect plants from herbivory by producing certain compounds that will prevent animals from further grazing on the same plant and sometimes act as biocontrol agents. A large number of bioactive compounds produced by them not only are useful for plants but also are of economical importance to humans. They serve as antibiotics, drugs or medicines, or compounds of high relevance in research or as compounds useful to the food industry. This chapter provides an in-depth detail of occurrence, microbial biological by-products of endophytes, their mechanism, discovery, and significances and uses them to enhance plant health and human life.

2 Needs for New Medicines and Agrochemical Agents

To explore new chemotherapeutic agents, antibiotics and highly effective agrochemicals with low toxicity and less environmental effect is the need of the day. This research is accelerated by developing resistance against infectious microorganism (e.g. *Mycobacterium*, *Streptococcus* and *Staphylococcus*) to recent compounds and threatened naturally resistant organism present in the result of this search. Detection and development of new medicines to tackle new diseases such as SARS and AIDS in humans. New treatments are used as well as medications to treat illnesses such as AIDS and ancillary (due to weak immune systems). Unscrupulous pathogens (*Aspergillus* spp., *Cryptococcus* spp. and *Candida* spp.) usually attack more on an immunocompromised human population, which is another major risk to human life. For treating nematode infections (malaria, trypanosomiasis, leishmaniasis and filariasis) and parasitic protozoans, new and more drugs are required. Generally speaking, a single malaria can kill more lives every year among all the aforementioned diseases except AIDS virus and *Mycobacterium tuberculosis*

(National Institute of Health 2001). Alternative methods to control farm pathogens and pests are required to be searched to remove many synthetic agricultural agents from the market due to environmental and safety problems (Demain 2000), where, opportunities for innovative drugs and agrochemical discovery are possible due to novel natural products and organisms.

3 Natural Products and Their Traditional Approaches in Medicinal World

These are the naturally derived metabolites and by-products of plants, microorganisms and animals (Baker et al. 2000). China is the largest traditional medicine users producing plants of approximately 5000 and obtained products in their pharmacopoeia. Aspirin (salicylic acid) is considered the most known and widely used medicine in the world, reported in various plant species of different genera, i.e. *Populus* and *Salix*. Salicylic acid is naturally originated from glycoside salicin. Mayans nearly 3000 years ago treat intestinal ailments using fungi grown on roasted green corns which indicated the benefits of medicinal plants in now-extinct civilizations (Buss and Hayes 2000). Around 800 AD, the Benedictine monks use *Papaver somniferum* for a pain reliever, which was done way back by Greeks. In the past, products obtained from the mixture of leaf, stem and roots are considered very helpful to treat certain diseases, reduce pain and sufferings and provide relief and quality improvement in life, but understanding the complex nature and function of these bioactive compounds remained a mystery. The mystery was partly solved from Pasteur discovery (fermentation caused by living cells). By then people thought seriously to search the sources of these bioactive compounds which were microorganisms. Later, the discovery of penicillin from *Penicillium notatum* (a fungus) provided motivation and observance power to Flemings, who led the antibiotic era. By then scientists are busy to overcome plant and human pathogens by applying different microbial metabolites. Since many of the beneficial micro-organisms had been found, the work in medicine (immunosuppressing functions and anti-cancer) which is used to combat various harmful illnesses and therefore in agriculture, has been made simple, a revolutionary and sophisticated screening method developed in medicine as well as agriculture.

4 The Endomicrobiome

Microbial community could be studied within plants using next-generation sequencing (NGS) technology and is together termed as "endomicrobiome". Various factors like geographical location and different genotypes of plants, etc. are somewhat less diverse than rhizomicrobiome (Edwards et al. 2015). The mechanism for microbe acquisition is still ambiguous in a particular ecosystem. A reasonable supposition suggests a two-step acquisition of microbiomes. During the first step, microorganism is first introduced in the rhizosphere which is followed by entrance mechanism within root tissue. This entrance depends upon general factors and is species-specific (Bulgarelli et al. 2013). Based on the time-staged profiling experiments in rice plants, it was concluded that microbial colonization starts within a day and reaches a steady state within 2 weeks. This shows the fast-growing process of microbial colonization in the endosphere. Root wounds, lateral roots, root hairs, lenticels or leaf stomata and cracks are usually the entry points of bacteria (Edwards et al. 2015). It is reported that *Proteobacteria* are found more in bulk soil as compared to endosphere while there is a reverse followed by acid bacteria, and *Gemmatimonadetes* are more in endosphere than in bulk soil. *Pseudomonas*-like operational taxonomic unit (OTU) which is gammaproteobacterial is found to be approximately 34% in endophytic bacterial sequence of *Populus*. After the analysis of sequencing approaches of different plant parts, a similar trend was observed in the whole community, not only in plant roots. When tomato leaf was analysed using 16S-rRNA pyrosequencing, it was reported the predominance of *Actinobacteria* (1.5%), *Proteobacteria* (90%), *Verrucomicrobia* (1.1%), *Planctomycetes* (1.4%) and *Acidobacteria* (0.5%). The phylum *Proteobacteria* is reported to dominate about 98% among the microbial communities which mostly consist of *Enterobacteriales*, *Pseudomonadales*, *Flavobacteriales*, *Actinomycetales*, *Xanthomonadales*, *Sphingomonadales* and *Rhizobiales* (Shi et al. 2014). All these studies assumed the endosphere of most land plants.

5 Definition, Classification and Origin of Endophytes

Endophytes are defined as the organisms residing inside the plant's internal tissues in its whole life period, no matter whether it was unbiased, beneficial or harmful to its host plant (Bacon and White 2000). They reside within plants for some part of the life cycle without initiating any signs of diseases (Sikora et al. 2007). Thus, endophytic microbes are an ecology concept and are an essential part of a plant-microecology system (Li 2005). About 270,000–4,000,000 different kinds of endophytic fungi live in the intercellular spaces and microtubule plant cells. Furthermore, a high density of about 104–106 CFU/g of endophytic bacteria live in plant roots (Dreyfuss and Chapela 1994). Moreover, McInroy and Kloepper (1996) discovered that *Neotyphodium coenophialum* (endophytic fungi) produced a syndrome called "fescue toxicosis" in cattle eating the grass *Festuca arundinacea,* providing new insights in this field.

6 Types of Endophytes

Endophytes are classified based on microbes into endophytic fungi, bacteria and actinomyces (Strobel et al. 2004).

6.1 Endophytic Fungi

An endophytic fungus can multiply asymptomatically in the tissues of plants including stems, leaves, and roots. Bacon and White (2000) reported that an endophytic fungus lives in the mycelial form in the biological organization within the living plant, at least for some time. Endophytic fungi are found to associate with above ground tissue of liverworts, hornworts, mosses, lycophytes, equisetopsids, fern, and seed plants from the arctic to the tropics and from agriculture fields to the most diverse tropical forest (Arnold 2007). They found that endophytic fungi could produce tolerance against drought and saline conditions (Waller et al. 2005). They act as stimulator against stress conditions more rapidly in comparison to the non-symbiotic plants (Redman et al. 2002). Red and chilli peppers contain a bioactive compound, capsaicin, that has been used as a remedy against pain and different types of human cancers. The endophytic fungus named *Alternaria alternata*, separated from *Capsicum annuum* (chilli), produces capsaicin, while *Eurotium* sp. from the rhizome of *Curcuma longa* (turmeric) produces asparaginase which can usually be used as an anticancerogenic enzyme (Jalgaonwala and Mahajan 2014).

6.1.1 Plant-Associated Fungi that Produce Bioactive Compounds

To adopt sustainable agriculture by maintaining a healthy ecosystem and reduce the residual effect of inorganic fertilizer and pesticides, the use of beneficial microorganism, i.e. fungi, as a biocontrol agent and growth promoter can be an effective alternative to various chemical pesticides and controlling pathogens in plants (Vurukonda et al. 2018; Aswani et al. 2020). Fungi interact with plants in a different way, playing a key role in the maintenance of ecosystems' well-being while developing associations with various plant tissues positively or negatively. These metabolites adopt different protective measures in protecting plants from herbivores, inducing systematic resistance to pathogens, and stimulate the synthesis of phytohormones and nutrient and water transport efficiency during different stress conditions (Zeilinger et al. 2016). They increase resistance against stress conditions in the plants by producing bioactive compounds like *Paecilomyces formosus* LWL1, an endophytic fungus in japonica rice cultivar 'Dongjin' that produced secondary metabolites under heat stress and improves growth-related attributes (Waqas et al. 2015). They promote accumulation of different secondary metabolites in the host plants under stress conditions (Venieraki et al. 2017).

6.1.1.1 Mycorrhizal Fungi

Mycorrhiza" the term used to describe the symbiotic association between a fungus and a root of higher plant. Endomycorrhizal fungi are involved in colonization of roots causing significant changes in their chemicals, produced by roots that influence the health status of plant, their performance under competitive condition, soil aggregate formation, increasing resistance against any biotic or abiotic stresses and activation of stimulated response (Jamiołkowska et al. 2017). Endophytes provide extensive types of bioactive secondary metabolites with a selected structure including flavonoids, alkaloids, chinones, phenolic acids, steroids, quinones, tetralones, terpenoids, xanthones, and others. Bioactive secondary metabolites are also isolated from conifer-associated endophytic fungi which are having anti-inflammatory, antimicrobial, antiproliferative, or cytotoxic activity toward human cancer cell lines and activity against plant insect pests or plant pathogens [96]. Such bioactive metabolites find wide-ranging application as anticancer, antiparasitics, agrochemicals, antibiotic, immune suppressants, and antioxidant agents (Stierle and Stierle 2015). Also, mycorrhizal fungi produce various bioactive compounds such as glomalin as defensive strategies that perform different functions by immobilizing contaminants on the hyphal cell wall and reduces predator infection (Souza et al. 2012). Under drought conditions, the association of plants with mycorrhizal fungi withstand drought-induced oxidative stress by the increased production of antioxidant compounds that scavenge reactive oxygen species and activate the activities of antioxidant enzymes (Rapparini and Penuelas 2014).

6.1.1.2 Fungi-Promoting Plant Growth

Such types of fungi living inside the soil can colonize the roots of plants. These fungi function as biocontrollers and growth promoters while improving the development and growth, as these microbes destroy pathogenic microorganisms and produce substrates of minerals. Furthermore, a series of metabolic responses were also observed in the plants through the volatile organic compounds' (VOCs) production by these fungi (Naznin et al. 2013). Plant defense is then achieved by priming for enhanced expression of sequences regulated by the production of jasmonic acid, ethylene, or salicylic acid. In other cases, the functions of mycorrhizal fungi are to produce active VOCs and antibiotics, both in plants and soil, which can be helpful in the identification of active biomolecules against plant pathogens and enhanced vegetative and reproductive performance of the plant (Vurukonda et al. 2018).

6.2 Endophytic Bacteria

Almost a diverse array of endophytic bacteria have some beneficial effects, i.e. as biocontrol agent, and act as an enhancer of N_2 fixation, plant hormone production, phosphate solubilization and inhibitors of ethylene (C_2H_2) biosynthesis against different biotic and abiotic stresses, having biocontrol activity (Fig. 11.1). They get multiplied at low-density population compared to bacterial pathogens and rhizospheric bacteria (Rosenblueth and Martinez Romero 2004), providing better protection than rhizospheric bacteria against abiotic stress. They help in repressing

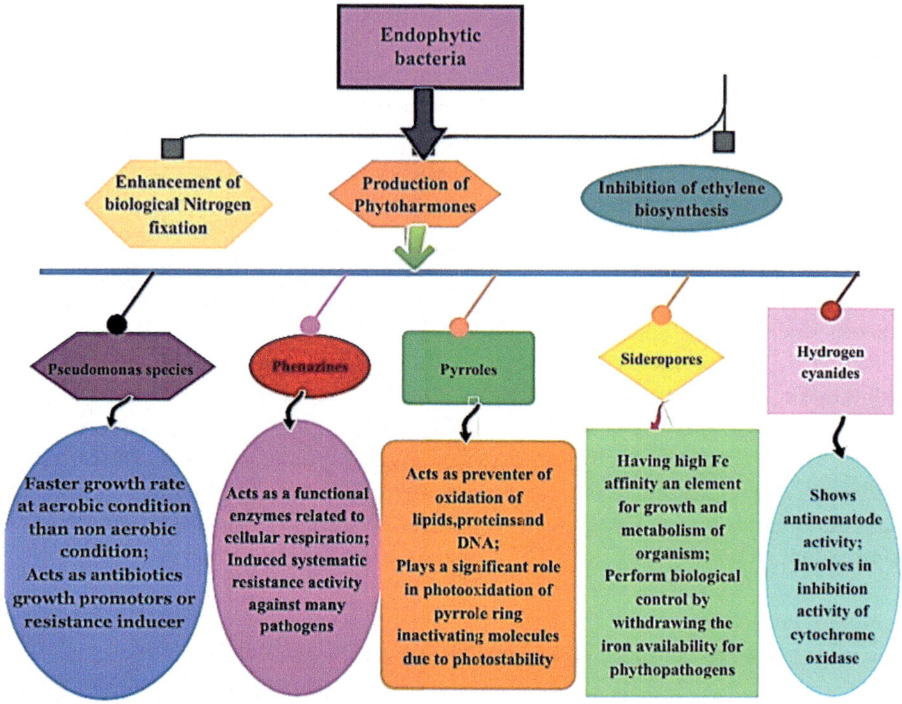

Fig. 11.1 Mechanism, classification and significance of endophytic bacteria

nematode proliferation in the rotation of other crops with other plant hosts (Sturz and Kimpinski 2004). They grow usually inside cellular space and plant vascular tissues. According to an estimation, about 129 or more endophytic bacterial types have been isolated from different kinds of plants, particularly, gram-positive and gram-negative bacteria representing about more than 54 genera. Gaiero et al. (2013) reported that bacterial endophytes promote the growth of plant, but have not obtained productive success to influence the growth of plants in the field conditions. Endophytes act as plant growth promoters, suppress pathogens, remove contaminants and help in solubilizing phosphate or contributing to plant nitrogen assembly (Rosenblueth and Martínez-Romero 2006).

6.2.1 Bacteria Associated with Plants Produce Bioactive Compounds

Bacterial endophytes have several potential applications in drug discovery and pharmaceutical (Tang et al. 2008) and serve as a key source of natural products for application in oxidative stress and as new bioactive agents (Nongkhlaw and Joshi 2015). They also act as antimicrobial agents that counteract the multidrug resistance in pathogenic microbes. The use of beneficial metabolites isolated from endophytes i.e. Amines and amides as the natural protective defense against pathogens which shows the toxic effect on insects but not mammals are increasing day by day and have shown a significant response compared to antimicrobial compounds that are environmentally friendly. Many endophytes adopt resistance mechanisms against plant pathogens by producing extracellular hydrolases, e.g. proteinase, cellulases, esterases and lipases establishing resistance against plant invasions (Tan and Zou 2001). The endophytic fungi enhance growth attributes of a dwarf mutant which is gas deficient, such as Dongjinbeyo and Waito-C rice through plant growth regulator production (Waqas et al. 2012).

6.2.1.1 Pseudomonas Species

They are facultative aerobic microorganisms and gram-negative bacteria; they can grow, both under aerobic and anaerobic conditions; however, their growth is fast in the aerobic condition under suitable nutritive ecological conditions. The physical appearance of this bacterium could be changed by using tools of molecular biology (Chin-A-Woeng et al. 2002). These bacteria with a minimum pathogenicity potential are saprophytic in nature, and show adaptation to changed environmental conditions, where they are found in all types of ecosystems, i.e. soil, water, plants and animals (Madigan et al. 2010).

6.2.1.2 Phenazines

These are heterocyclic compounds with aroma, produced by *Xanthomonas*, *Streptomyces*, *Mycobacterium*, *Burkholderia* and *Brevibacterium* (Pierson III and Pierson 2010), which could be extracted easily from microbial culture and have significant nitrogen and brightly coloured pigments. According to a hypothesis, as a subsistence capability approach, phenazines are used by bacteria for a competition of nutrients or improving the ability of survival (Laursen and Nielsen 2004) that is not easily understood. They also affect negatively electron flow and functional enzymes that play a vital role in the cellular respiration (Yu et al. 2018).

6.2.1.3 Pyrroles

Pyrrole pyrrolnitrin is considered to be a very prominent bacterial compound which is produced by *Burkholderia* and *Pseudomonas* in some of the bacteria in secondary metabolism (Kilani and Fillinger 2014). Pyrroles in general function in electron transportation to a respiratory chain of complex III in mitochondria and play a key role in preventing protein and lipid oxidation (Gomes 2012).

6.2.1.4 Siderophores

These are compounds diversely produced by genus *Pseudomonas* with low molecular weight and tendency of high affinity towards iron, helping in the metabolism and growth of organisms (Fedrizzi 2006). Pyoverdin and pyochelin have the ability to remove iron to perform the biological control of plant pathogens and decline in population (Scavino and Pedraza 2013).

6.2.1.5 Hydrogen Cyanide

It is a highly volatile compound which produces cyanide anions and is highly toxic when it comes in contact with air or water. A number of biocontrol PGPB have the ability to synthesize hydrogen cyanide (HCN). If the HCN produced by these bacteria were the only biocontrol mechanism being used in most instances, the low level of HCN would not be particularly effective at preventing the proliferation of most fungal phytopathogens. However, it is often the case that biocontrol PGPB that can produce HCN also synthesize some antibiotics or cell wall degrading enzymes (Fernando et al. 2005). Moreover, it has been observed that the low level of HCN synthesized by the bacterium improves the effectiveness of antifungals directed against fungal pathogens thereby ensuring that the fungi do not develop resistance to the particular antifungal in question. Thus, HCN synthesized by PGPB appears to act synergistically with other methods of biocontrol employed by the same bacterium (Fernando et al. 2005). It also stimulates ISR in plants and is also involved in the cytochrome oxidase inhibition playing a key role in electron transport during cellular respiration to prevent adenosine triphosphate production (Spence et al. 2014) and also acts as antinematode agent. The model nematode *Caenorhabditis elegans* was repelled by using HCN together with pyrrolnitrin (Nandi et al. 2015). Likewise, Kang et al. (2018) observed a positive correlation of HCN production with nematode biocontrol.

6.2.1.6 Bacillus Species

Biocontrol agents such as antibiotics are used to suppress pathogens that produced substances with the competition of antimicrobial activity, enhancing the growth of a plant and simulating resistance induction (Xu et al. 2013). They have an antimicrobial role formed by *Bacillus* sp. or ribosomal antibiotics like subtilin, chitinase, sublancin, Tas A and subtilisin A. They adopt environmental variations by forming endospore-resistant structures (Hoyles et al. 2012). Enabling polyketide synthases or NRPS formed various other compounds, e.g. mycobacillin, bacilysin, difficidin, chlorotetain, bacillaene, cyclic lipopeptides, and rhizocticins with a wide-ranging biotechnological significance (Shafi et al. 2017).

Iturins, consisting a subgroup of iturin, mycosubtilin and bacillomycin, through pore formation in the cytoplasmic membrane, affect cells (Gong et al. 2015). The isomers of iturin A have a high antifungal activity against several microorganisms (Ye et al. 2012). Bacillomycin belongs to the family iturin lipopeptide, with a ring structure. It helps in spore germination, mycelial growth inhibition, antifungal action and productivity and also has high ultrastructural variations, i.e. cell wall and membrane damage (Gu et al. 2017). Mycosubtilin considerably affected some of the crops like *F. oxysporum, B. cinerea, R. solani* and *Pythium* sp. by targeting cells of cytoplasmic membrane (Leclère et al. 2005). Fengycins, also called plastathins, consisted of fatty acids which are hydroxylated, having solid antifungal activity (Gong et al. 2015). Surfactins are hydroxylated fatty acids, which are not toxic fungi themselves. A combination of surfactins with iturin A shows the tendency for antifungal action (Deravel et al. 2014).

6.2.1.7 Actinobacteria, Beta- and Gammaproteobacteria

Actinobacteria are gram-positive bacteria generally found abundant in soil, water environment, animals and any other natural place and in anaerobic conditions with important morphological variations (El-Tarabily and Sivasithamparam 2006). The use of actinomycetes has been started by a human in the recent few decades assuring the quality of agricultural products. They have high potential as biological controlling agents of pathogens of plants. Antimicrobial compounds can be produced by actinomycetes acting as inhibition promoters against phytopathogens, as they also are producers of 70% of antibiotics (Lasudee et al. 2018).

Betaproteobacteria consisted of more than 400 species and 75 bacterial genera. They are mostly heterotrophic, while a few of them are known to be autotrophic and photo-heterotrophic, helping in soil pH maintenance and nitrate usage as a side acceptor of an electron. The majority of taxa within this class contain HPUT, which has been reported only for selected species of the gammaproteobacterial genus *Shewanella* and a *Colwellia* species, However, a few taxa have been reported to lack the *Betaproteobacteria*-specific HPUT (Ionczewski and Foster 2014).

However, *Gammaproteobacteria* consisted of a variety of polyamine patterns such as PUT, DAP, SPD and CAD in combinations and in a single form. Majority of *Shewanella* sp. contains HPUT and in common a diamine which is a clear seen trait for most of the beta bacteria. However, these species within the genus *Shewanella* did not show any relationship with this diamine. Species of the family *Pseudomonadaceae* consist of major polyamines PUT and SPD and often also CAD. Species of the genus *Aeromonas* contain DAP and PUT as major components, and significant amounts of CAD may also be present. The family *Enterobacteriaceae* requires separate consideration. Almost all species of the assigned genera contain the major diamine PUT, and the majority of them also contain major amounts of DAP and/or CAD and some also contain SPD. The physiological age of the biomass from which polyamines were extracted can greatly influence the resulting polyamine pattern. In late exponential phase cells of *K. pneumoniae*, PUT is predominant, and DAP is a second major polyamine. In contrast, CAD is predominant in late stationary phase cells, and PUT is a second major polyamine. These changes in the polyamine contents are reflected by a twofold reduction of relative amounts of DAP and PUT in late stationary phase cells, whereas the amount of CAD increases tenfold. Applying the same test conditions, DAP is the major polyamine in *E. cloacae*, and its relative amount is almost unaffected (reduced from 51 to 45%) by the growth phase of the cells. At the same time, CAD is increased from 2 to 22%, and relative amounts of PUT are unaffected as well. In late exponential phase cells of *Y. regensburgei*, PUT is the major polyamine (54%), whereas in stationary cells, PUT and CAD are predominant (33.5 + 1.3%) (Hamana and Kishimoto 1996).

7 Volatile Organic Compounds

Compounds which are organic, have high vapour pressure, and easily evaporate at room temperature and are produced by actinobacteria having great potential as biopesticides in the field of agriculture (Sharma and Salwan 2018) are known as volatile organic compounds (VOCs). They are also known as solvents having variable volatility and lipophilicity. A small amount of VOCs is present in correction fluid, graphics, copier and printer and carbonless copy paper. These are also found in craft material (adhesives, photographic solutions, glues and permanent markers) as well. These compounds have some major health concern from the last three decades due to their carcinogenic property (presence of predominant solvent, i.e. $CHCl_3$, trichloroethylene, tetrachloroethylene, benzene xylenes and ethylbenzene). VOCs volatilize during production, transport, storage and processing activities; hence their entry point to the environment is through evaporation process.

7.1 Ecological Role of VOCs and Interspecies Interactions

Loss in biodiversity and changes in ecosystem behaviour due to environmental pollution are major concerns to mankind causing different physiological disorders and diseases in human. Volatile organic compounds, with physical and chemical properties and mean life in the atmosphere, are introduced through biogenic and anthropogenic activity forming tropospheric ozone and less than 2.5 μm particles in big cities, degrading the quality of air and causing air pollution (Robinson 2005). According to World Health Organization, compounds with less than 250 °C boiling point (at a standard pressure of 101.3 kPa) are known as volatile organic compounds (Habre et al. 2014). Their life span is usually from few minutes to several months; hence transport through large distance from the emission source to the living body through air or skin causes several pathogenic diseases like atopic dermatitis, asthma, neurologic problem, etc. The International Agency for Research on Cancer (IARC) considered VOCs of group 1 as carcinogenic for humans (Rumana et al. 2014).

Direct and indirect interaction in community-wide scenario greatly depends on chemical traits of plants, which have a key role in running of these processes.

Plant-mediated interaction has several effects (allelopathy, natural communities, resource competition and facilitation) on plant community organization (Callaway 1995), while the structure of a community is due to mutualistic and antagonistic interaction of plants with other organisms. Such interactions directly alter physiology of another organism by plant traits or indirectly affect the third party (which is not involved in the interaction) due to the interaction of two organisms (Ohgushi et al. 2007). As a result, plant-mediated interaction adds complexity within community interactions and links potential organism of different trophic levels (Utsumi et al. 2010). It is well known that the effect of plant-mediated interaction on the agricultural system is of great importance. These effects include herbivore, pathogen and pollinators which affect plant productivity (Schiestl 2015).

7.2 Microbial VOCs (mVOCs) in Bacteria and Plant Interactions

Microbial interaction plays an important role within and outside kingdom interaction due to a variety of compounds and secondary metabolites released by these microorganisms. Plant and soil-related microbes produced a group of secondary metabolite which was reported; however, there are many more groups which are still unexplored. These compounds are small and odorous with high vapour pressure, low boiling point, a lipophilic moiety and low molecular weight, which has facilitated above and below ground diffusion and evaporation processes due to pore spaces in the ecological rhizosphere and soil (Effmert et al. 2012). The mVOCs produced belonging to several classes (ketones, alchohols, pyrazines, alkenes, sulfides, benzenoids, terpenes, etc.) are influenced by different factors which include nutrient and oxygen availability, the growth stage of microbes, temperature, soil moisture, pH, etc. (Schulz-Bohm et al. 2015). mVOCs benefit the plants in several ways which include induced resistance against plant pathogen, source of nutrients and plant secondary metabolite production and induce soil fungistasis and suppressiveness (Wintermans et al. 2016). There is a decrease in spore formation of *B. cinerea* and *Alternaria alternata*, and increase in plant defence reactions is due to a 6-pentyl-pyrone, a distinguishing compound of *Trichoderma asperellum* (Kottb et al. 2015). VOCs extracted from roots have multiple roles, i.e. as defence metabolites, chemical attractants, carbon sources, etc. (Van Dam et al. 2016).

7.3 Microbial VOCs in Fungi-Plant Interactions

Recent studies have shown the capability of soil fungi to produce volatile organic compounds that enhance growth (Lee et al. 2015). There are beneficial effects of *Trichoderma* strains found in root ecosystem and soil to enhance plant growth by mimicking themselves as secondary metabolites. This mimicry effect significantly improves biomass, plant size, chlorophyll concentration and root size of tomato and *Arabidopsis* (Lee et al. 2016). 1-Hexanol at low concentration had a growth-promoting effect on *Arabidopsis* while at high concentration it inhibited plant growth (Jelen et al. 2014) showing the change of volatile fungal profile with maturation and growth. Moreover, the survival of plants in certain habitats is also mediated by VOCs of endophytic fungi. VOCs enhance the growth of host by reducing the availability of nutrients to endophytic fungi showing the toxic effect of VOCs on endophytic fungi (Macias-Rubalcava et al. 2010).

7.4 Microbial VOCs in Fungi-Bacteria Interaction

There are different phenotypical responses in the interacting behaviour of fungi and bacteria due to fungal VOCs. Some of the recent studies showed the role of fungal VOCs in the suppression of bacterial growth, for example, VOCs formed by mushroom (oyster) have an inhibitory effect on *B. subtilis* and *B. cereus* (Werner et al. 2016). Transcriptomics and proteomics studies showed that there was a change in protein and gene expression (associated with energy metabolism, motility, secondary metabolite production, signal transduction and cell envelope biogenesis) of *S. plymuthica* when kept open to VOCs produced by *Fusarium culmorum*, a fungal pathogen (Schmidt et al. 2017). All the results indicate the significance of VOCs as molecules of signalling in bacterial and fungal interaction. In response to fungi, bacteria can also produce some VOCs which have an inhibitory effect on fungal growth. This phenomenon is known as soil fungistasis (Garbeva et al. 2011). VOCs produced by *Streptomyces* spp. have an inhibitory effect on the growth of a fungus *Rhizoctonia solani* thereby reducing the chances of diseases on the plant (Cordovez et al. 2015).

7.5 Endophytic Plants Secreting Microbial VOCs with Potential Aspects

The progress in biological, chemical and genomic analysis has led us to improve these mysterious natural volatile organic compounds produced by plant endophytes. We are at the beginning to explore the properties and nature of secondary metabolites, and by now several metabolites positively affect biocontrol, the stimulants of plant growth, biofuel and biopharmaceuticals. The significant volatile organic compounds produced organics of endophytes with the key role and their effect on socio-economic development (Table 11.1).

Table 11.1 Volatile organic compounds and their functions

Name of fungus	Compounds
Muscodor albus	Isoamyl acetate
Muscodor crispans	A mixture of antifungal and antibacterial volatile organic compounds
Daldinia concentrica	27 different compounds including alcohols, dienes, ketones, aldehydes and sesquiterpenes
Oxyporus latemarginatus	5-Pentyl-2-furaldehyde
Ascocoryne sarcoides	Hydrocarbons (preferentially produces several ketones and esters
Phomopsis sp.	15 volatile compounds including sesquiterpene with α-humulene or α-caryophyllene and several naphthalene derivatives
Myrothecium inundatum	Sabinene (monoterpene), 1-propanol, etc.
Trichoderma atroviride	Expresses biocontrol gene phlA that encodes 2,4-diacetylphloroglucinol
Pleurotus pulmonarius (oyster mushroom)	3-Octanone, 3-octanol, 1-octen-3-ol, benzaldehyde and unidentified trace components
Pseudomonas donghuensis	Dimethyl sulphide, *S*-methyl thioacetate, methyl thiocyanate, dimethyl trisulphide, 1-undecan and HCN
Bacillus subtilis GB03 (PGPR)	3-Hydroxy-2-butanone (acetoin), 2,3-butanediol, choline, and glycine betaine
Streptomyces sp.	Methyl 2-methylpentanoate, 1,3,5-trichloro-2

7.6 VOCs of Endophytes as a Plant Growth Stimulant

Plant endophytes ranging from bacteria to fungi produce a diverse class of volatile organic compounds (VOCs) that are important for the development of symbiotic relation under highly competitive environment with the host. They provide for an alternative to chemicals used to protect plants from pathogens and thus allow for better crop welfare. Microbial volatile organic compounds (mVOCs) act as a bio-control agent to control phytopathogens and as biofertilizers to promote plant growth. Various recent studies have proven the importance of mVOCs (eco-friendly) like a cost-effective sustainable strategy in the use of agriculture, which improves productivity, plant disease resistance and growth. Moreover, mVOCs can also be used as the substitutes to bactericides, fungicides and pesticides which are harmful (Ryu et al. 2003). It was evident that VOCs enhance plant nutrition, growth, health processes and resistance to stress, coined by a group of scientists who reported improvement in growth of *Arabidopsis thaliana* by volatile organic compounds released by *Bacillus subtilis* GB03. Furthermore, *Bacillus* species release volatiles that modify root architecture (Gutiérrez-Luna et al. 2010). Among the first volatiles produced by bacteria was 2,3-butanediol, which confers resistance in plants (Ryu et al. 2004).

7.7 VOCs of Endophytes as Aroma and Flavour Compounds

Some endophytes that live in aromatic plants are of commercial importance and produce abundant VOCs which produce aroma and fragrance. Terpenes, terpenoids and ester molecules are used in the preparation of beverages and food which has valuable aroma and flavour. Many fungal VOCs are found to be identical to natural flavouring and fragrance produced by plant molecules and are therefore of huge importance in chemical, feed, pharmaceutical, food and cosmetic industries. The fungal endophytes, which are volatile, produce a desirable aroma and flavour property which is used in many nonalcoholic beverages, jellies, backed goods, cheese, puddings, candies and other food products. The major component of rose oil, methyl eugenol (having a high demand in industries), has been identified in fungal endophytes *Aspergillus niger* and *Alternaria* sp. which were extracted from *Rosa damascena* (Abrahao et al. 2013). A remarkable molecule of terpene (β-caryophyllene) with spicy flavour has been reported in endophytic fungus volatile (*Phialocephala fortinii*) and extracted from *Pinus sylvestris* (Molina et al. 2012).

7.8 VOCs of Endophytes as Mycofumigation Agents and Biopharmaceuticals

Endophytic compounds are known to have anticancerogenic, antibacterial, immunosuppressant and antioxidant activities, reported from different researches of the last few decades. VOCs enhance plant defence and are discovered to have new antimicrobials to treat many diseases in medical science. Fungal endophytes produced different secondary metabolites used in pathogenic and pests attack control (Hung et al. 2015). *Muscodor albus* (an endophyte fungus) produce more than 25 volatile compounds extracted from cinnamon tree and are thought to have strategic defence against many pathogens (Stinson et al. 2003). The first ever fungus endophyte, *M. albus*, was thought of having potential antimicrobial function against humans and phytopathogens. The volatiles of *M. albus* are also used to treat different diseases like silver scurf, bacterial soft rot and dry rot in potatoes (*Solanum tuberosum*) inhibiting the three infectious fungi (*T. indica*, *T. tritici* and *Tilletia horrida*) causing many diseases in rice and wheat plants (Schalchli et al. 2016). Moreover, a special volatile, 2-phenylethanol, found in *Aspergillus niger* (endophyte fungi of rose) has great importance as preservatives in pharmaceuticals (Wani et al. 2010).

7.9 Significance of mVOC and Future Perspectives on Commercial Basis

Since the improvement in the analysis of gas-phase molecules, it can be observed that endophytic VOCs change biologically and chemically in a more active way. Gas chromatography-mass spectrometry (GC-MS) is known to be the most common and effective method that identifies volatile components but limited to the column used in this spectrometry. These columns used are selective for detecting some chemical groups of VOCs but not the total VOCs (Insam and Seewald 2010). Recently, quantitative analysis of VOCs becomes easy using a technique called proton transfer reaction-mass spectrometry (PTR-MS) that is a very sensitive method (Strobel et al. 2011). Hydronium ion in gas phase is used as a sourcing agent to monitor VOCs in ambient air. Nowadays, the most effective tool to detect and identify VOCs is the combination of PTR-MS and GC-MS (Insam and Seewald 2010).

Various environmental factors such as the composition of a microbial community, nutrient content, pH, humidity and temperature influence microbial volatile production (obtained as a complex mixture). These factors made it difficult to identify whether the effect is on an individual molecule and what is the mechanism. Hence, the commercial application of this volatiles is very limited as compared to the economic implications. Now it is well understood that there are varying differences of volatile compound effect from lab to field (Song and Ryu 2013). However, volatile compounds as a biocontrol and growth-promoting agent are effective against human and plant pathogens (Grimme et al. 2007). Endophyte developed strategies to overcome the challenges related to climate change (salinity, water and drought stress and high temperature) faced by agriculture crops. Moreover, the use of volatile compounds proves to be important in overcoming adversities on plant communities.

8 Signalling Pathway of Secondary Metabolism in Endophytes

To establish a stable biological community collaboration between plants and organisms is required. An ultimate model of studying the benefits of the interaction of fungi and plant is the relation between cool-season grasses and fungi (Schardl et al. 2013). The infection of endophytes and its effect in the light of expression profile relay on the sequencing of RNA. The reprogramming of infection of endophytes results in metabolism which makes secondary metabolism easier compared to primary metabolism. These types of infections can also produce variations in host development such as trichome formation and biogenesis of cell wall. The endophytic diazotrophic bacteria result in nitrogen signalling with endophytic bacteria. The diazotrophic bacteria help in growing a different variety of root associations and fixing N_2 to plant-available ammonium. The biogenesis pathway of swainsonine was reported to be beneficially important in the medical treatment of cancer and plays a significant role in anticancer activities and in regulating the immune system (Carvalho et al. 2014).

The mechanism for the signalling of ethylene reported that this signalling pathway helps in the production of endophytic fungus, the *Gilmaniella* sp. AL12, through induced production of ethylene in *Atractylodes lancea* (Yuan et al. 2016). Plantlet pretreatment with inhibitor aminooxyacetic acid (AOA) suppressed endophytic fungi-induced accumulation of sesquiterpenoids. The amino oxyacetic acid

Table 11.2 Mechanism of metabolite production by endophytic fungi

Host plant	Endophytic fungi	Mechanism	References
Anoectochilus formosanus	*Epulorhiza* sp.	Enhance enzyme activities of chitinase, β-1,3-glucase, phenylalanine ammonium lyase and polyphenol oxidase	Tang et al. (2008)
Atracty lancea	*Sclerotium* sp.	Increase cell protection from desiccation and leaf metabolic capability of host	Chen et al. (2008)
Cucumis sativus	*Penicillium* sp.	Secret phytohormones, viz. gibberellins and indoleacetic acid	Waqas et al. (2012)
Nicotiana attenuata	*Sebacina vermifera*	Enhance the absorption of nutrient and promote the growth and fitness by inhibiting ethylene signalling	Barazani et al. (2007)
Pecteilis susannae	*Epulorhiza* sp. *Fusarium* sp.	Enhance the absorption of N, P and K element in plant promoting the seed germination of host	Chutima et al. (2011)
Pedicularis sp.	Dark septate endophytic fungi	Increase their nutrient utilization efficiency	Li and Guan (2007)
Sesbania sesban	*Funneliformis mosseae, Rhizophagus intraradices* and *Claroideoglomus etunicatum*	Secrete plant hormones	Abd_Allah et al. (2015)

with ethylene inhibitor helps in the pretreatment of plantlets which inhibits the endophytic fungi (Ren and Dai 2012). The biosynthesis of sesquiterpenoid gives a theoretical base for active compound development in *A. lancea* and other compound biosynthesis like menthol, ginseng saponins, glycyrrhizic acid, artemisinin and paclitaxel. Jasmonic acid functions in the signalling pathway of fungal endophyte induced volatile oil for the plant *Atractylodes lancea.* Reports from research observed that jasmonic acid also helps in molecule signalling in mediated volatile of nitric oxide and hydrogen peroxidase by an endophytic fungus (Table 11.2).

9 Molecular and Metabolic Cooperation of Hosts and Endophytes

Many endophytes have the capability of producing different bioactive metabolites, which may be used as the agent for heals, either directly or indirectly, against a wide-ranging disease (Kharwar et al. 2011). Their vast biodiversity combined with the capability of biosynthesizing secondary metabolites has provided the impetus to many endophytic studies (Alvin et al. 2014). A symbiotic association between asexual endophytes of fungus and tall grasses from *Epichloe* exposes alkaloid biosynthesis that produces either beneficial or damaging effects (Ekanayake et al. 2017).

10 Uses and Importance of Endophytes in Plant Health

10.1 Antibiotics Prepared from Endophytic Microbes

Endophytes are a good source of antibiotics (organic natural products having low molecular weight) produced from active microorganisms. These natural products not only kill inclusive diversity of harmful pathogen (phytopathogen) but also those (bacteria, virus, protozoa and fungi) affecting humans and animals. The imperfect stage of *Pezicula cinnamomea* is *Cryptosporiopsis quercina*, known to be a fungus (isolated from an endophytic medicinal plant of Eurasia, i.e. *Tripterygium wilfordii*) which is associated with various deciduous species in European countries. Echinocandins, pneumocandins and antifungal are the major sources of bioactive compounds. A group of fluorescent bacteria (*Pseudomonas viridiflava*), mostly related to plants (linked with a portionof grass leaf present in or on the tissues), produce ecomycins (Strobel et al. 1999).

10.2 Antiviral Compounds

Another charming use of endophytic fungal antibiotic products is viruses' inhibition. sp. is an endophytic fungus, isolates two different cytomegalovirus protease inhibitors *Cytospora* (Ctyonic acid A and B) by solid-state fermentation process. Using mass spectrometry and NMR methods, structures of isomers can be fully elaborated. There is a still long way to detect the potential of endophytic compounds having antiviral activities. Inadequate screening systems of a virus limit the detection of antiviral compounds, but still some detected compounds have shown encouraging results against viruses.

10.3 Volatile Antibiotics from Endophytes

Muscodor albus (fungus), isolated from a cinnamon tree, is a newly studied fungus from endophytic group (Worapong et al. 2001). A fungus having no spores (xylariaceaous fungus) produces a mixture of volatile compounds (having antibiotic effect) that can alter different types of fungi and bacteria (Strobel et al. 2011). A non-*Muscodor* species, *Gliocladium* sp. (G. sp), for the first time has been discovered to be a producer of volatile antibiotics (different from volatile compounds of *M. albus* and *M. roseus*). Indeed, annulene could be found as the most abundant volatile inhibitor; previously, this was the first discovered natural product in an endophytic fungus and was used as rocket fuel (Stinson et al. 2003).

10.4 Biocontrol Activity of Endophytes

A large number of microorganisms are present inside plants producing microbe-plant interaction (some are destructive while others are beneficial). These microorganisms are rich sources of nutrients. Rhizobia, mycorrhiza and actinobacteria help the plants to get nutrients from the soil in a symbiotic interaction. Many bacterial species reduce the activity in the root system, stem, leaves and another plant organ by blocking plant tissues and vessels, but most of them are beneficial (metabolites producer) and help to increase plant defence mechanism against pathogens, nutrient uptake, growth promotion and hence crop productivity. *Streptomyces* belonging to actinomycetes are species-specific (having symbiotic relationship with plants) and are very much helpful to produce a variety of antibiotics. They protect the plants to fight against a pathogen, in response to boost up plant exudate production which is important for the growth of *Streptomyces* (El-Shanshoury 1991). Endophytic actinobacteria produce a chelated iron compound (siderophores), and chitinolytic enzymes have a supplemental role to hinder fungal growth. They also produce chitinase which damages fungal cell wall. About 90% of chitinolytic microorganisms are actinomycetes (Hastuti et al. 2012). A large number of bacteria (especially streptomycetes) obtain nutrients and degrade environmental chitin and soil-borne fungal cell wall by producing chitinases. Numerous bacteria, and especially streptomycetes, also form a variety of chitinases. Thus, selection and exploitation of chitinolytic mediators helps to control phytopathogenic fungi.

10.5 Endophytic-Mediated Plant Growth

Plants face hostile and unfavourable conditions in normal conditions, collectively called abiotic stresses which cause prevention in growth and homeostasis. Below or above optimum levels, severe ecological conditions often cause an effect on plant growth and development. Abiotic stress includes high or low temperature stress, nutrient stress, heavy metal stress, hunger stress, acidic, salt and drought stress that badly affect plant growth (Chaves and Oliveira 2004). Biotic stresses may consist of damage to plant caused by viruses, bacteria, fungi, pests, parasites, native or cultivated plants and weeds. Several microorganisms containing fungi, protozoa and bacteria make a symbiotic or beneficial association with plants, providing benefits to avoid various environmental stresses and support the development and growth of the plant as well (Shahzad et al. 2018). These endophytes contribute significantly to regulate many crucial physiological processes and enhance the overall growth and vigour of plants. For example, the endophytic fungi facilitate the cuticular cellulose degradation by improving the consequence of carbon absorption and promoting the germination of seed (Jerry 1994).

10.5.1 Production of Growth-Induced Compounds and Phytohormones

Plant growth, defence response and physiological processes are positively affected by phytohormones (regulatory molecules) (Egamberdieva et al. 2017). IAA homeostasis affects various physiological processes, comprising germination of a seed, cell differentiation, development of vascular tissues, vegetative growth, development and elongation of root, photosynthesis and pigmentation (Ahmad and Kibret 2013). Microbial representatives of this group enhance plant growth and development by producing a variety of proactive substances such as siderophores, 1- aminocyclopropane-1-carboxylate deaminase (ACC), phytohormones, e.g., indol acetic acid (IAA), gibberellic acid (GA), volatile organic compounds (VOCs), antibiotics, cyanides, and fungal cell-wall-degrading enzymes (Long et al. 2008). The enzyme ACC deaminase is thought to be a key trait in the arsenal that PGPB uses to promote plant growth. ACC and IAA deaminases produced by the rice plants cultivated in the fields of coastal areas recognized six endophytic bacteria in a study reported by Bal et al. (2013). Gibberellic acid-producing endophytic microorganisms often contribute to the improvement of the host plant yield.

Phoma herbarum (an endophytic fungus) obtained from soybean roots under salt stress, showed growth enhancing properties, leading to increased active GAs production and biomass (Hamayun et al. 2010). Strains (SF2, SF3 and SF4) of bacterial endophytes from sunflower under stress condition produced salicylic acid which was helpful to enhance plant growth (Forchetti et al. 2010). Root colonization by endophytic fungus Piriformospora indica caused stimulation in the growth and development of Arabidopsis due to the production of cytokinins (Vadassery et al. 2008).

10.5.2 Potential Role of Endophytes in the Acquisition of Nutrients

One of the key roles is the acquirement of plant nutrients from its natural habitation where most of the plants do not have the mechanism naturally to get vital nutrients. Nitrogen is essential for the plant growth and development but they can not obtain from the atmosphere, and dependent fertilizers containing nitrogen. Whereas, some other plants make a strong association with nitrogen-fixing bacteria, helping out the plants to consume atmospheric nitrogen. Others make symbiotic associations with the nitrogen-fixating bacteria, mostly seen in legumes, which help the plants to utilize the atmospheric nitrogen. However, the colonization of endophytes is markedly different than those of rhizobial nitrogen-fixating symbionts (Doty 2011) or an exchange offer by photosynthesis producing carbohydrates is given for available nitrogen. Through energy involvement and nitrogenase enzyme, ammonia is formed from atmospheric nitrogen by symbionts. *Herbaspirillum* spp., *Acetobacter* spp. and *Azoarcus* spp. help to fix nitrogen from the atmosphere in an association with the actinorhizal and rhizobial symbiosis of plant and bacteria. As like rhizobial bacteria, endophytic organisms adopt various strategies to protect nitrogenase, an enzyme that is sensitive to oxygen. In rhizobial condition, oxygen is usually at very low concentration, where, the haemoglobin in the legs provides help to clean free oxygen traces in the nodules. The endophytic associations between *Gluconacetobacter diazotrophicus* and sugarcane and pines are the well-studied symbiotic associations where the endophyte helps the host plant in nitrogen fixation (Hardoim et al. 2015). A high chelating iron compound, siderophores, functions in the absorption of iron (Johnson et al. 2013).

10.5.3 Endophytic-Mediated Tolerance to Abiotic Stress

Environmental stresses often disrupt the growth and development of plants. To overcome the challenges in such a situation, the endophytes present inside the host plant help out. Though endophytes have a very short life in comparison to its host, the short life cycle helps the host cope with its diversity. In association with plant endophytes, different strategies are then adopted to reduce the abiotic stresses emanating from the natural habitation of the host. Interaction between plants and microbes can be mostly classified as detrimental or neutral. In most of the cases, the interaction is considered as beneficial, because microbes consume the plants organic product for respiration and metabolism and at the same time help in nutrient recycling and tolerance against various stresses. Beneficial microbes encourage plant growth development and inhibit the plant diseases by enhancing different types of the mechanisms which mainly include production of growth regulators, hormones, and pathogen-inhibiting compounds (Lata et al. 2018). For example, *Phoma* spp. and *Penicillium* help to promote growth (uptake of nutrients and plant biomass) and overcome osmotic and drought stress caused by elevated polyethylene glycol and sodium levels (Waqas et al. 2012). Plants such as tomatoes and rice with useful endophytes could survive in water-deficient conditions, even exhibiting better growth potentials than plants which lack these endophytes (Lata et al. 2018).

Salt and drought stress mitigation is normally concerned with consequent scavenging and accumulation of reactive oxygen species (ROS) (Sekmen et al. 2007). Though reduced levels of ROS to plant may support various antioxidants, e.g. tocopherol and glutathione, the main ROS scavengers include glutathione reductases (GR), monodehydroascorbate reductase (MDHAR), and dehydroascorbate reductase (DHAR) (Rouhier et al. 2008). ROS accumulation in the cells of plants can be toxic, leading towards DNA, lipid and protein oxidative damages. ROS respond signalling cascades while acting as the preliminary plant stress event (Noctor et al. 2017). Pathogen-plant interaction observed the accumulation and production of ROS, similarly, leading salt and drought stress in association with successive scavenging of species of reactive oxygen (Sekmen et al. 2007). Whereas, ROS in low concentration is significant for the signalling and growth of pants, raised accumulation of ROS can create harmful effects. Endophytes residing within the plants benefit their host to manage the accumulation of ROS and, hence, protect them from the harmful effects of ROS. Various endophytic plants like those associated with roots are studied, showing the tolerance of host plant to the toxic level of heavy metals (Choo et al. 2015).

10.5.4 Endophytic-Mediated Response of Plant Defence

Plant growth and development is often compromised by the onset of several environmental stresses as plants prioritize resistance over growth. In this scenario, the endophytes living inside the host plants come in great support in overcoming the challenges. Although endophytes are very short-lived as compared to their host, their shorter life span helps in their rapid evolution in aiding the host toward tackling the diversities. The capacity of different endophytes providing resistance against these environmental stresses are exploited in modern sustainable agriculture (Zamioudis and Pieterse 2012) (Fig. 11.2). Moreover, the endophytic colonization in the plants induces a response to defence strategy while providing higher resistance to other pathogens of plants. Such idea in plants is known to be induced systematic resistance (ISR) which could be normally observed in endophytic association of plant and bacteria (Robert-Seilaniantz et al. 2011). ISR induction and pathogen defence enhanced repeatedly were studied in response to *Bacillus* spp. and *Pseudomonas* colonization. Endophytic bacteria can control plant defence manipulation and simulate the effect of primary defence against plant pathogens through ISR (Bae et al. 2011). Contrariwise, the endophytic fungus produces the chemical compound that inhibits growth, and these compounds function against offensive herbivores and invaded pathogens while protecting their host (Brader et al. 2014).

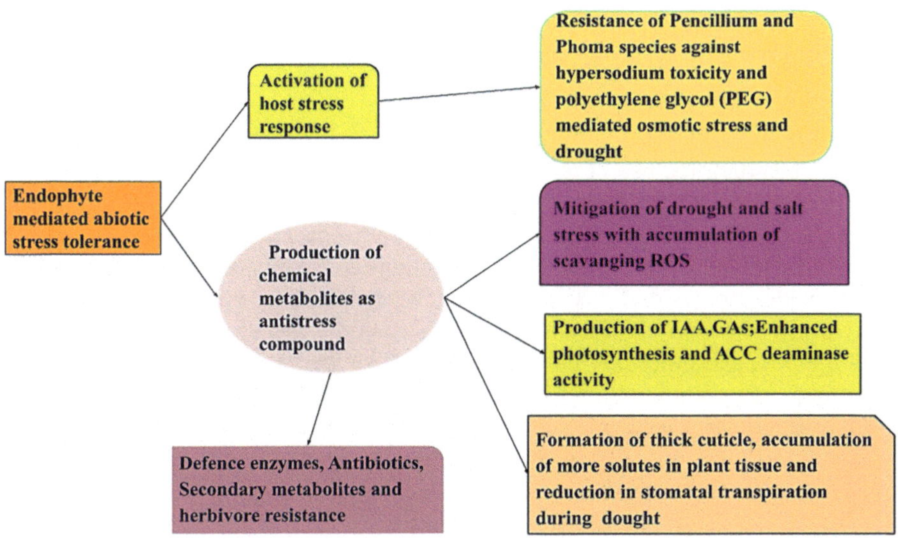

Fig. 11.2 Different mechanisms adopted by endophytes in promoting plant growth and mitigating different types of stress

11 Secondary Metabolite Production: Challenges in Endophytic Research

Although a lot of researches have been done on endophytes showing valuable sources of new metabolites, still, many features of endophytes are not explored. Endophytes have significant importance in the industry producing various enzymes which are helpful to speed up many processes, and also stresses in many plants are also relieved by endophytes. Although many endophytes are forming bacterial and fungal origin and affect many aspects of plant growth (growth, yield and bioactive metabolite production), in-depth understanding of secondary metabolites produced and a chemical released by endophytes in a host plant still needs to be explored. There is a need to investigate the role of endophytes in bioreactors although researches are there where some anticancer metabolites are produced by endophytes (Amna et al. 2006). A classical example of this might be *Entrophospora* infrequens (an endophytic fungus) that produced some anticancer alkaloid camptothecin in bioreactors. Nowadays, "omics" tools are used to better understand the host-endophyte niche. These omics tools consist of genome sequencing, next-generation sequencing, comparative genomics, microarray, metagenomics and metatranscriptome which help to recognize metabolic diversity and genetics of similar or related microbes. There is a lack of knowledge in the production of endophytes on a large scale for bioreactors to know the pathway shared by hosts and endophytes, an area of research for many scientists to explore and focus on.

12 Recent Developments in the Field of Microbiome Research

Studies regarding the use of microbiome have improved radically in recent years, due to the cost of analysis reduction and technological advancement. These researches have opened a gateway of data which has increased a significant amount of intuition to the scope of microbial populations consisting of interaction and their effect inside or outside of host as a particular portion of the ecological community. Keeping in view the significant role of microbiome including their combination with the host and other microbes provides a base for studying the engineering of new diagnostic techniques and strategies, which can be used in a diverse array of fields starting from ecology and agriculture to agriculture to medicine and from forensics to exobiology. The microbiome refers to a set of highly interactive microbial species that is shaped by the environment in which it exists, which includes hosts, and exogenous natural and human factors.

12.1 Interaction of Host with Microorganisms

The host along with its entire related microorganism is collectively called as "holobiont", while the study of host and microorganism genome is called "hologenome". According to researchers, the unit for natural selection is holobiont (Davenport et al. 2017). Every host can adopt two ways to procure microorganisms which are inherited from the parents (vertical) and taken from the environment (horizontal). A correlation of similar microbiome and host phylogeny is due to vertical transmission hence known as "phylosymbiosis"—however, it is important to bear in mind that the emergence of phylosymbiosis is irrespective of vertical transmission, e.g. contact of host species to other members (Groussin et al. 2017). Co-diversification (similar selective pressure or co-speciation results in microorganism with similar evolutionary histories) and co-speciation (host speciation results in microorganism speciation) are also the outcomes of vertical transmission (Davenport et al. 2017). In contrast to vertical transmission, horizontal transmission causes breakage of association with evolutionary histories, so mix them up. Hence, erode phylosymbiosis.

12.2 Interaction of Host and Microbiome with the Environment

12.2.1 Relationship of Microbiome with Environment and Ecology

Recent studies show the effect of microbiome on the different features of human health (Martí et al. 2017). However, generally speaking, interaction of microbiota with environment gives a clear picture of a healthy ecosystem and mankind. A healthy microbiome and environment results in healthy human microbiome and vice versa (Lloyd-Price et al. 2016). Therefore, it is very much important to study microbiome in ecosystem. The functional and structural richness of ecosystem communities determines the individual and populations of microbiome at various sides of biological organization (Rees et al. 2017).

12.2.2 Microbiome Ecology in a Population

There is a deep, empirical, computational and theoretical understanding of community ecology (a sub-branch of ecology) nowadays. Diversity determines a stable microbiome-related population health and microbiome itself (Coyte et al. 2015). The state of microbiome is determined by functional diversity (a more meaningful and fundamental feature) rather than taxonomic diversity (Li and Convertino 2019). Metacommunity approach (a useful tool to predict biodiversity assemblage) of microbiome is determined by alpha (diversity within), beta (diversity between) and gamma diversity (total diversity of microorganisms) that consists of multiple interacting communities. The scale for sharing fluctuation of information representing microorganism interdependencies greatly varies with biology, space and time (Leibold et al. 2004).

12.2.3 Nexus of Human, Microbiome and Environment

On a long and short timescale, microbiome research helps in making a positive relationship between human health and the environment. Efforts have been made to map microbiome of the globe for various habitats but the information regarding environment and microbiome population interaction is still lacking. Hence, a steady struggle for alternations in symptom-specific or disease analysis of microbiome to an outside environmental agent is the need of the day (Karkman et al. 2017; Mitmesser and Combs 2017). The noteworthy that targeted monitoring, models, and theory guides this ecological examination need no in-depth health analysis of microbiomes but time, space pattern establishing an ecological state of the co-evolving microbiomes Parfrey and Knight (2012) such as the pattern in biodiversity (Ochman 2016) and other services relating to the socio-ecological ecosystem.

13 Conclusion

Bioactive compounds, normally, can be used in controlling various diseases of plants biologically. The biological production of such antimicrobial bioactive compounds depends specifically on the resources and space competition. As a natural derivative metabolite, bioactive compounds played a havoc role in the world of pharmaceuticals and agrochemicals to combat against various diseases in plants and play a key role in human welfare. A huge number of biological antimicrobials are formed as a result of biological secondary metabolism providing benefits to the plants. Such bioactive metabolites have great potential use in the agriculture industry, specifically in controlling pathogens, and concerning the sustainability of the environment. Where the endophytes are known as biological endosymbiotic microbes found almost everywhere in the ecosystem, specifically distributed in a wide range in many plants, possess a long-term beneficial association with the host plant, combat against biotic and abiotic stresses and help in metabolism and stimulate immunity responses. Further, distributed in endophytic bacteria, fungi and actinomycetes in association with plants produce bioactive compounds. *Actinobacteria*, Beta- and *Gammaproteobacteria* function differently, i.e. *Actinobacteria* found abundantly in soil and other natural spaces act as a biological controlling agent against pathogens, *Betaproteobacteria* function as pH moderator and *Gammaproteobacteria* containing DAP work in the growth phase of cells. Where volatile organic compounds (VOCs) play a significant role in carbon sources, defence metabolites and chemical attractants, various endophytes can produce bioactive compounds/metabolites that can be used against many diseases either directly or indirectly. Studies regarding the use of microbiome have improved radically in recent years, due to the cost of analysis reduction and technological advancement. There is a great need for new bioactive compound production to replace agrochemicals used in controlling plant diseases, and a vast research study is needed to be carried out globally.

References

Abrahao MR, Molina G, Pastore GM (2013) Endophytes: recent developments in biotechnology and the potential for flavor production. Food Res Int 52(1):367–372

Abd_Allah EF, Hashem A, Alqarawi AA, Bahkali AH, Alwhibi MS (2015) Enhancing growth performance and systemic acquired resistance of medicinal plant Sesbania sesban (L.) Merr using Arbuscular mycorrhizal fungi under salt stress. Saudi J Biol Sci 22:274–283

Ahmad M, Kibret M (2013) Mechanism and applications of plant growth-promoting rhizobacteria: current perspective. J King Saud Univ Sci 26:1–20

Alvin A, Miller KI, Neilan BA (2014) Exploring the potential of endophytes from medicinal plants as sources of antimycobacterial compounds. Microbiol Res 169:483–495

Aminkhani A, Sharifi S, Ekhtiyari S (2020) Achillea filipendulina Lam. Chemical constituents and antimicrobial activities of essential oil of stem, leaf, and flower. Chem Biodivers e2000133:17

Amna T, Puri SC, Verma V, Sharma JP, Khajuria RK, Musarrat J, Spiteller M, Qazi GN (2006) Bioreactor studies on the endophytic fungus *Entrophospora infrequens* for the production of an anticancer alkaloid camptothecin. Can J Microbiol 52:189–196

Arnold AE (2007) Understanding the diversity of foliar endophytic fungi: progress, challenges, and frontiers. Fungal Biol Rev 21:51–66

Aswani R, Vipina Vinod T, Ashitha J (2020) Benefits of plant–endophyte interaction for sustainable agriculture. In: Kumar A, Radhakrishnan E (eds) Microbial endophytes: functional biology and applications. Elsevier, Amsterdam, pp 35–55

Atanasov AG, Waltenberger B, Pferschy-Wenzig EM, Linder T, Wawrosch C, Uhrin P et al (2015) Discovery and resupply of pharmacologically active plant-derived natural products: review. Biotechnol Adv 33:1582–1614

Bacon CW, White JF (2000) Microbial endophytes. Marcel Dekker, New York

Bae H, Roberts DP, Lim H-S, Strem MD, Park S-C, Ryu C-M, Melnick RL, Bailey BA (2011) Endophytic Trichoderma isolates from tropical environments delay disease onset and induce resistance against Phytophthora capsici in hot pepper using multiple mechanisms. Mol Plant-Microbe Interact 24:336–351. https://doi.org/10.1094/MPMI-09-10-0221

Baker D, Mocek U, Garr C (2000) Natural products vs. combinatorials: a case study. In: Wrigley SK, Hayes MA, Thomas R, Chrystal EJT, Nicholson N (eds) Biodiversity: new leads for pharmaceutical and agrochemical industries. The Royal Society of Chemistry, Cambridge, pp 66–72

Bal HB, Nayak L, Das S, Adhya TK (2013) Isolation of ACC deaminase PGPR from rice rhizosphere and evaluating their plant growth promoting activity under salt stress. Plant Soil 366:93–105

Barazani O, von Dahl CC, Baldwin IT (2007) Sebacina vermifera promotes the growth and fitness of Nicotiana attenuata by inhibiting ethylene signaling. Plant Physiol 144:1223–1232

Brader G, Compant S, Mitter B, Trognitz F, Sessitsch A (2014) Metabolic potential of endophytic bacteria. Curr Opin Biotechnol 27:30–37

Bulgarelli D, Schlaeppi K, Spaepen S, van Themaat EVL, Schulze-Lefert P (2013) Structure and functions of the bacterial microbiota of plants. Annu Rev Plant Biol 64:807–838

Buss T, Hayes MA (2000) Mushrooms, microbes and medicines. In: Wrigley SK, Hayes MA, Thomas R, Chrystal EJT, Nicholson N (eds) Biodiversity: new leads for pharmaceutical and agrochemical industries. The Royal Society of Chemistry, Cambridge, pp 75–85

Callaway RM (1995) Positive interactions among plants. Bot Rev 61:306–349

Cammack R, Atwood T, Campell P, Parish H, Smith A, Vella F, Stirling J (2006) Oxford dictionary of biochemistry and molecular biology, 2nd edn. Oxford University Press, Oxford, pp 74–75

Card S, Johnson L, Teasdale S, Caradus J (2016) Deciphering endophyte behaviour: the link between endophyte biology and efficacious biological control agents. FEMS Microbiol Ecol 92(8):fiw114

Carvalho ETL, Balsemão-Pires G, Saraiva RM, Ferreira PCG, Hemerly AS (2014) Nitrogen signalling in plant interactions with associative and endophytic diazotrophic bacteria. J Exp Bot 65(19):5631–5642

Chang CL, Lin Y, Bartolome AP, Chen YC, Chiu SC, Yang WC (2013) Herbal therapies for type 2 diabetes mellitus: chemistry, biology, and potential application of selected plants and compounds. Evid Base Complement Alternat Med 2013:378–657

Chaves MM, Oliveira MM (2004) Mechanisms underlying plant resilience to water deficits: prospects for water-saving agriculture. J Exp Bot 55:2365–2384

Chen JX, Dai CC, Li X, Tian LS, Xie H (2008) Endophytic fungi screening from Atracty lancea and inoculating into the host plantlet. Guihaia 28:256–260

Chin-A-Woeng TF, Bloemberg GV, Mulders IH, Dekkers lC, Lugtenberg BJ (2002) Root colonization by phenazine-1-carboxamide-producing bacterium *Pseudomonas chlororaphis* PCL1391 is essential for biocontrol of tomato foot and root rot. Mol Plant-Microbe Interact 13:1340–1345

Choo J, Sabri NBM, Tan D, Mujahid A, Müller M (2015) Heavy metal resistant endophytic fungi isolated from Nypa fruticans in Kuching Wetland National Park. Ocean Sci J 50:445–453

Chowdhury SP, Hartmann A, Gao XW, Borriss R (2015) Biocontrol mechanism by root associated Bacillus amyloliquefaciens FZB42—a review. Front Microbiol 6:780

Chutima R, Dell B, Vessabutr S, Bussaban B, Lumyong S (2011) Endophytic fungi from Pecteilis susannae (L.) Rafin (Orchidaceae), a threatened terrestrial orchid in Thailand. Mycorrhiza 21:221–229

Cordovez V, Carrion VJ, Etalo DW, Mumm R, Zhu H, van Wezel GP et al (2015) Diversity and functions of volatile organic compounds produced by Streptomyces from a disease suppressive soil. Front Microbiol 6:1081

Coyte KZ, Schluter J, Foster KR (2015) The ecology of the microbiome: networks, competition, and stability. Science 350:663–666

Davenport ER, Sanders JG, Song SJ, Amato KR, Clark AG, Knight R (2017) The human microbiome in evolution. BMC Biol 15:127

Demain AL (2000) Microbial natural products: a past with a future. In: Wrigley SK, Hayes MA, Thomas R, Chrystal EJT, Nicholson N (eds) Biodiversity: new leads for pharmaceutical and agrochemical industries. The Royal Society of Chemistry, Cambridge, pp 3–16

Deravel J, Lemière S, Coutte F, Krier F, Hese NV, Béchet M, Sourdeau N, Höfte M, Leprêtre A, Jacques P (2014) Mycosubtilin and surfactin are efficient, low ecotoxicity molecules for the biocontrol of lettuce downy mildew. Appl Microbiol Biotechnol 98:6255–6264

Doty SL (2011) Nitrogen-fixing endophytic bacteria for improved plant growth. In: Bacteria in agrobiology: plant growth responses. Springer, Berlin, pp 183–199

Dreyfuss M, Chapela I (1994) Potential of fungi in the discovery of novel, low-molecular weight pharmaceuticals. In: Gullo VP (ed) The discovery of natural products with therapeutic potential, vol 6. Elsevier, Newnes, pp 49–80

Edwards J, Johnson C, Santos-Medellín C, Lurie E, Podishetty NK, Bhatnagar S, Eisen JA, Sundaresan V (2015) Structure, variation and assembly of the root-associated microbiomes of rice. Proc Natl Acad Sci 112(8):911–920

Effmert U, Kalderas J, Warnke R, Piechulla B (2012) Volatile mediated interactions between bacteria and fungi in the soil. J Chem Ecol 38:665–703

Egamberdieva D, Wirth SJ, Alqarawi AA, Abd Allah EF, Hashem A (2017) Phytohormones and beneficial microbes: essential components for plants to balance stress and fitness. Front Microbiol 8:2104

Ekanayake PN, Kaur J, Tian P, Rochfort SJ, Guthridge KM, Sawbridge TI, Spangenberg GC, Forster JW (2017) Genomic and metabolic characterisation of alkaloid biosynthesis by asexual Epichloë fungal endophytes of tall fescue pasture grasses. Genome 60:496–509. NRC Research Press

Ek-Ramos MJ, Gomez-Flores R, Orozco-Flores AA, Rodríguez-Padilla C, González-Ochoa G, Tamez-Guerra P (2019) Bioactive products from plant-endophytic gram-positive bacteria. Front Microbiol 10:463. https://doi.org/10.3389/fmicb.2019.00463

El-Beltagi HS, Mohamed HI, Safwat G, Megahed BMH, Gamal M (2018) Evaluation of some chemical constituents, antioxidant, antibacterial and anticancer activities of *Beta vulgaris* L. root. Fresenius Environ Bull 27(9):6369–6378

El-Beltagi HS, Mohamed HI, Abdelazeem AS, Youssef R, Safwat G (2019a) GC-MS analysis, antioxidant, antimicrobial and anticancer activities of extracts from *Ficus sycomorus* fruits and leaves. Notulae Botanicae Hortic Agrobotanici Cluj-Napoca 47(2):493–505

El-Beltagi HS, Mohamed HI, Elmelegy AA, Eldesoky SE, Safwat G (2019b) Phytochemical screening, antimicrobial, antioxidant, anticancer activities and nutritional values of cactus (*Opuntia Ficus Indica*) pulp and peel. Fresenius Environ Bull 28(2A):1534–1551

El-Beltagi HS, Mohamed HI, Safwat G, Gamal M, Megahed BMH (2019c) Chemical composition and biological activity of *Physalis peruviana* L. Gesunde Pflanzen 71:113–122

El-Shanshoury AR (1991) Biosynthesis of indole-3-acetic acid in Streptomyces atroolivaceus and its changes during spore germination and mycelial growth. Microbiol Lett 67:159–164

El-Tarabily KA, Sivasithamparam K (2006) Non streptomycete actinomycetes as biocontrol agents of soil borne fungal plant pathogens and as plant growth promoters. Soil Biol Biochem 38:1505–1520

Fedrizzi SMG (2006) Produção de metabolitos antimicrobianos e sideróforos de isolados provenientes de Terra Preta Antropogênica da Amazônia Ocidental. Dissertation, Universidade de São Paulo

Fernando WGD, Nakkeeran S, Zhang Y (2005) Biosynthesis of antibiotics by PGPR and its relation in biocontrol of plant diseases. In: Siddiqui ZA (ed) PGPR: biocontrol and biofertilization. Springer, Dordrecht, pp 67–109

Forchetti G, Masciarelli O, Izaguirre MJ, Alemano S, Alvarez D, Abdala G (2010) Endophytic bacteria improve seedling growth of sunflower under water stress, produce salicylic acid, and inhibit growth of pathogenic fungi. Curr Microbiol 61:485–493

Fouda A, Hassan SED, Eid AM, Ewais EE (2019) The interaction between plants and bacterial endophytes under salinity stress. In: Jha S (ed) Endophytes and secondary metabolites, reference series in phytochemistry. Springer, Switzerland. https://doi.org/10.1007/978-3-319-90484-915

Gaiero JR, McCall CA, Thompson KA, Day NJ, Best AS, Dunfiel KE (2013) Inside the root microbiome: bacterial root endophytes and plant growth promotion. Am J Bot 100:1738–1750

Garbeva P, Hol WHG, Termorshuizen AJ, Kowalchuk GA, de Boer W (2011) Fungistasis and general soil biostasis - a new synthesis. Soil Biol Biochem 43:469–477

Gomes F (2012) A síntese de coenzima Q e a estabilidade de DNA mitocondrial em Saccharomyces cerevisiae. Dissertation, Universidade de São Paulo

Gong A-D, Li H-P, Yuan Q-S et al (2015) Antagonistic mechanism of iturin A and plipastatin A from Bacillus amyloliquefaciens S76-3 from wheat spikes against Fusarium graminearum. PLoS One 10:e0116871. https://doi.org/10.1371/journal.pone.0116871

Grimme E, Zidack N, Sikora R, Strobel G, Jacobsen B (2007) Comparison of Muscodor albus volatiles with a biorational mixture for control of seedling diseases of sugar beet and root-knot nematode on tomato. Plant Dis 91(2):220–225

Groussin M, Mazel F, Sanders JG, Smillie CS, Lavergne S, Thuiller W et al (2017) Unraveling the processes shaping mammalian gut microbiomes over evolutionary time. Nat Commun 8:14319

Gu Q, Yang Y, Yuan Q, Shi G, Wu L, Lou Z, Huo R, Wu H, Borriss R, Gao X (2017) Bacillomycin D produced by Bacillus amyloliquefaciens is involved in the antagonistic interaction with the plant pathogenic fungus Fusarium graminearum. Appl Environ Microbiol. https://doi.org/10.1128/AEM.01075-17

Gutiérrez-Luna F, López-Bucio J, Altamirano-Hernandez J, Valencia-Cantero E, Reyez H, Macías-Rodríguez L (2010) Plant growth- promoting rhizobacteria modulate root- system architecture in Arabidopsis thaliana through volatile organic compound emission. Symbiosis 51:75–83

Habre R, Coull B, Moshier E, Godbold J, Grunin A, Nath A et al (2014) Sources of indoor air pollution in New York city residences of asthmatic children. J Expo Sci Environ Epidemiol 24:269–278

Hamana K, Kishimoto N (1996) Polyamine distribution patterns in C, compound-utilizing eubacteria and acidophilic eubacteria. J Gen Appl Microbiol 42:431–437

Hamayun M, Khan SA, Khan AL (2010) Gibberellin production and plant growth promotion from pure cultures of Cladosporium sp. MH-6 isolated from cucumber (*Cucumis sativus* L.). Mycologia 102:989–995

Hamed MM, Abd El-Mobdy MA, Kamel MT, Mohamed HI, Bayoumi AE (2019) Phytochemical and biological activities of two asteraceae plants *Senecio vulgaris* and *Pluchea dioscoridis* L. Pharmacology 2:101–121

Hardoim PR, van Overbeek LS, Berg G et al (2015) The hidden world within plants: ecological and evolutionary considerations for defining functioning of microbial endophytes. Microbiol Mol Biol Rev 79(3):293–320. https://doi.org/10.1128/MMBR.0005014

Hastuti RD, Yulin L, Antonius S, Rasti S (2012) Endophytic *Streptomyces* sp. as biocontrol agents of Rice bacterial leaf blight pathogen (*Xanthomonas oryzae* pv. *oryzae*) HAYATI. J Biosci 19:155–162. https://doi.org/10.4308/hjb.19.4.155

Hoyles L, Honda H, Logan NA, Halket G, La Ragione RM, McCartney AL (2012) Recognition of greater diversity of Bacillus species and related bacteria in human faeces. Res Microbiol 163:3–13

Hung R, Lee S, Bennett JW (2015) Fungal volatile organic compounds and their role in ecosystems. Appl Microbiol Biotechnol 99(8):3395–3405

Insam H, Seewald MS (2010) Volatile organic compounds (VOCs) in soils. Biol Fertil Soils 46(3):199–213

Ionczewski JL, Foster JW (2014) Microbiology: an evolving science, 3rd edn. W. W. Norton, New York, pp 742–743. ISBN 9780393123678

Jalgaonwala RE, Mahajan RT (2014) Production of anticancer enzyme asparaginase from endophytic *Eurotium* sp. isolated from rhizomes of *Curcuma longa*. Eur J Exp Biol 4:36–43

Jamiołkowska A, Księżniak A, Hetman B, Kopacki M, Skwaryło-Bednarz B, Gałązka A, Thanoon AH (2017) Interactions of arbuscular mycorrhizal fungi with plants and soil microflora. Acta Sci Pol Hortorum Cultus 16:89–95

Jelen H, Błaszczyk L, Chełkowski J, Rogowicz K, Strakowska J (2014) Formation of 6-npentyl 2H-pyran-2-one (6-PAP) and other volatiles by different Trichoderma species. Mycol Prog 13(3):589–600

Jerry B (1994) A role of endophytic fungi in regulating nutrients and energy in plants within a desert ecosystem. In: International symposium and workshop on desertification in developed countries

Johnson LJ, Koulman A, Christensen M, Lane GA, Fraser K, Forester N, Johnson RD, Bryan GT, Rasmussen S (2013) An extracellular siderophore is required to maintain the mutualistic interaction of Epichloë festucae with Lolium perenne. PLoS Pathog 9:1003332

Kang BR, Anderson AJ, Kim YC (2018) Hydrogen cyanide produced by Pseudomonas chlororaphis O6 exhibits nematicidal activity against Meloidogyne hapla. Plant Pathol J 34:35–43

Karkman A, Lehtimaki J, Ruokolainen L (2017) The ecology of human microbiota: dynamics and diversity in health and disease. Ann N Y Acad Sci 1399:78–92

Kharwar RN, Verma VC, Kumar A, Gond SK, Harper JK, Hess WM, Lobkovosky E, Ma C, Ren Y, Strobel GA (2011) Javanicin, an antibacterial naphthaquinone from an endophytic fungus of neem, Chloridium sp. Curr Microbiol 58:233–238

Kilani J, Fillinger S (2014) Phenylpyrroles: 30 years, two molecules and (nearly) no resistance. Front Microbiol 7:1–10

Kottb M, Gigolashvili T, Grosskinsky DK, Piechulla B (2015) Trichoderma volatiles effecting Arabidopsis: from inhibition to protection against phytopathogenic fungi. Front Microbiol 6:995

Lasudee K, Tokuyama S, Lumyong S, Pathomaree W (2018) Actinobacteria associated with arbuscular mycorrhizal Funneliformis mosseae spores, taxonomic characterization and their beneficial traits to plants: evidence obtained from mung bean (*Vigna radiata*) and Thai jasmine rice (*Oryza sativa*). Front Microbiol 9:12–47

Lata R, Chowdhury S, Gond SK, White JF Jr (2018) Induction of abiotic stress tolerance in plants by endophytic microbes. Lett Appl Microbiol 66:268–276. https://doi.org/10.1111/lam.12855

Laursen JB, Nielsen J (2004) Phenazine natural products: biosynthesis, synthetic analogues, and biological activity. Chem Rev 104:1663–1685

Leclère V, Béchet M, Adam A, Guez JS, Wathelet B, Ongena M, Thonart P, Gancel F, CholletImbert M, Jacques P (2005) Mycosubtilin overproduction by Bacillus subtilis BBG 100 enhances the organism's antagonistic and biocontrol activities. Appl Environ Microbiol 8:4577–4584

Lee S, Hung R, Yap M, Bennett JW (2015) Age matters: the effects of volatile organic compounds emitted by Trichoderma atroviride on plant growth. Arch Microbiol 197(5):723–727

Lee S, Yap M, Behringer G, Hung R, Bennett JW (2016) Volatile organic compounds emitted by Trichoderma species mediate plant growth. Fungal Biol Biotechnol 3(1):7

Leibold MA, Holyoak M, Mouquet N, Amarasekare P, Chase JM, Hoopes MF et al (2004) The metacommunity concept: a framework for multiscale community ecology. Ecol Lett 7:601–613

Li AR, Guan KY (2007) Mycorrhizal and dark septate endophytic fungi of Pedicularis species from northwest of Yunnan Province, China. Mycorrhiza 17:103–109

Li WK (2005) Endophytes and natural medicines. Chin J Nat Med (In Chinese) 3:193–199

Li J, Convertino M (2019) Optimal microbiome networks: macroecology and criticality. Entropy 21:506. https://doi.org/10.3390/e21050506

Liarzi O, Bar E, Lewinsohn E, Ezra D (2016) Use of the endophytic fungus Daldinia cf. concentrica and its volatiles as bio-control agents. PLoS One 11(12):e0168242

Lloyd-Price J, Abu-Ali G, Huttenhower C (2016) The healthy human microbiome. Genome Med 8:51

Long HH, Schmidt DD, Baldwin IT (2008) Native bacterial endophytes promote host growth in a species-specific manner; phytohormone manipulations do not result in common growth responses. PLoS One 3:2702

Macias-Rubalcava ML, Hernández-Bautista BE, Oropeza F, Duarte G, González MC, Glenn AE, Hanlin RT, Anaya AL (2010) Allelochemical effects of volatile compounds and organic extracts from Muscodor yucatanensis, a tropical endophytic fungus from Bursera simaruba. J Chem Ecol 36(10):1122–1131

Madigan MT, Martinko JM, Dunlap PV, Clark DP (2010) Microbiologia industrial. In: Madigan MT, Martinko JM, Dunlap PV, Clark DP (eds) Microbiologia de Brock. Porto Alegre, Artmed, pp 734–760

Martí JM, Martínez-Martínez D, Rubio T, Gracia C, Peña M, Latorre A et al (2017) Health and disease imprinted in the time variability of the human microbiome. mSystems 2:e00144–e00116

McInroy JA, Kloepper JW (1996) Survey of indigenous bacterial endophytes from cotton and sweet corn. Plant Soil 173:337–342

Mitmesser S, Combs M (2017) Chapter 23 - Prebiotics: inulin and other oligosaccharides A2. In: Floch MH (ed) The microbiota in gastrointestinal pathophysiology. Dimensions, London

Molina G, Pimentel MR, Bertucci TC, Pastore GM (2012) Application of fungal endophytes in biotechnological processes. Chem Eng Trans 27(6):289–294

Nandi M, Selin C, Brassinga AKC, Belmonte MF, Fernando WD, Loewen PC, De Kievit TR (2015) Pyrrolnitrin and hydrogen cyanide production by Pseudomonas chlororaphis strain PA23 exhibits nematicidal and repellent activity against Caenorhabditis elegans. PLoS One 10:e0123184

National Institutes of Health (2001) NIAID global health research plan for HIV/AIDS, malaria and tuberculosis. U.S. Department of Health and Human Services, Bethesda
Naznin HA, Kimura M, Miyazawa M, Hyakumachi M (2013) Analysis of volatile organic compounds emitted by plant growth promoting fungus Phoma sp. GS8-3 for growth promotion effects on tobacco. Microbes Environ 28:42–49
Noctor G, Reichheld JP, Foyer CH (2017) ROS-related redox regulation and signaling in plants. Semin Cell Dev Biol 80:3–12
Nongkhlaw FMW, Joshi SR (2015) Investigation on the bioactivity of culturable endophytic and epiphytic bacteria associated with ethnomedicinal plants. J Infect Dev Ctries 9:954–961. https://doi.org/10.3855/jidc.4967
Ochman H (2016) Realizing microbial evolution. Cold Spring Harb Perspect Biol 8:a018101
Ohgushi T, Craig TP, Price PW (2007) Indirect interaction webs: an introduction. In: Ohgushi T, Craig TP, Price PC (eds) Ecological communities: plant mediation in indirect interaction webs. Cambridge University Press, Cambridge, pp 3–15
Parfrey LW, Knight (2012) Spatial and temporal variability of the human microbiota. Clin Microbiol Infect 18(suppl 4):8–11
Pierson LS III, Pierson EA (2010) Metabolism and function of phenazines in bacteria: impacts on the behavior of bacteria in the environment and biotechnological processes. Appl Microbiol Biotechnol 86:1659–1670
Rapparini F, Penuelas J (2014) Mycorrhizal fungi to alleviate drought stress on plant growth. In: Miransari M (ed) Use of microbes for the alleviation of soil stresses. Springer, NewYork, pp 21–42
Redman RS, Sheehan KB, Stout RG, Rodrigues RJ, Henson JM (2002) Thermotolerance conferred to plant host and fungal endophyte during mutualistic symbiosis. Science 298:158
Rees CM, Hall NJ, Fleming P, Eaton S (2017) Probiotics for the prevention of surgical necrotising enterocolitis: systematic review and metaanalysis. BMJ Paediatr Open 1:2066
Ren C-G, Dai C-C (2012) Jasmonic acid is involved in the signaling pathway for fungal endophyte-induced volatile oil accumulation of Atractylodes lancea plantlets. BMC Plant Biol 12:128
Robert-Seilaniantz A, Grant M, Jones JDG (2011) Hormone crosstalk in plant disease and defense: more than just jasmonate-salicylate antagonism. Annu Rev Phytopathol 49:317–343
Robinson DL (2005) Air pollution in Australia: review of costs, sources and potential solutions. Health Promot J Austr 16:213–220
Rosenblueth M, Martinez Romero E (2004) Rhizobium etli maize populations and their competitiveness for root colonization. Arch Microbiol 181:337–344
Rosenblueth M, Martínez-Romero E (2006) Bacterial endophytes and their interactions with hosts. Mol Plant-Microbe Interact 19:827–837
Rouhier N, San Koh C, Gelhaye E, Corbier C, Favier F, Didierjean C, Jacquot JP (2008) Redox based anti-oxidant systems in plants: biochemical and structural analyses. Biochim Biophys Acta 1780:1249–1260
Rumana HS, Sharma RC, Beniwal V, Sharma AK (2014) A retrospective approach to assess human health risks associated with growing air pollution in urbanized area of Thar Desert, Western Rajasthan, India. J Environ Health Sci Eng 12:23
Ryu C-M, Farag MA, Hu C-H, Reddy MS, Wei H-X, Paré PW, Kloepper JW (2003) Bacterial volatiles promote growth in Arabidopsis. Proc Natl Acad Sci 100(8):4927–4932
Ryu C-M, Farag MA, Hu C-H, Reddy MS, Kloepper JW, Paré PW (2004) Bacterial volatiles induce systemic resistance in Arabidopsis. Plant Physiol 134(3):1017–1026
Scavino AF, Pedraza RO (2013) The role of siderophores in plant growth promoting bacteria. In: Maheshwari DK, Saraf M, Aeron A (eds) Bacteria in agrobiology: crop productivity. Springer, Berlin, pp 265–286
Schalchli H, Tortella G, Rubilar O, Parra L, Hormazabal E, Quiroz A (2016) Fungal volatiles: an environmentally friendly tool to control pathogenic microorganisms in plants. Crit Rev Biotechnol 36(1):144–152
Schardl CL, Young CA, Hesse U, Amyotte SG, Andreeva K, Calie PJ, Fleetwood DJ, Haws DC, Moore N, Oeser B et al (2013) Plant–symbiotic fungi as chemical engineers: multi genome analysis of the Clavicipitaceae reveals dynamics of alkaloid loci. PLoS Genet 9:e1003323

Schiestl FP (2015) Ecology and evolution of floral volatile-mediated information transfer in plants. New Phytol 206:571–577

Schmidt R, Jager V, Zuhlke D, Wolff C, Bernhardt J, Cankar K et al (2017) Fungal volatile compounds induce production of the secondary metabolite Sodorifen in Serratia plymuthica PRI-2C. Sci Rep 7:862

Schulz-Bohm K, Zweers H, de Boer W, Garbeva P (2015) A fragrant neighborhood: volatile mediated bacterial interactions in soil. Front Microbiol 6:1212

Sekmen AH, Turkan I, Takio S (2007) Differential responses of antioxidative enzymes and lipid peroxidation to salt stress in salt-tolerant Plantagomaritima and salt-sensitive Plantago media. Physiol Plant 131:399–411

Shafi J, Tian H, Ji M (2017) Bacillus species as versatile weapons for plant pathogens: a review. Biotechnol Biotechnol Equip 31:446–459

Shahzad R, Waqas M, Khan AL, Al-Hosni K, Kang SM, Seo CW, Lee IJ (2018) Indole acetic acid production and plant growth promoting potential of bacterial endophytes isolated from rice (*Oryza sativa* L.) seeds. Acta Biol Hung 68:175–186

Sharma V, Salwan R (2018) Biocontrol potential and applications of Actinobacteria in agriculture. In: Singh B, Gupta V, Passari A (eds) Actinobacteria: diversity and biotechnological applications, new and future developments in microbial biotechnology and bioengineering, I edn. Elsevier, San Diego, pp 93–108

Shi Y, Yang H, Zhang T, Sun J, Lou K (2014) Illumina-based analysis of endophytic bacterial diversity and space-time dynamics in sugar beet on the north slope of Tianshan mountain. Appl Microbiol Biotechnol 98:6375–6385

Sikora RA, Schäfer K, Dababat AA (2007) Modes of action associated with microbially induced in planta suppression of plant-parasitic nematodes. Australas Plant Pathol 36:124–134

Singh M, Kumar A, Singh R, Pandey KD (2017) Endophytic bacteria: a new source of bioactive compounds. Biotech 7:315. https://doi.org/10.1007/s13205-017-0942-z

Song GC, Ryu C-M (2013) Two volatile organic compounds trigger plant self-defense against a bacterial pathogen and a sucking insect in cucumber under open field conditions. Int J Mol Sci 14(5):9803–9819

Souza CS, Menezes RSC, Sampaio EVSB, Lima FS (2012) Glomalin: characteristics, production, limitations and contribution to soils. Semina: Ciências Agrárias 33(1):3033–3044

Spence CA, Raman V, Donofrio NM, Bais HP (2014) Global gene expression in rice blast pathogen Magnaporthe oryzae treated with a natural rice soil isolate. Planta 239:171–185

Stinson A, Zidack N, Strobel G, Jacobsen B (2003) Mycofumigation with Muscodor albus and Muscodor roseus for control of seedling diseases of sugar beet and Verticillium wilt of eggplant. Plant Dis 87(11):1349–1354

Strobel GA, Miller RV, Martinez Miller C, Condron MM, Teplow DB, Hess WM (1999) Cryptocandin a potent and antimycotic from the endophytic fungus Cryptosporiopsis quercina. Microbiology 145:1919–1926

Strobel G, Daisy B, Castillo U, Harper J (2004) Natural products from endophytic, microorganisms. J Nat Prod 67:257–268

Strobel G, Singh SK, Riyaz-Ul-Hassan S, Mitchell AM, Geary B, Sears J (2011) An endophytic/pathogenic Phoma sp. from creosote bush producing biologically active volatile compounds having fuel potential. FEMS Microbiol Lett 320(2):87–94

Stierle AA, Stierle D (2015) Bioactive secondary metabolites produced by the fungal endophytes of conifers. Nat Prod Commun 10:1671–1682

Sturz A, Kimpinski J (2004) Endoroot bacteria derived from marigolds (Tagetes spp.) can decrease soil population densities of root lesion nematodes in the potato root zone. Plant Soil 262:241–249

Tan RX, Zou WX (2001) Endophytes: a rich source of functional metabolites. Nat Prod Rep 18:448–459. https://doi.org/10.1039/b100918o

Tang MJ, Meng ZX, Guo SX, Chen XM, Xiao PG (2008) Effects of endophytic fungi on the culture and four enzyme activities of Anoectochilus roxburghii. J Chin Pharm 43:890–893

Utsumi S, Ando Y, Miki T (2010) Linkages among trait-mediated indirect effects: a new framework for the indirect interaction web. Popul Ecol 52:485–497

Vadassery J, Ritter C, Venus Y, Camehl I, Varma A, Shahollari B, Novák O, Strnad M, Ludwig Müller J, Oelmüller R (2008) The role of auxins and cytokinins in the mutualistic interaction between Arabidopsis and Piriformospora indica. Mol Plant-Microbe Interact 21:1371–1383

Van Dam NM, Weinhold A, Garbeva P (2016) Calling in the dark: the role of volatiles for communication in the rhizosphere. In: Blande JD, Glinwood R (eds) Deciphering chemical language of plant communication. Springer, Cham, pp 175–210

Venieraki A, Dimou M, Katinakis P (2017) Endophytic fungi residing in medicinal plants have the ability to produce the same or similar pharmacologically active secondary metabolites as their hosts. Hellenic Plant Prot J 10:51–66

Villaescusa BP, Rangel-Huerta OD, Aguilera CM, Gil A (2015) A systematic review of the efficacy of bioactive compounds in cardiovascular disease: carbohydrates, active lipids and nitrogen compounds. Ann Nutr Metab 66:168–181. https://doi.org/10.1159/000430960

Vurukonda SSKP, Giovanardi D, Stefani E (2018) Plant growth promoting and biocontrol activity of Streptomyces spp. as endophytes. Int J Mol Sci 19:1–2

Waller F, Achatz B, Baltruschat H, Fodor J, Becker K, Fischer M, Heier T, Hückelhoven R, Neumann C, von Wettstein D, Franken P, Kogel KH (2005) The endophytic fungus Piriformospora indica reprograms barley to salt stress tolerance, disease resistance, and higher yield. Proc Natl Acad Sci 102:13386–13391

Wani MA, Sanjana K, Kumar DM, Lal DK (2010) GC–MS analysis reveals production of 2 phenylethanol from Aspergillus niger endophytic in rose. J Basic Microbiol 50(1):110–114

Waqas M, Khan AL, Kamran M, Hamayun M, Kang SM, Kim YH et al (2012) Endophytic fungi produce gibberellins and indoleacetic acid and promotes host-plant growth during stress. Molecules 17:10754–10773

Waqas M, Khan AL, Shahzad R, Ullah I, Khan AR, Lee J (2015) Mutualistic fungal endophytes produce phytohormones and organic acids that promote japonica rice plant growth under prolonged heat stress. J Zhejiang Univ-Sci B (Biomed & Biotechnol) 16:1011–1018

Werner S, Polle A, Brinkmann N (2016) Belowground communication: impacts of volatile organic compounds (VOCs) from soil fungi on other soil-inhabiting organisms. Appl Microbiol Biotechnol 100:8651–8665

Wintermans PCA, Bakker PAHM, Pieterse CMJ (2016) Natural genetic variation in Arabidopsis for responsiveness to plant growth-promoting rhizobacteria. Plant Mol Biol 90:623–634

Worapong J, Strobel GA, Ford EJ, Li JY, Baird G, Hess WM (2001) *Muscodor albus* gene sp. nov., an endophyte from *Cinnamomum zeylanicum*. Mycotaxon 79:67–79

Xu Z, Shao J, Li B, Shen Q, Zhang R (2013) Bacillomycin D in Bacillus amyloliquefaciens SQR9 contributes to antifungal activity and biofilm formation. Appl Environ Microbiol 79:755–781

Ye Y-F, Li Q-Q, Fu G, Yuan G-Q, Miao J-H, Lin W (2012) Identification of antifungal substance (Iturin A2) produced by Bacillus subtilis B47 and its effect on southern corn leaf blight. J Integr Agric 11(1):90–99

Yu JM, Wang D, Pierson LSIII, Pierson EA (2018) Effect of producing different phenazines on bacterial fitness and biological control in Pseudomonas chlororaphis 30-84. Plant Pathol J 34:44–58

Yuan J, Sun K, Deng-Wang M-Y, Dai C-C (2016) The mechanism of ethylene signaling induced by endophytic fungus Gilmaniella sp. AL12 mediating sesquiterpenoid biosynthesis in Atractylodes lancea, froniters. Plant Sci 7:361

Zamioudis C, Pieterse CMJ (2012) Modulation of host immunity by beneficial microbes. Mol Plant-Microbe Interact 25:139–150

Zeilinger S, Gupta VK, Dahms TES, Silva RN, Singh HB, Upadhyay RS, Gomes EV, Tsui CM, Chandra Nayak S, van der Meer JR (2016) Friends or foes? Emerging insights from fungal interactions with plants. FEMS Microbiol Rev 40(2):182–207

Chapter 12
Biosynthesis of Nanoparticles by Microorganisms and Applications in Plant Stress Control

Khaled M. A. Ramadan and Hossam S. El-Beltagi

Contents

K. M. A. Ramadan (✉)
King Faisal University, Central Laboratories, Riyadh, Saudi Arabia

Agricultural Biochemistry Department, Faculty of Agriculture, Ain-Shams University, Cairo, Egypt
e-mail: kramadan@kfu.edu.sa

H. S. El-Beltagi
Agricultural Biotechnology Department, College of Agriculture and Food Sciences, King Faisal University, Al-Ahsa, Saudi Arabia

Biochemistry Department, Faculty of Agriculture, Cairo University, Giza, Egypt

© The Author(s), under exclusive license to Springer Nature Switzerland AG 2021
H. I. Mohamed et al. (eds.), *Plant Growth-Promoting Microbes for Sustainable Biotic and Abiotic Stress Management*,
https://doi.org/10.1007/978-3-030-66587-6_12

Abbreviations

AgNPs	Silver nanoparticles
AuNPs	Gold nanoparticles
BacMPs	Bacterial magnetic particles
BMs	Bacterial magnetosomes
BRECs	Bovine retinal endothelial cells
CdS NPs	CdS nanoparticles
CSE	Cell-soluble extract
GTPase	Guanosine triphosphatase
HRP	Horseradish peroxidase
mAbs	Monoclonal antibodies
MRI	Magnetic resonance imaging
MTB	Magnetotactic bacteria
PHB	Polyhydroxybutyrate
TEM	Transmission electron microscope

1 Introduction

Nanotechnology's future applications and advantages in agriculture are immense. This involves the treatment of insect pests by new nanomaterial insecticide formulations (Ragaei and Sabry 2014). One nanometer is understood to be a milliard of a micrometer or a million of a micron. That is around 1/80,000 of human hair diameter or ten times hydrogen atom diameter. American scientists assert that "There is plenty of space at the bottom," which was also held as a way of paying attention to the nanotechnological field. Feynman (1960) discovered technique through which it is possible to manipulate single atoms and molecules, utilizing series with specialized instruments to construct and manage a limited range of necessary scales, etc. In this context, Feynman suggested that the shift in magnitude would lead to scaling problems in various physical phenomena: gravity became less relevant, and surface tension and the attraction of van der Waals might be more relevant. Many experiments on nanoparticles have shown their efficacy toward plant diseases, insects, or other threats. Therefore, such nanoparticles were still only used to repel insects, but also to prepare new products, such as pesticides and insecticides (Prasad et al. 2017a, b). But safety for plants to plants for metal-based nanostructures with far larger volume-to-volume particle size and with specific antimicrobials compared with their bulk materials is one of the latest with the rapid advancement of nanotechnology, and their special properties expand the use of a range of carbon nanomaterials (CNMs). The use of a buckyball molecule fullerene (C60) is, for example, commonly available in computers and aircraft airframes and as drug delivery carriers in the form of biomedicine and carbon nanotubes (CNTs) (Ngan et al. 2015; Liu et al. 2015). These have thoroughly studied interactions between CNMs and plants. In 30-day experiments with hydroponic tension, for instance, graphene concentrations ranging from 250 to 1500 mg/L inhibited wheat growth (Zhang et al. 2016). A great number of physical, electronic, biological, or hybrid methods depend on the fabrication of various classes of nanoparticles. Although organic compounds are most common throughout the production of nanoparticles, the use of dangerous substances severely restricts their medicinal use, especially in medical practice (Liu et al. 2011). Hence, it is of utmost importance that to extend their biomedical applications, healthy, nontoxic, and environmentally friendly approaches are developed for the production of nanomaterials. Synthesizing microorganisms with nanoparticles is one of the choices. The nanoparticles generated by biogenic enzyme process greatly outweigh those generated by chemical processes in many respects. Although the latter is capable of producing large amounts of nanoparticles of given size and shape in a reasonably short period, they become complex, obsolete, expensive, and ineffective and produce dangerous radioactive waste that is dangerous not only to the environment but also to public health. Usage of costly chemicals is avoided via an enzyme solution, and most suitable "green" pathway wasn't as energy-intensive and environmentally friendly as chemical route. A biogenic method is again confirmed by the fact that in varying temperature, pH, and pressure conditions, most bacteria exist. These procedures provide greater catalytic reaction, increased surface area, and enhanced interaction among enzyme and metal ion as a result of the bacterial cell membrane (Bhattacharya and Mukherjee 2008). Nanoparticles are biosynthesized as microorganisms take target ions out of the atmosphere and then transform metal into elemental metal by enzymes formed by cell activity. Depending on where nanoparticles are made, intracellular and extracellular synthesis can be categorized. Throughout the existence of enzymes, the intracellular process is the transport for ions to produce nanoparticles by bacterial cell. Extracellular nanoparticle synthesis includes capturing metal ions on the cell surface and decreasing the amount of ions when enzymes are present (Zhang et al. 2011). To biosynthesize nanoparticles, a number of applications have been used, like selective drug carriers, cancer treatment, gene therapy and DNA sequencing, antiviral activities, biosensors, reaction-enhancing rates, and isolation monitoring.

The objectives of this chapter highlight the extensive properties of inorganic nanoparticles and the synthesis of metal, oxide, sulfide, and other conventional nanoparticles among different species of microorganisms. It will also discuss the proposed pathways for the biosynthesis of inorganic nanoparticles. Size/shape and stabilization of synthesized nanoparticles were affected. Pharmaceutical formulations include such nanoparticles, crop protection, and antibacterial agents. Synthesized biometallic nanoparticles are also investigated by manipulating *Penicillium* species and their uses in pharmaceutical applications (Fig. 12.1).

2 Metallic Nanoparticles

Table 12.1 summarizes several standard nanoparticles made through microorganisms.

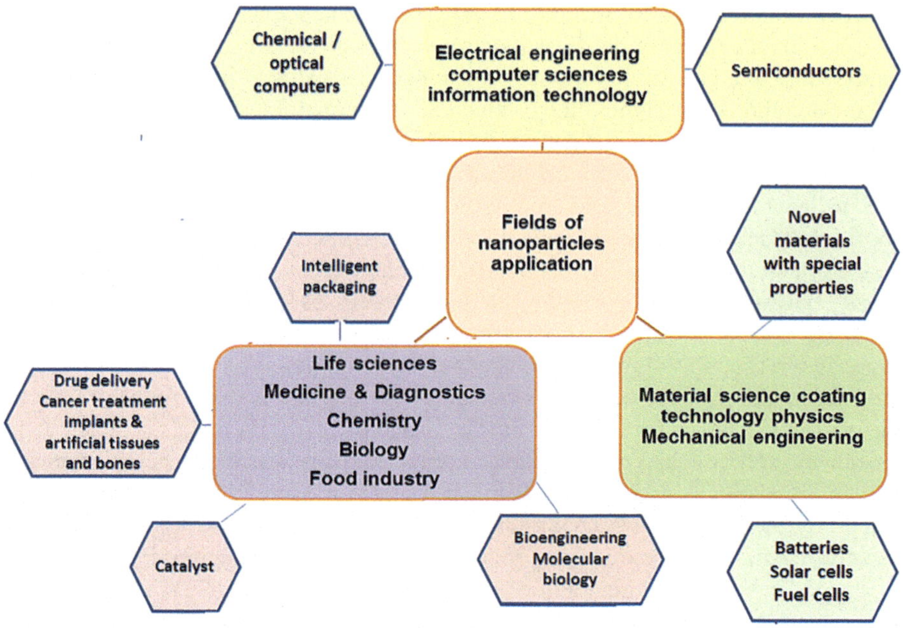

Fig. 12.1 Fields of application of nanoparticles

2.1 Gold Nanoparticles

In chemistry, Au nanoparticles get a long and glorious background to Roman times, wherein they were being used for aesthetic reasons to dye glasses. AuNPs were already used centuries earlier for the treatment of different diseases. Previous study recorded that colloidal gold substances had distinct characteristics than mass gold, which launched the modern era of AuNP synthesis (Hayat 1989). Because of the increasing need to improve environmentally sustainable material synthesis technologies, nanoparticles have received considerable attention as evolving bionanotechnology (overpass of nanotechnology and biotechnology). Extracellular production by *Fusarium oxysporum* fungus and actinomycete sp. with gold nanoparticles has been documented in previous research. Intracellular synthesis of *Verticillium* sp. fungal gold nanoparticles has been reported (Ahmad et al. 2003a). Southam and Beveridge (1996) showed nanoscale gold particles could be readily caused inside microbes by cells with Au^{3+} ions. The gold monodisperse nanoparticles were synthesized with *Rhodococcus* sp. alkalotolerant within extreme biological regulation, like alkaline conditions and environments with marginally greater temperatures (Ahmad et al. 2003b). Lengke et al. (2006a, b) have submitted Au complexes to synthesize filamentous cyanobacteria in various shapes, including spherical, cubic, and octahedral, and to research the mechanisms of nanostructure formation. There have been studies of the development of nanocrystals and nanoalloys using *Lactobacillus* (Nair and Pradeep 2002). Table 12.1 summarizes some other typical microorganism-formed gold nanoparticles (Konishi et al. 2007a; Singaravelu et al. 2007).

Table 12.1 Metal nanoparticles synthesized by microorganisms

Microorganisms	Products	Culturing temperature (°C)	Size (nm)	Shape	Location	References
Sargassum wightii	Au	Not available	8–12	Planar	Extracellular	Singaravelu et al. (2007)
Rhodococcus sp.	Au	37	5–15	Spherical	Intracellular	Ahmad et al. (2003a)
Shewanella oneidensis	Au	30	12 ± 5	Spherical	Extracellular	Suresh et al. (2011)
Plectonema boryanum	Au	25–100	<10–25	Cubic	Intracellular	Lengke et al. (2006a)
Plectonema boryanum UTEX 485	Au	25	10 nm–6 μm	Octahedral	Extracellular	Lengke et al. (2006b)
Escherichia coli	Au	37	20–30	Triangles, hexagons	Extracellular	Du et al. (2007)
Yarrowia lipolytica	Au	30	15	Triangles	Extracellular	Agnihotri et al. (2009)
Pseudomonas aeruginosa	Au	37	15–30	Not available	Extracellular	Husseiny et al. (2007)
Pseudomonas rhodesiae	Ag	37	20–100	Spherical	Extracellular	Hossain et al. (2019)
Pseudomonas sp. and *Achromobacter* sp.	Ag	37	20–50	Spherical		Kaur et al. (2018)
Rhodopseudomonas capsulate	Au	30	10–20	Spherical	Extracellular	He et al. (2007)
Shewanella algae	Au	25	10–20	Not available	Intracellular	Konishi et al. (2007a, b)
Brevibacterium casei	Au, Ag	37	10–50	Spherical	Intracellular	Kalishwaralal et al. (2010)
Trichoderma viride	Ag	27	5–40	Spherical	Extracellular	Fayaz et al. (2010)
Bacillus licheniformis	Ag	37	50	Not available	Extracellular	Kalimuthu et al. (2008)
Bacillus siamensis	Ag	37	25–50	Spherical	Extracellular	Ibrahim et al. (2019)
Escherichia coli	Ag	37	50	Not available	Extracellular	Gurunathan et al. (2009)
Shewanella loihica PV-4	Au	30	10–16	Spherical	Extracellular	Lv et al. (2018)
Corynebacterium glutamicum	Ag	30	5–50	Irregular	Extracellular	Sneha et al. (2010)
Trichoderma viride	Ag	10–40	2–4	Not available	Extracellular	Fayaz et al. (2009)
Ureibacillus thermosphaericus	Au	60–80	50–70	Not available	Extracellular	Juibari et al. (2011)

(continued)

Table 12.1 (continued)

Microorganisms	Products	Culturing temperature (°C)	Size (nm)	Shape	Location	References
Bacillus cereus	Ag	25	18–39 l	Spherical	Extracellular	Ahmed et al. (2020)
Aspergillus fumigatus	Ag	25	5–25	Spherical	Extracellular	Bhainsa et al. (2006)
Aspergillus niger	Ag	25	10–100	Spherical	Extracellular	Al-Zubaidi et al. (2019)
Verticillium sp.	Ag	25	25 ± 8	Spherical	Extracellular	Senapati et al. (2005)
Fusarium graminearum	Ag	25	20–45	Spherical	Extracellular	Ibrahim et al. (2020)
Fusarium oxysporum	Ag	25	5–50	Spherical	Extracellular	Senapati et al. (2005)
Trichoderma harzianum	Ag	25	11–13	Spherical	Extracellular	El-Moslamy et al. (2017)
Trichoderma hamatum	Au	25	5–30	Spherical, pentagonal, and Hexagonal	Extracellular	Abdel-Kareem and Zohri (2018)
Streptomyces griseus	Cu	25	5–50	Spherical	Extracellular	Ponmurugan et al. (2016)
Neurospora crassa	Au, Au/Ag	28	32, 20–50	Spherical	Intracellular Extracellular	Castro-Longoria et al. (2011)
Shewanella algae	Pt	25	5	Not available	Intracellular	Konishi et al. (2007a, b)
Enterobacter sp.	Hg	30	2–5	Spherical	Intracellular	Sinha and Khare (2011)
Shewanella sp.	Se	30	181 ± 40	Spherical	Extracellular	Lee et al. (2007)
Escherichia coli	CdTe	37	2.0–3.2	Spherical	Extracellular	Bao et al. (2010)
Yeast	Au/Ag	30	9–25	Irregular polygonal	Extracellular	Zheng et al. (2010)
Fusarium oxysporum	Au-Ag alloy	25	8–14	Spherical	Extracellular	Senapati et al. (2005)
Penicillium duclauxii	Ag	25	3–32	Spherical	Extracellular	Almaary et al. (2020)
Setosphaeria rostrata	Ag	25	2–50	Spherical	Extracellular	Akther and Hemalatha (2019)
Pyrobaculum islandicum	U(VI), Tc(VII), Cr(VI), Co(III), Mn(IV)	100	N/A	Spherical	Extracellular	Kashefi and Lovley (2000)
Desulfovibrio desulfuricans	Pd	25	50	Spherical	Extracellular	Lloyd et al. (1998)

2.2 Silver Nanoparticles

Ag nanoparticles exhibit Gram-positive bacteria with effective antimicrobial activity, particularly multiresistant strains such as *Staphylococcus aureus* which is resistant to methicillin, as its bulk counterpart (Panacek et al. 2006). The secrets of nature have contributed to the production of advanced nanoparticles through biomimetic approaches. Researchers have long made efforts to use microorganisms to manufacture as many silver nanoparticles as possible to create eco-friendly nanofactories. Various microbes are recognized as reducing Ag^{+} ions in silver nanoparticles, and most are spherical particles (Fayaz et al. 2010). Klaus et al. (1999) showed that when *Pseudomonas* bacterium is extracted from silver mine, while put within a solution containing aqueous silver nitrate, *stutzeri* AG259 played a significant function throughout the decrease of Ag^{+} ions as well as in production with well-defined silver nanoparticles and separate topography of bacteria within periplasmic space. AgNPs were produced as a film or formed in liquid or collected onto their cell surface when fungi *Verticillium* or *Fusarium oxysporum* were used (Jain et al. 2011). Table 12.1 lists some other microorganism-developed silver nanoparticles (Kalimuthu et al. 2008; Gurunathan et al. 2009; Sneha et al. 2010; Fayaz et al. 2009; Kalishwaralal et al. 2010; Castro-Longoria et al. 2011; Juibari et al. 2011). Synthesized AgNPs by Hamouda et al. (2019) demonstrated good antibacterial activity toward multidrug-resistant bacteria (*Bacillus cereus*, *Escherichia coli*) and anticancer activity toward cell lines of human (breast, colon, liver). Low concentrations of hemolytic activity of AgNPs have been studied and reported as nontoxic to human RBCs. Furthermore, the dynamics of absorption and cytotoxicity of these AgNPs have been studied in the cell lines of breast cancer, enabling them to be shown to be good antibacterial agents, with further proof of the different behavior of AgNPs to cause toxicity in cells and bacteria when collected at pH 7 or 8. Moreover, the theoretically unlimited source of the reducing agent (i.e., leaf extract obtained from agricultural processing waste) and its negligible environmental impact constitute another strength of this method (De Matteis et al. 2019; Tanase et al. 2019). It has been shown that the combination of $AgNP_{bio}$ and simvastatin may be a great future option for bacterial infection control, where lower doses of $AgNP_{bio}$ with the same antibacterial activity are needed when combined with simvastatin (Figueiredo et al. 2019). Also, the synthesized silver nanoparticles had a strong antibiofilm property and were also found to be biocompatible with the red blood cell lysis assay and their association with peripheral mononuclear blood cells and 293 cells of the human embryonic kidney. *Mesoflavibacter zeaxanthinifaciens* is therefore found to be an excellent source of exopolysaccharide synthesis that assists in production of silver nanoparticles (Oves et al. 2019).

2.3 Alloy Nanoparticles

Using alloy nanoparticles in catalytic reactions, electronics, and optical substances and coatings is of great interest. *Fusarium oxysporum* production of bimetallic Au-Ag alloy and argued that secreted NADH cofactor is a significant determinant of the composition of Au-Ag nanoparticles (Senapati et al. 2005). Au-Ag metal nanoparticles, biosynthesized by yeast cells, have been studied (Zheng et al. 2010). Nanoparticles of the Au-Ag alloy were commonly produced by extracellular phase, microscopically characterized by fluorescence and electron microscopic transmission, or generally existed as irregular polygonal nanoparticles. Electrochemical research has shown vanillin sensors have been able to enhance electrochemical reaction of vanillin at least five times by changing glass carbon electrodes based on Au-Ag metal nanoparticles. Au-Ag alloy nanoparticles from fungal strains have been used in *Fusarium semitectum* core-shell synthesis of nanoparticles and been very stable for several weeks (Sawle et al. 2008).

2.4 Other Metallic Nanoparticles

It is understood that heavy metals are life-threatening to microorganisms. Microbial tolerance to many other toxic metals is in nature due to its chemical detoxification or even cell-dependent ion excretion by protein complexes acting as ATPase, chemical cations, or anti-transporter protons. Solubility changes play a crucial role as well in resistant bacteria. Konishi et al. (2007b) studied the use of *Shewanella* algae, a metal ion-reducing bacterium, to obtain platinum nanoparticles. In most cells of *Shewanella* by time lactate was delivered as an electron donor, aqueous $PtC_{16}b_2$ ions in elemental platinum were reduced to room temperature and neutral pH within 60 min. Platinum nanoparticles of about 5 nm were found in periplasm. Sinha and Khare have shown that *Enterobacter* sp. can synthesize mercury nanoparticles (Sinha and Khare 2011). Cultivation conditions (pH 8.0 and lower mercury concentrations) facilitate the synthesis of uniformly sized, spherical, and monodispersed 2–5 nm intracellular mercury nanoparticles. Many of heavy metals with hydrogen as an electron donor of the anaerobic hyperthermophilic microorganism *Pyrobaculum islandicum*, like U(VI), Tc(VII), Cr(VI), Co(III), and Mn(IV), have been reported to be reduced (Kashefi and Lovley 2000). In palladium nanoparticles, sulfate-reducing bacteria, *Desulfovibrio desulfuricans*, or metal ion-reducing bacteria *sulfur* can be synthesized. Table 12.1 also lists some other nanoparticles formed by microorganisms (DeWindt et al. 2005; Lee et al. 2007; Bao et al. 2010).

3 Oxide Nanoparticles

Oxide nanoparticles are an essential type of microbial compound nanoparticles. The biosynthesized oxide nanoparticles from both sides have been investigated in this section: magnetic oxide nanoparticles or nonmagnetic oxide nanoparticles. In Table 12.2, many examples of magnetotactic bacteria (MTB) shown in development of nanoparticles of magnetic oxide and biological systems for the production of nanoparticles of nonmagnetic oxide are summarized.

3.1 Magnetic Nanoparticles

Owing to its peculiar microstructure and properties, such as magnetic nanoparticles, strong forces, and its potential to widespread implementation in fields of biological isolation and biomedicine, superparamagnetic nanoparticles become new materials discovered. It is known that magnetic nanoparticles are Fe_3O_4 (magnetite) and Fe_2O_3 (maghemite). Targeted treatment of cancer (magnetic hyperthermia), stem cell filtering and manipulation, drug delivery guidance, gene therapy, DNA sequencing, and magnetic resonance imaging (MRI) have been actively investigated (Fan et al. 2009). Magnetotactic bacteria produce intracellular magnetic particles containing iron oxide, iron sulfides, or either. To differentiate between them and artificially synthesized magnetic particles (AMPs), these particles were pointed as bacterial magnetic particles (BacMPs) (Arakaki et al. 2008). Its associations with bacterial links are presumed to function like biological compass points that allow bacteria to move to oxygen gradients in aquatic environments under geomagnetic field of Earth (Blakemore 1975). BacMPs, as they can be surrounded through biological membranes composed primarily of lipids and proteins, could be quickly spread into aqueous media. In addition, individual BacMPs with better magnetic characteristics involve individual magnetic field or magnetite (Thornhill et al. 1995). Since the first magnetotactic bacteria study in 1975, numerous morphological forms have been described and observed in numerous aquatic environments, including cocci, spirals, vibrants, ovoid bacteria, and multicellular bacteria, with specific characteristics (Spring and Schleifer 1995). For example, magnetotactic cocci showed a high diversity and distribution and were often found on aquatic sediment surfaces. Identification of such type of bacteria shows that it is microaerophilic, including the coccus strain cultivated by magnetic MC-1. In the case of *Vibrio* bacteria, three optional anaerobic marine vibrating forms were extracted from freshwater salt marshes. As part of *Alphaproteobacteria*, these bacteria are known to belong to *Rhodospirillaceae* family, and truncated hexoctahedron-type BacMPs have been synthesized to evolve heterotrophically and organically with chemo. On the other side, parts of the *Magnetospirillaceae* family are present in sediments containing fresh water. In this family, significant amounts of previously isolated magnetotactic bacteria have been detected by utilizing culture medium and magnetic isolation methods. The first family member was isolated from strain MS-1 of *Magnetospirillum magnetotacticum*, while the physiological and genetic features of strain MSR-1 of *Magnetospirillum gryphiswaldense* were also well studied. AMB-1 was discretionary magnetotactic anaerobic spirilla, separated by Arakaki et al. (2008). After 2000, several new magnetotactic bacteria were discovered in different ecological settings. Several of freshly described magnetotactic bacteria were recorded in Table 12.2. Uncultured magnetotactic bacteria were found in distinct environments (Lefevre et al. 2010a). Mesophilic bacteria are the most common cultivated magnetotactic bacteria, which appear to grow less than 30 °C. The majority of uncultivated magnetotactic bacteria is 30 °C and below. Thermophilic magnetotactic bacteria are described in only few studies. Each of magnetotactic bacteria known as HSMV-1 is identified in samples

Table 12.2 Oxide nanoparticles synthesized by microorganisms

Microorganisms	Products	Culturing temperature (°C)	Size (nm)	Shape	Location	References
Shewanella oneidensis	Fe_3O_4	28	40–50	Rectangular, rhombic, hexagonal	Extracellular	Perez-Gonzalez et al. (2010)
QH-2	Fe_3O_4	22–26	81 ± 23 × 58 ± 20	Rectangular	Intracellular	Zhu et al. (2010)
Recombinant AMB-1	Fe_3O_4	28	20	Cuboctahedral	Intracellular	Amemiya et al. (2007)
Yeast cells	Fe_3O_4	36	Not available	Wormhole-like	Extracellular	Zhou et al. (2009a)
Yeast cells	$FePO_4$	36	Not available	Nanopowders	Extracellular	Zhou et al. (2009b)
WM-1	Fe_3O_4	28	54 ± 12.3 × 43 ± 10.9	Cuboidal	Intracellular	Li et al. (2007)
Shewanella oneidensis MR-1	Fe_2O_3	25	30–43	Pseudohexagonal/irregular or rhombohedral	Intracellular	Bose et al. (2009)
HSMV-1	Fe_3O_4	63	113 ± 34 × 40 ± 5	Bullet-shaped	Intracellular	Lefevre et al. (2010a)
Saccharomyces cerevisiae	Sb_2O_3	25–60	2–10	Spherical	Intracellular	Jha et al. (2009)
Lactobacillus sp.	$BaTiO_3$	25	20–80	Tetragonal	Extracellular	Jha et al. (2010a)
Lactobacillus sp.	TiO_2	25	8–35	Spherical	Extracellular	Jha et al. (2010b)
Fusarium oxysporum	TiO_2	300	6–13	Spherical	Extracellular	Bansal et al. (2005)
Fusarium oxysporum	$BaTiO_3$	25	4–5	Spherical	Extracellular	Bansal et al. (2006)
Fusarium oxysporum	ZrO_2	25	3–11	Spherical	Extracellular	Bansal et al. (2004)
Streptomyces spp.	CuO	25	78–80	Spherical	Extracellular	Hassan et al. (2019)

of springs in which temperatures varied between 32 and 63 °C (Lefevre et al. 2010b). TEM images of the untouched HSMV-1 cell discovered single polar flagellum and single bullet-shaped magnetosome string. The average number per cell of magnetosome crystals is 12 ± 6 and 113 ± 34 nm by 40 ± 5 nm. Report's findings indicate that certain magnetotactic bacteria may at least indicate mild thermophilicity. Under conditions where magnetotactic bacteria are present and are expected to develop as high as 63 °C and where *Magnetosome magnetitis* (Magnetosomes are membranous structures present in magnetotactic bacteria) is deposited, maximum temperature level has been extended (Lefevre et al. 2010b). The use of yeast cells as a template has been reported to synthesize magnetic Fe_3O_4 materials with a mesoporous structure (Zhou et al. 2009a, b). Table 12.2 (Amemiya et al. 2007; Li et al. 2007; Bose et al. 2009; Perez-Gonzalez et al. 2010; Zhu et al. 2010;) mentions several other magnetic oxide nanoparticles.

3.2 Nonmagnetic Oxide Nanoparticles

Many oxide nanoparticles, including TiO_2, Sb_2O_3, SiO_2, $BaTiO_3$, and ZrO_2 nanoparticles, were also investigated in addition to magnetic oxide nanoparticles (Jha et al. 2009). A green, cheap-cost, repeatable biosynthesis induced by Sb_2O_3 nanoparticles of *Saccharomyces cerevisiae* has been described (Jha and Prasad 2010). The synthesis was carried out in compliance with room temperature. Analysis has shown that the Sb_2O_3 device is a 2–10 nm spherical aggregate (Jha et al. 2009). For processing of SiO_2 and TiO_2 nanoparticles of soluble SiF62- and TiF62-anionic complexes, *Fusarium oxysporum* (Fungus) is used. *F. oxysporum* 4–5 and 3–11 nm were also prepared from tetragonal $BaTiO_3$ and quasispheric ZrO_2 nanoparticles in size (Bansal et al. 2004, 2005, 2006).

4 Sulfide Nanoparticles

As quantum dot fluorescent biomarker and cell marking agent, sulfide nanoparticles have been strongly bounded to fundamental and technological research for its fascinating, innovative, optical, and electronic characteristics, in addition to oxide nanoparticles (Yang et al. 2005). Microorganisms have nanocrystal CdS synthesized, and it constitutes one typical form of sulfide nanoparticle. It was found that *Clostridium thermoaceticum* would aggregate CdS both on cell surface and in $CdCl_2$ media in existence of cysteine hydrochloride in raising environment, most likely serving as a sulfide source (Cunningham and Lundie Jr 1993). *Klebsiella pneumoniae* was reported to create CdS (20–200) nm of on cell surface, exposing growth environment to Cd^{2+} ions. Intercellular nanocrystals, consisting of rootite chrystal phase were formed, while *E. coli* incubates $CdCl_2$ and Na_2SO_4 (Sweeney et al. 2004). Depending on cell growth process, nanocrystal formation differs greatly and increases by approximately 20 *Escherichia coli* cultivated in stationary stage relative to that produced in retard logarithmic period. *S. pombe*, *C. pombe*, and *S. glabrata* (yeasts) were used in the production of CdS nanoparticles with intracellular cadmium mixture. PbS and ZnS nanoparticles have been designed and synthesized using biological systems. ZnS with 2–5 and 8 nm mean diameter intracellular nanoparticles were used with *Desulfobacter* and *R. sphaeroides* (Bai et al. 2006). The use of *Rhodobacter sphaeroides*, whose diameters are regulated by culture time, was also used to synthesize PbS nanoparticles (Bai and Zhang 2009). For extracellular development of sulfide metal nanoparticles, eukaryotic organisms like fungi have been reported for being ideal candidates (Ahmad et al. 2002). Certain stabilized metal-metal sulfide nanoparticles like CdS, ZnS, PbS, and MoS_2 may be formed extracellularly by fungus *Fusarium oxysporum* when exposed to aqueous metal sulfate solution. Quantum dots were produced from Cd^{2+} ion interaction to sulfide ions supplied via reduction of sulfide ions. Other types of sulfide nanoparticles were magnetic Fe_3S_4 or FeS nanoparticles. Uncultured magnetotactic bacteria have documented the development of Fe_3S_4 (Bazylinski et al. 1995). A sediment sample of magnetotactic bacteria was analyzed, and about 105 cells are collected the following purification by racetrack treatment. In uncultured cells, magnetosomes showed extended rectangular shapes. The overall amount of magnetosomes in each cell was around 40, and they have been usually observed with big groups of cells. Magnetosomes forming a chain-like structure were detected alongside major clusters. Sulfate reduction bacteria may generate magnetic FeS nanoparticles (Watson et al. 1999). Table 12.3 shows many sulfide nanoparticles formed via microorganisms.

5 Other Nanoparticles

A broad range of species from organic/inorganic composites in biological systems, are utilizing biopolymers, like microbial cells and protein, with organized structures. In addition to the above mentioned nanoparticles, microbe synthesis has been reported as $SrCO_3$, $PbCO_3$, $CdCO_3$, PHB, CdSe, and $Zn_3(PO_4)_2$ (Table 12.4). $SrCO_3$ crystals were produced with ionic Sr^{2+} ions while incubating demanding fungi (Rautaray et al. 2004). Researchers assume even through fungal development of *Fusarium oxysporum* in higher cognitive superstructures, protein excretion modulated the morphology and hierarchical assembly of strontianite crystals. Through yeast biotemplates, zinc phosphate nanopowder was produced (Pandian et al. 2009). Production of $Zn_3(PO_4)_2$ particles with a butterfly-like microstructure between 10–80 nm diameter and 80–200 nm in length was shown. It has been demonstrated that *Fusarium oxysporum* in extremely luminescent room temperature would synthesize CdSe quantum dots (Yan et al. 2009).

Table 12.3 Sulfide nanoparticles synthesized by microorganisms

Microorganisms	Products	Culturing temperature (°C)	Size (nm)	Shape	Location	References
Multicellular Prokaryotes	Fe_3S_4	25	Not available	Not available	Intracellular	Lefevre et al. (2010b)
Uncultured Magnetotactic Bacterium	Probably polyphosphate	Not available	Not available	Rectangular	Extracellular	Arakaki et al. (2010a, b)
Rhodopseudomonas palustris	CdS	30	8	Cubic	Intracellular	Bai et al. (2009)
Coriolus versicolor	CdS	25	100–200	Spherical	Extracellular	Sanghi and Verma (2009)
Lactobacillus	CdS	25–60	4.9 ± 0.2	Spherical	Intracellular	Prasad et al. (2010)
Yeast 1	CdS	25–60	3.6 ± 0.2	Spherical	Intracellular	Sweeney et al. (2004)
E. coli	CdS	25	2–5	Wurtzite crystal	Intracellular	Sweeney et al. (2004)
Rhodobacter sphaeroides	ZnS	Not available	10.5 ± 0.15	Spherical	Extracellular	Bai et al. (2009)
Sulfate-reducing bacteria	FeS	Not available	2	Spherical	Extracellular	Watson et al. (1999)

Table 12.4 Other miscellaneous nanoparticles synthesized by microorganisms

Microorganisms	Products	Culturing temperature (°C)	Size (nm)	Shape	Location	References
Fusarium oxysporum	$PbCO_3$, $CdCO_3$	27	120–200	Spherical	Extracellular	Sanyal et al. (2005)
Fusarium oxysporum	$SrCO_3$	27	10–50	Needlelike	Extracellular	Rautaray et al. (2004)
Brevibacterium casei	PHB	37	100–125	Not available	Intracellular	Pandian et al. (2009)
Yeasts	$Zn_3(PO_4)_2$	25	10–80 × 80–200	Rectangular	Extracellular	Yan et al. (2009)
Fusarium oxysporum	CdSe	10	9–15	Spherical	Extracellular	Kumar et al. (2007)

6 Mechanism of Nanoparticle Synthesis by Microbes

Different microorganisms have numerous pathways of nanoparticle creation. Nanoparticles, though, are usually shaped as follows: metal ions first were trapped in microbial cells or on the surface. Then, trapping metal ions in existence of enzymes was limited to nanoparticles. In fact, in two distinct ways, microorganisms affect mineral formation. At any point, you can change a solution's composition to oversaturate it or undersaturate it. Another way for microorganisms to affect mineral formation is through organic polymers that could affect nucleation by encouraging (or preventing) stabilization of first mineral seeds (Benzerara et al. 2010). Potential mechanisms for the production of some common nanoparticles were discussed in this section: gold and silver, heavy metals, and magnetic and sulfide nanoparticles. The basic process for intracellular creation of silver and gold nanoparticles from *Verticillium* sp. or algal biomass has not been entirely known. However, the observation in which nanoparticles have grown on mycelium surface rather than in the solution supports the following hypothesis: first electrostatic interactions of ions with the overlooked cell wall of carboxylated groups of enzymes have captured fungal cells on the surface. The metal ions were then reduced to nuclei of gold or silver, which were then produced further by reduction and aggregation (Sneha et al. 2010). It was suggested that nitrate reductase enzyme can synthesize nanoparticles of B silver (Kalishwaralal et al. 2008). Nitrate ions activate this enzyme and silver ions are reduced into silver. Reducing enzyme metals in electron shuttles is a potential way of minimizing silver ions. Nitrate reductase enzymes based on NADH and NADH-reliant enzymes are the essential factors for metal nanoparticle formation. NADH and NADH-reliant enzymes, especially nitrate reductase, are considered to be secrets for *Bacillus licheniformis*, which may be essential for biosynthesis of Ag^+ to Ag^0 or continued development of silver nanoparticles (Husseiny et al. 2007). Molecular and proteomic response to hazardous conditions in metalloplastic microorganisms can lead to the development of heavy metal nanoparticles (Reith et al. 2007). Toxic effect of the microorganisms on its survival is caused by strong metal ions like Ag^+, Cd^{2+}, Co^{2+}, CrO_4^{2+}, Cu^{2+}, Hg^{2+}, Pb^2, Ni^{2+}, and Zn^{2+}. To counteract certain impact or precisely control metal metabolism, microorganisms develop molecular and proteomic reactions (Nies 1999). Microbes have many essential genes of metal tolerance that allow cell removal through a range of techniques, including complexity, excretion, or limitation of precipitation. In conditions that require large amounts for moving ions of heavy metal, as mine waste dumps and metalworking plant flows including natural sedimentary areas, metallophilic microbes thus flourish (Tang et al. 2005). A multistage method is thought to be a molecular mechanism of BacMP biomineralization. First sage is cytoplasmic membrane invagination, which is a predecessor to BacMP membrane (Arakaki et al. 2008). The mechanism for envelope formation remains unknown. Vesicular pathways for magnetotactic bacteria were more likely similar to other eukaryotes, or precipitation is controlled by particular GTPase. In a linear cytoskeletal filament chain, vesicles which were formed were then assembled. Aggregation of iron ions in vesicles is the second stage in BacMP biomineralization. The movement of foreign iron is internalized by proteins and siderophores. An oxidation-reduction mechanism strictly controls internal iron. Closely bound BacMP proteins activate and/or regulate magnetized nucleation of crystal in the final step. Magnetite generation functional roles can be performed by different membrane proteins of BacMP. This requires iron supersaturation deposition, preservation of conditions of reduction, and iron oxidation to reduce or dehydrate ferrihydrate to magnetite (Arakaki et al. 2008). This implies mineralization. Perez-Gonzalez and the staff recently suggested a new possible *Magnetitis synthesization* method that uses both passive and active *Shewanella oneidensis* (Spring and Schleifer 1995). Secondly, Fe^{2+} activity occurs as a terminal electron admitter, as bacteria use ferrihydrite, and the cell pH value may be increased by the amino acid bacterial metabolism. Localized accumulations of Fe^{2+} and Fe^{3+} on a network, bacterial surface wall, cell compositions, or cell particles allow a passive mechanism to be precipitated by magnetite system to supersaturate magnetite process. It was proposed that the production of CdS NP was due to disulfide (cystine) bridges that could be related to slashing of S–H bonds or creation of new nanoparticle surface bonds, namely, Cd-thiolate (Cd–S–CH_2COOH) S–Cd-bond complex (Sanghi and Verma 2009). Cadmium thiolate group CoOH interacts with hydrogen bond, not with NH_2 protein. CdS-capped nanoparticles also bind to hydrogen bond groups of NH_2 (Tang et al. 2005). A coordinated link between oxygen Cd^{2+} ion atom was created by one of the carboxylic oxygen group atoms, COOH, thus competing with the thiol group to construct surfaces with CdS nanoparticles (Lover et al. 1997). In general, microbes synthesize nanoparticles by implanting metal ions, followed by enzyme reduction, on cell surfaces (extracellular) or in cells (intracellular). Using fungal cellular structure and cell membrane sugars, these metal ions can be absorbed and reduced. With different microorganisms, mechanisms of synthesis of nanoparticles differ. Three options, for example, consist of an extracellular synthesis of nanoparticles, i.e., action by both electron shuttle quinones or nitrate reductase. *Penicillium* and many other fungal species have initiated the synthesis of nitrate reductase (Deepa and Panda 2014). Nitrate reductase activity was conducted using 2,3-diaminophthalene nitrites (Kumar et al. 2007). Oxysporum is associated with quinone extracellular shuttle, NADPH-dependent reductases, and nitrate reductase. Studies have shown AgNP production is generated earlier with 33 kDa protein and then with protein capping agent (free amine groups and cysteine) that maintains NPs of *Aspergillus flavus* (Soni and Prakash 2011). Metal ions were trapped firstly in the cell surface of fungi by electrostatic activity by intracellular synthesis and later reduced with enzymes inside the cell wall, contributing to NP construction and production (Singh et al. 2014). Silver nanoparticles involved in nitrate reductase enzyme *Bacillus licheniformis* are synthesized. NADH and NADH-based enzymes essential for Ag^+ bioreduction and subsequent production of AgNPs secrete *Bacillus licheniformis* (Husseiny et al. 2007). Reduction of Ag + requires a process of reducing electron shuttle enzyme to metallic silver by convincing nitrate ions and silver ions. Strong metal nanoparticles (Co^{2+}, $CrO_4{}^{2+}$, Pb^{2+}, Zn^{2+}, Hg^{2+}, Cd^{2+}) synthesize genetic and proteomic reactions that specifically control metal homeostasis and fight harmful effects (Reith et al. 2007). *Shewanella oneidensis* synthesis, moreover, involves active and passive pathways. Owing to

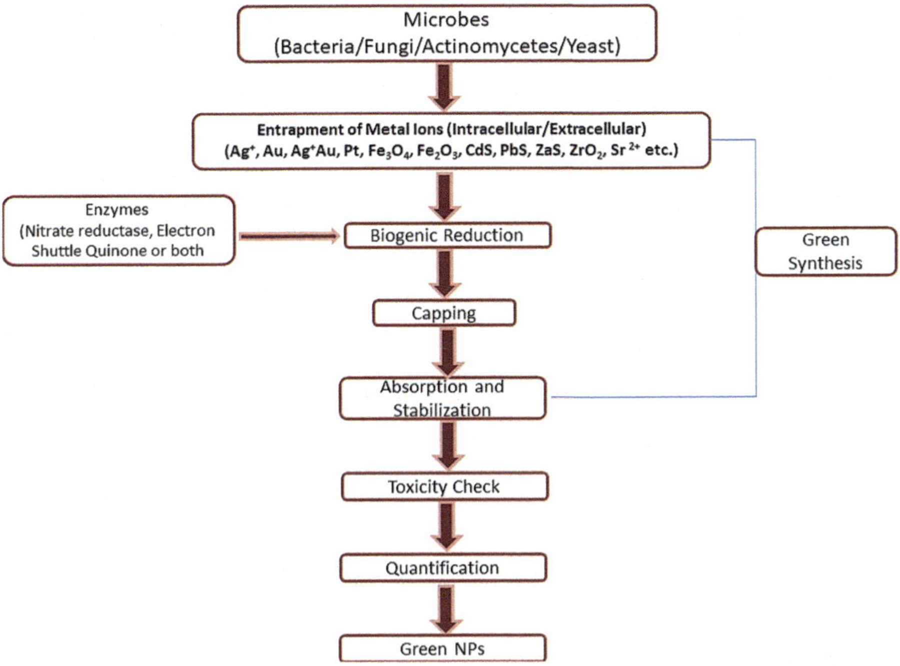

Fig. 12.2 Microbial synthesis of nanoparticles

amino acid metabolism and efficient Fe^{2+} growth, pH value rises, accompanied with active Fe^{2+} or Fe^{3+} levels that enable magnetite process to aggregate, *if* ferrihydrite is used by bacteria. The research was conducted on the production of disulfide (cysteine) cross-section CdS NPs that cause S–H bond divide and the new model nanoparticle complex (Cd–S–CH_2COOH) (Sanghi and Verma 2009). Acid carboxylic COOH groups with a hydrogen bond resulted in CdS nanoparticle capping bonds with NH_2 groups (Tang et al. 2005), cadmium-thiolate complex reaction. A coordination connection between Cd^{2+} and oxygen atoms has been generated by one carboxylic atom (–COOH) that competes to thiol for building nanoparticles on CdS surfaces (Li et al. 2007). Covalent binding to nanoparticles of carboxylic acids while still inhibiting the growth of surface oxides that minimize the magnetic characteristic of cobalt can induce biocompatibility. For the rational design of such entities, recognizing the origin of acid-metal interaction is important, but possibly most experimentally a difficult stage (Farkas et al. 2020) (Fig. 12.2).

7 Regulation of Nanoparticle Size and Morphology

It's so well established that electronic and optical characteristics of nanoparticles depend enormously on their size and shape. Significant attention was paid to monitoring the scale, shape, and media support for nanoparticles. Special emphasis has recently been put in the form regulation, as it also allows properties to be optimized to the highest degree of versatility, which gives particles their distinctive character. Although physical and chemical techniques are capable of generating, over a short time, significant quantities of nanoparticles of certain size and shape, these techniques are complex and present certain disadvantages, such as the development of radioactive waste that is hazardous not just to the environment but even to public health. Microbes that are considered to have been efficient green nanofabrics can regulate the size and shape of biological nanoparticles. Two fungal cultures of gold nanoparticles of different morphologies and sizes, *Verticillium luteoalbum* and one labeled isolate 3–6 (Gericke and Pinches 2006), were found to have an intracellular synthesis. Particle formation rate and particle size may be manipulated to a certain degree by manipulating parameters such as exposure times to pH, temperature, gold, and $AuCl_4$. As demonstrated by electron microscopy scans, numerous morphologies of particles were present, including circular, triangular, hexagonal, and other shapes. Shape and size of particles ranged dramatically from several nanometers to around 100 nm. Their observations often found that particles of spheres seemed to be lower than particles of triangles and hexagons. During the study, screened bacterial cultures appeared to intracellularly synthesize thin, nearly homogenous gold nanoparticles. Particles were mainly noticed in the cell cytoplasm, with most spherically shaped particles. Gurunathan et al. (2009) investigated optimal process requirements to complete AgNP production and particle size reduction. In a synthesis of AgNPs, process temperatures and pH values have been used to detect optimum conditions, various mediums, and media of varying $AgNO_3$ concentrations. A nitrate medium with a 5 mM $AgNO_3$, a reaction temperature of 60 °C, and a pH of 10 was described as the maximum synthesis subject. It took only 30 min to achieve more than 95% conversion using *Escherichia coli* supernatant culture under these optimum conditions. The rate of synthesis of identical particles obtained using chemical methods is comparable or faster. Average particle size can be tuned by varying the $AgNO_3$ concentration, temperature of reactions and pH from 10–90 nm. During the synthesis of the Pt nanoparticles, the cell-soluble extract (CSE) might decrease the Pt(IV) into nanoparticles that were stable by means of binding protein and exhibit both g in solution. Strong initial Pt(IV) levels seemed to have led to more regular and geometric particles. More hydrochloride (pH to 4) was produced inside the system at high initial amounts of Pt(IV), leading to precipitation of biocomposites of nanoparticle proteins and consequently a reduction in the level of soluble particle size in colloids. Besides, without cellular restrictions, high size and type variations of protein-stabilized biogenic Pt(0) nanoparticles can be synthesized. Magnetotactic bacteria create uniform size and morphological iron oxide magnetic particles. Magnetite shaped by magnetotactic bacteria takes different forms such as cuboid, rhombic, and rectangular shape of a bullet. A high degree of biological regulation has been observed in various species-dependent crystal morphologies and structures (Amemiya et al. 2007). It is discovered that Mms6 is a big protein closely linked to *Magnetospirillum magneticum* AMB-1, the surface of bacterial magnetites (Arakaki et al. 2010a). With a uniform cuboctahedral morphology, protein was shown to intercede the creation of magnetite crystals. Formation of magnetite with synthetic peptides imitating Mms6 protein was examined. A spherical structure of 0.70–0.90, similar to one of the bacterial magnetites and particulate matter formed by the Mms6 protein, was demonstrated by particles synthesized with short peptides comprising the Mms6 C-terminal acid region. Also, if other peptides are added in production, rectangular morphology was observed with circularities of 0.60–0.85 (Arakaki et al. 2010b). The same group developed an additional method for highly controlled synthesis of magnetite crystals using the recombinant magnetotactic bacterial protein Mms6 in aqueous solutions at reduced temperatures. Crystallographic study of magnetite crystals reveals that Mms6 mediates the development of a peculiar crystal shape of magnetite particles with narrow-scale distribution close to that seen in magnetic bacteria. Mms6 aggregates have a high affinity for iron ions in aqueous solution and have motif sequence in many biomineralization scaffold proteins, close to other organisms. If compared to Mms6, crystals have identical sizes (20 nm) and morphologies (cuboctahedral). This means that Mms6 has a direct impact through the synthesis process on size and shape of nanoparticles (Amemiya et al. 2007). Particle size control for other nanoparticles has also been seen. For instance, Yan et al. (2009) find that yeast induction is an efficient way of achieving a small diameter distribution of zinc phosphate powders. To prevent the large accumulation of $Zn_3(PO_4)_2$ particles to completely control particle size and shape, their method used the yeast feature in reaction mechanism.

8 Nanoparticle Applications

Nanomedicine is a booming scientific area with a vast potential to improve human disease diagnosis and care (Fadeel and Garcia-Bennett 2010). The most widely used nanomedicine nanoparticles are fluorescent biologic labeling, drug/molecular delivery agents, as well as tissue engineering (Tian et al. 2008), heat tumor destruction (hyperthermia), MRI contrast enhancement, and phagokinetic analysis (Parak et al. 2002). Many reviews and research articles have been published that analyze nanoparticles' applications in biomedicine (Piao et al. 2011). Though biosynthesized nanoparticles are relatively new, research has been initiated on applications in drug delivery, cancer care, genetic modification and DNA sequencing, antimicrobials, biomaterials, and response enhancement.

8.1 Antibacterial Agent

Silver-based antiseptics were stressed in recent times due to proliferation and rise of microorganism resistance to various antibiotics. The use of *Trichoderma viride* fungus in silver nanoparticles was biosynthesized (Fayaz et al. 2010). Aqueous silver ($Ag+$) ions were found to be decreased in solution when exposed to *Trichoderma viride* filtrate, resulting in production of pretty stabilized AgNPs. Nanoparticles have also been tested with multiple antibiotics for increased antimicrobial activity toward Gram (positive and negative) bacteria. With the existence of AgNPs, antibacterial efficacy of erythromycin, chloramphenicol, ampicillin, and kanamycin toward test strains has been improved. Strongest enhancement effect of ampicillin against test strains was detected. Results showed greater antimicrobial effects in combination with antibiotics with AgNPs and offered valuable insight into the production of new antibacterial agents. Duran et al. (2007) have demonstrated that extracellularly generated silver nanoparticles utilizing *F. oxysporum* could be integrated through woven materials in an effort to avoid or decrease contamination with infective bacteria like *S. aureus*. Silver nanoparticles of *Acalypha wilkesiana* (AW-AgNPs) demonstrated substantial repression toward dominant Gram-negative and Gram-positive selected bacteria. Therefore, AW-AgNPs may be suggested as a potential antimicrobial and therapeutic agent against multidrug-resistant pathogens (Dada et al. 2019). The key components of AgNPs, CuONPs, AuNPs, and ZnONPs have been updated and commonly used for therapeutic and medicinal purposes (e.g., as antibacterial, antifungal, antiviral, anti-amebial, anticancer, anti-angiogenic, anti-inflammatory factors). These particles were suggested as alternatives to standard antibiotics to overcome bacterial resistance due to their excellently described antibacterial activity toward Gram (positive and negative) bacteria. Nanoparticles utilize mechanisms involved that differ from traditional therapies, with the benefit of becoming effective toward antibiotic resistance bacteria which have already formed, as well as by attacking several biomolecules that compromise resistant strain growth (Sánchez-López et al. 2020).

9 BM-NPs: Synthesized as Antimicrobial, Antiviral, and Scolicidal Potential from *Penicillium* Species

There have been studies of silver nanoparticle (AgNPs) biosynthesis caused by *Penicillium citrinum* (Yassin et al. 2017). Biogenic AgNPs toward aflatoxinic *A. flavus* were also tested. Biogenic AgNPs toward aflatoxinic A. flavus var. columnaris isolated from sorghum seeds were also tested for antifungal activity (Fig. 12.3). They showed that action of AgNPs toward *Aspergillus flavus* varied from 20.28 to 50.00%, and 224.5 to 4001.8 ppm were calculated at ED50 and ED95, respectively. Such antifungal activity was linked to the cell membrane and cytoplasm modification, membrane permeability, and DNA energy depletion. In extracellular biomimetic synthesis, AgNPs induced by *Penicillium chrysogenum* strain FGCC/BLS1 have been reported (Saxena et al. 2017). Their analysis showed potent antibacterial activity of AgNP at 100 ppm and antifungal activity at 100 ppm toward *E. coli*, *K. pneumoniae*, and *S. aureus* against phytopathogenic fungi *sclerotiorum*. In hemolytic test with a dose of 10 ppm in red blood cells, no cytotoxicity was observed. Exceptionally, biogenic synthesis of gold nanoparticles in an extracellular approach with *P. funicular* BL1 in 18–28 nm range has been documented

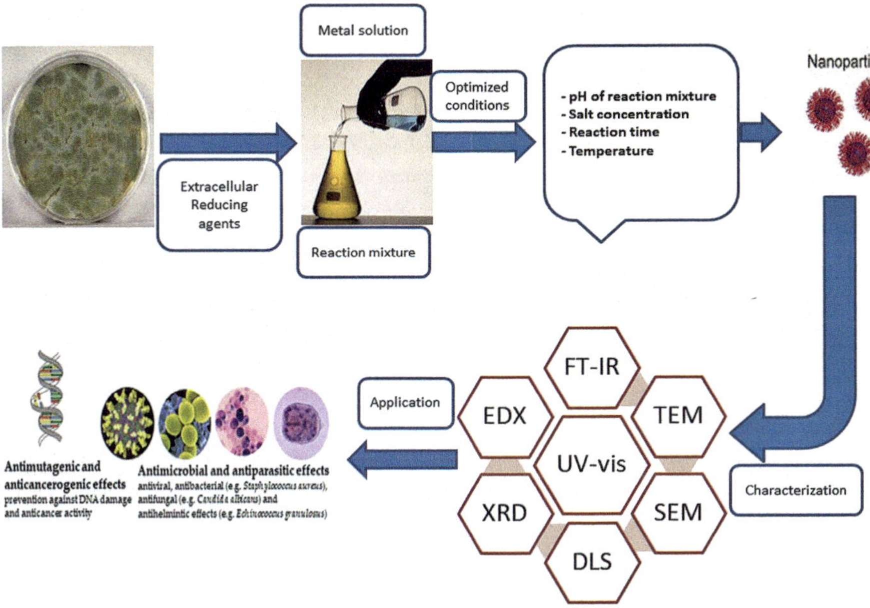

Fig. 12.3 A modern version of pharmaceutical nanobiotechnology and the interface of nanotechnology, bacteria, and pharmaceutical ability

(Maliszewska et al. 2017). They demonstrated a photodynamic inactivation of *Candida albicans* planktonic and biofilm cells in combination with synthesized biogenic AuNP exposure to rose bengal (RB). AuNPs showed no unusual murder of Xe lamp glare exposure to *Candida albicans*. However, killing was shown to be a fair efficiency of *Candida albicans* when RB and biogenic NPs are administered together like photosensitizing agent. Combination of RB and AuNP showed that 4.7 log10 and 4.89 log10 had decreased CFUs, which were 99.91 and 99.99%, while 98.21 and 99.37% were killed by RB alone after the same time. Furthermore, by using Penicillium spp. biosynthesized AgNPs. in an extracellular way (Verma et al. 2013). Maximum antibacterial activity in AgNPs was observed in *Bacillus* and *Pseudomonas* spp., accompanied by *E. coli* and *Salmonella* spp. at concentrations of 1 mg/mL if used in conjunction with tetracycline, and maximum inhibition was observed in *Salmonella*, *Pseudomonas*, and *Escherichia coli*. A research was performed using a disc diversion approach for *Pseudomonas aeruginosa*, *Escherichia coli*, *Bacillus subtilis*, *Staphylococcus aureus*, and *Candida albicans* to determine antimicrobial activities of biofabricated AgNPs of *Penicillium aculeatum* Su1. In either study, 200 μg/mL AgNPs had strongest antibacterial effect on all listed strains compared to 100 μg/mL AgNPs with a big variation relative to 50 and 200 μg/mL $AgNO_3$ (Osman et al. 2015). Notably, Solanki et al. (2016) extracellularly synthesized AgNPs *using Penicillium brevicompactum* between 6.28 and 15.12 nm. All through research, antimicrobial activity of biofabricated AgNPs has been evaluated utilizing disc-diffusion methods for clinically isolated pathogenic bacteria such as

E. coli, *S. aureus*, and *P. aeruginosa*. They found that regardless of whether AgNP concentration improved, a dose-dependent zone of inhibition often increased. The inhibition zone for the 10 μL concentration between 7 and 16 mm was found in depth, while for the 20 μL concentration, the inhibition region was significantly found between 9 and 28 mm. In addition, Khan and Jameel (2016) extracellularly biosynthesized AgNPs with *Penicillium fellutanum* within a domain of 10–100 nm. Antifungal activity was assessed through the use of discharge assays against *Candida glabrata*, *Candida albicans*, and *Candida tropicalis*, though $AgNO_3$ solution was not found to inhibit the region. Ammar and El-Desouky (2016) have also documented biosynthesis induced by HA_2N *Penicillium* expansion between 14 and 25 nm. For *A. ochraceus* and *A. niger* with disc-diffusion process, researchers even searched for an antifungal role for biogenic AgNPs. In particular, at concentration of 9 μg AgNPs in A, maximum inhibition level was observed in *Aspergillus niger*. Moreover, AgNPs with culture medium concentration of 220 μg/100 mL were found to cause, with 52.18% decrease percentage, the most important mycotoxin produced by *Aspergillus ochratoxin*, called *Aspergillus ochraceus*. Majeed et al. (2016) have documented an extracellular approach of biomimetic synthesis of AgNPs ranging from 30 to 60 nm. Appraised antibacterial activity of AgNPs using *Proteus vulgaris*, *Staphylococcus aureus*, *Escherichia coli*, and *Vibrio cholera* diffusion methods. For disc-diffusion research, every disc was saturated for 20 μg/mL of AgNPs. Antibiotics such as amoxicillin, carbenicillin, cefixime, ofloxacin, and piperacillin were contrasted with AgNPs. Antimicrobial activity of Ag nanoparticles recorded strong via a zone of inhibition for *E. coli*, *V. cholera*, *P. vulgaris*, and *S. aureus*. Amusingly, Ag nanoparticles strengthened their antibacterial efficacy in combination with the aforementioned antibiotics. Moreover, Sarsar et al. (2015) recorded biogenic AgNP production utilizing 5–25 nm range of *Penicillium atramentosum* KM filtrate extract. *Aeromonas hydrophila*, *Bacillus cereus*, *Enterobacter aerogenes*, *Micrococcus luteus*, *Staphylococcus aureus*, and *Salmonella typhimurium* disc-diffusion process tested antibacterial activity. Significant antimicrobial activity toward *Bacillus cereus* has been observed. A considerable surface area was provided as AgNPs, contributing to its connection to the cell wall, increasing the integrity of cell membranes causing apoptosis, and the authors advocate it for stronger bacterial communication. It also showed a substantial increase of antibacterial activity of microgravity-synthesized AgNPs than of usual gravity-synthesized AgNPs (Sheet et al. 2017). A research was carried out by Ali et al. (2014) that otherwise recorded antimicrobial activity for AgNP extracellular/intracellular production using *Pseudomonas citreonigrum* with micro-dilution technique toward *B. subtilis*, *S. aureus*, *S. typhimorium*, *E. coli*, and *P. aeruginosa* and demonstrated antifungal effect toward *Aspergillus* utilizing micro-dilution technique. In this research, the antiviral effect toward type 2 herpes virus and the cytotoxicity toward three cancer cell lines were also seen. Significant antiviral activity at concentrations of 50 μg/mL, medium antiviral activity at concentrations of 25 μg/mL, or poor performance at concentrations of 12.5 μg/mL has been seen in extracellular environment-generated AgNPs, while far poorer results were found in intracellular AgNPs at concentrations of 50 and 25 μg/mL. Authors proposed throughout viral membrane whether disulfide linking areas in the glycoprotein subunit would interact with AgNPs smaller than 10 nm in size because of their surface plasmon vibration and broad efficient dispersion cross-section including its individual AgNPs. It is important to remember that *P. aculeatum* used a mean diameter of about 60 nm and good scolicidal effect toward *Echinococcus granulosus* protoscolices. Extracellular biosynthesis of AuNPs is documented (Barabadi et al. 2017). Their results show that after 120 min of exposure, the scolicidal behavior of AuNPs was equal to that of AgNP, selenium NPs, 20% $AgNO_3$ at 20 min, and isotonic saline at 20%.

Synthesis of extracellular AgNP has been recently documented by Sheet et al. (2017) to assess its biological and physicochemical role, using microgravity and ordinary conditions. Findings indicate cytotoxic effects of microgravity-synthesized ANPs on cancer cells are much greater than standard severity-synthesized ANPs. In the range of 4–55 nm of exploited *Penicillium aculeatum* Su1, extracellular biosynthesis of AgNPs was stated (Ma et al. 2017). This research revealed that biosynthesized AgNPs are far more biocompatible with human bronchial epithelial cells than $AgNO_3$ and were substantially dose-determined toxic to A549 cells via IC_{50} of 48.73 μg/mL, reflecting a potential impact on human pulmonary adenocarcinoma cell proliferation. Moreover, cytotoxic activity of AgNPs was biosynthesized with the use of *Penicillium* spp. in vitro in a sample. Cell lines with human colon adenocarcinoma (HT-29) ranging from 5 to 100 μg/mL were tested in contrast to normal Vero cell lines. Findings showed that AgNPs of IC50 had a cytotoxic effect of 30 μg/mL to HT-29, while IC50 was anticipated to be far greater than 50 μg/mL for the standard Vero cell line (Verma et al. 2013). Also, a research study found that biogenic AgNPs provided cytotoxic effects on the A549 cancer cell line, whereas their toxicity was significantly lower at the same level as the usual Vero cell line. Expansion of AgNPs by active oxygen species, which causes oxidative damage that induces higher levels of necrosis at higher levels and not just affects critical enzymes, was explained by researchers (Majeed et al. 2016). Ali et al. (2014) also reported intracellular/extracellular AgNP biosynthesis by using *P. citreonigrum* throughout the order of 10–50 nm. AgNPs were tested for cytotoxicity on (breast, colon, liver) cell lines. In dramatic terms, extracellular AGNPs showed significantly greater inhibition effect of three cancer cell lines than intracellular NPs. For this relation, researchers indicated that interruptions of AgNPs in the mitochondrial breathing chain might contribute to ROS, which interrupts ATP production and leads directly to DNA damage. Furthermore, Vazquez-Muñoz et al. (2019) provide a deeper understanding of the complementary mechanism of AgNPs and antibiotics to effectively fight antimicrobial pathogens to alleviate current crises due to antibiotic resistance, particularly those with multidrug-resistant microorganisms.

10 Microbial-Based Crop Safety Nanoparticle Applications

Through the manufacture of nanomaterials, the distribution of inorganic fertilizers and biopesticides to agriculture or a fully qualified approach to gene transfer, nanobiotechnologies, including detection and control for phytopathogens and food safety against infections, can be widely used (Fig. 12.4). Nanoparticle crop protection applications are considered effective if they stay active in extreme conditions like temperature variations, target pathogen penetration, tolerance to phytopathogens, cheap cost of formulation preferably in advanced mode of action, and social and economic advantages (Smith et al. 2008). In growing effectiveness and stabilization of utilized cells and enzymes, nanoparticles play a pivotal role. Nanomaterials result from biomolecular integration (enzymes, metabolites, etc.) or full cell hybrid systems with different agricultural uses (Bailey et al. 2010). Microbe-integrated nanoparticles gain from improved biological efficacy, fast fixation over the wide surface region, increased bioavailability and versatility, reduced toxicity, and improved mass delivery systems. Next NPs are trapped and nanomaterials are fused, and active ingredient is released in a controlled manner. The use of NP aids would involve a tailored distribution strategy based on the actions and environmental conditions of phytopathogens. For instance, DNA-coated AuNPs have been utilized as a shot to bombard plant and tissue cells to induce gene transfer in gene gun protocol (Vijayakumar et al. 2010). Microbes (bacteria, fungi) and its metabolites (enzymes, inhibitors, antibiotics, toxins) have been able to use biocontrol factors to protect plants or to improve the productivity of plants for years.

Coating of polymeric NPs provided advanced pathways for improving efficiency and stability of biocontrol agents, as gravity preparations for formulations supplied to targeted pathogens with a directed distribution system. Besides, trapped nanomaterial products can support the growth of soil and plants (Peteu et al. 2010). Fungal biological control factors are highly precise and are widely available without ingestion, for mass manufacturing by contact. Many fungal genotypes (*Beauveria, Nomuraea, Verticillium*) spread infection via conidia, requiring humidity to allow host pathogenesis to germinate (Kulkarni et al. 2008). To stabilize *Myrothecium* complex enzymes, nanoformulation with chitosan and montmorillonite clay NPs was produced and demonstrated for *Fusarium* spp. *Gossyphilous Phenacoccus* and biocontrol, with a sluggish discharge of enzymes (cotton mealybug). Antifungal hydrolases and enhanced chitina and chitosanase enzymes are induced by Chito nanoparticles handled with curcuma plants to protect plant host that have made them resistant to turmeric red *Pythium aphanidermatum* rhizome (Anusuya and Sathiyabama 2013). Silica-based NPs (60 nm) packed with fluorescent dye and covalently linked with microbe surface antigen-specific antibodies are sensitive. Copper is converted through metal NPs by popular plant species (*Phragmites australis* and *Iris pseudacorus*) if produced using endomycorrhizal fungi in polluted soil (Manceau et al. 2008). The inhibition efficacy of Ag_2S nanocrystals and $ZnTiO_3$ was higher. In corn-treated plants by silica NPs, greater tolerance to *F. oxysporum* and *A. niger* has been exhibited (Suriyaprabha et al. 2014). TiO_2 NPs have improved

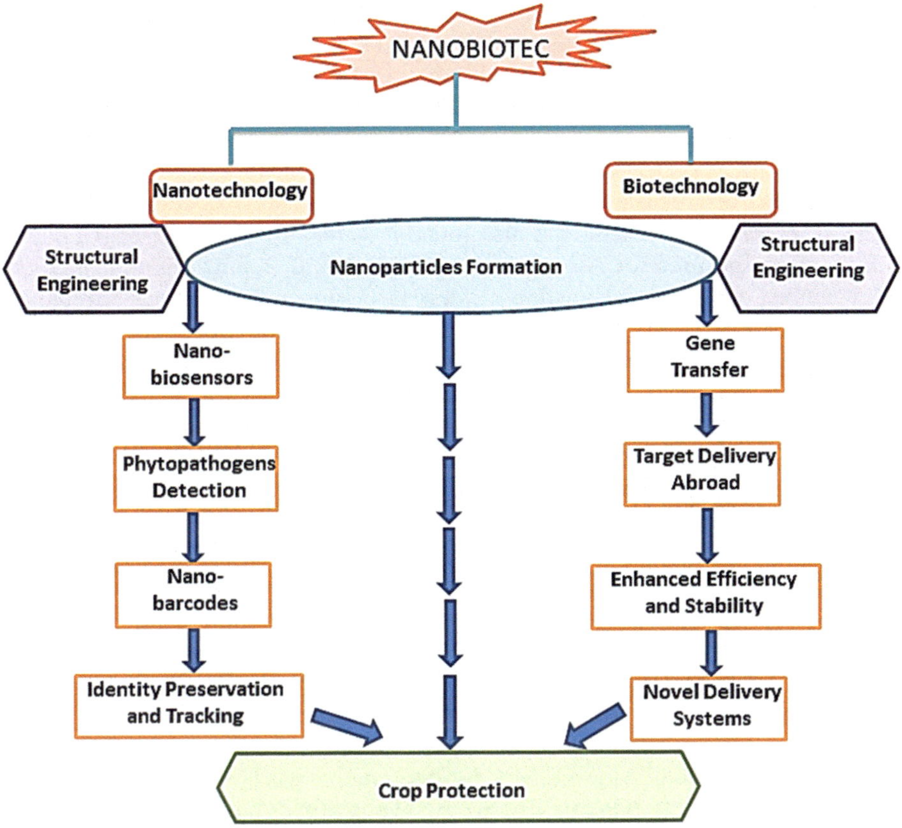

Fig. 12.4 Overview of the impact of nanoparticles on crop protection

and provided defense toward *Alternaria brassicae* (*Bacillus amyloliquefaciens*) in *Brassica napus* rhizosphere (Palmqvist et al. 2015). Zinc oxide NPs showed that conidium and *Penicillium expansum* conidium were inhibited, resulting in fungal mat absence (He et al. 2010). Magnetic reverse of nanoparticles is an extremely precise and sensitive approach. To detect *Prunus necrotic ringspot virus* promptly, reverse transcription loop-mediated isothermal amplification (RTLAMP) was established (Zong et al. 2014). Incubated under atmospheric conditions with combination of $CdCl_2$ + $SeCl_4$ and $CdCl_2$ + $TeCl_2$ by electron transmission microscopy (TEM) and electron diffraction under specific conditions, high fluorescence CdSe QDs and CdTe QDs are metabolized by *F. oxysporum* (Shaligram et al. 2009). Yeast cells have also been used for nanoparticle cadmium telluride (CdTe) biosynthesis QD of tunable fluorescence emission (Nayak et al. 2010). To reduce time to classify unique phytopathogens, the nucleic acid sensor bound to quartz crystal microbiological sensor surface could be coupled with rapid PCR protocols (Maliszewska et al. 2013). Through the use of AgNPs, nanobiotechnology has lately become more effective toward multiple phytopathogens. AgNP interaction with microbes increases because of a higher surface-to-volume proportion and hence greater permeability (Kim et al. 2008). This reduced solution results in the production of highly stable AgNPs with sizes of 5–40 nm when aqueous silver (Ag^+) ions are treated with *Trichoderma viride* filtrate (Fayaz et al. 2010). Antibiotic mixture with AgNPs has been tested to have a stronger antimicrobial effect on many types of bacteria (Aziz et al. 2014, 2015, 2016). Infection of *S. aureus* pathogens in textiles for extracellularly formed AgNPs containing *F. oxysporum* was reduced (Duran et al. 2007). Highest inhibition of disease was also found in *Colletotrichum* species (*C. acutatum*, *C. gloeosporioides*, *C. higginsianum*, *C. nigrum*, *C. orbiculare*, *C. dematium*) or cucumber, pumpkin, and powdery mildew. DNA-directed AgNPs can be removed by *Xanthomonas perforans* leaf spot disease (Ocsoy et al. 2013).

In other studies, biogenic silver nanoparticles have impregnated and reported superior antibiotic disc activity (chloramphenicol) with two pathogenic bacteria *Abelmoschus esculentus* and *Citrullus lanatus* (*Citrobacter freundii* and *Erwinia cacticida*) diseases (Paulkumar et al. 2014). Substantial antifungal effect toward spot blotching disease in wheat induced by *Bipolaris sorokiniana* has been metabolized and illustrated (Mishra et al. 2014). X*anthomonas axonopodis* fluorescent silica nanoparticles (FSNP) were correctly demonstrated in tomatoes and peppers in conjunction with antibody molecules to prevent *vesicatoria* that cause bacterial spot disease (Mishra et al. 2010). Nanoparticles include antibodies used to detect *Xanthomonas axonopodis* (Yao et al. 2009). Ag nanoparticles increasingly attracted researchers worldwide for their antimicrobial agents so their production is more cost-effective and competitive for plant disease control. If utilized in consortiums with several other nanocrystals, numerous studies have shown powerful effects on AgNPs. With the use of Ag-SiO_2 NPs, *Botrytis cinerea* has been reduced by significant antifungal activity (Oh et al. 2006). Ag nanoparticles have been tested toward *Phoma glomerata*, *Phoma herbarum*, *Fusarium semitectum* for antifungal activity with fluconazole spp., *Trichoderma*, and *C. albicans* through disc-diffusion method (Gajbhiye et al. 2009). Throughout the type of *Colletotrichum gloeosporioides* (competence of anthracnosis), *B. sorokiniana*, *M. grisea*, and *S. cepivorum*, sclerotium-forming phytopathogenic fungi, the existence of AgNPs has been significantly inhibited. AgNP fungistatic and fungicidal action against Ambrosian fungus *Raffaelea* spp. and *Fusarium culmorum* was examined, as well as some pathogenic yeasts (*Candida albicans*, *Candida parapsilosis*, *Candida tropicalis*) (Kasprowicz et al. 2010). Inhibition effect has shown to be 15 mg of AgNP toward *Alternaria alternata*, *Botrytis cinerea*, *Curvularia lunata*, *Macrophomina phaseolina*, *Sclerotinia sclerotiorum*, and *Rhizoctonia solani*.

11 Conclusion

Nanomedicine is a thriving scientific area with enormous potential for human diseases to be properly diagnosed and treated. Biological synthesis of microbial nanoparticles for "green chemistry" is considered safe, nontoxic, and environmentally acceptable. Depending on the location of intracellular and extracellular production of nanoparticles, microorganisms, like bacteria, leaves, fungi, and actinomycetes, may be used. Shape and size of nanoparticles in intracellular particle form could be manipulated to a certain degree using control factors like pH, temperature, substrate concentration, and exposure time. The study is presently being performed to monitor molecular and proteomic microorganisms. These techniques and their industrial use in medicine and health care are expected to be applied on a large scale in the next few years, with latest developments and ongoing attempts to increase the efficiency of particulate synthesis and to explore biomedical applications. Over the last decade, there have been huge advances in the field of nanoparticles developed by the microorganism and their applications. However, to improve synthesis and track size and morphology of particles, a lot of work needs to be done. Compared with the physical and chemical process, it is recognized that production of nanoparticles with microbes (several hours, even some days) is a really slow process. Reducing time of production would make this path even more appealing. Particle size and monodisperse particles are two main concerns in the assessment of nanoparticle synthesis. Efficient particle size and monodisperse regulation must therefore be thoroughly examined. Several studies have shown that after a certain period, nanoparticles produced by microorganisms can decompose. The stability of biological nanoparticles therefore needs further research and should be improved. Because particle shape control in the physical and chemical production of nanoparticles is indeed research subject, biological mechanisms with the ability to specifically regulate particle shape would seem to have significant benefits. Adequate control of particle size and monodisperse particle may be given with varying conditions like microorganism type, microbial growth phase, growth medium, synthesis, pH, substratum concentrations, target nanoparticles' origin compound, temperature, process period, and nontarget ion addition. Biosynthesis methods are also beneficial, as nanoparts are mostly covered by lipid molecules, which give biological stability and solubility, which is important for biomedical applications and other synthetic processes for bottling. Research is currently being conducted to control genomic and proteomic cells. Shorter response period and high composition efficiency are being achieved with a deeper understanding of the system of molecular and cellular synthesis, particularly separation and characterization for those molecules responsible for nanoparticle depletion.

References

Abdel-Kareem MM, Zohri AA (2018) Extracellular mycosynthesis of gold nanoparticles using Trichoderma hamatum: optimization, characterization and antimicrobial activity. Lett Appl Microbiol 67:465–475

Agnihotri M, Joshi S, Kumar AR, Zinjarde S, Kulkarni S (2009) Biosynthesis of gold nanoparticles by the tropical marine yeast *Yarrowia lipolytica* NCIM 3589. Mater Lett 63(15):1231–1234

Ahmad A, Mukherjee P, Mandal D et al (2002) Enzyme mediated extracellular synthesis of CdS nanoparticles by the fungus, *Fusarium oxysporum*. J Am Chem Soc 124(41):12108–12109

Ahmad A, Senapati S, Khan MI, Kumar R, Sastry M (2003a) Extracellular biosynthesis of monodisperse gold nanoparticles by a novel extremophilic actinomycete, *Thermomonospora* sp. Langmuir 19(8):3550–3553

Ahmad A, Senapati S, Khan MI et al (2003b) Intracellular synthesis of gold nanoparticles by a novel alkalotolerant actinomycete, Rhodococcus species. Nanotechnology 14(7):824–828

Ahmed T, Shahid M, Noman M, Niazi MBK, Mahmood F, Manzoor I, Zhang Y, Li B, Yang Y, Yan C et al (2020) Silver nanoparticles synthesized by using *Bacillus cereus* SZT1 ameliorated the damage of bacterial leaf blight pathogen in rice. Pathogens 9:160

Akther T, Hemalatha S (2019) Mycosilver nanoparticles: synthesis, characterization and its efficacy against plant pathogenic Fungi. BioNanoScience 9:296–301

Ali FT, El-Sheikh HH, El-Hady MM, Elaasser MM, El-Agamy DM (2014) Silver nanoparticles synthesized by Penicillium Citreonigrum and Fusarium moniliforme isolated from El-Sharkia, Egypt. Int J Sci Eng Res 5(4):181–186

Almaary KS, Sayed SRM, Abd-Elkader OH, Dawoud TM, El Orabi NF, Elgorban AM (2020) Complete green synthesis of silver-nanoparticles applying seed-borne Penicillium duclauxii. Saudi J Biol Sci 27:133–1339

Al-Zubaidi S, Alayafi AA, Abdelkader HS (2019) Biosynthesis, characterization and antifungal activity of silver nanoparticles by *Aspergillus niger* isolate. J Nanotechnol Res 1:23–36

Amemiya Y, Arakaki A, Staniland SS, Tanaka T, Matsunaga T (2007) Controlled formation of magnetite crystal by partial oxidation of ferrous hydroxide in the presence of recombinant magnetotactic bacterial protein Mms6. Biomaterials 28:5381–5389

Ammar HAM, El-Desouky TA (2016) Green synthesis of nanosilver particles by *Aspergillus terreus* HA1N and *Penicillium expansum* HA_2N and its antifungal activity against mycotoxigenic fungi. J Appl Microbiol 121:89

Anusuya S, Sathiyabama M (2013) Effect of chitosan on rhizome rot disease of turmeric caused by Pythium aphanidermatum. ISRN Biotechnol 305349:1–5

Arakaki A, Nakazawa H, Nemoto M, Mori T, Matsunaga T (2008) Formation of magnetite by bacteria and its application. J R Soc Interface 5(26):977–999

Arakaki A, Masuda F, Amemiya Y, Tanaka T, Matsunaga T (2010a) Control of the morphology and size of magnetite particles with peptides mimicking the Mms6 protein from magnetotactic bacteria. J Colloid Interf Sci 343(1):65–70

Arakaki A, Shibusawa M, Hosokawa M, Matsunaga T (2010b) Preparation of genomic DNA from a single species of uncultured magnetotactic bacterium by multiple-displacement amplification. Appl Environ Microbiol 76(5):1480–1485

Aziz N, Fatma T, Varma A, Prasad R (2014) Biogenic synthesis of silver nanoparticles using Scenedesmus abundans and evaluation of their antibacterial activity. J Nanoparticles 2014:689419. https://doi.org/10.1155/2014/689419

Aziz N, Faraz M, Pandey R, Sakir M, Fatma T, Varma A, Barman I, Prasad R (2015) Facile algae-derived route to biogenic silver nanoparticles: synthesis, antibacterial and photocatalytic properties. Langmuir 31:11605–11612

Aziz N, Pandey R, Barman I, Prasad R (2016) Leveraging the attributes of Mucor hiemalis-derived silver nanoparticles for a synergistic broad-spectrum antimicrobial platform. Front Microbiol 7:1984. https://doi.org/10.3389/fmicb.2016.01984

Bai HJ, Zhang ZM (2009) Microbial synthesis of semiconductor lead sulfide nanoparticles using immobilized *Rhodobacter sphaeroides*. Mater Lett 63(9–10):764–766

Bai HJ, Zhang ZM, Gong J (2006) Biological synthesis of semiconductor zinc sulfide nanoparticles by immobilized *Rhodobacter sphaeroides*. Biotechnol Lett 28(14):1135–1139

Bai HJ, Zhang ZM, Guo Y, Yang GE (2009) Biosynthesis of cadmium sulfide nanoparticles by photosynthetic bacteria *Rhodopseudomonas palustris*. Colloids Surf B Biointerfaces 70(1):142–146

Bailey KL, Boyetchko SM, Langle T (2010) Social and economic drivers shaping the future of biological control: a Canadian perspective on the factors affecting the development and use of microbial biopesticides. Biol Control 52:221–229

Bansal V, Rautaray D, Ahmad A, Sastry M (2004) Biosynthesis of zirconia nanoparticles using the fungus *Fusarium oxysporum*. J Mater Chem 14(22):3303–3305

Bansal V, Rautaray D, Bharde A et al (2005) Fungus-mediated biosynthesis of silica and titania particles. J Mater Chem 15(26):2583–2589

Bansal V, Poddar P, Ahmad A, Sastry M (2006) Room-temperature biosynthesis of ferroelectric barium titanate nanoparticles. J Am Chem Soc 128(36):11958–11963

Bao H, Lu Z, Cui X et al (2010) Extracellular microbial synthesis of biocompatible CdTe quantum dots. Acta Biomater 6(9):3534–3541

Barabadi H, Honary S, Mohammadi MA, Ahmadpour E, Rahimi MT, Alizadeh A, Naghibi F, Saravanan M (2017) Green chemical synthesis of gold nanoparticles by using *Penicillium aculeatum* and their scolicidal activity against hydatid cyst protoscolices of *Echinococcus granulosus*. Environ Sci Pollut Res Int 24:5800

Bazylinski DA, Frankel RB, Heywood BR et al (1995) Controlled biomineralization of magnetite (Fe_3O_4) and greigite (Fe_3S_4) in a magnetotactic bacterium. Appl Environ Microbiol 61(9):3232–3239

Benzerara K, Miot J, Morin G, Ona-Nguema G, Skouri-Panet F, Ferard C (2010) Significance, mechanisms and environmental implications of microbial biomineralization. Compt Rendus Geosci 343(2–3):160–167

Bhainsa KC, D'Souza SF (2006) Extracellular biosynthesis of silver nanoparticles using the fungus Aspergillus fumigatus. Colloids Surf. B Biointerfaces 47:160–164

Bhattacharya R, Mukherjee P (2008) Biological properties of naked metal nanoparticles. Adv Drug Deliv Rev 60(11):1289–1306

Blakemore R (1975) Magnetotactic bacteria. Science 190(4212):377–379

Bose S, Hochella MF, Gorby YA et al (2009) Bioreduction of hematite nanoparticles by the dissimilatory iron reducing bacterium *Shewanella oneidensis* MR-1. Geochim Cosmochim Acta 73(4):962–976

Castro-Longoria E, Vilchis-Nestor AR, Avalos- Borja M (2011) Biosynthesis of silver, gold and bimetallic nanoparticles using the filamentous fungus *Neurospora crassa*. Colloids Surf B Biointerfaces 83(1):42–48

Cunningham DP, Lundie LL Jr (1993) Precipitation of cadmium by *Clostridium thermoaceticum*. Appl Environ Microbiol 59(1):7–14

Dada AO, Adekola FA, Dada FE, Adelani-Akande AT et al (2019) Silver nanoparticle synthesis by *Acalypha wilkesiana* extract: phytochemical screening, characterization, influence of operational parameters, and preliminary antibacterial testing. Heliyon 5:10

De Matteis V, Rizzello L, Ingrosso C, Liatsi-Douvitsa E, De Giorgi ML, De Matteis G, Rinaldi R (2019) Cultivar-dependent anticancer and antibacterial properties of silver nanoparticles synthesized using leaves of different *Olea europaea* trees. Nano 9:1544

Deepa K, Panda T (2014) Synthesis of gold nanoparticles from different cellular fractions of *Fusarium oxysporum*. J Nanosci Nanotechnol 14:345–3463

DeWindt W, Aelterman P, Verstraete W (2005) Bioreductive deposition of palladium (0) nanoparticles on *Shewanella oneidensis* with catalytic activity towards reductive dichlorination of polychlorinated biphenyls. Environ Microbiol 7(3):314–325

Du L, Jiang H, Liu X, Wang E (2007) Biosynthesis of gold nanoparticles assisted by Escherichia coli DH5α and its application on direct electrochemistry of hemoglobin. Electrochem Commun 9(5):1165–1170

Duran N, Marcato PD, De Souza GIH, Alves OL, Esposito E (2007) Antibacterial effect of silver nanoparticles produced by fungal process on textile fabrics and their effluent treatment. J Biomed Nanotechnol 3(2):203–208

El-Moslamy SH, Elkady MF, Rezk AH, Abdel-Fattah YR (2017) Applying Taguchi design and large-scale strategy for mycosynthesis of nano-silver from endophytic *Trichoderma harzianum* SYA.F4 and its application against phytopathogens. Sci Rep 7:45297

Fan TX, Chow S.K, Zhang D (2009) Biomorphic mineralization: from biology to materials. Progress in Materials Science 54(5): 542–659

Fadeel B, Garcia-Bennett AE (2010) Better safe than sorry: understanding the toxicological properties of inorganic nanoparticles manufactured for biomedical applications. Adv Drug Deliv Rev 62(3):362–374

Farkas B, Terranova U, de Leeuw NH (2020) Binding modes of carboxylic acids on cobalt nanoparticles. Phys Chem Chem Phys 22:985–996

Fayaz AM, Balaji K, Kalaichelvan PT, Venkatesan R (2009) Fungal based synthesis of silver nanoparticles—an effect of temperature on the size of particles. Colloids Surf B Biointerfaces 74(1):123–126

Fayaz AM, Balaji K, Girilal M, Yadav R, Kalaichelvan PT, Venketesan R (2010) Biogenic synthesis of silver nanoparticles and their synergistic effect with antibiotics: a study against gram-positive and gram-negative bacteria. Nanomed Nanotechnol Biol Medicine 6(1):e103–e109

Feynman RP (1960) There's plenty of room at the bottom. Eng Sci 23:22–36

Figueiredo EP, Ribeiro JM, Nishio EK, Scandorieiro S, Costa AF et al (2019) New approach for simvastatin as an antibacterial: synergistic effect with bio-synthesized silver nanoparticles against multidrug-resistant bacteria. Int J Nanomedicine 14:7975–7985

Gajbhiye M, Kesharwani J, Ingle A, Gade A, Rai M (2009) Fungus mediated synthesis of silver nanoparticles and their activity against pathogenic fungi in combination with fluconazole. Nanomedicine 5:382–386

Gericke M, Pinches A (2006) Biological synthesis of metal nanoparticles. Hydrometallurgy 83(1–4):132–140

Gurunathan S, Kalishwaralal K, Vaidyanathan R et al (2009) Biosynthesis, purification and characterization of silver nanoparticles using *Escherichia coli*. Colloids Surf B Biointerfaces 74(1):328–335

Hamouda RA, Hussein MH, Abo-elmagd RA, Bawazir SS (2019) Synthesis and biological characterization of silver nanoparticles derived from the cyanobacterium Oscillatoria limnetica. Sci Rep 9:1–17

Hassan SE, Fouda A, Radwan AA, Salem SS, Barghoth MG, Awad MA, Abdo AM, El-Gamal MS (2019) Endophytic actinomycetes *Streptomyces* spp mediated biosynthesis of copper oxide nanoparticles as a promising tool for biotechnological applications. J Biol Inorg Chem 24:377–393

Hayat MA (1989) Colloidal gold: principles, methods, and applications. Academic Press, San Diego

He S, Guo Z, Zhang Y, Zhang S, Wang J, Gu N (2007) Biosynthesis of gold nanoparticles using the bacteria Rhodopseudomonas capsulate. Mater Lett 61(18):3984–3987

He L, Liu Y, Mustapha A, Lin M (2010) Antifungal activity of zinc oxide nanoparticles against *Botrytis cinerea* and *Penicillium expansum*. Microbiol Res 166:207–215

Hossain A, Hong X, Ibrahim E, Li B, Sun G, Meng Y, Wang Y, An Q (2019) Green synthesis of silver nanoparticles with culture supernatant of a bacterium *Pseudomonas rhodesiae* and their antibacterial activity against soft rot pathogen *Dickeya dadantii*. Molecules 24:2303

Husseiny MI, El-Aziz MA, Badr Y, Mahmoud MA (2007) Biosynthesis of gold nanoparticles using *Pseudomonas aeruginosa*. Spectrochim Acta A 67(3–4):1003–1006

Ibrahim E, Fouad H, Zhang M, Zhang Y, Qiu W, Yan C, Li B, Mo J, Chen J (2019) Biosynthesis of silver nanoparticles using endophytic bacteria and their role in inhibition of rice pathogenic bacteria and plant growth promotion. RSC Adv 9:29293–29299

Ibrahim E, Zhang M, Zhang Y, Hossain A, Qiu W, Chen Y, Wang Y, Wu W, Sun G, Li B (2020) Green-synthesization of silver nanoparticles using endophytic Bacteria isolated from garlic and its antifungal activity against wheat Fusarium head blight pathogen *Fusarium graminearum*. Nanomaterials (Basel) 10:219

Jain N, Bhargava A, Majumdar S, Tarafdar JC, Panwar J (2011) Extracellular biosynthesis and characterization of silver nanoparticles using Aspergillus flavus NJP08: a mechanism perspective. Nanoscale 3(2):635–641

Jha AK, Prasad K (2010) Ferroelectric $BaTiO_3$ nanoparticles: biosynthesis and characterization. Colloids Surf B Biointerfaces 75(1):330–334

Jha AK, Prasad K, Prasad K (2009) A green low-cost biosynthesis of Sb_2O_3 nanoparticles. Biochem Eng J 43(3):303–306

Jha AK, Prasad K (2010a) Ferroelectric $BaTiO_3$ nanoparticles: biosynthesis and characterization. Colloids Surf. B Biointerfaces 75(1): 330–334

Jha AK, Prasad K (2010b) Synthesis of $BaTiO_3$ nanoparticles: A new sustainable green approach. Integrated Ferroelectrics 117(1): 49–54

Juibari MM, Abbasalizadeh S, Jouzani GS, Noruzi M (2011) Intensified biosynthesis of silver nanoparticles using a native extremophilic *Ureibacillus thermosphaericus* strain. Mater Lett 65(6):1014–1017

Kalimuthu K, Suresh Babu R, Venkataraman D, Bilal M, Gurunathan S (2008) Biosynthesis of silver nanocrystals by *Bacillus licheniformis*. Colloids Surf B Biointerfaces 65(1):150–153

Kalishwaralal K, Deepak V, Ramkumarpandian S, Nellaiah H, Sangiliyandi G (2008) Extracellular biosynthesis of silver nanoparticles by the culture supernatant of *Bacillus licheniformis*. Mater Lett 62(29):4411–4413

Kalishwaralal K, Deepak V, Ramkumarpandian S et al (2010) Biosynthesis of silver and gold nanoparticles using *Brevibacterium casei*. Colloids Surf B Biointerfaces 77(2):257–262

Kashefi K, Lovley DR (2000) Reduction of Fe(III), Mn(IV), and toxic metals at 100°C by *Pyrobaculum islandicum*. Appl Environ Microbiol 66(3):1050–1056

Kasprowicz MJ, Kozio M, Gorczyca A (2010) The effect of silver nanoparticles on phytopathogenic spores of *Fusarium culmorum*. Can J Microbiol 56:247–253

Kaur P, Thakur R, Duhan JS, Chaudhury A (2018) Management of wilt disease of chickpea in vivo by silver nanoparticles biosynthesized by rhizospheric microflora of chickpea (Cicer arietinum). J Chem Technol Biotechnol 93:3233–3243

Khan JN, Jameel N (2016) Antifungal activity of silver nanoparticles produced from fungus, *Penicillium fellutanum* at different pH. J Microb Biochem Technol 8:440

Kim KJ, Sung WS, Moon SK, Choi JS, Kim JG, Lee DG (2008) Antifungal effect of silver nanoparticles on dermatophytes. J Microbiol Biotechnol 18:1482–1484

Klaus T, Joerger R, Olsson E, Granqvist CG (1999) Silverbased crystalline nanoparticles, microbially fabricated. Proc Natl Acad Sci USA 96(24):13611–13614

Konishi Y, Ohno K, Saitoh N et al (2007a) Bioreductive deposition of platinum nanoparticles on the bacterium Shewanella algae. J Biotechnol 128(3):648–653

Konishi Y, Tsukiyama T, Tachimi T, Saitoh N, Nomura T, Nagamine S (2007b) Microbial deposition of gold nanoparticles by the metal-reducing bacterium Shewanella algae. Electrochim Acta 53(1):186–192

Kulkarni SA, Ghormade V, Kulkarni G, Kapoor M, Chavan SB, Rajendran A et al (2008) Comparison of Metarhizium isolates for biocontrol of Helicoverpa armigera (Lepidoptera: Noctuidae) in chickpea. Biocontrol Sci Tech 18:809–828

Kumar SA, Ansary AA, Abroad A, Khan MI (2007) Extracellular biosynthesis of CdSe quantum dots by the fungus, *Fusarium oxysporum*. J Biomed Nanotechnol 3(2):190–194

Lee JH, Han J, Choi H, Hur HG (2007) Effects of temperature and dissolved oxygen on Se(IV) removal and Se(0) precipitation by Shewanella sp. HN-41. Chemosphere 68(10):1898–1905

Lefevre CT, Abreu F, Lins U, Bazylinski DA (2010a) Nonmagnetotactic multicellular prokaryotes from low-saline, nonmarine aquatic environments and their unusual negative phototactic behavior. Appl Environ Microbiol 76(10):3220–3227

Lefevre CT, Abreu F, Schmidt ML et al (2010b) Moderately thermophilic magnetotactic bacteria from hot springs in Nevada. Appl Environ Microbiol 76(11):3740–3743

Lengke MF, Fleet ME, Southam G (2006a) Morphology of gold nanoparticles synthesized by filamentous cyanobacteria from gold(I)-thiosulfate and gold(III)-chloride complexes. Langmuir 22(6):2780–2787

Lengke MF, Ravel B, Fleet ME, Wanger G, Gordon RA, Southam G (2006b) Mechanisms of gold bioaccumulation by filamentous cyanobacteria from gold(III)-chloride complex. Environ Sci Technol 40(20):6304–6309

Li W, Zhou L, Yu P, Zhu M (2007) A Magnetospirillum strain WM-1 from a freshwater sediment with intracellular magnetosomes. World J Microbiol Biotechnol 23(10):1489–1492

Liu J, Qiao SZ, Hu QH, Lu GQ (2011) Magnetic nanocomposites with mesoporous structures: synthesis and applications. Small 7(4):425–443

Liu Y, Page Z, Ferdous S, Liu F, Kim P, Emrick T, Russell T (2015) Dual functional zwitterionic fullerene interlayer for efficient inverted polymer solar cells. Adv Energy Mater 5(14):1500405

Lloyd JR, Yong P, Macaskie LE (1998) Enzymatic recovery of elemental palladium by using sulfate-reducing bacteria. Appl Environ Microbiol 64(11):4607–4609

Lover T, Henderson W, Bowmaker GA, Seakins JM, Cooney RP (1997) Functionalization and capping of a cds nanocluster: a study of ligand exchange by electrospray mass spectrometry. Chem Mater 9(8):1878–1886

Lv Q, Zhang B, Xing X, Zhao Y, Cai R, Wang W, Gu Q (2018) Biosynthesis of copper nanoparticles using Shewanella loihica PV-4 with antibacterial activity: novel approach and mechanisms investigation. J Hazard Mater 347:141–149

Ma L, Su W, Liu JX, Zeng XX, Huang Z., Li W, Liu ZC, Tang JX (2017) Optimization for extracellular biosynthesis of silver nanoparticles by Penicillium aculeatum Su1 and their antimicrobial activity and cytotoxic effect compared with silver ions. Materials Science and Engineering C: Mat Biol Appl 77:963–971

Majeed S, Abdullah MS, Dash GK, Ansari MT, Nanda A (2016) Biochemical synthesis of silver nanoparticles using filamentous fungi *Penicillium decumbens* (MTCC-2494) and its efficacy against A-549 lung cancer cell line. Chin J Nat Med 14:615

Maliszewska I, Juraszek A, Bielska K (2013) Green synthesis and characterization of silver nanoparticles using ascomycota fungi *Penicillium nalgiovense* AJ12. J Clust Sci 25:989–1004

Maliszewska I, Lisiak B, Popko K, Matczyszyn K (2017) Enhancement of the efficacy of photodynamic inactivation of candida albicans with the use of biogenic gold nanoparticles. J Photochem Photobiol 93:1081

Manceau A, Nagy K, Marcus M, Lanson M, Geoffroy N, Jacquet T et al (2008) Formation of metallic copper nanoparticles at the soil-root Interface. Environ Sci Technol 42:1766–1772

Mishra AN, Bhadauria S, Gaur MS, Pasricha R (2010) Extracellular microbial synthesis of gold nanoparticles using fungus *Hormoconis resinae*. J Microbiol 62:45–48

Mishra S, Singh BR, Singh A, Keswani C, Naqvi AH, Singh HB (2014) Biofabricated silver nanoparticles act as a strong fungicide against *Bipolaris sorokiniana* causing spot blotch disease in wheat. PLoS One 9(5):e97881

Nair B, Pradeep T (2002) Coalescence of nanoclusters and formation of submicron crystallites assisted by lactobacillus strains. Cryst Growth Des 2(4):293–298

Nayak RR, Pradhan N, Behera D, Pradhan KM, Mishra S, Sukla LB, Mishra BK (2010) Green synthesis of silver nanoparticle by *Penicillium purpurogenum* NPMF, the process and optimization. J Nanopart Res 13:3129–3137

Ngan CL, Basri M, Tripathy M, Karjiban RA, Abdul-Malek E (2015) Skin intervention of fullerene-integrated nanoemulsion in structural and collagen regeneration against skin aging. Eur J Pharm Sci 70:22–28

Nies DH (1999) Microbial heavy-metal resistance. Appl Microbiol Biotechnol 51(6):730–750

Ocsoy I, Paret ML, Ocsoy MA, Kunwar S, Chen T, You M, Tan W (2013) Nanotechnology in plant disease management: DNA directed silver nanoparticles on graphene oxide as an antibacterial against *Xanthomonas perforans*. ACS Nano 7:8972–8980. https://doi.org/10.1021/nn4034794

Oh SD, Lee S, Choi SH, Lee IS, Lee YM, Chun JH, Park HJ (2006) Synthesis of ag and Ag-SiO2 nanoparticles by y-irradiation and their antibacterial and antifungal efficiency against *Salmonella enterica* serovar Typhimurium and *Botrytis cinerea*. Colloids Surf A 275:228–233

Osman ME, Eid MM, Khattab OH, Abd-El All SM, El-Hallouty SM, Mahmoud DA (2015) Spectroscopic characterization of the effect of gamma radiation on the physical parameters of biosynthesized silver/chitosan nano-particles and their antimicrobial activity. J Chem Biol Phys Sci 5:2643

Oves M, Rauf MA, Hussain A, Qari AH, Parwaz Khan AA et al (2019) Antibacterial silver nanomaterial synthesis from mesoflavibacter zeaxanthinifaciens and targeting biofilm formation. Front Pharmacol 10:80

Palmqvist NGM, Bejai S, Meijer J, Seisenbaeva GA, Kessler VG (2015) Nano titania aided clustering and adhesion of beneficial bacteria to plant roots to enhance crop growth and stress management. Sci Rep 5:10146

Panacek A, Kvitek L, Prucek R et al (2006) Silver colloid nanoparticles: synthesis, characterization, and their antibacterial activity. J Phys Chem B 110(33):16248–16253

Pandian SRK, Deepak V, Kalishwaralal K, Muniyandi J, Rameshkumar N, Gurunathan S (2009) Synthesis of PHB nanoparticles from optimized medium utilizing dairy industrial waste using Brevibacterium casei SRKP2: a green chemistry approach. Colloids Surf B Biointerfaces 74(1):266–273

Parak WJ, Boudreau R, Le Gros M et al (2002) Cell motility and metastatic potential studies based on quantum dot imaging of phagokinetic tracks. Adv Mater 14(12):882–885

Paulkumar K, Gnanajobitha G, Vanaja M, Rajeshkumar S, Malarkodi C, Pandian K, Annadurai G (2014) Piper nigrum leaf and stem assisted green synthesis of silver nanoparticles and evaluation of its antibacterial activity against agricultural plant pathogens. Scientific World J. https://doi.org/10.1155/2014/829894

Perez-Gonzalez T, Jimenez-Lopez C, Neal AL et al (2010) Magnetite biomineralization induced by *Shewanella oneidensis*. Geochim Cosmochim Acta 74(3):967–979

Peteu SF, Oancea F, Sicuia OA, Constantinescu F, Dinu S (2010) Responsive polymers for crop protection. Polymers 2:229–251

Piao MJ, Kang KA, Lee IK et al (2011) Silver nanoparticles induce oxidative cell damage in human liver cells through inhibition of reduced glutathione and induction of mitochondria-involved apoptosis. Toxicol Lett 201(1):92–100

Ponmurugan P, Manjukarunambika K, Elango V, Gnanamangai BM (2016) Antifungal activity of biosynthesized copper nanoparticles evaluated against red root-rot disease in tea plants. J Exp Nanosci 11:1019–1031

Prasad R, Bhattacharyya A, Nguyen QD (2017a) Nanotechnology in sustainable agriculture: recent developments, challenges, and perspectives. Front Microbiol 8:1014. https://doi.org/10.3389/fmicb.2017.01014

Prasad R, Gupta N, Kumar M, Kumar V, Wang S, Abd-Elsalam KA (2017b) Nanomaterials act as plant defense mechanism. In: Prasad R, Kumar V, Kumar M (eds) Nanotechnology. Springer, Singapore, pp 253–269

Ragaei M, Sabry KH (2014) Nanotechnology for insect pest control. Int J Sci Environ Technol 3(2):528–545

Rautaray D, Sanyal A, Adyanthaya SD, Ahmad A, Sastry M (2004) Biological synthesis of strontium carbonate crystals using the fungus *Fusarium oxysporum*. Langmuir 20(16):6827–6833

Reith F, Lengke MF, Falconer D, Craw D, Southam G (2007) The geomicrobiology of gold. ISME J 1(7):567–584

Sánchez-López E, Gomes D, Esteruelas G, Bonilla L et al (2020) Metal-based nanoparticles as antimicrobial agents: an overview. Nanomaterials 10(292):1–39. https://doi.org/10.3390/nano10020292

Sanghi R, Verma P (2009) A facile green extracellular biosynthesis of CDs nanoparticles by immobilized fungus. Chem Eng J 155(3):886–891

Sanyal A, Rautaray D, Bansal V, Ahmad A, Sastry M (2005) Heavy-metal remediation by a fungus as a means of production of lead and cadmium carbonate crystals. Langmuir 21(16):7220–7224

Sarsar V, Selwal MK, Selwal KK (2015) Biofabrication, characterization and antibacterial efficacy of extracellular silver nanoparticles using novel fungal strain of *Penicillium atramentosum* KMJ. Saudi Chem Soc 19:682

Sawle BD, Salimath B, Deshpande R, Bedre MD, Prabhakar BK, Venkataraman A (2008) Biosynthesis and stabilization of Au and Au-Ag alloy nanoparticles by fungus, *Fusarium semitectum*. Sci Technol Adv Mater 9(3):1–6. Article ID 035012

Saxena J, Sharma P, Singh A (2017) Biomimetic synthesis of silver nanoparticles from *Penicillium chrysogenum* strain FGCC/BLS1 by optimizing physico-cultural conditions and assessment of their antimicrobial potential. IET Nanobiotechnol 11:576

Senapati S, Ahmad A, Khan MI, Sastry M, Kumar R (2005) Extracellular biosynthesis of bimetallic Au-Ag alloy nanoparticles. Small 1(5):517–520

Shaligram NS, Bule M, Bhambure R, Singhal RS, Singh SK, Szakacs G, Pandey A (2009) Biosynthesis of silver nanoparticles using aqueous extract from the compactin producing fungal strain. Process Biochem 44:939–943

Sheet S, Sathishkumar Y, Sivakumar AS, Shim KS, Lee YS (2017) Low-shear-modeled microgravity-grown *Penicillium chrysogenum*-mediated biosynthesis of silver nanoparticles with enhanced antimicrobial activity and its anticancer effect in human liver cancer and fibroblast cells. Bioprocess Biosyst Eng 40:1529–1542

Singaravelu G, Arockiamary JS, Kumar VG, Govindaraju K (2007) A novel extracellular synthesis of monodisperse gold nanoparticles using marine alga, Sargassum wightii Greville. Colloids Surf B 57(1):97–101

Singh D, Rathod V, Ninganagouda S, Hiremath J, Singh AK, Mathew J (2014) Optimization and characterization of silver nanoparticle by endophytic fungi Penicillium sp. isolated from *Curcuma longa* (turmeric) and application studies against MDR *E. coli* and *S. aureus*. Bioinorg Chem Appl 2014:408021. https://doi.org/10.1155/2014/408021

Sinha A, Khare SK (2011) Mercury bioaccumulation and simultaneous nanoparticle synthesis by Enterobacter sp. cells. Bioresour Technol 102:4281–4284

Smith K, Evans DA, El-Hiti GA (2008) Role of modern chemistry in sustainable arable crop protection. Phil Trans R Soc B 363:623–637

Sneha K, Sathishkumar M, Mao J, Kwak IS, Yun YS (2010) *Corynebacterium glutamicum*-mediated crystallization of silver ions through sorption and reduction processes. Chem Eng J 162(3):989–996

Solanki BD, Ramani HR, Garaniya NH, Parmar DV (2016) Biosynthesis of silver nanoparticles using fungus Penicillium Brevicompactum and evaluation of their anti-bacterial activity against some human pathogens. Res J Biotechnol 11:44

Soni N, Prakash S (2011) Factors affecting the geometry of silver nanoparticles synthesis in *Chrysosporium tropicum* and *Fusarium oxysporum*. Am J Nanotechnol 2(1):112–121

Southam G, Beveridge TJ (1996) The occurrence of sulfur and phosphorus within bacterially derived crystalline and pseudocrystalline octahedral gold formed in vitro. Geochim Cosmochim Acta 60(22):4369–4376

Spring S, Schleifer KH (1995) Diversity of magnetotactic bacteria. Syst Appl Microbiol 18(2):147–153

Suresh AK, Pelletier DA, Wang W et al (2011) Biofabrication of discrete spherical gold nanoparticles using the metal reducing bacterium *Shewanella oneidensis*. Acta Biomater 7(5):2148–2152

Suriyaprabha R, Karunakaran G, Kavitha K, Yuvakkumar R, Rajendran V, Kannan N (2014) Application of silica nanoparticles in maize to enhance fungal resistance. IET Nanobiotechnol 8(3):133–137

Sweeney RY, Mao C, Gao X et al (2004) Bacterial biosynthesis of cadmium sulfide nanocrystals. Chem Biol 11(11):1553–1559

Tanase C, Berta L, Coman NA, Ros ca I, Man A, Toma F, Mocan A, Nicolescu A, Jakab-Farkas L, Biró D et al (2019) Antibacterial and antioxidant potential of silver nanoparticles biosynthesized using the spruce bark extract. Nanomaterials 9:11

Tang H, Yan M, Zhang H, Xia M, Yang D (2005) Preparation and characterization of water-soluble CdS nanocrystals by surface modification of ethylene diamine. Mater Lett 59(8–9):1024–1027

Thornhill RH, Burgess JG, Matsunaga T (1995) PCR for direct detection of indigenous uncultured magnetic cocci in sediment and phylogenetic analysis of amplified 16S ribosomal DNA. Appl Environ Microbiol 61(2):495–500

Tian F, Prina-Mello A, Estrada G et al (2008) A novel assay for the quantification of internalized nanoparticles in macrophages. Nanotoxicology 2(4):232–242

Vazquez-Muñoz R, Meza-Villezcas AP, Fournier GJ, Soria-Castro E et al (2019) Enhancement of antibiotics antimicrobial activity due to the silver nanoparticles impact on the cell membrane. PLoS One 14(11):e0224904. https://doi.org/10.1371/journal.pone.0224904

Verma S, Abirami S, Mahalakshmi V (2013) Anticancer and antibacterial activity of silver nanoparticles biosynthesized by Penicillium spp. and its synergistic effect with antibiotic. Microbiol Biotechnol Res 3:54

Vijayakumar PS, Abhilash OU, Khan BM, Prasad BLV (2010) Nanogold-loaded sharp-edged carbon bullets as plant-gene carriers. Adv Funct Mater 20:2416–2423

Watson JHP, Ellwood DC, Soper AK, Charnock J (1999) Nanosized strongly-magnetic bacterially-produced iron sulfide materials. J Magn Magn Mater 203(1–3):69–72

Yan S, He W, Sun C et al (2009) The biomimetic synthesis of zinc phosphate nanoparticles. Dyes Pigments 80(2):254–258

Yang H, Santra S, Holloway PH (2005) Syntheses and applications of Mn-doped II-VI semiconductor nanocrystals. J Nanosci Nanotechnol 5(9):1364–1375

Yao KS, Li SJ, Tzeng KC, Cheng TC, Chang CY, Chiu CY, Liao CY, Hsu JJ, Lin ZP (2009) Fluorescence silica nanoprobe as a biomarker for rapid detection of plant pathogens. Adv Mater Res 79:513–516

Yassin MA, El-Samawaty A, Dawoud TM, Abd-Elkader OH, Al Maary KS, Hatamleh AA, Elgorban AM (2017) Characterization and anti-*Aspergillus flavus* impact of nanoparticles synthesized by *Penicillium citrinum*. Saudi J Biol Sci 24:1243

Zhang X, Yan S, Tyagi RD, Surampalli RY (2011) Synthesis of nanoparticles by microorganisms and their application in enhancing microbiological reaction rates. Chemosphere 82(4):489–494

Zhang P, Zhang R, Fang X, Song T, Cai X, Liu H, Du S (2016) Toxic effects of graphene on the growth and nutritional levels of wheat (*Triticum aestivum* L.): short- and long-term exposure studies. J Hazard Mater 317:543–551

Zhou W, He W, Zhang X et al (2009a) Biosynthesis of iron phosphate nanopowders. Powder Technol 194(1–2):106–108

Zhou W, He W, Zhong S et al (2009b) Biosynthesis and magnetic properties of mesoporous Fe_3O_4 composites. J Magn Magn Mater 321(8):1025–1028

Zhu K, Pan H, Li J et al (2010) Isolation and characterization of a marine magnetotactic spirillum axenic culture QH-2 from an intertidal zone of the China Sea. Res Microbiol 161(4):276–283

Zheng D, Hu C, Gan T, Dang X, Hu, S (2010) Preparation and application of a novel vanillin sensor based on biosynthesis of Au-Ag alloy nanoparticles. Sensors and Actuators B: Chemical 148: 247–252

Zong X, Wang W, Wei H, Wang J, Chen X, Xu L, Zhu D, Tan Y, Liu Q (2014) Rapid detection of Prunus necrotic ringspot virus using magnetic nanoparticle-assisted reverse transcription loop-mediated isothermal amplification. J Virol Methods 208:85–89

Chapter 13
Nano-enabled Approaches for the Suitable Delivery of Fertilizer and Pesticide for Plant Growth

Santwana Padhi and Anindita Behera

Contents

1 Introduction

Food remains as the mainstay for all living creatures who are predominantly dependent on agriculture. In today's scenario, agriculture faces the greatest challenge with pests, changes in the climate and decrease in the utilization of essential nutrients

S. Padhi
KIIT Technology Business Incubator, KIIT Deemed to be University, Bhubaneswar, Odisha, India

A. Behera (✉)
School of Pharmaceutical Sciences, Siksha 'O' Anusandhan Deemed to be University, Bhubaneswar, Odisha, India

© The Author(s), under exclusive license to Springer Nature Switzerland AG 2021

H. I. Mohamed et al. (eds.), *Plant Growth-Promoting Microbes for Sustainable Biotic and Abiotic Stress Management*,
https://doi.org/10.1007/978-3-030-66587-6_13

(Kagan 2016). Globally about 22,000 different types of plant pathogens, weeds, insects and mites affect farming (Zhang 2018). The challenge in plant disease management is the timely identification and availability of limited options of management (Adisa et al. 2019). Hence, conventional methods are adopted to manage the diseases by developing host-resistant crop variants (Servin et al. 2015). All crop plants do not have inherent resistant genes against pathogenic diseases, so their requirement is more significant as compared to genetically modified crops. Micronutrients like copper (Cu), manganese (Mn) and zinc (Zn) are responsible for the initiation of enzyme activities and generation of biomolecules which participate in defence of plants. Therefore, the search for a more sustainable alternative remains as the most challenging area in agriculture for improvement in crop production and management of plants from diseases and pest attack (Adisa et al. 2019). Engineered nanomaterial (ENMs) has gained much interest in the management of plant diseases and enhancement of soil fertility. Much research efforts have been conducted and reported for ENMs with potential for improvement for growth of healthy plants and crops with an increase in efficiency of nutrient use and defence for suppression of diseases in plants (Dimkpa and Bindraban 2017; Elmer and White 2018).

Rodrigues et al. worked on various nanotechnological techniques for improving sustainable agro-food systems. The study searched for improved techniques for delivery of nanofertilizers and nanopesticides ensuring controlled release and targeted delivery to control the pathogens and pests safeguarding food safety and security. The advances in detection of pathogen and toxins were also reported (Rodrigues et al. 2017). The efficiency of crop production may be increased by incorporating the engineered nanomaterial into traditional fertilizers and pesticides. The ENMs may be included in bulk or as the sole active ingredient (Dimkpa and Bindraban 2017; Prasad et al. 2017).

The essential nutrients for plant growth may be supplied in a nanoform for improved release with increased efficiency. This contributes to greater enhancement in the growth of plant resisting the environmental stress conditions. Fertilizers are most essential for the development and growth of plants. The use of nanofertilizers has been proved to be more advantageous as compared to conventional fertilizers. Nanofertilizers are the ENMs which provide essential nutrients to crops. They also increase the performance, accessibility or use of conventional fertilizers incorporated with ENMs (Liu and Lal 2015). The specific property of nanofertilizers, i.e. the small particle dimension of nanoparticles (NPs), has more surface area which augments absorption and utilization of nanofertilizers. The use of nanofertilizers prevents the loss of fertilizer by the processes like leaching and emissions and enduring assimilation by soil microflora (Liu et al. 2006; DeRosa et al. 2010). To maintain the soil fertility for extended period, nanofertilizers help by releasing at slower rates and decreasing the extent of toxicity due to the application of conventional fertilizers in large quantity (Suman et al. 2010).

As compared to nanofertilizers, nanopesticides help in the prevention or suppression of severity of infections to the plants by fungi, bacteria or oomycete diseases. Due to the nano-property of nanopesticides, they are found to be more potent, lower dose is required for application, and they maintain the productivity as compared to the conventional pesticides of chemical resemblance (Adisa et al. 2019). Nanopesticides reduce the frequency of application of pesticide, thereby reducing the wash-off and environmental pollution with reduced economic input cost by the farmer (Chhipa 2017).

This chapter emphasizes the mechanism of different nano-enabled fertilizers and pesticides which improves growth, crop production, crop quality, seed germination, seedling vigour, initiation of a root and photosynthesis to flowering under different biotic and abiotic stress conditions. The chapter focuses mainly on the biotic and abiotic stress conditions that affect the growth or development of plants and weaken the plant defence mechanism. The chapter elaborates the different types and the mechanism involved in the action of nanofertilizers and nanopesticides.

2 Conditions Affecting the Plant Growth

As plants are quiescent, they have to face all environmental changes like drought, flood, salinity in the soil, extreme variation of temperature, ultraviolet radiations, etc. and attack of pathogens, commonly categorized as abiotic and biotic stress conditions (Fig. 13.1). These abiotic stress conditions develop different reactive oxygen species (ROS) like hydrogen peroxide (H_2O_2), hydroxyl radical (OH^-), superoxides ($O_2^{\bullet-}$), singlet oxygen species ($^1O_2$) and hydroperoxy radical ($HO_2^{\bullet}$). These ROS accumulate in the plants and subsequently cause destructions in membrane biomolecules (Foyer and Noctor 2000). ROS initiates the noxious injury to cells and genes (Shen et al. 2010a, b; Yadav et al. 2014), thereby interfering the plant growth (Begum and Fugetsu 2012). Other than these deteriorating effects, ROS also facilitates different defence systems by activation of a cascade of cell signalling and

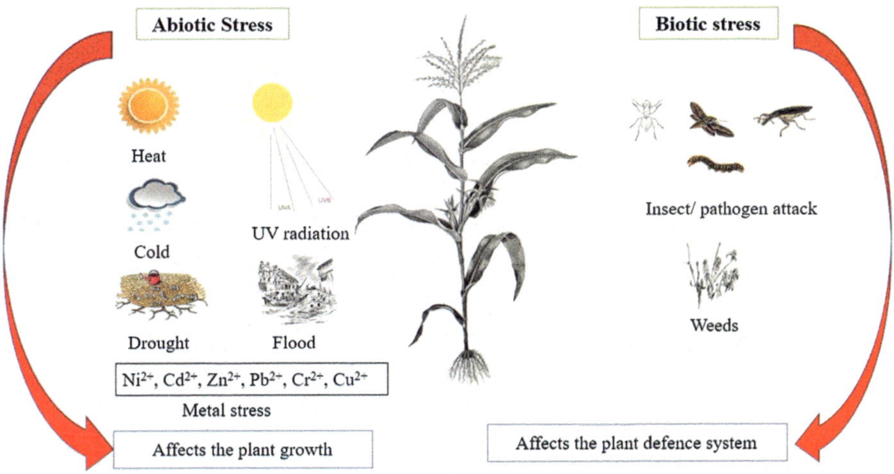

Fig. 13.1 Different abiotic and biotic stress conditions affecting the growth and defence system of plant

encouraging or suppression of many gene expressions (Hancock et al. 2001). Floras possess enzymatic antioxidants to counteract the oxidative stress like superoxide dismutase (SOD), catalase (CAT), ascorbate peroxidase (APX), glutathione reductase (GR) and some non-enzymatic antioxidants like glutathione and ascorbate that help in scavenging the ROS continuously. Plants fight the stress through osmosis by increasing the deposition of different organic osmolytes which helps in maintaining the normal hydration level. Similarly, hypoxia (deficiency of oxygen supply) results in a decrease in energy level, so the plants alter their metabolic processes to maintain the energy and replace the metabolism of carbohydrate with fermentation (Banti et al. 2013). Stress due to metal ions can be overcome by the plants, by synthesizing metal-ligand chelates, organic acids and polyphosphates. These metal chelates are restricted and confiscated in apoplasm or symplasm. However, the effect of ENMs depends on the plant varieties and the nature of nanomaterial used (Lin and Xing 2007). Most of the nanomaterials are responsible for alteration in gene expression as a response to abiotic and biotic stress conditions. They also affect the biosynthetic processes of the cell, cell organization, electron transport and energy pathways (Landa et al. 2012; Kaveh et al. 2013; Aken 2015).

Of all these stress conditions, most affected physiology of plants is photosynthesis. As photosynthesis is the only physiological process to maintain the strength of a plant, ENMs protect the plant from different stress conditions by improving the rate of photosynthesis, the conductance of stomata, the rate of transpiration, the efficiency of water use and the content of chlorophyll and the proline and carbonic anhydrase activity (Haghighi and Pessarakli 2013; Siddiqui et al. 2014). Advantages of ENMs are that its use in low concentrations is very potent in improving different abiotic traumas and improved growth and development of plant (Mahajan et al. 2011; Amira et al. 2015).

2.1 *Abiotic Stress*

Abiotic stress conditions like drought, flood, salinity, extreme temperature and metal stress affect and cause a huge loss in crop production globally by a reduction in crop quantity and quality (Wu and Ma 2015). These stress conditions alone or in combination affect the morphology and physiology of the plant negatively, and also the changes in biochemical and molecular level decrease the crop yield (Rao et al. 2016):

1. *Abiotic stress due to drought*: Stress due to drought is one type of anthropogenic climate change which restricts the production and distribution of crop (Khan et al. 2017).
2. *Abiotic stress due to soil salinity*: Salinity in the soil is the result of excess deposition of sodium chloride which causes stress to the crop plants due to osmosis and ions. In osmotic stress conditions, the uptake of water and nutrients by the plants is reduced. However, in ionic stress condition, there is an excess deposi-

tion of sodium ion (Na^+) in the cytosol of the cells, causing the lowering of potassium/sodium (K^+/Na^+) ratio (Khan et al. 2012). This imbalance results in excess production of ROS which causes damage to biomolecules and outflow of electrolytes and distresses different physiological processes in cytosol (Khan et al. 2010; Sharma et al. 2012; Ismail et al. 2014). The presence of excess sodium ion (Na^+) and chloride ion (Cl^-) induces salt stress and damage to the plant. So, to defend it, the plants reduce the uptake of ions from soil or deposition in the vacuoles (Khan et al. 2017).

3. *Abiotic stress due to temperature*: Stress due to temperature influences the sustainable growth and yield of plants. Temperature stress can be considered by two headings, i.e. high-temperature (HT) stress and low-temperature (LT) stress. Low-temperature stress may also be termed as cold stress or chilling stress or freezing stress. High-temperature stress causes damage to the plant irreversibly affecting the growth and development (Wahid 2007). Cold stress is generally referred to as the temperature from 0 to 15 °C in which the plant is injured without deposition of ice crystals in the tissues of plants, whereas freezing stress is generally temperature below 0 °C, and ice crystals are deposited into the tissues of plant (Hasanuzzaman et al. 2013). Cold stress causes loss of flexibility in the cell membrane and leakage of solutes. Plants under cold stress have slow development and seed germination and decreased production (Suzuki et al. 2008). Photosynthesis is the most affected physiological process of plant under cold stress as it decreases chlorophyll (Chl) content, carbon dioxide assimilation, transpiration rate and deterioration of enzymes (Yordanova and Popova 2007; Liu et al. 2012). Similarly, HT stress facilitates more generation of ROS, and the oxidative stress is responsible for the degeneration of lipids of membrane and biomolecules and leakage of electrolytes (Moller et al. 2007; Savicka and Skute 2010; Karuppanapandian et al. 2011). HT stress reduces the content of Chl and rate of photosynthesis (Prasad et al. 2011).
4. *Abiotic stress due to metals:* Metal stress (MS) is one of the main abiotic stresses inhibiting the plant growth and causing phytotoxicity (Chibuike and Obiora 2014). Metal stress causes metal toxicity-associated inhibition of plant growth by suppressing the activities of vital enzymes and obstruction in the uptake of essential nutrients causing deficiency symptoms (Capuana 2011). Metal stress facilitates excessive production of ROS and the oxidative stress along with the damage of cell and biomolecules of a cell (Rascio and Navari-Izzo 2011; Sharma et al. 2012).
5. *Abiotic stress due to ultraviolet-B radiation*: Ultraviolet-B (UV-B) radiation of wavelength 280–315 nm causes an increase in ROS level in the plant cells (Mackerness et al. 2001) which causes damages to the DNA and the structure of chloroplast and interferes with numerous cellular physiological developments comprising photosynthesis (Chen et al. 2011; Hideg et al. 2013). The enzyme and non-enzyme antioxidant system of plants deposit phenolic compounds (glycerol, inositols, sorbitols, etc.) which absorb harmful UV radiations (Shen et al. 2010a, b).
6. *Abiotic stress due to flood*: Flood causes a deficiency of oxygen to the plant due to water-clogging. Oxygen diffuses 10^4-fold slower in the aqueous medium than air (Armstrong and Drew 2002). Hypoxia or deficiency of oxygen causes a

decrease in energy level, inhibits respiration and upregulates the expression of genes responsible for ethylene synthesis (Komatsu et al. 2009). Stress due to flood hampers the seed germination, vegetative and reproductive growth along with root growth and hypocotyl pigmentation (Hou and Thseng 1991; Linkemer et al. 1998; Visser et al. 1997; Komatsu et al. 2012).

7. *Abiotic stress after harvesting and storage*: Plants also suffer abiotic stress due to transport to long-distance places and various post-harvest storage conditions. During transport and storage, darkness is also responsible for post-harvest stress. Dark stress declines chlorophyll content, rate of photosynthesis and activities of enzymes against oxidative stress and escalates production of ethylene, generation of ROS and damage of cell membrane (Prochazkova and Wilhelmova 2007). These changes in cellular contents reduce the shelf life and marketable price of the plant and plant products, whereas in horticultural harvest, these post-harvest stresses cause increased dehydration, oxidation, respiration and lipid peroxidation, so this affects the net mass and nutritious value of the product (Ouzounidou and Gaitis 2011).

2.2 Biotic Stress

Apart from abiotic stress, biotic stress is caused by pathogens or pests like bacteria, fungi, viruses, insects, arachnids and weeds. The causative pathogens or pests divert the uptake of nutrients by the host plant leading to the death of the plant due to deprivation of nutrition. Biotic stress leads to both pre- and post-harvest losses. The host plants can counteract the biotic stress by using their inherent genetic system or the genetic code present. In the presence of these biotic stress conditions, the resistant genes present in the plant genome are encoded by multiple orders (Gull et al. 2019). Among the causative pathogens, fungal infections are the most than bacterial infections. Very few viruses cause biotic stress conditions. Microbes lead to plant droop, spots on the leaves, rotting of the roots and damage to the seeds, whereas insects lead to physical damage to the plants and act like carriers or vectors of the viruses or bacteria, spreading the disease from infected plant to a healthy plant. Weeds are unwanted plants which interfere with the growth of the plants by competing for space and nutrients. Weeds grow very faster as compared to the desired crop, so they dominate the growth and development of the required crop (Trueman 2020).

3 Role of Nanofertilizers in the Plant Growth

Nanofertilizers act similarly to conventional fertilizers using nanotechnology. Nanofertilizers mainly provide nutrients to plants. The use of nanofertilizers in place of conventional fertilizers allows the controlled release of nutrients and water flow to enhance the agricultural gain (Moaveni and Kheiri 2011; Naderi and

Danesh-Shahraki 2013). Misuse or overuse of fertilizers leads to negative effects like declination in fertilization response ratios and incidents of multi-micronutrient scarcities in soils with a decrease in soil mineral content. This fails to maintain the optimum fertility standards of the soil affecting the crop yield (Kalia and Kaur 2019).

3.1 Advantages of Nanofertilizers over Conventional Fertilizers

Nanofertilizers which are preferred over the conventional fertilizers have the following advantages (Fig. 13.2):

1. Primarily nanofertilizers are used as they release the required micronutrient, macronutrient and non-nutrient in a controlled manner as compared to conventional fertilizers. The nutrients are lost as a response to abiotic and biotic stress conditions. Nitrogen-containing fertilizers volatilize as gaseous ammonia and nitrous oxides. Similarly, phosphorous-containing fertilizers precipitate as calcium, iron and aluminium phosphates. Micronutrients are more available in nanoformulation easing the uptake by plants (Kalia and Kaur 2019; Dimkpa and Bindraban 2017).
2. Nanoparticle-embedded nanofertilizers can have higher uptake by plants as the nanosize formulations can pass through the pores and by molecular transporters. Various ion channels are utilized by the nanofertilizers for higher absorption and nutrient uptake (Aamir Iqbal 2019).
3. ENMs can be used as soil binders (Majeed and Taha 2013) or land reclamation agents (Liu and Lal 2012). The soil binding prevents soil erosion. Polymeric nanoparticles (NPs) can absorb and store a high amount of water and simultane-

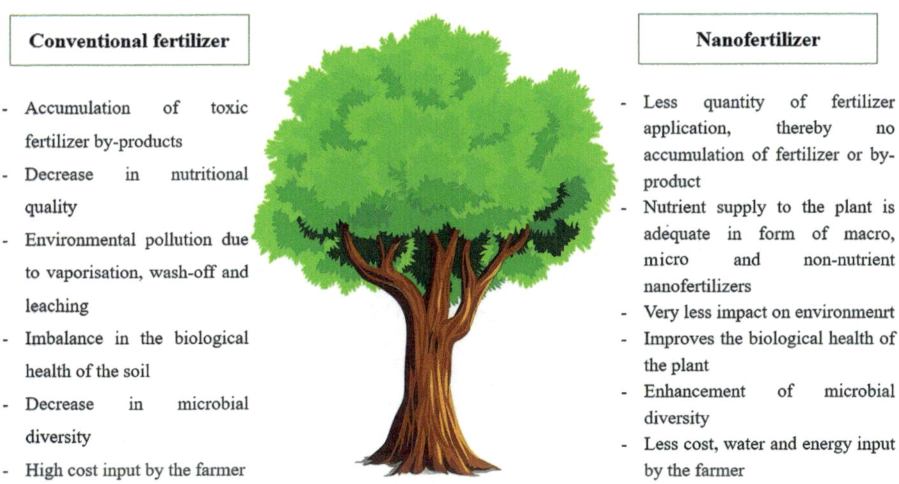

Fig. 13.2 Comparison of conventional fertilizer to nanofertilizers

ously decontaminate the pesticide residues, heavy metals and other toxic substances for the plants, animals and human beings (Kalia and Kaur 2019).
4. Loss of nanofertilizers by volatilization, emission or leaching is very less, so the quantity and frequency of application are less. It also prevents environmental pollutions (Aamir Iqbal 2019).
5. Nanotechnology opens the scope for development of nanodelivery systems for application of fertilizers. The ecological benefit can be achieved by reusing and recycling the agri-biowastes to fabricate the nano-products to carry the nutrient elements of plants (Wanyika et al. 2012).

3.2 Types of Nanofertilizers

The precision of agriculture can be enhanced by increasing the nutrient use efficiency using nanofertilizers. Nowadays, nanofertilizers are considered as smart technology to make sustainable agriculture (Khot et al. 2012; Davarpanah et al. 2016). The use of bulk conventional fertilizers is replaced with nanofertilizers, and that leads to declination in soil and water pollution due to excessive nutrients (Dimkpa and Bindraban 2017). Different nanoformulations like metal oxide nanoparticles are mainly used techniques for nanofertilizers (Davarpanah et al. 2017). The physicochemical properties of the ENMs depend on the type of method of synthesis and their origin. They may be prepared from inorganic metals or organic substances. Inorganic metal oxides like zinc oxide, magnesium oxide, titanium oxide and silver oxide are of more use as nanofertilizers, whereas polymers, lipids and carbon nanostructures are also a source of ENMs. As the primary objective of nanofertilizers is the adequate source of nutrients for growth and development of a plant, they are classically categorized into three types, i.e. micronutrient nanofertilizers, macronutrient nanofertilizers and non-nutrient nanofertilizers. Further another class is also emerging in the field of nanofertilizers, i.e. nano-biofertilizers produced by biological synthesis or green synthesis (Feregrino-Perez et al. 2018; Liu and Lal 2015):

1. *Micronutrient nanofertilizers*: Micronutrients are required in a very minute or trace quantity by plants for their growth and development. These micronutrients act as cofactors and are very essential for different physiological and biochemical processes. Other than this, they also help in the conjugation of different enzymes and macromolecules of primary or secondary importance. Supply of these micronutrients in the form of nanofertilizers helps in seed germination and root and shoot growth due to the increase in planta concentration of indole acetic acid [IAA]. This enriches the crop product with micronutrients mainly like zinc, boron, iron and manganese (Kalia and Kaur 2019). Zinc (Zn) affects plant growth as it is present in the structure of proteins or acts as a cofactor of different enzymes of physiological importance (Noreen et al. 2018). Zn mainly participates in carbohydrate synthesis, protein metabolism and regulation of a plant growth hormone, i.e. auxins. It also protects the plants from pathogen attack

(Broadley et al. 2007). ZnO NPs induce the morphological growth and development of pomegranate (*Punica granatum* L.) in in vitro condition by elevating the chlorophyll content and reduction of proline accumulation (El-Mahdy and Elazab 2020). Boron (B) is another important micronutrient which involves in the photosynthesis and other physiological processes. B plays a significant role in biosynthesis and lignification of cell walls (Navarro-León et al. 2016). Iron (Fe) is also equally important as zinc and boron in various physiological processes, and its deficiency mainly affects the yield (Palmqvist et al. 2017). Ngan et al. reported the effect of iron nanoparticles (FeNPs) on the formation of root and growth of carnation (*Dianthus caryophyllus*) plantlets in vitro and microponic system. FeNPs also affected the uptake of other nutrients like Ca, K, Mg and Fe and helped in the generation of antioxidant enzymes (Ngan et al. 2020). Manganese (Mn) is another micronutrient that participates in the photosynthesis as a cofactor of various enzymes and other physiological processes like biosynthesis of adenosine triphosphate (ATP), chlorophyll, fatty acids, proteins and different secondary metabolites like flavonoids (Palmqvist et al. 2017).

2. *Macronutrient nanofertilizers*: For appropriate growth and metabolism of plants, macronutrients (MNs) are required in large quantity. Mainly nine elements are considered as macronutrients. Carbon (C), hydrogen (H) and oxygen (O) are derivatives of air and water. Elements that are derivative of soil are nitrogen (N), phosphorous (P) and potassium (K) and considered as primary MNs, whereas secondary or tertiary MNs are calcium (Ca), magnesium (Mg) and sulphur (S). MNs are essential for more than 95% of the biomass for the plants (Kalia and Kaur 2019).
3. *Non-nutrient nanoparticulate nanofertilizers*: Non-nutrient nanofertilizers don't contribute for the nutrients but effectively deliver the required fertilizers in nanodelivery systems like carbon-based nanotubes, fullerenes, graphenes and quantum dots. Some inorganic nanoparticles of lithium (Li), silver (Ag) and gold (Au) or metal oxide nanoparticles like titanium dioxide (TiO_2), cerium oxide (CeO_2), bimetallic nanoparticles (CdSe) or nanoclay mineral particles are used for delivery of non-nutrient nanoparticulate nanofertilizers (Kalia and Kaur 2019). Xie et al. reported the co-activity of graphene oxide and indole acetic acid (IAA) on the growth of *Brassica napus* L. by multiple phytohormone pathways. Graphene oxide or IAA alone inhibits the root growth, but when they are co-administered, the growth of the root and shoot is facilitated by phytohormone pathways like abscisic acid (ABA), IAA, gibberellin (GA), cytokinin (CTK), brassinolide (BR) and salicylic acid (SA) (Xie et al. 2020).

3.3 Synthesis of Nanofertilizers

The nanomaterial to be used as nanofertilizers can be synthesized by any one of three methods, i.e. top-down method, bottom-up method or biological synthesis (Behera et al. 2020). The top-down method is a physical method by which the particle size is reduced to nanoscale in such a way to obtain uniform-sized assemblies. But this method has the limitation of non-uniform-sized nanoparticles with impurities, whereas the bottom-up method starts with an atomic or molecular level to produce nanoparticles with controlled size using different chemical methods. In this method, the size of the nanoparticles can be controlled easily with less impurity (Singh and Rattanpal 2014; Pradhan and Mailapalli 2017). Nanomaterials can be synthesized biologically, the so-called biosynthesis approach. In this approach, different natural sources are used, derived from plants or microorganisms. The advantage of the biosynthetic method is the better regulation of the toxicity and particle size (El-Ramady et al. 2018; Yadav et al. 2012).

4 Mechanism of Action of Nanofertilizers for Plant Growth in Stress Conditions

The defence system of the plant protects the plant from the damages caused by different stress conditions if they respond before the stimuli of stress. Cellular machinery and the defence system are the crucial mechanisms in the protection of plants from abiotic and biotic stress conditions. When the ENMs interact with plants under abiotic stress conditions, ROS generation is the most common reaction. The ROS triggers the defence system of the plants and exacerbates the damages of the plant cells (Dat et al. 2000). Though ENMs encourage ROS generation (Qi et al. 2013; Oukarroum et al. 2012; Ma et al. 2010; Simon et al. 2013), it also participates in scavenging the ROS by acting similarly to enzymes for antioxidation (Rico et al. 2013a, b; Wei and Wang 2013). The actual mechanism of nanofertilizers can be understood by two approaches, i.e. proteomic and genomic approach. These two mechanisms explain possible mechanisms of action of nanofertilizers in the presence of abiotic stress conditions.

4.1 A Proteomic Approach for a Mechanism of Action of Nanofertilizers

1. Vannini et al. reported a proteomic study on roots of *Eruca sativa* with two different forms of silver, i.e. silver nanoparticle and silver nitrate. The application changes the proteins involved in the metabolism of sulphur and redox regulation. This effect was owing to the distinctive physicochemical properties of silver nanoparticles (Vannini et al. 2013).
2. Abiotic and biotic stress condition elevates the cytosolic calcium ion (Ca^{2+}) in plant cells and triggers the generation of nitric oxide (NO) (Khan et al. 2012; Corpas et al. 2006; Del Rio et al. 2004). Mirzajani et al. (2014) studied the influence of silver nanoparticle on the roots of *Oryza sativa* assuming a proteomic

approach. He found that silver nanoparticles are involved in oxidative stress reaction pathway, regulation of Ca^{2+} and signalling, transcription and denaturation of protein, cell metabolisms and apoptosis. Mirzajani et al. followed the findings of Goyer (1995) and assumed that the silver nanoparticles or silver ions released from the nanoparticles interfere with the metabolism of cell by interacting with the second messenger like Ca^{2+} sensing receptors, voltage- or ligand-gated calcium ion channels and calcium-sodium (Ca^{2+}/Na^{+}) exchanger (Mirzajani et al. 2014).
3. On exposure to the stress condition, the defence mechanism of the plant is activated by a series of a response to signalling network. Calcium ion (Ca^{2+}) plays a vital role in signalling as a second messenger. Stress stimulus translocates the calcium ion from its store to cytosol by calcium channels so elevating the cytosolic Ca^{2+} level. The increased Ca^{2+} is detected by Ca^{2+}-binding proteins (CaBPs) which downregulate the alteration in expression of gene and plants' adaptations to stress conditions (Khan et al. 2014).
4. Miao et al. reported that the fullerene C60 nanocrystals suspended in aqueous medium caused functional variation of the Ca^{2+}/calmodulin-dependent protein kinase II (CaMKII) (Miao et al. 2014). Later the findings were reinforced by Marmiroli et al., (2015) and they experimented on *Arabidopsis thaliana* with cadmium sulphide (CdS) quantum dots. They found the overexpression of a calcium-binding protein CML45 and calcium-dependent protein kinase 23 (Marmiroli et al. 2015). These calcium-binding proteins control the stress responses, and their overexpression develops the increased resistance of plant towards different abiotic stresses (Xu et al. 2011; Boudsocq and Sheen 2013).
5. Improved activity of nitrate reductase is responsible for the increase in NO level in plants due to an increase in NO synthesis (Carpenter et al. 2012). ENMs elevate the enzyme activity (Shahrokh et al. 2014). So the elevated NO levels counteract the nanomaterial-induced phytotoxicity and also facilitate the genes responsible for antioxidant activity and suppress the generation of ROS and lipid peroxidation (Chen et al. 2015).

4.2 Genomic Approach for a Mechanism of Action of Nanofertilizers

1. Frazier et al. reported the interaction of miRNA, a small non-coding RNA, with a nanomaterial in abiotic and biotic stress conditions. miRNA controls different physiological processes of the plant in response to stress conditions. Nano-titanium dioxide (TiO_2) and aluminium oxide (Al_2O_3) NMs, when applied on tobacco plants, upregulated the miRNA expression in the protection of plants contrary to metal stress (Frazier et al. 2014).
2. Kim et al. performed the experimentation of nano-zerovalent iron (nZVI) on *Arabidopsis* species. nZVI improved the tolerance of the plant by expressing a

gene AHA_2 involved in stomatal opening in drought condition. The treatment decreased apoplastic pH, increased leaf area and widened stomatal opening. The AHA_2 gene enhanced the H^+-ATPase by fivefold in nZVI-treated plants than the control plants, which is responsible for stomatal opening (Kim et al. 2015).
3. ENMs behave similar to Ca^{2+} and interact with CABP and trigger the cascades of stress-responsive genes. So the enhanced gene expression facilitates cell division, cell elongation and tolerance against the stress conditions (Almutairi 2016).

Different nanomaterials are used as nanofertilizers to overcome different abiotic stress conditions. Table 13.1 illustrates different reported literature for nanomaterial to be used in different plant species.

5 Role of Nanopesticides in the Plant Growth

ENMs containing pesticides as an active constituent with antimicrobial activity are termed as nanopesticides (Kah and Hofmann 2014; Iavicoli et al. 2017). The main objective of nanopesticide is to protect the crop or plants from infections caused by different pathogens like fungi, bacteria, viruses and oomycete diseases. The modification of pesticides into nanopesticides brings about a sustainable amendment in the field of agriculture for better quality and quantity of yield. The modification to nanostructure changes and improves the properties of the nanomaterial which make them more potent; thereby low-dose application are required to get good results. ENMs as nanopesticides work promptly at lower dose, so the adverse effects to an environment like run-off and environmental pollution are avoided. The advantages of nanopesticides are also the conservation of energy and water inputs required for crop production and lower the economic cost of the pesticides (Chhipa 2017).

So many approaches were adopted to develop nano-enabled versions of conventional pesticides. They may be organic or inorganic polymeric nanomaterials with variations in their property related to shape, size and other morphological features (Kah and Hofmann 2014; Yin et al. 2012). Biodegradable polymers are used to formulate so many nanoformulations like nanospheres, nanocapsules, nanogels and nanofibres. In nanospheres, the active constituents are distributed uniformly, whereas nanocapsules contain the active ingredient in the core part enclosed within the polymeric matrix (Iavicoli et al. 2017). Nanogels are formulated by incorporating within the cross-linked biopolymer networks containing the active ingredient within the pores (Kah and Hofmann 2014; Yin et al. 2012).

Advantages of Nanopesticides over Conventional Pesticides

1. As compared to the conventional pesticides, carbon nanotubes were found to be more effective for germination of seed and growth of plant (Khodakovskaya et al. 2013; Zheng et al. 2005).
2. The small size and high-surface-to-volume ratio property of nanomaterials allow them to enter into intercellular space by apoplastic pathway or across the cell wall. Once they get inside the apoplast, they can easily come into epidermal and

Table 13.1 Different nanomaterial used for different abiotic stress conditions

Abiotic stress condition	Nanomaterial	Plant species	Effect on the plant	References
Drought	Titanium dioxide (TiO_2)	*Triticum aestivum* L.	Improves the growth of plant with high crop yield. The gluten and starch contents were higher in nano-TiO_2 plants	Jaberzadeh et al. (2013)
		Linum usitatissimum L.	Increase in content of chlorophyll and carotenoids with an increase in crop growth and yield, decrease in hydrogen peroxide (H_2O_2) and malondialdehyde (MDA) level	Aghdam et al. (2015)
		Ocimum basilicum L.	The adverse effects due to stress of drought got improved	Kiapour et al. (2015)
	Zinc oxide (ZnO)	*Glycine max*	Improvement in seed germination rate and percentage with declination in residual seed mass	Sedghi et al. (2013)
		Zea mays L.	Drought-stressed maize plants were exposed with nano-ZnO which improved the rate of photosynthesis, stomatal translocation and water use efficiency with drought tolerance. The ZnO NPs accelerated the biosynthesis of starch and sucrose by increasing the activity of UDP-glucose pyrophosphorylase, phosphoglucose isomerase and cytoplasmic invertase. Drought tolerance was improved by modulating the primary metabolic processes of carbohydrate	Sun et al. (2020)
	Analcite	*Triticum aestivum* L., *Zea mays* L.	Deposition of flavonoids and carotenoids and catalase (CAT) enzyme activity increased with decrease in deposition of proline. Seed germination, photosynthetic pigment concentration and biomass accumulation improved	Zaimenko et al. (2014)

(continued)

Table 13.1 (continued)

Abiotic stress condition	Nanomaterial	Plant species	Effect on the plant	References
	Zerovalent iron (Fe^0)	*Carthamus tinctorius* L.	Improvement of crop yield with a reduction of the impact of drought stress	Zareii et al. (2014)
		Arabidopsis thaliana	Expression of AHA_2, a gene responsible for the opening of stomata, thereby elevating the level H^+-ATPase in plasma membrane. This causes more opening of the stomata with elevated chlorophyll content and plant biomass. This helps in maintenance of usual drought sensitivity and increases CO_2 utilization	Kim et al. (2015)
	Silicon dioxide (SiO_2)	*Crataegus* sp.	Increase in rate of photosynthesis, stomatal conductance and plant biomass with very less impact on the content of chlorophyll and carotenoid	Ashkavand et al. (2015)
	Maghemite (γ-Fe_2O_3)	*Helianthus annuus* L.	The plant could fight with drought stress, and the amino acid and proline content and mobilization of trace constituents remain unaltered	Martínez-Fern-andez et al. (2015)
	Iron, copper, cobalt and zinc oxide (Fe, Cu, Co and ZnO) NPs	*Glycine max* (L.) Merrill	Overexpression of drought-responsive marker genes enhanced the level of drought tolerance by the soybean (*Glycine max* (L.) Merrill) plants. The physiological indicators of drought stress analysed were relative water content, drought tolerance index and reduction rate of biomass. The marker genes selected for the study were *GmRD20A*, *GmDREB2*, *GmERD1*, *GmFDL19*, *GmNAC11*, *GmWRKY27*, *GmMYB118* and *GmMYB174*	Linh et al. (2020)

Salinity	Silicon dioxide (SiO_2)	*Lycopersicon esculentum*	Concentration-dependent activity of nano form of SiO_2 on germination of seeds, root length growth and dry weight. Lower level favours whereas higher concentration suppress the seed germination	Haghighi et al. (2012)
		Solanum lycopersicum L.	SiO_2 improved the effects of salinity, so increasing fresh weight, chlorophyll content, rate of photosynthesis and water content in leaves	Haghighi and Pessarakli (2013)
		Ocimum basilicum	Improvement in chlorophyll and proline content and fresh and dry weight of crop yield	Kalteh et al. (2014)
		Lens culinaris Medik	Accelerated seed physiology like sprouting and sapling growth	Sabaghnia and Janmohammadi (2014)
		Cucurbita pepo L.	Upgraded germination and growth of seed with less levels of MAD (malondialdehyde) and H_2O_2 (hydrogen peroxide). It reduced the leakage of electrolytes with lesser chlorophyll degradation and damage due to oxidation accelerating antioxidative enzymes and an increase in the rate of photosynthesis	Siddiqui et al. (2014)
		Vicia faba L.	Seed germination is promoted with enhanced growth, accelerated antioxidant enzymes' activities and increase in total harvest and water content comparatively	Qados and Moftah (2015), Qados (2015)
		Solanum lycopersicum L.	Overexpression of four genes, AREB, TAS14, NCED3 and CRK1, in response to salt stress and downregulation of six genes, RBOH1, APX2, MAPK2, ERF5, MAPK3 and DDF2, and decreases the outcome of salinity on seed sprouting, length of root and fresh weight	Almutairi (2016)

(continued)

Table 13.1 (continued)

Abiotic stress condition	Nanomaterial	Plant species	Effect on the plant	References
	Zinc oxide and iron oxide (ZnO and Fe_3O_4)	*Moringa peregrina*	Reduces the sodium and chloride ion content and increase in nitrogen, phosphorous, iron, zinc and ions of potassium, calcium and magnesium. Also increases total chlorophyll, carotenoids and proline content with improvement in carbohydrate, protein and antioxidants	Soliman et al. (2015)
	Zinc oxide (ZnO)	*Helianthus annuus* L.	Improves development and speed of carbon dioxide assimilation and carbon dioxide concentration in stomata. Increases the chlorophyll content and zinc content and decreases the sodium concentration	Torabian et al. (2016)
	Selenium (Se)	*Hordeum vulgare* L.	Green synthesized Se NPs were applied exogenously on the leaves of barley plant (*Hordeum vulgare* L.) in hydroponic conditions. The Se NPs increased total contents of phenolic compounds and decreased the malondialdehyde significantly, increasing the shoot dry weight	Habibi and Aleyasin (2020)
	Titanium dioxide (TiO_2)	*Dracocephalum moldavica* L.	The activity of antioxidative enzymes was decreased so reducing the level of hydrogen peroxide significantly. The concentrations of essential oils like gernial, z-citral, geranyl acetate and geraniol were found to be maximum by application of TiO_2 NP	Gohari et al. (2020)
Cold	Selenium (Se)	*Cucumis sativus* L.	The physiological responses to the chilling stress condition got modified. This causes elevated proline content and reduction in lipid peroxidation, but resistance capability of the plant didn't alter so much in lower temperature	Hawrylak-Nowak et al. (2010)

	Titanium dioxide (TiO_2)	*Cicer arietinum* L.	Antioxidant enzyme activity was accelerated with reduction in generation of ROS like hydrogen peroxide and prevents the leakage of electrolytes	Mohammadi et al. (2013, 2014)
			Overexpression of genes of binding proteins for RuBisCO and chlorophyll. The content of hydrogen peroxide decreased with improved activity of phosphoenolpyruvate carboxylase	Hasanpour et al. (2015)
	Silicon dioxide (SiO_2)	*Agropyron elongatum* L.	Improved the seed germination with less seed dormancy, increased the mass of seedling	Azimi et al. (2014)
	Silver (Ag)	*Arabidopsis thaliana*	Antioxidant activity of genes were improved in the presence of silver nanoparticles and cold stress	Kohan-Baghkheirati and Geisler-Lee (2015)
Heat	Multiwalled carbon nanotubes (MWCNTs)	*Lycopersicum esculentum*	Overexpression of stress responsive genes including HSP90	Khodakovskaya et al. (2011)
	Titanium dioxide (TiO_2)	*Lycopersicon esculentum* Mill.	Photosynthesis was increased with controlled utilization of energy and cooling of leaves resulted to facilitated opening of stomata	Qi et al. (2013)
Heat and low temperature	Sodium selenate (Na_2SeO_4)	*Lycopersicon esculentum* Mill.	Enhancement in growth of plant with increase in chlorophyll and water content in the leaves	Haghighi et al. (2014)
Metal stress				
Cadmium	Titanium dioxide (TiO_2)	*Chlamydomonas reinhardtii*	Cadmium toxicity was declined	Yang et al. (2012)
		Glycine max L.	TiO_2 helped in more uptake of cadmium with less toxicity. Levels of chlorophyll and leaf water content increased with an increase in the rate of photosynthesis and growth. Lipid peroxidation decreased with decrease in proline content	Singh and Lee (2016)

(continued)

Table 13.1 (continued)

Abiotic stress condition	Nanomaterial	Plant species	Effect on the plant	References
		Vigna unguiculata (L.) Walp.	The foliar application of TiO_2 improved the chlorophyll b content and reduced the Cd content in roots, shoots and grains. Exposure to TiO_2 increased the activity of stress enzymes with elevation in mineral levels of Zn, Mn and Co in the seeds	Ogunkunle et al. (2020)
Copper, zinc and lead	Magnetic maghemite		Adsorption was increased, and Cu^{2+}, Zn^{2+} and Pb^{2+} were removed from water	Roy and Bhattacharya (2012)
Copper and lead	Carboxyl-cadmium selenide/zinc sulphide (CdSe/ZnS) quantum dots	*Chlorella kessleri* (cell-walled strain), *Chlamydomonas reinhardtii* (wall-less and a walled strain)	Copper and lead concentrations decreased in the intracellular contents of strains with cell wall, whereas it increased in strain without cell wall	Worms et al. (2012)
Nickel, cadmium, zinc, lead and chromium	Graphite oxide (GO), silicon dioxide (SiO_2)		About 90% of metal ions are removed by graphite oxide, and nanocomposites of silica and graphite oxide at 2:3 ratio were found to be the most effective adsorbent	Sheet et al. (2014)
Cadmium	Hydroxyapatite	*Brassica chinensis* L.	Enhancement in the quantity of chlorophyll and vitamin C with augmented biomass and antioxidative enzymes like SOD, CAT and POD with reduced concentration of malondialdehyde	Li and Huang (2014)
	Graphene oxide (GO)	*Microcystis aeruginosa*	Symptoms of cadmium toxicity increased	Tang et al. (2015)
		Lactuca sativa L.	Foliar application of graphene oxide reduced the toxicity due to Cd with an increase in rate of photosynthesis and transpiration, conductance of stomata and content of chlorophyll. The yield was improved with improvement in electron transport system and concentration of carboxylase and oxidase enzymes	Gao et al. (2020)

	Zinc oxide (ZnO)	*Lycopersicon esculentum*	ZnO NPs on foliar application to tomato plants (*Lycopersicon esculentum*) decreased the negative effects of Cd stress and elevated the protein content with increase in activities of enzymes like nitrate reductase and carbonic anhydrase. Microscopic study revealed the improvement of an aperture of stomata by decreasing the levels of malondialdehyde and superoxide radical	Faizan et al. (2020)
Chromium (VI)	Sodium metasilicate (Na_2SiO_3)	*Pisum sativum* L.	Pea seedlings were protected against phytotoxicity due to chromium (VI); uptake of Cr (VI) was reduced with oxidative stress. The antioxidative defence system was upregulated and more accumulation of nutrients enhancing the growth	Tripathi et al. (2015)
Chromium in tannery waste water	Lysine chelated Zn (Zn-Lys)	*Zea mays* L.	Tannery wastewater is used for crop irrigation in many parts of the world. This water contains different levels of metal ions which facilitate the crop growth, but high levels of metal ion concentration hamper the growth due to metal stress. So foliar spray application of Zn-Lys on maize plants (with chromium stress) improved with growth and biomass. Zn is a micronutrient, and lysine is an essential amino acid, and the Zn-Lys improved the growth by reducing the oxidative stress by a decrease in generation of hydrogen peroxide and malondialdehyde	Ahmad et al. (2020)

(continued)

Table 13.1 (continued)

Abiotic stress condition	Nanomaterial	Plant species	Effect on the plant	References
Arsenic	Zinc oxide (ZnO)	*Oryza sativa* L.	Metal stress due to As retarded the seed germination rate, length of seedling, dry weight of seedling and chlorophyll content. Application with ZnO increased the germination rate, weight of shoot and root, chlorophyll content, activity of SOD and CAT with reduced content of malondialdehyde	Wu et al. (2020)
UV-B radiation	Anatase-TiO_2	*Spinacia oleracea*	Increase in antioxidation process leading to decrease in the level of ROS and MDA	Lei et al. (2008)
	Cerium oxide (CeO_2)	*Chlorella vulgaris*	Proficient absorption of ultraviolet radiations with visible radiation and very less damage due to stress of oxidation	Sicard et al. (2011)
	Cadmium (Cd)-telluride quantum dots	*Triticum aestivum* L.	Root and shoot growth was repressed by automatic cell death and DNA laddering	Chen et al. (2014)
UV radiation	Copper oxide (CuO)	*Elodea nuttallii*	Reduction in chlorophyll content and photosynthesis, enhanced copper accumulation from CuO nanoparticles and activity of peroxidase	Regier et al. (2015)
Flooding	Silver (Ag)	*Crocus sativus*	Ethylene signalling was blocked promoting the growth of root	Rezvani et al. (2012)
		Glycine max	Production of toxic by-products of glycolysis was reduced with enhancement in levels of stress-related proteins and sapling growth	Mustafa et al. (2015a)
	Aluminium oxide (Al_2O_3)	*Glycine max*	Energy metabolism and cell death were regulated improving the plant growth	Mustafa et al. (2015b)

Post-harvest stress	Copper (Cu)	*Petroselinum crispum* Mill.	Weight loss and lipid peroxidation were checked to maintain higher levels of ascorbic acid enhancing the worth of parsley (*P. crispum* Mill.)	Ouzounidou and Gaitis (2011)
	Silver (Ag)	*Chrysanthemum morifolium* L.	Endurance and tenderness of cut flowers were improved, reduced loss of fresh weight and bacteria colonies in the stem part	Kazemipour et al. (2013)
Dark stress	Silver (Ag)	*Pelargonium zonale*	Activities of enzymes for antioxidation were Increased with improvement in leaf Chlorophyll and carotenoids content. lipid peroxidation and petal abscission were reduced with increased petal longevity	Hatami and Ghorbanpour (2013, 2014), Ghorbanpour and Hatami (2014)

cortical cells and reach the endodermis and get accumulated consistently (Larue et al. 2012; Zhao et al. 2013).

3. Rico et al. postulated another alternate approach of symplastic (through cytoplasm) route. The symplastic route is a more controlled pathway for movement of nanomaterials into plant cells. In this route, the nanomaterial binds with the carrier protein, favours entry into the cell and easily interchanges through ion channels, aquaporins and endocytosis (Rico et al. 2011).

6 Mechanism of Action of Nanopesticides for Growth of Plants in Biotic and Abiotic Stress Conditions

The efficacy of nanomaterial on plants is affected by abiotic and biotic stress conditions. In biotic stress, the efficacy of nanomaterial for the plant response especially depends on the properties of the ENMs and the species of the plant. El-Argawy et al. reported probable mechanisms of actions of ENMs against the pathogens causing the biotic stress. The study suggested that the cationic silver (Ag^{+}), copper (Cu^{2+}), zinc (Zn^{2+}) and titanium (Ti^{4+}) released from ENMs bind to the anionic sulfhydryl or other functional groups in proteins (El-Argawy et al. 2017). The ionic interaction of these functional groups with the metal ions causes an alteration in activity and functions of the proteins leading to cell membrane disruption. Sometimes these released ions may be toxic to plant genes causing interference in the electron transport chain (ETC) and alteration in the structure and function of DNA. All these destroy the integrity of the cell wall leading to the death of the pathogen.

ENMs can diminish the biotic stress in plants by two pathways, direct antimicrobial effect and an indirect effect by stimulation of nutrition-induced defence system, affecting the metabolic processes in the plant (Servin et al. 2015). In the direct pathway, biocidal effect of nanopesticides occurs by penetration and accumulation in the cell membrane, consequently leading to cell lysis (Salem et al. 2011). The ENMs can accelerate the generation of ROS (Lamsal et al. 2011a, b), and the increased level of ROS causes disturbances in the cellular homeostasis of microbes, and the oxidative stress damages the cell at a different level and causes automatic cell death (apoptosis) (Gill et al. 2005). Excessive ROS also destroys the DNA and enhances lipid peroxidation, enzyme inhibition and ultimately apoptosis, whereas ROS at lesser or moderate level acts as a secondary messenger in different physiological processes and develops the tolerability of plant towards biotic and abiotic stress conditions. This tolerability can be understood by the intracellular hormone-mediated signalling sequences like apoptosis, closure of the stomata, gravitropism and responses of plant towards different stress conditions (Mittler 2002; Yan et al. 2007; Joo et al. 2001; Miller et al. 2008):

1. *Nanopesticides containing silver*: Nanopesticides containing silver ion act by inducing toxicity to the pathogen. Silver ions (Ag^+) are released from the nanomaterial and recognized to be extremely toxic, disrupting the cell membrane of the pathogen by interacting with membrane proteins with cysteine moiety (Servin et al. 2015). Mishra et al. suggested that Ag^+ penetrates and accumulates destroying pathogen due to the sclerotial rind disruption (Mishra et al. 2017).
2. *Nanopesticides containing copper*: Cu-based NMs act similar to silver as antimicrobial nanopesticide (Adisa et al. 2019). Oussou-Azo et al. reported the antifungal activity study on *Colletotrichum gloeosporioides* using Cu NPs, CuO NPs and CuO. The antifungal activity of the Cu NPs was found to be maximum as compared to other forms. Cu NP prevents hyphal growth and spore germination (Oussou-Azo et al. 2020).
3. *Nanopesticides containing zinc*: Zinc oxide (ZnO) nanopesticides inactivate the pathogens by severely damaging the microbial cell wall in the presence of ROS. If the toxicity effect of Ag-, Cu- and Zn-containing NMs towards the plant cell is considered, then Zn has the least toxicity to the plant cells. So they are more preferred as nanopesticide having very less negative impact on the environment (Adisa et al. 2019).
4. *Nanopesticides containing chitosan polymer and another engineered nanomaterial (NM)*: The nanomaterial containing chitosan (CNMs) prevent the in vitro growth of pathogenic fungi and improve the in vivo seed germination and biomass (Sathiyabama and Parthasarathy 2016). It also causes the leakage of potassium ion (K^+) by disrupting the cell membrane and necrotic cell death. CNMs are positively charged superficially as amino groups are present on the surface which is responsible for antimicrobial activity of the nanopesticides.

The positive charge on the surface of chitosan increases its affinity towards negatively charged surfaces on the microbial cell membrane. It also enhances interaction with metals present in the cell. The electrostatic interaction between polycationic CNMs and anionic components of the pathogens can disrupt permeability of cell membrane, an outflow of intracellular materials and ultimately cell lysis (Xing et al. 2018).

Chitosan NMs can be used against the fungal infection as they disrupt the integrity of the cell wall or cell membrane similarly by electrostatic interaction (He et al. 2011). They inhibit the glucan-synthesizing enzymes which biosynthesize glucans, i.e. a significant biomolecule providing strength to the cell wall of the fungi (Aranda-Martinez et al. 2016). Other mechanisms involve the interaction of the CNMs with the fungal DNA and further interfering with protein synthesis by inhibiting the mRNA. The cationic amino group of CNMs interacts with an anionic phosphate group and amino acids of DNA, thereby inactivating the enzymes involved in protein synthesis (Kheiri et al. 2016).

Table 13.2 illustrates the reported nanopesticides to fight the biotic and abiotic stress.

Table 13.2 Nanopesticides for biotic and abiotic stress

Nanomaterial	Pathogen	Plant	Effects	References
Silver (Ag)	*Alternaria solani*	Tomato	The silver nanoparticles reduced the early blight disease with an increase in fresh weight and chlorophyll content	Kumari et al. (2017)
	Golovinomyces cichoracearum or *Sphaerotheca fusca*	Cucumber and pumpkin	Powdery mildew disease was reduced by 25% in cucumber and pumpkin leaves at the highest concentration with distortion of growth of mycelia and conidia as compared to commercial pesticides	Lamsal et al. (2011a)
	Colletotrichum sp.	Pepper	Distortion of growth of mycelia. Pepper anthracnose was inhibited in the field more effectively as compared to conventional fungicides	Lamsal et al. (2011b)
	Bipolaris sorokiniana	Wheat	Inhibition of spot blotch disease with 100% suppression of conidia germination as compared to untreated control group	Mishra et al. (2014)
	Fusarium incarnatum (*Desm.*) Sacc.	*Crossandra* spp.	Controls the disease occurrence from 75 to 55%, as compared to untreated control group	Mallaiah (2015)
	R. solani	Rice	Development of lesion on the leaves was decreased efficiently with increase in fresh and dry weight of the rice plant as compared to untreated control group	Nejad et al. (2016)
DNA-directed silver (Ag) on graphene oxide composite	*Xanthomonas perforans*	Tomato	Efficient inhibition of bacterial spot disease by 32%	Ocsoy et al. (2013)
Silver-sodium (Ag-Na) tallowate	*Phytophthora infestans* and *A. solani*	Tomato	Leaf surface area was increased with inhibition of pathogens and increase in yield of tomato by 35% and level of antioxidant enzymes	Zakharova et al. (2017)
Copper or oxide forms ($CuO/Cu/Cu_2O$)	*Phytophthora infestans*	Tomato	CuO nanoparticles decreased the leaf lesions by ~40% and ~61% after application of 3 and 7 days, respectively, as compared to untreated control group	Giannousi et al. (2013)

Copper oxide (CuO)	*Fusarium oxysporum* f. sp. *niveum*	Watermelon	Occurrence of *Fusarium* wilt was reduced by 25% with increase in yield by 21–53% as compared to conventional fungicides	Elmer et al. (2018)
	Poria hypolateritia	Tea plant	Root rot disease in tea plants was decreased by 80%, and the total leaf yield was increased by 30% as compared to untreated infected control	Ponmurugan et al. (2016)
	Fusarium oxysporum	Tomato	Fusarium wilt disease was inhibited with an increase in yield more effective than ZnO and MnO nanoparticles	Elmer and White (2016)
	Verticillium dahliae	Eggplant	Verticillium wilt disease was inhibited with an increase in yield more effective than ZnO and MnO nanoparticles	Elmer and White (2016)
	Pyricularia grisea	Finger millet	The occurrence of blast disease was reduced by 75% with an increased yield of leaves. Enhancement of defence enzyme activities	Sathiyabama and Manikandan (2018)
	Ralstonia solanacearum	Tobacco	Reduction in bacterial wilt disease by preventing biofilm formation and disturbing ATP production. CuO NPs caused downregulation of genes responsible for pathogenesis	Chen et al. (2019)
Copper NP	*Xanthomonas axonopodis*	Pomegranate	Foliar application of Cu NPs reduced the bacterial blight disease and bacterial colonization on the leaves	Chikte et al. (2019)
Copper composite	*Xanthomonas* spp.	Tomato	Bacterial spot disease in tomato plants was reduced more effectively than copper nanoparticles	Strayer-Scherer et al. (2018)
Copper chitosan	*Curvularia lunata*	Maize	The occurrence of leaf spot disease was reduced with increase in yield of grain, shoot length and chlorophyll content. Enhancement of defence enzyme activities	Choudhary et al. (2017)
Zinc oxide (ZnO)	*Fusarium graminearum*	Wheat	Incidence of disease was decreased significantly with reduced content of mycotoxin in the grain	Savi et al. (2015)

(continued)

Table 13.2 (continued)

Nanomaterial	Pathogen	Plant	Effects	References
Zinc oxide (ZnO) NM formulations (pale-like Zinkicide SG4 and particulate-like Zinkicide SG6)	*F. oxysporum* f. sp., *betae*, *S. rolfsii* and *R. solani*	Sugar beet	Severity of root rot disease was reduced by 86% with an increase in the plant growth, sugar content and PPO activity	El-Argawy et al. (2017)
Titanium dioxide/zinc (TiO_2/Zn)	*Xanthomonas citri* subsp. *citri*	Grapefruit trees	Reduced citrus canker lesion disease prevalence	Graham et al. (2016)
	Foliar Xanthomonas sp.	Rose	Bacterial leaf spot disease on rose was decreased by 99% after15 days of application	Paret et al. (2013)
Titanium dioxide (TiO_2)	*Cercospora beticola*	Sugar beet	Significantly reduced the disease	El-Argawy et al. (2017)
	F. oxysporum f. sp., *betae*, *S. rolfsii* and *R. solani*	Sugar beet	Root rot fungal disease was decreased by 95% with an increase in the growth of plant, the sugar content and the PPO activity	Hamza et al. (2016)
Cerium dioxide (CeO_2)	*Fusarium oxysporum*	Tomato	*Fusarium* wilt disease was reduced with improved yield and alteration in the defence and enzyme activities for stress tolerance	Adisa et al. (2018)
Magnesium oxide (MgO)	*Ralstonia solanacearum*	Tomato	Significantly reduced the disease by 30%	Imada et al. (2015)
Selenium (Se)/*Trichoderma asperellum*	*Sclerospora graminicola*	Pearl millet	Reduced the downy mildew disease in pearl millet with improvement in plant growth	Nandini et al. (2017)
Iron oxide/titanium dioxide (Fe_2O_3/TiO_2) or multiwalled carbon nanotubes/fullerene MWCNTs/C_{60}	Turnip mosaic virus	Tobacco plant (*Nicotiana benthamiana*)	Virus growth was reduced with an increase in biomass by 50%	Hao et al. (2018)
β-D-glucan	*Pythium aphanidermatum*	Turmeric plant	Reduced the rhizome rot disease with increased enzyme activities of defence system	Anusuya and Sathiyabama (2015)

Chitosan	*Fusarium graminearum*	Maize (*Zea mays*)	Resistance against the infection was developed with decreased mycotoxin in the maize grain	Kalagatur et al. (2018)
	Fusarium graminearum	Wheat	Reduced the pathogen attack	Kheiri et al. (2016)
	Pyricularia grisea	Rice	Suppressed the occurrence of disease by 100% on the 10th day of application	Manikandan and Sathiyabama (2016)
	Pyricularia grisea	Finger millet	The blast disease was decreased by 64% after 50th day of inoculation	Sathiyabama and Manikandan (2016)
Nano-emulsion containing polyethylene glycol (PEG)-coated NMs embedded in garlic essential oils	*Tribolium castaneum*		Significantly increased the insecticidal efficacy of PEG against *Tribolium castaneum* up to 80%	Yang et al. (2009)
Nanostructured alumina	*S. oryzae* and *R. dominica*	Wheat	Alumina NM was more effecting in controlling the pests as compared to commercial insecticidal dust	Stadler et al. (2010)
	Acromyrmex lobicornis		Nanostructured alumina was more toxic against *Acromyrmex lobicornis* as compared with diatomaceous earth	Buteler et al. (2018)
Silica NM	*Callosobruchus maculatus*, *R. dominica* F. and *Tribolium confusum* Jacquelin du Val	Pigeon pea (*Cajanus cajan*), horse gram (*Macrotyloma uniflorum*), black gram (*Vigna mungo*), green gram (*Vigna radiata*), chickpea (*Cicer arietinum*), cowpea (*Vigna unguiculata*), wheat (*Triticum aestivum* L.) and barley (*Hordeum vulgare*)	The nanomaterial had released the active ingredient slowly, thereby guarding the seeds against insect attacks. The silica NMs was absorbed into the cuticular lipid layer of the insect, thereby disrupting the cell membrane and cell death	Rouhani et al. (2013), Ziaee and Ganji (2016), Arumugam et al. (2016), Barik et al. (2008), Rai and Ingle (2012)

(continued)

Table 13.2 (continued)

Nanomaterial	Pathogen	Plant	Effects	References
Chitosan-based metal NMs including silver, aluminium oxide, titanium dioxide and zinc oxide (Ag, Al_2O_3, TiO_2 and ZnO)	Rice weevil (*Sitophilus oryzae*)		The application of metal-based chitosan NMs boosted the seed germination, plant growth and yield	Kumaraswamy et al. (2018)
Nanostructured copper oxide and calcium oxide (CuO and CaO)	Cotton leaf worm (*Spodoptera littoralis*)		The mortality rate was higher for nanostructured CuO and CaO	Ayoub et al. (2018)

7 Conclusion

Much progress persists along the prevailing pathways, and agrochemicals are quick to switch into or at least consider themselves as the next new genre of contaminants related to the farming activities. In such a context, nanotechnology could potentially prove as an element of burgeoning strategies to counteract pesticide and fertilizer contamination. Boosting collaborations among varied disciplines involved at all stages of the production and assessment of agrochemicals enable the production of products which integrate within the various limitations of the agrochemical industry framework, and this would possibly bring added value compared to the existing ones. With the prevalence of piling regulatory pressure, there is a need of opportunities to encourage creativity that can drive the development of smarter solutions for nutrition and plant protection. Fostering more collaboration across various scientific domains would guarantee consumer acceptance. Beginning on a wider nano-enabled development framework and building on experiences from other related sectors would be of greater benefit in promoting more sustainable agrochemical production.

References

Aamir Iqbal M (2019) Nano-fertilizers for sustainable crop production under changing climate: a global perspective. In: Hasanuzzaman M, Filho MCMT, Fujita M, Nogueira TAR (eds) Sustainable crop production. InTech Open, London

Adisa IO, Pullagurala VL, Rawat S, Hernandez-Viezcas JA, Dimkpa C, Elmer WH, Gardea-Torresdey JL (2018) Role of cerium compounds in Fusarium wilt suppression and growth enhancement in tomato (*Solanum lycopersicum*). J Agric Food Chem 66:5959–5970

Adisa I, Pullagurala V, Peralta-Videa J, Dimkpa C, Elmer W, Gardea-Torresdey J, White J (2019) Recent advances in nano-enabled fertilizers and pesticides: a critical review of mechanisms of action. Environ Sci Nano 6:2002–2030

Aghdam M, Mohammadi H, Ghorbanpour M (2015) Effects of nanoparticulate anatase titanium dioxide on physiological and biochemical performance of *Linum usitatissimum* (Linaceae) under well-watered and drought stress conditions. Braz J Bot 39:139–146

Ahmad R, Ishaque W, Khan M, Ashraf U, Riaz MA, Ghulam S, Ahmad A, Rizwan M, Ali S, Alkahtani S, Abdel-Daim MM (2020) Relief role of lysine chelated zinc (Zn) on 6-week-old maize plants under tannery wastewater irrigation stress. Int J Environ Res Public Health 17:5161

Aken BV (2015) Gene expression changes in plants and microorganisms exposed to nanomaterials. Curr Opin Biotechnol 33:206–219

Almutairi ZM (2016) Effect of nano-silicon application on the expression of salt tolerance genes in germinating tomato (*Solanum lycopersicum* L.) seedlings under salt stress. Plant Omics J 9:106–114

Amira S, Souad A, Essam D (2015) Alleviation of salt stress on *Moringa peregrina* using foliar application of nanofertilizers. J Hortic For 7:36–47

Anusuya S, Sathiyabama M (2015) Foliar application of β-D glucan nanoparticles to control rhizome rot disease of turmeric. Int J Biol Macromol 72:1205–1212

Aranda-Martinez A, Lopez-Moya F, Lopez-Llorca LV (2016) Cell wall composition plays a key role on sensitivity of filamentous fungi to chitosan. J Basic Microbiol 56(10):1059–1070

Armstrong W, Drew MC (2002) Root growth and metabolism under oxygen deficiency. In: Waisel Y, Eshel A, Kafkafi U (eds) Plant roots: the hidden half. Marcel Dekker, New York, pp 729–761

Arumugam G, Velayutham V, Shanmugavel S, Sundaram J (2016) Efficacy of nanostructured silica as a stored pulse protector against the infestation of bruchid beetle, *Callosobruchus maculatus* (Coleoptera: Bruchidae). Appl Nanosci 6(3):445–450

Ashkavand P, Tabari M, Zarafshar M, Tom-a-skov-a I, Struve D (2015) Effect of SiO_2 nanoparticles on drought resistance in hawthorn seedlings. Lesne Prace Badawcze 76:350–359

Ayoub HA, Khairy M, Elsaid S, Rashwan FA, Abdel-Hafez HF (2018) Pesticidal activity of nanostructured metal oxides for generation of alternative pesticide formulations. J Agric Food Chem 66(22):5491–5498

Azimi R, Borzelabad MJ, Feizi H, Azimi A (2014) Interaction of SiO_2 nanoparticles with seed prechilling on germination and early seedling growth of tall wheatgrass (*Agropyron elongatum* L.). Pol J Chem Technol 16:25–29

Banti V, Giuntoli B, Gonzali S, Loreti E, Magneschi L, Novi G, Paparelli E, Parlanti S, Pucciariello C, Santaniello A, Perata P (2013) Low oxygen response mechanisms in green organisms. Int J Mol Sci 14:4734–4761

Barik TK, Sahu B, Swain V (2008) Nanosilica—from medicine to pest control. Parasitol Res 103(2):253

Begum P, Fugetsu B (2012) Phytotoxicity of multi-walled carbon nanotubes on red spinach (*Amaranthus tricolor* L) and the role of ascorbic acid as an antioxidant. J Hazard Mater 243:212–222

Behera A, Mittu B, Padhi S, Patra N, Singh J (2020) Bimetallic nanoparticles: green synthesis, applications, and future perspectives. In: Abd-Elsalam KA (ed) Multifunctional hybrid nanomaterials for sustainable agri-food and ecosystems. Elsevier, Amsterdam, pp 639–681

Boudsocq M, Sheen J (2013) CDPKs in immune and stress signaling. Trends Plant Sci 18:30–40

Broadley MR, White PJ, Hammond JP, Zelko I, Lux A (2007) Zinc in plants. New Phytol 173:677–702

Buteler M, Garcia G, Stadler T (2018) Potential of nanostructured alumina for leaf-cutting ants *Acromyrmex lobicornis* (Hymenoptera: Formicidae) management. Aust Entomol 57(3):292–296

Capuana M (2011) Heavy metals and woody plants biotechnologies for phytoremediation. J Biogeo Sci For 4:7–15

Carpenter AW, Worley BV, Slomberg DL, Schoenfisch MH (2012) Dual action antimicrobials: nitric oxide release from quaternary ammonium-functionalized silica nanoparticles. Biomacromolecules 13:3334–3342

Chen H, Du M, Han R (2011) Influence of enhanced UV-B radiation on F - actin in wheat division cells. Plant Diver Resour 33:306–310

Chen H, Gong Y, Han R (2014) Cadmium telluride quantum dots (CdTe-QDs) and enhanced ultraviolet-B (UV-B) radiation trigger antioxidant enzyme metabolism and programmed cell death in wheat seedlings. PLoS One 9:e110400

Chen J, Liu X, Wang C, Yin SS, Li XL, Hu WJ, Simona M, Shen ZJ, Xiao Q, Chu CC, Peng XX, Zheng HL (2015) Nitric oxide ameliorates zinc oxide nanoparticles-induced phytotoxicity in rice seedlings. J Hazard Mater 297:173–182

Chen J, Mao S, Xu Z, Ding W (2019) Various antibacterial mechanisms of biosynthesized copper oxide nanoparticles against soilborne Ralstonia solanacearum. RSC Adv 9:3788–3799

Chhipa H (2017) Nanofertilizers and nanopesticides for agriculture. Environ Chem Lett 15(1):15–22

Chibuike GU, Obiora SC (2014) Heavy metal polluted soils: effect on plants and bioremediation methods. Appl Environ Soil Sci 2014:1–13

Chikte RG, Paknikar KM, Rajwade JM, Sharma J (2019) Nanomaterials for the control of bacterial blight disease in pomegranate: quo vadis? Appl Microbiol Biotechnol 103:4605–4621

Choudhary RC, Kumaraswamy RV, Kumari S, Sharma SS, Pal A, Raliya R, Saharan V (2017) Cu-chitosan nanoparticle boost defence responses and plant growth in maize (*Zea mays* L.). Sci Rep 7(1):9754

Corpas FJ, Barroso JB, Carreras A, Valderrama R, Palma JM, Leon AM, Sandalio L, Del Río LA (2006) Constitutive arginine-dependent nitric oxide synthase activity in different organs of pea seedlings during plant development. Planta 224:246–254

Dat J, Vandenabeele S, Vranova E, Van Montagu M, Inze D, Van Breusegem F (2000) Dual action of the active oxygen species during plant stress responses. Cell Mol Life Sci 57:779–795

Davarpanah S, Tehranifar A, Davarynejad G, Abadía J, Khorasani R (2016) Effects of foliar applications of zinc and boron nano-fertilizers on pomegranate (*Punica granatum* cv. Ardestani) fruit yield and quality. Sci Hortic. 210:57–64

Davarpanah S, Tehranifar A, Davarynejad G, Aran M, Abadía J, Khorassani R (2017) Effects of foliar nano-nitrogen and urea fertilizers on the physical and chemical properties of pomegranate (*Punica granatum* cv. Ardestani) fruits. Hortic Sci 52(2):288–294

Del Rio LA, Corpas FJ, Barroso JB (2004) Nitric oxide and nitric oxide synthase activity in plants. Phytochemistry 65:783–792

DeRosa MC, Monreal C, Schnitzer M, Walsh R, Sultan Y (2010) Nanotechnology in fertilizers. Nat Nanotechnol 5:91–94

Dimkpa C, Bindraban P (2017) Nanofertilizers: new products for the industry? J Agric Food Chem 66:6462–6473

El-Argawy E, Rahhal M, El-Korany A, Elshabrawy EM, Eltahan RM (2017) Efficacy of some nanoparticles to control damping-off and root rot of sugar beet in El-Behiera Governorate. Asian J Plant Pathol 11:35–47

El-Mahdy MT, Elazab DS (2020) Impact of Zinc Oxide nanoparticles on pomegranate growth under *in vitro* conditions. Russ J Plant Physl 67(1):162–167

Elmer WH, White JC (2016) The use of metallic oxide nanoparticles to enhance growth of tomatoes and eggplants in disease infested soil or soilless medium. Environ Sci Nano 3(5):1072–1079

Elmer W, White J (2018) The future of nanotechnology in plant pathology. Annu. Rev. Phytopathol 56:111–133

Elmer W, Torre-Roche R, Pagano L, Majumdar S, Zuverza-Mena N, Dimkpa C, White JC (2018) Effect of metalloid and metal oxide nanoparticles on fusarium wilt of watermelon. Plant Dis 102:1394–1401

El-Ramady H, Abdalla N, Alshaal T, El-Henawy A, Elmahrouk M, Bayoumi Y, Shalaby T, Amer M, Shehata S, Fári M, Domokos-Szabolcsy E, Sztrik A, Prokisch J, Pilon-Smits EAH, Pilon M, Selmar D, Haneklaus S, Schnug E (2018) Plant nano-nutrition: perspectives and challenges. In: Gothandam K, Ranjan S, Dasgupta N, Ramalingam C, Lichtfouse E (eds) Nanotechnology, food security and water treatment. Environmental chemistry for a sustainable world. Springer, Berlin, pp 129–161

Faizan M, Faraz A, Mir AR, Hayat S (2020) Role of zinc oxide nanoparticles in countering negative effects generated by cadmium in *Lycopersicon esculentum*. J Plant Growth Regul. https://doi.org/10.1007/s00344-019-10059-2

Feregrino-Perez A, Magaña-López E, Guzmán C, Esquivel K (2018) A general overview of the benefits and possible negative effects of the nanotechnology in horticulture. Sci Hortic 238:126–137

Foyer CH, Noctor G (2000) Oxygen processing in photosynthesis: regulation and signaling. New Phytol 146:359–388

Frazier TP, Burklew CE, Zhang B (2014) Titanium dioxide nanoparticles affect the growth and microRNA expression of tobacco (Nicotiana tabacum). Funct Integr Genom 14:75–83

Gao M, Chang X, Yang Y, Song Z (2020) Foliar graphene oxide treatment increases photosynthetic capacity and reduces oxidative stress in cadmium-stressed lettuce. Plant Physiol Biochem 154:287–294

Ghorbanpour M, Hatami M (2014) Spray treatment with silver nanoparticles plus thidiazuron increases anti-oxidant enzyme activities and reduces petal and leaf abscission in four cultivars of geranium (*Pelargonium zonale*) during storage in the dark. J Hort Sci Biotechnol 89:712–718

Giannousi K, Avramidis I, Dendrinou-Samara C (2013) Synthesis, characterization and evaluation of copper based nanoparticles as agrochemicals against *Phytophthora infestans*. RSC Adv 3(44):21743–21752

Gill SR, Fouts DE, Archer GL, Mongodin EF, DeBoy RT, Ravel J, Dodson RJ (2005) Insights on evolution of virulence and resistance from the complete genome analysis of an early methicillin-resistant *Staphylococcus aureus* strain and a biofilm-producing methicillin-resistant *Staphylococcus epidermidis* strain. J Bacteriol 187(7):2426–2438

Gohari G, Mohammadi A, Akbari A, Panahirad S, Dadpour MR, Fotopoulos V, Kimura S (2020) Titanium dioxide nanoparticles (TiO_2 NPs) promote growth and ameliorate salinity stress effects on essential oil profile and biochemical attributes of *Dracocephalum moldavica*. Sci Rep. 10:912

Goyer RA (1995) Nutrition and metal toxicity. Am J Clin Nutr 61:646S–650S

Graham JH, Johnson EG, Myers ME, Young M, Rajasekaran P, Das S, Santra S (2016) Potential of Nanoformulated zinc oxide for control of citrus canker on grapefruit trees. Plant Dis 100(12):2442–2447

Gull A, Lone AA, Wani NUI (2019) Biotic and abiotic stresses in plants. In: De Oliveira A (ed) Abiotic and biotic stress in plants. Intech Open, London. https://doi.org/10.5772/intechopen.85832

Habibi G, Aleyasin Y (2020) Green synthesis of Se nanoparticles and its effect on salt tolerance of barley plants. Int J Nano Dimens. 11(2):145–157

Haghighi M, Pessarakli M (2013) Influence of silicon and nano-silicon on salinity tolerance of cherry tomatoes (*Solanum lycopersicum* L.) at early growth stage. Sci Hortic 161:111–117

Haghighi M, Afifipour Z, Mozafarian M (2012) The effect of N-Si on tomato seed germination under salinity levels. J Biol Environ Sci 6:87–90

Haghighi M, Abolghasemi R, Teixeira JA (2014) Low and high temperature stress affect the growth characteristics of tomato in hydroponic culture with Se and nano-Se amendment. Sci Hortic 178:231–240

Hamza A, El-Mogazy S, Derbalah A (2016) Fenton reagent and titanium dioxide nanoparticles as antifungal agents to control leaf spot of sugar beet under field conditions. J Plant Prot Res 56(3):270–278

Hancock JT, Desikan R, Neill SJ (2001) Role of reactive oxygen species in cell signalling pathways. Biochem Soc Trans 29:345–350

Hao Y, Yuan W, Ma C, White J, Zhang Z, Adeel M, Zhou T, Rui Y, Xing B (2018) Engineered nanomaterials suppress turnip mosaic virus infection in tobacco (*Nicotiana benthamiana*). Environ Sci Nano 5:1685–1693

Hasanpour H, Maali-Amiri R, Zeinali H (2015) Effect of TiO_2 nanoparticles on metabolic limitations to photosynthesis under cold in chickpea. Russ J Plant Physiol 62:779–787

Hasanuzzaman M, Nahar K, Fujita M (2013) Extreme temperature responses, oxidative stress and antioxidant defense in plants. In: Vahdati K, Leslie C (eds) Abiotic stress - plant responses and applications in agriculture. InTech Open, London

Hatami M, Ghorbanpour M (2013) Effect of nanosilver on physiological performance of pelargonium plants exposed to dark storage. J Hortic Res 21:15–20

Hatami M, Ghorbanpour M (2014) Defence enzyme activities and biochemical variations of Pelargonium zonale in response to nanosilver application and dark storage. Turk J Biol 38:130–139

Hawrylak-Nowak B, Matraszek R, Szyma-nska M (2010) Selenium modifies the effect of short-term chilling stress on cucumber plants. Biol Trace Elem Res 138:307–315

He L, Liu Y, Mustapha A, Lin M (2011) Antifungal activity of zinc oxide nanoparticles against *Botrytis cinerea* and *Penicillium expansum*. Microbiol Res 166(3):207–215

Hideg E, Jansen MA, Strid A (2013) UV-B exposure, ROS and stress: inseparable companions or loosely linked associates? Trends Plant Sci 18:107–115

Hou FF, Thseng FS (1991) Studies on the flooding tolerance of soybean seed: varietal differences. Euphytica 57:169–173

Iavicoli I, Leso V, Beezhold DH, Shvedova AA (2017) Nanotechnology in agriculture: opportunities, toxicological implications, and occupational risks. Toxicol Appl Pharmacol 329:96–111

Imada K, Sakai S, Kajihara H, Tanaka S, Ito S (2015) Magnesium oxide nanoparticles induce systemic resistance in tomato against bacterial wilt disease. Plant Pathol 65(4):551–560

Ismail A, Takeda S, Nick P (2014) Life and death under salt stress: same players, different timing? J Exp Bot 65:2963–2979

Jaberzadeh A, Moaveni P, Reza H, Moghadam T, Zahedi H (2013) Influence of bulk and nanoparticles titanium foliar application on some agronomic traits, seed gluten and starch contents of wheat subjected to water deficit stress. Not Bot Hortic Agrobot 41:201–207

Joo JH, Bae YS, Lee JS (2001) Role of auxin-induced reactive oxygen species in root gravitropism. Plant Physiol 126(3):1055–1060

Kagan CR (2016) At the nexus of food security and safety: opportunities for nanoscience and nanotechnology. ACS Nano 10:2985–2986

Kah M, Hofmann T (2014) Nanopesticide research: current trends and future priorities. Environ Int 63:224–235

Kalagatur NK, Ghosh O, Sundararaj N, Mudili V (2018) Antifungal activity of chitosan nanoparticles encapsulated with *Cymbopogon martinii* essential oil on plant pathogenic fungi *Fusarium graminearum*. Front Pharmacol 9:610

Kalia A, Kaur H (2019) Nanofertilizers: an innovation towards new generation fertilizers for improved nutrient-use efficacy and environmental sustainability. In: Singh B, Katare OP, Souto EB (eds) Nano agroceuticals & nano phytochemicals. CRC Press, Taylor & Francis, Boca Raton, pp 45–62

Kalteh M, Alipour ZT, Ashraf S, Aliabadi MM, Nosratabadi AF (2014) Effect of silica nanoparticles on basil (*Ocimum basilicum*) under salinity stress. J Chem Health Risks 4:49–55

Karuppanapandian T, Wang HW, Prabakaran N, Jeyalakshmi K, Kwon M, Manoharan K, Kim W (2011) 2,4-Dichlorophenoxyacetic acid-induced leaf senescence in mung bean (*Vigna radiata* L. Wilczek) and senescence inhibition by co-treatment with silver nanoparticles. Plant Physiol Biochem 49:168–217

Kaveh R, Li YS, Ranjbar S, Tehrani R, Brueck CL, Aken BV (2013) Changes in *Arabidopsis thaliana* gene expression in response to silver nanoparticles and silver ions. Environ Sci Technol 47:10637–10644

Kazemipour S, Hashemabadi D, Kaviani B (2013) Effect of silver nanoparticles on the vase life and quality of cut chrysanthemum (*Chrysanthemum morifolium* L.) flower. Eur. J Exp Biol 3:298–302

Khan MN, Siddiqui MH, Mohammad F, Naeem M, Khan MMA (2010) Calcium chloride and gibberellic acid protect Linseed (*Linum usitatissimum* L.) from NaCl stress by inducing antioxidative defence system and osmoprotectant accumulation. Acta Physiol Plant 32:121–132

Khan MN, Siddiqui MH, Mohammad F, Naeem M (2012) Interactive role of nitric oxide and calcium chloride in the tolerance of plants to salt stress. Nitric Oxide 27:210–218

Khan MN, Mohammad F, Mobin M, Saqib MA (2014) Tolerance of plants to abiotic stress: A role of nitric oxide and calcium. In: Khan MN, Mobin M, Mohammad F, Corpas FJ (eds) Nitric oxide in plants: metabolism and role. Springer, Cham

Khan MN, Mobin M, Abbas Z, AlMutairi K, Siddiqui Z (2017) Role of nanomaterials in plants under challenging environments. Plant Physiol Biochem 110:194–209

Kheiri A, Jorf SM, Malihipour A, Saremi H, Nikkhah M (2016) Application of chitosan and chitosan nanoparticles for the control of Fusarium head blight of wheat (*Fusarium graminearum*) in vitro and greenhouse. Int J Biol Macromol 93:1261–1272

Khodakovskaya MV, de Silva K, Nedosekin DA, Dervishi E, Biris AS, Shashkov EV, Galanzha EI, Zharov VP (2011) Complex genetic, photothermal, and photoacoustic analysis of nanoparticle-plant interactions. Proc Natl Acad Sci U S A 108:1028–1033

Khodakovskaya MV, Kim BS, Kim JN, Alimohammadi M, Dervishi E, Mustafa T, Cernigla C (2013) Carbon nanotubes as plant growth regulators: effects on tomato growth, reproductive system, and soil microbial community. Small 9:115–123

Khot LR, Sankaran S, Maja JM, Ehsani R, Schuster EW (2012) Applications of nanomaterials in agricultural production and crop protection: a review. Crop Prot 35:64–70

Kiapour H, Moaveni P, Habibi D, Sani B (2015) Evaluation of the application of gibberellic acid and titanium dioxide nanoparticles under drought stress on some traits of basil (*Ocimum basilicum* L.). Int J Agron Agric Res 6:138–150

Kim JH, Oh Y, Yoon H, Hwang I, Chang YS (2015) Iron nanoparticle-induced activation of plasma membrane H^+-ATPase promotes stomatal opening in *Arabidopsis thaliana*. Environ Sci Technol 49:1113–1119

Kohan-Baghkheirati E, Geisler-Lee J (2015) Gene expression, protein function and pathways of Arabidopsis thaliana responding to silver nanoparticles in comparison to silver ions, cold, salt, drought, and heat. Nanomaterials 5:436–467

Komatsu S, Yamamoto R, Nanjo Y, Mikami Y, Yunokawa H, Sakata K (2009) A comprehensive analysis of the soybean genes and proteins expressed under flooding stress using transcriptome and proteome techniques. J Proteome Res 8:4766–4778

Komatsu S, Hiraga S, Yanagawa Y (2012) Proteomics techniques for the development of flood tolerant crops. J Proteome Res 11:68–78

Kumaraswamy RV, Kumari S, Choudhary RC, Pal A, Raliya R, Biswas P, Saharan V (2018) Engineered chitosan based nanomaterials: bioactivity, mechanisms and perspectives in plant protection and growth. Int J Biol Macromol 113:494–506

Kumari M, Pandey S, Bhattacharya A, Mishra A, Nautiyal CS (2017) Protective role of biosynthesized silver nanoparticles against early blight disease in Solanum lycopersicum. Plant Physiol Biochem 121:216–225

Lamsal K, Kim SW, Jung JH, Kim YS, Kim KS, Lee YS (2011a) Inhibition effects of silver nanoparticles against powdery mildews on cucumber and pumpkin. Mycobiology 39(1):26–32

Lamsal K, Kim SW, Jung JH, Kim YS, Kim KS, Lee YS (2011b) Application of silver nanoparticles for the control of Colletotrichum species in vitro and pepper anthracnose disease in field. Mycobiology 39(3):194–199

Landa P, Vankova R, Andrlova J, Hodekd J, Marsik P, Storchov H, White JC, Vanek T (2012) Nanoparticle-specific changes in *Arabidopsis thaliana* gene expression after exposure to ZnO, TiO_2, and fullerene soot. J Hazard Mater 241–242:55–62

Larue C, Laurette J, Herlin-Boime N, Khodja H, Fayard B, Flank A, Brisset F, Carriere M (2012) Accumulation, translocation and impact of TiO_2 nanoparticles in wheat (*Triticum aestivum* spp.): influence of diameter and crystal phase. Sci Total Environ 431:197–208

Lei Z, Mingyu S, Xiao W, Chao L, Chunxiang Q, Liang C, Hao H, Xiaoqing L, Fashui H (2008) Antioxidant stress is promoted by nano-anatase in spinach chloroplasts under UV-B radiation. Biol Trace Elem Res 121:69–79

Li Z, Huang J (2014) Effects of nanoparticle hydroxyapatite on growth and antioxidant system in pakchoi (*Brassica chinensis* L.) from cadmium-contaminated soil. J Nanomater 2014:1–7

Lin D, Xing B (2007) Phytotoxicity of nanoparticles: inhibition of seed germination and root growth. Environ Pollut 150:243–250

Linh TM, Mai NC, Hoe PT, Lien LQ, Ban NK, Hien LTT, Chau NH, Van NT (2020) Metal-based nanoparticles enhance drought tolerance in soybean. J Nanomater 2020:4056563

Linkemer G, Board JE, Musgrave ME (1998) Waterlogging effects on growth and yield components in late-planted soybean. Crop Sci 38:1576–1584

Liu R, Lal R (2012) Nanoenhanced materials for reclamation of mine lands and other degraded soils: a review. J Nanotechnol 461468:18

Liu R, Lal R (2015) Potentials of engineered nanoparticles as fertilizers for increasing agronomic productions. Sci Total Environ 514:131–139

Liu F, Wen LX, Li ZZ, Yu W, Sun HY, Chen JF (2006) Porous hollow silica nanoparticles as controlled delivery system for water soluble pesticide. Mater Res Bull 41:2268–2275

Liu YF, Qi MF, Li TL (2012) Photosynthesis, photoinhibition, and antioxidant system in tomato leaves stressed by low night temperature and their subsequent recovery. Plant Sci 196:8–17

Ma X, Geiser-Lee J, Deng Y, Kolmakov A (2010) Interactions between engineered nanoparticles (ENPs) and plants: phytotoxicity, uptake and accumulation. Sci Total Environ 408:3053–3061
Mackerness S, John CF, Jordan B, Thomas B (2001) Early signalling components in ultraviolet-B responses: distinct roles for different reactive oxygen species and nitric oxide. FEBS Lett 489:237–242
Mahajan P, Dhoke SK, Khanna AS (2011) Effect of nano-ZnO particle suspension on growth of mung (*Vigna radiata*) and gram (*Cicer arietinum*) seedlings using plant agar method. J Nanotechnol 696535:1–7
Majeed ZH, Taha MR (2013) A review of stabilization of soils by using nanomaterials. Aust J Basic Appl Sci 7:576–581
Mallaiah B (2015) Integrated approaches for the management of crossandra (*Crossandra infundibuliformis* L. Nees) wilt caused by Fusarium incarnatum (Desm.) Sacc, Doctoral dissertation, Tamil Nadu Agricultural University, Madurai
Manikandan A, Sathiyabama M (2016) Preparation of chitosan nanoparticles and its effect on detached rice leaves infected with *Pyricularia grisea*. Int J Biol Macromol 84:58–61
Marmiroli M, Imperiale D, Pagano L, Villani M, Zappettini A, Marmiroli N (2015) The proteomic response of arabidopsis thaliana to cadmium sulfide quantum dots, and its correlation with the transcriptomic response. Front Plant Sci 6:1140
Martínez-Fern-andez D, Vítkov-a M, Bernal MP, Kom-arek M (2015) Effects of nano-maghemite on trace element accumulation and drought response of *Helianthus annuus* L. in a contaminated mine soil. Water Air Soil Pollut 226:101
Miao Y, Xu J, Shen Y, Chen L, Bian Y, Hu Y, Zhou W, Zheng F, Man N, Shen Y, Zhang Y, Wang M, Wen L (2014) Nanoparticle as signalling protein mimic: robust structural and functional modulation of CaMKII upon specific binding to fullerene C60 nanocrystals. ACS Nano 8:6131–6144
Miller G, Shulaev V, Mittler R (2008) Reactive oxygen signalling and abiotic stress. Physiol Plant 133(3):481–489
Mirzajani F, Askari H, Hamzelou S, Schober Y, Römpp A, Ghassempour A, Spengler B (2014) Proteomics study of silver nanoparticles toxicity on Oryza sativa L. Ecotoxicol Environ Saf 108:335–339
Mishra S, Singh BR, Singh A, Keswani C, Naqvi AH, Singh HB (2014) Biofabricated silver nanoparticles act as a strong fungicide against *Bipolaris sorokiniana* causing spot blotch disease in wheat. PLoS One 9(5):e97881
Mishra S, Singh BR, Naqvi AH, Singh HB (2017) Potential of biosynthesized silver nanoparticles using Stenotrophomonas sp. BHU-S7 (MTCC 5978) for management of soil-borne and foliar phytopathogens. Sci Rep 7:45154
Mittler R (2002) Oxidative stress, antioxidants and stress tolerance. Trends Plant Sci 7(9):405–410
Moaveni P, Kheiri T (2011) Second international conference on agricultural and animal science, Maldives, 25–27 Nov 2011. IACSIT Press, Singapore, vol 22, pp 160–163
Mohammadi R, Maali-Amiri R, Abbasi A (2013) Effect of TiO_2 nanoparticles on chickpea response to cold stress. Biol Trace Elem Res 152:403–410
Mohammadi R, MaaliAmiri R, Mantri N (2014) Effect of TiO_2 nanoparticles on oxidative damage and antioxidant defence systems in chickpea seedlings during cold stress. Russ J Plant Physiol 61:768–775
Moller IM, Jensen PE, Hansson A (2007) Oxidative modifications to cellular components in plants. Ann Rev Plant Biol 58:459–481
Mustafa G, Sakata K, Hossain Z, Komatsu S (2015a) Proteomic study on the effects of silver nanoparticles on soybean under flooding stress. J Proteom 122:100–118
Mustafa G, Sakata K, Hossain Z, Komatsu S (2015b) Proteomic analysis of flooded soybean root exposed to aluminum oxide nanoparticles. J Proteom 128:280–297
Naderi MR, Danesh-Shahraki A (2013) Nanofertilizers and their roles in sustainable agriculture. Int J Agri Crop Sci 5(19):2229–2232

Nandini B, Hariprasad P, Prakash HS, Shetty HS, Geetha N (2017) Trichogenic-selenium nanoparticles enhance disease suppressive ability of Trichoderma against downy mildew disease caused by *Sclerospora graminicola* in pearl millet. Sci Rep 7(1):2612

Navarro-León E, Albacete A, Torre-González A, Ruiz JM, Blasco B (2016) Phytohormone profile in *Lactuca sativa* and *Brassica oleracea* plants grown under Zn deficiency. Phytochemistry 130:85–89

Nejad MS, Bonjar G, Khatami M, Amini A, Aghighi S (2016) In vitro and in vivo antifungal properties of silver nanoparticles against *Rhizoctonia solani*, a common agent of rice sheath blight disease. IET Nanobiotechnol 11(3):236–240

Ngan HTM, Tung HT, Le BV, Nhut DT (2020) Evaluation of root growth, antioxidant enzyme activity and mineral absorbability of carnation (*Dianthus caryophyllus* "Express golem") plantlets cultured in two culture systems supplemented with iron nanoparticles. Sci Hortic-Amsterdam 272:109612

Noreen S, Fatima Z, Ahmad S, Ashraf M (2018) Foliar application of micronutrients in mitigating abiotic stress in crop plants. In: Plant nutrients and abiotic stress tolerance. Springer, Singapore, pp 95–117

Ocsoy I, Paret M, Ocsoy M, Kunwar S, Chen T, You M, Tan W (2013) Nanotechnology in plant disease management: DNA-directed silver nanoparticles on graphene oxide as an antibacterial against *Xanthomonas perforans*. ACS Nano 7(10):8972–8980

Ogunkunle CO, Odulaja DA, Akande FO, Varun M, Vishwakarma V, Fatoba PO (2020) Cadmium toxicity in cowpea plant: Effect of foliar intervention of nano-TiO_2 on tissue Cd bioaccumulation, stress enzymes and potential dietary health risk. J Biotechnol 310:54–61

Oukarroum A, Bras S, Perreault F, Popovic R (2012) Inhibitory effects of silver nanoparticles in two green algae, Chlorella vulgaris and Dunaliella tertiolecta. Ecotoxicol Environ Saf 78:80–85

Oussou-Azo AF, Nakama T, Nakamura M, Futagami T, Vestergaard MCM (2020) Antifungal potential of nanostructured crystalline copper and its oxide forms. Nanomaterials 10:1003

Ouzounidou G, Gaitis F (2011) The use of nano-technology in shelf life extension of green vegetables. J Innov Econ Manag 2:163–171

Palmqvist NGM, Seisenbaeva GA, Svedlindh P, Kessler VG (2017) Maghemite nanoparticles acts as nanozymes, improving growth and abiotic stress tolerance in Brassica napus. Nanoscale Res Lett 12:631

Paret ML, Palmateer AJ, Knox GW (2013) Evaluation of a light-activated nanoparticle formulation of titanium dioxide with zinc for management of bacterial leaf spot on rosa 'Noare'. Hortscience 48(2):189–192

Ponmurugan P, Manjukarunambika K, Elango V, Gnanamangai BM (2016) Antifungal activity of biosynthesised copper nanoparticles evaluated against red root-rot disease in tea plants. J Exp Nanosci 11(13):1019–1031

Pradhan S, Mailapalli DR (2017) Interaction of engineered nanoparticles with the agrienvironment. J Agric Food Chem 65:8279–8294

Prasad P, Pisipati S, Momčilović I, Ristic Z (2011) Independent and combined effects of high temperature and drought stress during grain filling on plant yield and chloroplast EF-Tu expression in spring wheat. J Agron Crop Sci 197(6):430–441

Prasad R, Bhattacharyya A, Nguyen QD (2017) Nanotechnology in sustainable agriculture: recent developments, challenges, and perspectives. Front Microbiol 8:1014

Prochazkova D, Wilhelmova N (2007) Leaf senescence and activities of the antioxidant enzymes. Biol Plant 51:401–406

Qados A (2015) Mechanism of nanosilicon-mediated alleviation of salinity stress in faba bean (Vicia faba L.) plants. Am J Exp Agric 7:78–95

Qados A, Moftah AE (2015) Influence of silicon and nano-silicon on germination, growth and yield of faba bean (*Vicia faba* L.) under salt stress conditions. Am J Exp Agric 5:509–524

Qi M, Liu Y, Li T (2013) Nano-TiO_2 improve the photosynthesis of tomato leaves under mild heat stress. Biol Trace Elem Res 156:323–328

Rai M, Ingle A (2012) Role of nanotechnology in agriculture with special reference to management of insect pests. Appl Microbiol Biotechnol 94(2):287–293

Rao NS, Laxman RH, Shivashankara KS (2016) Physiological and morphological responses of horticultural crops to abiotic stresses. In: Abiotic stress physiology of horticultural crops. Springer, New Delhi, pp 3–17

Rascio N, Navari-Izzo F (2011) Heavy metal hyperaccumulating plants: how and why do they do it? And what makes them so interesting? Plant Sci 180:169–181

Regier N, Cosio C, Moos N, Slaveykova VI (2015) Effects of copper-oxide nanoparticles, dissolved copper and ultraviolet radiation on copper bioaccumulation, photosynthesis and oxidative stress in the aquatic macrophyte *Elodea nuttallii*. Chemosphere 128:56–61

Rezvani N, Sorooshzadeh A, Farhadi N (2012) Effect of nano-silver on growth of saffron in flooding stress. Int J Biol Biomol Agric Food Biotechnol Eng 6:11–16

Rico CM, Majumdar S, Duarte-Gardea M, Peralta-Videa JR, Gardea- Torresdey JL (2011) Interaction of nanoparticles with edible plants and their possible implications in the food chain. J Agric Food Chem 59:3485–3498

Rico CM, Hong J, Morales MI, Zhao L, Barrios AC, Zhang JY, Peralta-Videa JR, Gardea-Torresdey JL (2013a) Effect of cerium oxide nanoparticles on rice: a study involving the antioxidant defence system and in vivo fluorescence imaging. Environ Sci Technol 47:5635–5642

Rico CM, Morales MI, McCreary R, Castillo-Michel H, Barrios AC, Hong J, Tafoya A, Lee WY, Varela-Ramirez A, Peralta-Videa JR, Gardea-Torresdey JL (2013b) Cerium oxide nanoparticles modify the antioxidative stress enzyme activities and macromolecule composition in rice seedlings. Environ Sci Technol 47:14110–14118

Rodrigues SM, Demokritou P, Dokoozlian N, Hendren CO, Karn B, Mauter MS, Sadik OA, Safarpour M, Unrine JM, Viers J, Welle P, White J, Wiesner M, Lowry G (2017) Nanotechnology for sustainable food production: promising opportunities and scientific challenges. Environ Sci Nano 4(4):767–781

Rouhani M, Samih MA, Kalantari S (2013) Insecticidal effect of silica and silver nanoparticles on the cowpea seed beetle, *Callosobruchus maculatus* F. (Col.: Bruchidae). J Entomol Res 4(4):297–305

Roy A, Bhattacharya J (2012) Removal of Cu (II), Zn (II) and Pb (II) from water using microwave-assisted synthesized maghemite nanotubes. Chem Eng J 211–212:493–500

Sabaghnia N, Janmohammadi M (2014) Effect of nanosilicon particles application on salinity tolerance in early growth of some lentil genotypes. Ann UMCS Biol 69:39–55

Salem HF, Kam E, Sharaf MA (2011) Formulation and evaluation of silver nanoparticles as antibacterial and antifungal agents with a minimal cytotoxic effect. Int J Drug Delivery 3(2):293

Sathiyabama M, Manikandan A (2016) Chitosan nanoparticle induced defence responses in finger millet plants against blast disease caused by *Pyricularia grisea* (Cke.) Sacc. Carbohydr Polym 154:241–246

Sathiyabama M, Manikandan A (2018) Application of copper-chitosan nanoparticles stimulate growth and induce resistance in finger millet (*Eleusine coracana* Gaertn.) plants against blast disease. J Agric Food Chem 66(8):1784–1790

Sathiyabama M, Parthasarathy R (2016) Biological preparation of chitosan nanoparticles and its in vitro antifungal efficacy against some phytopathogenic fungi. Carbohydr Polym 151:321–325

Savi GD, Piacentini KC, de Souza SR, Costa ME, Santos CM, Scussel VM (2015) Efficacy of zinc compounds in controlling Fusarium head blight and deoxy nivalenol formation in wheat (*Triticum aestivum* L.). Int J Food Microbiol 205:98–104

Savicka M, Skute N (2010) Effects of high temperature on malondialdehyde content, superoxide production and growth changes in wheat seedlings (Triticum aestivum L.). Ekologija 56:26–33

Sedghi M, Hadi M, Toluie SG (2013) Effect of nano zinc oxide on the germination parameters of soybean seeds under drought stress. Ann West Univ Timisoara Ser Biol 16:73–78

Servin A, Elmer W, Mukherjee A, Torre-Roche R, Hamdi H, White J, Bindraban P, Dimkpa C (2015) A review of the use of engineered nanomaterials to suppress plant disease and enhance crop yield. J Nanopart Res 17(2):92

Shahrokh S, Hosseinkhani B, Emtiazi G (2014) The impact of silver nanoparticles on bacterial aerobic nitrate reduction process. J Bioprocess Biotechnol 4:152

Sharma P, Jha AB, Dubey RS, Pessarakli M (2012) Reactive oxygen species, oxidative damage, and antioxidant defence mechanisms in plants under stressful conditions. J Bot 2012:1–26

Sheet I, Kabbani A, Holail H (2014) Removal of heavy metals using nanostructured graphite oxide, silica nanoparticles and silica/graphite oxide composite. Energy Proc 50:130–138

Shen CX, Zhang QF, Li J, Bi FC, Yao N (2010a) Induction of programmed cell death in Arabidopsis and rice by single-wall carbon nanotubes. Am J Bot 97:1–8

Shen X, Zhou Y, Duan L, Li Z, Eneji AE, Li J (2010b) Silicon effects on photosynthesis and antioxidant parameters of soybean seedlings under drought and ultraviolet-B radiation. J Plant Physiol 167:1248–1252

Sicard C, Perullini M, Spedalieri C, Coradin T, Brayner R, Livage J, Jobbagy M, Bilmes SA (2011) CeO2 nanoparticles for the protection of photosynthetic organisms immobilized in silica gels. Chem Mater 23:1374–1378

Siddiqui M, Al-Whaibi M, Faisal M, Al Sahli A (2014) Nano-silicon dioxide mitigates the adverse effects of salt stress on *Cucurbita pepo* L. Environ Toxicol Chem 33:2429–2437

Simon DF, Domingos RF, Hauser C, Hutchins CM, Zerges W, Wilkinson KJ (2013) Transcriptome sequencing (RNA-seq) analysis of the effects of metal nanoparticle exposure on the transcriptome of Chlamydomonas reinhardtii. Appl Environ Microbiol 79:4774–4785

Singh J, Lee BK (2016) Influence of nano-TiO_2 particles on the bioaccumulation of Cd in soybean plants (Glycine max): a possible mechanism for the removal of Cd from the contaminated soil. J Environ Manag 170:88–96

Singh G, Rattanpal H (2014) Use of nanotechnology in horticulture: a review. Int J Agric Sci Vet Med 2:34–42

Soliman S, El-feky SA, Darwish E (2015) Alleviation of salt stress on *Moringa peregrina* using foliar application of nanofertilizers. J Hortic For 7:36–47

Stadler T, Buteler M, Weaver DK (2010) Novel use of nanostructured alumina as an insecticide. Pest Manage Sci 66(6):577–579

Strayer-Scherer A, Liao Y, Young M, Ritchie L, Vallad G, Santra S, Paret ML (2018) Advanced copper composites against copper-tolerant *Xanthomonas perforans* and tomato bacterial spot. Phytopathology 108(2):196–205

Suman PR, Jain VK, Varma A (2010) Role of nanomaterials in symbiotic fungus growth enhancement. Curr Sci 99:1189–1191

Sun L, Song F, Zhu X, Liu S, Liu F, Wang Y, Li X (2020) Nano-ZnO alleviates drought stress via modulating the plant water use and carbohydrate metabolism in maize. Arch Agron Soil Sci. https://doi.org/10.1080/03650340.2020.1723003

Suzuki K, Nagasuga K, Okada M (2008) The chilling injury induced by high root temperature in the leaves of rice seedlings. Plant Cell Physiol 49:433–442

Tang Y, Tian J, Li S, Xue C, Xue Z, Yin D, Yu S (2015) Combined effects of graphene oxide and Cd on the photosynthetic capacity and survival of *Microcystis aeruginosa*. Sci Total Environ 532:154–161

Torabian S, Zahedi M, Khoshgoftar AH (2016) Effects of foliar spray of two kinds of zinc oxide on the growth and ion concentration of sunflower cultivars under salt stress. J Plant Nutr 39:172–180

Tripathi DK, Singh VP, Prasad SM, Chauhan DK, Dubey NK (2015) Silicon nanoparticles (SiNp) alleviate chromium (VI) phytotoxicity in *Pisum sativum* (L.) seedlings. Plant Physiol Biochem 96:189–198

Trueman S (2020). Plant stresses: abiotic and biotic stresses. https://www.thoughtco.com/plant-stresses-abiotic-and-biotic-stresses-419223

Vannini C, Domingo G, Onelli E, Prinsi B, Marsoni M, Espen L, Bracale M (2013) Morphological and proteomic responses of Eruca sativa exposed to silver nanoparticles or silver nitrate. PLoS One 8:e68752

Visser E, Nabben R, Blom C, Voesenek L (1997) Elongation by primary lateral roots and adventitious roots during conditions of hypoxia and high ethylene concentrations. Plant Cell Environ 20:647–653

Wahid A (2007) Physiological implications of metabolites biosynthesis in net assimilation and heat stress tolerance of sugarcane (*Saccharum officinarum*) sprouts. J Plant Res 120:219–228

Wanyika H, Gatebe E, Kioni P, Tang Z, Gao Y (2012) Mesoporous silica nanoparticles carrier for urea: potential applications in agrochemical delivery systems. J Nanosci Nanotechnol 12:2221–2228

Wei H, Wang E (2013) Nanomaterials with Enzyme-Like Characteristics (Nanozymes): Next-Generation Artificial Enzymes. Chem Soc Rev 42:6060–6093

Worms I, Boltzman J, Garcia M, Slaveykova VI (2012) Cell-wall-dependent effect of carboxyl-CdSe/ZnS quantum dots on lead and copper availability to green microalgae. Environ Pollut 167:27–33

Wu W, Ma B (2015) Integrated nutrient management (INM) for sustaining crop productivity and reducing environmental impact: a review. Sci Total Environ 512–513:415–427

Wu F, Fang Q, Yan S, Pan L, Tang X, Ye W (2020) Effects of zinc oxide nanoparticles on arsenic stress in rice (*Oryza sativa* L.): germination, early growth, and arsenic uptake. Environ Sci Pollut Res 27:26974–26981

Xie L, Chen F, Du H, Zhang X, Wang X, Yao G, Xu B (2020) Graphene oxide and indole-3-acetic acid co-treatment regulates the root growth of *Brassica napus* L. via multiple phytohormone pathways. BMC Plant Biol 20:101

Xing K, Xing Y, Liu Y, Zhang Y, Shen X, Li X, Qin S (2018) Fungicidal effect of chitosan via inducing membrane disturbance against Ceratocystis fimbriata. Carbohydr Polym 192:95–103

Xu GY, Rocha P, Wang ML, Xu ML, Cui YC, Li LY, Zhu YX, Xia X (2011) A novel rice calmodulin-like gene, OsMSR2, enhances drought and salt tolerance and increases ABA sensitivity in Arabidopsis. Planta 234:47–59

Yadav TP, Yadav RM, Singh DP (2012) Mechanical milling: a top down approach for the synthesis of nanomaterials and nanocomposites. Nanosci Nanotechnol 2:22–48

Yadav T, Mungray AA, Mungray AK (2014) Fabricated nanoparticles: current status and potential phytotoxic threats. In: Whitacre DM (ed) Reviews of environmental contamination and toxicology. Springer, Cham

Yan J, Tsuichihara N, Etoh T, Iwai S (2007) Reactive oxygen species and nitric oxide are involved in ABA inhibition of stomatal opening. Plant Cell Environ 30(10):1320–1325

Yang FL, Li XG, Zhu F, Lei CL (2009) Structural characterization of nanoparticles loaded with garlic essential oil and their insecticidal activity against *Tribolium castaneum* (Herbst) (Coleoptera: Tenebrionidae). J Agric Food Chem 57(21):10156–10162

Yang WW, Miao AJ, Yang LY (2012) Cd^{2+} toxicity to a green alga *Chlamydomonas reinhardtii* as influenced by its adsorption on TiO_2 engineered nanoparticles. PLoS One 7:e32300

Yin YH, Guo QM, Yun H, Wang LJ, Wan SQ (2012) Preparation, characterization and nematicidal activity of lansiumamide B nano-capsules. J Integr Agric 11(7):1151–1158

Yordanova R, Popova L (2007) Effect of exogenous treatment with salicylic acid on photosynthetic activity and antioxidant capacity of chilled wheat plants. Gen Appl Plant Physiol 33:155–170

Zaimenko NV, Didyk NP, Dzyuba OI, Zakrasov OV, Rositska NV, Viter AV (2014) Enhancement of drought resistance in wheat and corn by nanoparticles of natural mineral analcite. Ecol Balk 6:1–10

Zakharova OV, Gusev AA, Zherebin PM, Skripnikova EV, Skripnikova MK, Ryzhikh VE, Krutyakov YA (2017) Sodium tallow ampho polycarboxy glycinate stabilized silver nanoparticles suppress early and late blight of *Solanum lycopersicum* and stimulate the growth of tomato plants. Bionanoscience 7(4):692–702

Zareii FD, Roozbahani A, Hosnamidi A (2014) Evaluation the effect of water stress and foliar application of Fe nanoparticles on yield, yield components and oil percentage of safflower (*Carthamus tinctorius* L.). Int J Adv Biol Biomed Res 2:1150–1159

Zhang W (2018) Global pesticide use: profile, trend, cost/benefit and more. Proc Int Acad Ecol Environ Sci 8(1):1–27

Zhao L, Sun Y, Hernandez-Viezcas JA, Servin AD, Hong J, Niu G, Peralta-Videa J, Duarte-Gardea M, Gardea-Torresdey J (2013) Influence of CeO_2 and ZnO nanoparticles on cucumber physiological markers and bioaccumulation of Ce and Zn: a life cycle study. J Agric Food Chem 61:11945–11951

Zheng L, Hong F, Lu S, Liu C (2005) Effect of nano-TiO_2 on strength of naturally aged seeds and growth of spinach. Biol Trace Elem Res 104:83–91

Ziaee M, Ganji Z (2016) Insecticidal efficacy of silica nanoparticles against *Rhyzopertha dominica* F. and *Tribolium confusum* Jacquelin du Val. J Plant Prot Res 56(3):250–256

Chapter 14
Production of Plant Hormones from Algae and Its Relation to Plant Growth

Sanaa M. Shanab and Emad A. Shalaby

Contents

S. M. Shanab
Botany and Microbiology Department, Faculty of Science, Cairo University, Giza, Egypt

E. A. Shalaby (✉)
Biochemistry Department, Faculty of Agriculture, Cairo University, Giza, Egypt

© The Author(s), under exclusive license to Springer Nature Switzerland AG 2021

H. I. Mohamed et al. (eds.), *Plant Growth-Promoting Microbes for Sustainable Biotic and Abiotic Stress Management*,
https://doi.org/10.1007/978-3-030-66587-6_14

1 Introduction

Plant growth regulators (such as auxins, cytokinins, gibberellin, abscisic acid, and ethylene) play an important role in mediating growth of different plant species at very low concentrations as well as signaling environmental alterations or changes, initiating stress responses (biotic and abiotic) and indicator molecules in the regulation of almost all phases of plant growth and development (maturation) from embryogenesis to senescence (Li et al. 2010).

Seaweed and microalgae extracts are used as growth stimulants or growth regulators in cultivation of agricultural species due to its content from the plant growth regulator (hormones) concentrations (Stirk and Van Staden 2006).

In various algal species, phytohormones were recorded in significant amount when compared with their amount in plants, and the various biological activities of algal hormones corresponded to the functions of hormones in higher plant (Tarakhovskaya et al. 2007).

El Shoubaky and Salem (2009) investigated green macroalgae (*Ulva lactuca* and *Enteromorpha clathrate*) as biofertilizers due to their high concentrations of inorganic nutrients in addition to organic compounds and plant growth hormones. Phytohormones of microalgae and macroalgae are exogenous growth regulators, affecting the tolerance ability to different factors of various (abiotic and biotic) stress conditions (Romanenko et al. 2015).

Hormonal level may undergo changes when the alga is exposed to an alteration in natural environmental or laboratory conditions such as light (quantity, quality, and duration), temperature, salinity, etc. Moreover, Nimura and Mizuta (2002) reported that the endogenous abscisic acid (ABA) content of laminaria sporophyte increased as a result of its transition from the vegetative growth stage to the reproductive state. Also, in some microalgae, ABA was increased under salt stress or lowered moisture content. Polyamines and betaines are important active signal factors required for different processes in plant and algal development and participate in biotic and abiotic stress responses as illustrated by Kusano et al. (2008). These factors were recorded in different stressed microalgal and seaweed species (Mackinnon et al. 2010; Gebser and Pohnert 2013).

In the following sections, we will summarize the major plant hormones and its analogues present in algal species and its physiological functions and methods of extractions and determinations and if any abiotic stress factors can affect the accumulation of these regulators inside the algal species.

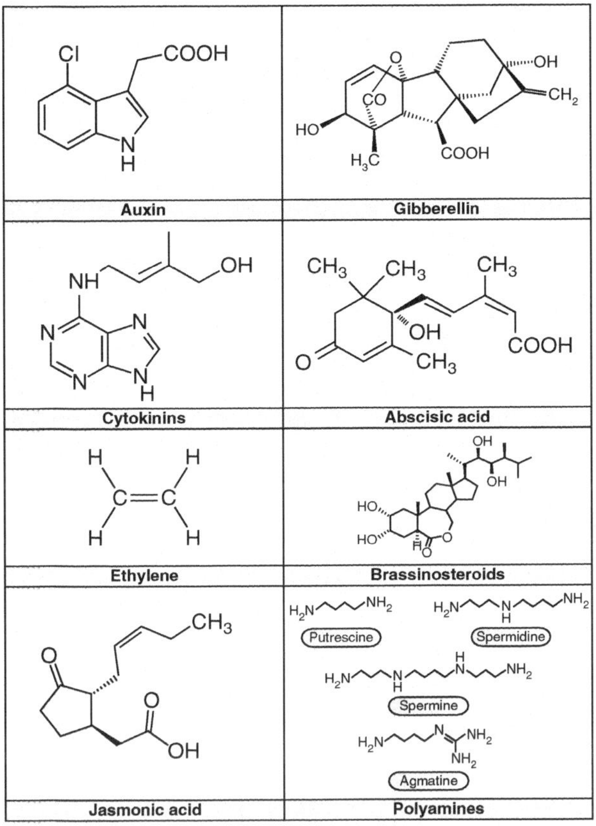

The chemical structure of major phytohormones and other regulators produced by algae

2 Algal Hormones

2.1 Auxins

It is interesting to know that not only the hormonal substances produced by the highly evolved terrestrial plants are already produced by the lower primitive thallophytic algae, but they also function similarly. Regarding auxins, there have been numerous investigations that dealt with auxin production (especially IAA) in many algal species (microalgae, macroalgae, and cyanobacteria).

Starting with Du Buy and Olson (1937) who reported the presence of auxin in the tissue of *Fucus vesiculosus*, occasionally, other scientists recorded the presence of auxin in *Bryopsis muscosa*. Also, indole acetic acid (IAA) were found in the brown seaweeds *Fucus*, *Macrocystis*, and *Desmarestia* sp. Few years later (Skibola 2004; Tarakhovskaya et al. 2007; Li et al. 2007), a growth substance similar to IAA was found in *Laminaria agardhii*. These previous investigations proved their findings by the use of bioassay of Avena coleoptile curvature.

This was followed by a lot of investigations which dealt with the presence of auxinic substance in many algal species as well as the isolation and identification of indolic substance using different analytical methods. The previously recorded studies confirmed the production of auxins in various algal species belonging to different divisions, but they also proved that the hormonal function and its catabolism followed the same pathway as that in angiosperms (Sitnik et al. 2003; Stirk et al. 2009). Researches continued in this field till now, recording the presence of one or more plant hormones in different algal species, identifying its (or their) chemical structure by chromatographic analysis, and confirming its hormonal properties by specific plant bioassays (Table 14.1 and Figs. 14.1, 14.2, and 14.3). Auxins in algal thalli varied from season to season, and developmental stage and highest concentration were recorded especially in summer season and in vegetative tissues (EL Shoubaky and Salem 2016; Mori et al. 2017).

The effect of different culture conditions especially a biotic stress (such as concentration of L-tryptophan, acidity degree, and light conditions) on the synthesis of indole by *Spirulina* sp. was reported by Ahmed et al. (2010). It was found that the formation of IAA in *Spirulina* sp. was organized by 1.5 μg/mL L-tryptophan concentration. Moreover, the height amount of IAA was found at pH 6 in light-dark cycle 8:16 h. However, in the dark, auxin synthesis was not observed.

IAA is the naturally occurring growth regulator in the kingdom of plant. It is present in very low concentration (0.5–15 μg/kg) and is in equilibrium with bound forms (such as glucose, ester, aldehyde, and more complex forms as glucobrassicin).

IAA is synthesized in different plants from the tryptophan (amino acid) by different biosynthetic routes.

2.1.1 Physiological Properties of Auxins

1. Initiation of root formation.
2. Apical dominance.
3. Tropisms.
4. Differentiation of phloem elements.
5. Induction of elongation.

2.1.2 Separation and Detection

Auxins can be separated by paper chromatography (PC) using the mobile phase/isopropanol/ammonium hydroxide/water (8:1:1).

It can be separated by thin layer chromatography (TLC F_{254}) using the mobile phase/chloroform/ethyl acetate/formic acid (5:4:1); standard indoles are used for comparison.

Table 14.1 Seaweed species producing auxins

Algal species	Division
Fucus vesiculosus	Phaeophyta
Macrocystis sp.	
Desmarestia sp.	
Fucus sp.	
Ascophyllum sp.	
Ascophyllum nodosum	
Laminaria agardhii	
Laminaria sp.	
Undaria pinnatifida	
Pylaiella littoralis	
Fucus vesiculosus	
Ecklonia maxima	
Macrocystis pyrifera	
Dictyota humifusa	
Sargassum heterophyllum	
Laminaria japonica	
Valonia utricularis	Chlorophyta
Valonia macrophysa	
Bryopsis muscosa	
Acetabularia sp.	
Cladophora sp.	
Ulva pertusa	
Enteromorpha compressa	
Caulerpa paspaloides	
Enteromorpha prolifera	
Ulva fasciata	
Ulva lactuca	
Ulva rigida	
Gelidium amansii	Rhodophyta
Eisenia bicyclis	
Ceramium rubrum	
Botryocladia sp.	
Porphyra sp.	
Pyropia yezoensis	
Bangia fuscopurpurea	
Sarconema filiforme	
Nemalion multifidum	
Furcellaria fastigiata	

To detect the separated spots, they can be sprayed with the coloring reagent DMAC (0.1 g of *p*-dimethylaminocinnamaldehyde in 10 mL concentrated HCl, then diluting to 200 mL with acetone). In case of using PC for separation, the PC paper will be dipped in the reagent, dried, and then heated at 65 °C for 2.5 min.

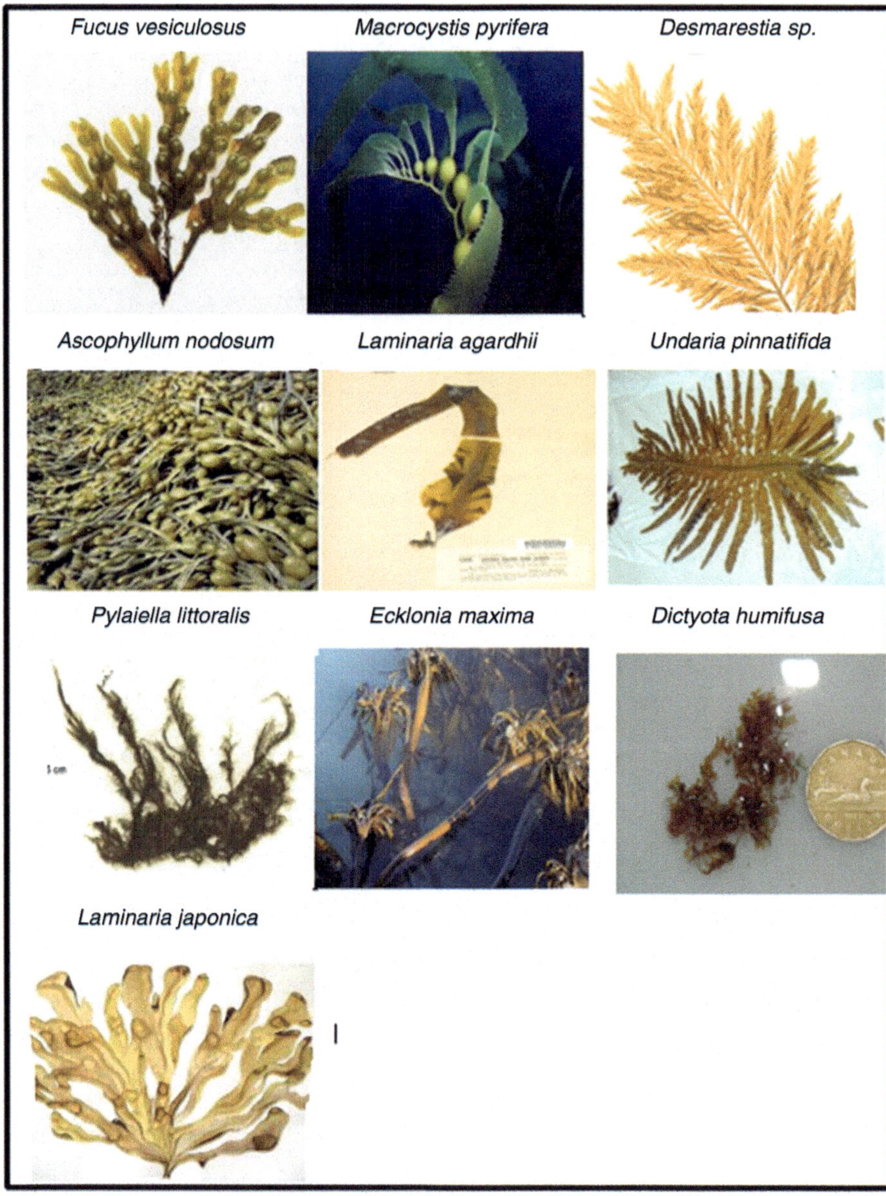

Fig. 14.1 Some Phaeophyta species producing auxins

Fig. 14.2 Some Chlorophyta species producing auxins

Another coloring reagent can be used in case of PC which is called Salkowski reagent (0.001 M ferric chloride in 5% perchloric acid), giving pink spots with $FeCl_3$/perchloric acid/ethanol reagent. Using TLC for separation of the indolic extract, the spray reagent is used as 0.25% of DMCA in ethanol/conc. HCl (1:1) and the color will develop over night at room temperature (purple spot).

2.1.3 Identification and Determination

For identification of indoles, spectral measurements in methanol must be at wavelength 220–320 nm. Indole acetic acid has fluorescence peak of 365 nm and activity peak of 285 nm. Identification of auxins are performed by LC/MS, GLC/MS, HPLC/MS, and GC/MS, which identify the indolic compound(s) (compared with

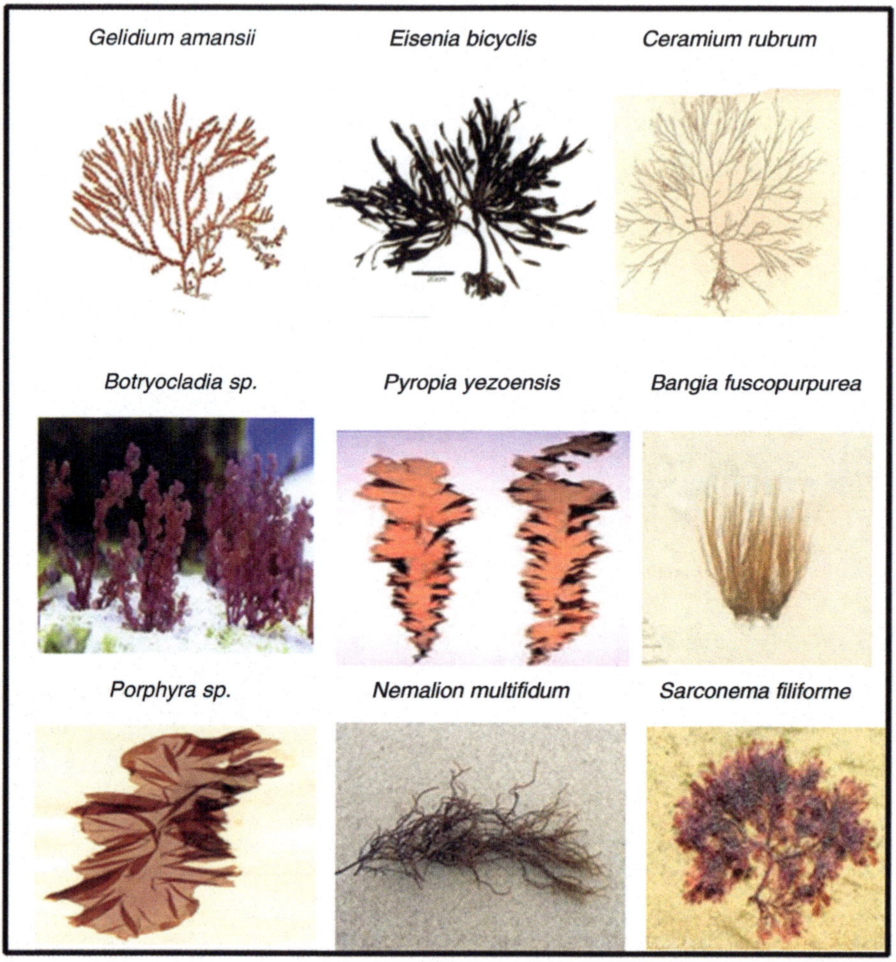

Fig. 14.3 Some Rhodophyta species producing auxins

either standard or not) at its specific retention time, and recording its chemical structure and formula.

2.1.4 Measurement of Its Characteristic Biological Activity

1. Using suitable plant bioassays (for each hormone compared to the control of synthetic standard).
2. Avena coleoptile curvature test.
3. Elongation of wheat (or barley) coleoptile sections.
4. Induction of rooting in cutting stem of mung bean.

2.2 Cytokinins

They are group of plant growth regulators consist of purines substituted in the six position. The first naturally occurring cytokinin was zeatin (from *Zea mays*) which was commonly found as riboside forms. Cytokinin was found to initiate all division in plant tissue culture during growth. It interferes with auxins in many developmental stages controlled by the balance on the ratios of cytokinin to indoles which may influence shoot and root differentiation and growth of lateral buds or remain at the undifferentiated callus stage. Cytokinins are synthesizing from adenosine-5-monophosphate producing iso-pentenyladenine (ip) which was believed to be the precursor of all other naturally occurring cytokinins. Different cytokinin-like substances were reported in different algal species belonging to various algal groups, as zeatin, dihydrozeatin, iso-pentenyladenine (ip), *N*6-methylaminopurine, and *N*6,8,8-dimethyl-allylaminopurine.

In Phaeophyta, Stirk et al. (2003) recorded cytokinin-like activity in *Fucus serratus*, *Ascophyllum nodosum*, *Ecklonia maxima*, *Laminaria saccharina*, *Fucus vesiculosus*, *Dictyota* sp., and *Sargassum heterophyllum* (coincide with the release of gametes), while Stirk and Van staden (1997) reported the cytokinin activity in the green seaweeds: *Ulva* sp., *Cladophora contexta*, *Codium capitatum*, *C. extricatum*, *Halimeda cuneata*, *Caulerpa racemosa*, and *Valonia macrophysa*.

Detection of cytokinin-like activity in the red seaweeds (Rhodophyta) was achieved by Yokoya et al. (2010) and Mori et al. (2017), who reported that this activity was demonstrated in *Galaxaura diesingiana*, *Gelidium amansii*, *Amphiroa bowbankii*, *A. ephedraea*, *Arthrocardia* sp., *Cheilosiphorum sagiltatum*, *Jania crassa*, *Plocamium corallorhiza*, *Hypnea rosea*, *H. spicifera*, *Spyridia hypnoides*, *Pyropia yezoensis*, and *Bangia fuscopurpurea*. In various algal groups (Stirk et al. 2009; Mori et al. 2017), aromatic cytokinins (topolins) were recorded as shown in Table 14.2 and Figs. 14.4, 14.5, and 14.6.

Regarding the response of algal cells to abiotic stress condition and its relation to phytohormone accumulation, Maršálek et al. (1992) found that during cultivation of microalgae species, the concentration of ABA in mother algal cultures was three times folded when compared with new culture. An increase in ABA level was observed during the first 24 h of microalgae cultivation in the absence of light condition and decreased in the following 24 h; at the same time, the amount of ABA was decreased gradually in light and dark environment conditions (14:10) (Stirk et al. 2014). Also, ABA softened the effect of various oxidative stress conditions, with positive relation to the activity of AO enzymes such as glutathione *S*-transferase, peroxidase, and catalase (Yoshida et al. 2003).

2.2.1 Separation and Detection

Separate the ammoniacal fraction by descending paper chromatography using propanol/ammonia/water (10:1:1) as described by Stirk and Van staden (1997) and using standard cytokinin (as kinetin) for comparison. *t*-Butanol/conc. NH_4OH/H_2O (3:1:1) or *n*-butanol/acetic acid/water (4:1:1) can be used as mobile systems. The separated cytokinin appears as dark spots in short UV light. Using TLC of alumina G, separation of cytokinin (as zeatin) can be performed using butanone saturated with water or EA saturated with H_2O or by chloroform/EthOH (9:1). Detection done using Dische reagent (spraying with 0.5 g cysteine hydrochloride in $3MH_2SO_4$ giving a pink color after 20 min).

Table 14.2 Seaweed species producing cytokinins

Algal species	Division
Fucus vesiculosus	Phaeophyta
Desmarestia sp.	
Undaria pinnatifida	
Bifurcaria brassicaeformis	
Ascophyllum nodosum	
Sargassum muticum	
Laminaria japonica	
L. pallida	
Macrocystis pyrifera	
Ecklonia maxima	
Laminaria saccharina	
Dictyota sp.	
Dictyota humifusa	
Sargassum heterophyllum	
Macrocystis angustifolia	
Splachnidium rugosum	
Ulva sp.	Chlorophyta
Ulva fasciata	
Cladophora coelothrix	
Codium capitatum	
C. extricatum	
Halimeda cuneata	
Caulerpa racemosa	
Caulerpa filiformis	
Valonia macrophysa	

(continued)

Table 14.2 (continued)

Algal species	Division
Amphiroa bowbankii	Rhodophyta
A. ephedraea	
Arthrocardia sp.	
Plocamium corallorhiza	
Cheilosporum sp.	
Hypnea spicifera	
Galaxaura diesingiana	
Gelidium amansii	
Cheilosporum sagittatum	
Jania crassa	
Hypnea rosea	
Griffithsia pacifica	
Pyropia yezoensis	
Bangia fuscopurpurea	
Aeodes orbitosa	
Gigartina clathrata	
Gigartina polycarpa	
Sarcothalia scutellata	
Hymenena venosa	
Nothogenia erinacea	
Plocamium corallorhiza	
Carradoeriella virgata	
Porphyra capensis	
Sarcothalia stiriata	
Suhria vittata	
Amphiroa bowerbankii	
Arthrocardia sp.	
Cheilosporum sp.	
Jania sp.	

2.2.2 Identification and Determination

Occur by using HPLC.

2.2.3 Bioassay/Biological Activity

The ability of cytokinin to promote growth of secondary phloem of carrot. The effect of cytokinin on barley germination.

Fig. 14.4 Some Phaeophyta species producing cytokinins

The cytokinin-like activity promoting cell division can be assayed by soybean callus culture.

2.2.4 Physiological Properties of cytokinins

1. Shoot and Root differentiation in tissue culture.
2. Growth of lateral buds and leaf expansion.
3. Chloroplast development.
4. Leaf senescence.
5. Morphogenesis in cultured tissues.

Fig. 14.5 Some Chlorophyta species producing cytokinins

2.3 *Gibberellins (GAs)*

It is clear from literatures that gibberellin-like substances are synthesized by different types of macroalgae. Bently (1960) suggested the presence of GA_3 like substances in microalgae and macroalgae. She revealed that there are two unknown components in acidic extracts of phytoplankton which have some growth stimulatory characteristics. Meanwhile, Stirk et al. (2013a, b) extracted gibberellin-like substance from *Fucus vesiculosus*, purified it by PC and identified one or an analogue of GA_1, GA_3, and GA_6 (~10 μg/kg F.wt) ($R_f = 0.3–0.4$). Furthermore, other data found gibberellin-like activity in *Fucus spiralis* (GA_1 or GA_3).

Detection of gibberellins in other brown seaweeds was recorded by many investigations (EL shoubaky and Salem 2016). Moreover, these scientists recorded gibberellin in the red seaweeds *Hypnea musciformis*, *Gracilaria corticata*, and *Porphyra leucostricta*. Table 14.3 and Figs. 14.7, 14.8, and 14.9 recorded some algal species producing gibberellins. Researches continued till now searching for plant hormones in seaweed species and applying recent techniques for their extraction, separation, identification, and determination (Stirk et al. 2013a, b). Gibberellins are group of hormones (belongs to diterpenoids) which stimulate plant growth and are widespread in plants and algae.

Fig. 14.6 Some Rhodophyta species producing cytokinins

In fact, more than 40 compounds of gibberellin structure have been recognized till now. The most familiar is gibberellic acid (GA_3).

2.3.1 Physiological Properties of Gibberellins

1. It promotes seed germination and organ differentiation.
2. It stimulates stem elongation and shoot growth.
3. It interferes with leaf expansion, development, and fruit maturity (Yamaguchi 2008; Sun 2010). It has positive effect with IAA to differentiation of cell and elongation of root but has negative effect with abscisic acid on growth and ger-

Table 14.3 Seaweed species producing gibberellins (GAs)

Algal species	Division
Cystoseira sticta	Phaeophyta
Fucus vesiculosus	
Fucus spiralis	
Sargassum plagiophyllum	
Ecklonia radiata	
Ascophyllum nodosum	
Enteromorpha flexuosa	Chlorophyta
Oedogonium cardiacum	
Caulerpa prolifera	
Codium fragile	
Enteromorpha prolifera	
Ulva lactuca	
Ulva rigida	
Gracilaria corticata	Rhodophyta
Porphyra leucostricta	
Hypnea musciformis	
Sarconema filiforme	

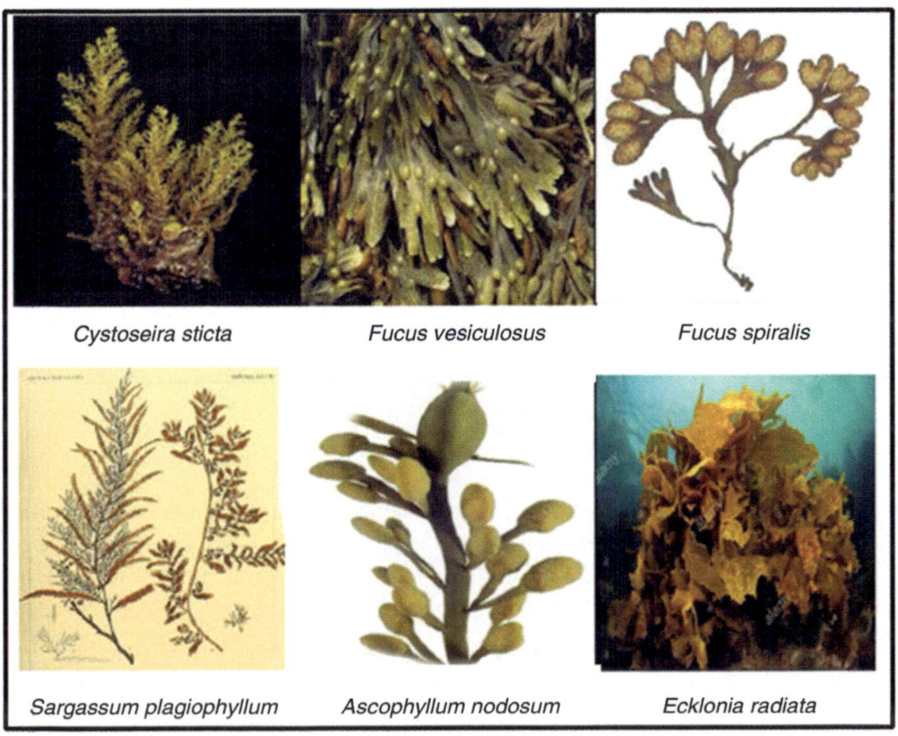

Fig. 14.7 Some Phaeophyta species producing gibberellins

Fig. 14.8 Some Chlorophyta species producing gibberellins

mination. Negative and positive effects depend on environmental conditions and factors with stress-related ethylene and negative effects with cytokinin concentration (Weiss and Ori 2007; Yamaguchi 2008). Gibberellins were synthesized from glyceraldehyde 3-phosphate in young shoot tissues and developing seeds.

Fig. 14.9 Some Rhodophyta species producing gibberellins

2.3.2 Separation and Detection

Gibberellins contain more than 40 chemically closely related compounds which are difficult to separate and distinguish. Separation of gibberellins can be performed on column of 5% OV-22, on DMCS-treated chromosorb W.

Gibberellins are separated on silica gel plates with the solvent system benzene/butanol/acetic acid (70:25:5) and benzene/acetic acid/water (50:19:31). Detection was carried out by H_2SO_4/water (7:3) spraying the plate and then exposure at 120 °C; GA_3 appear as spots with yellow-green.

2.3.3 Identification and Determination

The most satisfactory method for gibberellin determination is by GC/MS; using GLC, gibberellins converted first to their methyl ester (by Methylation) or TMS esters (silylation).

2.3.4 Plant Bioassays (Measurement of Gibberellin Biological Activity)

1. Lettuce hypocotyl.
2. α-amylase,
3. Dwarf rice leaf.

2.4 Abscisic Acid (ABA) and Lunularic Acid

It is a sesquiterpenoid growth inhibitor (inhibitor of elongation), but it may significantly stimulate maize root elongation at some concentrations. ABA is present in several plant species in root tips and root caps. White light and stresses (mineral starvation and leaf dehydration) induce high content of ABA in leaves. In the brown algae of genus *Ascophyllum* (*A. nodosum*) and some species of *Laminaria* (Nimura et al. 2002), a hormone was detected which suppressed plant growth in bioassay.

In various algal groups, the growth-inhibiting complex includes lunularic acid and abscisic acid, and other undifferentiated biologically active compounds were recorded.

Lunularic acid is a kind of growth inhibitor which was detected in liverworts. Its structure, activity, and metabolism resemble those of ABA. It suppressed the growth of cut discs from cultured *Laminaria japonica* and induced reproductive tissue formation at the same concentration used in higher plants (10^{-6} to 10^{-4} M). Also, ABA induced the morphogenesis of *Hypnea pluvialis* cells to form cysts. In some microalgae, the endogenous ABA content increased under stress conditions (salinity, light intensity, drought, etc.).

El shoubaky and Salem (2016) recorded ABA in the green seaweeds *Ulva rigida* and *Ulva lactuca* as well as in the red *Sarconema filiforme*, and higher concentration of ABA was recorded in *U. lactuca* where the ABA profile (by GC/MS) contained *cis*-, *trans*-ABA-L-alanine methyl ester, *cis*-, *trans*-ABA-L-valine, and *cis*-, *trans*-ABA-L-alanine. Stirk et al. (2009) detected endogenous ABA in the green *Ulva fasciata* and the brown *Dictyota humifusa*. It was also detected in different seaweed commercial extracts as Kelpak R from *Ecklonia maxima*. Also, red seaweeds were found to produce ABA as in the case of *Bangia fuscopurpurea* and *Pyropia yezoensis* by Mori et al. (2017), as well as red algae of Brazil (Yokoya et al. 2010), as illustrated in Table 14.4 and Figs. 14.10 and 14.11.

2.4.1 Separation, Detection, and Determination

Using paper chromatography for separation of ABA in the algal extract and standard ABA (at conc. 10^{-6} M) for comparison. Detect the isolated spots by UV light of 254 nm as dark absorbed spots. The most widely used techniques for quantification are GC/MS, GLC/MS, and HPLC/MS.

2.4.2 The Biological Activity of ABA Using Specific Bioassays

1. Inhibition of elongation of wheat coleoptile sections (ABA of more than 10^{-8} M).
2. Induction of stomatal closure.

Table 14.4 Seaweed species producing abscisic acid (ABA) and lunularic acid

Algal species	Division
Ascophyllum nodosum	Phaeophyta
Laminaria japonica	
Dictyota humifusa	
Laminaria sp.	
Ascophyllum sp.	
Laminaria digitata	
Ulva rigida	Chlorophyta
Ulva lactuca	
Enteromorpha compressa	
Ulva fasciata	
Sarconema filiforme	Rhodophyta
Bangia fuscopurpurea	
Pyropia yezoensis	

2.5 *Ethylene*

Plants and macroalgae produce a range of volatile compounds, such as alcohols, alkane, alkenes, esters, etc.

These volatile compounds are produced in response to biotic or abiotic stimuli. These compounds have several biological roles in higher plants such as promotion of seed germination, inhibition of the stem, root elongation, ripening of fruits, senescence of leaves and flowers, and sex determination as reported by Bleecker and Kenode (2000), Klee (2004), Grennan (2008), Holopainen and Gershenzon (2010), and Loreto and Schnitzler (2010). Most of the biosynthetic pathways of volatile compound production depend on *S*-adenosylmethionine compound which may act as a substrate for the enzyme reactions or as a source of methyl group (as in the synthesis of jasmonates, salicylates, and brassinosteroids). Ethylene and dimethyl sulfide are examples of etherial compounds which are produced from the red alga *Gelidium* sp.

The concentrations and types of these compounds were affected by various abiotic stress factors as salinity, light quality, and exogenous ethylene. The period of light and darkness causes the production of amines and methyl alkyl compounds.

Reaction oxygen species (ROS) act as a secondary messenger initiating a signal cascade which stimulate ethylene synthesis (Mackerness 2000). Accumulation of volatile compounds was recorded after the exposure to red light and application of exogenous ethylene.

The level of dimethyl sulfide (DMS) which emitted in all conditions didn't increase after incubation with ethylene (they appear to be not coordinated as reported in the red alga *G. arbuscula*. In *Acetabularia mediterranea*, the rate of algal development decreases with increase of ethylene production.

In *Enteromorpha intestinalis*, reduction of chlorophyll content below that of control occurred on the addition of ethephon which decompose to generate ethylene as

Fig. 14.10 Some (**a**) Phaeophyta and (**b**) Chlorophyta species producing abscisic acid and lunularic acid

reported by Garcia-Jimenez et al. (2013). Moreover, ethylene was involved in growth of the red alga *Pterocladiella capillacea* (Garcia-Jimenez and Robaina 2012).

When the acclimatized *Ulva intestinalis* to low light intensity was transferred to high light condition, ethane level was increased causing an inhibition of chlorophyll

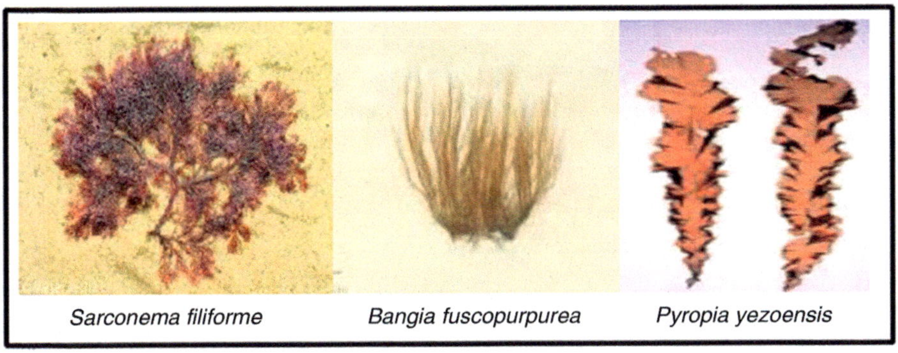

Fig. 14.11 Rhodophyta species producing abscisic acid and lunularic acid

Table 14.5 Seaweed species producing ethylene

Algal species[a]	Division
Ulva intestinalis	Chlorophyta
Acetabularia mediterranea	
Pterocladiella capillacea	Rhodophyta

[a]No Phaeophyta species were recorded

content (by 30%). Table 14.5 and Fig. 14.12a, b recorded some algal species producing ethylene.

2.5.1 Gas Chromatography/MS Analysis of the Released Volatiles (GC/ MS)

Volatile compounds are analyzed using Varian 431GC/210MS with capillary column and He as a carrier gas.

2.5.2 Physiological Properties of Ethylene

Ethylene production increased during leaf abscission, flower senescence, and fruit ripening.

Physiological stresses and wounding induce ethylene biosynthesis.

During storage of fruits, vegetables, and flowers, an effective ethylene absorbent is used ($KMnO_4$, pot. permanganate) to reduce ethylene concentration in the storage area (extending the storage life of the fruits).

Inhibition of Ethylene Action.

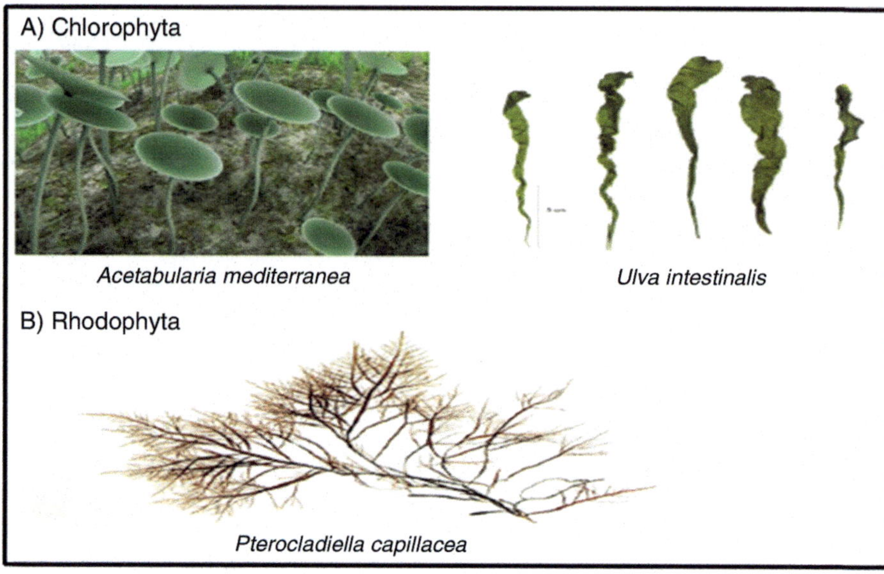

Fig. 14.12 Seaweed species producing ethylene. (**a**) Chlorophyta and (**b**) Rhodophyta

Ethylene effects can be antagonized by silver ion Ag^+ [in the form of $AgNO_3$ or silver thiosulfate $Ag(S_2O_3)_2$]. Also, CO_2 at high concentrations (5–10%) inhibit many ethylene effects.

3 Growth Substances (Growth Regulators)

There are five classical phytohormones detected in angiosperms and in the lower plants as algae. They are auxins, cytokinins, gibberellins, abscisic acid, and ethylene which control different physiological and developmental processes. Their extraction, separation, detection, and determination by various methods are well known. Also, their synthetic pathway(s) and biological functions and bioassays have long been documented. Different chemical compounds were reported to control growth and ameliorate plant (or algal) defensive system against biotic or abiotic stresses. These compounds are termed growth substances or growth regulators; they include brassinosteroids, jasmonic acid, salicylic acid, polyamines, and betaines. Higher plants and lower plants (as algae) were reported to produce different growth regulators (Mikami et al. 2016; Mori et al. 2017).

3.1 Brassinosteroids (BRs)

They are group of polyhydroxylated steroid growth regulators which have a remarkable role in various biochemical and development processes in different organisms such as plants and algae, including elongation, reproduction, and cell division, in stems and roots, stress responses, leaf senescence, and photomorphogenesis. The most active components of brassinosteroids are termed brassinolide and castasterone which are widely reported in various tissues of seeds, flowers, leaves, pollens, stems, and roots (Bajguz and Hayat 2009). The precursor of brassinolide is the campesterol (C28-sterol) by oxidation at C6 and addition of OH groups into the β-ring. Brassinazole is an inhibitor of brassinosteroid biosynthesis.

Brassinosteroids have been recorded in *Hydrodictyon reticulatum* by Bajguz and Hayat (2009) as well as in angiosperms, gymnosperms, the pteridophyte *Equisetum arvense*, and the bryophyte *Marchantia polymorpha* (Bajguz and Tretyn 2003). Table 14.6 and Fig. 14.13 recorded some seaweed species producing growth substances.

Secondary internode bioassay of beans is used, where brassins cause both cell elongation and cell division as well as bending, swelling, and splitting of the second internode. Brassinosteroids act locally near their site of synthesis and transported in the xylem.

Brassinosteroids have usually effect on the transport of auxin by indirect way (Symons et al. 2008), increase the percentage of ethylene accumulation, and have an additive effect with GA_3, in addition to its synergistic effect with auxins (IAA). Moreover, Brassinosteroids have effects on increase cytokinin and jasmonic acid production and decrease ABA responses. In the seaweed extract of *Ecklonia maxima* (Phaeophyta) called Kelpak (Stirk et al. 2013a, b), auxins, cytokinin, GAs, ABA, and brassinosteroids were detected.

Table 14.6 Seaweed species producing growth substances (growth regulators)

Algal species	Algal group
Brassinosteroids	
Hydrodictyon reticulatum	Chlorophyta
Jasmonic acid	
Fucus vesiculosus	Phaeophyta
Gelidium sp.	Rhodophyta
Polyamines	
Dictyota dichotoma	Phaeophyta
Ulva rigida	Chlorophyta
Gelidium canariensis	Rhodophyta
Grateloupia doryphora	
Cyanidium caldarium	
Rhodomorphin	
Griffithsia pacifica	Rhodophyta

Fig. 14.13 Seaweed species producing growth substances (growth regulators). (**a**) Jasmonic acids. (**b**) Polyamines

3.2 Jasmonic Acid (JA)

Jasmonic acid (is a fatty acid) and its volatile methyl ester (jasmonate) were detected in some seaweed's species. Also, a hydroxylated compound called tuberonic acid, its ME, and its glucosides control potato tuberization development.

Jasmonic acid is produced from the fatty acid linolenic acid (18:3) and plays an important role in plant defense by inducing the synthesis of proteinase inhibitors.

Jasmonates inhibit seed and growth germination and promote abscission, fruit pigmentation, and ripening formation. Jasmonic acid and methyl jasmonate were detected in many species of microalgae and cyanobacteria. Moreover, it was observed in the red seaweed *Gelidium latifolium* and in brown seaweeds (oxilipins and lipoxygenases) as reported by Arnold et al. (2001) shown in Table 14.6 and Fig. 14.13.

3.3 Polyamines (Aliphatic Amines)

Different seaweed species produced polyamines in the red macroalgae *Gelidium canariensis*, *Grateloupia doryphora*, and *Cyanidium caldarium*, in the brown *Dictyota dichotoma*, as well as in the green *Ulva rigida* (Table 14.6 and Fig. 14.13). The content as well as the biosynthesis of polyamines in algae doesn't differ from that in higher plants (50–150 μg/g F.wt).

Polyamine content in macroalgae changes with seasons and developmental stage (Marián et al. 2000; Sacramenta et al. 2004; Alcazar et al. 2010). Polyamines belong to the putrescine group (putrescine, spermine, and spermidine). It derived from the carboxylation of the amino acids, arginine and ornithine, putrescine (diamine) → spermidine (triamine) → spermine (quadramine).

It exerts regulatory control on the development and growth at very low concentration (especially cell division and morphology). In carrot tissue culture, when polyamine content is low, callus growth only occurs, but at higher concentration, the embryo is formed (polyamines are released to the outer growth media). They aren't recorded in the commercial seaweed products till now.

3.4 Salicylic Acid (SA)

It is recognized recently as potential regulatory compound. It is produced from phenylalanine (AA).

Salicylic acid plays a role in the pathogen's resistance. It was reported that SA enhance flower longevity, inhibit biosynthesis of ethylene and seed germination, and reverse the effect of ABA.

Salicylic acid was detected in the red seaweeds *Pyropia yezoensis* and *Bangia fuscopurpurea* (Mori et al. 2017). The precursors of SA are cinnamic acid and benzoic acid. To quantify the amount of SA in an extract, methyl salicylate (*ortho*-anisic acid) can be used as an internal standard (HPLC analysis) (Forcat et al. 2008).

3.5 Signal Peptides

Small molecular weight peptides (as systemin) were found to have regulatory properties in plants which travel in phloem from attacked leaves by herbivore insects to the distant leaves to protect them from insect attack. The traveled signal peptides induce an increase in the production of jasmonic acid and proteinase in the distant leaves for protection.

Signal peptides produced by plants many have a role in:

1. Activation of defensive responses.
2. Cell proliferation promotion.
3. Nodule formation (in legumes).

No known recorded studies that extracted, separated, and identified such signal peptides in seaweeds till now.

3.6 Small RNA Molecules

Recently, many small RNA molecules of single-stranded RNA that consist of 21–22 nucleotides have been identified in plant phloem which may act as transportable signals that regulate gene expression involved in plant defense against viruses. Many of these micro-RNAs (miRNAs) have been identified which means that they may represent a more general means of regulating gene expression.

No such RNA molecules have been identified in seaweeds.

3.7 Rhodomorphin

This regulator was detected in *Griffithsia pacifica* (red alga) following morphogenetic effects in this alga (Table 14.6 and Fig. 14.13). When an intercalary cell in the filament is removed, the basal cell of the filament starts to secrete the regulator rhodomorphin which increases the formation of reparatory cell. Further studies showed that rhodomorphin is a glycoprotein with molecular weight of 14 kDa. Similar glycoproteins were recorded in the green alga *Volvox* sp. where it acts as a pheromone facilitating the gamete adhesion and fusion. So, the function of these glycoproteins in algae was to provide adhesion and fusion of gametes during sexual reproduction.

3.8 Commercial Seaweed Concentrates (or Extracts)

Many reports were published in literature concerning the presence of plant growth hormones in brown, green, and red seaweeds (Crouch and Van Staden 1993; Stirk et al. 2013a, b; Tuhy et al. 2013). Their presence was determined and confirmed by plant bioassays and chromatographic analysis (TLC/LC/MS, HPLC/MS, and GC/MS).

A commercial seaweed concentrate was firstly prepared from the water zone occupied by *Fucus* and *Ascophyllum* sp. which was tested to have cytokinin-like activity due to the presence of isopentenyladenine (ipA).

During the past two decades, utilization of commercial seaweed products increased as natural sources of fertilizers, biostimulants, and soil ameliorants. It improves plant growth is relatively cheap and easy to apply either as soil manure or as foliar spray. It was suggested that bioactive organic compounds in the seaweed concentrate are responsible for the increase in crop yield (Crouch et al. 1992).

Recent researches showed that macroalgae have recorded the ability as a source for products that contain growth regulators and plant hormones and many of the observed effects to treated crops are now attributed to these constituents. The chemical composition of seaweeds revealed that all the major plant nutrients and trace elements are present in marine seaweeds.

Many investigators recently reviewed the presence of antibiotic, antiviral, antibacterial, and antioxidant activities due to the active substances obtained from marine algae. These substances may be responsible for the reduced harmful effects of some plant pathogens (Hamed et al. 2018).

Some of commercial seaweed products (have commercial names):

1. Maxi-crop/Seasol.
2. Algifert/Algimex/Algistim.
3. SM3/Seamac/Marinure.
4. SWC (Kelpak66).

They contain gibberellin-like activity, betaines, ABA, ethylene, and cytokinins. The commercial seaweed extracts use the seaweeds *Ascophyllum nodosum*, *Ecklonia maxima*, *Enteromorpha compressa*, *Durvillaea potatorum*, *Fucus serratus*, *Porphyra perforata*, *Sargassum muticum*, *Laminaria japonica*, *Macrocystis pyrifera*, *Ectocarpus confervoides*, and *Pylaiella littoralis*.

4 Conclusion

From the obtained data, we can conclude that algae (micro and macro) extracts are rich with plant hormones and other growth regulator substances. So we can use these species commercially as growth stimulants in different agricultural sectors. Some of these algal species can be used as organic fertilizer and biofertilizers due to its ability for nitrogen fixation (in case of species with heterocysts) and its content from inorganic chemical substances (such as phosphorus and potassium), in addition to organic substances and hormones. The wider distribution, high adaptability to different cultural factor conditions (biotic and abiotic stresses), and high growth rates led algae to be considered as an attractive feedstock for developing fertilizer and biorefinery products, in addition to the ability of these algal species to increase the accumulation of phytohormones when exposed to different abiotic stress conditions such as salinity, drought, light intensity, chemical substances, etc.

References

Ahmed M, Stal LJ, Hasnain S (2010) Production of indole-3-acetic acid by the cyanobacterium Arthrospira platensis strain MMG-9. J Microbiol Biotechnol 20(9):1259–1265

Arnold TM, Targett NM, Tanner CE, Halch WI, Ferrari KE (2001) Evidence for methyl Jasmonate induced phlorotannin production in Fucus vesiculosus (Phaeophyceae). J Phycol 37:1026–1029

Alcazar R, Altabbella T, Marco F, Bortolotti C, Reymond M, Koncz C, Carrasco P, Tiburcio AF (2010) Polyamines: molecules with regulatory functions in plant abiotic stress tolerance. Planta 231:1237–1249
Bajguz A, Hayat S (2009) The effect of brassinosteroids on the plant responses to environmental stresses. Plant Physiol Biochem 47:1–8
Bajguz A, Tretyn A (2003) The chemical characteristic and distribution of brassinosteroids in plants. Phytochemistry 62:1027–1046
Bently JA (1960) Role of plant hormones in algal metabolism and ecology. Nature 181:1499–1502
Bleecker AB, Kenode H (2000) Ethylene: a gaseous signal molecule in plants. Annu Rev Cell Dev Biol 16:1–18
Crouch IJ, Van Staden J (1993) Evidence for the presence of plant growth regulators in commercial seaweed products. Plant Growth Regul 13:21–29
Crouch IJ, Smith MT, Van Staden J, Lewis MJ, Hoad GV (1992) Identification of auxins in a commercial seaweed concentrate. J Plant Physiol 138:590–594
Du Buy HG, Olson RA (1937) The presence of growth regulators during the early development of *Fucus*. Am J Bot 24:609–611
El Shoubaky GA, Salem EA (2009) Biodiversity in Timsah Lake as a biofertilizer source to the economic plants. Egypt J Bot 49:53–69
El Shoubaky GA, Salem EA (2016) Effect of abiotic stress on endogenous phytohormones profile in some seaweeds. Int J Pharm Phytochem Res 8(1):124–134
Forcat S, Bennett MH, Mansfield JW, Grant MR (2008) A rapid and robust method for simultaneously measuring changes in the phytohormones ABA, JA and SA in plants following biotic and abiotic stress. Plant Methods 4:16–23
Garcia-Jimenez P, Robaina RR (2012) Effects of ethylene on tetrasporogenesis in *Pterocladiella capillacea* (Rhodophyta). J Phycol 48:710–715
Garcia-Jimenez P, Brito-Romano O, Robaina RR (2013) Production of volatiles by the red seaweed *Gelidium arbuscula* (Rhodophyta): emission of ethylene and dimethyl sulfide. J Phycol 49:661–669
Gebser B, Pohnert G (2013) Synchronized regulation of different zwitterionic metabolites in the osmoadaption of phytoplankton. Mar Drugs 11:2168–2182
Grennan AK (2008) Ethylene response factors in Jasmonate signaling and defense response. Plant Physiol 146:1457–1458
Hamed SM, Abd EL-Rhman AA, Abdel-Raouf N, Ibraheem IBM (2018) Role of marine macro algae in plant protection and improvement for sustainable agriculture technology. Beni-Suef Univ J Basic Appl Sci 7:104–110
Holopainen IK, Gershenzon J (2010) Multiple stress factors and emission of plant VOC_s. Trends Plant Sci 15:176–184
Klee H (2004) Ethylene signal transduction. Moving beyond *Arabidopsis*. Plant Physiol 135:660–667
Kusano T, Berberich T, Tateda C, Takahashi Y (2008) Polyamines: essential factors for growth and survival. Planta 228:367–381
Li T, Wang CG, Miao J (2007) Identification and quantification of indole-3- acetic acid in the kelp Laminaria japonica Areschoug and its effect on growth of marine micro algae. J Appl Phycol 19:479–484
Li XG, Su YH, Zhao XY, Li W, Gao XQ, Zhang XS (2010) Cytokinin over production caused alteration of flower development is partially mediated by CUC2 and CUC3 in Arabidopsis. Gene 450:109–120
Loreto F, Schnitzler JP (2010) Abiotic stress and induced BVOCs. Trends Plant Sci 15:154–166
Mackerness SAH (2000) Plant responses to ultraviolet-B (UV-B:280-320nm) stress. What are the key regulators? Invited review. Plant Growth Regul 32:27–39
Mackinnon SA, Craft CA, Hilty D, Ugate R (2010) Improved method of analysis for betaines in Ascophyllum nodosum and its commercial seaweed extracts. J Appl Phycol 22:489–494
Marián FD, Garcia-Jiménez P, Robaina RR (2000) Polyamines in marine macro algae: levels of putrescine, spermidine and spermine in the thalli and changes in their concentration during glycerol-induced cell growth *in vitro*. Physiol Plant 110:530–534

Maršálek B, Zahradníčková H, Hronková M (1992) Extracellular abscisic acid produced by cyanobacteria under salt stress. J Plant Physiol 139(4):506–508
Mikami K, Mori IC, Matsuura T, Ikeda Y, Kojima M, Sakakibara H, Hirayama T (2016) Comprehensive quantification and genome survey reveal the presence of novel phytohormone action modes in red seaweeds. J Appl Phycol 28:2539–2548
Mori IC, Ikeda Y, Matsuura T, Hirayama T, Mikami K (2017) Phytohormones in red seaweeds : a technical review of methods for analysis and a consideration of genomic data. Bot Mar 60(2):153–170
Nimura K, Mizuta H (2002) Inducible effects of abscisic acid on sporophyte discs from *Laminaria japonica* Areschoug (Laminariales, Phaeophyceae). J Appl Phycol 14:159–163
Romanenko EA, Kosakovskaya IV, Romanenko PA (2015) Phytohormones of microalgae: biological role and involvement in the regulation of physiological processes. Pt I. auxins, abscisic acid, ethylene. Int J Algae 17(3):275–289
Sacramenta AT, Garcia-Jimenez P, Alcazar R, Tiburcio AF, Robaina RR (2004) Influence of polyamines on the sporulation of *Grateloupia* (Halymeniaceae, Rhodophyta). J Phycol 40:887–894
Sitnik KM, Musatenko LI, Vosyuk VA, Vedenicheva NP, Generalova VM, Martin GG, Nesterova AN (2003) Gormonal'nii Kompleks roslin i gribiv (Hormonal complex in plants and fungi). Akademperiodika, Kiev
Skibola CF (2004) The effect of *Fucus vesiculosus*, an edible brown seaweed, upon menstrual cycle length and hormonal status in three pre-menopausal women: a case report. BMC Complement Altern Med 4:10–18
Stirk WA, Van Staden J (1997) Comparison of cytokinin and auxin-like activity in some commercially used seaweed extracts. J Appl Phycol 8:503–508
Stirk WA, Van Staden J (2006) Seaweed products as biostimulants in agriculture. In: Critchley AT, Ohno M, Largo DB (eds) World seaweed resources. ETI Information Services Ltd., Wokingham, pp 1–32, (DVD ROM)
Stirk WA, Novák O, Strnad M, Van Staden J (2003) Cytokinins in macroalgae. Plant Growth Regul 41:13–24
Stirk WA, Novák O, Hradecká V, Pěnčik A, Rolčik J, Strnad M, Van Standen J (2009) Endogenous cytokinins, auxins and abscisic acid in *Ulva fasciata* (Chlorophyta) and *Dictyota humifusa* (Phaeophyta): towards understanding their biosynthesis and homoeostasis. Eur J Phycol 44(2):231–240
Stirk WA, Balint P, Tarkowská D, Novák O, Strnad M, Ordog V, Van Staden J (2013a) Hormone profiles in microalgae: gibberellins and brassinosteroids. Plant Physiol Biochem 70:348–353
Stirk WA, Tarkowská D, Turecová V, Strnad M, Van Staden J (2013b) Abscisic acid, gibberellins and brassinosteroids in Kelpak R, a commercial seaweed extract made from Ecklonia maxima. J Appl Phycol 26(1):561–567
Stirk WA, Tarkowska D, Turecová V, Strnad M, Van Staden I (2014) Abscisic acid, gibberellins and brassinosteroids in Kelpak®, a commercial seaweed extract made from *Ecklonia maxima*. J Appl Phycol 26:561–567
Sun TP (2010) Gibberellin-GIDI-DELLA: a pivotal regulatory module for plant growth and development. Plant Physiol 154:567–570
Symons GM, Ross JJ, Jager CE, Reid JB (2008) Brassinosteroid transport. J Exp Bot 59:17–24
Tarakhovskaya ER, Maslov YI, Shishova MF (2007) Phytohormones in algae. J Plant Physiol 54:163–170
Tuhy L, Chowanska J, Chojnacka K (2013) Seaweed extracts as bio-stimulants of plant growth: review. Chemik 67(7):636–641
Weiss D, Ori N (2007) Mechanisms of cross talk between gibberellins and other hormones. Plant Physiol 144:1240–1246
Yamaguchi S (2008) Gibberellin metabolism and its regulation. Annu Rev Plant Biol 59:225–251
Yokoya NS, Stirk WA, Van staden J, Novák O, Tureckova V, Pencik A, Strnad M (2010) Endogenous cytokinins, auxins and abscisic acid in red algae from Brazil. J Phycol 46:1198–1205
Yoshida K, Igarashi E, Mukai M, Hirata K, Miyamoto K (2003) Induction of tolerance to oxidative stress in the green alga Chlamydomonas reinhardtii by abscisic acid. Plant Cell Environ 26:451–457

Chapter 15
Role of *Trichoderma* in Agriculture and Disease Management

Varucha Misra and Mohammad Israil Ansari

Contents

1 Introduction

Trichoderma spp., free-living saprophytic fungi, is found commonly in the soil where plant roots sustain especially in intercellular spaces. This fungus is known to be highly interactive in three different environments, viz., soil, root, and foliar (Singh et al. 2006). The first description of this fungus was recorded in Germany in the year 1791. In 1927, four species of this fungus is identified based on color, conidial shape, and colony appearance by Gilman and Abbott. There are two major species, i.e., *T. lignorum* (due to conidial globose structure) and *T. koningii* (due to conidial oblong structure), which are mostly known. In 1932, Weindling has shown its capability as an effective biocontrol agent toward pathogen, *Rhizoctonia solani*. Harman et al. (2004) had revealed this fungus to be opportunistic and avirulent symbiont, and at times, it also possesses parasitic capability. Several *Trichoderma* species such as *T. harzianum*, *T. viride*, *T. hamatum*, *T. koningii*, and *T.*

V. Misra
ICAR-Indian Institute of Sugarcane Research, Lucknow, India

M. I. Ansari (✉)
Department of Botany, University of Lucknow, Lucknow, India
e-mail: ansari_mi@lkouniv.ac.in

© The Author(s), under exclusive license to Springer Nature Switzerland AG 2021

H. I. Mohamed et al. (eds.), *Plant Growth-Promoting Microbes for Sustainable Biotic and Abiotic Stress Management*,
https://doi.org/10.1007/978-3-030-66587-6_15

longibrachiatum have phytopathogenic property against a number of fungi like *Pythium ultimum*, *Fusarium oxysporum*, *Sclerotinia sclerotiorum*, etc. (Manczinger et al. 2002).

Trichoderma is one recognized fungus which is being used as a biocontrol agent since 1920 (Samuels 1996). They are known to improve plant health along with their natural capability to degrade the toxic compounds produced by the plants. It is important for the overall growth of the plant, and its function is not limited to disease control particularly to soil-borne diseases (Zaidi et al. 2014). *Trichoderma* is a ubiquitous genus which grows in wider habitats and at high population densities (Chaverri et al. 2011). This could be proved through its diverse applications and role. The fungus have increase reproductive ability. It is known to survive under abiotic stress conditions and compete with other pathogens for the uptake of nutrients for their survival, augmenting the plant defense system (Tripathi et al. 2013; Daguerre et al. 2014; Keswani et al. 2014). Certain species of this fungus also have multiple interactions with crop plants, for example, *Trichoderma harzianum* strain T22 and *Trichoderma atroviride* strain P1 (Woo et al. 2006). This chapter focuses on the role of *Trichoderma* in agriculture and disease management.

2 Characteristic Features of *Trichoderma*

Increased growth rate, bright green conidia in major strains of this fungus, and repetitively branched structure of conidiophore are the main characteristics of this fungus (Gams and Bissett 1998). This fungus is known to be a flourishing colonizer of their habitat. It can be indicated by the way it utilizes the substrate and secretes enzymes and antibiotic compounds irrespective of the environmental condition, whether the condition is like that of tropical rainforest or of biotechnological fermentor (Schuster and Schmoll 2010). In *Trichoderma* colonization, the fungus identifies and adheres to root via hydrophobins or expansin-like proteins through which it penetrates in the tissues of the plant. Hydrophobins are small proteins which are hydrophobic, and it coats the cell wall of the fungus, whereas swollenin is also protein molecule that is known to break the cell wall of the plant (composed of crystalline cellulose structure) due to the carrier of cellulose-binding molecule which assists in the expansion of cell wall of root cells and root hairs (Brotman et al. 2008). For instance, *T. asperellum* produces TasHyd 1 (belonging to class I hydrophobin) and swollenin TasSwo (belonging to expansin-like proteins) that helps in protecting its hyphal tips and root colonization (Viterbo and Chet 2006; Brotman et al. 2008). Druzhinina et al. (2011) had revealed that due to an increase in root surface area by swollenin molecule, *Trichoderma* takes extra benefit during its establishment in the rhizosphere. The plant-derived sucrose is an important resource by which *Trichoderma* cells assist three aspects, i.e., root colonization, synchronization of defense mechanisms, and improved photosynthetic rate (Vargas et al. 2009). In root colonization process, *Trichoderma* swaps molecular messages and also causes fungal deposition by elicitors in apoplastic cells of roots

(Contreras-Cornejo et al. 2014; Gupta et al. 2014). Shoresh and Harman (2008) had shown that though *T. harzianum* Rifai strain 22 (T22) resides in roots, only their role during colonization is prominent as it stimulates impactful alterations in proteome of corn shoot seedlings. Morán-Diez et al. (2009) had also revealed that *T. harzianum* secretes endopolygalacturonase, ThPG 1 (plant cell wall-degrading enzymes) during active root colonization. Furthermore, Chacón et al. (2007) had illustrated that after 72 h of colonization of roots with *Trichoderma*, cell walls of plant epidermis and the cortex are much stronger than nontreated plants, and even they possess cellular deposition (consists of an abundance of callose) which acts as a barrier for the pathogens.

3 Role of *Trichoderma* in Agriculture

Trichoderma is a well-known fungus for its diverse uses in agriculture. Some strains of this fungus cause a direct impact on the plant by enhancing their growth and uptake of nutrients (Table 15.1). The nutrient uptake by *Trichoderma* causes the secretion of organic acids which help in dissolving many minerals and trigger the uptake of nutrients from soil. This in turn led to consumption and movement of nutrients. Besides, the involvement of *Trichoderma* in the soil causes expansion in the area of rhizosphere and rise in secretion of organic acids and extracellular enzymes (phosphatase, urease, etc.) due to its ability of colonization. This will result in an improvement of cycling of nutrients and enzymatic activity. Harman (2011) and Khan et al. (2017) had revealed that this fungus helps in the conversion of nutrients into useful nutrients as required by the plant. This was also supported by Mbarki et al. (2016) who suggested that rise in nutrient and enzymatic activity helps in improving the quality of soil and enhancing the growth of a plant. Different species of *Trichoderma* are also known to break down N compounds into available N by releasing nitrous oxide (Maeda et al. 2015). Soil-borne diseases are known to arise due to the discrepancy in soil microbes, and *Trichoderma* is effective in controlling soil-borne diseases due to its property of rapid growth and vitality as it covers the space where microbes develop and even uptake the nutrients which otherwise could be used up by the microbes causing soil-borne diseases (Zhang 2015). *Trichoderma* besides increasing nutrient uptake also promotes the growth of beneficial microbes and their biomass (Wagner et al. 2016). Hyperparasitism is another property of this fungus in which there is a secretion of cell wall-degrading enzymes, such as xylanases, cellulases, etc., that helps in good growth and development. Besides, higher-use efficiency of fertilizer, seed germination rate, and plant defense system are also having a strong positive impact of this fungus (Shoresh et al. 2010). *Trichoderma* is also playing an effective role in unraveling the mysteries of the molecular biology of plants. A significant rise in height and weight of dwarf tomato plants has been reported after treatment with *T. viride* by 28% and 8%, respectively (Lindsey and Baker 1967). This was also seen in other plant species too such as pepper, chrysanthemum, and periwinkle where this fungus (*Trichoderma*

Table 15.1 Role of certain *Trichoderma* species in plant growth promotion and nutrient uptake

Trichoderma species	Plants	Role of the fungus on plant	References
Plant growth promotion			
T. harzianum	Pepper, chrysanthemum, and periwinkle	• Improved the germination. • Flowering incidence and occurrence. • Height and fresh weight.	Chang et al. (1986)
T. viride	Tomato	Height and weight of dwarf tomato plants	Lindsey and Baker (1967)
T. harzianum and *Trichoderma koningii*	Corn, tomato, tobacco, and radish	Increased germination rates, emergence, and dry weights	Windham et al. (1986)
Trichoderma harzianum T22	Crack willow (*Salix fragilis*)	Shoots and roots that were 40% longer and more than double the dry biomass of controls	Adams et al. (2007)
Trichoderma asperellum PR11	Cacao seedlings	Significantly increased plant height, fresh root, and shoot weight against control	Tchameni et al. (2011)
Nutrient uptake			
Trichoderma asperellum PR11	Cacao seedlings	Increase in acid phosphatase activity and phosphorus uptake	Tchameni et al. (2011)
T. harzianum T447	Tomato seedling	Increase in calcium, magnesium, phosphorus, and potassium concentration	Azarmi et al. (2011)
Trichoderma virens As19-1 (T.v7)	Soya bean	Fe uptake is increased up to 77%	Entesari et al. (2013)
Trichoderma asperellum CHF 78	Tomato	Increase dry weight of plant	Li et al. (2018)

harzianum) improved the germination and flowering incidence and occurrence, besides height and fresh weight of plant. Furthermore, Windham et al. (1986) had also revealed that in corn, radish, tomato, and tobacco, *T. harzianum* and *T. koningii* play an important part in augmenting the germination rate of the plant along with its emergence and dry weight.

The following are the major role of *Trichoderma* (Fig. 15.1) in agriculture:

1. *Bio-fertilization*: *Trichoderma* plays an efficient role in improving plant health even when there is no pathogen present. This fungus shows its maximum production in acidic soil as it creates favorable conditions for itself by secreting organic acids which in turn gives additional benefit to the crop grown in such soils. This fungus helps in dissolving mineral ions (Fe, Mn, and Mg) and phosphate ions present in the soil that cause the crop to absorb these nutrients in an easier and better way in which in general condition may not be sufficiently available.
2. *Plant defense system*: This fungus secretes a number of lytic and proteolytic enzymes as well as volatile and secondary metabolites (Table 15.2) for surviving

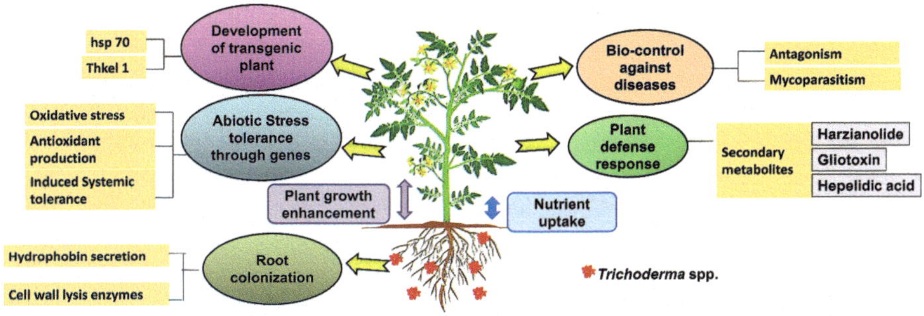

Fig. 15.1 Different responses of *Trichoderma* on plant and its panoply mechanism. *Trichoderma* works in a plant through five ways, i.e., as developer of transgenic plant, biocontroller against diseases, abiotic stress alleviator, root colonizer, and plant defense response

Table 15.2 Some of the secondary metabolites secreted by different species of *Trichoderma*

Trichoderma species	Secondary metabolites	References
T. harzianum	Azaphilone	Vinale et al. (2006)
	Butenolide	
	Harzianolide	Almassi et al. (1991), Claydon et al. (1991), Ordentlich et al. (1992)
	Harzianic acid	Vinale et al. (2009)
	Trichorzianines	Hajji et al. (1987)
	Harzianopyridone	Cutler and Jacyno (1991)
	Dehydroharzianolide	Almassi et al. (1991)
T. viride, *T. atroviride*, *T. harzianum*, *T. koningii*	6-Pentyl-α-pyrone	Vinale et al. (2014), Marra et al. (2006)
T. virens	Gliotoxin	Rajasekaran and Murugesan (2005)
	Heptelidic acid	Pachauri et al. (2020)
T. viridens	Viridian	Awad et al. (2018)
T. viridens	Viridiol	Moffatt et al. (1969)
T. koningii	Koninginin A 1	Harman (2000)
T. koningii	Trichoviridin	Nobuhara et al. (1976)
	Cyclonerodiol	Cutler et al. (1991)
T. cerinum	Cerinolactone	Cutler et al. (1986)

against pathogens present in the same environment. These secondary metabolites are known to be produced at minimal nutrition requirements and are even used in various purposes due to its beneficial properties (Khan et al. 2020). The antifungal activities exhibited by this fungus are known against many fungal pathogens (Vizcaino et al. 2005) wherein secondary metabolites are being involved (Vinale et al. 2008). Besides, it also secretes hydrolytic enzymes such as chitinases, proteases, and glucanases, which are the bases of its relationship with pathogens. This relation is known as mycoparasitism.

3. *As plant survivor under abiotic and biotic stress*: *Trichoderma* fungus is also being used for coping out the plant from abiotic and biotic stress conditions. The interaction of *Trichoderma* and plants exposed to biotic and abiotic stress with pathogenic microbes particularly nematode and fungus is antagonistic (Singh et al. 2004). This antagonistic activity helps in enhancing plant growth, root growth, and resistance to many diseases and abiotic stress (Lorito et al. 2010; Bae et al. 2011; Harman 2000; Shoresh et al. 2010), nitrogen use efficiency, P solubilization, availability of nutrients, and humic acid content (due to organic matter decomposition) (Harman 2011a; Harman and Mastouri 2010; Shoresh et al. 2010). The abiotic stress includes salt stress, high temperatures, and drought (Shoresh et al. 2010). Zaidi et al. (2014) showed that the use of this fungus helps in declining the use of nitrogen efficiency by 30% in certain crops without affecting the crop yields. Such application of this fungus has repercussion in agriculture.
4. *Development of transgenic plants*: Several studies had illustrated that in transgenic plants in which overexpression of genes isolated from *Trichoderma* occurs is a new approach to overcome the situation of adverse condition. For example, development of transgenic plants such as *Nicotiana tabacum* and *Solanum tuberosum* using genes isolated from *T. harzianum* revealed to be tolerant to diseases like *Alternaria*, *Botrytis*, or *Rhizoctonia* (Lorito et al. 1998), and overexpression of chitinases in the same plants were tolerant to abiotic (salt stress and heavy metals) and biotic stress (diseases including fungal and bacterial). Montero-Barrientos et al. (2010) had revealed that cloning of heat-shock protein, HSP 70 gene, from *T. harzianum* in *Arabidopsis* resulted in providing tolerance to heat stress and other associated stresses like salt, osmotic, and oxidative stress. Another gene encoding protein, Thkel 1, from *T. harzianum* showed regulation in glucosidase activity which helped in improving plant growth in *Arabidopsis* plant by providing tolerance against salt and osmotic stress (Hermosa et al. 2011). Studies had also shown that there are many proteins isolated from *Trichoderma*, like small protein 1 (Sm1), PKS/NRPS hybrid enzyme, etc., which are useful in bestowing resistance against various pathogens either soil-borne or foliar (Howell et al. 2000; Perazzoli et al. 2012; Viterbo et al. 2005).

4 Property and Mechanism of *Trichoderma* in Disease Management

Trichoderma strains have long back identified as a biological agent that helps the plant to improve its growth and productivity (Ansari 2017; Singh et al. 2006). It is considered as one of the best biocontrol agents known so far and has attracted the interest of many scientists as a promising substitute to chemical fungicides against several disease-causing pathogenic organisms (Kubicek et al. 2001). Among many of the species identified in *Trichoderma*, five species are known as biological agents (Rifai 1969; Benitez et al. 2004). These are *T. harzianum*, *T. asperellum*, *T.*

atroviride, *T. virens*, and *T. reesei*. These strains can curtail disease severity by inhibiting pathogens which attack the plant either through soil or through roots. They do so by their antagonistic and mycoparasitic property (Viterbo and Horwitz 2010). This fungus stimulates the release of many compounds which provide resistance either in localized or in a systemic manner. In induced systemic resistance (ISR), certain strains of this fungus affect the growth, development, and biochemistry of plant as the fungus colonizes and penetrates inside the root of the plant reaching to its tissues. This helps the plant to defend against many pathogens attacking it (Shoresh et al. 2010; Lorito et al. 2010). Kubicek et al. (2011) had shown its mycoparasitism capability in two species, viz., *T. atroviride* and *T. virens*. Moreover, Druzhinina et al. (2011) had illustrated that there are two aspects which attracted this fungus to grow in rhizosphere, one being the presence of the organism on which it can feed and another being the available nutrients in the root zone of the plants. Both these aspects also help this fungus to improve the growth of the plant. Several studies have reported its role in controlling pathogens of the plant either by elicitation or by developing resistance toward the pathogen (Harman et al. 2004). In addition to this, one of the major mechanisms used in *Trichoderma* for acting as biocontrol agent is its capability of competition for space, nutrients, and formation of volatile compounds (enzymes and antibiotics) against other microbes. The hydrolytic enzymes secreted by this fungus degrade partially the cell wall of pathogen and cause parasitization on the attacked pathogen (Kubicek et al. 2001).

Trichoderma spp. is also known to decline the incidence and severity of disease through plant-mediated mechanism. This mechanism is alike to systemic acquired resistance (SAR) on the phenotypic basis and is known as induced resistance which is mainly concerned with plant parts above the ground and gets activated by this fungus (Singh et al. 2011; Harman 2011). This induced systemic resistance (ISR) is newly discovered in *Trichoderma* and is now attaining much more importance. When roots were inoculated in cucumber plant (of age 7 days), *T. harzianum* helped in increasing plant defense system by increasing activities of peroxidase and chitinase enzyme along with cellulose and cellobiose wall deposition (Yedidia et al. 2001).

5 Interaction of *Trichoderma* spp. with Other Microbes

Trichoderma is an antagonist microorganism that causes a reduction in the growth of the pathogens, and their survival gets difficult by the various mechanisms this fungus adopt (Fig. 15.2) such as enzyme secretion, competition, antibiosis, interactions of its hyphae with another fungus, mycoparasitism, etc. (Singh et al. 2006). During the competition process, this fungus suppresses the growth and survivability of pathogen through its antagonistic property. For instance, 80–85% of collar rot disease in elephant foot yam plant is effectively controlled by *T. harzianum* (Singh et al. 2006). Another important aspect is mycoparasitism where *Trichoderma* attacks the target organism physically not only by acting as a parasite but also by producing toxic chemicals. Some of these chemicals are volatile, like trichothecine,

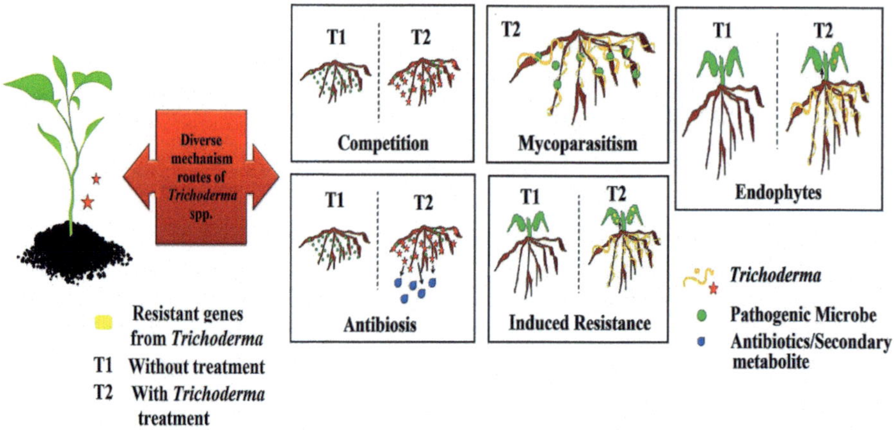

Fig. 15.2 Interaction of *Trichoderma* spp. with pathogens in the plant. *Trichoderma* adopts diverse mechanisms to enhance the overall growth and productivity of the plant. In competition mode, it competes with pathogenic microbe by growing faster and dominating over others. In mycoparasitism, it feeds on to other microbes present in the rhizospheric region of the plant. In antibiosis, secretion of antibiotics or secondary metabolites from this fungus helps in inhibiting the growth of other microbes. In induced resistance, it secretes chemicals which protect the plants from pathogens, and even genes isolated from this fungus provide resistance to biotic and biotic stress. In endophytes, *Trichoderma* can grow in a plant as endophytes, thereby benefiting the plant development

sesquiterpene, etc., and may travel via air. Chitinases and antibiotics secreted by *Trichoderma* spp. work in synergistic manner, causing a relatively stronger impact on the target organisms. The mechanism involves three stages wherein the first stage comprises of the interaction of chemical stimulus of a pathogen with antagonistic nature of *Trichoderma* that results in a chemotropic response, the second stage comprises of identifying and recognizing the pathogen and antagonistic fungi through lectins, and the third stage is the interaction of hyphae of *Trichoderma* with hyphae of pathogen fungi where the hyphae of *Trichoderma* coils around the hyphae of pathogen and secretes enzymes such as pectinase and chitinase. This could be seen in interaction of *Trichoderma* with pathogens like *Fusarium roseum*, *Phytophthora colocasiae*, *F. solani*, etc. (Singh et al. 2006). Besides, this fungus has a characteristic feature to take up nutrients from the source and survive effectively in comparison to other microbes as it can break down chitin component of other fungi or cellulose of plants which are generally difficult to break down by other microbes due to their complexity.

Besides this, some strains of this fungus can even bind with ions of iron present in soil to produce siderophore (Leong 1986), for example, *Serpula lacrymans*. This specialized compound is difficult to be uptake by other microbes, and so it results in unavailability of iron uptake to the microbes present in the same environmental condition. This causes the target organisms not to become resistant toward it as in doing so the organism needs to be resistant to many mechanism routes involved in the mode of action of *Trichoderma*.

6 Role of *Trichoderma* in Management of Viral, Fungal, and Bacterial Pathogens

Trichoderma is well known for its biocontrol activity against several crucial plant pathogens like virus, fungi, and bacteria causing severe diseases (Madan et al., 2000; Al-Ani 2018). As a biocontroller against many fungal infections in plants, several studies had reported that this fungus works either by inhibiting or by parasitizing the pathogen mycelial growth by production of certain enzymes like chitinases, permeases, etc. and thus helps in controlling the disease-causing pathogen to proliferate (Table 15.3). *Trichoderma* was also found to be effective in red rot disease of sugarcane, the most damaging disease (Madan et al. 1997; Ansari et al. 2008; Ansari 2012). In viral infections, Luo et al. (2010) had showed that *T. pseudokoningii* SMF2 have antimicrobial peptaibols referred to as trichokonin which increased upregulation of genes governing plant defense and are being used against tobacco mosaic virus (TMV) infection for coping out the plant from the disease with increased reactive oxygen species (ROS) and phenolic compounds. Cucumber mosaic virus (CMV) also showed effective results in its management by the use of this fungus (Sachdev and Singh 2020). Elsharkawy et al. (2013) had illustrated that *T. asperellum* SKT-1 showed increased levels of genes associated with salicylic acid, jasmonic acid, and ethylene in leaves by inducing resistance in plants with this disease. However, in the case of pretreatment of this fungus in *Arabidopsis* plant against this disease, the defense mechanism gets activated against this disease. In *Solanum lycopersicum*, defense response is induced by *T. harzianum* T-22 strain against CMV disease (Vitti et al. 2015). In bacterial diseases, Al-Ani (2018) had showed that *T. asperellum* T203 gives a protective effect against *Pseudomonas syringae* pv. *lachrymans* in cucumber plants. Studies had revealed another strain of *Trichoderma*, *T. pseudokoningii* SMF2, possessing antibacterial property against a wide range of Gram-positive and Gram-negative bacteria (Bora et al. 2020; Shi et al. 2012; Li et al. 2014). *Pectobacterium carotovorum* ssp. *carotovorum* causing disease of soft rot in Chinese cabbage was able to manage by this *Trichoderma* strain through the production of trichokonins which inhibited bacterial growth by increasing production of PR-1a gene, ROS, and SA (Li et al. 2014). Khalili et al. (2016) had also illustrated that in charcoal rot of soybean, *Trichoderma* acts as an effective biocontrol agent. Studies have also reported that *T. harzianum* also proved to be a positive controller of wilt diseases caused by *Ralstonia solanacearum* in a number of crops such as chili, brinjal, ginger, tomato, etc. (Bora et al. 2013; Deuri 2013). The use of *T. viride* in lettuce plant had reported to effectively manage the disease caused by *R. solanacearum* and *F. oxysporum* f. sp. *lactucae* (Khan et al. 2018).

Table 15.3 Different strains of *Trichoderma* controlling fungal infection and mode of action

Trichoderma species	*Trichoderma* strain	Fungal pathogen	Mode of action	References
T. koningii	MTCC 796	*Macrophomina phaseolina*	Parasitize fungal mycelia growth	Gajera et al. (2012)
T. harzianum	T12			Khalili et al. (2016)
T. harzianum	*Foc*TR4	*Fusarium oxysporum* f. subspecies *cubense*	Restrain the growth of mycelium	Al-Ani et al. (2013)
T. atroviride	Tveg1 and TR10			
T. asperellum	CCTCC-RW0014	*F. oxysporum* f. subspecies *cucumerinum*	Increasing production of protease, cellulose, and chitinase	Saravanakumar et al. (2016)
Trichoderma asperellum strain	T34	*F. oxysporum* f. sp. *lycopersici*	Competition for iron and form siderophores	Segarra et al. (2010)
T. hamatum	URM 6656	*F. solani*	Production of chitinases	da Silva et al. (2016)
T. harzianum	T3	*Ceratocystis radicicola*	Lysis of hyphae, phialoconidia, and aleurioconidia	Al-Naemi et al. (2016)
T. atroviride	T17	*Guignardia citricarpa*	Antagonistic activity by secreting proteins such as chitinase, mutanase, a-1,2-mannosidase, α-galactosidase, a-1,3-glucanase, neutral protease, carboxylic hydrolase ester, etc.	de Lima et al. (2016)
T. harzianum	T39	*Gliocladium virens*	Inhibit growth of fungal mycelia	Bora and Deka (2007)
	CICR G	*S. sclerotiorum*		Mukherjee et al. (2014)
T. atroviride	P1	*Phytophthora cinnanerium*		Olabiyi and Ruocco (2013)
T. viride	T30, T31	*R. solani*		
T. harzianum	T22	*Botrytis cinerea*		

7 Conclusion

Trichoderma is a free-living soil fungus that is frequently seen in the soil and rhizospheric region of the plant. This fungus is known for its many characteristics and peculiar properties which benefit the plant in its growth and development. It is being known worldwide for its protectant activity and growth enhancement. Different strains of *Trichoderma* produce compounds that elicit the plant defense responses. These compounds include low-molecular-weight compounds, proteins, and peptides. *Trichoderma* also have many potential abilities such as tolerant capability against a number of biotic and abiotic stresses, enhancement in nutrient uptake activity of plant, and augmentation in nitrogen use efficiency and even in photosynthetic activity. A large number of genes are known to over express in *Trichoderma* species that helps in abiotic stress tolerance to plants. Some antibiotic substances are also being secreted by this fungus to dominate and kill other fungal pathogens, thereby maintaining its colonization where it uses its hyphae to adhere to plant roots through hydrophobins or swollenin. *Trichoderma* is a renowned biocontrol agent that helps manage the diseases occurring in the plants.

References

Adams P, De-Leij FAAM, Lynch JM (2007) *Trichoderma harzianum Rifai* 1295-22 mediates growth promotion of crack willow (*Salix fragilis*) saplings in both clean and metal-contaminated soil. Microb Ecol 54(2):306–313

Al-Ani LKT (2018) Trichoderma: beneficial role in sustainable agriculture by plant disease management. In: Egamberdieva D, Ahmad P (eds) Plant microbiome: stress response, microorganisms for sustainability. Springer, Berlin, pp 105–126

Al-Ani LKT, Salleh B, Ghazali AHA (2013) Biocontrol of Fusarium wilt of banana by Trichoderma spp. In: 8th PPSKH colloquium, Pust Pengajian Sains Kajihayat/School of Biological Sciences, USM, June 5-6

Almassi F, Ghisalberti EL, Narbey MJ, Sivasithamparam K (1991) New antibiotics from strains of *Trichoderma harzianum*. J Nat Prod 54:396–402

Al-Naemi FA, Ahmed TA, Nishad R, Radwan O (2016) Antagonistic effects of *Trichoderma harzianum* isolates against *Ceratocystis radicicola*: pioneering a biocontrol strategy against black scorch disease in date palm trees. J Phytopathol 164(7–8):464–475

Ansari MI, Madan VK, Arya N, Lal RJ (2008) Cloned DNA probes for identification of red rot pathogen, Colletotrichum falcatum infection in sugarcane at early stages. Green Farming (An International Journal of Agricultural Sciences 1(5):41–43

Ansari MI (2012) Colletotrichum falcatum, a causal organism of sugarcane red rot disease. Trends in Biosciences 5(2):95–96

Ansari MI (2017) Red Rot: The Cancer of Sugarcane. Lambert Academic Publishing, Germany. Pp 56. ISBN: 978-3-330-34604-8.

Awad NE, Kaseem HA, Hamed MA, Elfeky AM, Elnaggar MAA, Mahmoud K, Ali MA (2018) Isolation and characterization of the bioactive metabolites from the soil-derived fungus *Trichoderma viride*. Mycology 9(1):70–80

Azarmi R, Hajighrari B, Giglou A (2011) Effect of *Trichoderma* isolates on tomato seedling growth response and nutrient uptake. Afr J Biotechnol 10(31):5850–5855

Bae H, Roberts DP, Strem M, Lim HS, Park SC, Ryu CM, Melnick R, Bailey BA (2011) Endophytic *Trichoderma* isolates from tropical environments delay disease and induce resistance against *Phytophthora capsici* in hot pepper using multiple mechanisms. Mol Plant-Microbe Interact 24(3):336–351

Benitez T, Rincon AM, Limon MC, Antonia C (2004) Biocontrol mechanisms of *Trichoderma* strains. Int Microbiol 7:249–260

Bora LC, Deka SN (2007) Wilt disease suppression and yield enhancement in tomato (*Lycopersicon esculentum*) by application of *Pseudomonas fluorescens* based biopesticide (Biofor-Pf) in Assam. Indian J Agric Sci 77(8):490–494

Bora LC, Sarkar, R, Kataky, L (2013) Genomic characterization of microbial antagonists, their interactive effects and utility in management of bacterial wilt of Bhut Jolokia (*Capsicum chi-

nense Jacq). In: Acta phytopathologica sinica, 10th international congress of plant pathology ICPP 2013, 20-30 Aug 2013, Beijing, China

Bora LC, Bora P, Gogoi M (2020) Potential of Trichoderma spp. for pest management and plant growth promotion in NE India. In: Sharma AK, Sharma P (eds) Trichoderma, rhizosphere biology. Springer, Berlin, pp 205–220

Brotman Y, Briff E, Viterbo A, Chet I (2008) Role of swollenin, an expansin-like protein from *Trichoderma*, in plant root colonization. Plant Physiol 147:779–789

Chacón MR, Rodríguez-Galán O, Benítez T, Sousa S, Rey M, Llobell A, Delgado-Jarana J (2007) Microscopic and transcriptome analyses of early colonization of tomato roots by *Trichoderma harzianum*. Int Microbiol 10:19–27

Chang YC, Baker R, Kleifeld O, Chet I (1986) Increased growth of plants in the presence of the biological-control agent *Trichoderma harzianum*. Plant Dis 70(2):145–148

Chaverri P, Gazis RO, Samuels GJ (2011) *Trichoderma amazonicum*, a new endophytic species on *Hevea brasiliensis* and *H. guianensis* from the Amazon basin. Mycologia 103(1):139–151

Claydon N, Hanson JR, Truneh A, Avent AG (1991) Harzianolide, a butenolide metabolite from cultures of *Trichoderma harzianum*. Phytochemistry 30:3802–3803

Contreras-Cornejo HA, Macías-Rodríguez LI, Alfaro-Cuevas R, Lopez-Bucio J (2014) *Trichoderma* improves growth of *Arabidopsis* seedlings under salt stress through enhanced root development, osmolite production and Na+ elimination through root exudates. Mol Plant Microbe Interact 27:503–514

Cutler HG, Jacyno JM (1991) Biological activity of (-)-harziano-pyridone isolated from *Trichoderma harzianum*. Agric Biol Chem 55(10):2629–2631

Cutler HG, Cox RH, Crumley FG, Cole PD (1986) 6-Pentyl-a-pyrone from *Trichoderma harzianum*: its plant growth inhibitory and antimicrobial properties. Agricult Biol Chem 50:2943–2945

Cutler HG, Jacyno JM, Phillips RS, von Tersch RL, Cole PD, Montemurro N (1991) Cyclonerodiol from a novel source, *Trichoderma koningii*: plant growth regulatory activity. Agric Biol Chem Tokyo 55:243–244

da Silva JAT, de Medeiros EV, da Silva JM, Tenório DA, Moreira KA, Nascimento TCE, Souza-Motta C (2016) *Trichoderma aureoviride* URM 5158 and *Trichoderma hamatum* URM 6656 are biocontrol agents that act against cassava root rot through different mechanisms. J Phytopathol 164(11–12):1003–1011

Daguerre Y, Siegel K, Edel-Hermann V, Steinberg C (2014) Fungal proteins and genes associated with bio-control mechanisms of soil borne pathogens: a review. Fungal Biol Rev 28:97–125

de Lima FB, Félix C, Osório N, Alves A, Vitorino R, Domingues P, Correia A, Ribeiro RTS, Esteves AC (2016) Secretome analysis of *Trichoderma atroviride* T17 biocontrol of *Guignardia citricarpa*. Biol Control 99:38–46

Deuri D (2013) Bio-intensive approach for management of bacterial wilt of ginger (*Zingiber officinale*). MSc (Agri) thesis, Assam Agricultural University, Jorhat, Assam

Druzhinina IS, Seidl-Seiboth V, Herrera-Estrella A, Horwitz BA, Kenerley CM, Monte E, Mukherjee PK, Zeilinger S, Grigoriev IV, Kubicek CP (2011) *Trichoderma*: the genomics of opportunistic success. Nat Rev Microbiol 9:749–759

Elsharkawy MM, Shimizu M, Takahashi H, Ozaki K, Hyakumachi M (2013) Induction of systemic resistance against cucumber mosaic virus in *Arabidopsis thaliana* by *Trichoderma asperellum* SKT-1. Plant Pathol J 29(2):193–200

Entesari M, Sharifzadeh F, Ahmadzadeh M, Farhangafar M (2013) Seed bio-priming with *Trichoderma* species and *Pseudomonas fluorescent* n growth parameters, enzymes activity and nutritional status of soya bean. Int J Agron Plant Prod 4(4):610–619

Gajera HP, Bambharolia RP, Patel SV, Khatrani TJ, Goalkiya BA (2012) Antagonism of *Trichoderma* spp. against *Macrophomina phaseolina*: evaluation of coiling and cell wall degrading enzymatic activities. J Plant Pathol Microbiol 3:7

Gams W, Bissett J (1998) Morphology and identification of *Trichoderma*. In: Harmann GE, Kubicek CP (eds) Trichoderma and Gliocladium. Taylor and Francis, London, pp 3–34

Gupta KJ, Mur LA, Brotman Y (2014) *Trichoderma asperelloides* suppresses nitric oxide generation elicited by *Fusarium oxysporum* in *Arabidopsis* roots. Mol Plant-Microbe Interact 27:307–314

Hajji ME, Rebuffat S, Lecommandeur D, Bodo B (1987) Isolation and sequence determination of Trichorzianines A antifungal peptides from *Trichoderma harzianum*. Int J Pept Protein Res 29(2):207–215

Harman GE (2000) Myths and dogmas of biocontrol: changes in perceptions derived from research on *Trichoderma harzianum* T-22. Plant Dis 84:377–393

Harman GE (2011) Multifunctional fungal plant symbionts: new tools to enhance plant growth and productivity. New Phytol 189:647–649

Harman GE (2011a) *Trichoderma*—not just for biocontrol anymore. Phytoparasitica 39:103–108

Harman GE, Mastouri F (2010) Enhancing nitrogen use efficiency in wheat using *Trichoderma* seed inoculants. Int Soc Plant Microbe Interact 7:1–4

Harman GE, Howell CR, Viterbo A, Chet I, Lorito M (2004) *Trichoderma* species—opportunistic, avirulent plant symbionts. Nat Rev Microbiol 2(1):43–56

Hermosa R, Botella L, Keck E, Jiménez JA, Montero-Barrientos M, Arbona V, Gómez-Cadenas A, Monte E, Nicolás C (2011) The overexpression in *Arabidopsis* thaliana of a *Trichoderma harzianum* gene that modulates glucosidase activity, and enhances tolerance to salt and osmotic stresses. J Plant Physiol 168:1295–1302

Howell CR, Hanson LE, Stipanovic RD, Puckhaber LS (2000) Induction of terpenoid synthesis in cotton roots and control of *Rhizoctonia solani* by seed treatment with *Trichoderma virens*. Phytopathology 90:248–252

Keswani C, Singh SP, Singh HB (2014) A superstar in biocontrol enterprise: *Trichoderma* spp. Biotechnol Today 3(2):27–30

Khaleil M, El-Mougith A, Hashem H, Lokma N (2016) Biocontrol potential of entomopathogenic fungus, *Trichoderma hamatum* against the cotton aphid, Aphis Gossypii. J Environ Sci Toxicol Food Technol 10(5):11–20

Khan MY, Haque MM, Molla AH, Rahman M, Alam MZ (2017) Antioxidant compounds and minerals in tomatoes by, Trichoderma-enriched biofertilizer and their relationship with the soil environments. J Integr Agric 16:691–703

Khan P, Bora LC, Bora P, Talukdar K, Kataky L (2018) Efficacy of microbial consortia against bacterial wilt caused by *Ralstonia solanacearum* in hydroponically grown lettuce plant. Int J Curr Microbiol Appl Sci 7(6):3046–3055

Khan RAA, Najeeb S, Hussain S, Xie B, Li Y (2020) Bioactive secondary metabolites from *Trichoderma* spp. against Phytopathogenic fungi. Microorganisms 8:817–838

Kubicek CP, Mach RL, Peterbauer CK, Lorito M (2001) *Trichoderma*: from genes to biocontrol. J Plant Pathol 83:11–23

Kubicek CP, Herrera-Estrella A, Seidl-Seiboth V, Martinez DA, Druzhinina IS, Thon M, Zeilinger S, Casas-Flores S, Horwitz BA, Mukherjee PK, Mukherjee M, Kredics L, Alcaraz LD, Aerts A, Antal Z, Atanasova L, Cervantes-Badillo MG, Challacombe J, Chertkov O, McCluskey K, Coulpier F, Deshpande N, Dohren HV, Ebbole DJ, Esquivel-Naranjo EU, Fekete E, Flipphi M, Glaser F, Gomez-Rodriguez EY, Gruber S, Han C, Henrissat B, Hermosa R, Hernandez-Onate M, Karaffa L, Kosti I, Crom SL, Lindquist E, Lucas S, Lubeck M, Lubeck PS, Margeot A, Metz B, Misra M, Nevaaleinen H, Omann M, Packer N, Perrone G, Uresti-Riveria EE, Salamov A, Schmoll M, Seiboth B, Shapiro H, Sukno S, Tamayo-Ramoos JA, Tisch D, Wiest A, Wilkinson HH, Zhang M, Coutinho PM, Kenerley CM, Monte E, Baker SE, Grigoriev IV (2011) Comparative genome sequence analysis underscores mycoparasitism as the ancestral life style of *Trichoderma*. Genome Biol 12:R40

Leong J (1986) Siderophores: their biochemistry and possible role in the biocontrol of plant pathogens. Annu Rev Phytopathol 24:187–209

Li Y-H, Luo Y, Zhang X-S, Shi W-L, Gong Z-T, Shi M, Chen L-L, Chen X-L, Zhang Y-Z, Song X-Y (2014) Trichokonins from *Trichoderma pseudokoningii* SMF2 induce resistance against

Gram-negative *Pectobacterium carotovorum* subsp. carotovorum in Chinese cabbage. FEMS Microbiol Lett 354:75–82
Li YT, Hwang SG, Huang YM, Huang CH (2018) Effects of *Trichoderma asperellum* on nutrient uptake and *Fusarium* wilt on tomato. Crop Protect 110:275–282
Lindsey DL, Baker R (1967) Effect of certain fungi on dwarf tomatoes grown under gnotobiotic conditions. Phytopathology 57:1262–1263
Lorito M, Woo SL, Fernández IG, Colucci G, Harman GE, Pintor-Toro JA, Filippone E, Muccifora S, Lawrence CB, Zoina A, Tukun S, Scala F (1998) Genes from mycoparasitic fungi as a source for improving plant resistance to fungal pathogens. Proc Natl Acad Sci U S A 95:7860–7865
Lorito M, Woo SL, Harman GE, Monte E (2010) Translational research on *Trichoderma*: from 'omics to the field. Annu Rev Phytopathol 48:395–417
Luo Y, Zhang D-D, Dong X-W, Zhao P-B, Chen L-L, Song X-Y, Wang X-J, Chen X-L, Shi M, Zhang Y-Z (2010) Antimicrobial peptaibols induce defense responses and systemic resistance in tobacco against tobacco mosaic virus. FEMS Microbiol Lett 313:120–126
Madan VK, Bikash M, Ansari MI, Anjani S, Soni N, Solomon S, Agnihotri VP (2000) RAPD-PCR analysis of molecular variability in the red rot pathogen (Colletotrichum falcatum) of sugarcane. Sugar Cane International, 5–8
Madan VK, Mandal B, Misra SR, Ansari MI, Srivastava A, Agnihotri VP (1997) Isolation and characterization of total and plasmid DNA from Colletotrichum falcatum causing red dot of sugarcane. Sugar Cane (United Kingdom) 6:9–11
Maeda K, Spor A, Edel-Hermann V, Heraud C, Breuil MC, Bizouard F, Toyoda S, Yoshida N, Steinberg C, Philippot L (2015) N_2O production, a widespread trait in fungi. Sci Rep 5:96–97
Manczinger L, Antal Z, Kredics L (2002) Ecophysiology and breeding of mycoparasitic *Trichoderma*. Acta Microbiol Immunol Hung 49:1–25
Marra R, Ambrosino P, Carbone V, Vinale F, Woo SL, Ruocco M, Ciliento R, Lanzuise S, Ferraioli S, Soriente I, Gigante S, Turrà D, Fogliano V, Scala F, Lorito M (2006) Study of the three-way interaction between *Trichoderma atroviride*, plant and fungal pathogens by using a proteomic approach. Curr Genet 50:307–321
Mbarki S, Cerdà A, Brestic M, Mahendra R, Abdelly C, Pascual JA (2016) Vineyard compost supplemented with Trichoderma harzianum t78 improve saline soil quality. Land Degrad Dev 28:1028–1037
Moffatt JS, BuLock JD, Yuen TH (1969) Viridiol, a steroid like product from *Trichoderma viride*. J Chem Soc D Chem Commun 14:839
Montero-Barrientos M, Hermosa R, Cardoza RE, Gutierrez S, Nicolás C, Monte E (2010) Transgenic expression of the *Trichoderma harzianum* HSP70 gene increases *Arabidopsis* resistance to heat and other abiotic stresses. J Plant Physiol 167:659–665
Morán-Diez E, Hermosa R, Ambrosino P, Cardoza RE, Gutiérrez S, Lorito M, Monte E (2009) The ThPG1 endopolygalacturonase is required for the *Trichoderma harzianum*–plant beneficial interaction. Mol Plant-Microbe Interact 22:1021–1031
Mukherjee AK, Kumar AS, Kranthi S, Mukherjee PK (2014) Biocontrol potential of three novel Trichoderma strains: isolation, evaluation and formulation. 3 Biotech 4:275–281
Nobuhara M, Tazima H, Shudo K, Itai A, Okamoto T, Iitaka Y (1976) A fungal metabolite, novel isocyano epoxide. Chem Pharm Bull 24:832–834
Olabiyi TI, Ruocco M (2013) *In-vitro* competition bio-assay experiment on the effect of *Trichoderma* species and some crop pathogenic fungi. J Biol Agric Healthc 3(12):2224–3208
Ordentlich A, Wiesman Z, Gottlieb HE, Cojocaru M, Chet I (1992) Inhibitory furanone produced by the biocontrol agent *Trichoderma harzianum*. Phytochemistry 31:485–486
Pachauri S, Gupta GD, Mukherjee PK, Kumar V (2020) Expression of a heptelidic acid insensitive recombinant GAPDH from *Trichoderma virens* and its biochemical and biophysical characterization. Protein Expr Purif 175:105697
Perazzoli M, Moretto M, Fontana P, Ferrarini A, Velasco R, Moser C, Delledonne M, Pertot I (2012) Downy mildew resistance induced by *Trichoderma harzianum* T39 in susceptible grapevines partially mimics transcriptional changes of resistant genotypes. BMC Genomics 13:660

Rajasekaran A, Murugesan K (2005) Production of gliotoxin on natural substrates by *Trichoderma virens*. J Basic Microbiol 45(1):12–19

Rifai MA (1969) Revision of genus *Trichoderma*. Mycol Pap 116:1–56

Sachdev S, Singh RP (2020) *Trichoderma*: a multifaceted fungus for sustainable agriculture. In: Bauddh K et al (eds) Ecological and practical applications for sustainable agriculture. Springer, Berlin, pp 261–304

Samuels GJ (1996) *Trichoderma*: a review of biology and systematics of the genus. Mycol Res 100:923–935

Saravanakumar K, Yu C, Dou K, Wang M, Li Y, Chen J (2016) Synergistic effect of *Trichoderma* derived antifungal metabolites and cell wall degrading enzymes on enhanced biocontrol of *Fusarium oxysporum* f. sp. *cucumerinum*. Biol Control 94:37–46

Schuster A, Schmoll M (2010) Biology and biotechnology of *Trichoderma*. Appl Microbiol Biotechnol 87:787–799

Segarra G, Casanova E, Avilés M, Trillas I (2010) *Trichoderma asperellum* strain T34 controls *Fusarium* wilt disease in tomato plants in soilless culture through competition for iron. Fungal Microbiol 59:141–149

Shi M, Chen L, Wang X-W, Zhang T, Zhao P-B, Song X-Y, Sun C-Y, Chen X-L, Zhou B-C, Zhang Y-Z (2012) Antimicrobial peptaibols from *Trichoderma pseudokoningii* induce programmed cell death in plant fungal pathogens. Microbiology 158:166–175

Shoresh M, Harman G (2008) The molecular basis of shoot responses of maize seedlings to *Trichoderma harzianum* T22 inoculation of the root: a proteomic approach. Plant Physiol 147:2147–2156

Shoresh M, Harman GE, Mastouri F (2010) Induced systemic resistance and plant responses to fungal biocontrol agents. Annu Rev Phytopathol 48:21–43

Singh US, Zaidi NW, Joshi D, John D, Khan T, Bajpai A (2004) *Trichoderma*: a microbe with multifaceted activity. Annu Rev Plant Pathol 3:33–75

Singh RN, Saurabh A, Nedunchezhiyan M (2006) Use of *Trichoderma* in disease management. Orissa Review, September-October: 68–70

Singh BN, Singh SP, Singh A, Singh HB (2011) Reprogramming of oxidant and antioxidant metabolites in root apoplast of sunflower by *Trichoderma harzianum* NBRI-1055 against *Rhizoctonia solani*. Eur J Plant Pathol 131:121–134

Tchameni SN, Ngonkeu MEL, Begoude BAD, Nana LW, Fokom R, Owona AD, Mbarga JB, Tchana T, Tondje PR, Etoa EX, Kuate J (2011) Effect of *Trichoderma asperellum* and arbuscular mycorrhizal fungi on cacao growth and resistance against black pod disease. Crop Prot 30(10):1321–1327

Tripathi P, Singh PC, Mishra A, Chauhan PS, Dwivedi S, Bais RT, Tripathi RD (2013) *Trichoderma*: a potential bioremediator for environmental clean up. Clean Technol Enviro Policy 15(4):541–550

Vargas WA, Mandawe JC, Kenerley CM (2009) Plant-derived sucrose is a key element in the symbiotic association between *Trichoderma virens* and maize plants. Plant Physiol 151:792–808

Vinale F, Marra R, Scala F, Ghisalberti EL, Lorito M, Sivasithamparam K (2006) Major secondary metabolites produced by two commercial *Trichoderma* strains active against different phytopathogens. Lett Appl Microbiol 43:143–148

Vinale F, Sivasithamparam K, Ghisalberti EL, Woo SL, Nigro M, Marra R (2008) *Trichoderma* secondary metabolites active on plants and fungal pathogens. Open Mycol J 8:127–139

Vinale F, Flematti G, Sivasithamparam K, Lorito M, Marra R, Skelton BW, Ghisalberti EL (2009) Harzianic acid, an antifungal and plant growth promoting metabolite from *Trichoderma harzianum*. J Nat Prod 72:2032–2035

Vinale F, Sivasithamparam K, Ghisalberti EL, Woo SL, Nigro M, Marra R, Lombardi N, Pascale A, Ruocco M, Lanzuise S, Manganiello G, Lorito M (2014) *Trichoderma* secondary metabolites active on plants and fungal pathogens. Open Mycol J 8(Suppl-1, M5):127–139

Viterbo A, Chet I (2006) TasHyd1, a new hydrophobin gene from the biocontrol agent Trichoderma asperellum, is involved in plant root colonization. Mol Plant Pathol 7:249–258

Viterbo A, Horwitz BA (2010) Mycoparasitism. In: Borkovich KA, Ebbole DJ (eds) Cellular and molecular biology of filamentous fungi, vol 42. American Society for Microbiology, Washington, DC, pp 676–693
Viterbo M, Harel B, Horwitz A, Chet I, Mukherjee PK (2005) *Trichoderma* mitogen-activated protein kinase signaling is involved in induction of plant systemic resistance. Appl Environ Microbiol 71:6241–6246
Vitti A, Monaca EL, Sofo A, Scopa A, Cuypers A, Nuzzaci M (2015) Beneficial effects of *Trichoderma harzianum* T-22 in tomato seedlings infected by cucumber mosaic virus (CMV). BioControl 60:135–147
Vizcaino JA, Sanz L, Cardoza RE, Monte E, Gutierrez S (2005) Detection of putative peptide synthetase genes in *Trichoderma* species. Application of this method to the cloning of a gene from *T. harzianum* CECT 2413. FEMS Microbiol Lett 244:139–148
Wagner K, Apostolakis A, Daliakopoulos I, Tsanis I (2016) Can tomato inoculation with *Trichoderma* compensate yield and soil health deficiency due to soil salinity? In: Proceedings of the EGU general assembly conference abstracts, Vienna, Austria, 17–22 April 2016
Windham MT, Elad Y, Baker R (1986) A mechanism for increased plant-growth induced by *Trichoderma* spp. Phytopathology 76(5):518–521
Woo SL, Scala F, Ruocco M, Lorito M (2006) The molecular biology of the interactions between *Trichoderma* spp., phytopathogenic fungi and plants. In: Symposium on the nature and application of biocontrol microbes II: *Trichoderma* spp, vol 96, no 2, pp 181–185
Yedidia I, Srivastava AK, Kapulnik Y, Chet I (2001) Effect of *Trichoderma harzianum* on micro element concentration and increased growth of cucumber plants. Plant Soil 235:235–242
Zaidi NW, Dar MH, Singh S, Singh US (2014) *Trichoderma* species as abiotic stress relievers in plants. In: Biotechnology and biology of *Trichoderma*. Elsevier, Amsterdam, pp 515–524
Zhang FG (2015) The a effects and mechanisms of Puta five *Trichoderma harzianum* mutant and ITS bio-organic fertilizer on growth of cucumber. Nanjing Agricultural University, Nanjing, pp 15–18

Chapter 16
Production of Antibiotics from PGPR and Their Role in Biocontrol of Plant Diseases

Saima Hamid, Rafiq Lone, and Heba I. Mohamed

Contents

1 Introduction

The symbiotic bacteria present around the roots of plants are called rhizobacteria which are free-living unswervingly correlated with root surface or dwell inter alia on the roots such as endophytic bacteria without adding any value to the soil (Kloepper and Beauchamp 1992). When rhizobacteria help plants to grow, they are defined as plant growth-promoting rhizobacteria (PGPRs) (Kloepper et al. 2004).

S. Hamid · R. Lone (✉)
Centre of Research for Development, University of Kashmir,
Srinagar, Jammu and Kashmir, India

H. I. Mohamed
Department of Biological and Geological Sciences, Faculty of Education, Ain Shams University, Cairo, Egypt

© The Author(s), under exclusive license to Springer Nature
Switzerland AG 2021

H. I. Mohamed et al. (eds.), *Plant Growth-Promoting Microbes for Sustainable Biotic and Abiotic Stress Management*,
https://doi.org/10.1007/978-3-030-66587-6_16

Bacteria must be competent for rhizosphere, i.e., capable of interacting with rhizosphere-based nutrients secreted from the root or from sites which can be occupied on the root, to exert their beneficial effects in the root system (Hao et al. 2012; Kim et al. 2012). Also, the characteristic inherence of PGPR is that it interacts with other microbes, such as arbuscular mycorrhizal fungi (AMF), to promote plant growth. In addition to soil-based microbes other than AMF, the plant-AMF relationship is mostly manipulated through indirect mechanisms by the increased availability of soil nutrients (Ghignone et al. 2012; Pii et al. 2015), while its impact directly is still under debate for plant-based growth (Glick 2012). A very favorable habitat for the growth of microorganisms in the rhizosphere, which covers a volume of root soil, which is chemically and physically affected by the plant root, can have a potential effect on plant health and soil fertility (Sorensen 1997). The microorganisms colonizing at root may be free, parasitic, and saprophytic, and their diversity remains varied because population and species abundance often change amino acid, monosaccharide, and organic acids into primary sources of nutrients released from a root environment which support the dynamic increase and activities of different microorganisms (Kunc and Macura 1988). It has been observed that in various plant species with aid of PGPR, plant growth has been improved in terms of an increase in seedlings, biomass, vigor, root system proliferation, and production. During the past 30 decades, various reports have been performed at a very exponential rate to identify PGPR in different agricultural systems and agroecological regions, as they are an important component of the root-colonizing microorganism (Podile 2006) (Table 16.1).

The protective effect of PGPR inoculated to seedlings was observed against soil-borne pathogens (Manjula and Podile 2001; Guo et al. 2004). Therefore, the role of PGPR as defense products for soil pathogens has been increased. However, in recent years, PGPR has once again been discovered as biofertilizers, and organic farming has become more important with minimum to no input. The requirement for a threshold point to sustain plant development for the initial bacterial inoculum indicates the quorum sensing of bacteria in plant-PGPR interactions plays a significant

Table 16.1 Commercially accessible PGPR strains, which are primarily assisted by mechanisms for direct plant production

PGPR strain	Trade name	Manufacturer	Recommended application
Azotobacter spp.	Bioplin	Kumar Krishi Mitra Bioproducts Pvt. Ltd., Pune, India	Soil drenching for sunflower, tomato, and another vegetable crops
Bacillus subtilis	Kodiak	GB03 Gustafson, LLC, Dallas, TX	Seed treatment in fruits and vegetables
Bacillus spp.	Bioyield	Gustafson, LLC, Plano, TX	Seed treatment in tomato, tobacco, cucumber, and pepper
Bacillus, Pseudomonas and *Streptomyces* spp.	Compete	Plant Health Care BV, CA Vught	Soil drenching for turfgrass, nursery, and greenhouse plantations

role (Teplitski et al. 2000). With understanding and knowledge of genetics, biochemical and physiological pathways aimed to help as to how PGPR can be used for plant growth promotion and disease control, hence with the goal of choosing and improving potential strains for crop improvement. PGPR innovation and distribution systems in various crop systems increase the rapid acceptance of strains and satisfy farmers by reducing costs in respect of chemical fertilizer.

2 Mechanism of Action of PGPR

PGPR-mediated growth in plants is promoted as per the reports of Kloepper and Schroth (1981), with the alteration in the rhizosphere niche of the entire microbial community through the formation of different compounds (Kloepper and Schroth 1981). In general, PGPR promotes plant growth through promoting either the production or regulation of the hormone levels of plants or indirectly interfering with the rhizosphere, by fixing nitrogen, solubilized phosphorus, and potassium, or the production of siderophore. Other biocontrol mechanisms, such as antibiotics (Chin-A-Woeng et al. 1998) and CNN (competition for nutrients and niches) (Validov et al. 2009), have been certainly required to create root colonization over the current years. The development of exopolysaccharides may be one of the potential explanations. Thus, produced exopolysaccharides reduce Na uptake by binding them and also by forming biofilms (Qurashi and Sabri 2012). Although there are two mechanisms involved for PGPR, mostly studied is the direct one which is also discussed below.

3 Direct Mechanisms Involved in PGPR

In the absence of pathogens, direct PGPR promotes plant production. According to Vessey (2003), plant rhizosphere soil bacterial species growing in, on, or around plant tissue enhance plant development and growth through a multitude of processes. In addition to supplying mechanical assistance and supporting water and nutrient absorption, rhizosphere microbial behavior influences the habits of rooting and the availability of nutrients to plants.

3.1 Nutrient Acquisition

A part of these organically grown plants are additionally metabolized by nearby microorganisms as carbohydrate and nitrogen sources and replanted for the growth and processing by certain microbiological molecules (Kang et al. 2010).

3.2 Nitrogen Fixation

Certain microorganisms are in a position to transform nitrogen to ammonia through the process of fixing nearly two-thirds of the global amount of nitrogen by means of complex enzyme mechanism known as nitrogenase (Kim and Rees 1994). There are two groups of microbes which fix atmospheric nitrogen into a usable form: (a) symbiotic nitrogen-fixing bacteria (Ahemad and Khan 2011) and *Frankia* (nonleguminous tree) and (b) nonsymbiotic nitrogen-fixing form such as cyanobacteria (Bhattacharyya and Jha 2012). Host plant, which is associated with nonsymbiotic nitrogen-fixing bacteria, fixes a minimal amount of nitrogen (Glick 2012). Diazotrophs are nitrogen-fixing microbes and contain molybdenum nitrogenase which is responsible for biological nitrogen fixation and other related activities (Bishop and Jorerger 1990). A variety of free-living bacteria, such as *Azospirillum* spp., in addition to *Rhizobia* spp., can also fix and distribute nitrogen to the plants (Wisniewski-Dyé et al. 2015). However, the bacteria which live freely produce only a small number of fixed nitrogen required by the bacterial host plant. Nitrogenases (nif) are also important for the fixation of nitrogen into structural genes, iron-protein activation genes, molybdenum cofactor genes, electron donations, and regulatory genes required for the synthesis and action of enzymes (Bruto et al. 2014). As with the NIF genes, they usually occur in a group of 7 operons between the dimensions of 10 and 20 kb, encoding 20 proteins (Glick 2012).

3.3 Phosphate Solubilization

Phosphorus (P) is the second most important nutrient-restricting plant growth in soils, in both organic and inorganic forms, following nitrogen (Khan et al. 2009). The phosphorus mass of soil is found as an insoluble form when only the monobasic ions (H_2PO_4) and the diabasic ions are taken into consideration when absorbing plants in two soluble forms, although phosphorus is available to plants in minimal amounts (Bhattacharyya and Jha 2012). Phosphorus deficits are frequently used in soil fields because plants absorb lower phosphatic fertilizers and the remaining complexes quickly become insoluble when a reaction to other soil component phosphatic fertilizers is carried out (Mckenzie and Roberts 1990). However, routine treatments of phosphate fertilizers are both costly and unnecessary (Kaur and Reddy 2014). This led to the search for environmentally sustainable and affordable alterna Pyoluteorin was first isolated in tive to grow crops in low phosphorus soils.

In this respect, a viable substitute for the chemical phosphatic fertilizers is provided by the phosphorus sources used by the plant (Khan et al. 2007). The microorganisms which can solubilize the phosphate are called phosphate solubilizing microorganism (PSM). Although the most potential biofertilizer of different PSMs inhabiting the rhizosphere was the use of phosphate-solubilizing bacteria (PSB), plants that can easily absorb via biological routes obtain a good amount of phosphorus (Zaidi et al. 2009; Yadav et al. 2014). Kumar et al. (2001) put forward that many crops like radish, potato, tomato, and wheat are associated with microbial species which solubilizes phosphorus.

Therefore, it is very relevant for agricultural microbiologist that PGPRs can solubilize the mineral phosphate since they can boost phosphorus availability for effective plant development. PGPRs for the solubilization of precipitated plant phosphates were registered as a possible plant growth support mechanism in field conditions (Verma et al. 2001; Guo et al. 2015). The reason for the solubilization of inorganic phosphorus might be the organic acid synthesis by rhizospheric microbes (Barea and Richardson 2015). The commercial use of PGPB phosphate solubilization was unfortunately limited due to variable results (Ghosh et al. 2014).

Biological disease control is an attractive alternative strategy for the control of plant diseases. Meanwhile, it also provides practices compatible with the goal of a sustainable agricultural system. Understanding the mechanisms of biological control of plant diseases through the interactions between antagonists and pathogens may allow us to select and construct the more effective biocontrol agents and to manipulate the soil environment to create a conducive condition for successful biocontrol. Many factors have to be considered in deciding whether a biological system is feasible for the control of a particular pathogen. Of prime importance is the availability of a suitable antagonist capable of maintaining itself on the host plant. The environment under which the crop is grown will play a significant part in determining whether effective population levels of an antagonist can be established in competition with the existing microflora. Environment may also govern the choice of antagonist; for example, yeasts can survive on leaves more readily than non-spore-forming bacteria under adverse humidity conditions. It is essential that the primary mechanism by which antagonism is brought about should be known. A variety of biological controls are available for use, but further development and effective adoption will require a greater understanding of the complex interactions among plants, people and the environment (Nega 2014). Currently, agriculture faces challenges, such as soil fertility reductions, changes in climate, and increased pathogen attacks (Gopalakrishnan et al. 2015). In this way, our future main priorities are environmentally sustainable plant conservation strategies. There are growing questions about the use of chemical and synthetic fertilizers and pesticides and environmentally sustainable and effective approaches to crop growth and development. The sustainability and safety of the horticulture industry depend on eco-adaptation methods such as biopesticides, biofertilizers, and crop residues. PGPR is a big part of the conservation of crops, the development of growth, and the improvement of soil health (Beneduzi et al. 2012; Liu et al. 2017). Some of the exceptional PGPR strains that play a large part in inhibiting or destroying pathogens by making unique antibiotic mixtures are *Pseudomonas*, *Bacillus*, *Azospirillum*, and *Rhizobium*. In addition to chemical pesticides, the microbial antagonist is another way to suppress plant pathogen in crops. A broad variety of pathogens are regulated by PGPR, including bacteria, fungi, viruses, and nematic diseases (Liu et al. 2017).

4 Major Antibiotics of PGPR

In the management of plant diseases, PGPR plays a vital role in the production of antibiotics, and the system is called pathogenic microbes' inhibition or suppression. PGPR such as *Bacillus* species and fluorescent *Pseudomonas* help in the destruction of pathogens, generating inhibitory, antagonistic metabolites in their defense mechanism against harmful strains of microbes. Furthermore, in plant induced systemic resistance mechanism (ISR) antibiotics play a critical role in direct antagonistic action. Specific microorganisms can produce a range of antibiotic products, for example, PGPR produce multiple antibiotics (Table 16.2).

Antibiotic is defined as a heterogeneous community of low-molecular organic complexes that harm the production or metabolism of various microorganisms (Kumar et al. 2015). In vitro and in situ, the development of the target pathogen was smothered more effectively with the help of antibiotics. The formation of one or more antibiotics is the most crucial aspect of plant growth that promotes rhizopathological bacteria and promotes resistance to other pathogens (Glick et al. 2007). Moreover, the antibiotics are classified as volatile and nonvolatile, as aldehydes, alcohols, sulfides, ketones, and hydrogen cyanide come under the category of

Table 16.2 Antibiotics produced by PGPR

PGPR	Antibiotics
Pseudomonas sp.	Antifungal antibiotics
	Phenazines
	Phenazine-1-carboxylic acid
	Phenazine-1-carboxamide
	Pyrrolnitrin
	Pyoluteorin
	Cepaciamide A
	Oomycin A
	Viscosinamide
	Pyocyanin
	Antibacterial antibiotics
	Pseudomonic acid
	Azomycin
	Antitumor antibiotics
	FR901463
	Cepafungins
	Antiviral antibiotic
	Karalicin
Bacillus sp.	Kanosamine
	Zwittermicin A
	Iturin A (cyclopeptide)
	Bacillomycin
	Plipastatins A and B

volatile antibiotics, while the nonvolatile antibiotics include heterocyclic nitrogenous compound (Gouda et al. 2017; Fernando et al. 2018). Antibiotics promote plant growth and possess other potentially beneficial properties like antimicrobial, antiviral, and antioxidant (Ulloa-Ogaz et al. 2015; Fernando et al. 2018).

The antibiotics that play a critical role when plant pathogens are suppressed are classified into two groups: volatile and nonvolatile antibiotic products.

4.1 Nonvolatile Antibiotics

4.1.1 Polyketides (2,4-Diacetylphloroglucinol (DAPG or Phl))

DAPG or Phl is a phenolic polyketide compound which is obtained from fluorescent pseudomonas with antibacterial, antifungal, and antioxidant activities (Gaur 2002). Phl is a key determinant of plant growth-enhancing rhizobacteria's biocontrol activity. *Gaeumannomyces graminis* var. *tritici* is responsible to cause take-all diseases in wheat which can be suppressed by 2,4-DAPG antibiotic to act as take-all decline (TAD) which is produced from strains of *P. fluorescens* (Weller et al. 2007). The diseases are caused by some soil-borne pathogens and are prevented by some strains of *P. fluorescens* which also have nematicidal activity (McSpadden Gardener 2007; Meyer et al. 2009). As per reports of Dwivedi and Johri (2003), Phl's mode of action remains uncertain, although the interaction between root-associated Phl-producing microorganisms and pathogens is considered to be a significant cause of disease suppression. So, in plants, Phl elicits ISR microorganisms and, therefore, can serve as unique elicitors in plant disease management of the development of phytoalexins or other related molecules.

4.1.2 Pyoluteorin (Plt)

Pyoluteorin (Plt) is a natural antibiotic that is biosynthesized from a hybrid nonribosomal peptide synthetase (NRPS) and polyketide synthase (PKS) pathway (Fermando et al. 2005). Pyoluteorin was first isolated in the 1950s from *Pseudomonas aeruginosa* strains T359 and IFO 3455 and was found to be toxic against oomycetes, bacteria, fungi, and against certain plants (Kraus and Loper 1995). Plt inhibited most pathogens of oomycete, like *Pythium ultimum*. The severity of *Pythium* damping decreased when seeds are applied with *Pseudomonas* Plt producing, reported by now by Nowak-Thompsan et al. (1999). Hassan et al. (2011) put forward that in sugarcane, *Glomerella tucumanensis* is responsible to cause disease, namely, red root rot, but pyoluteorin produced by *P. putida* has been found to be effective against this disease.

4.1.3 Heterocyclic Nitrogenous Compounds

Heterocyclic nitrogen pigments called phenazines, which are low-molecular-weight compounds, were developed by a small group of bacterial species including *Pseudomonas*, *Burkholderia*, *Brevibacterium*, or *Streptomyces*, since more than 50 phenazine compounds occurring naturally were examined. Some bacterial strains will generate blends of different phenazine derivatives simultaneously (Guttenberger et al. 2017; Dasgupta et al. 2015). Like phenazine-1-carboxylic acid (PCA) found in *P. fluorescens* 2–79, however, *P. aureofaciens* 30–84 has been identified as a mixture of PCA along with a minimum amount of 2-hydroxyphenazine.

Several PGPR pseudomonad strains have antibiotic and antitumor features and are active in their ability to suppress pathogenic plant fungi and nematodes (Cezairliyan et al. 2013; Zhou et al. 2016). The disease caused by *G. graminis* var. *tritici* in wheat has been biocontrolled by compound known as phenazine-1-carboxylic acid (PCA) which is produced by *P. fluorescens* 2–79 and *P. aureofaciens* 30–84 (Thomashow and Weller 1988; Chin-A-Woeng et al. 2000; Shanmugaiah et al. 2010; Ju et al. 2018). *P. aeruginosa* PNA1 (wild-type) phenazine-1-carboxylic acid and phenazine-1-carboxamide are important in controlling cocoyam root rot caused by *P. myriotylum* (Tambong and Hofte 2001). *P. aeruginosa* is known to produce pyocyanin and phenazine-1-carboxylic acid which are having antagonistic activity against *F. oxysporum*, *Aspergillus niger*, and other various pathogens (Rane et al. 2007; Abo-Zaid 2014). In *P. chlororaphis*, 30–84 phenazine derivatives have to be developed to prevent plant pathogens (Ju et al. 2018). Several volatile antibiotics, such as hydrogen cyanide, aldehydes, alcohols, ketones, and sulfides, are present in this region, but hydrogen cyanide is the most important metabolite (Yu et al. 2018).

4.2 *Volatile Antibiotics*

4.2.1 Hydrogen Cyanide (HCN)

Various Gram-negative bacteria, namely, *Chromobacterium violaceum*, *P. aeruginosa*, and *P. fluorescens*, produce cyanide as their secondary metabolite (Hass and Defago 2005). It has been reported by many workers that hydrogen cyanide (HCN) showed the nematicidal activity against *Meloidogyne hapla* as produced from the bacterial strain, namely, *P. chlororaphis* O6 (Kang et al. 2018). Sarhan and Shehata (2014) reported that in alfalfa, infection caused by *F. solani* can be stopped by generation of HCN from *F. solani*. Hydrogen cyanide (HCN) production is an essential determinant of biocontrol (Anderson and Kim 2018). The characterized hcnABC gene set was found to be responsible in Q2-87 and CHA0 for biosynthesis of HCN (Hass and Defago 2005).

4.2.2 Aldehydes, Alcohols, Ketones, and Sulfides

Mycelium formation, ascosporous germination, and survival of sclerotia were entirely impeded by these substances. These volatiles come directly into contact with sclerotial structures that lead to a reduction in inoculum capacity, preventing the occurrence of the disease (Fernando et al. 2004). The pathogen *Erwinia carotovora* has been inhibited by bacterial volatiles such as 2,3-butadienol (Ryu et al. 2003).

5 Biostimulants of PGPR

Plant growth regulators or phytostimulants which include auxin (indole-3-acetic acid (IAA)), gibberellic acid (GA), cytokinins (CK), and ethylene are organic chemical compounds that are known to regulate plant growth and development. Throughout the years, these chemical molecules became known as the main biochemical, physiological, and morphological hormones required for growth. PGPR species of the genera *Azospirillum*, *Pseudomonas*, *Xanthomonas*, *Rhizobium*, and *Bradyrhizobium* can form phytohormones (Mohamed and Gomaa 2012).

Auxin is a vital hormone, which controls most plant processes directly or indirectly. Being the first phytohormone identified in the *Phalaris canariensis* seeds by Darwin (1887), it has since paved the way for further exploration leading to the detection of the most active and prominent plant hormones in the auxin community, namely, indole-3-acetic acid (IAA). No matter how the plants can synthesize this chemical compound (endogen supply), their success still depends entirely on external (exogenous) supply. PGPR is mainly supervised and is correlated with soil bacteria in this external gathering (Khalid et al. 2006). The cell function of auxin ranges from distinguishing the vascular tissue, initiating lateral and adventitious roots, stimulating the division of cells, and elongating the growth of the shoots and roots (Glick 1995). PGPR is significant in the development of the stage cum availability of nutrients in the rhizosphere for more efficient IAA production, considering the type of species and strain it cultivates, the condition, and the development (Ashrafuzzaman et al. 2009). While plants have now recognized other auxins including indole-3-butyric acid (IBA) and phenylacetic acid (PAA) (Normanly 1997), researchers also need to learn their structure mechanism of action and functioning. In comparison, in the soil-plant auxin pool and L-tryptophan (L-TRP) as a substitute for the production of the auxin, IAA producers are found to be more prevalent. The results indicate a rise in the L-tryptophan level that raises the biochemical and metabolic activities of bacterial BIPs or APBs, with subsequent root length reactions and root architecture modifications. Tryptophol, tryptamine, indole-3-pyruvic acid (IPA), and indole-3-acetamide are the primary metabolic pathways (Bartel 1997). Emergent evidence demonstrates that species that produce low auxins due to the lack of L-tryptophan are likely to grow high auxins when increased by L-tryptophan, especially in the presence of a viable strain of *Rhizobium* (Zahir et al. 2010). Importantly, it is important to notice that plant-based indigenous auxin (IAA) might still not be automatically adequate to achieve maximum plant performance but should contribute to plant growth (Pilet and Saugy 1987). Therefore, it is important to explain the chemical 130 messengers (IAA produced by PGPR) with an exogenous need to bring about optimum plant development, growth, and adaptation to the stressful setting.

It is not well known yet what exact pathways PGPR stimulate to promote plant growth through the synthesis of gibberellic acid (Kang et al. 2009). GA is a group of diterpenes which greatly affect the processes of sprouting, leaf growth, elongation of the root, extension of the lateral root, fruit development, flowering, and initiation of trichomes (Yamaguchi 2008). Gibberellins and genera are the primary targets during environmental stress conditions because of the important role played by them in improving effective photosynthetic processes in plants, and they are a major plant growth biological regulator, which can enhance stress tolerance in many crops. The exogenous application of these growing hormones can be useful in soil shift and crop production improvements (Iqbal et al. 2011). Gibberellins are essentially interested in the alteration of plant morphology and promote the production of an aerial component, (Van Loon 2007) and has also been given to their effect in increasing tolerance of abiotic and biotic stresses. At the cellular level, the growth rate is regulated by the combined activity of two processes: cell proliferation and expansion. Gibberellins (GA) are plant specific hormones that play a central role in the regulation of growth and development with respect to environmental variability. It is well established that GA promotes growth through cell expansion by stimulating the destruction of growth-repressing DELLA proteins (DELLAs) and promotes chloroplast biogenesis, shoot proliferation, senescence, apical dominance, development of anthocyanin, and photomorphogens (Davies 2004). This also contributes to the susceptibility to vascular changes, proliferation of root hair, and suppression of the development of lateral root and main elongation (Aloni et al. 2006), and this molecule can be obtained by either plants or PGPR in an endogenous and exogenous way.

Plants improve the absorption through biosynthesis of endogenous cytokinin (Pospíšilová 2003). Studies have shown that cytokinin perfectly regulates plant adaptation, especially in salt-exposed areas, during plant growth (Hadiarto and Tran 2011). Through a biochemical test, cytokinin is a major antagonist of abscisic acid (ABA), resulting in certain phytohormone regulation (Pospíšilová 2003). The cytokinin content of the plant declines significantly during water scarcity, resulting in a favorable rise in ABA concentration. The evaluation of the development in broth media for plant hormones by various streptomycin strains indicates that cytokinin and gibberellin are synthesized by both strains (Mansour et al. 1994). While essential to phyto-development, it does not yet have a well-defined mechanism of action. The cytokinin in the receptor gene is regulated by changes in osmotic conditions (Merchan et al. 2007). Various studies have shown that the plants are immune to environmental stress by inoculating seedlings with cytokinin strains of *Bacillus subtilis*.

Ethylene which is a special phytohormone has a wide spectrum of chemical activity as at low concentrations the useful function of this biomolecule is better reported. This impairs certain significant developmental features, e.g., root elongation, defoliation, and other cellular processes, which lead to reduced crop production at high concentrations (Bhattacharyya and Jha 2012). An enzyme 1-aminocyclopropane-1-carboxylic acid (ACC) deaminase is required to resolve these troubling effects. The biocatalyst's function is to degenerate the ACC plant which is the direct precursor to α-ketobutyrate and ammonium for ethylene synthesis on the plant (Glick et al. 2007). The decay results from the decrease of plant production of ethylene by a variety of pathways, while PGPR producing ACC deaminase controls the amount of ethylene of plants and stops high levels of ethylene from inhibiting development (Noumavo et al. 2016). However, this vaporous hormone also governs the initiation, maturation, and germination of the seeds and abscission of the leaf and wilting (Kaur et al. 2016).

6 Role of PGPR in Biocontrol of Plant Disease

The greatest danger to food security worldwide is the loss of crops from plant diseases. The losses vary from small reductions in plant growth to major damage resulting in plant death and reduced yields (Savary et al. 2012). Many methodologies were studied to avoid or control these pathogens, including the production of resistant varieties by plant breeding, the production of GMO plants, as well as the chemical enrollments such as fungicides. Furthermore, there could be a detrimental effect on the health of humans through the presence of pesticide and fungicide leftovers. Due to the imperatives on antibiotic development in standard environments,

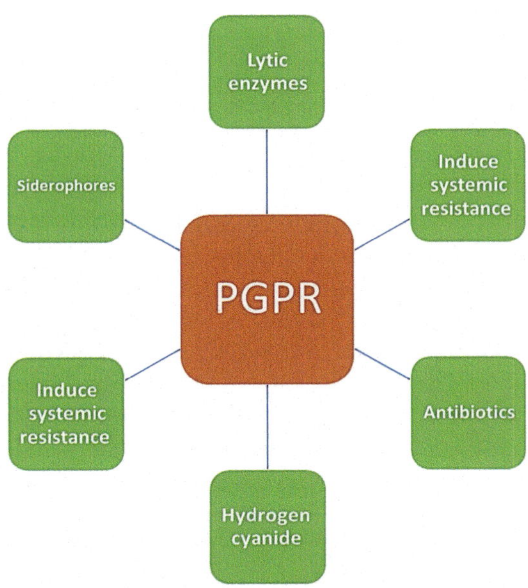

Fig. 16.1 Model to illustrate the role of PGPR in plant protection

the role of antibiotics in biocontrol and microbial antagonism has been discussed. PGPR is a biocontrol agent with the ability to kill a large variety of potential species with plant disease. PGPR must use one of the following mechanisms to be an effective biocontrol agent against pathogenic microbes: antibiotic formation, systemic resistance induction, hydrogen cyanide formation, and lytic enzyme formation (Fig. 16.1) (Lugtenberg and Kamilova 2009). According to reports by Junaid et al. (2013), key organisms which attack the plants typically include, in host plants, bacteria, fungus, and nematodes which cause bad diseases. Thereby, rhizobacteria or their metabolites are known to function as a sort of protection against disease.

6.1 Antibiotic Production

Antibiotic production by PGPR is one of the essential components for the promotion of plant growth and antimicrobial activity (Table 16.3). These antibiotics have been shown to play a part in disease concealment through mutant study and biochemical exams using distilled antibiotics in various biocontrol frameworks. These antimicrobial mixes can track pathogenic plant microbes or their growth by inhibiting the germination of spores and fungal mycelia lysis (Adhya et al. 2018; Ulloa-Ogaz et al. 2015). PGPR is known as a biocontrol agent due to the generation of antibiotics which includes known examples, i.e., DAPG, phenazine, cyclic lipopeptides, and amphisin (Loper and Gross 2007), while there is certain list of antibiotics which includes zwittermicin A, oligomycin A, xanthobaccin, and kanosamine known to be generated by *Pseudomonas strains*, *Bacillus*, *Streptomyces*, and *Stenotrophomonas* sp. (Compant et al. 2005). However, these biochemicals are found to be regulated by abiotic, biotic, and other environmental factors, and diseases caused by pathogens can be suppressed by low-weight-molecular compounds known as antibiotics as various good known drugs from PGPR as biocontrol agents which have been utilized for the disease control include 2-hexyl-5-propyl resorcinol (HPR), 2-hydroxymethyl-chroman-4-one, D-gluconic acid, hydrogen cyanide (HCN), and phloroglucinols (Phl) (Cazorla et al. 2006). To maintain other microbes in the soil as niche competition in the field will have passed, and the fundamental path to decreasing the incidence of disease in plants will be followed by *Rhizobacteria* with a view to nutrient supply and spatial abundance (Kamilova et al. 2005, b). If an association of competent microbes flourishes in the rhizosphere and affects the radically colonized PGPR by releasing noxious metabolites or compounds, it thus impedes the root absorption capacity to assimilate growing and developing nutrients. Apart from the ability to survive in the nutrients of PGPR, flagellum, lipopolysaccharide, chemotaxis, and root exudate secretion enhance its longevity (Lugtenberg and Kamilova 2009). According to Saraf et al. (2011), it is important in heme growth the reduction of ribotide precursors of DNA and ATP synthesis that siderophores are synthesized in PGPR with iron chelation when not present in pathogenic fungal species of plant. Space exposure is thus a significant factor in the thriving and dominant role of PGPR over pathogens in niche competition, and the rhizosphere role plays a vital role in supplying plant nutrient exposure (Heydari and Pessarakli 2010).

Table 16.3 Generation of antibiotics for soil-borne diseases via the PGPR microorganism

Antibiotics/functions	PGPR	Pathogen/disease	References
Phenazine, 2,4-diacetylphloroglucinol (DAPG)	*Pseudomonas fluorescens*	*Meloidogyne incognita* *Fusarium oxysporum*	Meyer et al. (2016)
Surfactin Iturin Fengycin	*Bacillus velezensis*	*Ralstonia solanacearum* *Fusarium oxysporum*	Cao et al. (2018)
Volatile antibiotics	*Bacillus amyloliquefaciens* subsp. *plantarum XH-9*	*Fusarium oxysporum*	Wang et al. (2018)
Bacilysin	*B. subtilis*	*Phytophthora infestans*	Caulier et al. (2017)
Hydrogen cyanide Phenazine	*Fluorescent pseudomonads*	*Pythium aphanidermatum*	Prabhukarthikeyan and Raguchander (2016)
Pyrrolnitrin	*Pseudomonas aeruginosa*	*Rhizopus microsporus*, *Fusarium*	Uzair et al. (2018)
Bacillus Peptide Antibiotics	*Bacillus*	*Fusarium graminearum*	Khan et al. (2017)
Surfactin Iturin A Iturin D Fengycin Bacillomycin D	*Bacillus subtilis*	Wilt and root rot	Smitha et al. (2017)
Bacillomycin D Fengycin A	*B. subtilis*	*Sclerotinia sclerotiorum*	Abdeljalil et al. (2016)
Pyrrolnitrin Hydrogen cyanide	*Pseudomonas chlororaphis*	*Sclerotinia sclerotiorum*	Nandi et al. (2015)
Triterpenoid soyasapogenol	*Rhizobium leguminosarum bv. viciae*	*Didymella pinodes*	Ranjbar Sistani et al. (2017)
Fengycin	*Bacillus amyloliquefaciens* subsp. *plantarum*	*Rhizomucor variabilis*	Zihalirwa Kulimushi et al. (2017)
Iturin Bacilysin Bacillomycin Surfactin Subtilin Subtilosin	*B. amyloliquefaciens*	*Sclerotinia sclerotiorum*	Vinodkumar et al. (2017)

(continued)

Table 16.3 (continued)

Antibiotics/functions	PGPR	Pathogen/disease	References
DAPG	*Pseudomonas* sp. LBUM300	*Clavibacter michiganensis* subsp. *michiganensis*	Lanteigne et al. (2012)
2,4-diacetylphloroglucinol (2,4-DAPG), pyoluteorin (PLT) pyrrolnitrin (PRN)	*Pseudomonas fluorescens*	*Botrytis cinerea Monilinia fructicola*	Zhang et al. (2020)

6.2 *Induced Systemic Resistance (ISR)*

To combat pathogenic bacteria, fungi, and viruses, PGPR activates some form of protection mechanism. This will improve and adapt the plant much better (Van Loon 2007). The gene and gene products have not been well established for this form of biological control phenomenon. Unlike systemic acquired resistance (SAR), a protection state is triggered in the entire plant following primary pathogen infections (Bakker et al. 2013). To act against plant pathogens, a mechanism called induced systemic resistance (ISR) uses plant hormones like jasmonic acid (JA), salicylic acid (SA), and ethylene and other organic acids for the stimulation and signaling in host plant for the defense purpose (Pieterse et al. 2000). This mechanism is mediated through JA, ethylene, and SA biosynthesis pathways (Dempsey and Klessig 2012). The interaction of these hormones is either antagonistic or synergistic to change the mechanism of defense (Nassem and Dandekar 2012). A large number of secondary metabolites that have antibiotic activity (phenolic, flavonoids, alkaloids, cyanide glycosides, etc.) were identified as an ISR mechanism in noninfected crops following receipt of chemical signals from infected plants, with volatile methyl salicylic acid as a signal (Dempsey and Klessig 2012). Antimicrobial active ingredients, such as phenols, can inhibit microbial development, and different phenolic metabolic cells that are less harmful to plant cells accumulate in the cells than aglycones. After infection, aglycone is released by hydrolysis which is toxic to both plant cells and microbes (Kenawy 2016). The defense response in the plant system can cause cell wall thickening and lignification, callus deposition, a buildup of phytoalexins, and synthesis of many lytic enzymes (Sticher et al. 1997).

According to Labuschagne et al. (2010), to cope up with environmental stress, PGPR reaction toward ISR can be achieved through adjustment of physical and biochemical reaction to environmental stress and also by increasing physical and mechanical vigor of the cell wall, and it has been observed that certain molecules such as lipopolysaccharide, *N*-acyl homoserine lactone (AHL), salicylic acid, etc. are antibiotic forms of ISR in PGPR (Van Loon 2007). There are certain bacterial species which are found to be involved in the process to biocontrol including *Pseudomonas* sp., *Bacillus pumilus*, and *Enterobacteriaceae* (Jourdan et al. 2009). Zehnder et al. (2001) found that ISR has wider scope when applied PGPR strain is used as a seed coat against *Pseudomonas syringae* causing angular leaf spot, *Colletotrichum lagenarium* causing anthracnose in cucumber, and *Erwinia tracheiphila* leading to bacterial wilt.

Besides, *P. fluorescens* has protected tomatoes from wilt diseases and may serve as an ISR signal to cause DAPG pools in tomato root rhizosphere (Haas and Keel 2003).

7 Conclusion

Over the last century, the effective application of organic fertilizers, herbicides, and pesticides should not be overlooked in an agricultural environment. They help plant growth initially while having a long-term negative impact. This practice not only affects the land and its inhabitants but also threatens people's lives through the food chain. The soil has become extremely infertile and unproductive due to the rise in soil pollution, condition of climate, soil pathogens, and extensive land overuse. Food insecurity and the increasing population are evident at the low agro-yield. To achieve auto-sufficiency, a wide understanding of the microbial interaction and its mechanism of action must be made, particularly in the tropic world, to be essential to scientific knowledge. Not only does this lead to bumpers but also keeps the ground healthy and safe. Although the PGPR campaign has been in progress for decades, in Africa, it has not been adopted due to a lack of understanding and governmental policies. Nonetheless, efforts will be based on the replacement of bioproduct agrochemicals such as biofertilizers, bioinsecticides, and bioherbicides by a supportive PGPR consortium. To boost crop yield while preserving the soil conditions, farmers must carefully define and recognize the benefits of these bioinoculants in terms of improved plant nutrients and biocontrol through the introduction of systemic resistance and nutrients or space rivalry. This approach is to mitigate soil degradation, habitat change, and land flora and fauna loss by genetically modified processing of PGPR as an essential compound of modern food production. Finally, this technology, especially in developed countries, must be used and implemented to curb the possible humanitarian (famine) crisis in areas ravaged by war and terrorism, thus stimulating the production of food and improving our community's environmental safety..

References

Abdeljalil NOB, Vallance J, Gerbore J, Bruez E, Martins G (2016) Characterization of tomato-associated rhizobacteria recovered from various tomato-growing sites in Tunisia. J Plant Pathol Microbiol 7(351):2

Abo-Zaid GA (2014) Scaling-up production of biocontrol agents from *Pseudomonas* spp. Faculty of Agriculture, Alexandria University, Alexandria

Ashrafuzzaman M, Hossen FA, Ismail MR, Hoque MA, Islam ZM, et al. (2009) Efficiency of Plant Growth-Promoting Rhizobacteria (PGPR) for the enhancement of rice growth. Afr J Biotech 8:1247–1252

Adhya TK, Lal B, Mohapatra B, Paul D, Das S (2018) Advances in soil microbiology: recent trends and future prospects. Springer, Singapore

Ahemad M, Khan MS (2011) Toxicological assessment of selective pesticides towards plant growth promoting activities of phosphate solubilizing Pseudomonas aeruginosa. Acta Microbiol Immunol Hung 58:169–187

Aloni R, Aloni E, Langhans M, Ullrich CI (2006) Role of cytokinin and auxin in shaping root architecture: regulating vascular differentiation, lateral root initiation, root apical dominance and root gravitropism. Ann Bot 97(5):883–893

Anderson AJ, Kim YC (2018) Biopesticides produced by plant-probiotic Pseudomonas chlororaphis isolates. Crop Prot 105:62–69

Bakker PA, Doornbos RF, Zamioudis C, Berendsen RL, Pieterse CMJ (2013) Induced systemic resistance and the rhizosphere microbiome. Plant Pathol J 29(2):136–143

Barea JM, Richardson AE (2015) Phosphate mobilisation by soil microorganisms. In: Lugtenberg B (ed) Principles of plant-microbe interactions. Springer, Heidelberg, pp 225–234

Bartel B (1997) Auxin biosynthesis. Annu Rev Plant Physiol Plant Mol Biol 48:51–66 . PMID: 15012256. https://doi.org/10.1146/annurev.arplant.48.1.51

Beneduzi A, Ambrosini A, Passaglia LMP (2012) Plant growth-promoting rhizobacteria: their potential as antagonists and biocontrol agents. Genet Mol Biol 35(4):1044–1051

Bhattacharyya PN, Jha DK (2012) Plant growth-promoting rhizobacteria (PGPR): emergence in agriculture. World J Microbiol Biotechnol 28:1327–1350

Bishop PE, Jorerger RD (1990) Genetics and molecular biology of an alternative nitrogen fixation system. Plant Mol Biol 41:109–125

Bruto M, Prigent-Combaret C, Muller D, Moënne-Loccoz Y (2014) Analysis of genes contributing to plant-beneficial functions in plant growth-promoting rhizobacteria and related Proteobacteria. Sci Rep 4:6261

Cao Y, Pi H, Chandrangsu P, Li Y, Wang Y, Zhou H, Cai Y (2018) Antagonism of two plant-growth promoting *Bacillus velezensis* isolates against *Ralstonia solanacearum* and *Fusarium oxysporum*. Sci Rep 8(1):4360

Caulier S, Gillis A, Colau G, Licciardi F, Liépin M, Desoignies N, Bragard C (2017) Versatile antagonistic activities of soil-borne Bacillus spp. and *Pseudomonas* spp. against *Phytophthora infestans* and other potato pathogens. Front Microbiol 9:143

Cazorla FM, Duckett SB, Bergstro€m ET, Noreen S, Odijk R, Lugtenberg BJJ, Thomas-Oates J, Bloemberg GV (2006) Biocontrol of avocado dematophora root rot by antagonistic Pseudomonas fluorescens PCL1606 correlates with the production of 2-hexyl 5-propyl resorcinol. Mol Plant Microbe Interact 19:418–428

Cezairliyan B, Vinayavekhin N, Grenfell-Lee D, Yuen GJ, Saghatelian A, Ausubel FM (2013) Identification of *Pseudomonas aeruginosa* phenazines that kill *Caenorhabditis elegans*. PLoS Pathog 9:e1003101

Chin-A-Woeng TF, Bloemberg GV, Van der Bij AJ, Van der Drift KM, Schripsema J, Kroon B, de Bruijn FJ (1998) Biocontrol by phenazine-1-carboxamide-producing *Pseudomonas chlororaphis* PCL1391 of tomato root rot caused by *Fusarium oxysporum* f. sp. *radicis-lycopersici*. Mol Plant 11(11):1069–1077

Chin-A-Woeng TF, Bloemberg GV, Mulders IH, Dekkers LC, Lugtenb BJ (2000) Root colonization by the phenazine-1-carboxamide producing bacterium *Pseudomonas chlororaphis* PC3L1391 is essential for biocontrol of tomato foot and root rot. Am Phytopathol Soc 13:1340–1345

Compant S, Duffy B, Nowak J, Clément C, Barka EA (2005) Use of plant growth-promoting bacteria for biocontrol of plant diseases: principles, mechanisms of action, and future prospects. Appl Environ Microbiol 71:4951–4959

Darwin F (1887) The life and letters of Charles Darwin, London: John Murray. Volume 3.

Dasgupta D, Kumar A, Mukhopadhyay B, Sengupta TK (2015) Isolation of phenazine 1,6-dicarboxylic acid from *Pseudomonas aeruginosa* strain HRW.1-S3 and its role in biofilm-mediated crude oil degradation and cytotoxicity against bacterial and cancer cells. Appl Microbiol Biotechnol 99:8653–8665

Davies PJ (2004) Plant hormones: biosynthesis, signal transduction, action. Kluwer Academic Publishers, Dordrecht

Dempsey DA, Klessig DF (2012) SOS—too many signals for systemic acquired resistance? Trends Plant Sci 17(9):538–545

Dwivedi D, Johri BN (2003) Antifungal from fluorescent pseudomonads: biosynthesis and regulation. Curr Sci 85:1693–1703

Fernando WGD, Nakkeeran S, Zhang Y (2004) Ecofriendly methods in combating *Sclerotinia sclerotiorum* (Lib.) de Bary. Rec Res Dev Environ Biol 1:329–347

Fernando W, Nakkeeran S, Zhang Y, Savchuk S (2018) Biological control of Sclerotinia sclerotiorum(lib.) de Bary by *Pseudomonas* and *Bacillus* species on canola petals. Crop Prot 26:100–107

Fernando WDG, Nakkeeran S, Zhang Y (2005) Biosynthesis of antibiotics by PGPR and its relation in biocontrol of plant diseases. In: Siddiqui ZA (ed) PGPR: Biocontrol and Biofertilizer. Springer, Dordrecht, pp 67–109

Gaur R (2002) Diversity of 2,4-diacetylphloroglucinol and 1-aminocyclopropane 1-carboxylate deaminase producing rhizobacteria from wheat rhizosphere. PhD thesis, G.B. Pant University of Agriculture and Technology, Pantnagar

Ghignone S, Salvioli A, Anca I, Lumini E, Ortu G, Petiti L, Bonfante P (2012) The genome of the obligate endobacterium of an AM fungus reveals an interphylum network of nutritional interactions. ISME J 6(1):136–145

Ghosh UD, Saha C, Maiti M, Lahiri S, Ghosh S, Seal A, Mitra Ghosh M (2014) Root associated iron-oxidizing bacteria increase phosphate nutrition and influence root to shoot partitioning of iron in tolerant plant *Typha angustifolia*. Plant Soil 381(1–2):279–295

Glick BR (1995) The enhancement of plant growth by free-living bacteria. Can J Microbiol 41:109–117

Glick BR (2012) Plant growth-promoting bacteria: mechanisms and applications. Scientifica 2012:963401

Glick BR, Cheng Z, Czarny J, Duan J (2007) Promotion of plant growth by ACC deaminase- producing soil bacteria. Eur J Plant Pathol 119:329–339

Gopalakrishnan S, Sathya A, Vijayabharathi R, Varshney RK, Gowda CL, Krishnamurthy L (2015) Plant growth-promoting rhizobia: challenges and opportunities. 3 Biotech 5(4):355–377

Gouda S, Kerry RG, Das G, Paramithiotis S, Shin HS, Patra JK (2017) Revitalization of plant growth-promoting rhizobacteria for sustainable development in agriculture. Microbiol Res 206:131–140

Guo JH, Qi HY, Guo YH, Ge HL, Gong LY, Zhang LX, Sun PH (2004) Biocontrol of tomato wilt by plant growth-promoting rhizobacteria. Biol Cont 29:66–72

Guo JK, Ding YZ, Feng RW, Wang RG et al (2015) *Burkholderia metalliresistens* sp. nov., a multiple metal-resistant and phosphate-solubilising species isolated from heavy metal-polluted soil in Southeast China. Antonie van Leeuwenhoek 107(6):1591–1598

Guttenberger N, Blankenfeldt W, Breinbauer R (2017) Recent developments in the isolation, biological function, biosynthesis, and synthesis of phenazine natural products. Bioorg Med Chem S0968–S0896:31180–31844

Haas D, Keel C (2003) Regulation of antibiotic production in root-colonizing Pseudomonas spp. and relevance for biological control of plant disease. Annu Rev Phytopathol 41:117–153

Hadiarto T, Tran LS (2011) Progress studies of drought-responsive genes in rice. Plant Cell Rep 30:297–310

Hao X, Xie P, Johnstone L, Miller SJ, Rensing C, Wei G (2012) Genome sequence and mutational analysis of plant-growth promoting bacterium Agrobacterium tumefaciens CCNWGS0286 isolated from a zinc-lead mine tailing. Appl Environ Microbiol 78:5384–5394

Hass D, Defago G (2005) Biological control of soil-borne pathogens by fluorescent pseudomonads. Nat Rev Microbiol 3:307–319

Hassan MN, Afghan S, Hafeez FY (2011) Biological control of red rot in sugarcane by native pyoluteorin-producing *Pseudomonas putida* strain NH-50 under field conditions and its potential modes of action. Pest Manag Sci 67:1147–1154

Heydari A, Pessarakli M (2010) A review on biological control of fungal plant pathogens using microbial antagonists. J Biol Sci 10:273–290

Iqbal N, Nazar R, Iqbal MRK, Masood A, Nafees AK (2011) Role of gibberellins in regulation of source sink relations under optimal and limiting environmental conditions. Curr Sci 100:998–1007

Jourdan E, Henry G, Duby F, Dommes J, Barthelemy JP, Thonart P, Ongena M (2009) Insights into the defense-related events occurring in plant cells following perception of surfactin-type lipopeptide from *Bacillus subtilis*. Mol Plant Microbe Interact 22:456–468

Ju M, Wang D, Pierson L, Pierson E (2018) Disruption of MiaA provides insights into the regulation of phenazine biosynthesis under suboptimal growth conditions in *Pseudomonas chlororaphis* 30-84. Microbiology 163:94–108

Junaid JM, Dar NA, Bhat TA, Bhat AH, Bhat MA (2013) Commercial biocontrol agents and their mechanism of action in the management of plant pathogens. Int J Mod Plant Anim Sci 1:39–57

Kang S, Joo GJ, Hamayun M (2009) Gibberellin production and phosphate solubilization by newly isolated strain of Acinetobactercalcoaceticus and its effect on plant growth. Biotech Lett 31: 277–281

Kang BR, Anderson AJ, Kim YC (2018) Hydrogen cyanide produced by Pseudomonas chlororaphis o6 exhibits nematicidal activity against Meloidogyne hapla. Plant Pathol J 34:35–43

Kamilova F, Validov S, Azarova T, Mulders I, Lugtenberg B (2005) Enrichment for enhanced competitive plant root tip colonizers selects for a new class of biocontrol bacteria. Environ Microbiol 7:1809–1817

Kang BG, Kim WT, Yun HS, Chang SC (2010) Use of plant growth-promoting rhizobacteria to control stress responses of plant roots. Plant Biotechnol Rep 4(3):179–183

Kaur G, Reddy MS (2014) Influence of P-solubilizing bacteria on crop yield and soil fertility at multilocational sites. Eur J Soil Biol 61:35–40

Kaur H, Kaur J, Gera R (2016) Plant growth promoting rhizobacteria: a boon to agriculture. Int J Cell Sci Biotechnol 5:17–22

Kenawy AMA (2016) characterization of two udp glycosyltransferase genes from hybrid poplar, MSc thesis, Faculty of Forestry, The University of British Columbia, Vancouver, Canada

Khalid A, Akhtar MJ, Mahmood MH, Arshad M (2006) Effect of substrate-dependent microbial ethylene production on plant growth. Microbiology 75:231–236

Khan MS, Zaidi A, Wani PA, Oves M (2009) Role of plant growth-promoting rhizobacteria in the remediation of metal contaminated soils. Environ Chem Lett 7:1–19

Khan N, Maymon M, Hirsch AM (2017) Combating *Fusarium* infection using bacillus-based antimicrobials. Microorganisms 5(4):75

Khan MS, Zaidi A, Wani PA (2007) Role of phosphate-solubilizing microorganisms in sustainable agriculture - A review. Agron Sustain Dev 27: 29–43

Kim J, Rees D (1994) Nitrogenase and biological nitrogen fixation. Biochemistry 33:389–397

Kim BK, Chung JH, Kim SY, Jeong H, Kang SG, Kwon SK, Kim JF (2012) Genome sequence of the leaf-colonizing bacterium Bacillus sp. strain 5B6, isolated from a cherry tree. J Bacteriol 194(14):3758–3759

Kloepper JW, Beauchamp CJ (1992) A review of issues related to measuring colonization of plant roots by bacteria. Can J Microbiol 38(12):1219–1232

Kloepper JW, Schroth MN (1981) Plant growth-promoting rhizobacteria and plant growth under gnotobiotic conditions. Phytopathology 71:642–644

Kloepper JW, Ryu CM, Zhang S (2004) Induced systemic resistance and promotion of plant growth by Bacillus spp. Phytopathology 94:1259–1266

Kraus J, Loper JE (1995) Characterization of a genomic locus required for the production of the antibiotic pyoluteorin by the biological control agent *Pseudomonas fluorescens* Pf-5. Appl Environ Microbiol 61:849–854

Kumar V, Behl RK, Narula N (2001) Establishment of phosphate solubilizing strains of *Azotobacter chroococcumin* the rhizosphere and their effect on wheat cultivars under greenhouse conditions. Microbiol Res 156:87–93

Kumar A, Vandana RS, Singh M, Pandey KD (2015) Plant growth-promoting rhizobacteria (PGPR). A promising approach to disease management. Microbes and environmental management. Studium Press, New Delhi, pp 195–209

Kunc F, Macura J (1988) Mechanisms of adaptation and selection of microorganisms in the soil. In: Vancura V, Kunc F (eds) Soil microbial associations. Amsterdam, Elsevier, pp 281–299

Labuschagne N, Pretorius T, Idris AH (2010) Plant Growth Promoting Rhizobacteria as Biocontrol Agents Against Soil-Borne Plant Diseases. In: Plant Growth and Health Pro- Moting Bacteria, Microbiology Monographs, Maheshwari, D.K. (Ed.), SpringerVerlag Berlin Heidelberg, pp:211–230

Lanteigne C, Gadkar VJ, Wallon T, Novinscak A, Filion M (2012) Production of DAPG and HCN by *Pseudomonas* sp. LBUM300 contributes to the biological control of bacterial canker of tomato. Phyto Pathol 102:967–973

Liu K, McInroy JA, Hu C-H, Kloepper JW (2017) Mixtures of plant-growth-promoting rhizobacteria enhance biological control of multiple plant diseases and plant-growth promotion in the presence of pathogens. Department of Entomology and Plant Pathology, Auburn University, Auburn

Lugtenberg B, Kamilova F (2009) Plant-growth-promoting bacteria. Annu Rev Microbiol 63:541–555

Loper JE, Gross H (2007) Genomic analysis of antifungal metabolite production by Pseudomonas fluorescens Pf-5. Eur J Plant Pathol 119: 265–278

Manjula K, Podile AR (2001) Chitin-supplemented formulations improve biocontrol and plant growth promoting efficiency of Bacillus subtilis AF 1. Can J Microbiol 47:618–625

Mansour FA, Ildesuguy HS, Hamedo HA (1994) Studies on plant growth regulator sand enzyme production by some bacteria. J. Qatar Univ Sci 14:81–288

McKenzie RH, Roberts TL (1990) Soil and fertilizers phosphorus update. In: Proceedings of Alberta soil science workshop proceedings, Edmonton, Alberta, Feb 20–22, pp 84–104

McSpadden Gardener BB (2007) Diversity and ecology of biocontrol *Pseudomonas* in agricultural systems. Phytopathology 97:221–226

Merchan F, de Lorenzo L, González-Rizzo S, Niebel A, Megías M, Frugier F, Sousa C, Crespi M (2007) Analysis of regulatory pathways involved in the reacquisition of root growth after salt stress in *Medicago truncatula*. Plant J 51:1–17

Meyer SLF, Halbrendt JM, Carta LK, Skantar AM, Liu T, Abdelnabby HME, Vinyard BT (2009) Toxicity of 2,4-diacetylphloroglucinol (DAPG) to plant-parasitic and bacterial-feeding nematodes. J Nematol 41:274–280

Meyer SL, Everts KL, Gardener BM, Masler EP, Abdelnabby HM, Skantar AM (2016) Assessment of DAPG-producing *Pseudomonas fluorescens* for management of *Meloidogyne* incognita and *Fusarium oxysporum* on watermelon. J Nematol 48(1):43

Mohamed HI, Gomaa EZ (2012) Effect of plant growth promoting *Bacillus subtilis* and *Pseudomonas fluorescens* on growth and pigment composition of radish plants (*Raphanus sativus*) under NaCl stress. Photosynthetica 50(2):263–272

Nandi M, Selin C, Brassinga AKC, Belmonte MF, Fernando WD, Loewen PC, De Kievit TR (2015) Pyrrolnitrin and hydrogen cyanide production by *Pseudomonas chlororaphis* strain PA23 exhibits nematicidal and repellent activity against *Caenorhabditis elegans*. PLoS One 10:e0123184

Nega A (2014) Review on Concepts in Biological Control of Plant Pathogens. J Biol Agri Healthcare 4(27): 33–54

Nassem M, Dandekar T (2012) The role of auxin-cytokinin antagonism in plant pathogen interactions. PLoS Pathog 8:1–4

Normanly J (1997) Auxin metabolism. Physiol Plant 100:431–442

Noumavo PA, Agbodjato NA, Baba-Moussa F, Adjanohoun A, Baba-Moussa L (2016) Plant growth-promoting rhizobacteria: beneficial effects for healthy and sustainable agriculture. Afr J Biotechnol 15(27):1452–1463

Nowak-Thompsan B, Chancey N, Wing JS, Gould SJ, Loper JE (1999) Characterization of a pyoluteorin biosynthetic gene cluster of *Pseudomonas fluorescens* Pf-5. J Bact 181:2166–2174

Pieterse CMJ, Van Pelt JA, Ton J, Parchmann S, Mueller MJ, Buchala AJ, Métraux JP, Van Loon LC (2000) Rhizobacteria-mediated induced systemic resistance (ISR) in *Arabidopsis* requires sensitivity to jasmonate and ethylene but is not accompanied by an increase in their production. Physiol Mol Plant Pathol 57:123–134

Pii Y, Mimmo T, Tomasi N, Terzano R, Cesco S, Crecchio C (2015) Microbial interactions in the rhizosphere: beneficial influences of plant growth-promoting rhizobacteria on nutrient acquisition process—a review. Biol Fertil Soil 51:403–415

Pilet PE, Saugy M (1987) Effect of root growth of endogenous and applied IAA and ABA. A critical reexamination. Plant Physiol 83:33–38

Podile AR (2006) Seed bacterization with Bacillus subtilis AF 1 enhances seedling emergence, growth and nodulation of pigeonpea. Indian J Microbiol 35:199–204

Pospíšilová J (2003) Interaction of cytokinins and abscisic acid during regulation of stomatal opening in bean leaves. Photosynthetica 41:49–56

Prabhukarthikeyan SR, Raguchander T (2016) Antifungal metabolites of *Pseudomonas fluorescens* against Pythium aphanidermatum. J Pure Appl Microbiol 10(1):579–585

Qurashi AW, Sabri AN (2012) Bacterial exopolysaccharide and biofilm formation stimulate chickpea growth and soil aggregation under salt stress. Braz J Microbiol 11:83–91

Rane MR, Sarode PD, Chaudhari BL, Chincholkar SB (2007) Detection, isolation and identification of phenazine-1-carboxylic acid produced by biocontrol strains of *Pseudomonas aeruginosa*. J Sci Ind Res 66:627–631

Ranjbar Sistani N, Kaul HP, Desalegn G, Wienkoop S (2017) Rhizobium impacts on seed productivity, quality, and protection of *Pisum sativum* upon disease stress caused by *Didymella pinodes*: phenotypic, proteomic, and metabolomic traits. Front Plant Sci 8:1961

Ryu CM, Farag MA, Hu CH, Reddy MS, Wei HX, Pare PW, Kloepper JW (2003) Volatiles produced by PGPR elicit plant growth promotion and induced resistance in *Arabidopsis*. In: Proceedings of the 6th international workshop on plant growth promoting rhizobacteria, pp 436–443

Saraf M, Jha CK, Patel D (2011) The role of ACC deaminase producing PGPR in sustainable agriculture. In: Maheshwari DK (ed) Plant growth and health promoting bacteria microbiology. Steinbuchel A (series ed) Monographs, vol 18. Springer, Berlin, pp 365–386

Sarhan EAD, Shehata HS (2014) Potential plant growth-promoting activity of *Pseudomonas* spp. and *Bacillus* spp. as biocontrol agents against damping-off in alfalfa. Plant Pathol J 13:8–17

Savary S, Ficke A, Aubertot JN, Hollier C (2012) Crop losses due to diseases and their implications for global food production losses and food security. Food Secur 4:519

Shanmugaiah V, Mathivanan N, Varghes B (2010) Purification, crystal structure and antimicrobial activity of phenazine-1-carboxamide produced by a growth-promoting biocontrol bacterium, *Pseudomonas aeruginosa* MML2212. J Appl Microbiol 108:703–711

Smitha K, Mohan R, Devadason A, Raguchander T (2017) Exploiting novel rhizosphere Bacillus species to suppress the root rot and wilt pathogens of chickpea. Afr J Microbiol Res 9:1098–1104

Sorensen J (1997) The rhizosphere as a habitat for soil microorganisms. In: van Elsas JD, Trevors JT, Welington EMH (eds) Modern soil ecology. Marcel Dekker, New York, pp 21–46

Sticher L, Mauch-Mani B, Metraux JP (1997) Systemic acquired resistance. Annu Rev Phytopathol 35:235–270

Tambong JT, Hofte M (2001) Phenazines are involved in biocontrol of *Pythium myriotylum* on cocoyam by *Pseudomonas aeruginosa* PNA1. Eur J Plant Pathol 107:511–521

Thomashow LS, Weller DM (1988) Role of a phenazine antibiotic from Pseudomonas fluorescens in biological control of Gaeumannomyces graminis var. tritici. J Bact 170:3499–3508

Teplitski M, Robinson JB, Bauer WD (2000) Plants secrete substances that mimic bacterial nacyl homoserine lactone signal activities and affect population density-dependent behaviour in associated bacteria. Mol Plant Microbe Interact 13:637–648

Ulloa-Ogaz AL, Muñoz-Castellanos LN, Nevárez-Moorillón GV (2015) Biocontrol of phytopathogens: antibiotic production as a mechanism of control. The battle against microbial pathogens: basic science, technological advances, and educational programmes. Formatex Research Center, Badajoz, pp 305–309

Uzair B, Kausar R, Bano SA, Fatima S, Badshah M, Habiba U, Fasim F (2018) Isolation and molecular characterization of a model antagonistic *Pseudomonas aeruginosa* divulging in vitro plant growth-promoting characteristics. Biomed Res Int 2018:6147380

Validov SZ, Kamilova F, Lugtenberg BJ (2009) *Pseudomonas putida* strain PCL1760 controls tomato foot and root rot in stonewool under industrial conditions in a certified greenhouse. Biol Control 48(1):6–11

Van Loon LC (2007) Plant responses to plant growth-promoting rhizobacteria. Eur J Plant Pathol 119:243–254

Verma SC, Ladha JK, Tripathi AK (2001) Evaluation of plant growth-promoting and colonization ability of endophytic diazotrophs from deep water rice. J Biotechnol 91(2):127–141

Vessey JK (2003) Plant growth-promoting rhizobacteria as biofertilizers. Plant Soil 255:571–586

Vinodkumar S, Nakkeeran S, Renukadevi P, Malathi VG (2017) Biocontrol potentials of antimicrobial peptide producing *Bacillus* species: multifaceted antagonists for the management of stem rot of carnation caused by *Sclerotinia sclerotiorum*. Front Microbiol 8:446

Wang X, Wang C, Ji C, Li Q, Zhang J, Song X, Jun Kang S, Liu Z, Liu X (2018) Isolation and characterization of antagonistic bacteria with the potential for biocontrol of soil-borne wheat diseases. bioRxiv preprint first posted online 18 May 2018

Weller DM, Landa BB, Mavrodi OV, Schroeder KL, De La Fuente L, Bankhead SB, Molar RA, Bonsall RF, Mavrodi DV, Thomashow LS (2007) Role of 2,4-diacetylphloroglucinol-producing fluorescent *Pseudomonas* spp. in the defense of plant roots. Plant Biol 9:4–20

Wisniewski-Dyé F, Vial L, Burdman S, Okon Y, Hartmann A (2015) Phenotypic variation in Azospirillum spp and other root-associated bacteria. In: De Bruijn FJ (ed) Biological nitrogen fixation. Wiley, Hoboken, pp 1047–1054

Yadav J, Verma JP, Jaiswal DK, Kumar A (2014) Evaluation of PGPR and different concentration of phosphorus level on plant growth, yield and nutrient content of rice (*Oryza sativa*). Ecol Eng 62:123–128

Yamaguchi S (2008) Gibberellin metabolism and its regulation. Annu Rev Plant Physiol 59:225–251

Yu JM, Wang D, Pierson LS, Pierson EA (2018) Effect of producing different phenazines on bacterial fitness and biological control in *Pseudomonas chlororaphis* 30–84. Plant Pathol J34:44–58

Zahir ZA, Shah MK, Naveed M, Akhtar MJ (2010) Substrate dependent auxin production by Rhizobium phaseoli improve the growth and yield of Vigna radiate L. under salt stress conditions. J Microbiol Biotechnol 20:1288–1294

Zaidi A, Khan MS, Ahemad M, Oves M (2009) Plant growth promotion by phosphate solubilizing bacteria. Acta Microbiol Immunol Hung 56:263–284

Zehnder GW, Murphy JF, Sikora EJ, Kloepper JW (2001) Application of rhizobacteria for induced resistance. Eur J Plant Pathol 107:39–50

Zhang QX, Kong XW, Li SY, Chen XJ, Chen XJ (2020) Antibiotics of *Pseudomonas protegens* FD6 are essential for biocontrol activity. Aust Plant Pathol 49:307–317

Zhou L, Jiang HX, Sun S, Yang DD, Jin KM, Zhang W, He YW (2016) Biotechnological potential of a rhizosphere *Pseudomonas aeruginosa* strain producing phenazine-1-carboxylic acid and phenazine-1-carboxamide. World J Microbiol Biotechnol 32:50

Zihalirwa Kulimushi P, Argüelles Arias A, Franzil L, Steels S, Ongena M (2017) Stimulation of fengycin-type antifungal lipopeptides in *Bacillus amyloliquefaciens* in the presence of the maize fungal pathogen *Rhizomucor variabilis*. Front Microbiol 8:850

Chapter 17
Role of Phosphate-Solubilising Microorganisms in Agricultural Development

Ghulam Jilani, Dongmei Zhang, Arshad Nawaz Chaudhry, Zahid Iqbal, Muhammad Ikram, and Muneeb Bashir

Contents

G. Jilani (✉) · A. N. Chaudhry · Z. Iqbal · M. Ikram · M. Bashir
Institute of Soil Science, PMAS Arid Agriculture University, Rawalpindi, Pakistan
e-mail: jilani@uaar.edu.pk

D. Zhang
College of Environmental Science and Engineering, Guangdong University of Petrochemical Technology, Maoming, Guangdong, China

© The Author(s), under exclusive license to Springer Nature Switzerland AG 2021

H. I. Mohamed et al. (eds.), *Plant Growth-Promoting Microbes for Sustainable Biotic and Abiotic Stress Management*,
https://doi.org/10.1007/978-3-030-66587-6_17

1 Introduction

Bioavailability of essential nutrients is generally low in cultivated and infertile soils due to resource competition, especially phosphorus (P), which limits the growth of plants (Dubey et al. 2020). Biologically, it is required in plants for their cell division, synthesis of cell organelles, energy and cellular metabolism, synthesis of starch and amino and fatty acids, and N fixation. So, it contributes to root and stalk development, reproduction, fruit/seed quality, disease resistance, and eventually production. Phosphorus is among the less-abundant macronutrients (except N) in the soil (about 0.1% of all elements). It is crucially required by all microorganisms for cell synthesis and metabolism, energy transfer, and signalling (Bünemann et al. 2011).

Contents of bioavailable P in agricultural lands are very little (<0.01–3.07 mg L^{-1}), due to firstly poor P content of parent materials and secondly its high reactivity causing fixation in the mineral matrix (Sharma et al. 2013). This small content fulfils only a little portion of plants' requirement, so the rest has to be obtained via biotic and abiotic processes for which P-solubilising microflora could be quite helpful. Thus soil microorganisms have developed diverse strategies to enhance P bioavailability. Plants can take up only inorganic P (viz. HPO_4^{2-}, $H_2PO_4^-$), while bacteria and fungi also have the capability of consuming low-molecular-weight (LMW) compounds of organic P (Schwöppe et al. 2003). However, protozoa could also take up high-molecular-weight (HMW) compounds of organic P. It infers that a little fraction of organic P pools remains microbially unavailable, so diverse sources of phosphorus in soil could provide ecological niches for various species (Jones and Oburger 2011).

Phosphorus bioavailability in the soil is associated with reversible processes of immobilisation-mineralisation (biological), sorption-desorption (physical), and dissolution-precipitation (chemical). Unfortunately, most of the native P in soil and applied through fertiliser become immobile or fixed via reactions with Al^{3+} and Fe^{3+} in low-pH soils and with Ca^{2+} in alkaline soils (Khan et al. 2015). Therefore, fertiliser's phosphate use efficiency rarely exceeds 30% with its soluble concentration in the soil around 1.0 mg kg^{-1} (Mengel et al. 2001) if the total P ranges 500–800 mg kg^{-1} in soil. Total P content in surface soils (0–15 cm) falls in the range of 50–3000 mg kg^{-1} contingent upon the type of parent material/soil, land management, and vegetation cover (Sims and Pierzynski 2005). Phosphorus fixation phenomena prevail extensively in the soil as hardly 0.1% of the entire P pool, viz. 0.05% (w/w) is bioavailable, which renders it inaccessible to plants, so its deficiency impedes the growth and yields of plants.

Under this scenario, sustainable crop production and long-term agricultural development demand for exploration of natural processes and biological entities to mobilise the large resource of fixed P accumulated in soil (Dixon et al. 2020). Heterotrophic microbes mainly govern the soil P solubilisation by excreting organic acids and enzymes for P supply to plant roots. Phosphate-solubilising microorganisms (PSM) comprising the P-solubilising bacteria (PSB) as well as P-solubilising fungi (PSF) are used as biofertiliser for P release from immobilised organic and fixed mineral forms in soil (Khan et al. 2014). Sections proceeding below encompass all types of PSM existing in soil, forms of phosphates available for use by PSM, mechanisms of phosphate utilisation, and potentials of native soil P for agricultural production.

2 Phosphate-Solubilising Microorganisms

A larger proportion of total P in the soil is organically bound; therefore, microbes contribute enormously in P turnover. Microorganisms produce carbon dioxide, protons, and secondary metabolites (viz. amino acids, starch, organic acid anions, enzymes, siderophores, phenols, etc.), which catalyse the processes of phosphate solubilisation (Jones and Oburger 2011). Several species of microbes possess great capability of enhancing the organic P cycling through solubilisation of bound organic and mineral P. Research on the PSM spans over a century witnessing the superiority of bacteria (1–50% of total soil bacteria) with greater potential of P solubilisation than fungi (only 0.1–0.5% of total soil fungi) classified as PSM (Chen et al. 2006). Within the total PSM population/species in soil, the PSB outnumber the PSF greatly; however, fungal strains possess higher P-solubilising capability (Gyaneshwar et al. 2002).

More diversified populations of PSM proliferate physically and remain active metabolically in the rhizospheric soils than in other environments. More frequently studied PSM species among bacteria are *Bacillus*, *Burkholderia*, *Enterobacter*, and *Pseudomonas*, while that of fungi are *Aspergillus*, *Penicillium*, and *Trichoderma* (Bononi et al. 2020). Microorganisms responsible for P acquisition also include ectomycorrhizal and endomycorrhizal fungi. Greater number of metabolically active PSM is found in the rhizosphere than in other ecologies. They exist ubiquitously in forms and population in almost all types of soils depending on the cultural activities, physicochemical properties, organic matter, and phosphate minerals, eventually with their highest populations in cultivated and range lands (Khan et al. 2015).

2.1 *Bacteria*

Phosphate-solubilising bacteria (PSB) produce organic acids and phosphatases, which mineralise the P-bearing organic materials present in soil (Rodríguez and Fraga 1999). A number of bacteria in fertile soils range from 10^1 to 10^{10}, and their live biomass could be around 2000 kg ha^{-1}. Structural forms of bacteria found in soil are spherical (cocci, 0.5 μm), rod-shaped (bacilli, 0.5–0.3 μm), or spiral (1–100 μm). Bacilli are the most common type in soil, while spirilla exist scarcely in the natural environments (Baudoin et al. 2002). Relatively more efficient PSB communities in soil are *Bacillus*, *Enterobacter*, *Pseudomonas*, *Burkholderia*, *Azotobacter*, and *Rhizobium* (Jones and Oburger 2011). Strains of *Serratia marcescens* have been suggested as supersolubiliser for biofertiliser preparation (Ben Farhat et al. 2009).

Multiple P solubilisation mechanisms operating simultaneously have been found in some bacteria as in the case of *Gluconacetobacter diazotrophicus* (Intorne et al. 2009). Thus, P-solubilising indole acetic acid-producing rhizobacteria (PSIRB) could perform more efficiently than P-solubilising rhizobacteria (PSRB) or indole acetic acid-producing rhizobacteria (IRB) individually (Hariprasad et al. 2009). Similarly, several P-solubilising bacteria proliferating on the outer membrane of mycorrhizal hyphae proliferating in the soil (hyphosphere) add up indirectly to P uptake of mycorrhizae and eventually the plants (Gonzalez-Chavez et al. 2008). Such PSB species may colonise in the mucilage of hyphae, on hyphoplane, among the walls of hyphal layers or within the hyphae (Mansfeld-Giese et al. 2002).

2.2 Fungi

Several non-mycorrhizal fungi isolated from agricultural soils exhibit P-solubilising capability. Among them, inoculation with *Aspergillus*, *Mucor*, *Penicillium*, and *Trichoderma* species has shown 5–20% improvement in the production of crops (Gunes et al. 2009). Like some ectomycorrhizal fungi, the non-mycorrhizal fungi (e.g. *Arthrobotrys*, *Emericella*, *Penicillium*) could use one or more of these mechanisms for P solubilisation, viz. soil acidification, production of organic acid anions (e.g. citrate, oxalate, gluconate), and release of acid and alkaline phosphatases/phytases (Xiao et al. 2009). Phytase enzyme-producing fungi efficiently hydrolyze the phytases/inositol phosphates, which have the main share in the organic P of soil.

Chaetomium globosum is efficient in the production of phosphatase and phytase for mobilising organic P and has great potential to produce citric, formic, lactic, and malic acid, which are important for about one unit reduction in soil pH (Tarafdar 2019). *Aspergillus* is known to be the most efficient fungus for producing both types of phosphatases (acid and alkaline). The minimum concentration of fungal released organic acids required to solubilise the phosphates is in the range of 0.2–0.5 mM (Tarafdar 2019). In PSF-treated crops, fungal share in the P uptake is usually greater than that of plants themselves. Extracellular enzymes produced by fungi are more efficient than their intracellular counterparts for P solubilisation.

2.3 Actinomycetes

General characteristics of actinomycetes found in soil include their capability of existing in extreme environmental conditions (e.g. water and salinity stress), production of antibiotics and phytohormones, plant growth promotion, and P solubilisation by some strains (Hamdali et al. 2010). Several species of P-solubilising actinomycetes (PSA) isolated from the rhizosphere soil have been recognised to enhance P use efficiency and stimulate the plant growth when reinoculated. Around 20% genera (including *Streptomyces* and *Micromonospora*) among all the actinomycetes are capable of performing P solubilisation process.

Opposite to most fungi, the majority of the PSA does not possess acidifying characteristics. Rather, they release the citrate, formiate, lactate, malate, and succinate anions from respective organic acid and other organic substances associated with P dissolution (Hoberg et al. 2005). They store phosphorus as polyphosphate in their mycelia (Hamdali et al. 2010). Due to their thermotolerance, actinomycetes could be employed preferentially to accelerate the P-release process during compost production (Chang and Yang 2009).

2.4 *Mycorrhizae*

Mycorrhizal symbiosis among the plant roots and fungal species in soil is usually established in the agricultural and forest ecosystems. The presence of arbuscular mycorrhizal fungi (AMF) is very common in the surface and subsoils, where they establish symbiotic as well as mutualistic associations as found in several plant types (Yang et al. 2018). The rhizosphere is the main habitat for a diversity of beneficial microbes including AMF, which improve P bioavailability and increase the growth of inhabited plants (Zhang et al. 2014). These fungi release protons as well as extend their hyphae for the acquisition of soluble and/or insoluble forms of soil P and then share it with plants especially on P-deficient soils (Smith and Smith 2011). Interaction of such AMF and plant growth-promoting rhizobacteria (PGPR) boosts their activities in the rhizosphere, also to enhance P uptake by the colonised roots of plants (Pierre et al. 2014). By having long aerial mycelia, the AMF transports phosphate from long distances unreachable by plant roots.

The AMF may also release the organic acids and phosphatases that could help solubilise the native P sources unavailable to plants (Tarafdar 2019). Thus, plant root infection with AMF enhances their nutrient absorption efficiency, improves the growth of AMF-infected plants, and influences their root morphology depending upon the mycorrhizal density. The AMF enhance the nodulation in legumes and increase the plant root surface area to approach and uptake more nutrients by the plants. Thus, nitrogen fixation in legumes being dependent on P supply improves with AMF colonisation. The mycorrhizal hyphae also combine the mineral particles of soil and organic materials to make macroaggregates, which join to make more stable macroaggregates (Bibi et al. 2018). The AMF application could reduce the use of P fertiliser.

3 Phosphate Sources vs. Microbial P Solubilisation

Soil microbes have enormous potential for mineralising and solubilising the organic as well as inorganic phosphate compounds, respectively. Various kinds of P minerals as contained in the phosphate rocks exhibit variable solubility. However, the forms of P found in different soils may not be similar to that in phosphate rocks. In the soil's solid phase, physical and chemical forms of phosphates strongly regulate the efficiency of PSM to mobilise the bound P (Jones and Oburger 2011). As the PSM display differential response to various forms of phosphates, so their inocula are developed according to their ability to dissolve the particular forms of phosphates under field conditions. The following paragraphs comprehend the most dominant P forms in soil:

3.1 Mineralisation of Organic Phosphates

The proportion of organic phosphates in soil varies greatly ranging from 4% (in sandy soils) to 90% (in organic soils) of the total phosphorus contents, but generally it's 30–65% in mineral soils (Jones and Oburger 2011). Abundant kinds of organic P present in soil are mainly inositol phosphates (dominant, ≥80%), phospholipids (0.5–7.0%), and nucleic acids which account for ≤3% (Quiquampoix and Mousain 2005). Some organic P compounds being less abundant are sugar P, monophosphorylated carboxylic acids, and teichoic acids. Inositol phosphates are component of the insoluble complexes or polymers containing proteins and lipids and have high acidity. Their stability is related to phosphate group counts, which render the higher-order esters being stronger recalcitrant liable to biodegradation and thus higher in abundance. Phospholipids are mostly in the form of phosphoglycerides. Nucleic acids and their derivatives are quickly mineralised, resynthesised, and incorporated into microbial biomass or soil constituents.

Organic soil amendments, viz. municipal biosolids, compost, crop residues, and animal manures, also contribute to P nutrition of plants and soil microorganisms. Nevertheless, P bioavailability from them depends on the P forms present therein and their interaction with soil. Therefore, it has created great interest in the interactive effects of PSM inocula with organic amendments for providing nutrients to the crops. Precise analysis reveals that biosolids contain mainly inorganic P forms, like variscite (Al-P, containing 86% of total P) and less-soluble hydroxylapatite (Ca-P, having 14% of total P), while manures contain 12–68% each of dicalcium phosphate dehydrate and struvite (magnesium ammonium phosphate), 0–18% variscite, and 20–70% organic P as calcium phytate (Ajiboye et al. 2007). Inorganic P component of compost mostly binds to calcium forming the minerals apatite or octacalcium phosphate. Distribution of inorganic P among Al, Fe, and Ca fractions in the compost is also dependent upon the type of additives, e.g. lime, metal salts, etc., which reduce the P solubility and immobilisation (Maguire et al. 2006). Phosphate solubility in organic additives is mostly affected by the equilibrium soil solution and its pH, as at lower pH Fe and Al phosphates render stronger recalcitrant, while Ca phosphate at acidic pH is less recalcitrant. Nonetheless, in addition to phosphate, organic additives supply considerable amounts of carbon and nitrogen, which enhance microbial activities, like respiration, mineralisation, turnover, and biomass build-up (Saha et al. 2008), and accelerate C, N, and P cycling. Increased microbial activity enhances not only the solubilisation of organic and inorganic P found in the organic amendments but also the solubilisation of originally existing P forms in soil too.

Considerable amount in the pool of organic P contributed through the biomass of soil microorganisms. Nevertheless, the contents (mg kg^{-1}) of microbial P present in different soil types range widely from 0.75 (in sandy soils) to 106 (in grasslands) and 169 (in forest litter), which could constitute 0.51–26% of the total P therein (Oberson and Joner 2005). Various phosphate-containing compounds in microbes (as % of the total microbial P) include nucleic acids (30–65%), phospholipids, phosphate esters, and phosphorylated coenzymes (15–20%), along with some P-storage compounds, viz. polyphosphates and teichoic acid found only in Gram-positive bacteria (Bünemann et al. 2011). Phosphorus immobilisation by microorganisms depends more upon C than P limitation; thus P contents in microbial biomass are related to soil C dynamics (Achat et al. 2009). Seasonal variations leading to dry periods, increased soil depth, decreased organic matter, and P fertilisation reduce the biomass P content of microorganisms (Chen et al. 2003).

Nearly 50% among the microbial communities associated with soil and plant root system perform P mineralisation through phosphatase enzymes, e.g. acidic/alkaline phosphatases and phytases (Zineb et al. 2020). Phosphatase enzymes mineralise their substrate of organic phosphate and yield inorganic forms of phosphorus. Major mechanisms for the mineralisation or hydrolysis of organic phosphates and residues to make them bioavailable involve organic anions/acids, siderophores, and phosphatase enzymes produced largely by the microbial population and partially by plant roots (Dodor and Tabatabai 2003). Some microbes, for instance, *Enterobacter agglomerans*, can perform both functions, viz. hydrolysis of organic P compounds and solubilisation of inorganic P minerals like hydroxyapatite.

3.2 Solubilisation of Phosphate Minerals

Within the growing season, only a small portion (around 1%) from the total soil P assimilates into vegetation biomass, which reflects little P bioavailability to plants (Quiquampoix and Mousain 2005). A fraction of inorganic P ranges 35–70% of the entire soil P being related to the parent material, pH, vegetation, and pedogenesis (Sims and Pierzynski 2005). The pool of organic P rises through soil development processes; however, it declines in greatly weathered and past developed soils. Therefore, soil development processes, P allocation between organic and inorganic P pools, and P forms greatly influence the P accessibility to microbial community and eventually effectiveness of PSM to promote crop growth in the field (Jones and Oburger 2011).

Phosphatic minerals are subjected to solubilisation with several species of saprophytic bacteria and fungi mostly through chelation by both organic and inorganic acids produced by them. Hydroxyl and carboxyl ions from these acids effectively chelate the cations (Al, Fe, Ca) while lower down the pH under basic conditions; resultantly several phosphate compounds are solubilised (Stevenson 2005). Organic acids produced by these microbes are mostly low molecular weight, e.g. gluconic and ketogluconic acids (Deubel and Merbach 2005). The pH mainly in rhizosphere reduces with the release of protons/bicarbonates (anion/cation balance) as well as with the gaseous (O_2/CO_2) exchange. Thus, organic acids contribute to phosphate solubilisation through pH reduction, cation chelation, and competition with phosphate to find adsorption sites in the soil. Generally, organic acids are more efficient than inorganic acids to solubilise the phosphates if compared to the same level of pH.

3.2.1 Solubilisation of Ca-Bound Phosphates

Sources of primary P minerals in less weathered and unweathered soils having neutral or alkaline pH are calcium phosphates (various types of apatites), e.g. fluorapatite, hydroxyapatite, and francolites (Benmore et al. 1983). Acidification through lowering of soil pH by PSM inocula solubilises Ca phosphates and releases inorganic P. For this purpose, several types of acidifying PSM are employed to enhance the dissolution of phosphate rocks before incorporation into the soil, via inoculation of individual PSM or compost enrichment with microbial consortia (Aria et al. 2010).

Under alkaline conditions, phosphate minerals present in soil as apatites and phosphate from P fertilisers are fixed as phosphates with calcium like $Ca_3(PO_4)_2$. These compounds and rock phosphates (fluorapatite, francolite) exhibit low solubility rate in soil releasing very little concentration of inorganic P being insufficient to support the normal plant growth. Phosphate solubilisation in alkaline soils undergoes with the joint influence of pH reduction and the release of organic acids (e.g. carboxylic acid). Both these mechanisms operated by soil microorganisms dissociate the bound forms of phosphorus (Stephen and Jisha 2009). Reduced pH or excretion of H^+ around microbial cells releases phosphate from P-fixed minerals through proton substitution (with more absorption of cations than anions) or production of Ca^{2+} (Villegas and Fortin 2002). However, an opposite reaction takes place when anion uptake exceeds that of cations/H^+ due to excretion of OH^-/HCO_3^- (Tang and Rengel 2003).

Carboxylic anions released from PSM show greater affinity to Ca, and thus it solubilises more P than the acidification alone. Being an important P solubilisation mechanism, complexing of cations is mainly through pH decrease by organic acids and influenced by nutrition, physiology, growth, and metabolites of the PSM (Reyes et al. 2007). Organic anions and associated protons are important for solubilisation of precipitated P compounds. They would chelate the metal ions attached with complexed P compounds or could release the adsorbed P via ligand exchange reactions. Thus Ca-P releases through joint mechanisms of pH reduction and carboxylic acid production, as the proton release mechanism alone cannot proceed this process (Deubel and Merbach 2005).

3.2.2 Solubilisation of Al and Fe Phosphates

In lower pH and highly weathered soils, the dominant P minerals are Fe and Al phosphates and inorganic P bound and/or occluded by Fe and Al oxy(hydr)oxides (Sims and Pierzynski 2005). Under neutral and acidic soil conditions, Al and Fe oxides/hydroxides greatly influence the P availability, due to rare occurrence of various Fe and Al phosphates, e.g. wavellite, variscite, and strengite. With decreasing pH, positive surface charge of Fe and Al oxides is increased, and strong covalent bonds (chemisorption) are developed through negatively charged P, which renders it recalcitrant to exchange reactions. However, low-molecular-weight (LMW) organic anions (e.g. gluconate and oxalate) excreted from PSM could compete with inorganic P for sorption sites. Further, pH dynamics may influence the surface potential of oxides, resulting in the solubility of inorganic P (Jones and Oburger 2011).

Iron- and aluminium-associated phosphates are solubilised through proton produced by PSM via reducing the negative charge on adsorbing sites that ultimately enhances sorption of negatively charged P ions. Release of protons may also reduce the P sorption due to acidification that increases $H_2PO_4^-$ as compared to HPO_4^{2-} exhibiting greater affinity to the reactive sites on soil. Carboxylic acids mostly solubilise the Al-P, while Fe-P is solubilised via direct dissolution of phosphate mineral due to anion exchange of PO_4^{3-} by acid anion, which could chelate both Fe and Al ions attached to phosphate (Henri et al. 2008). Root-associated pseudomonas strains have high-affinity Fe-uptake system depending upon release of Fe^{3+}-chelating agents, viz. siderophores (Khan et al. 2007). Further, carboxylic anions replace the PO_4^{3-} anions from sorption complexes through ligand exchange, thus chelating both Fe and Al ions attached to phosphate, which after transformation releases bioavailable phosphate for plants. The capability of organic acids for chelating the metal cations is highly affected by these acids' molecular structure, principally by the abundance of carboxyl and hydroxyl ions.

3.3 Release of Immobilised P from PSM Biomass

It is a general understanding that phosphorus released by PSM is consumed mainly by the plants and soil organisms. Conversely, the fact is that inevitably the greater portion of released P gets assimilated by the PSM biomass itself. Normally, the release of PSM's immobilised P takes place after their cell death with environmental changes, starvation, or predation by microflora and microfauna. Fluctuation in soil conditions, e.g. drying-rewetting and/or freezing-thawing, results in higher rates of microbial cell lysis (breakdown) causing flush events, which witness a sudden rise of bioavailable P in soil solution (Butterly et al. 2009). Approximately, 30–45% of microbial P (constituting 0.8–1 mg kg^{-1}) is mineralised within a day during the first flush event after drying-rewetting cycles (Grierson et al. 1998). Nevertheless, P bioavailability proceeding the flush events is mainly relying on the P-sorption capability of soil, as the major part of released P could subsequently be immobilised on the solid phase.

Grazing of microorganisms by microbivores (e.g. nematodes, protozoa) also releases microbial P. During a preliminary study, the presence of bacterial grazers caused substantial P mineralisation within a week, while in their absence, vigorous P immobilisation continued beyond 3 weeks without any P release (Cole et al. 1978). Similarly, the presence of organic matter and its C/P ratio render a substantial influence on microbial P immobilisation-mineralisation dynamics (Silvan et al. 2003). Inputs of easily available C sources as fresh organic materials improve the microbial P with subsequent decrease and rise in soil P on the depletion of a substrate (Jones and Oburger 2011). Nevertheless, substrate quality and soil characteristics determine the time passing between P immobilisation and remineralisation, as the dynamics is smaller for stronger recalcitrant organic materials (Oehl et al. 2001).

4 Mechanisms of Phosphate Solubilisation

Phosphate-solubilising efficiency is the ability of PSM to produce organic acids, whose hydroxyl as well as carboxyl ions chelate cations associated with phosphate, so bringing them to soluble state. Phosphate solubilisation in the global P cycling undergoes several mechanisms, which also include organic acids and/or proton release attributed to soil microorganisms. Phosphorus assimilated in the microbial biomass is immobilised for a shorter time, but remineralisation or turnover by microorganisms transforms it in a bioavailable form after some time depending upon the soil conditions. Therefore, P-solubilising microbes are the key players in all the three components of P cycle being operated in soil, viz. mineralisation-immobilisation, dissolution- precipitation, and sorption-desorption. Bioavailability of inorganic P from the P-containing minerals is largely governed by their dissolution properties, which are influenced mainly by the pH and equilibrium reactions in soil solution (viz. sorption and desorption). Whereas, the P bioavailability from organic P materials entirely depends upon the activities of soil microorganisms, e.g. mineralisation, enzymatic hydrolysis, etc. Therefore, various factors and mechanisms are involved in the solubilisation of organic and mineral phosphates in soil as detailed in the following paragraphs:

4.1 Phosphate Release Through pH Dynamics

Microorganisms release protons or hydroxide ions, which change the pH of soil solution as well as mineral nutrient bioavailability. Although phosphate solubilisation via alkalinisation is rarely reported, P solubilisation through microbial acidification rendered by numerous species of bacteria and fungi is well recognised (Ben Farhat et al. 2009) especially if phosphate is associated with calcium. Release of protons sometimes relates to production of organic acid anions, which is enhanced with NH_4^+ supply (rather than NO^{3-}), and decrease in pH resulting to more P solubilisation (Sharan et al. 2008). *Penicillium rugulosum* with the assimilation of amino acids as a sole N source also decreased pH in external medium and thus enhanced P mobilisation. Contrastingly in *Pseudomonas fluorescens*, C source (e.g. glucose vs. fructose) but not N source (e.g. NH_4^+ vs. NO_3^-) imparts more impact on proton release (Park et al. 2009).

It reflects that in various microbial species, dissimilar strategies operate in proton release, influenced somewhat by NH_4^+. Although pH dynamics is a potential P solubilisation mechanism, nevertheless, situations in the field (against in vitro) might not be favourable for enough acidification due to insufficient labile N and C as limiting factors for microbial activity in the bulk soil (Jones and Oburger 2011). Also, especially the calcareous soils have strong pH buffering capacity that might reduce the P solubilisation and reduce the growth of PSM.

4.2 Phosphate Release Via Organic Acid Anions

Just only acidification may not be enough to understand the process of P mineral solubilisation. The LMW organic acid anions (carboxylates) produced from microorganisms are also involved in the solubilisation of inorganic P (Patel et al. 2008). Frequently observed organic acid anions released from PSM are citric, gluconic, glycolic, 2-ketogluconic, lactic, malic, malonic, oxalic, succinic, and tartaric acids (Gyaneshwar et al. 2002). Secretion of protons (rather than organic anions) compensates the loss of negative charge, which reduces the pH of soil. On the other hand, organic anions influence P solubilisation through their negative charges or metal complexation properties. So, inorganic P is mobilised from the metal oxide surface through ligand exchange or solubilisation of iron or aluminium oxides and calcium phosphates, and adsorption/chelation of organic anion liberates the occluded P due to weakening of mineral bonds (Jones and Oburger 2011). Further, adsorption of organic anions on metal oxides reduces positive surface potential that also facilitates the release of adsorbed P.

Organic acids mostly released by bacteria are gluconic and 2-ketogluconic acid, while that by fungi include citric, gluconic, and oxalic acid (Khan et al. 2009). Tricarboxylic acid anions (e.g. citrate) have a greater potential of inorganic P solubilisation due to mineral dissolution mechanism than that of dicarboxylic acids (e.g. gluconate, oxalate), whereas oxalate is more efficient for P mobilisation in calcareous soils due to greater affinity for making Ca precipitates (Ström et al. 2005). Phosphorus mobilisation by organic anions is influenced mainly through soil characteristics (e.g. sorption sites, pH) and properties/quantity of PSM-released organic acid anions, differing greatly from a few micromolars to 100 mM (Gyaneshwar et al. 2002; Patel et al. 2008). The P-solubilising property of organic acid anions mostly declines in soils with higher contents of carbonate and Fe or Al (hydr)oxides (Ström et al. 2005; Oburger et al. 2009).

The LMW carboxylates released from microorganisms as well as roots of a plant are used by microbes as labile C substrate and being removed from the solution; thus their P-mobilisation potential is reduced. For continuous P dissolution during the crop season, organic acid anions must be released by PSM regularly, as their half-life is very short, viz. 0.5–12 h (Jones et al. 2003). Importantly, within high-sorbing soils, breakdown of organic acid anions by microorganisms is greatly reduced (Oburger et al. 2009). In addition to enhancing the growth of microbes and solubilisation of inorganic P, organic acid anions increase the solubility of organic P to make it more prone to enzymatic hydrolysis (Tang et al. 2006).

4.3 Phosphate Release by Enzymes

Phosphorus demand mostly provokes the release of enzymes required for the breakdown of organic P, which is catalysed by phosphatases produced by PSM present in the soil. Usually, extracellular phosphatases instead of intracellular ones release larger amounts of phosphates in soil solution (Nannipieri et al. 2011). Phosphatases or phosphohydrolases represent the large category of enzymes, which catalyse the breakdown of both esters and anhydrides of H_3PO_4 (Dodor and Tabatabai 2003). Their activities are inhibited at higher contents of orthophosphate (end product), other polyvalent anions (e.g. MoO_4^{2-}, AsO_4^{3-}), and some metals, while lower contents of divalent cations (e.g. Ca, Mg, Zn, Co) activate these enzymes (Quiquampoix and Mousain 2005). Moreover, adsorption on soil minerals or organominerals may change enzymes' conformation and activities. Sorption to solid phase decreases enzymatic activity, but it shields enzymes from microbial decay or thermal inactivation. Clay particles most strongly hold the phosphatases, cluing that soil characteristics (e.g. minerals, SOM, pH) influence PSM-released enzymes, and their activity is not only depending on release rate (Jones and Oburger 2011).

Among the several classes of phosphatase enzymes produced by PSM, phosphatases are the most abundant ones and are categorised as acid and alkaline phosphatases depending upon their pH optima and external conditions (Jorquera et al. 2008). Thus, acid phosphatases are more abundant in low-pH soils, and alkaline phosphatases predominate in neutral- to high-pH soils. The plant roots mostly release acid phosphatases, but rarely alkaline phosphatases, so this could be a niche for PSM. It is very exhaustive to determine the difference among root- and PSM-produced phosphatases; however, microbial phosphatases exhibit higher affinity to organic P compounds as compared to those coming from plant roots (Richardson et al. 2009). Reports on both positive and negative correlations between phosphatase activity and inorganic P concentration in soil highlight the uncertainty and interactive complexity of biochemical processes of P mobilisation (George et al. 2002; Ali et al. 2009).

4.4 Phosphate Release by Siderophores

Siderophores are biochemical complexing agents having a greater affinity for iron, and they are produced by most of soil microbes in response to Fe deficiency. About 500 siderophores have been recognised, and the majority is used by several microbes and plants, while some are utilised by the producing microbes themselves (Crowley 2007). Production of siderophores by PSM is well documented, but not widely known for P solubilisation mechanism (Hamdali et al. 2010). Due to dominance of mineral dissolution against the ligand exchange by organic acid anions as P-solubilising mechanism, siderophores might also be considered for enhancing P bioavailability.

In spite of extensive evidence of Fe mobilisation by siderophores, only one study reported the impact of microbial siderophores on P bioavailability (Jones and Oburger 2011). Improved Fe and P diffusion of two siderophores (desferrioxamine B, desferriferrichrome) and iron-chelating agent EDDHA if compared with water through root simulation method was found long before by Reid et al. (1985). Further, desferriferrichrome enhanced the P diffusion 13-folds against water, while desferrioxamine B rendered very little impact. By keeping in view the large reserves of Fe phosphates in soil, greater P-sorption capacity of Fe (hydr)oxides, and Fe requirements of microbes, the role of siderophore-enhanced P solubilisation is quite obvious.

4.5 Phosphate Release Mediated by Exopolysaccharides

Microbes in soil produce exopolysaccharides (EPS) and biosurfactants mostly to cope with biofilm formation and stress conditions. Recently, nonenzymatic high-molecular-weight (HMW) microbial exudates (viz. mucilage, EPS, etc.) are also being investigated for their effectiveness in P solubilisation from soil components. Gaume et al. (2000) reported that maize root mucilage if adsorbed onto synthetic ferrihydrite reduced the P adsorption continuously, but this mucilage couldn't mobilise the pre-adsorbed P in a significant amount. Nonetheless, the indirect effect of microbial mucilages has been observed on the P availability via increased soil aggregation and pore connectivity in soil, which facilitates the soil moisture retention and movement (Ionescu and Belkin 2009).

It has been reported that microbially produced EPS can make complexes with the metals in soil variably (Ochoa-Loza et al. 2001), which indicates that they could have some influence on the P solubility in soil. Microbial EPS and organic acid anions produced in pure culture have been found to enhance the dissolution of tricalcium phosphate in a synergistic manner (Yi et al. 2008). The microbial EPS production could be favoured under P-deficient soil conditions, thus being more favoured with N supply rather than available P (Wielbo and Skorupska 2008). Moreover, the rate of phosphate solubilisation depends upon the microbial population/source and EPS contents in soil.

5 Interactive Effects of PSM on Plants

The PSM might also come in competition with growing plants for the uptake of released P from any source. Phosphorus in the soil solution increases under the situation when (a) active P solubilisation from the soil minerals containing large P contents and (b) sum of the SOM mineralisation and remineralisation of organic P detained in microbial biomass exceed P immobilisation (via P uptake and its assimilation in plants/microbial biomass) and sorption onto the surface of soil minerals. These processes involved in P cycling are driven by several physicochemical soil characteristics (e.g. mineral contents, organic matter, texture, structure, temperature, moisture percentage) and vegetation properties, which collectively influence the P bioavailability in soil from PSM inoculation (Jones and Oburger 2011). Since the microbial populations and activities are greater in the rhizosphere, so the combined efforts of microbes and plant roots in proton (or hydroxide) extrusion could enhance the P bioavailability to both. Further, respiration by plant roots and microbes would increase CO_2 concentration in the rhizosphere and might cause the pH to decrease.

Tarafdar (2019) mentioned that co-inoculation with compatible fungi could mobilise greater amount of soil P for better plant growth; for instance, the AMF *Glomus mosseae* combined with *Aspergillus fumigatus* had a greater activity of phytase enzymes. Plants and PSM have a synergistic association, where microbes provide the soluble P and plant roots supply the carbon compounds (mostly sugars), which are metabolised for microbial proliferation (Pérez et al. 2007). Thus, the presence of PSM in the rhizosphere is highly beneficial for improving crop production. Combined inoculation of *Rhizobium* and PSM or AMF renders better plant growth than inoculation of each microbe alone in P-deficient soil (Zaidi and Khan, 2006). Positive interactive effects on plant growth through simultaneous application of PSB with AMF or with N-fixing bacteria, e.g. *Azospirillum* and *Azotobacter*, have been investigated extensively (Figueiredo et al. 2017; Wahid et al. 2020).

6 Extent of Phosphate Solubilisation in Soil

Contribution of PSM for enhancing the plant growth is influenced mainly by microbial P- mobilisation activities, viz. P uptake followed by its release and P redistribution throughout the soil mass. Numerous studies at the greenhouse and field level have been undertaken to assess the number of phosphates solubilised through inoculation of PSM, and an increase of crop yields up to 70% has been reported (Kumar et al. 2016). The PSB species of *P. striata* and *B. polymyxa* mobilised correspondingly 156 and 116 mg P L^{-1} (Rodríguez and Fraga 1999). Similarly, *P. fluorescens* released 100 mg P L^{-1} from $Ca_3(PO_4)_2$, 92 mg P L^{-1} from $AlPO_4$, and 51 mg P L^{-1} from $FePO_4$ (Henri et al. 2008). Acid-producing PSM also improve the solubilisation of phosphatic rocks (Gyaneshwar et al. 2002).

The PSB strains have the capabilities to solubilise the inorganic P from 53 to 42 μg P mL^{-1} and mineralise organic P ranging 8–18 μg P mL^{-1} (Tao et al. 2008). Seed inoculation of *C. globosum*, *P. purpurogenum*, and *E. rugulosa* could mobilise 45–60 kg P from soil, rendering 416–25% improvement in the production of various crops (Tarafdar 2019). The PSB applied along with SSP fertiliser and rock phosphate decreased the P fertiliser rate by 25% and 50%, respectively (Sundara et al. 2002). The PSB strains of *P. putida*, *P. fluorescens*, and *P. fluorescens* solubilised 51%, 29%, and 62% phosphate, correspondingly (Ghaderi et al. 2008).

Zineb et al. (2020) reported that *Bacillus*, *Burkholderia*, *Pseudomonas*, and *Serratia* strains inoculated to rock phosphate solubilised up to 600 mg P mL^{-1} by producing phytases (16.1–24.8 U mL^{-1}), IAA (up to 39.6 μg mL^{-1}), and siderophores (9–81.1%). The use of PSM inoculum can benefit equivalent to 100–150 kg P ha^{-1} the fields growing horticultural crops (Gunes et al. 2009). The PSM have a daily potential of mineralising 1–4 mg P kg^{-1} in the soil; but without any distinction between the enzymatic (biochemical) and biological (microbial turnover) strategies of mineralisation (Oehl et al. 2001).

7 Contribution of PSM in Crop Production

The worldwide consensus is evolving extensively to encourage the adoption of sustainable practices for the management of both agroecosystems and environment. Among them, the high emphasis has been put on the use of beneficial/effective microbes, referred to as biofertilisers or inoculants. These active biological agents containing beneficial microorganisms drive the biogeochemical nutrient cycles. The PSM, mainly the bacteria *Bacillus*, *Pseudomonas*, *Burkholderia*, and *Enterobacter* and the fungi *Trichoderma*, *Aspergillus*, and *Penicillium*, including ectomycorrhizae and endomycorrhizae have been found beneficial for enhancing the bioavailable P in the cultivated lands as well as improving the production of crops (Bononi et al. 2020). The PSM solubilise the precipitated soil P and fertiliser P contributing significantly to meet P deficiency and enhance crop yields (Sharma et al. 2013).

The combined use of PSB and AMF renders better P uptake both from soil and rock phosphates applied in the field (Cabello et al. 2005). Not only the PSM enhance plant growth by P solubilisation, but they could also increase the N fixation undertaken by crop plants (Ponmurugan and Gopi 2006). The PSB strains of *Pseudomonas* sp. have been reported to enhance the number and mass of nodules, growth attributes, grain yield, nutrient bioavailability, and their uptake in the soybean crop (Son et al. 2006). In another study, seedling length of *Cicer arietinum* was increased by PSB application (Sharma et al. 2007). Co-inoculation of PSB and PGPR decreased the P application rate up to 50% in maize (Yazdani et al. 2009). Inoculation with PSB alone raised the biological yield, whereas co-inoculation of the same PSB along with AMF gave the highest yield of barley grains (Mehrvarz et al. 2008). The PSB application improved sugarcane production by 12.6% (Sundara et al. 2002). Inoculation of alpine *Carex* with *Pseudomonas fortinii* significantly improved the weight of fresh roots and foliage and P content in shoots (Bartholdy et al. 2001).

Application of PSB in addition to P fertiliser produced 30–40% higher yield of wheat grains than with sole P fertiliser, while inoculation without P fertiliser enhanced 20% yield over control (Afzal and Bano 2008). *Pseudomonas putida* and AMF co-inoculation in barley also improved the content of leaf chlorophyll (Mehrvarz et al. 2008). With combined inoculation, *Bradyrhizobium*, *G. fasciculatum*, and *B. subtilis* interacted positively for improving plant growth and N and P uptake of green gram, and seed yield was increased by 24% over control (Zaidi and Khan 2006). The PSB strains of *Bacillus*, *Burkholderia*, *Pseudomonas*, and *Serratia* inoculated to *Medicago truncatula* increased the dry shoot weight in the range of 40–134% and 13–87% in two soils, and the best results were obtained with their consortium (Zineb et al. 2020).

Currently, Wahid et al. (2020) reported the potential of AMF inoculum containing six species (viz. *G. microaggregatum*, *F. geosporum*, *C. etunicatum*, *F. mosseae*, *R. intraradices*, and *G. claroideum*) and PSB strain *Bacillus* sp. PIS7 along with phosphate rock on field-grown maize followed by wheat in alkaline soil. Their combined application significantly enhanced the grain yield of crops and P uptake as compared to control and sole applications. In legumes, co-inoculation of *Rhizobium* and PSM demonstrates great potential in terms of enhancing the nodulation, crop growth and nutrient uptake from chemical fertilisers, e.g. 30% yield improved in soybean (Govindan and Thirumurugan 2005).

Although the strong buffering capacity of soil suppresses the solubilisation of bound P by native microorganisms, efficient PSM inoculants could enhance the microbial activity of P solubilisation contributing significantly in agricultural production. Phosphorus bioavailability in soil depends upon the natural processes of sorption-desorption and immobilization-mineralization. Soil microorganisms contribute enormously in supplying soil phosphorus to the plants through solubilisation of inorganic compounds and mineralization of organic materials. These microorganisms operate two mechanisms in soil, viz., lowering of soil pH via production of organic acids and their anions to solubilise mineral phosphates, and mineralization of organic phosphates via acid phosphatases. Soil enriched with phosphate solubilisers increases the phosphorus bioavailability to the crops. Better efficiency is achieved by co-inoculation of phosphorus solubilising bacteria with other beneficial bacteria, fungi and mycorrhizae. Hence, exploitation of PSM through biofertilisers bears great prospective for utilisation of fixed soil P present hugely in the soil. Similarly, bio-mineralisation of phosphate rocks by the PSM could be an eco-friendly alternative to mineral fertilisers, especially in alkaline soils. So, this chapter concludes that PSM exhibit high potential for the development of a safe biofertiliser product, which could improve the P bioavailability in soil and enhance the plant growth and crop yields to achieve sustainable agricultural production.

References

Achat DL, Bakker MR, Morel C (2009) Process-based assessment of phosphorus availability in a low phosphorus sorbing forest soil using isotopic dilution methods. Soil Sci Soc Am J 73:2131–2142

Afzal A, Bano A (2008) *Rhizobium* and phosphate solubilizing bacteria improve the yield and phosphorus uptake in wheat (*Triticum aestivum* L.). Int J Agric Biol 10:85–88
Ajiboye B, Hu Y, Flaten DN (2007) Phosphorus speciation of sequential extracts of organic amendments using nuclear magnetic resonance and X-ray absorption near-edge structure spectroscopies. J Environ Qual 36:1563–1576
Ali MA, Louche J, Legname E, Duchemin M, Plassard C (2009) Pinus pinaster seedlings and their fungal symbionts show high plasticity in phosphorus acquisition in acidic soils. Tree Physiol 29:1587–1597
Aria MM, Lakzian A, Haghnia GH, Berenji AR, Besharati H, Fotovat A (2010) Effect of *Thiobacillus*, sulfur, and vermicompost on the water-soluble phosphorus of hard rock phosphate. Bioresour Technol 101:551–554
Bartholdy BA, Berreck M, Haselwandter K (2001) Hydroxamate siderophore synthesis by *Phialocephala fortinii*, a typical dark septate fungal root endophyte. Bio Metals 14:33–42
Baudoin E, Benizri E, Guckert A (2002) Impact of growth stages on bacterial community structure along maize roots by metabolic and genetic fingerprinting. Appl Soil Ecol 19:135–145
Ben Farhat M, Farhat A, Bejar W, Kammoun R, Bouchaala K, Fourati A, Antoun H, Bejar S, Chouayekh H (2009) Characterization of the mineral phosphate solubilizing activity of *Serratia marcescens* CTM 50650 isolated from the phosphate mine of Gafsa. Arch Microbiol 191:815–824
Benmore RA, Coleman ML, McArthur JM (1983) Origin of sedimentary francolite from its sulphur and carbon composition. Nature 302:516–518
Bibi Z, Ding W, Jilani G, Khan NU, Saleem H (2018) Soil amendments delineate amino acid biomarkers modulating macro-aggregation and carbon stabilization. Pak J Agric Sci 55(4):809–818
Bononi L, Chiaramonte JB, Pansa CC, Moitinho MA, Melo IS (2020) Phosphorus-solubilizing *Trichoderma* spp. from Amazon soils improve soybean plant growth. Sci Rep 10:2858
Bünemann EK, Prusisz B, Ehlers K (2011) Characterization of phosphorus forms in soil microorganisms. In: Bünemann E, Oberson A, Frossard E (eds) Phosphorus in action: biological processes in soil phosphorus cycling. Soil biology, vol 26. Springer, Heidelberg, pp 37–58
Butterly CR, Bünemann EK, McNeill AM, Baldock JA, Marschner P (2009) Carbon pulses but not phosphorus pulses are related to decrease in microbial biomass during repeated drying and rewetting of soils. Soil Biol Biochem 41:1406–1416i
Cabello M, Irrazabal G, Bucsinszky AM, Saparrat M, Schalamuck S (2005) Effect of an arbuscular mycorrhizal fungus, *G. mosseae* and a rock-phosphate-solubilizing fungus, *P. thomii* in *Mentha piperita* growth in a soilless medium. J Basic Microbiol 45:182–189
Chang CH, Yang SS (2009) Thermo-tolerant phosphate-solubilizing microbes for multi-functional biofertilizer preparation. Bioresour Technol 100:1648–1658
Chen CR, Condron LM, Davis MR, Sherlock RR (2003) Seasonal changes in soil phosphorus and associated microbial properties under adjacent grassland and forest in New Zealand. For Ecol Manag 117:539–557
Chen YP, Rekha PD, Arunshen AB, Lai WA, Young CC (2006) Phosphate solubilizing bacteria from subtropical soil and their tricalcium phosphate solubilizing abilities. Appl Soil Ecol 34:33–41
Cole CV, Elliot ET, Hunt HW, Coleman DC (1978) Trophic interactions in soil as they affect energy and nutrient dynamics. V. Phosphorus transformations. Microb Ecol 4:381–387
Crowley DE (2007) Microbial siderophores in the plant rhizosphere. In: Barton LL, Abadia J (eds) Iron nutrition in plants and rhizospheric microorganisms. Springer, Dordrecht, pp 169–198
Deubel A, Merbach W (2005) Influence of microorganisms on phosphorus bioavailability in soils. In: Buscot F, Varma A (eds) Microorganisms in soils: roles in genesis and functions. Soil Biology, vol 3. Springer, Berlin, pp 177–191
Dixon M, Simonne E, Obreza T, Liu G (2020) Crop response to low phosphorus bioavailability with a focus on tomato. Agronomy 10:617
Dodor DE, Tabatabai AM (2003) Effect of cropping systems on phosphatases in soils. J Plant Nutr Soil Sci 166:7–13

Dubey RK, Tripathi V, Prabha R, Chaurasia R, Singh DP, Rao CS, El-Keblawy A, Abhilash PC (2020) Belowground microbial communities: key players for soil and environmental sustainability. In: Unravelling the soil microbiome: perspectives for environmental sustainability. Springer Nature Switzerland AG, Basel, pp 5–22

Figueiredo GGO, Lopes VR, Fendrich RC, Szilagyi-Zecchin VJ (2017) Interaction between beneficial bacteria and sugarcane. In: Singh DP, Singh HB, Prabha R (eds) Plant-microbe interactions in agro-ecological perspectives, vol 2. Springer Nature Singapore Pte Ltd, Singapore, pp 1–27

Gaume A, Weidler PG, Frossard E (2000) Effect of maize root mucilage on phosphate adsorption and exchangeability on a synthetic ferrihydrite. Biol Fertil Soils 31:525–532

George TS, Gregory PJ, Robinson JS, Buresh RJ (2002) Changes in phosphorus concentrations and pH in the rhizosphere of some agroforestry and crop species. Plant Soil 246:65–73

Ghaderi A, Aliasgharzad N, Oustan S, Olsson PA (2008) Efficiency of three *Pseudomonas* isolates in releasing phosphate from an artificial variable-charge mineral (iron III hydroxide). Soil Environ 27:71–76

Gonzalez-Chavez MDA, Newsam R, Linderman R, Dodd J, Valdez-Carrasco JM (2008) Bacteria associated with the extraradical mycelium of an arbuscular mycorrhizal fungus in an As/Cu polluted soil. Agrociencia 42:1–10

Govindan K, Thirumurugan V (2005) Synergistic association of rhizobium with phosphate-solubilizing bacteria under different sources of nutrient supply on productivity and soil fertility in soybean (*Glycine max*). Indian J Agron 50:214–217

Grierson PF, Comerford NB, Jokela EJ (1998) Phosphorus mineralization kinetics and response of microbial phosphorus to drying and rewetting in a Florida Spododol. Soil Biol Biochem 30:1323–1331

Gunes A, Ataoglu N, Turan M, Esitken A, Ketterings QM (2009) Effects of phosphate-solubilizing microorganisms on strawberry yield and nutrient concentrations. J Plant Nutr Soil Sci 172:385–392

Gyaneshwar P, Kumar GN, Parekh LJ, Poole PS (2002) Role of soil microorganisms in improving P nutrition of plants. Plant Soil 245:83–93

Hamdali H, Smirnov A, Esnault C, Ouhdouch Y, Virolle MJ (2010) Physiological studies and comparative analysis of rock phosphate solubilization abilities of Actinomycetales originating from Moroccan phosphate mines and of Streptomyces lividans. Appl Soil Ecol 44:24–31

Hariprasad P, Navya HM, Nayaka SC, Niranjana SR (2009) Advantage of using PSIRB over PSRB and IRB to improve plant health of tomato. Biol Control 50:307–316

Henri F, Laurette NN, Annette D, John Q, Wolfgang M, François-Xavier E, Dieudonné N (2008) Solubilization of inorganic phosphates and plant growth promotion by strains of *Pseudomonas fluorescens* isolated from acidic soils of Cameroon. Afr J Microbiol Res 2:171–178

Hoberg E, Marschner P, Lieberei R (2005) Organic acid exudation and pH changes by Gordonia sp and Pseudomonas fluorescens grown with P adsorbed to goethite. Microbiol Res 160:177–187

Intorne AC, de Oliveira MVV, Lima ML, da Silva JF, Olivares FL, de Souza GA (2009) Identification and characterization of Gluconacetobacter diazotrophicus mutants defective in the solubilization of phosphorus and zinc. Arch Microbiol 191:477–483

Ionescu M, Belkin S (2009) Overproduction of exopolysaccharides by an Escherichia coli K-12 rpoS mutant in response to osmotic stress. Appl Environ Microbiol 75:483–492

Jones DL, Oburger E (2011) Solubilization of phosphorus by soil microorganisms. In: Bünemann E, Oberson A, Frossard E (eds) Phosphorus in action: biological processes in soil phosphorus cycling. Soil biology, vol 26. Springer, Heidelberg, pp 169–198

Jones DL, Dennis PG, Owen AG, van Hees PAW (2003) Organic acid behaviour in soils: misconceptions and knowledge gaps. Plant Soil 248:31–41

Jorquera MA, Hernandez MT, Rengel Z, Marschner P, Mora MD (2008) Isolation of culturable phosphobacteria with both phytate-mineralization and phosphate-solubilization activity from the rhizosphere of plants grown in a volcanic soil. Biol Fertil Soils 44:1025–1034

Khan MS, Zaidi A, Wani PA (2007) Role of phosphate-solubilizing microorganisms in sustainable agriculture—a review. Agron Sustain Dev 27:29–43

Khan AA, Jilani G, Akhtar MS, Naqvi SMS, Rasheed M (2009) Phosphorus solubilizing bacteria: occurrence, mechanisms and their role in crop production. J Agric Biol Sci 1(1):48–58
Khan MS, Zaidi A, Ahmad E (2014) Mechanism of phosphate solubilization and physiological functions of phosphate-solubilizing microorganisms. In: Khan MS, Zaidi A, Musarrat J (eds) Phosphate solubilizing microorganisms. Principles and application of microphos technology. Springer International Publishing Switzerland, Basel, pp 31–62
Khan AA, Jilani G, Akhtar MS, Islam M, Naqvi SMS (2015) Potential of phosphorus solubilizing microorganisms to transform soil P fractions in sub-tropical Udic Haplustalfs soil. J Biodivers Environ Sci 7(3):220–227
Kumar M, Singh DP, Prabha R, Rai AK, Sharma L (2016) Role of microbial inoculants in nutrient use efficiency. In: Singh DP, Singh HB, Prabha R (eds) Microbial inoculants in sustainable agricultural productivity. Functional applications, vol 2. Springer India, New Delhi, pp 133–142
Maguire RO, Hesterberg D, Gernat A, Anderson K, Wineland M, Grimes J (2006) Liming poultry manures to decrease soluble phosphorus and suppress the bacteria population. J Environ Qual 35:849–857
Mansfeld-Giese K, Larsen J, Bodker L (2002) Bacterial populations associated with mycelium of the arbuscular mycorrhizal fungus Glomus intraradices. FEMS Microbiol Ecol 41:133–140
Mehrvarz S, Chaichi MR, Alikhani HA (2008) Effects of phosphate solubilizing microorganisms and phosphorus chemical fertilizer on yield and yield components of barely (*Hordeum vulgare* L.). Am Euras J Agric Environ Sci 3:822–828
Mengel K, Kirkby EA, Kosegarten H, Appel T (2001) Phosphorus. In: Mengel K, Kirkby EA, Kosegarten H, Appel T (eds) Principles of plant nutrition. Springer, Dordrecht, pp 453–479
Nannipieri P, Giagnoni L, Landi L, Renella G (2011) Role of phosphatase enzymes in soil. In: Bünemann E, Oberson A, Frossard E (eds) Phosphorus in action: biological processes in soil phosphorus cycling, Soil biology, vol 26. Springer, Heidelberg, pp 215–243
Oberson A, Joner EJ (2005) Microbial turnover of phosphorus in soil. In: Turner BL, Frossard E, Baldwin DS (eds) Organic phosphorus in the environment. CABI, Wallingford, pp 133–164
Oburger E, Kirk GJD, Wenzel WW, Puschenreiter M, Jones DL (2009) Interactive effects of organic acids in the rhizosphere. Soil Biol Biochem 41:449–457
Ochoa-Loza FJ, Artiola JF, Maier RM (2001) Stability constants for the complexation of various metals with a rhamnolipid biosurfactant. J Environ Qual 30:479–485
Oehl F, Oberson A, Sinaj S, Frossard E (2001) Organic phosphorus mineralization studies using isotopic dilution techniques. Soil Sci Soc Am J 65:780–787
Park KH, Lee CY, Son HJ (2009) Mechanism of insoluble phosphate solubilization by *Pseudomonas fluorescens* RAF15 isolated from ginseng rhizosphere and its plant growth-promoting activities. Lett Appl Microbiol 49:222–228
Patel DK, Archana G, Naresh Kumar G (2008) Variation in the nature of organic acid secretion and mineral phosphate solubilization by Citrobacter sp. DHRSS in the presence of different sugars. Curr Microbiol 65:168–174
Pérez E, Sulbarán M, Ball MM, Yarzabál LA (2007) Isolation and characterization of mineral phosphate-solubilizing bacteria naturally colonizing a limonitic crust in the south-eastern Venezuelan region. Soil Biol Biochem 39:2905–2914
Pierre MJ, Sopan Bhople DB, Kumar DA, Erneste H, Emmanuel B, Singh YN (2014) Contribution of arbuscular mycorrhizal fungi (AM fungi) and rhizobium inoculation on crop growth and chemical properties of rhizospheric soils in high plants. IOSR J Agric Vet Sci 7:45–55
Ponmurugan P, Gopi C (2006) Distribution pattern and screening of phosphate solubilizing bacteria isolated from different food and forage crops. J Agron 5:600–604
Quiquampoix H, Mousain D (2005) Enzymatic hydrolysis of organic phosphorus. In: Turner BL, Frossard E, Baldwin DS (eds) Organic phosphorus in the environment. CABI, Wallingford, pp 89–112
Reid RK, Reid CPP, Szaniszlo PJ (1985) Effects of synthetic and microbially produced chelates on the diffusion of iron and phosphorus to a simulated root in soil. Biol Fertil Soils 1:45–52

Reyes I, Valery A, Valduz Z (2007) Phosphate solubilizing microorganisms isolated from rhizospheric and bulk soils of colonizer plants at an abandoned rock phosphate mine. In: Velázquez E, Rodríguez-Barrueco C (eds) First international meeting on microbial phosphate solubilization, pp 69–75

Richardson AE, Hocking PJ, Simpson RJ, George TS (2009) Plant mechanisms to optimize access to soil phosphorus. Crop Pasture Sci 60:124–143

Rodríguez H, Fraga R (1999) Phosphate solubilizing bacteria and their role in plant growth promotion. Biotechnol Adv 17:319–339

Saha S, Prakash V, Kundu S, Kumar N, Mina BL (2008) Soil enzymatic activity as affected by long term application of farm yard manure and mineral fertilizer under a rainfed soybean-wheat system in N-W Himalaya. Eur J Soil Biol 44:309–315

Schwöppe C, Winkler HH, Neuhaus HE (2003) Connection of transport and sensing by UhpC, the sensor for external glucose-6-phosphate in Escherichia coli. Eur J Biochem 270:1450–1457

Sharan A, Shikha, Darmwal NS (2008) Efficient phosphorus solubilization by mutant strain of *Xanthomonas campestris* using different carbon, nitrogen and phosphorus sources. World J Microbiol Biotechnol 24:3087–3090

Sharma K, Dak G, Agrawal A, Bhatnagar M, Sharma R (2007) Effect of phosphate solubilizing bacteria on the germination of *Cicer arietinum* seeds and seedling growth. J Herb Med Toxicol 1:61–63

Sharma SB, Sayyed RZ, Trivedi MH, Gobi TA (2013) Phosphate solubilizing microbes: sustainable approach for managing phosphorus deficiency in agricultural soils. Springerplus 2:587

Silvan N, Vasander H, Karsisto M, Laine J (2003) Microbial immobilisation of added nitrogen and phosphorus in constructed wetland buffer. Appl Soil Ecol 24:143–149

Sims JT, Pierzynski GM (2005) Chemistry of phosphorus in soil. In: Tabatabai AM, Sparks DL (eds) Chemical processes in soil, SSSA book series 8. SSSA, Madison, pp 151–192

Smith SE, Smith FA (2011) Roles of arbuscular mycorrhizas in plant nutrition and growth: new paradigms from cellular to ecosystem scales. Annu Rev Plant Biol 62:227–250

Son TTN, Diep CN, Giang TTM (2006) Effect of bradyrhizobia and phosphate solubilizing bacteria application on soybean in rotational system in the Mekong delta. Omonrice 14:48–57

Stephen J, Jisha MS (2009) Buffering reduces phosphate solubilizing ability of selected strains of bacteria. World J Agric Sci 5:135–137

Stevenson FJ (2005) Cycles of soil: carbon, nitrogen, phosphorus, sulfur, micronutrients. Wiley, New York

Ström L, Owen AG, Godbold DL, Jones DL (2005) Organic acid behaviour in a calcareous soil: implications for rhizosphere nutrient cycling. Soil Biol Biochem 37:2046–2054

Sundara B, Natarajan V, Hari K (2002) Influence of phosphorus solubilizing bacteria on the changes in soil available phosphorus and sugarcane and sugar yields. Field Crops Res 77:43–49

Tang C, Rengel Z (2003) Role of plant cation/anion uptake ratio in soil acidification. In: Rengel Z (ed) Handbook of soil acidity. Marcel Dekker, New York, pp 57–81

Tang J, Leung A, Leung C, Lim BL (2006) Hydrolysis of precipitated phytate by three distinct families of phytases. Soil Biol Biochem 38:1316–1324

Tao G, Tian S, Cai M, Xie G (2008) Phosphate solubilizing and mineralizing abilities of bacteria isolated from soils. Pedosphere 18:515–523

Tarafdar JC (2019) Fungal inoculants for native phosphorus mobilization. In: Giri B, Prasad R, Wu Q-S, Varma A (eds) Biofertilizers for sustainable agriculture and environment, Soil biology, vol 55. Springer Nature, Switzerland, Basel, pp 21–40

Villegas J, Fortin JA (2002) Phosphorus solubilization and pH changes as a result of the interactions between soil bacteria and arbuscular mycorrhizal fungi on a medium containing NO_3 as nitrogen source. Can J Bot 80:571–576

Wahid F, Fahad S, Danish S, Adnan M, Yue Z, Saud S, Siddiqui MH, Brtnicky M, Hammerschmiedt T, Datta R (2020) Sustainable management with mycorrhizae and phosphate solubilizing bacteria for enhanced phosphorus uptake in calcareous soils. Agriculture 10:334

Wielbo J, Skorupska A (2008) Influence of phosphate and ammonia on the growth, exopolysaccharide production and symbiosis of *Rhizobium leguminosarum* bv. Trifolii TA1 with clover (*Trifolium pratense*). Acta Biol Hung 59:115–127

Xiao C, Chi R, He H, Qiu G, Wang D, Zhang W (2009) Isolation of phosphate-solubilizing fungi from phosphate mines and their effect on wheat seedling growth. Appl Biochem Biotechnol 159:330–342

Yang H, Schroeder-Moreno M, Giri B, Hu S (2018) Arbuscular mycorrhizal fungi and their responses to nutrient management. In: Giri B, Prasad R, Varma A (eds) Root biology, Soil biology, vol 52, pp 429–449

Yazdani M, Bahmanyar MA, Pirdashti H, Esmaili MA (2009) Effect of phosphate solubilization microorganisms (PSM) and plant growth promoting rhizobacteria (PGPR) on yield and yield components of corn (*Zea mays L.*). Proc World Acad Sci Eng Technol 37:90–92

Yi YM, Huang WY, Ge Y (2008) Exopolysaccharide: a novel important factor in the microbial dissolution of tricalcium phosphate. World J Microbiol Biotechnol 24:1059–1065

Zaidi A, Khan MS (2006) Co-inoculation effects of phosphate solubilizing microorganisms and *Glomus fasciculatum* on green gram—*Bradyrhizobium* symbiosis. Turk J Agric 30:223–230

Zhang L, Fan J, Ding X, He X, Zhang F, Feng G (2014) Hyphosphere interactions between an arbuscular mycorrhizal fungus and a phosphate solubilizing bacterium promote phytate mineralization in soil. Soil Biol Biochem 74:177–183

Zineb AB, Trabelsi D, Ayachi I, Barhoumi F, Aroca R, Mhamdi R (2020) Inoculation with elite strains of phosphate-solubilizing bacteria enhances the effectiveness of fertilization with rock phosphates. Geomicrobiol J 37:22–30

Chapter 18
Cyanobacteria as Biofertilizer and Their Effect Under Biotic Stress

Nihal Gören-Sağlam

Contents

1 Introduction

Cyanobacteria, also known as blue-green algae, were the first organisms that created molecular oxygen and transformed the biosphere from anaerobic to largely aerobic. Many cyanobacteria have a very wide distribution. Thanks to these features, they are considered as a model organism that enables us to learn about microbial biogeography and evolution (Gupta et al. 2013; Prasanna et al. 2009; Ahmed et al. 2010).

Cyanobacteria have been identified as important inhabitants of many agricultural soils that potentially contribute to biological nitrogen fixation, phosphate dissolution, mineral release to increase soil fertility, and crop productivity (Singh 2014). They produce and secrete a variety of biologically active substances, such as proteins, vitamins, carbohydrates, amino acids, polysaccharides, and phytohormones, which act as signal molecules to support plant growth. So, they protect plants against environmental stress. It is determined that the related bacteria are also found

N. Gören-Sağlam (✉)
Faculty of Science, Department of Biology, Istanbul University, Istanbul, Turkey
e-mail: gorenn@istanbul.edu.tr

© The Author(s), under exclusive license to Springer Nature Switzerland AG 2021

H. I. Mohamed et al. (eds.), *Plant Growth-Promoting Microbes for Sustainable Biotic and Abiotic Stress Management*,
https://doi.org/10.1007/978-3-030-66587-6_18

in cultivated fields. Identification of dominant strains effective in plant growth was found important for plant production (Osman et al. 2010; Prasanna et al. 2009).

Cyanobacteria show antagonistic activity against many plant pathogenic fungi. The application of cyanobacteria as biological fertilizers reduced the disease severity caused by the pathogen in many plants (Küçük and Sezen 2019).

2 General Features of Cyanobacteria

Cyanobacteria members are the oldest oxygen-producing photoautotrophs on earth. Plant chloroplasts evolved from cyanobacteria through the process of endosymbiosis. Cyanobacteria are known as blue-green algae, which is commonly confused with algae because it shares traits with algae and bacteria, because of the C-fucocyanin, a blue-green pigment they contain (Yadav et al. 2017) (Table 18.1).

Their cell structures are simple, and individual cells can also exist as spheres, courses, or flat colonies. The most common form of the colonies is a filament. The colonies can contain several cells or several thousands of cells in a mucilage sheath. Threads of cyanobacteria are called “trichomes.” There is no organization or division of labor between cells in the threads. However, it is seen that some cells grow and take a homogeneous appearance, and structures called “heterocyst” occur. A thick wall, enriched with nutrients, surrounds some of the cells, and structures resistant to unfavorable conditions called “akinetes” are formed. In some cells, real branching is seen, while in other cells, false branching is also observed. In some species, it is seen that the trichome thinners from the bottom to the end and there is a heterocyclic at the bottom (Mishra et al. 2013).

Since cyanobacteria cells have a prokaryotic organization, they do not have any membrane organelles. The cell wall is similar in structure and function to Gram-negative bacteria (Whitton and Potts 2012; Mishra et al. 2013). The cytoplasm structure consists of two different layers, namely, chromoplasm and centroplasm.

Table 18.1 General characteristics of algae (on the left) and bacteria (on the right). Cyanobacteria have combination characteristics that come from algae and bacteria (middle column) (adapted from https://www.deq.ok.gov, DEQ n.d.)

Algae	*Cyanobacteria*	*Bacteria*
• Eukaryote • Photosynthetic • Unicellular and multicellular • Can be filamentous • Found only in aquatic environments • Does not produce toxins • Can form visible colonies in water	• Prokaryote • Photosynthetic • Unicellular and multicellular • Can be filamentous • Found in many diverse habitats • Capable of producing toxins • Can form visible colonies in water	• Prokaryote • Non-photosynthetic • Unicellular • Found in many diverse habitats • Capable of producing toxins • Can cause increased turbidity, not visible colonies

Chromoplasm is a colorful and networked structure with uncertain boundaries around the centroplasm. Generally, it does not have a vacuole and is immobile. As a chemical structure, RNA is dispersed, and assimilation pigments have a lamellar structure. However, they are not homogeneously disperplastics as plastids surrounded by a real membrane. Centroplasm is colourless and located in the centre. Its chemical structure consists of DNA; it contains elements in the form of a stick, reticular, or thread. All of these correspond to the nucleus and are called chromatin devices. There is no real nucleus (Shevela et al. 2013).

There is only chlorophyll-a from chlorophylls in cyanobacteria. Among the carotenoids, they contain β-carotene and E-carotene. Cyanobacteria often have all the types of xanthophylls and lutein. They contain C-fucocyanin and allophycocyanin, which are phycobilins. The color of *Cyanophyta* is mostly bluish green, olive green, and yellow brown. Cyanobacteria take the blue-green color from fucocyanin. There is also a small amount of phycoerythrin (Takaichi et al. 2009; Singh 2014).

Food storage substances in chromoplasma are glycogen, cyanophilin from proteins, and volutin. Nitrogen constitutes 8% of the dry weight of blue-green algae.

Reproduction in cyanobacteria occurs by dividing the cells into two, as in bacteria. Colony-forming species are seen cell division, and asexual reproduction occurs in a type of fragmentation. In some of the filamentous species, with the death of the cells in between, the thread breaks down into several cells. These parts are called "hormogonium." Hormogoniums occur in abnormal conditions and develop and form the thread when the conditions are favorable (Cohen and Meeks 1997).

2.1 *Ecology and Phylogeny of Cyanobacteria*

Cyanobacteria have spread to all parts of the earth. They live in freshwaters and seas. Some of their species are planktonic. Some species are benthic; they live on the grounds of streams, lakes, pond waters, and marshes. In suitable conditions and seasons, some of the planktonic species can over-proliferate and cause the death of fish and other aquatic organisms due to the toxic substances that appear. Some species of cyanobacteria are found in moist soils and on rocks that leak water as a blackish-mucilage cover. They also live on bare rocks on the shores of the seas, bark, and arctic regions (Nagarajan et al. 2011). In addition to their association with plants, they can develop epiphytically on bark, leaves, roots, and stems of submerged areas (Aguiar et al. 2008; Boopathi et al. 2013). They are the most abundant algae after diatoms on the soil surface and below. There are also species living in the dark cave walls as they show chromatic adaptation according to the light intensity. Some species live at 75–85 °C in hot water sources. There are also species living in deserts, poles, snow, rarely in salt waters, and oceans.

Cyanobacteria provide nitrogen for the growth of the plant partner. It has been explained that cyanobacteria can convert atmospheric nitrogen to ammonium form with nitrogenase enzyme, and ATP is used in this conversion (Magnuson 2019):

$$N_2 + 8H^+ + 8e^- + 16MgATP \rightarrow 2NH_3 + H_2 + 16MgADP + 16Pi$$

Species belonging to some blue-green algae genus (*Chroococcus*, *Gloeocapsa*, etc.) live symbiotically with fungi and form "lichens." Some species of *Anabaena* and *Nostoc* also live symbiotically with some species of ferns, Gymnosperm and Angiosperms. Cyanobacteria are known to affect tallus morphogenesis in lichens (Singh et al. 2016; Singh 2014). It is known that cyanobacteria, especially those that form symbiotic relationships with plants, secrete protein from carbohydrate-rich arabinogalactan. It has been found that these proteins act as signaling molecules which do not play an important role in the regulation of plant growth and development (Abdel-Raouf et al. 2012). The secretion of phytohormones by cyanobacteria begins with the formation of a symbiotic relationship (Singh et al. 2016).

Nitrogen fixation is an important feature of cyanobacteria. Various species can physiologically detect the free nitrogen of the air. Cyanobacteria are similar to bacteria in these aspects. Apart from cyanobacteria, no other algae group has this feature. The nitrogen-binding species in the structure of lichens give nitrogen they detected to the fungus (Zehr 2011; Stal 2013).

Base compositions of DNA molecules belonging to different cyanobacteria have been determined. GC rates of cyanobacteria with unicellular form vary between 35 and 71%. This ratio indicates that this group includes a very large group of organisms that have very few genetically related relationships. On the other hand, DNA ratios of DNA molecules of the cyanobacteria group that form the heterocysts very much less (between 38 and 46%). Cyanobacteria are grouped with their morphological lines as well as phylogenetic features. Unicellular cyanobacteria are very broad phylogenetic, and different representatives show phylogenetic relationship with different morphological groups (Yadav et al. 2017; Chittora et al. 2020).

3 Biofertilizers

Agricultural systems that use more inputs for high yields cause environmental problems and depletion of natural resources. The rapid production increase caused by the application of chemicals decreases gradually, and a healthy agriculture system becomes inevitable. The production of clean foods without agricultural chemicals is compulsory for the future of humanity and natural resources. Plant nutrients are essential for crop and healthy food production, given the growing population of the world. Today, agricultural strategies are mainly carried out on inorganic chemical-based fertilizers, which pose a serious threat to the environment and human health (Itelima et al. 2018). Biofertilizer is used as an alternative way to increase soil fertility and crop production in sustainable agriculture. The use of beneficial microorganisms as biofertilizers is crucial for the agricultural sector, given their potential in food safety and sustainable crop production (Vessey 2003). Research is ongoing to make biofertilizers an important component of nutritional management. According to a report by the FAO published in 2006, biofertilizer is a substance used for products containing microorganisms that fix atmospheric N or secrete growth-promoting substances that help dissolve soil nutrients (FAO 2006).

Nitrogen fixers (N-fixer), potassium and phosphorus solubilizers, plant growth-promoting rhizobacteria (PGPRs), endo- and ectomycorrhizal fungi, and cyanobacteria are commonly used as biofertilizer components (Fig. 18.1) (Ansari and Mahmood 2017; Zakeel and Safeena 2019). The use of biofertilizers provides improved nutrients and water intake, plant growth, and enhanced plant defense against abiotic and biotic stresses. These properties of biofertilizers play a very important role in soil fertility and environmental protection. Also, their low cost will benefit farmers economically (Itelima et al. 2018).

Biofertilizer is an alive, pure, or mixed microorganism formulation that, when applied to seed, plant surface, or soil, colonizes in the rhizosphere or enters the plant tissues, fixes atmospheric nitrogen, and increases soil uptake and plant nutrient uptake and vegetative growth (Chatterjee et al. 2017) (Fig. 18.2). Biofertilizers are cheaper than chemical fertilizers, do not show toxic effects to plants, do not pollute groundwater, do not increase soil acidity, and do not adversely affect plant development. The most prominent features of biofertilizers related to plant development are nitrogen fixation, making plant nutrients available, biological control of diseases, and secretion of plant growth stimulants. While a significant amount of fossil energy is used in chemical fertilizer production, energy is free in biological fertilization. The species that are active among the bacteria generally isolated from the rhizosphere are chosen by considering their adaptability to activity and environmental conditions and are stored for use in single or multiple species containing biological fertilizers. Reducing the use of excessive chemical fertilizers, potential nitrogen fixation and the use of phosphate-dissolving bacteria as biological fertilizers increases productivity in agricultural products. However, it is necessary to develop special plant-microorganism combinations that will show high efficiency in wide environmental conditions (Vessey 2003; Adesemoye and Kloepper 2009; Sinha et al. 2010; Khosro and Yosef 2012; Santos et al. 2012; Raja 2013; Youssef and Eissa 2014; Chun-Li et al. 2014).

As biofertilizers are living content and product content, quality of life and shelf life directly affect the availability or efficiency of biofertilizer.

Biofertilizer:

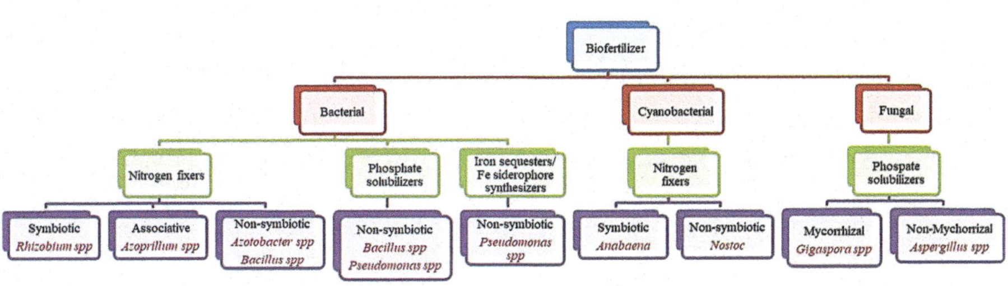

Fig. 18.1 Classification of biofertilizers. (Adapted from Zakeel and Safeena 2019)

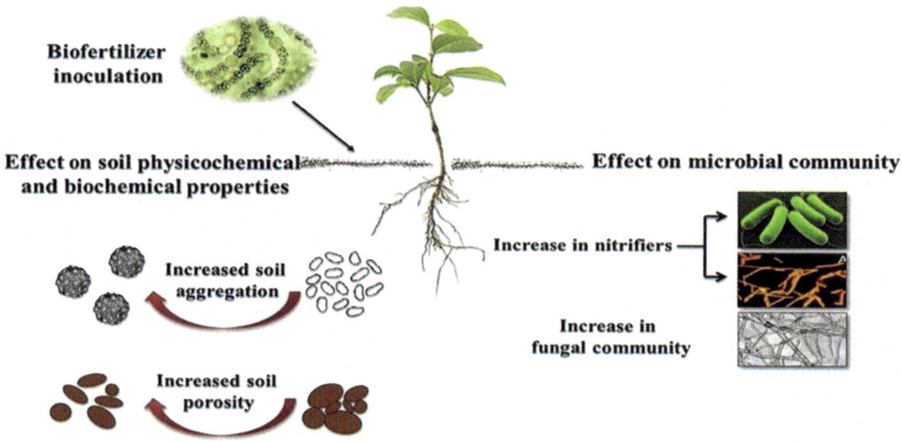

Fig. 18.2 Effects of biofertilizers on physiological and biochemical properties of soil

- Colonized in the rhizosphere when entering seed, plant surface, or soil or entering plant tissues.
- Fixing atmospheric nitrogen.
- A living, pure, or mixed microorganism formulation that increases soil.

These:

- Cheap cost.
- Do not show toxic effects to plants.
- Do not pollute groundwater.
- Do not increase soil acidity.
- Biologically controlling soil-borne diseases and secreting substances that stimulate plant growth (increase tolerance to environmental stresses) phosphorus, and uptake of plant nutrients and plant growth (Çakmakçı 2014).

Effective work of microorganisms occurs only when there are favorable and optimal conditions for them to metabolize their substrates. Some of these conditions are adequate water and oxygen (varies depending on whether microorganisms are aerobic or anaerobic), pH, and ambient temperature.

3.1 Types of Biofertilizers

According to the general classification in the FAO's report entitled "Plant Nutrition for Food Security" published in 2006, biofertilizers can be divided into four main categories:

1. **N-fixing biofertilizers**: These include *Rhizobium*, *Azotobacter*, *Azospirillum*, *Clostridium*, and *Acetobacter* bacteria; cyanobacteria; and fern *Azolla* (collaborating with cyanobacteria).
2. **P-solubilizer/activating biofertilizers:** Phosphate-solubilizing bacteria (PSB) and phosphate-solubilizing microorganisms (PSMs), for example, *Bacillus*, *Pseudomonas*, and *Aspergillus*. Mycorrhiza is a nutrient-activating fungus.
3. **Composting accelerators:** Cellulosic (*Trichoderma*) and ligninolytic (Humicola).
4. **Plant growth-promoting rhizobacteria (PGPRs)**: *Pseudomonas* species. PGPRs increase plant growth and performance.

Different types of biological fertilizers and related microorganisms are given in Table 18.2 (Itelima et al. 2018).

Among these, the groups of N-fixing organisms are the most important biological fertilizers used in plant growing. Another important biofertilizer is those containing P-dissolving organism cultures. Unlike industrial nitrogen fixation, biological nitrogen fixation involves the conversion of nitrogen (N2) to ammonia via microorganisms. Many microorganisms (e.g., *Rhizobium*, *Azotobacter*, and *Cyanobacteria*) reduce the atmospheric N_2 to ammonia (NH_3) using molecular N_2 with the help of nitrogen enzyme:

$$N_2 + 6H^+ + 6e^- \rightarrow 2NH_3$$

Biological nitrogen fixation is an important nitrogen source for plant life. Biological nitrogen fixation estimates range from 100 to 290 million tons N/year. It is estimated that 40–48 million tons of this total is biologically fixed in agricultural

Table 18.2 Types of biofertilizers and related microorganisms (Itelima et al. 2018)

Groups	Examples
	Nitrogen-fixing biofertilizers
Free-living	*Azotobacter*, *Beijerinckia*, *Clostridium*, *Klebsiella*, *Anabaena*, *Nostoc*
Symbiotic	*Rhizobium*, *Frankia*, *Anabaena*, *Azolla*
Associative symbiotic	*Azospirillum*
	Phosphate-solubilizing biofertilizers
Bacteria	*Bacillus megaterium* var. *phosphaticum*, *Bacillus subtilis*, *Bacillus circulans*
Fungi	*Penicillium* spp., *Aspergillus awamori*
	Phosphate-mobilizing biofertilizers
Arbuscular mycorrhiza	*Glomus* spp., *Gigaspora* spp., *Acaulospora* spp., *Scutellospora* spp., *Sclerocystis* spp.
Ectomycorrhiza	*Laccaria* spp., *Pisolithus* spp., *Boletus* spp., *Amanita* spp.
Ericoid mycorrhiza	*Pezizella ericae*
Orchid mycorrhiza	*Rhizoctonia solani*
	Plant growth-promoting rhizobacteria (PGPRs)
Pseudomonas	*Pseudomonas fluorescens*

crops and fields. Only nitrogen-fixing microorganisms supply an additional nutrient (N) to the soil plant system. Other biological fertilizers dissolve or activate the nutrients already in the soil. *Azolla* is an almost unique species when evaluated as a green fertilizer among nitrogen-fixing cyanobacteria. In this process, it does not only add the nitrogen it fixes biologically but also other nutrients it receives from the soil. While *Rhizobium* is specific to legumes, *Cyanobacteria* and *Azolla* are useful in increasing N supplies during flooded rice cultivation as they are abundant in wetlands (FAO 2006).

Some of the biofertilizers promote plant growth through the production of plant hormones. The production of hormones such as auxins, cytokinins, and giberellins has an effect on plant development and quality via direct and/or indirect mechanisms (Eşitken et al. 2003a, b; Elsheikh and Elzidany 1997).

Direct mechanisms:

- Biological nitrogen fixation.
- Reducing environmental stress.
- Harmony in a bacteria-plant relationship.
- Increasing the inorganic phosphorus solubility.
- Mineralization of organic phosphorus compounds.
- Increasing iron intake and increasing the ratio of some trace elements.
- Vitamin synthesis.
- Increasing root permeability.

Indirect mechanisms:

- Taking a role as biocontrol agents, reducing diseases with antibiotic production.
- In soils contaminated with various organic compounds, it is counted as protecting plants by breaking down barrier xenobiotics.

The main idea in biological fertilization is to reduce the use of chemicals to support agricultural sustainability, to protect natural resources and the environment, and to improve the quality. In its current state, biofertilizers cannot replace agricultural chemicals alone, but they reduce their usage rates and support ecological agriculture (Eşitken et al. 2003a, b; Elsheikh and Elzidany 1997).

4 Biotic Stress

Stress in plants is defined as all external factors that adversely affect the growth, development, or productivity. Plants are constantly subjected to environmental stresses due to their immobile structure. Stresses in plants cause a wide variety of events such as cellular metabolism, gene expression, changes in growth rates, crop yields, etc. Plants developed effective strategies and mechanisms to deal with environmental stresses. Stress response mechanisms contribute to stress resistance or stress tolerance at different morphological, biochemical, and molecular levels (Bakır 2020). The stresses to which plants are exposed are gathered under two

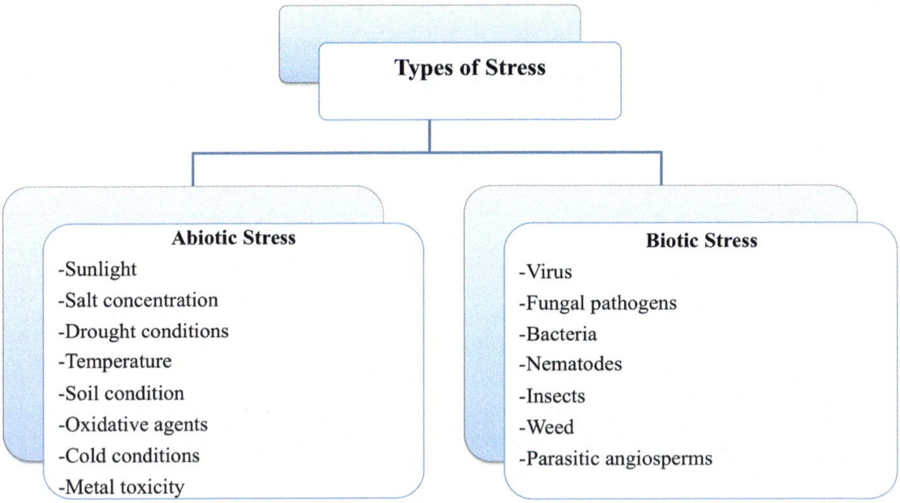

Fig. 18.3 Types of stress in plants

important topics. These are "abiotic" and "biotic" stresses (Fig. 18.3). Biotic factors are stresses caused by infection of microorganisms (fungi, bacteria, and virus) and attacks of harmful animals (Lichtenhaler 1996; Büyük et al. 2012). Abiotic stress factors are environmental factors including drought, cold, hot, salt, and nutritional deficiencies and are among the factors that decrease productivity in agricultural production. Biotic and abiotic stresses have been shown to reduce the average crop productivity by 65–87% depending on the crop type (Verma et al. 2013).

Viruses, bacteria, fungi, nematodes, insects, arachnids, and weeds are known as living organisms that cause biotic stress in plants. The organisms that cause biotic stress can lead to the death of plants by depriving their hosts of nutrients directly. Biotic stresses are very important for agriculture due to pre- and postharvest losses. Generally biotic stresses affect photosynthesis, because of chewing insects and virus infections, and reduce the rate of photosynthesis (Gull et al. 2019). The increase in the amount of pests and pathogens in nature can be caused by climate changes. For example, it is known that an increase in temperatures facilitates pathogen spread. At the same time, many abiotic stress conditions weaken the defense mechanisms of plants and thereby increase their susceptibility to pathogen infection (Suzuki et al. 2014).

Three different pathogen attack strategies have been defined (Koeck et al. 2011; Elad et al. 2011):

1. Necrotrophy: Plant cells are killed by pathogen infection (gray mold, *Botrytis cinerea*).
2. Biotrophy: In biotrophy the plant cells remain alive (powdery mildew, *Podosphaera aphanis*).

3. Semibiotrophy: The pathogen does not immediately kill the cells, causing them to die later in the infection, in this type (anthracnose, *Colletotrichum acutatum*).

Some pathogens that cause biotic stress in plants and their effects on the area they infect are given in the table below (Table 18.3) (Kanwar and Jha 2019).

Plants use highly complex defense systems against pathogen attacks. The defense mechanism has two types: innate and systemic plant response. However, the plant in two ways exhibits a natural defense: specific (specific to species/pathogen race) and nonspecific (non-host or general resistance). Nonspecific resistance is based on both structural barriers and inducible responses, including numerous proteins and other organic molecules produced before infection or during a pathogen attack. Structural defenses include morphological and structural barriers, chemical compounds, proteins, and enzymes. These compounds not only protect the plant from invasion but also give the plant strength and hardness, giving it tolerance or resistance to biotic factors (Onaga and Wydra 2016).

5 Usage of Cyanobacteria as Biofertilizer for Biotic Stress

Different microorganism groups associated with plants have been described to produce metabolites with beneficial effects on plants (Berendsen et al. 2012; Mendes et al. 2013). The harmful effects of pathogens on plants have been known for a long time. Studies reveal signals related to microorganisms promoting plant growth (PGPR = plant growth-promoting rhizobacteria), and plant communications have accelerated in recent years. PGPRs have been reported to release signaling compounds that can bind to receptor sites on the plasma membrane and cause activation of genes, leading to the synthesis of proteins and enzymes or secondary metabolites (Hussain et al. 2013). Many of the signaling compounds included in the phytochemical reaction belonging to the carbohydrate, lipid, glycolipid, or glycoprotein group have been identified (Yamaguchi and Huffaker 2011). Some of these compounds have been found to cause an increase in the accumulation of glucosinolates, alkaloids, polyphenols, flavonoids, flavonoid glycosides, saponins, terpenes, and phytoalexins, when applied to plants as spray or root treatments (Hussain et al. 2013; Rodriguez et al. 2006). These phytochemicals protect plants from biotic and abiotic stress and help plants develop resistance to these stresses (Shan et al. 2012; Sokolova et al. 2011).

When studies on microorganisms that support plant growth are examined, it has been determined that the most researched studies are rhizobacteria, symbiotic rhizobia, and mycorrhizal fungi. In recent studies, it is seen that another group of microorganisms that encourage plant development is cyanobacteria (Mendes et al. 2013; Willis et al. 2013). In recent studies, data affecting the gene expression of host plants have been obtained with the signals produced by cyanobacteria; thus it has been determined that various changes occur in the phytochemical structures of plants (Manjunath et al. 2010; Singh et al. 2016; Yadav et al. 2017). The

Table 18.3 Some biotic stresses and their effect in plants (Kanwar and Jha 2019)

Pathogen	Plant	Effect	References
Bacteria			
Pseudomonas syringae	Soybean	Reduced photosynthesis	Zou et al. (2005)
Xanthomonas campestris pv. vesicatoria	Tomato	Reduced photosynthesis	Kocal et al. (2008)
Pseudomonas syringae	Arabidopsis	Reduced photosynthetic rate at the infection site	Bonfig et al. (2006), Berger et al. (2007), de Torres Zabala et al. (2015)
Viruses			
Tobacco mosaic virus	Tobacco	Photo inhibition and photo oxidation of chlorophyll in infected cells	Balachandran et al. (1994)
Cucumber mosaic virus	*Cucurbita pepo*	Reduced photosynthesis, starch mobilization, and alteration in metabolism	Tecsi et al. (1996)
Potato virus Y	Tobacco	Accumulation of soluble sugars	Herbers et al. (2000)
Abutilon mosaic virus	*Abutilon striatum*	Carbohydrate accumulation in leaves during early symptom development	Lohaus et al. (2000)
Pepper mild mottle virus (PMMoV)-I	*Nicotiana benthamiana*	Increase in NPQ values of the areas invaded by the pathogen	Pérez-Bueno et al. (2006)
Rice stripe virus	Rice	Repression of genes related to photosynthesis	Cho et al. (2015)
Strawberry vein banding virus (SVBV)	*Fragaria vesca*	Altered photosynthesis	Chen et al. (2016)
Grapevine leafroll-associated virus 3 (GLRaV-3)	*Vitis vinifera*	Reduced photosynthesis and altered expression of genes related to sugar metabolism	Vega et al. (2011), Montero et al. (2016)
Bean common mosaic virus (BCMV)	*Phaseolus vulgaris*	Repression of genes related to photosynthesis and carbohydrate metabolism	Martin et al. (2016)
Herbivores attack or wounding			
Caterpillar	Wild parsnip	Reduced CO_2 assimilation in the attacked leaf is proportionally greater than the leaf area that is actually damaged	Zangerl et al. (2002)

(continued)

Table 18.3 (continued)

Pathogen	Plant	Effect	References
Manduca sexta	*Nicotiana attenuata*	Repression of genes related to photosynthesis, while induction of genes related to carbohydrate metabolism	Hui et al. (2003)
Mechanical wounding or (*Choristoneura occidentalis* or *Pissodes strobi*)	*Picea sitchensis*	Repression of genes related to photosynthesis	Ralph et al. (2006)
Trichoplusia ni	Arabidopsis	Reduced maximum quantum efficiency of photosystem II and increased dark respiration rates	Tang et al. (2006)
Mirid bug (*Tupiocoris notatus*)	*Nicotiana attenuata*	Increased photosynthesis	Halitschke et al. (2011)
Meloidogyne incognita	Tomato	Altered expression of genes related to primary metabolism	Shukla et al. (2017), Zhao et al. (2018)

development of phytochemicals has opened a new field of research that may have significant economic benefits for the agricultural industry. Studies on resistance induced to control plant diseases in laboratory, greenhouse, and field conditions enabled the commercialization of R&D products, thereby providing new-generation microbial fertilizers or product preservatives.

Bioactive compounds produced by cyanobacteria have been found to increase phytohormone levels, which are responsible for triggering the development of the subsoil and aboveground parts of the plant. It is also known that phytohormones regulate the enzymatic activities and metabolic changes that occur during plant growth. Therefore, the increase in the activity of peroxidase and phenylalanine ammonia-lyase enzymes from defense enzymes has also been linked to phytohormone levels (Tvorogova et al. 2013). The presence of jasmonic acid (JA) has been detected in cyanobacteria (Singh 2014). These bacteria have been reported to trigger the accumulation of abscisic acid (ABA), which ensures plant survival in stress conditions such as wilt, water stress, osmotic stress, and salt stress (Khan et al. 2012). Jasmonic acid and its various metabolites are known to be responsible for regulating plant development as well as plant reactions to abiotic and biotic stress (Khan et al. 2012). In addition, members of *Synechococcus*, *Anabaena*, *Nostoc*, *Calothrix*, *Scytonema*, and *Cylindrospermum* can produce ethylene (Singh et al. 2016). Flavonoids and phytohormones have been reported to aid plant-microorganism interactions (Jaiswal et al. 2018); these compounds increased root colonization of microorganisms (Kehr et al. 2011), providing an allelochemical effect on the population of other organisms (Khan et al. 2012). These also served as signal molecules (Kehr et al. 2011; Khan et al. 2012).

Cyanobacteria are used as biological fertilization of some rice cultures. It is known that over a hundred of cyanobacteria species fix N. Common cyanobacteria, *Nostoc*, *Anabaena*, *Aulosira*, *Tolypothrix*, and *Calothrix*, are used as biological fertilizers for rice (Chittora et al. 2020). Cyanobacteria also release plant growth substances such as IAA (indoleacetic acid) and GA (gibberellic acid) and improve polysaccharides that help bind soil particles (improving soil structure). These are also used as a soil conditioner and to protect the soil against erosion by entangled bulk formation (FAO 2006). The optimum temperature for cyanobacteria is about 30–35 °C. The pH of the soil is the most important factor in the growth of cyanobacteria and N fixation. The optimal pH for growth of cyanobacteria in the culture medium is 7.5–10, and the lower limit is around 6.5–7. The growth of cyanobacteria is better in neutral to alkaline soils under natural conditions. Cyanobacteria need all plant nutrients to grow and fix nitrogen (N). N-containing fertilizers often inhibit the growth and N fixation of cyanobacteria. Since phosphorus (P) increases the growth and N fixation of cyanobacteria, sufficient phosphorus must be present in irrigation water. Consequently, P deficiency causes a marked decrease in the growth of cyanobacteria and thus N fixation. Cyanobacteria vaccine can be prepared in the laboratory or open areas. The open-air soil culture method is simple, is less expensive, and can be easily adapted by farmers (FAO 2006).

Some cyanobacteria have been found to reduce the occurrence of a disease caused by plant pathogens in plants (Table 18.4), for example, culture filter and ethyl acetate extract of *Calothrix elenkinii* Kossinskaja; in pot experiments, it has been found that it decreases disease severity on *Pythium aphanidermatum* (Edson) Fitzp-infected soybean, tomato, and pepper seeds (Manjunath et al. 2010). It was investigated that damping-off disease in tomato seedlings inoculated with a group of fungal pathogens containing *Pythium debaryanum* R. Hesse, *Fusarium oxysporum* f. sp. *lycopersici* W.C. Snyder & H.N. Hansen, *Gibberella fujikuroi* (Sawada) Wollenw, and *Rhizoctonia solani* J.G. Kühn decreases with *Trichormus variabilis* (Kützing ex Bornet & Flahault) Komarek & Anagnostidis and *Anabaena oscillarioides* Bory ex Bornet & Flahault applications (Chaudhary et al. 2012). *Trichormus variabilis* and *A. laxa* A. Braun were found to produce a systemic defense response in tomato plants struggling with *Fusarium* sp. wilt. Some enzyme activities, phenylalanine ammonia-lyase, polyphenol oxidase, chitosanase, and β-1,3-glucanase, were found high in the tomato roots treated with cyanobacterial formulations. This situation revealed the importance of cyanobacterial interaction with tomato seedlings (Prasanna et al. 2013).

The use of bacteria that promote plant growth as biocontrol agents to be used against soil-borne plant pathogens has become very attractive in recent years for sustainable agriculture. These microorganisms reveal their induced systemic resistance (ISR), which strengthens the physical and mechanical of the cell wall and alters the synthesis of metabolites for defense against pathogens and the physiological and biochemical reaction of the host (Chaudhary et al. 2012).

6 Conclusions

Today, strategies that can help reduce chemicals used for agricultural products, a more economical product to be used instead of chemicals, and environmentally friendly agriculture are demanded. Various methods are tried to increase product

Table 18.4 Some cyanobacteria and their biocidal activities against plant pathogens (Kumar et al. 2019)

Cyanobacteria	Extract	Plant pathogens	References
Fischerella muscicola	Fischerellin	• *Uromyces appendiculatus* (brown rust) • *Erysiphe graminis* (powdery mildew) • *Phytophthora infestans* • *Pyricularia oryzae* (rice blast)	Hagmann and Juttner (1996)
Nostoc muscorum	Bis(2,3-dibromo-4,5-dihydroxybenzyl) (BDDE)	• *Sclerotinia sclerotiorum* (cottony rot of vegetables and flowers) • *Rhizoctonia solani* • *Candida albicans*	Borowitzka (1995)
Tolypothrix byssoidea	Antifungal peptides (dehydrohomoalanine, Dhha)	Antifungal activity against the yeast *Candida albicans*	Jaki et al. (2001)
Oscillatoria redekei syn. *Limnothrix redekei* HUB 051	Antibacterial fatty acids (α-dimorphecolic acid, a 9-hydroxy-10E,12Z-octadecadienoic acid (9-HODE), and coriolic acid)	Inhibited the growth of Gram-positive bacteria • *Bacillus subtilis* SBUG 14 • *Micrococcus flavus* SBUG 16 • *Staphylococcus aureus* SBUG11 and ATCC 25923	Mundt et al. (2003)
Nostoc sp.	Cryptophycin	Natural pesticides against the fungi, insects, and nematodes	Biondi et al. (2004)
Anabaena subcylindrica, *Nostoc muscorum*, *Oscillatoria angusta*	Efficient algal filtrate concentration (EAFC)	• *Alternaria alternata* • *M. phaseolina* • *F. saloni*	Abo-Shady et al. (2007)
Spirulina platensis, *Oscillatoria* sp., *Nostoc muscorum*		*Cercospora beticola* causing leaf spot of sugar beat	Mostafa et al. (2009)
Calothrix elenkenii	Ethyl acetate extract	*Pythium aphanidermatum*	Manjunath et al. (2010)
Lessonia trabeculata	Ethanolic extracts	Reduced number and size of the necrotic lesion in tomato leaves following infection with *Botrytis cinerea*	Jimenez et al. (2011)
Gracilaria chilensis (red algae)	Aqueous and ethanolic extracts	*Phytophthora cinnamomi*	Jimenez et al. (2011)
Durvillaea antarctica	Crude extracts	Tobacco mosaic virus (TMV) in tobacco leaves	Jimenez et al. (2011)

(continued)

Table 18.4 (continued)

Cyanobacteria	Extract	Plant pathogens	References
Anabaena variabilis RPAN59, *A. oscillarioides* RPAN69	Antifungal	• *Pythium debaryanum* • *Fusarium oxysporum lycopersici* • *F. moniliforme* • *Rhizoctonia solani*	Chaudhary et al. (2012)
Anabaena variabilis, *S. platensis*, *Synechococcus elongatus*	Butanol extract	• *Aspergillus niger* • *Alternaria solani*	Tiwari and Kaur (2014)
Nostoc muscorum, *Oscillatoria* sp.	Norharmane and α-isomethyl ionone	*Alternaria porri* (purple blotch of onion)	Abdel-Hafez et al. (2015)

yield. Cyanobacteria are abundant in agricultural areas and, especially in rice-cultivated soils, together with microalgae, are considered as microbial photosynthetic agents of the soil. Because of its important roles in nitrogen fixation, cyanobacteria are inevitable to be used in agriculture to increase vegetative production. Although there are several studies on nitrogen fixation abilities, their ecological roles are not fully understood. It has been determined that cyanobacterial inoculation in agricultural areas provides increased yield even in the presence of high doses of nitrogen fertilizers. In addition to increasing the nitrogen content of plants, cyanobacteria can be used to promote plant growth. For this reason, significant progress has been made in recent years in the development and application of cyanobacterial biofertilizers.

Biosynthesis of phytohormones, polysaccharides, vitamins, amino acids, and peptides is considered crucial for plant growth and development. Microorganisms release these active compounds in the rhizosphere where plant roots can absorb.

Cyanobacterial strains have been identified in studies that support the growth of the plant, usually by greenhouse and pot experiments performed under controlled conditions. New studies are needed to try cyanobacterial strains in field conditions. This chapter is expected to shed light on the work to be done in the application of cyanobacteria to agricultural fields.

References

Abdel-Hafez SI, Abo-Elyousr KA, Abdel-Rahim IR (2015) Fungicidal activity of extracellular products of cyanobacteria against *Alternaria porri*. Eur J Plant Pathol 50:239–245

Abdel-Raouf N, Al-Homaidan AA, Ibraheem IBM (2012) Agricultural importance of algae. Afr J Biotechnol 11:11648–11655

Abo-Shady AM, Al-ghaffar BA, Rahhal MMH, Abd-El Monem HA (2007) Biological control of faba bean pathogenic fungi by three cyanobacterial filtrates. Pak J Biol Sci 10:3029–3038

Adesemoye AO, Kloepper JW (2009) Plant–microbes interactions in enhanced fertilizer-use efficiency. Appl Microbiol Biotechnol 85:1–12

Aguiar R, Fiore MF, Franco MW, Ventrella MC, Lorenzi AS, Vanetti CA, Alfenas AC (2008) A novel epiphytic cyanobacterial species from the genus *Brasilonema causin* damage to *Eucalyptus* leaves. J Phycol 44:1322–1334
Ahmed M, Stal LJ, Hasnain S (2010) Association of non-heterocystous cyanobacteria with crop plants. Plant Soil 336:363–375
Ansari RA, Mahmood I (2017) Optimization of organic and bio-organic fertilizers on soil properties and growth of pigeon pea. Sci Hortic 226:1–9
Bakır Ö (2020) Abiotic stress-related miRNAs in Triticeae. Atatürk Univ J Agric Faculty 51:207–218
Balachandran S, Osmond CB, Daley PF (1994) Diagnosis of the earliest strain-specific interactions between tobacco mosaic virus and chloroplasts of tobacco leaves in vivo by means of chlorophyll fluorescence imaging. Plant Physiol 104:1059–1065
Berendsen RL, Pieterse CMJ, Bakker P (2012) The rhizosphere microbiome and plant health. Trends Plant Sci 17:478–486
Berger S, Benediktyová Z, Matous K et al (2007) Visualization of dynamics of plant-pathogen interaction by novel combination of chlorophyll fluorescence imaging and statistical analysis: differential effects of virulent and avirulent strains of P. syringae and of oxylipins on A. thaliana. J Exp Bot 58:797–806
Biondi N, Piccardi R, Margheri MC, Rodolfi L, Smith GD, Tredici MR (2004) Evaluation of *Nostoc* strain ATCC 53789 as a potential source of natural pesticides. Appl Environ Microbiol 70:3313–3320
Bonfig KB, Schreiber U, Gabler A et al (2006) Infection with virulent and avirulent P. syringae strains differentially affects photosynthesis and sink metabolism in Arabidopsis leaves. Planta 225:1–12
Boopathi T, Balamurugna V, Gopinath S, Sundararaman M (2013) Characterization of IAA production by the mangrove cyanobacterium *Phormidium sp.* MI405019 and its influence on tobacco seed germination and organogenesis. J Plant Growth 32:758–766
Borowitzka MA (1995) Microalgae as source of pharmaceuticals and other biologically active compounds. J Appl Phycol 7:3–15
Büyük İ, Soydam-Aydın S, Aras S (2012) Molecular responses of plants to stress conditions. Turk Hij Den Biyol Derg 69:97–110
Çakmakçı R (2014) Mikrobiyal gübre olarak kullanılabilecek mikroorganizmaların etki mekanizmaları ve özellikleri. In: Mikrobiyal Gübre çalıştayı, pp 5–18 (in Turkish)
Chatterjee A, Singh S, Agrawal C, Yadav S, Rai R, Rai L (2017) Role of algae as a biofertilizer. In: Algal green chemistry. Elsevier, Amsterdam, pp 189–200. https://doi.org/10.1016/b978-0-444-63784-0.00010-2
Chaudhary V, Prasanna R, Nain L, Dubey SC, Gupta V, Singh R, Bhatnagar AK (2012) Bioefficacy of novel cyanobacteria-amended formulations in suppressing damping off disease in tomato seedlings. World J Microbiol Biotechnol 28:3301–3310
Chen J, Zhang H, Feng M et al (2016) Transcriptome analysis of woodland strawberry (Fragaria vesca) response to the infection by strawberry vein banding virus (SVBV). Virol J 13:128–137
Chittora D, Meena M, Barupal T, Swapnil P, Sharma K (2020) Cyanobacteria as a source of biofertilizers for sustainable agriculture. Biochem Biophys Rep 22:100737
Cho WK, Lian S, Kim S-M et al (2015) Time-course RNA-Seq analysis reveals transcriptional changes in rice plants triggered by rice stripe virus infection. PLoS One 10:e0136736
Chun-Li W, Shiuan-Yuh C, Chiu-Chung Y (2014) Present situation and future perspective of biofertilizer for environmentally friendly agriculture. Annu Rep 21:1–5
Cohen MF, Meeks JC (1997) A hormogonium regulating locus, hrmUA, of the cyanobacterium Nostoc punctiforme strain ATCC 29133 and its response to an extract of a symbiotic plant partner Anthoceros punctatus. Mol Plant-Microbe Interact 10:280–289
de Torres Zabala M, Littlejohn G, Jayaraman S et al (2015) Chloroplasts play a central role in plant defence and are targeted by pathogen effectors. Nat Plants 1:1–10
DEQ (n.d.). https://www.deq.ok.gov. Accessed June 2020

Elad Y, Cytryn E, Harel YM, Lew B, Graber ER (2011) The biochar effect: plant resistance to biotic stresses. Phytopathol Mediterr 50:335–349
Elsheikh EA, Elzidany AA (1997) Effects of rhizobium inoculation, organic and chemical fertilizers on yield and physical properties of faba bean seeds. Plant Foods Hum Nutr 51:137–144
Eşitken A, Karlıdağ H, Ercişli S, Turan M, Şahin F (2003a) The effect of spraying a growth promoting bacterium on the yield, growth and nutrient element composition of leaves of apricot (Prunus armeniaca L. cv. Hacihaliloglu). Aust J Agric Res 54:377–380
Eşitken A, Ercişli S, Şevik İ, Şahin F (2003b) Effect of indole-3-butyric acid and different strains of agrobacterium rubi on adventive root formation from softwood and semi-hardwood wild sour cherry cuttings. Turk J Agric For 27:37–42
FAO (2006) Plant nutrition for food security. A guide for integrated nutrient management. FAO Fertilizer and Plant Nutrition Bulletin, Rome
Gull A, Lone AA, Wani NUI (2019) Biotic and abiotic stresses in plants. In: Abiotic and biotic stress in plants. IntechOpen, Rijeka
Gupta V, Ratha SC, Sood A, Chaudhary V, Prasanna R (2013) New insight into the biodiversity and applications of cyanobacteria (blue-green algae)-prospects and challenges. Algal Res 2:79–97
Hagmann L, Juttner F (1996) Fischerllin: a novel with the fungicide Diathane M45 on the control of photosystem II inhibiting allelochemical of the chocolate spot on leaf and pods spot on horse cyanobacterium *Fischerella muscicola* with beans. Agric Res Rev Cairo 53:123–134
Halitschke R, Hamilton JG, Kessler A (2011) Herbivore-specific elicitation of photosynthesis by mirid bug salivary secretions in the wild tobacco Nicotiana attenuata. New Phytol 191:528–535
Herbers K, Takahata Y, Melzer M et al (2000) Regulation of carbohydrate partitioning during the interaction of potato virus Y with tobacco. Mol Plant Pathol 1:51–59
Hui D, Iqbal J, Lehmann K et al (2003) Molecular interactions between the specialist herbivore Manduca sexta (lepidoptera, sphingidae) and its natural host Nicotiana attenuata: V. microarray analysis and further characterization of large-scale changes in herbivore-induced mRNAs. Plant Physiol 131:1877–1893
Hussain A, Hamayun M, Shah ST (2013) Root colonization and phytostimulation by phytohormones producing entophytic *Nostoc sp.* AH-12. Curr Microbiol 67:624–630
Itelima JU, Bang WJ, Onyimba IA, Oj E (2018) A review: biofertilizer; a key player in enhancing soil fertility and crop productivity. J Microbiol Biotechnol Rep 2:22–28
Jaiswal A, Das K, Koli DK, Pabbi S (2018) Characterization of cyanobacteria for IAA and siderophore production and their effect on rice seed germination. Int J Curr Microbiol Appl Sci 7:5212–5222
Jaki B, Zerbe O, Heilmann J, Sticher O (2001) Two novel cyclic peptides with antifungal activity from the cyanobacterium *Tolypothrix byssoidea* (EAWAG 195). J Nat Prod 64:154–158
Jimenez E, Dorta F, Medina C, Ramírez A, Ramírez I, Pena-Cortes H (2011) Anti-phytopathogenic activities of macro-algae extracts. Mar Drugs 9:739–756
Kanwar P, Jha G (2019) Alterations in plant sugar metabolism: signatory of pathogen attack. Planta 249:305–318
Kehr J, Picchi DG, Dittmann E (2011) Natural product biosynthesis in cyanobacteria: a treasure trove of unique enzymes. Beilstein J Org Chem 7:1622–1635
Khan MIR, Syeed S, Nazar R, Anjum NA (2012) An insight into the role of salicylic acid and jasmonic acid in salt stress tolerance. In: Phytohormones and abiotic stress tolerance in plants. Springer, Berlin, p 300
Khosro M, Yosef S (2012) Bacterial biofertilizers for sustainable crop production: a review. J Agric Biol Sci 7:307–316
Kocal N, Sonnewald U, Sonnewald S (2008) Cell wall-bound invertase limits sucrose export and is involved in symptom development and inhibition of photosynthesis during compatible interaction between tomato and Xanthomonas campestris pv vesicatoria. Plant Physiol 148:1523–1536
Koeck M, Hardham AR, Dodds PN (2011) The role of effectors of biotrophic and hemibiotrophic fungi in infection. Cell Microbiol 13:1849–1857

Küçük Ç, Sezen G (2019) Cyanobacteria that promote that plant growth and metabolites. Comm J Biol 3:117–123

Kumar G, Teli B, Mukherjee A, Bajpai R, Sarma BK (2019) Secondary metabolites from cyanobacteria: a potential source for plant growth promotion and disease management. In: Secondary metabolites of plant growth promoting rhizomicroorganisms. Springer, Singapore, pp 239–252

Lichtenhaler HK (1996) Vegetation stress: an introduction to the stress concept in plants. J Plant Physiol 148:4–14

Lohaus G, Heldt HW, Osmond CB (2000) Infection with phloem limited Abutilon mosaic virus causes localized carbohydrate accumulation in leaves of Abutilon striatum: relationships to symptom development and effects on chlorophyll fluorescence quenching during photosynthetic induction. Plant Biol 2:161–167

Magnuson A (2019) Heterocyst thylakoid bioenergetics. Life 9:13

Manjunath M, Prasanna R, Nain L, Dureja P, Singh R, Kumar A, Kaushik BD (2010) Biocontrol potential of cyanobacterial metabolites against damping off disease caused by Pythium aphanidermatum in solanaceous vegetables. Arch Phytopathol Plant Prot 43:666–677

Martin K, Singh J, Hill JH et al (2016) Dynamic transcriptome profiling of bean common mosaic virus (BCMV) infection in common bean (Phaseolus vulgaris L.). BMC Genomics 17:613–619

Mendes R, Garbeva P, Raaijmakers JM (2013) The rhizosphere microbiome: significance of plant beneficial, plant pathogenic, and human pathogenic microorganisms. FEMS Microbiol Rev 37:634–663

Mishra PK, Sailo JL, Mehta SK (2013) Structural, physiological, and ecological adaptations in cyanobacterial mats under stressful environment. In: Stress biology of cyanobacteria: molecular mechanisms to cellular responses. CRC Press, Boca Raton, pp 352–364

Montero R, Pérez-Bueno ML, Barón M et al (2016) Alterations in primary and secondary metabolism in Vitis vinifera "Malvasía de Banyalbufar" upon infection with grapevine leafroll-associated virus 3. Physiol Plant 157:442–452

Mostafa SM, Abdel El-All AAM, Hussien MY (2009) Bioactivity of algal extracellular byproducts on cercospora leaf spot disease, growth performance and quality of sugar beet. In: 4th conference on recent technologies in agriculture, Faculty of Agriculture, Cairo University

Mundt S, Kreitlow S, Jansen R (2003) Fatty acids with antibacterial activity from the cyanobacterium *Oscillatoria redekei* HUB 051. J Appl Phycol 15(2–3):263–267

Nagarajan M, Maruthanayagam V, Sundararamam M (2011) A review of pharmacological and toxicological potentials of marine cyanobacterial metabolites. J Appl Toxicol 32:153–185

Onaga G, Wydra K (2016) In: Abdurakhmonov IY (ed) Advances in plant tolerance to biotic stresses, plant genomics. IntechOpen, Rijeka. https://doi.org/10.5772/64351

Osman MEH, El-Sheekh MM, El-Naggar AH, Gheda SF (2010) Effect of two species of cyanobacteria as biofertilizers on some metabolic activities, growth, and yield of pea plant. Biol Fert Soils 46:861–875

Pérez-Bueno ML, Ciscato M, VandeVen M et al (2006) Imaging viral infection: studies on Nicotiana benthamiana plants infected with the pepper mild mottle tobamovirus. Photosyn Res 90:111–123

Prasanna R, Babu S, Rana A et al (2013) Evaluating the establishment and agronomic proficiency of cyanobacterial consortia as organic options in wheat–rice cropping sequence. Experimental Agriculture, 49(3), 416–434. https://doi.org/10.1017/S001447971200107X

Prasanna R, Jaiswal P, Nayak S, Sood A, Kaushik BD (2009) Cyanobacterial diversity in the rhizosphere of rice and its ecological significance. Indian J Microbiol 49:89–97

Raja N (2013) Biopesticides and biofertilizers: ecofriendly sources for sustainable agriculture. J Biofertil Biopest 4:112–115

Ralph SG, Yueh H, Friedmann M et al (2006) Conifer defence against insects: microarray gene expression profiling of Sitka spruce (Picea sitchensis) induced by mechanical wounding or feeding by spruce budworms (Choristoneura occidentalis) or white pine weevils (Pissodes strobi) reveals large-scale. Plant Cell Environ 29:1545–1570

Rodriguez AA, Stella AM, Storni MM, Zulpa G, Zaccaro MC (2006) Effects of cyanobacterial extracellular products and gibberellic acid on salinity tolerance in *Oryza sativa* L. Saline Syst 2:7–10

Santos VB, Araújo AS, Leite LF, Nunes LA, Melo WJ (2012) Soil microbial biomass and organic matter fractions during transition from conventional to organic farming systems. Geoderma 170:227–231

Shan X, Yan J, Xie D (2012) Comparison of phytohormone signaling mechanisms. Curr Opin Plant Biol 15:84–91

Shevela D, Pishchalnikov RY, Eichacker LA (2013) Oxygenic photosynthesis in cyanobacteria. In: Stress biology of cyanobacteria: molecular mechanisms to cellular responses. CRC Press, Boca Raton, pp 3–40

Shukla N, Yadav R, Kaur P et al (2017) Transcriptome analysis of root-knot nematode (Meloidogyne incognita)-infected tomato (Solanum lycopersicum) roots reveals complex gene expression profiles and metabolic networks of both host and nematode during susceptible and resistance responses. Mol Plant Pathol 19:615–633

Singh S (2014) A review on possible elicitor molecules of cyanobacteria: their role in improving plant growth and providing, tolerance against biotic abiotic stress. J Appl Microbiol 7:1–19

Singh JS, Kumar A, Rai AN, Singh DP (2016) Cyanobacteria: a precious bioresource in agriculture, ecosystem and environmental sustainability. Front Microbiol 7:1–19

Sinha RK, Valani D, Chauhan K, Agarwal S (2010) Embarking on a second green revolution for sustainable agriculture by vermiculture biotechnology using earthworms: reviving the dreams of Sir Charles Darwin. J Agric Biotechnol Sustain Dev 2:113–128

Sokolova MG, Akimova GP, Vaishlya OB (2011) Effect of phytohormones synthesized by rhizosphere bacteria on plants. Appl Biochem Microbiol 47:274–278

Stal JL (2013) Environmental factors regulating nitrogen fixation in heterocystous and non-heterocystous cyanobacteria. In: Stress biology of cyanobacteria: molecular mechanisms to cellular responses. CRC Press, Boca Raton, pp 291–306

Suzuki N, Rivero RM, Shulaev V, Blumwald E, Mittler R (2014) Abiotic and biotic stress combinations. New Phytol 203:32–43

Takaichi S, Maoka T, Mochimaru M (2009) Unique carotenoids in the terrestrial cyanobacterium *Nostoc commune* NIES-24: 2-hydroxymyxol 20-fucoside, nostoxanthin and canthaxanthin. Curr Microbiol 59:413–419

Tang JY, Zielinski RE, Zangerl AR et al (2006) The differential effects of herbivory by first and fourth instars of Trichoplusia ni (Lepidoptera: Noctuidae) on photosynthesis in Arabidopsis thaliana. J Exp Bot 57:527–536

Tecsi LI, Smith AM, Maule AJ, Leegood RC (1996) A spatial analysis of physiological changes associated with infection of cotyledons of marrow plants with cucumber mosaic virus. Plant Physiol 111:975–985

Tiwari A, Kaur A (2014) Allelopathic impact of cyanobacteria on pathogenic fungi. Int J Pure Appl Biosci 2(3):63–70

Tvorogova VY, Osipova MA, Doduyeva IY, Lutova LA (2013) Interactions between transcription factors and phytohormones in the regulation of plant meristem activity. Russ J Genet Appl Res 3:325–337

Vega A, Gutiérrez RA, Peña-Neira A et al (2011) Compatible GLRaV-3 viral infections affect berry ripening decreasing sugar accumulation and anthocyanin biosynthesis in Vitis vinifera. Plant Mol Biol 77:261–274

Verma S, Nizam S, Verma PK (2013) Biotic and abiotic stress signalling in plants. In: Sarwat M, Ahmad A, Abdin MZ (eds) Stress signaling in plants: genomics and proteomics perspective, vol 1. Springer, Berlin, pp 25–49

Vessey JK (2003) Plant growth promoting Rhizobacteria as bio-fertilizers. Plant Soil 225:571–586

Whitton BA, Potts M (2012) Introduction to the cyanobacteria. In: Ecology of cyanobacteria II. Springer, Dordrecht, pp 1–13

Willis BF, Rodrigues BF, Harris PJC (2013) The ecology of arbuscular mycorrhizal fungi. Crit Rev Plant Sci 32:1–20

Yadav S, Rai S, Rai R, Shankar A, Singh S, Rai LCR (2017) Cyanobacteria: role in agriculture, environmental sustainability, biotechnological potential and Agroecological impact. In: Singh DP, Singh HB, Prabha R (eds) Plant-microbe interactions in agro-ecological perspectives, Microbial interactions and agro-ecological impacts, vol 2. Springer, Berlin, p 277

Yamaguchi Y, Huffaker A (2011) Endogenous peptide elicitors in higher plants. Curr Opin Plant Biol 14:351–357

Youssef MMA, Eissa MFM (2014) Biofertilizers and their role in management of plant parasitic nematodes. A review. J Biotechnol Pharm Res 5:1–6

Zakeel MCM, Safeena MIS (2019) Biofilmed biofertilizer for sustainable agriculture. In: Plant health under biotic stress. Springer, Singapore, pp 65–82

Zangerl AR, Hamilton JG, Miller TJ et al (2002) Impact of folivory on photosynthesis is greater than the sum of its holes. Proc Natl Acad Sci U S A 99:1088–1091

Zehr JP (2011) Nitrogen fixation by marine cyanobacteria. Trends Microbiol 19:162–171

Zhao D, You Y, Fan H et al (2018) The role of sugar transporter genes during early infection by root-knot nematodes. Int J Mol Sci 19:302–310

Zou J, Rodriguez-Zas S, Aldea M et al (2005) Expression profiling soybean response to Pseudomonas syringae reveals new defense-related genes and rapid HR-specific downregulation of photosynthesis. Mol Plant-Microbe Interact 18:1161–1174

Chapter 19
Microorganism: A Potent Biological Tool to Combat Insects and Herbivores

Syed Tanveer Shah, Abdul Basit, Muhammad Sajid, and Heba I. Mohamed

Contents

S. T. Shah · A. Basit (✉) · M. Sajid
Department of Horticulture, Faculty of Crop Production Sciences, The University of Agriculture, Peshawar, Pakistan

H. I. Mohamed
Department of Biological and Geological Sciences, Faculty of Education, Ain Shams University, Cairo, Egypt

© The Author(s), under exclusive license to Springer Nature Switzerland AG 2021

H. I. Mohamed et al. (eds.), *Plant Growth-Promoting Microbes for Sustainable Biotic and Abiotic Stress Management*,
https://doi.org/10.1007/978-3-030-66587-6_19

1 Introduction

Insects and herbivores are the diverse living organism with millions of species around the globe (Behmer 2009). These insects and herbivores require an adequate amount of food and nutrients to fulfil their need to live (Wetzel et al. 2016). These food resources help them grow, develop and reproduce and continue their race (Moore et al. 2014). For such purpose, these organisms require the host to fulfil their need. Plants are the major source that provides nutrients and food supply to these insects and herbivores. The search for host plants greatly depends on their plant traits which help the insects and herbivores identify them. Therefore, insect/ herbivore-plant interaction needs to be explored (Bruce et al. 2005).

Plant-associated signals are helpful for herbivores to gather information related to plant location and defensive strategies and also identify nutritional quality during the process of food-seeking (Hassani et al. 2018). These signals (from sensory modalities) play a vital role to identify the initial location of the plants rather to individual plants or tissues during the process of host-plant selection (Biere and Bennett 2013). One of the criteria that help insects locate plants over long spaces is visual signals (Reeves 2011). On contrary, insects and herbivores use other signals like taste, smell, etc. that require plant contact which identifies suitable feeding tissues (Pan et al. 2015). These signals vary with varying insects and herbivores, but there are certain useful signals, i.e. soil environment, nocturnal, diurnal or crepuscular activity, which are common among different insects and herbivores to identify host plants (Bruce and Pickett 2011).

Plants live with insects and herbivores together in an ecosystem since very long from 350 million years. When they are living together so insects and herbivores feed these plants, in response plants also have developed a variety of defensive strategies ranging from morphological to biochemical defence to restrict these insects and herbivores to completely vanish the plants from the earth (Howe and Jander 2008). These strategies can identify nonselfed signals or molecules as nonfunctional cell same like animals and humans activating the immune system of plants against these herbivores and insects (War et al. 2013a, b). These defensive strategies (morphological/mechanical and biochemical) (War et al. 2012) involve the increase in latex deposition, thorns, spines, sclerophylly, thicker leaves, hairs, trichomes, etc.; toxic chemical production like alkaloids, terpenoids, quinones and certain secondary metabolites (anthocyanins, phenols); and proteins in plants, respectively, to delay or kill the development and growth of herbivores and insects (Hanley et al. 2007). Moreover, plants not only directly but also indirectly affect the biology of insects and herbivores. The direct effect may be the preference of host plant or reproductive success and survival where the strategies indirectly include the recruitment of another organism (which are enemies of those insects/pests and herbivores) and microorganisms which counter affect the harm created by these insects and herbivores. The indirect defence also includes blending of herbivore-induced plant volatiles (HIPVs) and extrafloral nectar that attracts enemy which increase the effectiveness in resistance against these insects and herbivores (Arimura et al. 2009). Induced resistance against the attack of insects and herbivores makes the plants phenotypically plastic and tissues less nutritious resulting to reduced attractive food choice and practically revolting insects/pests and herbivores. The induced resistance not only has the above-mentioned advantage; rather this resistance can be transferred to the next generation (transgenerational effect), making the plant more vigorous and minimize insect/pest or herbivores attack in the progeny (Karban 2011).

Furthermore, an additional layer of complexity in aging of insects/pests and herbivores arises from plant-herbivore interaction with microorganism (Hassani et al. 2018). Microorganism may be beneficial or pathogenic that indirectly affects the selection process by insects and herbivores. Additionally these microbes also help modify plant traits (Biere and Bennett 2013). Interactions between plants and microorganisms are universal. The role of plant-associated microbes are well understood modifying plant phenotypes and reshaping the interaction between insects/herbivores and plants (Biere and Bennett 2013). These microorganisms adjust plant-produced signals which greatly affect the foraging behaviour and oviposition of insects and herbivores (Kariyat et al. 2013). There are various visual, olfactory and gustatory signals which are associated with a beneficial and pathogenic microorganism that greatly affects the insect/herbivore-plant interaction (Bruce et al. 2005). This chapter emphasizes on the forging behaviour of herbivores and insects for the selection of host plants and their interaction with microorganism.

2 Ecological and Evolutionary Pattern of Host-Microbe Interaction.

Plants are chief facilitator between related microbe and insect interaction. The interaction of microorganism with their host ranged from parasitic to a mutualistic relationship; hence, it is engaged with insects, herbivores and plants in either a short term or an everlasting relationship. There are several factors which greatly affect the nature of this interaction which includes environmental and ecological effects of microbial communities (Sugio et al. 2014). For example, according to De Vries et al. (2004), trips are mostly infected by *Enterobacter Erwinia* sp. which has a beneficial effect to the host on which trips feed on. Additionally, various parasites have been reported to attack plants and insect communities; thus, under various conditions, the interaction of parasites, however, evolves to a useful relationship. The evolutionary period has a great role to change the parasitic to a mutualistic relationship, but this change may also be fast within 20 years (Sugio et al. 2014). For such purpose, Weeks et al. (2007) reported a 10% fertility increase in *Drosophila simulans* population by *Wolbachia* over uninfected females (due to endosymbiont genome).

A surging type of effect has been observed between interactions of two different plant species which greatly changes their modelling and community structure. The effect of the modelling of community structure can be seen indirectly within plant-associated insect communities (Colman et al. 2012). Therefore, there is not only a two-way interaction observed between the insect host and plants but also an interaction of three ways such as microbes, insects and plants which has great evolutionary and co-evolutionary effects (Biere and Bennett 2013). This three-way interaction was greatly explained by Jones et al. (2013). They reported that the interaction of aphids and barley is reliant on the different genotypes of rhizosphere bacteria and also the species that interact. Two basic pathways are involved in plant-insect interaction mediated by microbes. According to the first pathway, the suitability for the food resource for herbivorous insect by host plants is greatly affected by symbiont, pathogen and microbe interaction. These interactions greatly alter plant abundance, phenology, morphology, biochemistry, physiology and other aspects which affect herbivore population and community structure. Example for such type of interaction includes the role of some phytopathogens in inducing defences against herbivores. The second pathway includes insect microbial pathogen and symbiont interaction that affects the ability of their insect hosts to explore food plants. This influence greatly affects the performance and specialization of food plants. For instance, not only the plant sap-feeding lifestyle has evolved due to the acquirement of microbial nutritional endosymbionts but also exerts a greater shift from the current host to other crops. Ecological opportunities are brought in the hosts due to microbial activities which is a great innovation in them. Bacterial taxa change the sap-feeding and colony-making ability of insects in almost all plant species which is brought about by symbiotic associations between microbe, plants and insects. Sometimes one symbiont is required to fulfil the nutritional requirement of microbes, e.g. hemipteran group hosts specific bacteria (*Buchnera* for aphids and *Carsonella* for psyllids) (Hansen and Moran 2014), while some required two symbionts such as conifer aphid that acts as a host for *Cinara cedri* and *Serratia symbiotica* (Lamelas et al. 2011). Similarly, a leafhopper species *Homalodisca vitripennis* associates with *Sulcia muelleri* and *Baumannia cicadellinicola* (Toju and Fukatsu 2011). Genotype of the symbionts is the criteria to determine the performance of host plant for some herbivorous insects. For example, modification of pest status on soybean between two stinkbug species of genus *Megacopta* is due to the exchange of symbiont *Candidatus Ishikawaella capsulata* (Hosokawa et al. 2007). The driven forces that help insect-microbe, plant-microbe and plant-insect-microbe interaction are usually the transmission patterns of microbial communities. Regular infection from the ecosystem helps their hosts to acquire plant symbionts of which are mostly facultative (Sugio et al. 2014). For example, nitrogen-fixing rhizobia from the soil frequently attack new legume plants. Leaf nodules of certain plant species of genus Rubiaceae are inhibited by a bacteria, i.e. *Burkholderia* sp., in obligatory symbiosis. Symbiont transmission maintains symbioses through host generations and has a pivotal role in their evolution. Two fundamentally different modes of transmission can be distinguished: horizontal (that is, from an environmental, free-living symbiont source) and vertical (that is, inheritance of the symbiont from the mother or, more rarely, from both parents). However, there is great variation, and transmission can also be mixed, involving both vertical and horizontal transfers from the environment and intraspecific or interspecific host switching (Lemaire et al. 2012).

There is a greater role of plant-mediated microbes in structuring communities of herbivores. There is a greater effect of microbes that induced changes in plant traits on performance and behaviour of individual herbivore or their population. A bottom-up effect on the above-ground insect herbivores is created by below-ground microbes. This is done through increasing the nutritional quality and defence mechanism and also through an alternation of plant abundance, thus exerting a substantial influence on insects individually or at a community level (Sugio et al. 2014).

Another important evolutionary significance of microbe, insect and plant interaction is the transfer of lateral gene in between the host and symbionts or insects and microbes which are plant-mediated (Hansen and Moran 2014). This leads to a variation of traits either in a direct way or indirectly in various stress responses or in insect and plant nutrition. Incorporation of gene, *Bacillus mannanase* in the genome *Hypothenemus hampei* (an important coffee plant pest), allows coffee berries to exploit a new ecological niche (Acuna et al. 2012). The cellulase gene responsible for the degradation process is reported to originate from microbes in both plant nematodes and termites (Todaka et al. 2010). There is a possibility of gene transfer between plant and insect microbial companion. Furthermore, the limited proof has been reported for lateral gene transfer in plants' nuclear genome and chloroplast that alters insect traits or host plants, which may alter the microbe and host association (Richardson and Palmer 2007).

3 Microbial Diversity Associated with Herbivorous Insects and Plants

Advanced technologies focused on the overabundance of earlier unseen microbial associates and the well-studied microbe, insect and plant symbiosis. Many aspects of host ecology are mainly affected by microbial communities, but these do not contribute to host survival or reproduction. These aspects include utilization of plant, climate change response and defence against enemies (Oliver et al. 2010). Plant and insect microbiomes show microbial communities that are internally interacted with their hosts. The rising and fast growth of next-generation sequencing (NGS) technologies benefits the assessment of microbial communities related to plants and insects. This permits recognizing the taxonomic diversity of microbial communities in different environmental conditions and ecosystems. PCR-amplified taxonomic genes, whole genome sequencing and whole transcriptomics have been realized through NGS technologies for bacteria (16S rRNA gene) and fungi (16S rRNA gene). These whole genomic datasets help identify the existence and density of microbial communities relating to their host in a specific environment if combined with a specific database and modified bioinformatics. This (datasets of transcription) will also indirectly help assess the biological functions of microbes (Sugio et al. 2014).

There is a dire need to collect information about some important groups of herbivorous insects as the insect-associated bacterial diversity is fast growing. Although herbivorous insects have protected different microbial communities with limited diversity, still, it is let by some taxa (Jones et al. 2013; Colman et al. 2012). Not more than 30 different operational taxonomic units (OTUs) or taxa on the average basis are by protected by *Drosophila* species; in contrast, mammalian gut can accommodate more than 1000 taxa (Chandler et al. 2011a, b). Recent researches support the hypothesis of feeding habits and evolutionary history of insect microbial communities and conclude low diversity of bacterial species (an average of 10–15 OTUs per insect). For instance, xylophagous leaf feeders can harbour the communities of bacteria about 38 and 103 OTUs as per sample of communities of bacteria (Russell et al. 2013), while insects which are sap-feeding, e.g. whiteflies, aphids and psyllids, are having the poorest microbial diversity even not more than 3–7 OTUs per sample (Jing et al. 2014). Primary and secondary symbionts especially related to different sap feeder groups dominate most individuals of these insects.

Insect-derived structure hosts obligatory symbionts, while facultative symbionts can reside in various positions in the host like haemolymph and sheath cells, etc. Different bacterial associates can reside at different positions of insect gut and play direct or indirect roles in nutrition process (Dillon and Dillon 2004), while different other groups, i.e. *Spiroplasma* and *Wolbachia*, can alter host sex ratio by infecting the insect reproductive organs (Engelstadter and Hurst 2009) and in salivary glands of insect hosts (Kaiser et al. 2010). Plants can also host different diversity of microbes, i.e. pathogens, beneficial associates and commensals as compared to insects as well. The most common and well-established mutualistic association in plants is the mycorrhizal fungi and nitrogen-fixing bacteria. Apart from this, there are certain other bacteria and endophytic fungi that benefit the host plants with different functional biological ranges (Bulgarelli et al. 2013).

4 Mechanism of Insect-Plant Interaction Under the Influence of Microbes

4.1 Plant Hormones Mediating Responses of Plant Defences Against Herbivorous Insects

Hormones are used to facilitate the defence mechanism of plants against different insect herbivores. Plants also produce various chemicals and volatiles to protect themself from the attack of insect/pests. They also produce (when needed) such molecules that are toxic to insects. Plant defense against herbivore attack involves many signal transduction pathways that are mediated by a network of phytohormones. Plant hormones play a critical role in regulating plant growth, development, and defense mechanisms (Wu and Baldwin 2010). A number of plant hormones have been implicated in intraand inter-plant communication in plants damaged by herbivores. Most of the plant defense responses against insects are activated by signaltransduction pathways mediated by JA, SA, and ethylene (Zarate et al. 2007). Specific sets of defense related genes are activated by these pathways upon wounding or by insect feeding. These hormones may act individually, synergistically or antagonistically, depending upon the attacker. It also varies from species to species. Thus, manipulations in hormonal balance or signalling process manipulate the plant-insect interactions (Zarate et al. 2007).

4.2 Insect Symbionts Can Counteract Plant Defences

The consequences of plant-insect interactions are altered by insect-associated microbes which are location dependent. During the plant-insect-symbiont interaction, insect symbiont lives within an insect body, neutralizing plant defence responses by producing different enzymes required by insects. Due to climate change in North America, *Dendroctonus ponderosae* (mountain pine beetle) colonize and kill conifers, thus extending the range of its host trees (Raffa et al. 2013). One of the recent studies using metagenomic study on mountain pine beetle revealed that *D. ponderosae* live in association with those microbial communities which produce terpene-degrading genes as compared to microbial communities using plant biomass processing. Terpenoids are usually noxious to beetle and are being synthesized by pine trees (Adams et al. 2013). Moreover, *D. ponderosae*-associated bacteria also synthesize diterpene acids and monoterpenes showing the role of

D. ponderosae degrading terpenoid-based defences in plants (Boone et al. 2013). The western corn rootworm (WCR, Diabrotica virgifera virgifera LeConte) (Coleoptera: Chrysomelidae) is a major corn pest that has been controlled via annual rotation between corn (Zea mays) and nonhost soybean (Glycine max) in the United States. This practice selected for a "rotation-resistant" variant (RRWCR) with reduced ovipositional fidelity to cornfields. When in soybean fields, RR-WCRs also exhibit an elevated tolerance of antiherbivory defenses (i.e., cysteine protease inhibitors) expressed in soybean foliage. Here we show that gut bacterial microbiota is an important factor facilitating this corn specialist's (WCR's) physiological adaptation to brief soybean herbivory. Comparisons of gut microbiota between RR- and wild-type WCR (WT-WCR) revealed concomitant shifts in bacterial community structure with host adaptation to soybean diets. Antibiotic suppression of gut bacteria significantly reduced RR-WCR tolerance of soybean herbivory to the level of WT-WCR, whereas WT-WCR were unaffected. Our findings demonstrate that gut bacteria help to facilitate rapid adaptation of insects in managed ecosystems (Chu et al. 2013).

4.3 Insects Symbionts Can Change the Physio-Morphology of Plants

During the interaction of plants and insects, various symbionts of insects increase the supplement of nutrition to the insects, host in nature (Kaiser et al. 2010). Insects meet their nutritional requirement by altering the source-sink relationship and to avoid plant tissues from being senesced. They also protect themselves from the seasonal variation for nutrient provision (Giron and Huguet 2011). Several galling insects and leaf minor system are a clear example of such type of phenomena. *Phyllonorycter blancardella* (herbivorous leaf-mining moth) promote "green islands" (prompt green and active area in the yellow senesced leaf). This is made possible through altering the profile of plant cytokinins (Giron et al. 2007), manipulating protein-sugar content and overcoming the need for food which is unavailable under the condition of senescence. Induction of green islands and control of nutritional imbalances are made possible in the presence of symbiotic bacteria (Body et al. 2013; Kaiser et al. 2010). Not only cytokinins may be involved in promoting growth and source-sink relationship and reducing leaf senescence but are also involved to improve plant defence by mediating JA signalling (Erb et al. 2012). Thus, it is clear from the discussion that insect symbionts play a key role in the expression of genes and hormonal balance within the host plant that causes morphological and metabolic variations in plants (Giron and Glevarec 2014).

5 Biological Control Using Invertebrates and Microorganisms

The use of a population of one organism (living) to suppress the population of another organism (living) is known as biological control. The use of biological control is since long about 2000 years, but the contemporary use is underway from the nineteenth century (van Lenteren and Godfray 2005). There are four known classes of biological control, namely, natural, conservative, classical and augmentative. According to the Millennium Ecosystem Assessment, when a pest population is suppressed by a natural beneficial organism, it is known as a natural control. In conservative biological control, natural organism controls plant diseases in soil as well as

Table 19.1 Worldwide use of major augmentative biological control programs (after van Lenteren and Bueno 2003), with updates and supported with references when large differences in areas under control existed between 2003 and 2016

Natural enemy	Pest and crop	Area under control (in ha)
Trichogramma spp.	Lepidopteran pests in vegetables, cereals, cotton	10 million, former USSR
Trichoderma spp.	Soil diseases, various crops	5 million, Brazil, Europe
Trichogramma spp.	Lepidopteran pests in various crops, forests	4 million, China
Cotesia spp.	Sugarcane borers	3.6 million, South America, China
Metarhizium anisopliae	Lepidopteran pests in sugarcane	2 million, Brazil
Trichogramma spp.	Lepidopteran pests in corn, cotton, sugarcane, tobacco	1.5 million, Mexico
Trichogramma spp.	Lepidopteran pests in cereals, cotton, sugarcane, pastures	1.2 million, South America
AgMNPV	Soybean caterpillar in soybean	1 million, Brazil
Beauveria bassiana	Coffee berry borer in coffee, whitefly in several crops	1 million, Brazil
Entomopathogenic fungi	Coffee berry borer in coffee	0.5 million, Colombia
Trichogramma spp.	Lepidopteran pests in cereals and rice	0.3 million, Southeast Asia
Trichogramma spp.	Lepidopteran pests in sugarcane and tomato	0.3 million, Northeast Africa
Predatory mites	Spider mites in greenhouses, fruit orchards, tea and cotton	0.07 million China
Trichogramma spp.	*Ostrinia nubilalis* in corn	0.05 million, Europe
Orgilus sp.	Pine shoot moth, pine plantations	0.05 million, Chile
30 spp. of nat. Enemies	Many pests in greenhouses and interior plantscapes	0.05 million, worldwide
Egg parasitoids	Soybean stinkbugs in soybean	0.03 million, South America
Five spp. of nat. enemies	Lepidoptera, Hemiptera, spider mites in orchards	0.03 million, Europe

crop residues minimizing the infection of pest and pathogen. When natural enemies are collected from the area where the pest has originated and released to invaded pest areas, thereby permanently suppressing pest population is known as classical biological control (Cock et al. 2010). Augmentative biological control (ABC) is usually defined as the release of large-scale additional natural enemies when there are very few natural enemies present to effectively control a pest. Usually, large mass production of additional pests is prepared in the laboratory and growers to buy on a large scale to control pests (van Lenteren 2012). The modern use of ABC was initiated in Russia to control beetles in different crops using *Metarhizium anisopliae* (insect pathogens) by Metchnikoff during the 1880s. Currently, ABC is used in many integrated pest programs to control pest populations in different fruits, vegetables, field crops and greenhouse crops (Table 19.1). ABCs are reported to be economically and environmentally affective against chemical pest control (Cock et al. 2010).

5.1 Availability of Biological Control Agents in the Market

About 170 species of augmentative biological control agents (invertebrates) are used in Europe (Cock et al. 2010). Moreover, about 230 species of invertebrates (biological control agents) worldwide are used to control the pest population, but the list is still not complete (Van Lenteren 2012). There are many reasons for the non-provision of data by manufacturers of natural enemies which include market development, sales volumes and profit margins. A worth of about US$58.46 billion was recorded by global pesticide market in 2015 (Research and Markets 2016). Approximately US$1.7 was obtained from invertebrates and microorganisms (biological agents) in the global market during 2015 which is less than 2% of the pesticide market. The annual growth rate of biological control markets is 15% since 2005 which was 10% before 2005 showing the faster trend of biological markets as compared to synthetic pesticide markets (Dunham 2015). The use of commercial ABC is in garden crops (vegetables and ornamentals) and high-value outdoor crops (vineyards, strawberry, etc.), which involve about 15–20 different natural enemy species, contributing about 80% of invertebrate biological control agent markets. The remaining 20% of the natural enemy market is from a relatively cheap and simple application of biological control programs (use only biological control agent), e.g. *Cotesia* spp., against lepidopterans in sugarcane and *Trichogramma* spp. against lepidopterans in sugarcane and cereals. Predatory mites are reported to immensely contribute to the growth of invertebrate biological control agent markets (Messelink et al. 2014).

The problem of pests rose due to the creation of less stable natural ecosystem and simplified agroecosystems in evolutionary agricultural practices. Enormous losses in agriculture crops (about 40%) are reported in various parts of the world due to the attack of insects/pests mostly termites, grasshoppers, cattle ticks and locusts. Though there is a marginal increase in pesticide use, crop losses remained fairly stable. Developed countries nowadays are shifting strategies for pest management to transgenic plants that produce different resistance traits to fungi, viruses or insects. Farmers are still using some agrochemical pesticides to control insect/pests and diseases in agricultural practices, thus responsible to maintain quantity as well as quality of food globally. The scientists are forced to focus on alternative environmentally safe, cost-effective and reliable strategies due to the overuse of these chemical pesticides resulting in an adverse effect on non-targeted organisms, groundwater pollutions, the resistance of insects to chemicals, deposition on edible food crops and also a negative effect on human health (Chandler et al. 2011a, b).

6 Role of Biopesticide in Integrated Pest Management as a Biocontrol Agent

There is a very high demand of active compounds (biological origin and their synthetic derivatives) to protect crop against traditional pesticides. These synthetic compounds not only have minimized the accessibility and harmful toxicological issues but also improve crop resistance against pests (their survival). There is very much less burden of biologically originated insecticide (biopesticides) on the environment, as they are less noxious, required in very low amount, and decompose quickly. Furthermore, these are confined to specific target and do not affect other organs. There are many biological agents used in insect pest management (IPM) which include fungi, bacteria, virus, protozoans, botanicals (plants or products derived from plants), predator/pathogen system, plant-incorporated protectants (PIPs) and insect pheromones. The primarily and commercialized product developed as bioinsecticide nowadays is from *Bacillus thuringiensis*. Among very large number, a very few biologically originated compounds are commercially used (available and affordable to growers) (Singh et al. 2019a). Hynes and Boyetchko (2006) clarified the term "biopesticide" which is taken in a misleading sense. Biopesticide does not mean that microbes can destroy pest population, rather it means that it suppresses the pest population, and crop produce is not affected due to resistance against the toxic effect of pests. There are several classes of biopesticides: fungi (*B. bassiana*), entomopathogenic bacteria (*Bacillus thuringiensis*) or viruses (*Baculovirus*), protozoa and nematodes (entomopathogenic) as well. *Bacillus thuringiensis* belongs to family *Bacillaceae* that produces a toxin which is used against various classes of insects (Fisher and Garczynski 2012).

6.1 Pesticides Derived from a Microorganism

Biopesticides (derived from bacteria, fungi, viruses and nematodes) are environmentally friendly and do not have any toxic effect of non-targeted pests but very effective against species-specific pests. There are either beneficial or harmful microorganisms

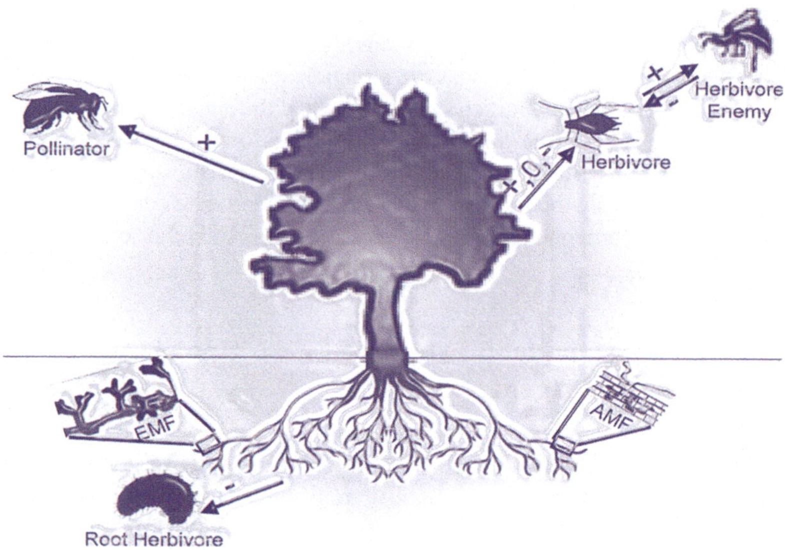

Fig. 19.1 An illustrative representation of insect and plant interaction as effected by mycorrhizal fungi. Roughly describe the *Eucalyptus* genus associating both arbuscular mycorrhizal (AM) and ectomycorrhizal (EM) fungi subjected to both pollination (i.e. bees and various other insects) and insect herbivory (subsequent of aphids and different other insects). Herbivores subjected to enemies on these plants are also linked with the mycorrhizal fungi. Different signs (+,−, 0) put on the arrows show plant effect either immunized with AM or EM

living in close vicinity of plants. There is a potential importance of beneficial microorganism to control pests and improve soil fertility in contrast to harmful microbes which cause losses in crop production. So, there is a need to isolate, test, facilitate and commercialize different types of beneficial microbes in agriculture (Fig. 19.1)

6.2 Biopesticides Derived from Bacteria

One of the cheap and widely used pest bioregulation means is bacterial-based pesticides (Sarwar 2015). There are so many bacterial species with insecticidal properties, but very few have reached to commercialization stage (Table 19.1). About 100 well-renowned *Bacillus thuringiensis* or Bt (well-known microbial pesticide) reported controlling insects/pests especially belonging to Coleoptera (Scarabaeidae or Chrysomelidae, Diptera (Nematocera) and Lepidoptera) (van Frankenhuyzen 2009). It is a spore-forming, gram-positive and facultative bacterium (Jurat-Fuentes and Jackson 2012) and contains potential properties of bio- as well as a chemical pesticide. It is inexpensive and formulated easily; action mode is quick with prolonged shelf life just like synthetic pesticide but is not harmful to the environment. Bt is sunlight sensitive which is only demerit reported and therefore needs to be applied at specific timings. Humans, beneficial organisms, vertebrates and environment are harmed with Bt formulations, but one should be selective and safe while applying Bt formulations on vegetables and fruits (Chandler et al. 2011a, b). *B. thuringiensis* and *B. thuringiensis* subsp. *kurstaki* are worldwidely used to check the attack of leaf rollers and defoliators (lepidopteran insects) in orchards and fields (Glare et al. 2012) especially pests of corns, cotton, legumes, crucifers, solanaceous vegetables and cucurbits (Kroschel and Lacey 2009).

6.3 Biopesticides Derived from Fungi

Mycoinsecticides/mycopesticide (class of microbial insecticides) are natural pathogen derived from entomopathogenic fungi, useful against many pests and Acari in agriculture. One of the many suitable characteristic features of fungi is that they are useful to the non-targeted organisms (predators and parasites of pests and bees) minimizing the risk for the growth and development of earthworms and collembolan (useful organisms), hence acting as a useful biocontrol agent (potential IPM agent and useful to safeguard biodiversity and long-term crop husbandry) (Koike et al. 2011).

Fungi in IPM exploit proper environmental conditions and ecological approaches that suppress the growth of pest (promotion and spread of pathogen within pest) (Lacey et al. 2015). *Metarhizium anisopliae* and *Beauveria bassiana* are the two ascomycetes used as commercial mycopesticide among the so many available commercial mycobiopesticides (*Metarhizium* spp., *Beauveria* spp., *Lecanicillium* spp. and *Isaria fumosorosea*) (Table 19.2). These are applied as mycelium or conidia that sporulate after their application. Luz et al. (2008) reported that *Aedes albopictus* and *A. aegypti* (adult mosquitoes) can be controlled using *M. anisopliae* (insect-pathogenic fungus). Moreover, locust control in Australia and Africa is also possible due to mycopesticide (Chandler et al. 2011a, b).

In recent time, it is also reported that apart from entomopathogenic role of fungi, they also act as endophytes that promote plant growth as the rhizosphere. Other studies also reported increased potentiality of fungus in integrated pest management due to the use of genetic, ecological and functional diversity. Using tools like genetic engineering, information regarding virulence and tolerance to adverse situations will initiate cost-effective applications of mycoinsecticides against pest control in the field of agriculture (Sharma et al. 2020).

6.4 Biopesticides Derived from Viruses

Apart from bacteria and fungi, viruses are also reported to be helpful against various insects/pests that cause severe losses in crops. Virus-based pesticides are effective against lepidopteran pests of rice, cotton, vegetables and plant-chewing insects.

Table 19.2 Bacterial and fungal biopesticides developed to control pest attack on various crop plants

Microorganism species	Type	Target pest
Bacteria	*Bacillus popilliae*	Members of Coleoptera
	Paenibacillus popilliae	Coleoptera: Scarabaeidae: *Popillia japonica*
	Bacillus thuringiensis var. *kurstaki*	Members of Lepidoptera and Coleoptera
	B. thuringiensis var. *aizawai*	Lepidoptera
	B. thuringiensis var. *galleriae*	*Helicoverpa armigera* and *Plutella xylostella*
	B. thuringiensis var. *israelensis*	Diptera: Culicidae, Simuliidae
	B. thuringiensis subspecies japonensis strain Buibui	Coleoptera: Scarabaeidae
	B. thuringiensis subspecies *tenebrionis*	Coleoptera: Chrysomelidae, predominantly
	Leptinotarsa Lysinibacillus sphaericus	Diptera: Culicidae
	Serratia entomophila	*Costelytra zealandica*
	Chromobacterium subtsugae	*Leptinotarsa decemlineata*, Hemiptera, Acarina
Fungi	*Aschersonia aleyrodis*	Hemiptera (Aleyrodidae)
	B. bassiana sensu lato	Acari, Diptera, Lepidoptera, Hemiptera, Isoptera Coleoptera, Diplopoda, Hymenoptera, Lepidoptera, Orthoptera, Siphonaptera, Thysanoptera
	B. bassiana	Coleoptera, Acari, Diptera, Orthoptera, Thysanoptera, Hymenoptera, Hemiptera
	Beauveria brongniartii	Coleoptera (Scarabaeidae)
	Conidiobolus thromboides	Acari Hemiptera, Thysanoptera
	Hirsutella thompsonii	Acari
	Isaria fumosorosea	Acari, Diptera, Coleoptera, Hemiptera, Thysanoptera
	Lagenidium giganteum	Diptera (Culicidae)
	Lecanicillium longisporum	Hemiptera
	Lecanicillium muscarium	Acari, Hemiptera, Thysanoptera
	Metarhizium anisopliae sensu lato	Acari, Blattoidea, Coleoptera, Diptera, Hemiptera, Isoptera, Lepidoptera, Orthoptera *Metarhizium acridum* Orthoptera
	Nomuraea rileyi	Lepidoptera
	Paecilomyces fumosoroseus	Hemiptera

Heliothis zea nucleopolyhedrosis is reported to be the first viral pesticide used against different insects/pests. Among the various entomopathogenic viruses (iridoviruses, poxviruses, nodaviruses baculoviruses (BVs), polydnaviruses, cytoplasmic polyhedrosis viruses, parvoviruses, picorna-like viruses, nucleopolyhedrosis viruses (NPVs), tetraviruses, granuloviruses (GVs), acoviruses and reoviruses), baculovirus (BV) is reported to be the most effective viral pesticide produced commercially (Moscardi et al. 2011) and reported infecting different species of *Heliothis* or *Helicoverpa* genera. HzSNPV (types of baculovirus) are noxious against the pest of sorghum, beans, soybean, tomato and maize. There is a limited use of non-BV viruses, i.e. cyprovirus, tetraviruses, etc., in protecting crops (Sarwar 2015).

7 Plant Attributes That Determine Herbivore

Several evidences are available which clearly indicate the role of plant traits in minimizing the growth, survival and feeding aspects of herbivores. These may be ecological traits (resistance) and evolutionary response (defence). The traits responsible for resistance in plants against herbivores are constitutive (sustainable at a steady level) and inducible (responsible for maintaining physical, allelochemical and/or nutritional traits of plants) (Karban and Myers 1989). Resistance traits can also be classified by tolerance (plant capability to regrow after herbivore attack) and avoidance mechanism (capability of plants to deter herbivore either through physical barriers, probability of being found or quality of plants as food). The following paragraphs shows the avoidance mechanism from herbivore attacks (Zamora et al. 1999).

7.1 Physical Barriers

Herbivores should overcome all the hurdles after searching for a host plant. In response to herbivore, plants also have a countless specialized structure to deter these herbivores which are leaf hairs and trichomes which prevent invertebrates (Bernays and Chapman 1994), while thorns, spines, scales and barbs are effective against mammals (Grubb 1992). Other plants secrete adhesive from glands which is not only effective in repelling small pests but also traps small arthropods which has a carnivory effect on plant. Herbivores also avoid sclerophylly which is a mechanical barrier to their digestive system (Turner 1994). The spine is also another defensive strategy used by plant to avoid the attack of herbivore. Spines negatively affect the performance of herbivores by reducing their rate of consumption (Gowada 1996).

7.2 *Possibility of Existence*

Initially, the herbivores need to find a host plant. In response to herbivore, plant has discovered certain traits through which herbivores are unable to find his host plant. For example, plants occupying free sites from an enemy, showing a good defence system, may lack odour for attracting insect or herbivore and may have damaged tissue and short life span; the production of the edible portion of plant does not synchronize with the presence of herbivore. Plants having longer life cycle are more prone to herbivore than those having a short life span. Another way for plants to protect themselves is "mass flowering or fruiting". In this strategy plants synchronize its production of tissues (flowers, leaves or fruits) as closely as possible which helps overcome the capability of herbivore to consume all the available tissues. In this way time and space effect is being multiplied (Kelly 1994).

7.3 *Quality of Plants as Food*

Herbivores have no problem in habitats where plants are abundant. The potential food for herbivores is the cell wall (cellulose, lignin and hemicellulose) and cytoplasm of plants. About 90 and 65% of the total biomass of trees and grasses, respectively, are cellulose, hemicellulose and lignin. Herbivore animals cannot be producing the decomposing cellulose enzymes and hence cannot consume the abounding food directly. Apart from the cell wall, the plant cytoplasm is rich in proteins, and lipids and starches are abundant in reproductive and photosynthetic tissue. Animals solve plant chemical defences by consuming the plant cytoplasm (Howe and Westley 1988).

Herbivores are not only confined to the available energy but also nutritional quality of plant tissues. Comparing animal and plant tissues shows clear differences. Plant tissues are rich in carbon with a lower quantity of N, P and S compared to animal tissues (Sterner and Hessen 1994). In general, the animal herbivores have nitrogen ten times more than the plants they ingest. There is a marked difference to characterize the boundary line of plants and animals in the biological make-up of resource and the consumer. Hence, herbivores are largely dependent specifically (for female reproduction living) on better quality and rare plants and plants' organ in "green deserts" (Moen et al. 1993). Furthermore, McNaughton (1998) found the distribution of African ungulates is largely associated with a level of minerals (Na, Mg and P), and herbivore density clearly describes selective ingestion and assimilation of essential limiting minerals. Several nitrogenous compounds, i.e. alkaloids and cyanogenic glycosides, may be poisonous (Bentley and Johnson 1992). Therefore, it is worth noted that all the nutritional minerals may not be present in the tissues of the plant; however, they are may be available equally to herbivores. There is an unbalanced chemical composition of herbivore diets which would cause a decrease in the efficiency of herbivores. The consequence of which will be reflected to a decreased production of all the trophic levels of the food chain. Therefore, herbivores are very much selective in searching for food hosts, but plants have also adapted certain strategies to overcome the attack of herbivores (White 1993).

8 Recent Advances in Plant-Herbivore Interaction

Plant-herbivore interaction has a crucial link between food webs and primary production and hence is considered important to understand community dynamics and functioning of ecosystem. They are also considered a backbone for multiple fields within ecology and ecosystem, i.e. co-evolution (Johnson et al. 2015), nutritional ecology (Wetzel et al. 2016), chemical ecology (Schuman and Baldwin 2016) and ecological stoichiometry (Lemoine et al. 2014a, b). Recent technological and statistical advances, i.e. phylogenetic and genetic basis of plant and herbivore interaction (Edger et al. 2015), remote sensing technology (RS) and global positioning system (GPS), statistical advances (Lemoine et al. 2016) and DNA barcoding to explain herbivore diets (Kartzinel et al. 2015), have rapidly increased the importance of this field. Plant-herbivore interaction is now considered a leading driving force to study various aspects of ecology and evolution. This topic highlights the importance of different areas of plant-herbivore interaction which is given as under.

8.1 Plant Defence Theory

One of the foundations laid for the field of ecology, evolution, a theory of co-evolution, chemical ecology and mechanism explaining the success of invasive species (Verhoeven et al. 2009) is plant defence against herbivores. It has been thought from long that success of plant invasion through enemy release hypothesis (enemy loss) is conflicting (Alofs and Jackson 2014), alternatively supported (Heckman et al. 2016), and ambivalent (Heger and Jeschke 2014) from the direct evidence produced in the recent studies which has become a challenge to some key models in these fields. Moreover, increased invasiveness and competitive effect of neighbouring plants are not associated with an enemy release. Based on recent studies findings, it could be observed that one should examine the interactions of released enemies with environmental carriers are whether through the availability of resources or disturbance (Gruntman et al. 2017). It should also be examined that plant-herbivore interactions alone are invasion resistance predictors as compared to the integrity or complexity of the whole food web (Smith-Ramesh et al. 2017). It is thought from long that plant chemical defence is regarded to be the primary source of defence against herbivores, whereas a weaker relationship between chemical defence and herbivory is observed at a community scale across sympatric species (Schuldt et al. 2012). However, recent studies at a comparative and community level re-emphasize the significance of diverse strategic defences against a wide range of community types (Moles et al. 2013) such as traits of life history, structural defences, nutrient quality and redistribution of above- and below-ground chemical defences that control herbivory (Mundim et al. 2017). Endara and Coley (2011) predict a linear reduction in herbivory due to low nutrient quality as compared to Wetzel et al. (2016) who reported that it is not the nutrient quality but variance in nutrient traits that determine herbivory performance. For example, insect performance is due to the high or low nutrient level and nutrient toxicity or nutrient-toxin interactions. The low nutrient level leads to better performance of insects which decreased when nutrient availability is more (Tao et al. 2014). Weak natural communities and crops are susceptible more to the eruption of insects in comparison to a varied system when they were provided with relatively homogenous nutrient levels (Dyer et al. 2012). Plant-herbivore interaction is situation dependent. For example, plant-herbivore interaction is controlled by neighbouring plant community (Barabas and D'Andrea 2016), by local predator community (Flagel et al. 2016), by plant-fungal and bacterial microbiomes (Christian et al. 2015) and by local nutrient condition (Burghardt 2016). All these factors sharpen the co-evolution of plants, and their herbivore gives rise to a highly complex "phytochemical landscape" (Glassmire et al. 2016). It is concluded that the functional traits of plants play an important role in understanding the interactions of food web ecological processes (Schmitz et al. 2015).

8.2 *Diversity of Herbivores and Role of Ecosystem*

The potential role of functional diversity and different species is widely recognized to maintain a resilient, healthy ecosystem (Lefcheck et al. 2015). The function of an ecosystem can also be maintained through consumer diversity even though the function of ecological diversity shows dominance in various specific studies of plants (Lefcheck and Duffy 2015). There are many effects of herbivore diversity on many aspects (primary production, consumption of producer biomass and plant diversity) of primary producer communities (Burkepile et al. 2016). The function of ecosystem is affected by loss in diversity of herbivores. The integration of various aspects of herbivory ecology (population growth, movement and predation risk) with diet is proven by recent studies (Adam et al. 2015) which results in more integrative knowing of herbivore complementarity.

Empirical work and synthetic analyses have shown multifunctionality of ecosystem due to biodiversity (Lefcheck and Duffy 2015). Besides primary production, biodiversity also affects various other processes of an ecosystem like nutrient cycling, decomposition rate, nutrient retention and many other interlinked processes. Lefcheck et al. (2015), in a multifunctionality experiment of species diversity on 94 ecosystem biodiversities, concluded that there was a consistent effect on aquatic and terrestrial habitat as the function grew stronger. Perhaps critically speaking, the diversity of herbivore influences numerous ecosystems positively at a high threshold level as compared to plant diversity at a higher threshold level which has a negative effect. Therefore, it is predicted that consumer diversity has a stronger influence on the ecosystem than plant diversity (Duffy 2003).

8.3 Interaction of Plants and Herbivores in Climate Change Era

There are intense consequences of climate change for life on earth and thought from the recent researches that climate change was 10 times faster in the last 65 million years (Diffenbaugh and Field 2013) than in any other time, but climate change varies across different ecosystems (Loarie et al. 2009). Other recent studies also concluded that there are fast evolutionary and environmental reactions of herbivore and plant interaction to our climate warming. For example, there is a movement of hundreds of pests and pathogen in the northern hemisphere at an average of 2.7 km/year, since the 1960s which matched with observed increased temperature. Herbivores, in most of the cases, respond much rapidly to climate change as compared to host plants (Lu et al. 2013). For example, *Aratus pisonii* (mangrove crab) move much faster at 6.2 km/year (Riley et al. 2014) surpassing the average movement rate of mangrove of 1.3–4.5 km/year (Williams et al. 2014).

Furthermore, climate change has created "oceanic hotspots" by strengthening the ocean flow and expanded the range of many species of tropical fish to temperate regions. The result of this tropicalization has led to overgrazing on the communities of temperate macroalgae in Japan, Australian Gulf, Mediterranean, South Africa and Mexico (Verges et al. 2014). Reduced snowpack due to warmer winter initiated a process "phonological mismatch" that has increased the attack of herbivory on woody species especially aspen (Brodie et al. 2012). It is worth noted that the increased amount of rains in the warmer winter has an opposite effect in the High Arctic and hence hardened the snowpack and reduced food availability for vertebrate herbivores in the winter (Hansen et al. 2014). Increased carbon dioxide and temperature (the main driving forces of climate change) greatly affect the physiology and metabolism of both plants and herbivores (DeLucia et al. 2012). According to the metabolism theory, high rate of metabolism in ectothermic consumers due to high temperature results in increased feeding rate (O'Connor et al. 2011), but there are irregularities in the feeding rate (decrease, increase or remain unchanged) of insect herbivores due to high temperature (Lemoine et al. 2014a, b). In some cases, high temperature has caused toxicity of compounds (Kurnath et al. 2016) but varies from species to species (Lemoine et al. 2013). Future work is suggested on plant-herbivore interaction to temperature and carbon dioxide and should observe the evolutionary consequences of climate change.

8.4 Neo-technologies for Considerate Diversity of Herbivores and Metabarcoding of DNA

It is a challenging and difficult job to determine diet breadth of herbivore and functional diversity, but recent technological advances made this challenge more manageable. For example, careful and time-consuming observation of feeding and behaviour of herbivore results in dietary characterization, but not always visual observation gives you the actual species of being consumed in a mixed population (Nash et al. 2016). To solve some of these problems, new and advanced technologies are required to give an exact and accurate measurement of diet physiologies of herbivore.

The ambiguous aspects of functional diversity, niche partitioning and complementarity have been solved by the development of DNA metabarcoding and helped to accurately measure herbivore gut content to resolve herbivore diet (Kress et al. 2015). DNA metabarcoding left behind all the previous traditional techniques to resolve herbivore diet identification and give quantitative measurements of relative consumption of food and capture rare diet items (Willerslev et al. 2014). Kartzinel et al. (2015) examined diet niche partitioning of seven mammalian herbivores (impala, buffalo, dik-dik, African savanna elephant, Kenyan savanna and two species of zebra), using metabarcoding technique, and concluded that there was a considerable difference in the diet of all herbivore at all comparison levels.

9 Interaction of Plants, Insects and Fungal Mycorrhiza

A symbiotic association between various plant species roots and fungi for the exchange of resources from fungi (soil resources) to plants (photosynthetic carbon) is known as mycorrhizas. This type of association not only improves the individual plant performance but also modifies plant productivity, nutrient cycling and plant community structure (Smith and Read 2008). Insects have a major role as herbivores, seed disperser, pollinator and parasitoids and affect individual plants, plant communities and even the whole ecosystem (Price 1997). Insects also associate with plants in most ecosystems, where they influence individual plants, plant communities, and ecosystems through their roles as herbivores, parasitoids, seed dispersers, and pollinators. Different research showed that insects and mycorrhizal fungi interact with one another in complex ways likely to be important to fungi, insects, and host plants. Several reviews have been published synthesizing many of these findings and proposing conceptual models by which the enormous variation in interaction outcomes can be understood (Gange 2007).

Table 19.3 Summary of studies examining the effects of mycorrhizal fungi on insect herbivores

Insect species	Mode of feeding	Host range	Plant species	AM/EM	Mycorrhizal fungi species	Effect on insect	References
Myzus ascalonicus	Phloem sap	Generalist	*Plantago lanceolata*	AM	*Glomus intraradices*	Positive	Gange et al. (1999)
Myzus persicae	Phloem sap	Generalist	*Plantago lanceolata*	AM	*Glomus intraradices*	Positive	Gange et al. (1999)
Polyommatus icarus Leaf chewing	Leaf chewing	Specialist	*Lotus corniculatus*	AM	3 unidentified *Glomus* spp.	Positive	Goverde et al. (2000)
Myzus persicae	Phloem sap	Generalist	*Plantago lanceolata*	AM	Field	Positive	Gange and West (1994)
Chromatomyia syngenesiae	Leaf miner	Generalist	*Leucanthemum vulgare*	AM	Field	Positive	Gange et al. (2005a)
Chromatomyia syngenesiae	Leaf miner	Generalist	*Leucanthemum vulgare*	AM	*Glomus fasciculatum* *Glomus caledonium*	Positive	Gange et al. (2005a)
Tephritis neesii	Seed		*Leucanthemum vulgare*	AM	*Glomus mosseae*	Positive	Gange et al. (2005a)
Scopula ornata	Leaf chewing	Specialist	*Clinopodium vulgare*	AM	*Glomus fasciculatum*	Positive	Gange et al. (2002)
Cryptomyzus ribis	Phloem sap	Specialist	*Stachys sylvatica*	AM	*Glomus fasciculatum*	Positive	Gange et al. (2002)
Myzus persicae	Phloem sap	Generalist	*Stachys sylvatica*	AM	*Glomus fasciculatum*	Positive	Gange et al. (2002)
Chaitophorus populicola	Phloem sap	Specialist	*Populus angustifolia* × *P. fremontii*	EM	*Pisolithus tinctorius*	Positive	Gehring and Whitham (2002)
Lygus rugulipennis	Meristem	Generalist	*Pinus sylvestris*	EM	Field	Variable[a]	Manninen et al. (2000)
Lymantria dispar	Leaf chewing	Generalist	*Castanea dentata*	EM	*Pisolithus tinctorius*	Positive[b]	Rieske et al. (2003)
Schizaphis graminum	Phloem sap	Generalist	*Sorghum* sp.	AM	*Glomus fasciculatum*	No effect	Pacovsky et al. (1985)
Anomala cupripes	Leaf chewing	Generalist	*Eucalyptus urophylla*	AM	*Glomus caledonium*	No effect	Gange et al. (2005b)
Strepsicrates spp.	Leaf chewing	Specialist	*Eucalyptus urophylla*	AM	*Glomus caledonium*	No effect	Gange et al. (2005b)
Matsucoccus acalyptus	Mesophyll	Specialist	*Pinus edulis*	EM	Field	No effect	Gehring et al. (1997)
Chromatomyia syngenesiae	Leaf miner	Generalist	*Leucanthemum vulgare*	AM	*Glomus mosseae*	No effect	Gange et al. (2005a)

(continued)

Table 19.3 (continued)

Insect species	Mode of feeding	Host range	Plant species	AM/EM	Mycorrhizal fungi species	Effect on insect	References
Ozirhincus leucanthemi	Seed		*Leucanthemum vulgare*	AM	*Glomus fasciculatum*	No effect	Gange et al. (2005a)
Tupiocoris notatus	Mesophyll	Specialist	*Nicotiana rustica*	AM	*Glomus etunicatum*	Negative	Wooley and Paine (2007)
Idaea aversata	Leaf chewing	Generalist	*Clinopodium vulgare*	AM	*Glomus fasciculatum*	Negative	Gange et al. (2002)
Udea prunalis	Leaf chewing	Generalist	*Origanum vulgare*	AM	*Glomus fasciculatum*	Negative	Gange et al. (2002)
Phlogophora meticulosa	Leaf chewing	Generalist	*Lolium perenne*	AM	*Glomus mosseae*	Negative	Vicari et al. (2002)
Heliothis zea	Leaf chewing	Generalist	*Glycine max*	AM	*Glomus fasciculatum*	Negative	Rabin and Pacovsky (1985)
Spodoptera frugiperda	Leaf chewing	Generalist	*Glycine max*	AM	*Glomus fasciculatum*	Negative	Rabin and Pacovsky (1985)

9.1 *Influence of Insects on Mycorrhizal Fungi*

Insect consumes fungal hyphae (mycophagy), feed host tissues and disperse mycorrhizal fungal propagules, hence influencing the performance of fungus directly or indirectly (Smith and Read 2008). The influence of insect herbivory on mycorrhizal fungi is generally on above-ground herbivore (Currie et al. 2006) which estimates how much a mycorrhizal fungus occupies the plant root system. There is a negative effect of insect herbivory on EM fungi colonization (Mueller et al. 2005). An increase in the colonization of mycorrhiza is found at the insect herbivory's early life stage or moderate or low level of insect herbivory (Kula et al. 2005). This may result from increased nutrient acquisition for regrowth after defoliation and increased root exudation by plants (Gange 2007).

9.2 *Influence of Mycorrhizal Fungi on Insects*

Many studies on plant-mycorrhizal fungi-insects reported the influence of mycorrhizal fungi on plant-herbivore interaction. Mycorrhizal fungi through changes in nutrient content not only increase the plant size the plant quality which may also change (Smith and Read 1997) but also affect numerous traits of plants, where mycorrhizal fungi also change inducible and constitutive defences and tolerance to herbivory during plant-herbivore interaction (Bennett et al. 2006). Mycorrhizal fungi have a positive as well as a negative effect on the above- and below-ground herbivores (Table 19.3, Fig. 19.1). In contrast, the attack of root herbivore has generally been reduced by mycorrhizal fungi (Gange 2007). It is predicted that through a quality change in plants, tolerance or defence quality and mycorrhizal fungi influence herbivore in abundance on host plant which results from cascade up to herbivore enemies (Fig. 19.1). Multiple mechanisms (as earlier discussed) have been adopted by arbuscular mycorrhizal (AM) and ectomycorrhizal (EM) fungi that affect plant herbivores. There are very few studies focusing on constitutive secondary compounds within plant tissues. Wurst et al. (2004) found that catalpol levels are reduced when AM fungi were inoculated, whereas Gange and West (1994) reported that the amount of aucubin and catalpol was minimum when fungicide has suppressed AM fungal community.

10 Mechanisms for Enzymatic Virulence of Entomopathogenic Fungus

The process in which death of insect occurs due to pathogenesis is known as virulence (Mondal et al. 2016). The virulence of entomopathogenic fungi is determined by the production of cuticle-degrading enzyme and varies from insect to insect due to the production of a variety of enzymes (proteases, lipase and chitinase) which is considered to be necessary causing death of insect during pathogenesis (Samuels et al. 2011). The main barrier for the entry of pathogenic fungus in the various stages of insect life cycle is usually cuticle of insect (very heterogeneous structure), which is composed of protein, chitin associated with lipids, wax and phenolic compounds (Ortiz-Urquiza and Keyhani 2015). These enzymes break down the insect cell wall's cuticular polymers that help the fungus pathogen enter the insect body, taking nutrients for its growth and taking control of insect, hence preventing the crops from damage and helping in the potential benefit to the growers (Petrisor and Stoian 2017). Many studies have been carried out on the extracellular activity of insect and cuticle-degrading enzymes produced by fungi and also biochemical characterization of these enzymes during the pathogenesis process and during the interaction of host and pathogens (Cristina and Stoian 2017).

10.1 Abiotic Stresses vs. Improved Virulence

The most noticeable damage to metabolic and molecular functions of entomopathogenic fungus is due to various abiotic stresses (high temperature, UV radiation and low water) that greatly limit fungal activities in the field. The virulence of *Beauveria*, *Metarhizium* and pathogenic fungi is increased by different metabolic pathway genes (HsPHR2 (CPD photolyase), heat shock protein 25 (HSP25), Try (tyrosinase), MrPhr1 (CPD photolyase), trxA (thioredoxin) and BbSOD1 (superoxide dismutase)) (Zhao et al. 2016). The overexpression in DNA repair photolyase is known to improve resistance in fungus to solar radiation (Fang and St. Leger 2012), while the tolerance level of *B. bassiana* is improved with the expression of thioredoxin (trxA) against UV-B irradiation, heat and oxidation (Ying and Feng 2011). Furthermore, increased fungal virulence in *M. anisopliae* against UV radiation is due to the integration of PKS gene cluster for melanin biosynthesis (Liao et al. 2014). Therefore, improvement in the stability of fungus can be achieved using a genetic engineering tool. Vigorous genetic management technologies (genome combination of several *Metarhizium* spp. and *Bassiana*) with being helpful to determine complete information about the host-specific genes and pathogenicity and new virulence techniques against insects can be achieved (Xu et al. 2014).

10.2 Strategies of Entomopathogenic Fungi in Integrated Pest Management

Chemical insecticides used to protect plants from noxious insects/pests may not only cause a negative effect on the ecosystem but also insect resistance to different chemical substances. Thus, scientists nowadays are compelled to look for new effective, eco-friendly methods to minimize insect/pest outbreak. In biological control or insect/pest suppression, special attention is given to the use of natural enemies of insects (entomopathogenic fungi) (Sahayaraj 2014). In both conventional and natural production systems, the most important method in IPM is entomopathogens. Several examples have shown the role of bioinsecticide having entomopathogens to control pests (Nana et al. 2015).

The use of entomopathogens depends on pest, environment or specific crop and may be used alone or in combination with botanical pesticides, chemicals or other entomopathogens. The performance of entomopathogenic fungi is mostly increased with formulation development. Potential formulations are developed from about 171 products globally since the 1960s, of which *Beauveria* and *Metarhizium* sp. contribute 33.9%, respectively, and *Isaria fumosorosea* and *B. brongniartii* contribute 5.8%, 4.1, respectively. These formulations are helpful in natural, classical and augmented biocontrol but do not harm the beneficial organism and hence are characterized as low-danger substance. The whole plant can adopt its defence system against a variety of insects/pests using a potent mechanism, i.e. induced system resistance (ISR), which is induced by beneficial entomopathogenic fungi (Pieterse et al. 2014).

11 Interaction of Insects and Plants Under the Influence of Bacteria

Plants and insects have been co-existing for more than 400 million years, leading to intimate and complex relationships. Throughout their own evolutionary history, plants and insects have also established intricate and very diverse relationships with microbial associates. Studies in recent years have revealed plant- or insect-associated microbes to be instrumental in plant-insect interactions, with important implications for plant defences and plant utilization by insects. Microbial communities associated with plants are rich in diversity, and their structure greatly differs between below- and above-ground levels. Microbial communities associated with insect herbivores generally present a lower diversity and can reside in different body parts of their hosts including bacteriocytes, haemolymph, gut, and salivary glands. Acquisition of microbial communities by vertical or horizontal transmission and possible genetic exchanges through lateral transfer could strongly impact on the host insect or plant fitness by conferring adaptations to new habitats. Recent developments in sequencing technologies and molecular tools have dramatically enhanced opportunities to characterize the microbial diversity associated with plants and insects and have unveiled some of the mechanisms by which symbionts modulate plant-insect interactions. These microbes may alter the metabolism in plant defence system and both in primary and secondary metabolites against insects benefiting either plants or insects.

11.1 Bacterial Diversity Specified in Insects and Plant Host

Further, recent researches of the well-studied symbioses in insects (i.e. *Buchnera aphidicola* and the obligatory symbiont of aphids) and plants (i.e. mycorrhizal and nitrogen-fixing bacterial mutualists) employ innovative technologies highlighting a surfeit of microbial associates hidden previously. In insects, this concerns the growing field of research on facultative symbionts (also referred to as secondary or accessory symbionts as opposed to obligatory or primary symbionts) and, to a lesser extent, the inhabitants of the digestive tract, referred to as gut associates. These microbial communities, although not essential for the host's reproduction and survival, may considerably affect many aspects of their host's ecology, behavior, and physiology, such as traits associated with plant utilization, protection against natural enemies, or responses to climate changes (Oliver et al. 2010).

11.2 Communities of Bacteria in Relation to Insects and Plant Herbivores

11.2.1 Bacteria Community Associated with Insects

A vast studied literature study has been observed on the diversity of bacteria associated with insects; however, more evidence for many groups of herbivorous insects (i.e. Lepidoptera) is required to the study. It is observed from literature studies that a few taxa have dominated the limited diversity of herbivorous insect microbial populations (Jones et al. 2013). Whereas approximately 1000 taxa inhibit mammalian guts in general, operational taxonomic units (OTU) or distinguished taxa, not exceeded to 30 sequences, are harboured by *Drosophila* spp. (Chandler et al. 2011a, b), and bacterial diversity of various species is not that much high (an average of 10–15 OTUs per insect) supporting the feeding habits and evolutionary history microbial community of insects (Colman et al. 2012). The richest gut bacterial communities are xylophagous insect harbour (as per sample of 103 OTUs), and the insects feeding on leaves are considered to be in the intermediate ones, as per each sample, 38 OTUs, while the poorest microbial diversity (not more than 3–7 OTUs per sample) has been observed in sap-feeding insects (i.e. whiteflies, aphids and psyllids). *Proteobacteria*, specifically, the *Enterobacteriaceae*, is known to be the biggest community of microbes from insecticidal herbivores. The two major bacterial groups are facultative and obligatory insects' symbionts in a wide range and the endosymbiont of psyllids (*Carsonella*); aphids (*Buchnera*); whiteflies (*Portiera*), the major endosymbiont (Moran et al. 2008); and rice weevil (*Sitophilus oryzae*). Facultative symbionts occurred in various parts of the host's cell sheath in primary bacteriocyte (insects' derived structures) and peripheral parts in secondary bacteriocytes, while obligatory symbionts are inhibited in bacteriocytes (Dillon and Dillon 2004). Some of the bacteria like *Spiroplasma* and *Wolbachia* cause infection in the reproductive organs of insects and change the ratio of host sex in males (Engelstadter and Hurst 2009).

11.2.2 Plant-Associated Bacterial Communities

Plants in comparison to insects are found to be significant diversity hosts of microbial enmities consisting of commensals, the actual valuable associates and pathogens. Distinguished examples of mutualist microbes are nitrogen-fixing bacteria and mycorrhizal fungi, while other endophytic bacteria and fungi benefit only their plant hosts in a wide array of biological purposes (Bulgarelli et al. 2013). Usually, plant-associated communities of bacteria are categorized into two; according to the location of bacteria, colonizing either beneath the layer/tissues of the ground (rhizosphere) or else above (phyllosphere, in an abundance of 106–107 cells of bacteria on the leaf area per centimetre), either live on or within the leaves (Humphrey et al. 2014). All the bacterial communities differ both in structure and diversity. The structure of the communities of bacteria of rhizosphere appears to be found specifically through mutual influence of exudates of roots which carry the soil biome differentiation through factors dependent on host genotype and in rhizosphere, whereas the phyllosphere communities could be seen on the surface of the leaves (Bulgarelli et al. 2013).

11.3 Mediation of Plant Defence Responses in Contrast to Herbivorous Insects by Plant Hormones

Plants produce different chemicals and volatiles repelling attackers. Elemental molecule production having toxicity against insects could be planted damaging and costly. Hence, plants protect themselves while inducing such molecules production at the time they needed it. On the attack of insects, plants produce defensive molecules that trigger a process providing complex signals bringing a reaction for resistance (Wu and Baldwin 2010). Plant defense responses are mainly regulated by three phytohormones: jasmonic acid (JA), salicylic acid (SA) and ethylene (ET). JA and ET are generally associated with plant defense responses against necrotrophic pathogens and herbivorous arthropods. In particular, activation of the JA signaling pathway is characterized by the induction of defensive compounds in vegetative tissues such as secondary metabolites (e.g. polyamines, quinones, terpenoids, alkaloids, phenylpropanoids, glucosinolates and antioxidants), proteins (e.g. polyphenol oxidases and proteinase inhibitors) and leaf trichomes (Escobar-Bravo et al. 2017). The pathway of signalling, opposed to ET and JA, signals pathway supported by salicylic acid and tempted by biotrophic microbes and more or fewer insects feeding on saps (hemipterans), i.e. aphids and whiteflies, requiring plants' living cells' active feeding process (Moran and Thompson 2001), whereas the signalling pathway of jasmonic acid initiates plant defence responses which are effective against few or more sap-feeding insects, i.e. leafhopper and whitefly (Zarate et al. 2007). The scenario of defence of plants counter to aphids is more complex. As revealed from different experiments, plant defence against aphids seems to be mediated by JA (De Vos et al. 2007) and further pathways of defence (Louis and Shah 2013).

11.4 Plant-Allied Bacteria May Interfere with Plant Defence Signalling

Plant roots, beneficial or pathogenic to plants, linked with various bacterial and microbial soil-borne communities (may be fungi as well), whereas few of the rhizobacteria which are growth promoting are known to produce induced systematic resistance in contrast to pathogenic microbes and insect herbivores and function as plant growth promoter (Pineda et al. 2010), where priming defence genes mediate ISR and more often contributes plant reaction in response to hormones of plants, i.e. ET and JA (Van der Ent et al. 2009). A stronger and faster expression of cellular defence is induced by ISR-primed plants upon insect and pathogen attacks, resulting in a better improved resistance level against attackers (Pozo et al. 2008).

12 Plant Defence Against Insect Herbivores

To counter the herbivore attack effects, a plant response is shown to herbivory through a different morphological, molecular and biochemical mechanism. The biochemical mechanism mediated both by direct and indirect defences is wide-ranging against the herbivores. Moreover, some unstable natural compounds are unconfined by plants attracting herbivores' natural competitors. Still, our approach to understanding these defensive mechanisms is very limited.

12.1 Herbivory and Shoot Morphology

Morphology of shoots has a great influence, the way how to keep themselves protected in response to photosynthetic tissue loss and nutrients which are valuable to herbivores (Sebata and Ndlovu 2012). Regarding shoot morphology, defences have been spread between the plants woody in nature in savanna of semiarid regions as they affect the susceptibility of plant parts to browsers (Scogings et al. 2004). Plants which are woody have separated into two different groups, i.e. shoot-dominated species (produce newly arisen leaves on new long shoots) and species of limited

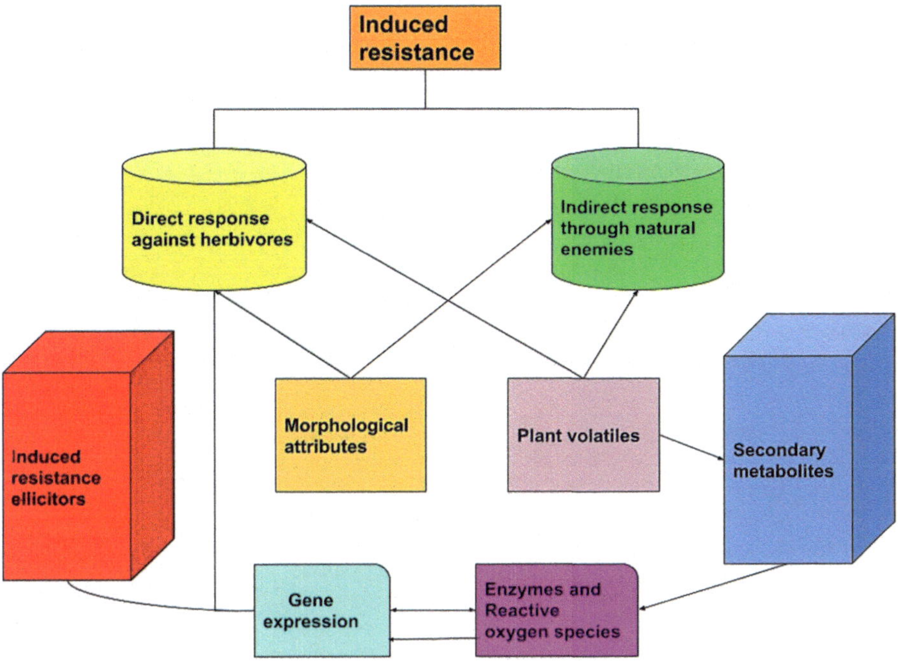

Fig. 19.2 Mechanism of induced resistance in plants

shoots (produce bunches of new leaves on the nodes of previous unbrowsable shoots), where the species with dominating branches are influenced by adding new leaf area and epical buds extending internodes and have higher nutrient concentration than species which are having limited shoots which add newly arisen area of leaf with no shoot elongation (Ganqa and Scogings 2007). The epical shoot meristem species are less susceptible to microbial herbivores as compared to the species which are dominated in shoots (Dziba et al. 2003), requiring improved defences against herbivory. The shoot-limited species depend on structural defences (thorns) as they have poor chemical defences where the goats have shown the preference to use their upper mobile lips and favour shoot-limited overshoot-dominant species (Sebata and Ndlovu 2012). Secondary compounds (i.e. compressed tannins and fibres) of plants are lesser in species which have limited shoots and replace its lost tissue rapidly by growing again (Scogings et al. 2004). Both of them adopt anti-herbivory defences.

Table 19.4 Defensive proteins of plants against pests

	Plant species	Insects species	References
Protease inhibitors	*Sorghum bicolor* Tomato *Gossypium hirsutum* *Solanum nigrum* *Nicotiana attenuata* *Arabidopsis* transgenic/rape seed oil Transgenic *Arabidopsis*/ tobacco	*Schizaphis graminum* *Manduca sexta* *Helicoverpa armigera* *Manduca sexta* *Spodoptera littoralis* *Spodoptera exigua* *Spodoptera exigua*	Zhu-Salzman et al. (2004) Chen et al. (2005) Dunse et al. (2010) Hartl et al. (2010) Steppuhn and Baldwin (2007) De Leo et al. (2001)
LOXs	*Cucumis sativus* *Nicotiana attenuata* *Alnus glutinosa* Wheat Tomato *Nicotiana attenuate*	*Spodoptera littoralis* *Bemisia tabaci* *Agelastica alni* *Sitobion avenae* *Macrosiphum euphorbiae* *Myzus persicae* *Myzus nicotianae*	Reymond et al. (2004) Kempema et al. (2007) Tscharntke et al. (2001) Zhao et al. (2009) Fidantsef et al. (1999) Voelckel et al. (2004)
Peroxidases	*Alnus glutinosa* *Arabidopsis* Buffalo grasses Poplar *Medicago sativa* Corn *Oryza sativa*/rice	*Agelastica alni* *Bemisia tabaci* (whitefly) *Blissus occiduus* *Lymantria dispar* *Aphis medicaginis* *Spodoptera littoralis* *Spodoptera frugiperda*	Tscharntke et al. (2001) Kempema et al. (2007) Heng-Moss et al. (2004) Gulsen et al. (2010) Barbehenn et al. (2009) Huang et al. 2007 Chen et al. (2009) Stout et al. (2009)
Hevein-like protein	*Arabidopsis*	*Bemisia tabaci*	Kempema et al. (2007)
Chitinase	*Sorghum bicolor*	*Schizaphis graminum*	Zhu-Salzman et al. (2004)
Catalase	Buffalo grass	*Blissus occiduus*	Heng-Moss et al. (2004)

12.2 Host Plant Defence Against Insects

Plants show a strong response to attack from herbivore through a forceful and complex defence system consisted of physical barriers and chemical toxin in nature. Plant defense can be divided into resistance and tolerance strategies. Plant traits that confer herbivore resistance typically prevent or reduce herbivore damage through expression of traits that deter pests from settling, attaching to surfaces, feeding and reproducing, or that reduce palatability. Plant tolerance of herbivory involves expression of traits that limit the negative impact of herbivore damage on productivity and yield. Identifying the defensive traits expressed by plants to deter herbivores or limit herbivore damage, and understanding the underlying defense mechanisms, is crucial for crop scientists to exploit plant defensive traits in crop breeding (Agrawal 2011). In the previous decades, a remarkable and well-established progress was observed finding the response in which plants induced to stress are not similar. Plants respond to herbivory through various morphological, biochemicals, and molecular mechanisms to counter/offset the effects of herbivore attack. The biochemical mechanisms of defense against the herbivores are wide-ranging, highly dynamic, and are mediated both by direct and indirect defenses. The defensive compounds are either produced constitutively or in response to plant damage, and affect the feeding, growth, and survival of herbivores. In addition, plants also release volatile organic compounds that attract the natural enemies of the herbivores. These strategies either act independently or in conjunction with each other. Host plant resistance to insects, particularly, induced resistance, can also be manipulated with the use of chemical elicitors of secondary metabolites, which confer resistance to insects (Steppuhn and Baldwin 2007).

12.3 Secondary Metabolites and Plant Defence

Such compounds which are not effecting normal plant development and growth are called secondary metabolites. These compounds decrease the lusciousness in tissues of plants where they have formed. Metabolites for defence are utilized in response to microbial or insect attacks or stored in inactive forms. The earlier is called phytoanticipins, initially activated by β-glucosidase the time herbivores mediate the release of various metabolites and aglycones (Barakat et al. 2010), and later are known to be phytoalexins.

12.4 Defensive Proteins of Plants

Insects' nutritional requirements are same as that of several other animals where insect always search for a healthy and true host to get food for the offsprings. Gene expression change with stresses includes the attack of insects leading to both quantitative and qualitative protein alterations that as a result play a significant role in oxidative defence and transduction of signals (Fig. 19.2) (Gulsen et al. 2010). Numerous proteins of vegetable taken by insects remain whole and stabilized within the middle intestine and there then move through the wall of the intestine into the haemolymph. Changes in the contents of amino acids or sequential proteins influence the role of these proteins. Similarly, the anti-insect activity of proteins (toxic) susceptible to proteolysis could increase by administrating a protease inhibitor (PI) that prevents toxic proteins from degrading allowing them to use protective functions (Table 19.4). Advances in proteomics and microarray approaches revealed that a vast spectrum of plant resistance protein (PRP) is included in defence of plants counter to herbivores (Chen et al. 2009). Due to arthropods' changed eating habits, several paths for signalling including ethylene, salicylic acid and jasmonic acid (JA) regulate the proteins induced by arthropods (Arimura et al. 2009).

13 Plant Defence Against Herbivory and Insect Adaptation

The arms race of evolution between insects and plants headed towards the huge defence system development in plants tending to identify signals from cells which are damaged and unnatural molecules, like animals, and trigger the immune system of plants in response to herbivores (Moore et al. 2014). Combating the attack of herbivores, plants produce some specific secondary metabolites, morphological structures and proteins having repellent, toxic and anti-feed effects on microbial herbivores (Bruce et al. 2005). Plant directly deals with herbivores, influencing the host plant survival or success of reproduction and in an indirect way by various other species as pests' natural enemies (Kariyat et al. 2013). The direct defence is influenced through the characteristics of plant-mediating biology of herbivores as protection surface mechanics in plants (i.e. thorns, spines, trichrome, thicker leaves and hairs) or toxic chemical production like alkaloids, phenols, terpenoids, quinones and anthocyanins, inhibiting or killing the herbivorous development (Reeves 2011), whereas the indirect defence against microbial insects is arbitrated through the release of mixtures and volatile attracting natural enemies of herbivores specifically or through the provision of food (i.e. nectar of flower) (Kariyat et al. 2013). Herbivores require developing pathways to fight with the protection of plant in a way using woody plants as slabs. Herbivores which feed on plants have a very small mouth to handle a difficult task of small leaves removed from thorny bushes (Belovsky et al. 1991). Majority of exploratory animals are having sprightly lips and tongues which let the animals to avoid thorns and select only leaves (Gordon and Illius 1988). More likely, the goats with narrow and mobile noses can give movement easily to their mouths in the thorny bush and pick only leaves (Shipley et al. 1999). Giraffe (*Giraffa camelopardalis*) having flexible long tongue is facilitated to feed on acacia trees (solid) (Hanley et al. 2007). Rumen development in ruminants could also be considered as an evolution to plant protection, like it lets ungulate digesting fibrous plant parts (Perez-Barberia et al. 2004).

14 Structural Traits of Plants and Their Role in Anti-herbivorous Defence

Plants having different resistance mechanisms to herbivores can generally be divided into two main groups: evasion and tolerance. Protection is considered to be in the concept of an umbrella including both tolerance and avoidance (Stowe et al. 2000), whereas some authors distinguish between plant tolerance to attack herbivores and avoid plant protection properties that appear through protection. Avoidance is deliberated to be chemical (i.e. phenol production which stop herbivores from nurturing early after the first bite) (Hanley and Lamont 2001) and structural (i.e. thorns surrounding the leaves) (Gowada 1996), or it is the herbivores' avoidance through synchronization of life cycle (Saltz and Ward 2000). The structural defence is a mechanism of avoidance based on structural properties, whether they are microscopic changes in the thickness of the cell wall or they may be visible plant prominences, a convenient structural protection definition (Boege and Marquis 2005), which could be an anatomical or morphological feature which gives an advantage of fitness for plants through direct prevention of herbivore feeding. Thus, we shortly suppose explanations of alternative adaption for structural defence deliberated below, considering crop protection as the only trait to be highlighted as a need; however, a complex formed from a group of related features and related structures adapts together (Agrawal and Fishbein 2006).

15 The Role of Plant-Associated Microbes in Mediating the Selection of Host Plants by Insect Herbivores

Plant microbial interactions are pervasive and may differ from beneficial parasites or pathogens. It is recognized that plant-related bacteria may play an important role modulating phenotypes of plants to form plant-insect interactions (Porter et al. 2019), i.e. increasing evidence shows that microbes change plant-produced cues which subsequently affect the foraging behaviour and oviposition of herbivorous insects (Eigenbrode et al. 2018).

16 Potential Role of Plant Growth Regulators in Plant Defence Against Pathogens and Insects

Phytohormones play a significant role in plant defence and growth regulation through mediating the processes of development and network signalling in response to plants in a vast range of biotic and parasitic interactions. Such plant hormones may be a target to pathogens and arthropods in the duration of arms race evolution between plants and their biological partners disrupting the metabolism of plants, controlling their morphology and physiology, and attack the plants successively. In arthropods, bacterial symbionts mediate the phenotypes induced by cytokines, producing intricate insects, microbes and plant interaction. In the previous years, the key model plants (i.e. *Arabidopsis*) and improved molecular techniques caused a great increase to understand the involved signalling cascades. The phytohormones, i.e. salicylic acid, ethylene and jasmonic acid, emerged swiftly as significant regulators in response to the specific defence gene activation (Pieterse et al. 2012). Their consequences and involvement for plant defence, fitness and survival were studied both in field and laboratory conditions; in open filed they were exposed to a wide range of natural ecological communities. Other plant hormones like auxins, gibberellin, cytokinins and abscisic acid were evolved as significant regulators for defence recently (Robert-Seilaniantz et al. 2011).

17 Potential Role of Secondary Metabolites in Defence Mechanism of Plants Against Herbivory and Insects

17.1 Secondary Metabolites

A huge and wide range of organics are produced by plants showing no role in the development and growth directly, i.e. these organic compounds usually function in photosynthesis, transportation of solutes, respiration process, differentiation and assimilation of nutrients (Hartmann 1991). A great variety of secondary metabolites or natural products are produced with a bulging function against microbial pathogens and predators on the base of repellence and toxic nature to microbes and herbivores, while some of the functions contrary to abiotic stress (i.e. to UV-B exposure) are important for communication of plants with several other organisms (Schafer and Wink 2009). The three principal compounds are phenolics, terpenes and other compounds containing N and S. Phenolic compounds are primarily produced from the shikimic acid pathway products, playing a key role in plants as defensive compounds, and terpenes consisted of units of 5-carbon isopentanoids, known as chains of feed and toxins to several herbivores. Moreover, the compounds containing N and S are mainly produced by normal amino acids (Van Etten et al. 2001).

17.1.1 Terpenes

These compounds are known to be the largest secondary metabolite class as combined by the biosynthetic basis of acetyl coA and intermediates of glycolytic sources (Grayson, 1998), where its subclass comprised of monoterpenes (C10) lies in the flowers and leaves of chrysanthemum species providing a strong response of insecticidal to different insects like bees, wasps, beetles and moths (Turlings et al. 1995). Sesquiterpenes (C15) are distinguished by a lactone ring of five members with a strong repellent feeding to several insect herbivores and mammals (Picman 1986). Diterpenes (C20) are found in leguminous trees and pines in the form of abietic acid (Bradley et al. 1992). Triterpenes (C30), e.g. limonoid, are found in citrus fruits in the form of a bitter substance which performs as anti-herbivore. Polyterpenes (C5), e.g. carotenoids, are primary tetraterpenes, from the pigment family. The second one, rubber, is found in vessels that are long, known as laticifers protecting wound healing and defence counter to insect herbivores (Eisner and Meinwald 1995).

17.1.2 Phenolic Compounds

A wide range of secondary products containing chemically heterogeneous phenol groups are produced by plants. It may be an important part of plant defence system (Wuyts et al. 2006). Phenolic compounds are secondary metabolites of plants. So far, more than 8000 phenolics have been found from natural sources and are

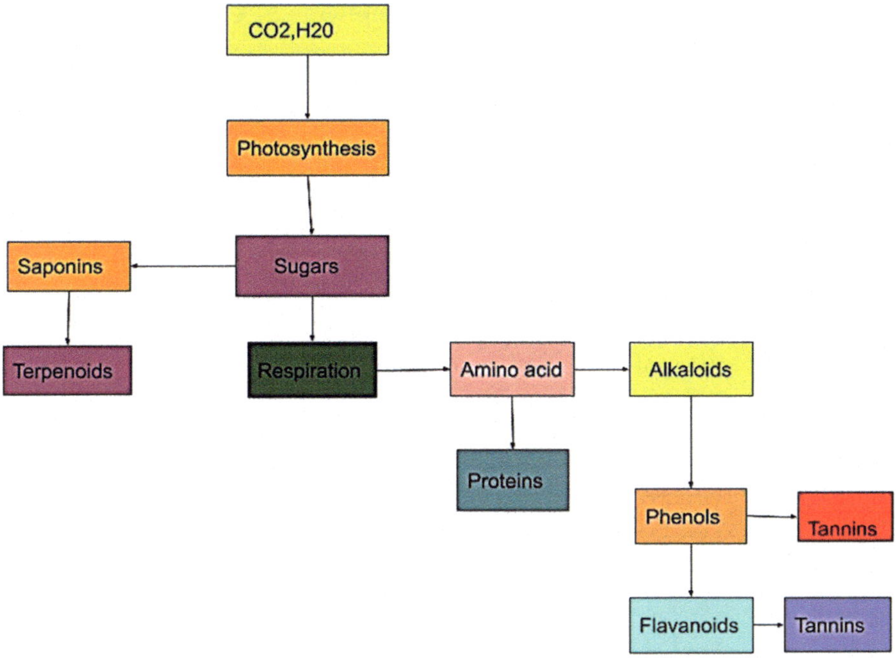

Fig. 19.3 Biosynthetic association between the primary and secondary metabolites

classified into phenolic acids, flavonoids, stilbenes, coumarins, lignins, and tannins. Phenolics play a crucial role in plants by controlling their growth as an internal physiological regulator and play a role in the defense mechanism of many plants against herbivorous insects and fungi. In addition, some of the coumarin derivatives are having a huge antifungal performance against pathogenic fungi that are soil-borne (Brooker et al. 2008). Furanocoumarins are found largely in members of Umbelliferae, i.e. parsnip, celery and parsley. And these are simply considered non-toxic until they are activated by UV-A (Rice 1984). Lignin are generally formed from different alcohols, i.e. coumaryl, coniferyl and sinapyl, known as a polymer of phenylpropanoids which are branched highly (Lewis and Yamamoto 1990). Flavonoids, considered to be one of the largest phenolic groups, vary plant functions like pigmentation and defence methods (Kondo et al. 1992). Isoflavonoids, usually secreted by legumes, play a key role in plant development and defence responses and also promote the formation of nodules fixing nitrogen by rhizobial symbiosis (Sreevidya et al. 2006). Tannins, a common toxin, significantly reduce the growth and survival of many herbivores and in a great diversity of animals act as a feeding repellent. They are known to cause an astringent sensation in the mouth of mammalian herbivores as a result of their salivary protein binding (Oates et al. 1980).

17.1.3 Secondary Metabolites Involving Sulphur

Such metabolites having GSH are found in the soluble fraction of plants in the form of organic S, playing a significant role as a reduced S mobile group in the regulation of plant development and growth, and in stress response, it acts as a cellular antioxidant (Kang and Kim 2007). It is accumulated rapidly after the attack of fungus, where it may act as a systematic messenger to carry some useful attack-related information to the tissues that are not infested, whereas the GSL is known as a group of N and S with low molecular mass and contains phytoglucoside, produced in higher plants enhancing its resistance in contrary to the negative effects of parasites and predators (De Vos and Jander 2009). Phytoalexins are produced in response to bacterial and fungal infections and many other stresses which help in limiting the invading pathogens spread (Grayer and Harborne 1994); defensins and thionins are non-stored, S-rich plant proteins which are produced and accumulated after the attack of microbes (Van Loon et al. 1994), involved in natural defence system. Thionins are found in infected spikes of the cell walls of resistive wheat germplasms, and they may act as a defence response in *Fusarium culmorum* spread and other infections (Kang and Buchenauer, 2003).

17.1.4 Secondary Metabolites Having Nitrogen

Such metabolites including alkaloids are known to be a family of secondary metabolites consisting of N and are found in higher plants specifically in about 20% of the plant species (Hegnauer 1988), mostly in herbaceous dicots and a very few in gymnosperms and monocots. This is re-believed to work as a defensive substance against predators (Hartmann 1991), and some of them cause interference with several components of the nervous system, i.e. chemical receivers/transmitters (Creelman and Mullet 1997), whereas the cyanogenic glucosides consist of N-containing group of protective compounds, which release the HCN poison, found in members of Rosaceae, Gramineae and Leguminosae families (Seigler 1981). They are not toxic themselves; instead, they are broken down to produce H2S and HCN like volatile toxic substances when the plant is crushed (Taiz and Zeiger 1995). Some unusual amino acids to be found in plants are non-proteinaceous, are present in free form, are incorporated in proteins and play a key role as substances involved in defence system (Johnson et al. 1989). Usually, plants synthesizing amino acids which are non-proteinaceous are toxicity resistant to these substances, though they enhance defensive system against insects, herbivorous animals and pathogenic microbes (Funck et al. 2009) (Fig. 19.3).

18 The Potential Anti-herbivory Defence Role of Microbial Organisms on Plant Thorns

Several anti-herbivorous defences adopted by plants are spines, thorns and prickles. It has recently been discovered that they are often aposematic (warning staining). However, the anti-herbivory physical defence performs some structures which are sharp and found to be a tip of the iceberg. It can be observed that thorns of various species of plants usually anchorage the aerobic and anaerobic bacterial amount, especially *Clostridium perfringens*, the potentially casual lethal agent of gangrene gas. Research work of medical studies showed that pathogenic fungi in animals and humans are introduced by spines, thorns and prickles (Halpern et al. 2007).

18.1 Bacterial Microbe and Thorns

Spines, thorns and prickles are fleeting antibacterials, mechanically protected in thousands of species of plants originated in different arid zones (Grubb 1992). A complete defensive story might not be of mechanical defence provided by thorns against herbivores. Injuries can cause thorns to infect bacteria and cause serious infections, which can be much more dangerous and painful than lone thorns alone. Recent publications have indicated that thorns found in *Crataegus* spp. (known as hawthorn commonly) and *Phoenix dactylifera* (commonly called date palm) have some bacterial pathogens. Pathogenic organisms and dead bacteria contain further bacterial hubs, which are inhabited by thorns and are divided with the greatest plausibility into preventing and damaging to the herbivores (Gowada 1996).

18.2 Pathogenic Fungal Microbes and Thorns

Bacteria not only result in the injury by plant thorns causing infected inflammation. Various medical studies indicated that thorns, spikes and prickles introduce fungal pathogens into humans and animals. Dermal infections of fungi which cannot penetrate into the skin are caused by dermatophytes, even though they may penetrate dermal tissues through a punctured wound (Willey et al. 2008). Chromoblastomycosis, a kind of dermal mycosis, is caused by demacia or pigmented saprophytic mould. Thorns of *Mimosa pudica* plant produce *Fonsecaea pedrosoi* (Salgado et al. 2004), and, sporotrichosis, a dermal mycosis, was found in *Sporothrix schenckii* fungus. This can be hazardous to the working gardeners, florists and forestry, which is also identified as a disease of a rose garden as it is generally spread by prickles of roses (Haldar et al. 2007). The skin eruption then spread to muscles and bones. This disease are categorized into actinomycetoma and eumycetoma commonly spread/caused by a filamentous fungus (Fahal 2004).

19 Role of Microorganism in Controlling Root-Knot Nematodes in Different Crops

Nematodes being severe pests to crops can be controlled by nematicides which are environmentally hazardous. Therefore, alternative methods could be developed to reduce the harmful effect of these nematicides to suppress nematodes. The most common traditional method to control nematodes is soil fumigants. Being a cost-effective technique but due to the chemical nature, a broad-spectrum range produces economic losses and harm to non-target species. Hence there should be an alternative way (plant-derived products) which is eco-friendly to alleviate nematode infestation in crops (Singh et al. 2019b). As root-knot nematode lives underground, hence the damage cannot be estimated which makes their control difficult. Parasitic nematodes present in the soil hinder the uptake of water and nutrients to different plant parts like bulbs, roots and tubers, thus causing nutrition and water deficiency symptoms leading to stunted plant growth. Nematodes initially attacked the root tip portion degrading the cell wall and moved towards the vascular cylinder leading to gall formation (giant cell). These structures draw off the nutrients and photoassimilates of the plants. Both the young (infection process is lethal) and mature plants (losses in yield) are affected by root-knot nematodes. Stunted growth, loss of yield and quality (changes in organic acids and amino acid levels and chlorophyll content) and reduced resistance against various biotic and abiotic stresses in fruits, vegetables and field crops are the damages related to the attack of root-knot nematodes (Kepenekci et al. 2018). Fruits and vegetables are badly affected by the attack of nematodes. Some of the horticulture crops affected by root-knot nematodes along with their management practices are given below.

Tomatoes, being the most important and popular horticulture crop worldwide, greatly suffer from biotic and abiotic stresses. Nematodes can cause several losses in yield and make the plant susceptible to bacterial and fungal attack (Zhou et al. 2016). About 10–30% losses annually in vegetables is due to root-knot nematode invasion. Crop management, resistant cultivars and chemical nematicides are some of the management strategies to control nematodes in tomato. Apart from these, biological control management is the best alternative. Among the biocontrol agents, *Trichoderma album*, *Bacillus megaterium*, *Ascophyllum nodosum*, *Trichoderma harzianum* (Radwan et al. 2012), *Streptomyces* (Ruanpanun et al. 2010) and arbuscular mycorrhiza (Sharma and Sharma 2017) having nematicidal properties are affective against root-knot nematodes in tomatoes.

Carrot, being a root crop, is among the ten vegetables grown worldwide in the tropical and subtropical areas. Six species of nematodes attack carrot plant. Among them, an endoparasitic nematode "*Meloidogyne hapla*" badly affects carrot crop. Severe losses (quantitative losses of about 24–55% and qualitative losses of 13–77%) have been observed due to the attack of root-knot nematodes. Poor or no tap root formation by the invasion of root-knot nematodes is due to giant cell "galls" produced in the growing root tips resulting in defective and forked carrots (Nagachandrabose 2018). *Pseudomonas fluorescens*, *Purpureocillium lilacinum*

(Killani et al. 2011), *B. subtilis* (Huang et al. 2010) and *Pochonia chlamydosporia* (Bontempo et al. 2014) are among the various biocontrol strategies adopted to suppress root-knot nematodes with improved growth and yield as well.

Chilli is among the commercially grown crops globally, recorded several losses due to parasitic nematodes. A root-knot nematode *Meloidogyne incognita* infests chillies resulting in less flower production, less yield and stunted growth. *Pseudomonas fluorescens* compared with pesticides has proven prolonged resistance against nematode attack (Khan et al. 2012).

Banana, one of the chief economic crops grown in the tropical areas of the world, faces severe losses due to nematodes, and about 132 species (belonging to 54 genera) are known to be present in rhizosphere of banana (Eissa et al. 2005). Mokbel et al. (2006) reported about 76% occurrences of nematodes in banana. The biocontrol agents used to control the attack of root-knot nematodes are *Glomus fasciculatum*, *Paecilomyces lilacinus*, *Penicillium* spp., *Bacillus subtilis* and *Trichoderma viride*, showing effective nematicidal activities (Esnard et al. 1998), hence resisting the plant against nematodes. Apart from these some alga species like *Laurencia obtusa*, *Sargassum vulgare*, *Jania rubens* and *Ulva lactuca* are also useful to control nematode activity.

20 Conclusions

It is essential to understand the ambiguous signalling molecules of herbivores, their mode of action, identification and transduction of signals. A single trait can influence the natural competitors, positively or negatively, associated with herbivores. There are many different factors either ecological or mutual community relationship or evolutionary effects can effect the microbial communities. Hence, it is essential to understand and know about the multitrophic interactions and significances of defensive attributes for herbivory insect and pest management, understanding the induced resistance in plants. Herbivores are confined to the available food and nutritional minerals, whether not all the minerals are present in the tissues of plants, but they may be available to herbivores equally. Also, there are fast evolutionary and environmental reactions of herbivore and plant interaction to climate change or warming. Pest population suppressed by a natural beneficial organism is known as a natural control. The scientists are forced to focus on alternative environmentally safe, cost-effective and reliable strategies due to the overuse of these chemical pesticides resulted in an adverse effect on non-targeted organisms, groundwater pollutions, the resistance of insects to chemicals, deposition on edible food crops and also negative effect on human health. Farmers are still using some agrochemical pesticides to control insects/pests and diseases in agricultural practices, thus responsible to maintain quantity as well as quality of food globally. There are many biological agents used in insect pest management (IPM) which include fungi, bacteria, virus, protozoans, botanicals (plants or products derived from plants), predator/pathogen system, plant-incorporated protectants (PIPs) and insect pheromones. But the main issue is that among the very large number, a very few biologically originated compounds are commercially used (available and affordable to growers) (Singh et al. 2019a; Hynes and Boyetchko 2006). It could be suggested as well that thorns, prickles and spines, by wounding pathogenic fungi or bacteria of plants inside the herbivorous body, also pass through the major defence line (the skin) and cause diseases. It is furtherly needed to collect information about some important groups of herbivorous insects as the insect-associated bacterial diversity is fast growing.

References

Acuna R, Padilla BE, Florez-Ramos CP et al (2012) Adaptive horizontal transfer of a bacterial gene to an invasive insect pest of coffee. Proc Natl Acad Sci U S A 109:4197–4202

Adam TC, Kelley M, Ruttenberg BI et al (2015) Resource partitioning along multiple niche axes drives functional diversity in parrotfishes on Caribbean coral reefs. Oecologia 179(4):1173–1185

Adams AS, Aylward FO, Adams SM, Erbilgin N, Aukema BH, Currie CR, Suen G, Raffa KF (2013) Mountain pine beetles colonizing historical and naive host trees are associated with a bacterial community highly enriched in genes contributing to terpene metabolism. Appl Environ Microbiol 79:3468–3475

Agrawal AA (2011) Current trends in the evolutionary ecology of plant defence. Funct Ecol 25:420–432. https://doi.org/10.1111/j.1365-2435.2010.01796

Agrawal AA, Fishbein M (2006) Plant defence syndromes. Ecology 87:S132–S149

Agrawal AA, Janssen A, Bruin J, Posthumus MA, Sabelis MW (2002) An ecological cost of plant defence: attractiveness of bitter cucumber plants to natural enemies of herbivores. Ecol Lett 5:377–385. https://doi.org/10.1046/j.1461-0248.2002.00325.x

Alofs KM, Jackson DA (2014) Meta-analysis suggests biotic resistance in freshwater environments is driven by consumption rather than competition. Ecology 95(12):3259–3270

Arimura GI, Matsui K, Takabayashi J (2009) Chemical and molecular ecology of herbivore-induced plant volatiles: proximate factors and their ultimate functions. Plant Cell Physiol 50(911):23; PMID: 19246460. https://doi.org/10.1093/pcp/pcp030

Bradley DJ, Kjellborn P, Lamb CJ (1992) Elicitor and wound induced oxidative cross linking of a proline rich plant cell protein: A novel rapid defence response. Cell, 70:21–30

Barabas G, D'Andrea R (2016) The effect of intraspecific variation and heritability on community pattern and robustness. Ecol Lett 19(8):977–986

Barakat A, Bagniewska-Zadworna A, Frost CJ, Carlson JE (2010) Phylogeny and expression profiling of CAD and CAD-like genes in hybrid Populus (P. deltoides x P. nigra): evidence from herbivore damage for subfunctionalization and functional divergence. BMC Plant Biol 10:100; PMID: 20509918. https://doi.org/10.1186/1471-2229-10-100

Barbehenn RV, Jaros A, Lee G, Mozola C, Weir Q, Salminen JP (2009) Hydrolyzable tannins as "quantitative defenses": limited impact against Lymantria dispar caterpillars on hybrid poplar. J Insect Physiol 55:297–304; PMID: 19111746

Behmer ST (2009) Insect herbivore nutrient regulation. Annu Rev Entomol 54:165–187

Belovsky GE, Schmitz OJ, Slade JB, Dawson TJ (1991) Effects of spines and thorns on Australian arid zone herbivores of different body masses. Oecologia 88:521–528

Bennett AE, Alers-Garcia J, Bever JD (2006) Three-way interactions among mutualistic mycorrhizal fungi, plants, and plant enemies: hypotheses and synthesis. Am Nat 167:141–152

Bentley BL, Johnson ND (1992) Plants as food for herbivores: the roles of nitrogen fixation and carbon dioxide enrichment. In: Price PW, Lweinsohn TM, Fernandes GW, Benson WW (eds) Plant-animal interaction: evolutionary ecology in tropical and temperate regions. Wiley, New York, pp 257–272

Bernays EA, Chapman RF (1994) Host plant selection by phytophagous insects. Chapman and Hall, New York

Biere A, Bennett AE (2013) Three-way interactions between plants, microbes and insects. Funct Ecol 27:567–573

Body M, Kaiser W, Dubreuil G, Casas J, Giron D (2013) Leaf-miners co-opt microorganisms to enhance their nutritional environment. J Chem Ecol 39:969–977

Boege K, Marquis RJ (2005) Facing herbivory as you grow up: the ontogeny of resistance in plants. Trends Ecol Evol 20:441–448

Bontempo AF, Fernandes RH, Lopes J, Freitas LG, Lopes EA (2014) Pochonia chlamydosporia controls Meloidogyne incognita on carrot. Aust Plant Pathol 43:421–424

Boone CK, Keefover-Ring K, Mapes AC, Adams AS, Bohlmann J, Raffa KF (2013) Bacteria associated with a tree-killing insect reduce concentrations of plant defense compounds. J Chem Ecol 39:1003–1006

Brodie J, Post E, Watson F et al (2012) Climate change intensification of herbivore impacts on tree recruitment. Proc Biol Sci 279(1732):1366–1370

Brooker N, Windorski J, Blumi E (2008) Halogenated coumarins derivatives as novel seed protectants. Commun Agric Appl Biol Sci 73(2):81–89

Bruce TJA, Pickett JA (2011) Perception of plant volatile blends by herbivorous insects. Finding the right mix. Phytochemistry 72:1605–1611

Bruce TJA, Wadhams LJ, Woodcock CM (2005) Insect host location: a volatile situation. Trends Plant Sci 10:269–274

Bulgarelli D, Schlaeppi K, Spaepen S, van Themaat EVL, SchulzeLefert P (2013) Structure and functions of the bacterial microbiota of plants. Annu Rev Plant Biol 64:807–838

Burghardt KT (2016) Nutrient supply alters goldenrod's induced response to herbivory. Funct Ecol 30(11):1769–1778

Burkepile DE, Fynn RW, Thompson DI, et al (2016) Herbivore size matters for productivity-richness relationships in African savannas. J Ecol 105: 674–686

Chandler D, Bailey AS, Tatchell GM, Davidson G, Greaves J, Grant WP (2011a) The development, regulation and use of biopesticides for integrated pest management. Philos Trans R Soc Lond B Biol Sci 366(1573):1987–1998

Chandler JA, Morgan Lang J, Bhatnagar S, Eisen JA, Kopp A (2011b) Bacterial communities of diverse Drosophila species: ecological context of a host-microbe model system. PLoS Genet 7:e1002272

Chen H, Wilkerson CG, Kuchar JA, Phinney BS, Howe GA (2005) Jasmonate-inducible plant enzymes degrade essential amino acids in the herbivore midgut. Proc Natl Acad Sci U S A 102:19237–19242; PMID: 16357201. https://doi.org/10.1073/pnas.0509026102

Chen Y, Ni X, Buntin GD (2009) Physiological, nutritional, and biochemical bases of corn resistance to foliage-feeding fall armyworm. J Chem Ecol 35:297–306; PMID: 19221843. https://doi.org/10.1007/s10886-009-9600-1.5

Christian N, Whitaker BK, Clay K (2015) Microbiomes: unifying animal and plant systems through the lens of community ecology theory. Front Microbiol 6:869

Chu CC, Spencer JL, Curzi MJ, Zavala JA, Seufferheld MJ (2013) Gut bacteria facilitate adaptation to crop rotation in the western corn rootworm. Proc Natl Acad Sci U S A 110:11917–11922

Cock MJW, van Lenteren JC, Brodeur J, Barratt BIP, Bigler F, Bolckmans K, Cônsoli FL, Haas F, Mason PG, Parra JRP (2010) Do new access and benefit-sharing procedures under the convention on biological diversity threaten the future of biological control? BioControl 55:199–218

Colman DR, Toolson EC, Takacs-Vesbach CD (2012) Do diet and taxonomy influence insect gut bacterial communities? Mol Ecol 21:5124–5137

Creelman RA, Mullet JE (1997) Biosynthesis and action of jasmonates in plants. Annu Rev Plant Physiol Plant Mol Biol 48:355–381

Cristina P, Stoian G (2017) The role of hydrolytic enzymes produced by entomopathogenic fungi in pathogenesis of insects. Rom J Plant Prot X:66–72

Currie AF, Murray PJ, Gange AC (2006) Root herbivory by *Tipula paludosa* larvae increases colonization of *Agrostis capillaris* by arbuscular mycorrhizal fungi. Soil Biol Biochem 34:1994–1997

De Leo F, Bonade-Bottino M, Ceci LR, Gallerani R, Jouanin L (2001) Effects of a mustard trypsin inhibitor expressed in different plants on three lepidopteran pests. Insect Biochem Mol Biol 31:593–602; PMID:11267898. https://doi.org/10.1016/S0965-1748(00)00164-8

De Vos M, Jander G (2009) *Myzus persicae* (Green peach aphid) salivary components induce defence responses in Arabidopsis thaliana. Plant Cell Environ 32(11):1548–1560

De Vos M, Kim JH, Jander G (2007) Biochemistry and molecular biology of Arabidopsis–aphid interactions. Bioessays 29:871–883

De Vries EJ, Jacobs G, Sabelis MW, Menken SBJ, Breeuwer JAJ (2004) Diet-dependent effects of gut bacteria on their insect host: the symbiosis of *Erwinia sp.* and western flower thrips. Proc R Soc B Biol Sci 271:2171–2178

DeLucia EH, Nabity PD, Zavala JA et al (2012) Climate change: resetting plant-insect interactions. Plant Physiol 160(4):1677–1685

Diffenbaugh NS, Field CB (2013) Changes in ecologically critical terrestrial climate conditions. Science 341(6145):486–492

Dillon RJ, Dillon VM (2004) The gut bacteria of insects: nonpathogenic interactions. Annu Rev Entomol 49:71–92

Duffy JE (2003) Biodiversity loss, trophic skew and ecosystem functioning. Ecol Lett 6(8):680–687

Dunham WC (2015) Evolution and future of biocontrol. Paper presented at the 10th annual biocontrol industry meeting (ABIM), Basel, Switzerland, October 20th, 2015. http://www.abim.ch/index.php?eID=tx_nawsecuredl&u=0&g=0&t=1489234639&hash=9a70d39f93f7e559c74c63844ae047a9aa3c37ea&file=fileadmin/abim/documents/presentations2015/Keynote_Dunham_ABIM_2015.pdf. Accessed 10 Mar 2017

Dunse KM, Stevens JA, Lay FT, Gaspar YM, Heath RL, Anderson MA (2010) Coexpression of potato type I and II proteinase inhibitors gives cotton plants protection against insect damage in the field. Proc Natl Acad Sci U S A 107:15011–15015; PMID: 20696895. https://doi.org/10.1073/pnas.1009241107

Dyer LA, Carson WP, Leigh EG (2012) Insect outbreaks in tropical forests: patterns, mechanisms, and consequences. In: Barbosa P, Letourneau DK, Agrawal A (eds) Insect outbreaks revisited. Wiley, Chichester

Dziba LE, Scogings PF, Gordon IJ, Raats JG (2003) Effects of season and breed on browse species intake rates and diet selection by goats in the False Thornveld of the Eastern Cape, South Africa. Small Ruminant Res 47:17–30

Edger PP, Heidel-Fischer HM, Bekaert M et al (2015) The butterfly plant arms-race escalated by gene and genome duplications. Proc Natl Acad Sci U S A 112(27):8362–8366

Eigenbrode SD, Bosque-Perez NA, Davis TS (2018) Insect-borne plant pathogens and their vectors: ecology, evolution, and complex interactions. Annu Rev Entomol 63:169–191

Eisner T, Meinwald J (eds) (1995) Chemical ecology: the chemistry of biotic interaction. National Academy Press, Washington, DC

Eissa MFM, El-Gindi AY, Abd-Elgawad MM, Ismail AE, El-Nagdi WA (2005) Application of some bioagents and oxamyl in controlling Meloidogyne incognita, *Helicotylenchus exallus* and Criconemoides spp. infesting banana cv. Williams. Pak J Biotechnol 2(1–2):70–79

Endara M, Coley PD (2011) The resource availability hypothesis revisited: a meta-analysis. Funct Ecol 25(2):389–398

Engelstadter J, Hurst GDD (2009) The ecology and evolution of microbes that manipulate host reproduction. Annu Rev Ecol Evol Syst 40:127–149

Erb M, Meldau S, Howe GA (2012) Role of phytohormones in insect-specific plant reactions. Trends Plant Sci 17:250–259

Esnard J, Marban MN, Zuckerman BM (1998) Effects of three microbial broth cultures and an organic amendment on growth and populations of free living and plant-parasitic nematodes on banana. Eur J Plant Pathol 104:457–463

Escobar-Bravo R, Klinkhamer PGL, Leiss KA (2017) Induction of Jasmonic Acid-Associated Defenses by Thrips Alters Host Suitability for Conspecifics and Correlates with Increased Trichome Densities in Tomato. Plant Cell Physiol 58(3):622–634

Fahal AH (2004) Mycetoma: a thorn in the flesh. Trans R Soc Trop Med Hyg 98:3–11

Fang W, St. Leger RJ (2012) Enhanced UV resistance and improved killing of malaria mosquitoes by photolyase transgenic entomopathogenic fungi. PLoS One 7:e43069

Fidantsef AL, Stout MJ, Thaler JS, Duffey SS, Bostock RM (1999) Signal interactions in pathogen and insect attack: expression of lipoxygenase, proteinase inhibitor II, and pathogenesis-related protein P4 in the tomato, *Lycopersicon esculentum*. Physiol Mol Plant Pathol 54:97–114. https://doi.org/10.1006/pmpp.1998.0192

Fisher TW, Garczynski SF (2012) Isolation, culture, preservation, and identification of entomopathogenic bacteria of the Bacilli. In: Lacey LA (ed) Manual of techniques in invertebrate pathology. Academic, London, pp 75–98

Flagel DG, Belovsky GE, Beyer DE Jr (2016) Natural and experimental tests of trophic cascades: gray wolves and white-tailed deer in a Great Lakes forest. Oecologia 180(4):1183–1194

Funck D, Stadelhofer B, Koch W (2009) Ornithine-delta-aminotransferase is essential for arginine catabolism but not for proline biosynthesis. BMC Plant Biol 8:40–45

Gange AC (2007) Insect-mycorrhizal interactions: patterns, processes, and consequences. In: Ohgushi T, Craig TP, Price PW (eds) Ecological communities: plant mediation in indirect interaction webs. Cambridge University Press, London, pp 124–143

Gange AC, West HM (1994) Interactions between arbuscular mychorrhizal fungi and foliar-feeding insects in Plantago lanceolata L. New Phytol 128:79–87

Gange AC, Bower E, Brown VK (1999) Positive effects of an arbuscular mycorrhizal fungus on aphid life-history traits. Oecologia (Berl.) 120:123–131

Gange AC, Stagg PG, Ward LK (2002) Arbuscular mycorrhizal fungi affect phytophagous insect specialism. Ecol Lett 5:11–15

Gange AC, Brown VK, Aplin DM (2005a) Ecological specificity of arbuscular mycorrhizae: evidence from foliar- and seed-feeding insects. Ecology 86:603–611

Gange AC, Gane DRJ, Chen YL, Gong MQ (2005b) Dual colonization of *Eucalyptus urophylla* ST Blake by arbuscular and ectomycorrhizal fungi affects levels of insect herbivore attack. Agric For Entomol 7:253–263

Ganqa NM, Scogings PF (2007) Forage quality, twig diameter, and growth habit of woody plants browsed by black rhinoceros in semi-arid sub-tropical thicket, South Africa. J Arid Environ 70:514–526

Gehring CA, Whitham TG (2002) Mycorrhizae herbivore interactions: population and community consequences. In: van der Heijden MGA, Sanders IR (eds) Mycorrhizal ecology. Springer, Berlin, pp 295–320

Gehring CA, Cobb NS, Whitham TG (1997) Three-way interactions among ectomycorrhizal mutualists, scale insects, and resistant and susceptible pinyon pines. Am Nat 149:824–841

Giron D, Glevarec G (2014) Cytokinin-induced phenotypes in plant-insect interactions: learning from the bacterial world. J Chem Ecol 40:826–835

Giron D, Huguet E (2011) A genomically tractable and ecologically relevant model herbivore for a model plant: new insights into the mechanisms of insect-plant interactions and evolution. Mol Ecol 20:990–994

Giron D, Kaiser W, Imbault N, Casas J (2007) Cytokinin-mediated leaf manipulation by a leafminer caterpillar. Biol Lett 3:340–343

Glare T, Caradus J, Gelernter W, Jackson T, Keyhani N, Köhl J, Marrone P, Morin L, Stewart A (2012) Have biopesticides come of age? Trends Biotechnol 30:250–258

Glassmire AE, Jeffrey CS, Forister ML et al (2016) Intraspecific phytochemical variation shapes community and population structure for specialist caterpillars. New Phytol 212(1):208–219

Gordon IJ, Illius AW (1988) Incisor arcade structure and diet selection in ruminants. Funct Ecol 2:15–22

Goverde M, van der Heijden MGA, Wiemken A, Sanders IR, Erhardt A (2000) Arbuscular mycorrhizal fungi influence life history traits of a lepidopteran herbivore. Oecologia (Berl.) 125:362–369

Gowada JH (1996) Spines of Acacia tortilis: what do they defend and how? Oikos 77:279–286

Grayer RJ, Harborne JB (1994) A survey of antifungal compounds from higher plants 1982–1993. Phytochemistry 37:19–42

Grayson DH (1998) Monoterpenoids. Nat Prod Rep 5:497–521

Grubb PJ (1992) A positive distrust in simplicity-lessons from plant defenses and from competition among animals. J Ecol 80:585–610

Gruntman M, Segev U, Glauser G et al (2017) Evolution of plant defenses along an invasion chronosequence: defence is lost due to enemy release—but not forever. J Ecol 105(1):255–264

Gulsen O, Eickhoff T, Heng-Moss T, Shearman R, Baxendale F, Sarath G et al (2010) Characterization of peroxidase changes in resistant and susceptible warm-season Turf grasses challenged by *Blissus occiduus*. Arthropod Plant Interact 4:45–55. https://doi.org/10.1007/s11829-010-9086-3

Haldar N, Sharma MK, Gugnani HC (2007) Sporotrichosis in north-east India. Mycoses 50:201–204

Halpern M, Raats D, Lev-Yadun S (2007) The potential anti-herbivory role of microorganisms on plant thorns. Plant Signal Behav 2(6):503–504

Hanley ME, Lamont BB (2001) Herbivory, serotiny and seedling defence in Western Australian Proteaceae species. Oecologia 126:409–417

Hanley ME, Lamont BB, Fairbanks MM, Rafferty CM (2007) Plant structural traits and their role in anti-herbivore defence. Perspect Plant Ecol Evol Syst 8:157–178

Hansen AK, Moran NA (2014) The impact of microbial symbionts on host plant utilization by herbivorous insects. Mol Ecol 23:1473–1496

Hansen BB, Isaksen K, Benestad RE et al (2014) Warmer and wetter winters: characteristics and implications of an extreme weather event in the High Arctic. Environ Res Lett 9(11):114021

Hartl M, Giri AP, Kaur H, Baldwin IT (2010) Serine protease inhibitors specifically defend Solanum nigrum against generalist herbivores but do not influence plant growth and development. Plant Cell 22:4158–4175; PMID: 21177479. https://doi.org/10.1105/tpc.109.073395

Hartmann T (1991) Alkaloids. In: Rosenthal GA, Berenbaum MR (eds) Herbivores; their interaction with secondary plant metabolites, Vol. I, The chemical participants, 2nd edn. Academic, San Diego, pp 33–85

Hassani MA, Duran P, Hacquard S (2018) Microbial interactions within the plant holobiont. Microbiome 6:58. https://doi.org/10.1186/s40168-018-0445-0

Heckman RW, Wright JP, Mitchell CE (2016) Joint effects of nutrient addition and enemy exclusion on exotic plant success. Ecology 97(12):3337–3345

Heger T, Jeschke JM (2014) The enemy release hypothesis as a hierarchy of hypotheses. Oikos 123(6):741–750

Hegnauer R (1988) Biochemistry, distribution and taxonomic relevance of higher plant alkaloids. Phytochemistry 27:2423–2427

Heng-Moss TM, Sarath G, Baxendale F, Novak D, Bose S, Ni X et al (2004) Characterization of oxidative enzyme changes in buffalo grasses challenged by *Blissus occiduus*. J Econ Entomol 97:1086–1095; PMID: 15279295. https://doi.org/10.1603/0022-0493(2004)097[1086:COOECI]2.0.CO;2

Hosokawa T, Kikuchi Y, Shimada M, Fukatsu T (2007) Obligate symbiont involved in pest status of host insect. Proc R Soc B Biol Sci 274:1979–1984

Howe GA, Jander G (2008) Plant immunity to insect herbivores. Annu Rev Plant Biol 59:41–66

Howe HF, Westley LC (1988) Ecological relationship of plants and animals. Oxford University Press, New York

Huang W, Zhikuan J, Qingfang H (2007) Effects of herbivore stress by *Aphis medicaginis* Koch on the malondialdehyde contents and activities of protective enzymes in different alfalfa varieties. Acta Ecol Sin 27:2177–2183. https://doi.org/10.1016/S1872-2032(07)60048-1
Huang Y, Xu C, Ma L, Zhang K, Duan C, MoM (2010) Characterization of volatiles produced from *Bacillus megaterium* YFM 3.25 and their nematicidal activity against Meloidogyne incognita. Eur J Plant Pathol 26:417–422
Humphrey PT, Nguyen TT, Villalobos MM, Whiteman NK (2014) Diversity and abundance of phyllosphere bacteria are linked to insect herbivory. Mol Ecol 23:1497–1515
Hynes RK, Boyetchko SM (2006) Research initiatives in the art and science of biopesticide formulations. Soil Biol Biochem 38:45–849
Jing X, Wong ACN, Chaston JM, Colvin J, McKenzie CL, Douglas AE (2014) The bacterial communities in plant phloem-sap-feeding insects. Mol Ecol 23:1433–1444
Johnson R, Narvaez J, An G, Ryan C (1989) Expression of proteinase inhibitors I and II in transgenic tobacco plants: effects on natural defence against *Manduca Sexta* larvae. Proc Natl Acad Sci U S A 86:9871–9875
Johnson MT, Campbell SA, Barrett SC (2015) Evolutionary interactions between plant reproduction and defense against herbivores. Annu Rev Ecol Evol Syst 46(1):191–213
Jones RT, Sanchez LG, Fierer N (2013) A cross-taxon analysis of insect-associated bacterial diversity. PLoS One 8:e61218
Jurat-Fuentes JL, Jackson TA (2012) Bacterial entomopathogens. In: Insect pathology. Academic, San Diego, pp 265–349
Kaiser W, Huguet E, Casas J, Commin C, Giron D (2010) Plant green-island phenotype induced by leaf-miners is mediated by bacterial symbionts. Proc R Soc B Biol Sci 277:2311–2319
Kang Z, Buchenauer H (2003) Immunocytochemical localization of cell wall-bound thionins and hydroxyproline-rich glycoproteins in *Fusarium culmorum*-infected wheat spikes. J Phytopathol 151(3):120–129
Kang SY, Kim YC (2007) Decursinol and decursin protect primary cultured rat cortical cells from glutamate-induced neurotoxicity. J Pharm Pharmacol 59(6):863–870
Karban R (2011) The ecology and evolution of induced resistance against herbivores. Funct Ecol 25:339–347. https://doi.org/10.1111/j.1365-2435.2010.01789.x
Karban R, Myer JH (1989) Induced plant responses to herbivory. Annu Rev Ecol Syst 20:331–348
Kariyat RR, Mauck KE, Balogh CM, Stephenson AG, Mescher MC, De Moraes CM (2013) Inbreeding in horsenettle (*Solanum carolinense*) alters night-time volatile emissions that guide oviposition by *Manduca sexta* moths. Proc R Soc B Biol Sci 280:20130020
Kartzinel TR, Chen PA, Coverdale TC et al (2015) DNA metabarcoding illuminates dietary niche partitioning by African large herbivores. Proc Natl Acad Sci U S A 112(26):8019–8024
Kelly D (1994) The evolutionary ecology of mast seeding. Trends Ecol Evol 9:465–470
Kempema LA, Cui XP, Holzer FM, Walling LL (2007) Arabidopsis transcriptome changes in response to phloem-feeding silverleaf whitefly nymphs. Similarities and distinctions in responses to aphids. Plant Physiol 143:849–865; PMID: 17189325. https://doi.org/10.1104/pp.106.090662
Kepenekci I, Hazir S, Oksal E, Lewis E (2018) Application methods of Steinernema feltiae, Xenorhabdus bovienii and Purpureocillium lilacinum to control root-knot nematodes in greenhouse tomato systems. Crop Prot 108:31–38. https://doi.org/10.1016/j.cropro.2018.02.009
Khan Z, Tiyagi SA, Mahmood I, Rizvi R (2012) Effect of N fertilisation, organic matter and biofertilisers on growth and yield of chilli in relation to management of plant-parasitic nematodes. Turk J Bot 36:73–81
Killani AS, Abaidoo RC, Akintokun AK, Abiala MA (2011) Antagonistic effect of indigenous Bacillus subtilis on root soil-borne fungal pathogens of cowpea. Researcher 3:11–18
Koike M, Shinya R, Aiuchi D, Mori M, Ogino R, Shinomiya H, Tani M, Goettel M (2011) Future biological control for soybean cyst nematode. In: El-Shemy HA (ed) Soybean physiology and biochemistry. Intech Open Access, Rijeka, pp 193–208

Kondo T, Yoshida K, Nakagawa A, Kawai T, Tamura H, Goto T (1992) Structural basis of blue-colour development in flower petals from *Commelina communis*. Nature 358:515–518

Kress WJ, Garcia-Robledo C, Uriarte M et al (2015) DNA barcodes for ecology, evolution, and conservation. Trends Ecol Evol 30(1):25–35

Kroschel J, Lacey LA (2009) Integrated pest management for the potato tuber moth, Phthorimaea operculella (Zeller)—a potato pest of global importance. In: Kroschel J, Lacey LA (eds) Tropical agriculture 20, advances in crop research 10. Margraf, Weikersheim, p 147

Kula AAR, Hartnett DC, Wilson GWT (2005) Effects of mycorrhizal symbiosis on tallgrass prairie plant-herbivore interactions. Ecol Lett 8:61–69

Kurnath P, Merz ND, Dearing MD (2016) Ambient temperature influences tolerance to plant secondary compounds in a mammalian herbivore. Proc Biol Sci 283(1822):pii: 20152387.

Lacey LA, Grzywacz D, Shapiro-Ilan DI, Frutos R, Brownbridge M, Goettel MS (2015) Insect pathogens as biological control agents: back to the future. J Invertebr Pathol 132:1–41

Lamelas A, Gosalbes MJ, Manzano-Marin A, Pereto J, Moya A, Latorre A (2011) *Serratia symbiotica* from the aphid *Cinara cedri*: a missing link from facultative to obligate insect endosymbiont. PLoS Genet 7:e1002357

Lefcheck JS, Duffy JE (2015) Multitrophic functional diversity predicts ecosystem functioning in experimental assemblages of estuarine consumers. Ecology 96(11):2973–2983

Lefcheck JS, Byrnes JE, Isbell F et al (2015) Biodiversity enhances ecosystem multifunctionality across trophic levels and habitats. Nat Commun 6:6936

Lemaire B, Janssens S, Smets E, Dessein S (2012) Endosymbiont transmission mode in bacterial leaf nodulation as revealed by a population genetic study of *Psychotria leptophylla*. Appl Environ Microbiol 78:284–287

Lemoine NP, Drews WA, Burkepile DE et al (2013) Increased temperature alters feeding behavior of a generalist herbivore. Oikos 122(12):1669–1678

Lemoine NP, Burkepile DE, Parker JD (2014a) Variable effects of temperature on insect herbivory. PeerJ 2:e376

Lemoine NP, Giery ST, Burkepile DE (2014b) Differing nutritional constraints of consumers across ecosystems. Oecologia 174(4):1367–1376

Lemoine NP, Burkepile DE, Parker JD (2016) Quantifying differences between native and introduced species. Trends Ecol Evol. 31(5):372–381

Lewis NG, Yamamoto E (1990) Lignin: occurrence, biogenesis and biodegradation. Annu Rev Plant Physiol Plant Mol Biol 41:455–496

Liao X, Lu HL, Fang W, St. Leger RJ (2014) Overexpression of a *Metarhizium robertsii* HSP25 gene increases thermotolerance and survival in soil. Appl Microbiol Biotechnol 98:777–783

Loarie SR, Duffy PB, Hamilton H et al (2009) The velocity of climate change. Nature 462(7276):1052–1055

Louis J, Shah J (2013) *Arabidopsis thaliana-Myzus persicae* interaction: shaping the understanding of plant defense against phloem-feeding aphids. Front Plant Sci 4:213

Lu X, Siemann E, Shao X et al (2013) Climate warming affects biological invasions by shifting interactions of plants and herbivores. Glob Chang Biol 19(8):2339–2347

Luz C, Tai, MHH, Santos AH, Silva HHG (2008) Impact of moisture on survival of Aedes aegypti eggs and ovicidal activity of Metarhizium anisopliae under laboratory conditions. Mem Inst Oswaldo Cruz 103: 214–215

Manninen AM, Holopainen T, Lyytikainen-Saarenmaa P, Holopainen JK (2000) The role of low-level ozone exposure and mycorrhizas in chemical quality and insect herbivore performance on Scots pine seedlings. Global Change Biol 6:111–121

McNaughton SJ (1998) Mineral nutrition and spatial concentration of African ungulates. Nature 334:343–345

Messelink GJ, Bennison J, Alomar O, Ingegno BL, Tavella L, Shipp L, Palevsky E, Wäckers FL (2014) Approaches to conserving natural enemy populations in greenhouse crops: current methods and future prospects. BioControl 59:377–393

Moen JH, Gardfjell L, Oksanen L, Ericson L, Ekerholm P (1993) Grazing by food-limited microtine rodents on a productive experimental plant community; does the "green desert" exist? Oikos 68:401–413

Mondal S, Baksi S, Koris A, Vatai G (2016) Journey of enzymes in entomopathogenic fungi. Pac Sci Rev 18:85–99

Mokbel AA, Ibrahim IKA, El-Saedy MAM, Hammad SE (2006) Plant-parasitic nematodes associated with some fruit trees and vegetable crops in northern Egypt. Egypt J Phytopathol 34(2):43–51

Moles AT, Peco B, Wallis IR et al (2013) Correlations between physical and chemical defenses in plants: tradeoffs, syndromes, or just many different ways to skin a herbivorous cat? New Phytol 198(1):252–263

Moore BD, Andrew RL, Kulheim C, Foley WJ (2014) Explaining intraspecific diversity in plant secondary metabolites in an ecological context. New Phytol 201:733–750

Moran PJ, Thompson GA (2001) Molecular responses to aphid feeding in Arabidopsis in relation to plant defense pathways. Plant Physiol 125:1074–1085

Moran NA, McCutcheon JP, Nakabachi A (2008) Genomics and evolution of heritable bacterial symbionts. Annu Rev Genet 42:165–190

Moscardi F, de Souza ML, de Castro MEB, Moscardi ML, Szewczyk B (2011) Baculovirus pesticides: present state and future perspectives. In: Ahmad I, Ahmad F, Pichtel J (eds) Microbes and microbial technology. Springer, New York, pp 415–445

Mueller RC, Martinez T, Sthultz CM, Gehring CA, Whitham TG (2005) The relationship between stem-galling wasps and the mycorrhizal colonization of *Quercus turbinella.* Can J Bot 83:1349–1353

Mundim FM, Alborn HT, Vieira-Neto EH et al (2017) A whole-plant perspective reveals unexpected impacts of above- and belowground herbivores on plant growth and defense. Ecology 98(1):70–78

Nagachandrabose S (2018) Liquid bioformulations for the management of root-knot nematode, Meloidogyne hapla that infects carrot. Crop Prot 114:155–161

Nana P, Nchu F, Ekesi S, Boga HI, Kamtchouing P, Maniania NK (2015) Efficacy of spot spray application of *Metarhizium anisopliae* formulated in emulsifiable extract of *Calpurnia aurea* in attracting and infecting adult *Rhipicephalus appendiculatus* ticks in semi-field experiments. J Pest Sci 88:613–619

Nash KL, Graham NA, Jennings S et al (2016) Herbivore cross-scale redundancy supports response diversity and promotes coral reef resilience. J Appl Ecol. 53:646–655

O'Connor MI, Gilbert B, Brown CJ (2011) Theoretical predictions for how temperature affects the dynamics of interacting herbivores and plants. Am Nat 178(5):626–638

Oates JF, Waterman PG, Choo GM (1980) Food selection by the south Indian leaf-monkey, *Presbytis johnii*, in relation to leaf chemistry. Oecologia 45:45–56

Oliver KM, Degnan PH, Burke GR, Moran NA (2010) Facultative symbionts in aphids and the horizontal transfer of ecologically important traits. Annu Rev Entomol 55:247–266

Ortiz-Urquiza A, Keyhani NO (2015) Action on the surface: entomopathogenic fungi versus the insect cuticle. Insects 4:357–374

Pacovsky RS, Rabin LB, Montllor CB, Waiss ACJ (1985) Host-plant resistance to insect pests altered by *Glomus fasciculatum* colonization, pp. 288. In: Molina R (ed) Proceedings of the 6th North American conference on mycorrhiza. Oregon State University, Corvallis

Pan H, Xiu C, Lu Y (2015) A combination of olfactory and visual cues enhance the behavioral responses of *Apolygus lucorum.* J Insect Behav 28:525–534

Perez-Barberia FJ, Elston DA, Gordon IJ, Illius AW (2004) The evolution of phylogenetic differences in the efficiency of digestion in ruminants. Proc R Soc Lond Ser B 271:1081–1090

Petrisor C, Stoian G (2017) The role of hydrolytic enzymes produced by entomopathogenic fungi in pathogenesis of insects. Rom J Plant Prot X:66–72

Picman AK (1986) Biological activities of sesquiterpene lactones. Biochem Syst Ecol 14:255–281

Pieterse CMJ, Van der Does D, Zamioudis C, Leon-Reyes A, Van Wees SCM (2012) Hormonal modulation of plant immunity. Annu Rev Cell Dev Biol 28:489–521
Pieterse CM, Zamioudis C, Berendsen RL, Weller DM, Van Wees SC, Bakker PA (2014) Induced systemic resistance by beneficial microbes. Annu Rev Phytopathol 52:34
Pineda A, Zheng SJ, van Loon JJA, Pieterse CMJ, Dicke M (2010) Helping plants to deal with insects: the role of beneficial soil-borne microbes. Trends Plant Sci 15:507–514
Porter SS, Bantay R, Friel CA, Garoutte A, Gdanetz K, Ibarreta K, Moore BM, Shetty P, Siler E, Friesen ML (2019) Beneficial microbes ameliorate abiotic and biotic sources of stress on plants. Funct Ecol 1365–2435:13499
Pozo MJ, Van Der Ent S, Van Loon LC, Pieterse CMJ (2008) Transcription factor Myc2 is involved in priming for enhanced defense during rhizobacteria-induced systemic resistance in Arabidopsis thaliana. New Phytol 180:511–523
Price PW (1997) Insect ecology, 3rd edn. Wiley, New York
Rabin LB, Pacovsky RS (1985) Reduced larva growth of two lepidoptera (Noctuidae) on excised leaves of soybean infected with a mycorrhizal fungus. J Econ Entomol 78:1358–1363
Radwan M, Farrag S, Abu-Elamayem M, Ahmed N (2012) Biological control of the root-knot nematode, Meloidogyne incognita on tomato using bioproducts of microbial origin. Appl Soil Ecol 56:58–62
Raffa KF, Powell EN, Townsend PA (2013) Temperature-driven range expansion of an irruptive insect heightened by weakly coevolved plant defenses. Proc Natl Acad Sci U S A 110:2193–2198
Reeves JL (2011) Vision should not be overlooked as an important sensory modality for finding host plants. Environ Entomol 40:855–863
Research and Markets (2016) Global pesticides market segmented by type, application area and geography. Trends and forecasts (2015-2020). Sustainability, regulation & competition. http://www.researchandmarkets.com/research/4hd338/global_pesticides. Accessed 10 Mar 2017
Reymond P, Bodenhausen N, Van Poecke RMP, Krishnamurthy V, Dicke M, Farmer EE (2004) A conserved transcript pattern in response to a specialist and a generalist herbivore. Plant Cell 16:3132–3147; PMID: 15494554. https://doi.org/10.1105/tpc.104.026120
Rice EL (1984) Allelopathy, 2nd edn. Academic, New York
Richardson AO, Palmer JD (2007) Horizontal gene transfer in plants. J Exp Bot 58:1–9
Rieske LK, Rhoades CC, Miller SP (2003) Foliar chemistry and gypsy moth, *Lymantria dispar* (L.), herbivory on pure Am. chestnut, Castanea dentata (Fam: Fagaceae), and a disease-resistant hybrid. Environ Entomol 32:359–365
Riley ME, Johnston CA, Feller IC et al (2014) Range expansion of *Aratus pisonii* (Mangrove tree crab) into novel vegetative habitats. Southeast Nat 13(4):N43–N48
Robert-Seilaniantz A, Grant M, Jones JDG (2011) Hormone crosstalk in plant disease and defense: more than just Jasmonate-Salicylate antagonism. Annu Rev Phytopathol 49:317–343
Ruanpanun P, Tangchitsomkid N, Hyde KD, Lumyong S (2010) Actinomycetes and fungi isolated from plant-parasitic nematode infested soils: screening of the effective biocontrol potential, indole-3-acetic acid and siderophore production. World J Microbiol Biotechnol 26:1569–1578
Russell JA, Weldon S, Smith AH, Kim KL, Hu Y, Lukasik P, Doll S, Anastopoulos I, Novin M, Oliver KM (2013) Uncovering symbiont driven genetic diversity across North American pea aphids. Mol Ecol 22:2045–2059
Sahayaraj K (ed) (2014) Basic and applied aspects of biopesticides. Springer, New Delhi, pp 31–46
Salgado CG, da Silva JP, Diniz JA, da Silva MB, da Costa PF, Teixeira C, Salgado UI (2004) Isolation of Fonsecaea pedrosoi from thorns of *Mimosa pudica*, a probable natural source of chromoblastomycosis. Rev Inst Med Trop Sao Paulo 46:33–36
Saltz D, Ward D (2000) Responding to a three-pronged attack: desert lilies subject to herbivory by *Dorcas gazelles*. Plant Ecol 148:127–138
Samuels RI, Santos AV, Silva CP (2011) Enzymology of entomopathogenic fungi. In: Borgio JF et al (eds) Microbial insecticides -principles and applications. Nova Science, New York, pp 71–92

Sarwar M (2015) Biopesticides: an effective and environmental friendly insect-pests inhibitor line of action. Int J Eng Adv Res Technol 1(2):10–15
Schafer H, Wink M (2009) Medicinally important secondary metabolites in recombinant microorganisms or plants: progress in alkaloid biosynthesis. Biotechnol J 4(12):1684–1703
Schmitz OJ, Buchkowski RW, Burghardt KT et al (2015) Chapter Ten—Functional traits and trait-mediated interactions: connecting community-level interactions with ecosystem functioning. In: Samraat Pawar GW, Anthony ID (eds) Advances in ecological research, vol 52. Academic, New York, pp 319–343
Schuldt A, Bruelheide H, Durka W et al (2012) Plant traits affecting herbivory on tree recruits in highly diverse subtropical forests. Ecol Lett 15(7):732–739
Schuman MC, Baldwin IT (2016) The layers of plant responses to insect herbivores. Annu Rev Entomol 61:373–394
Scogings PF, Dziba LE, Gordon IJ (2004) Leaf chemistry of woody plants in relation to season, canopy retention and goat browsing in a semiarid subtropical savanna. Aust Ecol 29:278–286
Sebata A, Ndlovu LR (2012) Effect of shoot morphology on browse selection by free ranging goats in a semi-arid savanna. Livestock Sci 144:96–102
Seigler DS (1981) Secondary metabolites and plant systematic. In: Conn EE (ed) The biochemistry of plants, Secondary plant products, vol 7. Plenum, New York, pp 139–176
Sharma IP, Sharma AK (2017) Co-inoculation of tomato with an arbuscular mycorrhizal fungus improves plant immunity and reduces root-knot nematode infection. Rhizosphere 4:25–28
Sharma A, Srivastava A, Shukla AK, Srivastava K, Srivastava AK, Saxena AK (2020) Entomopathogenic fungi: a potential source for biological control of insect pests. In: Solanki MK et al (eds) Phytobiomes: current insights and future vistas. Springer Nature Singapore Pte Ltd, Singapore, pp 225–250. https://doi.org/10.1007/978-981-15-3151-4_9
Shipley LA, Illius AW, Danell K, Hobbs NT, Spalinger DE (1999) Predicting bite size selection of mammalian herbivores: a test of a general model of diet optimization. Oikos 84:55–68
Singh A, Bhardwaj A, Singh IK (2019a) Biocontrol agents: potential of biopesticides for integrated pest management. In: Giri B et al (eds) Biofertilizers for sustainable agriculture and environment, Soil biology, vol 55. Springer Nature Switzerland AG, Basel. https://doi.org/10.1007/978-3-030-18933-4_19
Singh A, Sharma P, Kumari A, Kumar R, Pathak DV (2019b) Management of root-knot nematode in different crops using microorganisms. In: Varma A et al (eds) Plant biotic interactions. Springer Nature Switzerland AG, Basel, pp 85–99. https://doi.org/10.1007/978-3-030-26657-8_6
Smith SE, Read DJ (1997) Mycorrhizal symbiosis, 2nd edn. Academic, London
Smith SE, Read DJ (2008) Mycorrhizal symbiosis, 3rd edn. Academic, London
Smith-Ramesh LM, Moore AC, Schmitz OJ (2017) Global synthesis suggests that food web connectance correlates to invasion resistance. Glob Chang Biol 23(2):465–473
Sreevidya VS, Srinivasa RC, Rao C, Sullia SB, Ladha JK, Reddy PM (2006) Metabolic engineering of rice with soyabean isoflavone synthase for promoting nodulation gene expression in rhizobia. J Exp Bot 57(9):1957–1969
Steppuhn A, Baldwin IT (2007) Resistance management in a native plant: nicotine prevents herbivores from compensating for plant protease inhibitors. Ecol Lett 10:499–511; PMID: 17498149. https://doi.org/10.1111/j.1461-0248.2007.01045.x
Sterner RW, Hessen DO (1994) Algel nutrient limitation and nutrition of aquatic herbivores. Annu Rev Ecol Syst 25:1–29
Stout MJ, Riggio MR, Yang Y (2009) Direct induced resistance in *Oryza sativa* to Spodoptera frugiperda. Environ Entomol 38:1174–1181
Stowe KA, Marquis RJ, Hochwender CG, Simms EL (2000) The evolutionary ecology of tolerance to consumer damage. Annu Rev Ecol Syst 31:565–595
Sugio A, Dubreuil G, Giron D, Simon J (2014) Plant–insect interactions under bacterial influence: ecological implications and underlying mechanisms. J Exp Bot 66:435. https://doi.org/10.1093/jxb/eru435
Taiz L, Zeiger E (1995) Plant physiology. Panima Publishing Corporation, New Delhi

Tao L, Berns AR, Hunter MD (2014) Why does a good thing become too much?: interactions between foliar nutrients and toxins determine performance of an insect herbivore. Funct Ecol 28(1):190–196

Todaka N, Inoue T, Saita K, Ohkuma M, Nalepa CA, Lenz M, Kudo T, Moriya S (2010) Phylogenetic analysis of cellulolytic enzyme genes from representative lineages of termites and a related cockroach. PLoS One 5:e8636

Toju H, Fukatsu T (2011) Diversity and infection prevalence of endosymbionts in natural populations of the chestnut weevil: relevance of local climate and host plants. Mol Ecol 20:853–868

Tscharntke T, Thiessen S, Dolch R, Boland W (2001) Herbivory, induced resistance, and interplant signal transfer in *Alnus glutinosa*. Biochem Syst Ecol 29:1025–1047

Turlings TCJ, Loughrin JH, Mccall PJ, Roese USR, Lewis WJ, Tumlinson JH (1995) How caterpillar-damaged plants protect themselves by attracting parasitic wasps. Proc Natl Acad Sci U S A 92:4169–4174

Turner JM (1994) Scherophylly: primarily protective? Funct Ecol 8:669–675

Van der Ent S, Van Hulten M, Pozo MJ, Czechowski T, Udvardi MK, Pieterse CMJ, Ton J (2009) Priming of plant innate immunity by rhizobacteria and beta-aminobutyric acid: differences and similarities in regulation. New Phytol 183:419–431

Van Etten H, Temporini E, Wasmann C (2001) Phytoalexin (and phytoanticipin) tolerance as a virulence trait: why is it not required by all pathogens? Physiol Mol Plant Pathol 59:83–93

Van Frankenhuyzen K (2009) Insecticidal activity of *Bacillus thuringiensis* crystal proteins. J Invertebr Pathol 101(1):1–16

van Lenteren JC (2012) The state of commercial augmentative biological control: plenty of natural enemies, but a frustrating lack of uptake. BioControl 57:1–20

van Lenteren JC, Godfray HCJ (2005) European science in the enlightenment and the discovery of the insect parasitoid life cycle in The Netherlands and Great Britain. Biol Control 32:12–24

van Lenteren JC, Bueno VHP (2003) Augmentative biological control of arthropods in Latin America. BioControl 48:123–139

Van Loon LC, Pierpoint WS, Boller T, Conejero V (1994) Recommendations for naming plant pathogenesis-related proteins. Plant Mol Biol Rep 12:245–264

Verges A, Steinberg PD, Hay ME et al (2014) The tropicalization of temperate marine ecosystems: climate-mediated changes in herbivory and community phase shifts. Proc Biol Sci 281(1789):20140846

Verhoeven KJ, Biere A, Harvey JA et al (2009) Plant invaders and their novel natural enemies: who is naïve? Ecol Lett 12(2):107–117

Vicari M, Hatcher PE, Ayres PG (2002) Combined effect of foliar and mycorrhizal endophytes on an insect herbivore. Ecology 83:2452–2464

Voelckel C, Weisser WW, Baldwin IT (2004) An analysis of plant-aphid interactions by different microarray hybridization strategies. Mol Ecol 13:3187–3195; PMID: 15367131. https://doi.org/10.1111/j.1365-294X.2004.02297.x

War AR, Paulraj MG, Ahmad T, Buhroo AA, Hussain B, Ignacimuthu S, Sharma HC (2012) Mechanisms of plant defense against insect herbivores. Plant Signal Behav 7:1306–1320

War AR, Hussain B, Sharma HC (2013a) Induced resistance in groundnut by jasmonic acid and salicylic acid through alteration of trichome density and oviposition by *Helicoverpa armigera* (Hubner) (Lepidoptera: Noctuidae). AOB Plants 5:plt053. https://doi.org/10.1093/aobpla/plt053

War AR, Sharma HC, Paulraj MG, Hussain B, Buhroo AA, War MY, Ignacimuthu S, Sharma HC (2013b) Effect of plant secondary metabolites on *Helicoverpa armigera*. J Pest Sci 86:399–408

Weeks AR, Turelli M, Harcombe WR, Reynolds KT, Hoffmann AA (2007) From parasite to mutualist: rapid evolution of Wolbachia in natural populations of Drosophila. PLoS Biol 5:997–1005

Wetzel WC, Kharouba HM, Robinson M, Holyoak M, Karban R (2016) Variability in plant nutrients reduces insect herbivore performance. Nature 539:425–427

White TRC (1993) The inadequate environment: nitrogen and the abundance of animals. Springer, Berlin

Willerslev E, Davison J, Moora M et al (2014) Fifty thousand years of Arctic vegetation and megafaunal diet. Nature 506(7486):47–51

Willey JM, Sherwood LM, Woolverton CJ (2008) In: Willey JM, Sherwood LM, Woolverton CJ (eds) Prescot, Harley, and Klein's microbiology, 7th edn. McGraw Hill, New York, pp 1009–1011

Williams AA, Eastman SF, Eash-Loucks WE et al (2014) Record northernmost endemic mangroves on the United States Atlantic Coast with a note on latitudinal migration. Southeast Nat 13(1):56–63

Wooley SC, Paine TD (2007) Can intra-specific genetic variation in arbuscular mycorrhizal fungi (*Glomus etunicatum*) affect a mesophyll-feeding herbivore (*Tupiocoris notatus* Distant)? Ecol Entomol 32:428–434

Wu JQ, Baldwin IT (2010) New insights into plant responses to the attack from insect herbivores. Annu Rev Genet 44:1–24

Wurst S, Dugassa-Gobena D, Langel R, Bonkowski M, Scheu S (2004) Combined effects of earthworms and vesicular-arbuscular mycorrhizas on plant and aphid performance. New Phytol 163:169–176

Wuyts N, De waele D, Swennen R (2006) Extraction and partial characterization of polyphenol oxidase from banana (*Musa acuminate* grandrnaine) roots. Plant Physiol Biochem 44:308–314

Xu C, Zhang X, Qian Y, Chen X, Liu R, Zeng G et al (2014) A high-throughput gene disruption methodology for the entomopathogenic fungus *Metarhizium robertsii*. PLoS One 9:e107657

Ying SH, Feng MG (2011) Integration of *Escherichia coli* thioredoxin (trxA) into Beauveria bassiana enhances the fungal tolerance to the stresses of oxidation, heat and UV-B irradiation. Biol Control 59:255–260

Zamora R, Hodar JA, Gomez JM (1999) Plant herbivore interaction: beyond a binary vision. In: Pugnaire FI, Marcel VF (eds) Handbook of functional plant ecology. Dekker, New York, pp 681–683

Zarate SI, Kempema LA, Walling LL (2007) Silver leaf whitefly induces salicylic acid defenses and suppresses effectual jasmonic acid defenses. Plant Physiol 143:866–875

Zhao LY, Chen JL, Cheng DF, Sun JR, Liu Y, Tian Z (2009) Biochemical and molecular characterizations of Sitobion avenae-induced wheat defense responses. Crop Prot 28:435–442. https://doi.org/10.1016/j.cropro.2009.01.005

Zhao H, Lovettx B, Fang W (2016) Genetically engineering entomopathogenic fungi. Adv Genet 94:137–163

Zhou L, Yuen G, Wang Y et al (2016) Evaluation of bacterial biological control agents for control of root-knot nematode disease on tomato. Crop Prot 84:8–13

Zhu-Salzman K, Salzman RA, Ahn JE, Koiwa H (2004) Transcriptional regulation of sorghum defense determinants against a phloem-feeding aphid. Plant Physiol 134:420–431; PMID: 14701914. https://doi.org/10.1104/pp.103.028324

Chapter 20
Eco-Friendly Approaches for the Alleviation of Root-Knot Nematodes

Gufran Ahmad, Yasar Nishat, Moh. Sajid Ansari, Amir Khan, Mohammad Haris, and Abrar Ahmad Khan

Contents

1 Introduction

Phytopathogens are major biological barriers and a significant threat to food products (Hussain et al. 2020). Every pathogen in any community may inhibit the growth and development and sometime kill the whole plant. Root-knot nematodes (*Meloidogyne incognita*) are microscopic and eel-like roundworms. They proliferate many problems for different crops and ornamentals. Root-knot nematode lives and feeds in the root of various plants, and the most susceptible crop is brinjal. Although numerous saprophytic nematodes exist in soils, they differ from plant pathogenic nematodes in that the latter have styles that allow feeding on plants. Plant pathogenic nematodes can be ecto- or endoparasites depending on their feeding location in the root and sedentary or migratory depending on their movement during feeding. Nematode feeding on plants can cause mechanical injury and alter

G. Ahmad (✉) · M. S. Ansari · A. Khan · M. Haris · A. A. Khan
Section of Environmental Botany and Plant Nematology, Department of Botany, Aligarh Muslim University, Aligarh, India

Y. Nishat
Division of Plant Pathology and Nematology, Department of Botany, Aligarh Muslim University, Aligarh, India

© The Author(s), under exclusive license to Springer Nature Switzerland AG 2021

H. I. Mohamed et al. (eds.), *Plant Growth-Promoting Microbes for Sustainable Biotic and Abiotic Stress Management*,
https://doi.org/10.1007/978-3-030-66587-6_20

cell physiology because of enzymes in their saliva. Consequent symptoms on plants include yellowing, stunting, hyperplasia at feeding sites such as root knots and root galls, root lesions, etc. Root-knot nematodes (*Meloidogyne* spp.) are capable of damaging a wide range of plants, mainly vegetables, and cause yield losses in agriculture (Sikora and Fernandez 2005). Root-knot nematode most economically destructing species is on field, horticulture, crops (Ye et al. 2015), and vegetable crops losses around 10% (Koenning et al. 1999). *Meloidogyne* species are obligate parasites of the plant roots like monocots, dicots, herbs, shrubs, and woody plants. *Meloidogyne*'s infection symptoms include root galls (formation of galls due to damages in water and nutrient-conducting abilities of the roots), shoot chlorosis, deficiency of nutrient, stunted growth, and wilting (Hunt and Handoo 2009). Root-knot nematodes are controlled by various types of nematicides that have hazardous effects on the environment. Hence, the identification of new approaches alternate to harmful chemical nematicides could be effective in controlling root-knot nematodes.

On the other hand, many studies (Meyer and Roberts 2002; Ciancio and Mukerji 2007; Hashem and Abo-Elyousr 2011; Affokpon et al. 2011) reveal that some other methods of management of root-knot nematode by biocontrol (fungal, bacterial) and organic amendments are beneficial cultural practices in nematode management and suppress the nematode population.

In the past several years, a strong movement that uses biological control agents in the management of nematode and the momentum behind this due to nematicides is toxic, and biocontrol agents are eco-friendly. The use of biocontrol agents at the place of nematicides is an awareness of their danger. In the present scenario, we have limited cultivars, economic pressure, and limited land for agriculture. We cannot afford expensive nematicides available in the market and also hazardous for flora and fauna of agricultural soil.

Nematologists emphasize on integrated biological management strategies for nematode management. Various fungi, bacteria, and different sporozoans are biological tools called biocontrol to reduce nematode populations under the greenhouse conditions, laboratory, and agricultural fields. However, the field experiments have been inconclusive, and some studies were disappointing. According to the International Potato Center (CIP), Peru, the nematologists have discovered a fungus *Purpureocillium lilacinum* that manages the root-knot nematode population and *Meloidogyne* spp. (Franco et al. 1981). The International Potato Center (CIP) provides a new path and an alternate method of nematode management from chemical control to biological control.

Bacterial biocontrol agents also play a significant role in plant growth and health and reduce the nematode population. Rhizobacteria reduce pathogen growth and supply micro- and macronutrient to protect from diseases caused by a different type of pathogens (Kloepper et al. 1980; Griffin 1990). Various bacteria have the capability that causes infections in plant-parasitic nematode's different patterns like making colonies or trapping nematode by the spores. The primary bacteria studied as a biocontrol agent of the nematode are active in the rhizosphere of the plants and capable of invading the plant tissue. They are facultative endophytic bacteria like *Pseudomonas* species and *Bacillus* species. Some are obligatory parasitic bacteria, such as *Pasteuria* species. These are all the above bacteria that show exceptional destructive abilities to reduce the nematode (*Meloidogyne* spp.) population (Machado et al. 2012). Ambo et al. (2010) reveal that vermicompost and *Glomus aggregatum* alone and in combination with *Bacillus coagulans* observed an excellent plant growth and biomass of test plant and notice that the root-knot nematode population was decreasing. The seven strains of *Gluconacetobacter diazotrophicus* bacteria show antagonistic efficacy against *M. incognita* (Bansal et al. 2005). The growth of *Solanum lycopersicum*, *P. fluorescens*, *A. chroococcum*, and *A. brasilense* and composted organic fertilizers show a reduction of the multiplication of root-knot nematode and enhance plant health (Siddiqui 2004).

An organic amendment has been used from immemorial time when the farmers add different types of crop waste conveniently available. Sometimes observed by landholders, organic additive may be animal and plant origin, improve soil health, and increase crop yield. Organic wastes showed a beneficial effect and provide extra nourishment to the crop (Muller and Gooch 1982). In 1973 a summary published by Singh and Sitaramaiah focused on use of organic amendments for control nematode. In the literature, many trails of organic amendments like oil cake, crop waste, compost, manure, green manure, organic manure, poultry manure, different extracts, processed crop waste (husk, straw, etc.), agro-industrial waste (fly ash, charcoal, etc.), sawdust, and chopped leaves are used by different researchers as organic additives to improve crop yields and suppress the root-knot disease (Muller and Gooch 1982). The severity of *Meloidogyne incognita* was maximum reduced when chopped leaves of *Azadirachta indica* and *Ricinus communis* were used in soil (Akhtar and Mahmood 1993). Root-knot nematode, *Meloidogyne incognita*, population significantly reduced when treated with neem, mahua, and castor oil cake on the *Polianthes tuberosa*; oil cake also enhances the growth of plant (Jothi and Poornima 2017). Biocontrol agents *Paecilomyces lilacinus* and *Trichoderma viride* alone and in combination with mustard oil cake reduced root-knot disease caused by *M. incognita* (Goswami et al. 2006). Fungal biocontrol of *Trichoderma harzianum* and organic wastes, viz., cow urine and organic additives, farmyard manure, and vermicompost, was tested separately and in combination with control of *M. incognita* population in *Withania somnifera* (Pandey et al. 2011).

In the management of root-knot disease caused by *Meloidogyne* spp., the previous study reveals that this disease is reduced by biocontrol agents, organic amendments, and different eco-friendly organic wastes.

2 Fungi as a Biocontrol Agent Against Root-Knot Nematode

The use of chemical nematicides is being limited, given the growing concern for the umbworld and health of human being that has led to its ban. Control of chemically synthesized nematicides is one of the most trending methods in the present; however, the environment and health perilous by chemical nematicides have led to the gradual elimination of certain effective nematicides for certain crops. Biological control is a plausible option with different bacteria, fungi, and other predators (e.g.,

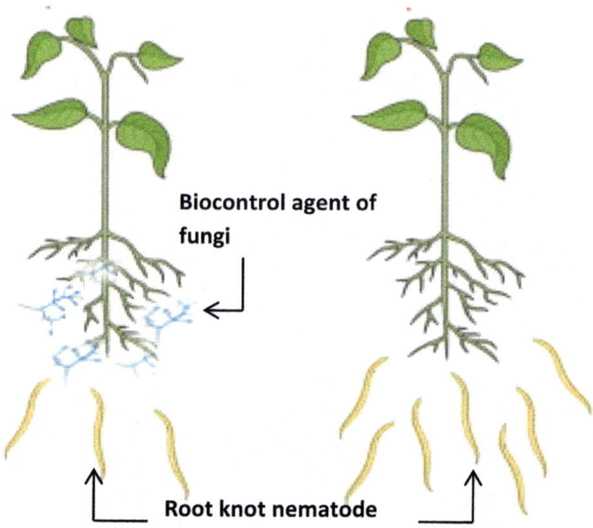

Fig. 20.1 Management of RKN by nematophagous fungi

mites). Management of RKN through biocontrol agents of fungi is the most diverse among all the enemies of nematodes, using a variety of mechanisms to hold and kill the nematodes (Fig. 20.1).

Different nematophagous fungi perform various mechanisms to arrest the nematode and kill them (Stirling 2014). Nematophagous fungi have more than 700 species which come under various phyla, such as *Zygomycota*, *Chytridiomycota*, *Basidiomycota*, and *Ascomycota*. Furthermore, organisms that belong to the phylum *Oomycota* have also shown nematicidal activities (Li et al. 2015).

Wei et al. (2009) utilized a detection strategy that depends on the production of chitinase and protease to recognize fungi with the greatest potential which suppress nematodes. Three isolates selected for this strategy, one *P. chlamydosporia* and two *P. lilacinus*, are responsible for reduced root galling of *Meloidogyne* sp. in tomato field from 48% to 61% and increase the yield through a similar percentage.

Nematophagous fungi are cosmopolitan in nature which are capable of modifying saprophytic behavior in carnivores, due to which they can feed on nematodes under harsh conditions for their nutrition. They are natural adversary of plant-parasitic nematodes that are developed in a highly sophisticated infection manner (Braga and Araújo 2014; Degenkolb and Vilcinskas 2016). The use of nematophagous fungi for all natural enemies of nematodes is an exciting and fast-developing area of research (Jyoti and Singh 2017). Many researchers have prepared bionematicides that are sold in the world market. The list of certain available fungi against RKNs acts as biocontrol (Table 20.1).

Overall, most fungal genera controlled RKNs which improved plant growth and yield performance. Some fungal genera such as *Paecilomyces* spp., *Pochonia* spp., and *Trichoderma* spp. are maximum studied against RKNs. *Trichoderma* spp.,

Table 20.1 Fungal species (biocontrol agents) against RKNs on different crops

Fungus	Nematode managed	Crops	References
Aspergillus niger	*Meloidogyne incognita*	Mung bean	Bhat and Wani (2012)
Aspergillus niger	*Meloidogyne* spp.	Tomato	Li et al. (2011)
Aspergillus niger	*Meloidogyne javanica*	Pigeon pea	Askary (2012)
Paecilomyces lilacinus	*Meloidogyne graminicola*	Rice	Narasimhamurthy et al. (2017a, b)
Paecilomyces lilacinus	*Meloidogyne incognita*	Black gram	Kumar et al. (2017)
Paecilomyces lilacinus	*Meloidogyne incognita*	Okra	Simon and Pandey (2010)
Paecilomyces lilacinus	*Meloidogyne incognita*	Tobacco	Ramakrishnan and Nagesh (2011)
Paecilomyces lilacinus	*Meloidogyne javanica*	Tomato	Ganaie and Khan (2010)
Paecilomyces lilacinus	*Meloidogyne incognita*	Pittosporum tobira (mock orange)	Baidoo et al. (2017)
Paecilomyces lilacinus	*Meloidogyne incognita*	Tobacco	Ramakrishnan and Rao (2013)
Paecilomyces lilacinus	*Meloidogyne incognita*	Brinjal	Nisha and Sheela (2016)
Paecilomyces lilacinus	*Meloidogyne incognita*	Tomato	Khalil et al. (2012a, b)
Pochonia chlamydosporia	*Meloidogyne incognita*	Tomato	Silva et al. (2017)
Pochonia chlamydosporia	*Meloidogyne javanica*	Lettuce	Viggiano et al. (2015)
Pochonia chlamydosporia	*Meloidogyne javanica*	Brinjal	Parihar et al. (2015)
Pochonia chlamydosporia	*Meloidogyne incognita*	Okra	Kumar and Jain (2010a)
Pochonia chlamydosporia	*Meloidogyne incognita*	Okra	Dhawan and Singh (2010)
Pochonia chlamydosporia	*Meloidogyne incognita*	Common bean	Sharf et al. (2014)
Trichoderma harzianum	*Meloidogyne javanica*	Tomato	Feyisa et al. (2016)
Trichoderma harzianum	*Meloidogyne incognita*	Pea	Brahma and Borah (2016)
Trichoderma harzianum	*Meloidogyne incognita*	Brinjal	Devi et al. (2016)
Trichoderma harzianum	*Meloidogyne incognita*	French bean	Gogoi and Mahanta (2013)

(continued)

Table 20.1 (continued)

Fungus	Nematode managed	Crops	References
Trichoderma harzianum	*Meloidogyne incognita*	Brinjal	Kumar and Chand (2015)
Trichoderma harzianum	*Meloidogyne incognita*	Green gram	Deori and Borah (2016)
Trichoderma harzianum	*Meloidogyne incognita*	Green gram	Singh and Mahanta (2013)
Trichoderma harzianum	*Meloidogyne javanica*	Tomato	Jamshidnejad et al. (2013)
Trichoderma harzianum	*Meloidogyne* spp.	Tomato	Khattak and Khattak (2011)
Trichoderma harzianum, Trichoderma viride	*Meloidogyne incognita*	Okra	Kumar and Jain (2010b)
Trichoderma harzianum, Trichoderma viride	*Meloidogyne javanica*	Tomato	Al-Hazmi and Javeed (2016)
Trichoderma harzianum and *Trichoderma viride*	*Meloidogyne* spp.	Round-leaf fountain palm	Jegathambigai et al. (2011)
Trichoderma viride	*Meloidogyne incognita*	Mulberry	Muthulakshmi et al. (2010)
Trichoderma viride	*Meloidogyne incognita*	Mulberry	Muthulakshmi and Devrajan (2015)
Trichoderma viride	*Meloidogyne graminicola*	Rice	Priya (2015)
Trichoderma viride	*Meloidogyne incognita*	Cowpea	Kumar et al. (2011)

Paecilomyces chlamydosporia, and *P. lilacinus* are common soil dwellers, and few strains are aggressive parasites of sedentary stage of nematode (Siddiqui and Mahmood 1996; Sharon et al. 2001). Some toxic metabolites are also released from *Trichoderma* spp. (Khan and Saxena 1997; Sharon et al. 2001). A field area infected by *M. incognita* and *T. harzianum* decreased galling in the roots of tomato approx. 47% compared with untreated field area (Goswami et al. 2008). Another eco-friendly biological control organism, arbuscular mycorrhizal fungi (AMF), is responsible for the management of PPN. In vitro, field and greenhouse experiments have shown protective effects against PPN with the help of AMF on plants like bananas, tomato, and coffee (Calvet et al. 2001; Vos et al. 2012; Koffi et al. 2013; Alban et al. 2013).

3 Bacteria as Biocontrol Against *Meloidogyne* spp.

Since microorganisms inside the rhizosphere are widespread, they are an abundant source of biologically active, beneficial compounds that can help farmers (Ramezani Moghaddam et al. 2014). Throughout the previous years, several reports have examined the impact of rhizobacteria on RKNs, like *Bacillus*, *Pseudomonas*, and *Pasteuria*. These are key genera of nematophagous soil microbe (Tian et al. 2007; Li et al. 2015) and have a significant potential since the first pathogen defense line targeting rhizosphere root plants could be named (Raddy et al. 2013; Tiwari et al. 2017). Such nematophagous bacteria exhibit diverse pathways to combat nematodes and overt antibiosis direct parasites (Lee and Kim 2016), reducing competition from the plant root for essential resources and ecological niches (Mendoza et al. 2008), and may cause systemic resistance to a host plant (Cawoy et al. 2011). Antibiosis for the production of reactive antibiotics, toxin, and volatile organic compounds (Rahman et al. 2018) is the most closely researched and commonly known pathway (Saraf et al. 2014). Certain rhizobacteria, such as *Pasteuria*, *Pseudomonas*, *Streptomyces*, and *Bacillus* (74%), are known as nematicidal traits as compared to *Pseudomonas* (54.77%) which had the largest decrease rate for *M. javanica* egg hatching (Turatto et al. 2017). In a research of the antagonistic ability, growth-enhancing strains of rhizobacteria adolescents of the second stage were *M. incognita*. The ability of three species of *Bacillus*, namely, *B. subtilis*, *B. coagulans*, and *B. firmus*, to combat *M. javanica* was investigated, and *B. subtilis* was identified as the cause of the highest decline in eggplant growth in nematode infection (Abbasi et al. 2014), while *B. coagulans* increased significantly plant production. Containing eight bacterial isolates, respectively, *B. subtilis*, the inhibitions of complete *M. incognita* egg hatching were found to be responsible. Thus, *Bacillus* genus members can be used as the natural biocontrol rhizobacteria (Metwally et al. 2015). B. subtilis produces various compounds with antimicrobial properties, including lipopeptides, exoenzymes, and volatile organic compounds and used in biological control and responsible for the development, under airborne conditions, of the most resistant endospores (Gao et al., 2016). One of the most studied and characterized secondary metabolites of B. subtilis is surfactin, a cyclic lipopeptide that has many important but distinct functions, such as signaling activities and reduction of surface tension (Sansinenea and Ortiz 2011). Such dormant spores are highly resistant to severe conditions like food scarcity, water shortages, elevated temperatures, and adverse pH levels (Cawoy et al. 2011). In the technical point of view, *Bacillus*' capacity to manufacture thermal-resistant and desiccator-prone endospores is extremely beneficial to formulate (Ongena and Jacques 2008) a healthy commodity which can be sold on the market under normal conditions (Padgham and Sikora 2007). This also helps keep bionematicides healthy when they are being processed (Rosas-García 2009). *Bacillus* spp.'s capacity to stimulate the growth of plants and antagonize plant pests, like root-knot nematode (RKNs) (Wu et al. 2015), exists concurrently and is therefore interlinked. Many advantageous characteristics would also not be isolated as goods that are sold by regulatory

Table 20.2 Application of various Bacterial spp. that have a nematicidal action against root-knot nematode species

Bacteria	Nematode control	Crop control	References
Pasteuria penetrans	*Meloidogyne javanica*	Tomato	Mankau and Prasad (1972)
Pasteuria penetrans	*Meloidogyne* spp.	Sugarcane	Spaull (1984)
Pasteuria penetrans	*Meloidogyne javanica*	Tomato, grape	Stirling (1984)
Pasteuria penetrans	*Meloidogyne incognita*	Soybean, hairy vetch, tobacco	Brown et al. (1985)
Pasteuria penetrans	*Meloidogyne incognita*	Tomato	Maheswari and Mani (1989)
	Meloidogyne javanica		
Pasteuria penetrans	*Meloidogyne species*	Mung bean and brinjal	Zaki and Maqbool (1990)
Pasteuria penetrans	*Meloidogyne arenaria*	Rye, peanut, and vetch	Oostendorp et al. (1991)
Pasteuria penetrans	*Meloidogyne incognita*	Tomato	Vargas et al. (1992)
Pasteuria penetrans	*Meloidogyne incognita*	Kiwi	Verdejo-Lucas (1992)
	Meloidogyne arenaria		
	Meloidogyne hapla		
Pasteuria penetrans	*Meloidogyne javanica*	Chickpea	Sharma (1992)
Pasteuria penetrans	*Meloidogyne arenaria*	Peanut	Chen et al. (1996, 1997)
Pasteuria penetrans	*Meloidogyne arenaria*	Tomato, oriental melon	Cho et al. (2000)
Pasteuria penetrans	*Meloidogyne incognita*	Tomato and banana	Jonathan et al. (2000)
Pasteuria penetrans	*Meloidogyne incognita*	Tomato	Chand and Gill (2002)
Pasteuria penetrans	*Meloidogyne javanica*	Brinjal	Kumar et al. (2005)
Pasteuria fluorescens	*Meloidogyne incognita*	Sugar beet	Kavitha et al. (2007)
Pasteuria penetrans	*Meloidogyne incognita*	Tomato	Ravichandra and Reddy (2008)
Pasteuria penetrans	*Meloidogyne incognita*	Tomato	Singh et al. (2008)
Pasteuria fluorescens	*Meloidogyne incognita*	Field pea	Siddiqui et al. (2009)

(continued)

Table 20.2 (continued)

Bacteria	Nematode control	Crop control	References
Pasteuria fluorescens	*Meloidogyne incognita*	Tomato	Abo-Elyousr et al. (2010)
Pseudomonas sp.	*Meloidogyne incognita*	Okra	Vetrivelkalai et al. (2010)
Pasteuria fluorescens	*Meloidogyne incognita*	Jasmine	Seenivasan and Poornima (2010)
Pasteuria fluorescens	*Meloidogyne incognita*	Mulberry	Muthulakshmi et al. (2010)
Pasteuria fluorescens	*Meloidogyne incognita*	Maize	Ashoub and Amara (2010)
Pasteuria fluorescens	*Meloidogyne incognita*	Okra	Kumar and Jain (2010a)
Pasteuria fluorescens	*Meloidogyne arenaria*	Groundnut	Kalaiarasan et al. (2010)
Pasteuria fluorescens	*Meloidogyne incognita*	Tomato	Singh and Siddiqui (2010)
Pseudomonas sp.	*Meloidogyne incognita*	Black pepper	Devapriyanga et al. (2012)
Pasteuria fluorescens	*Meloidogyne incognita*	Tomato	Khalil et al. (2012a, b)
Pasteuria fluorescens	*Meloidogyne incognita*	Black gram	Akhtar et al. (2012)
Pasteuria fluorescens	*Meloidogyne graminicola*	Rice	Anita and Samiyappan (2012)
Pasteuria fluorescens	*Meloidogyne incognita*	Chili	Wahla et al. (2012)
Pasteuria fluorescens	*Meloidogyne graminicola*	Rice	Seenivasan et al. (2012)
Pasteuria penetrans	*Meloidogyne javanica*	Tomato	Vikram and Walia (2014)
Pasteuria fluorescens	*Meloidogyne graminicola*	Rice	Priya (2015)
Pasteuria penetrans	*Meloidogyne incognita*	Tomato, cucumber	Kokalis-Burelle (2015)
Pasteuria penetrans	*Meloidogyne arenaria*	Snapdragon	Kokalis-Burelle (2015)
Pasteuria fluorescens	*Meloidogyne incognita*	Mulberry	Muthulakshmi and Devrajan (2015)
Pasteuria penetrans	*Meloidogyne javanica*	Okra, chickpea	Vikram and Walia (2015)
Pasteuria penetrans	*Meloidogyne graminicola*	Rice	Thakur and Walia (2016)
Pasteuria fluorescens	*Meloidogyne graminicola*	Rice	Narasimhamurthy et al. (2017a, b)

(continued)

Table 20.2 (continued)

Bacteria	Nematode control	Crop control	References
Pasteuria fluorescens	*Meloidogyne incognita*	Cowpea	Nama and Sharma (2017)
Pasteuria fluorescens	*Meloidogyne javanica*	Tomato	Eltayeb (2017)

authorities. According to Saraf et al. (2014), it points out that owing to their rapid growth, quick handling, and robust colonization of the rhizosphere and antagonistic bacteria, like *Bacillus*, are suitable for biocontrol (Table 20.2). The US Food and Drug Administration (USFDA) has shown effectiveness in the usage of *Bacillus* in bionematicides which provide the classification of *B. subtilis* as mainly considered healthy (Usta 2013).

4 Use of Organic Amendments for Root-Knot Nematode Management

The management of nematodes is a difficult task due to their wide occurrence, wide feeding habit, and simple dispersal. In this way, it is important to look through effective management practices to diminish and keep plant-parasitic nematode population underneath the harm level. The management of plant-parasitic nematodes is progressively troublesome in contrast with other microbes since they normally live in the soil and assault the underground parts (mainly roots) of the plant. Several methods are known to manage the root-knot nematodes which include the use of nematicides, organic matters, resistant cultivars, and biological control. Utilization of organic matters such as oil cakes, neem sawdust, and cleaved leaves is very helpful and gives numerous advantages, for example, the expansion in some natural enemies (Oka 2010) changes the physical and chemical properties of soil and conductivity of water which prompts better plant development (Akhtar and Malik 2000; Gonzalez et al. 2010).

Oil cakes like castor (*Ricinus communis*), mustard (*Brassica campestris*), *Jatropha* spp. flax (*Linum usitatissimum*), neem (*Azadirachta indica*), groundnut (*Arachis hypogea*), mahua (*Madhuca indica*), and sesame (*Sesamum indicum*) have been discovered as powerful in diminishing the augmentation of root-knot nematodes (Ansari and Azam 2010; Ganai et al. 2011; Rehman et al. 2011). Soil amended with oil cakes of neem, castor, mustard, and other plant items has been effectively utilized for the managing plant-parasitic nematodes (Mohan 2011; McGeehan 2012). Khan et al. (2011) found that neem oil cake was discovered essentially decreasing the number of *Meloidogyne incognita* and expanded yield of carrot. Youssef and El-Nagdi (2004) reported that sesame seed cake essentially decreased root galls and egg masses and prevented nematode build-up on squash plants. The impact of urea covered with nimin and oil cakes of neem, castor, and rocket plate of mixed greens/duan was discovered viable against the root-knot nematode, *M. incognita*, and increment development of mung bean (Wani and Bhat 2012; Gupta 2017).

Amended soil with neem cake, castor cake, groundnut cake, sunflower cake, and farmyard manure has fundamentally decreased the nematode population and expanded the plant development (Jagadeeswaran and Singh 2011). The utilization of neem cake and mustard cake was effective and keeps their population in soil below the threshold level (Adhikari et al. 2017). Kumar and Khanna (2008) observed that neem cake improves the growth of tomato and inhibits the development of *M. incognita*. Luma et al. (2003) found that neem cake and Rakshak gold (neem-based item) were malicious against the eggs of *M. incognita*. According to Saikia et al. (2007) and Seenivasan (2010), various plant products such as neem cake, vermicompost, neem seed kernel, sawdust, and carbofuran 3G essentially improved the plant development boundaries and yield of brinjal with comparison diminished in nematode population both in soil and roots.

Plants have been a reservoir of glorious and fascinating natural biopesticides. Addition of plant parts such as freshly chopped leaves to the soil is one of the traditional agriculture practices for the nematode management that improves the soil structure, texture, nutrient content, and soil flora and fauna. After a brief analysis of neem by an Indian pharmaceutical scientist in 1919, it has been found that neem oil contains an acidic compound named as "margosic acid." However, actual research on neem in 1942 isolated three active compounds such as nimbin, nimbidin, and nimbinene. From time immemorial, neem as a reservoir of various components has insecticidal, fungicidal, bactericidal, and nematicidal properties (Gajalakshmi and Abbasi 2004). According to Akhtar and Alam (1993), neem tree has various components that suppress the population of plant-parasitic nematode. In the big nations, for example, the USA, Canada, and Europe, neem is sold as insecticidal in the market. Azadirachtin, one of the potent parts of the neem tree, has been marketed. The toxicological profile of azadirachtin is commonly ideal (Stark 2007). Soil amended with various parts of neem (*A. indica*) has been seen as profoundly successful in diminishing the number of inhabitants in different phytonematodes (Rather and Siddiqui 2007). Freshly cleaved leaves and seeds of neem (*A. indica*) contain various terpenoids, some of which are perceived as having nematicidal impacts although the exact compounds remain uncertain (Akhtar 2000; Chitwood 2002). Enormous amounts of sawdust are formed as by-products in wood industries. It is used as modest fuel, fluid retentive, filling material, and so on. Since it is insufficient in nitrogen, it isn't supported as a soil amendment; however, a few reports demonstrate its potential for nematode control (Miller and Edgington 1962; Singh et al. 1967; Singh and Sitaramaiah 1971). Singh et al. (1967) suggest that sawdust amended soil also supplemented with nitrogenous fertilizers to improve its efficacy against soil microorganism as well as plant growth. Mian and Rodriguez-Kabana (1982) have identified that nematode-control properties of soil organic matters are directly related to the nitrogen content or inversely related to the C/N ratio, while according to Alam, M. M. (unpublished) in case of sawdust, not the C/N ratio, but carbon content or quantity of sawdust is related to nematode management. Be that as it may, in another report, Rodriguez-Kabana et al. (1987) communicated the assessment that the adequacy of a given organic matter relies upon its chemical composition and the types of microorganisms that create during its decomposition. To additionally check these outcomes, an endeavor has been made to analyze the viability of sawdust of two unique sources against plant-parasitic nematodes swarming certain vegetables. To further verify these results, an attempt has been made to compare the efficacy of sawdust of two different sources against plant-parasitic nematodes infesting certain vegetables.

5 Conclusions

The species of root-knot nematodes such as *Meloidogyne* sp. are responsible for causing a greater loss in crop production. Application of chemical pesticides causes a toxic effect to plant and consumer also. However, the use of organics and biocontrol agents replaces the chemically based control measures. The use of organic amendments and biological agents shows eco-friendly behavior to our environment rather than chemical control methods.

References

Abo-Elyousr, K. A, Khan, Z, & Abedel-Moneim, M. F (2010) Evaluation of plant extracts and Pseudomonas for control of root-knot nematode, Meloidogyne incognita on tomato. Nematropica, 289–299

Akhtar A, Hisamuddin, Robab MI, Abbasi SR (2012) Plant growth promoting Rhizobacteria: an overview. J Nat Prod Plant Resour 2(1):19–31

Anita, B, & Samiyappan, R (2012) Induction of systemic resistance in rice by Pseudomonas fluorescens against rice root knot nematode Meloidogyne graminicola. Journal of Biopesticides, 5, 53.

Ashoub, A.H, Amara, M.T, 2010. Biocontrol activity of some bacterial genera against root-knot nematode, Meloidogyne incognita. J. Am. Sci. 6, 321–328

Abbasi M, Ahmed N, Zaki M, Shuakat S, Khan D (2014) Potential of Bacillus species against *Meloidogyne javanica* parasitizing eggplant (*Solanum melongena* L.) and induced biochemical changes. Plant Soil 375:159–173

Adhikari N, Bahadur GKC, Shrestha SM, Khadka RB (2017) Eco-friendly management of root-knot nematodes (*Meloidogyne* spp.) in okra at farmers' field in chitwan, Nepal. Int J Agric Environ Res 3:4187–4197

Affokpon A, Coyne DL, Htay CC, Agbèdè RD, Lawouin L, Coosemans J (2011) Biocontrol potential of native Trichoderma isolates against root-knot nematodes in west African vegetable production systems. Soil Biol Biochem 43:600–608

Akhtar M (2000) Approaches to biological control of nematode pests by natural products and enemies. J Crop Prod 3:367–395

Akhtar M, Alam MM (1993) Utilization of waste materials in nematode control: review. Bioresour Technol 45:1–7

Akhtar M, Mahmood I (1993) Effect of *Mononchus aquaticus* and organic amendments on *Meloidogyne incognita* development of chilli. Nematol Medit 21:251–252

Akhtar M, Malik A (2000) Roles of organic amendments and soil organisms in the biological control of plant-parasitic nematode: a review. Bioresour Technol 74:35–47

Alban R, Guerrero R, Toro M (2013) Interactions between a root-knot nematode (*Meloidogyne exigua*) and arbuscular mycorrhizae in coffee plant development (Coffea arabica). Am J Plant Sci 4:19–23

Al-Hazmi AS, Javeed MT (2016) Effects of different inoculum densities of *Trichoderma harzianum* and *Trichoderma viride* against *Meloidogyne javanica* on tomato. Saudi J Biol Sci 23:288–292

Ambo PBN, Ethiopia EA, Serfoji P, Rajeshkumar S, Selvaraj T (2010) Management of root-knot nematode, *Meloidogyne incognita* on tomato cv Pusa Ruby. by using vermicompost, AM fungus, Glomus aggregatum and mycorrhiza helper bacterium, Bacillus coagulans. J Agric Sci Technol 6:37–45

Ansari N, Azam MF (2010) Studies on the management of disease complex caused by root-knot nematode, *Meloidogyne incognita* and wilt fungus, *Fusarium oxysporum* f. sp. *ciceri* on chickpea. In: Proceedings of the national seminar, plant biotechnology: advances, impact and relevance. Department of Botany, AMU, Aligarh, p 42

Askary TH (2012) Management of root-knot nematode *Meloidogyne javanica* in pigeonpea through seed treatment. Indian J Ecol 39:151–152

Baidoo R, Mengistu T, McSorley R, Stamps RH, Brito J, Crow WT (2017) Management of root-knot nematode (*Meloidogyne incognita*) on *Pittosporum tobira* under greenhouse, field, and on-farm conditions in Florida. J Nematol 49:133–139

Bansal RK, Dahiya RS, Narula N, Jain RK (2005) Management of Meloidogyne incognita in cotton, using strains of the bacterium Gluconacetobacter diazotrophicus. Nematol Medit 33:101–105

Bhat MY, Wani AH (2012) Bio-activity of fungal culture filtrates against root-knot nematode egg hatch and juvenile motility and their effects on growth of mung bean (*Vigna radiata* L. Wilczek) infected with the root-knot nematode, *Meloidogyne incognita*. Arch Phytopathol Plant Prot 45:1059–1069

Braga FR, Araújo JV (2014) Nematophagous fungi for biological control of gastrointestinal nematodes in domestic animals. Appl Microbiol Biotechnol 98:71–82

Brahma U, Borah A (2016) Management of *Meloidogyne incognita* on pea with bioagents and organic amendment. Indian J Nematol 46:58–61

Brown, S.M, Kepner, J.L, & Smart G.C (1985) Increase crop yield following application of Bacillus penetrans to field plots infested with M. incognita. Soil Bio,Biochem, 17: 483–486

Calvet C, Pinochet J, Hernández-Dorrego A, Estaún V, Camprubí A (2001) Field microplot performance of the peach-almond hybrid GF-677 after inoculation with arbuscular mycorrhizal fungi in a replant soil infested with root-knot nematodes. Mycorrhiza 10:295–300

Cawoy H, Wagner B, Fickers P, Ongena M (2011) Bacillus-based biological control of plant diseases. In: Stoytcheva M (ed) Pesticides in the modern world- pesticides use and management. InTech, Rijeka, pp 272–302

Chitwood DJ (2002) Phytochemical based strategies for nematode control. Annu Rev Phytopathol 40:221–249

Ciancio A, Mukerji KG (eds) (2007) Integrated management and biocontrol of vegetable and grain crops nematodes, vol 2. Springer Science & Business Media, Berlin

Chand, R. & Gill, J.S (2002) Evaluation ofvarious application methods of Pasteuria penetrans against Meloidogyne incognita in tomato. Indian Journal of Nematology 32: 23–25

Chen, Z. X, Dickson, D. W, McSorley, R, Mitchell, D. J, & Hewlett, T (1996) Suppression of Meloidogyne arenaria race 1 by soil application of endospores of Pasteuria penetrans. Journal of Nematology, 28(2): 159.

Chen, Z. X, Dickson, D. W, Mitchell, D. J, McSorley, R, & Hewlett, T (1997) Suppression mechanisms of Meloidogyne arenaria race 1 by Pasteuria penetrans. Journal of Nematology, 29(1):1

Cho, H. J, Farrand, S. K, Noel, G. R, & Widholm, J. M (2000) High-efficiency induction of soybean hairy roots and propagation of the soybean cyst nematode. Planta, 210(2):195–204

Devapriyanga, R, Jonathan, E. I, Meena, K. S, & Kavitha, P. G (2012) Bioefficacy of Pseudomonas and Bacillus isolates against root-knot nematode, Meloidogyne incognita in black pepper cv. Panniyur 1. Indian J Nematol, 42, 57–65

Degenkolb T, Vilcinskas A (2016) Metabolites from nematophagous fungi and nematicidal natural products from fungi as an alternative for biological control. Part I: metabolites from nematophagous ascomycetes. Appl Microbiol Biotechnol 100:3799–3812

Deori R, Borah A (2016) Efficacy of *Glomus fasciculatum*, *Trichoderma harzianum* for the management of *Meloidogyne incognita* and *Rhizoctonia solani* disease complex in green gram. Indian J Nematol 46:61–64

Devi TS, Mahanta B, Borah A (2016) Comparative efficacy of *Glomus fasciculatum*, *Trichoderma harzianum*, carbofuran and carbendazim in management of *Meloidogyne incognita* and *Rhizoctonia solani* disease complex on brinjal. Indian J Nematol 46:161–164

Dhawan SC, Singh S (2010) Management of root-knot nematode, *Meloidogyne incognita* using *Pochonia chlamydosporia* on okra. Indian J Nematol 40:171–178

Eltayeb, F. M. E (2017) The effects of Bacillus subtilis bacteria on Meloidogyne javanica (Nematode) infection and tomato plant growth. European Journal of Advanced Research in Biological and Life Sciences Vol, 5(2)

Feyisa B, Lencho A, Selvaraj T, Getaneh G (2016) Evaluation of some botanicals and *Trichoderma harzianum* against root-knot nematode *Meloidogyne incognita* (Kofoid and white) Chitwood in tomato. J Entomol Nematol 8:11–18

Franco J, Jatala P, Bocangel M (1981) Efficiency of *Paecilomyces lilacinus* as a biocontrol agent of *Globodera pallida*. J Nematol 13:438–439

Gao, H, Qi, G, Yin, R, Zhang, H, Li, C, & Zhao, X (2016) Bacillus cereus strain S2 shows high nematicidal activity against Meloidogyne incognita by producing sphingosine. Scientific reports, 6(1):1–11.

Gajalakshmi S, Abbasi SA (2004) Neem leaves as a source of fertilizer-cum-pesticide vermicompost. Bioresour Technol 92:291–296

Ganai MA, Rather AA, Siddiqui MA (2011) Pathogenicity of root-knot nematode, *Meloidogyne incognita* on okra and its management through botanicals. Arch Phytopathol Plant Prot 44:1683–1688

Ganaie MA, Khan TA (2010) Biological potential of *Paecilomyces lilacinus* on pathogenesis of *Meloidogyne javanica* infecting tomato plant. Eur J Appl Sci 2:80–84

Gogoi D, Mahanta B (2013) Comparative efficacy of *Glomus fasciculatum*, *Trichoderma harzianum*, carbofuran and carbendazim in management of *Meloidogyne incognita* and *Rhizoctonia solani* disease complex on French bean. Ann Plant Prot Sci 21:172–175

Gonzalez M, Gomez E, Comese R, Quesada M, Conti M (2010) Influence of organic amendments on soil quality potential indicators in an urban horticultural system. Bioresour Technol 101:8897–8901

Goswami BK, Pandey RK, Rathour KS, Bhattacharya C, Singh L (2006) Integrated application of some compatible biocontrol agents along with mustard oil seed cake and furadan on *Meloidogyne incognita* infecting tomato plants. J Zhejiang Univ Sci B 7:873–875

Goswami J, Pandey RK, Tewari JP et al (2008) Management of root-knot nematode on tomato through application of fungal antagonists, *Acremonium strictum* and *Trichoderma harzianum*. J Environ Sci Health B 43:237–240

Griffin GJ (1990) Importance of *Pythium ultimum* in a disease syndrome of cv. Essex soybean. Can J Plant Pathol 12:135–140

Gupta S (2017) Effect of organic amendment on organic metabolites in root-knot nematode (*Meloidogyne Incognita*) infested spinach. Pollution 3:377–382

Hashem M, Abo-Elyousr KA (2011) Management of the root-knot nematode Meloidogyne incognita on tomato with combinations of different biocontrol organisms. Crop Protect 30:285–292

Hunt D, Handoo ZA (2009) Taxonomy, identification and principal species. In: Perry RN, Moens M, Starr JL (eds) Root-knot nematodes. CABI, Wallingford, pp 55–97

Hussain T, Akhtar N, Aminedi R, Danish M, Nishat Y, Patel S (2020) Role of the potent microbial based bioagents and their emerging strategies for the ecofriendly management of agricultural phytopathogens. In: Natural bioactive products in sustainable agriculture. Springer, Berlin, pp 45–66. https://doi.org/10.1007/978-981-15-3024-1_4

Jagadeeswaran R, Singh RV (2011) Management of reniform nematode, *Rotylenchulus reniformis* on okra hybrid by organic amendments. Indian J Nematol 41:9–13

Jamshidnejad V, Sahebani N, Etebarian H (2013) Potential biocontrol activity of *Arthrobotrys oligospora* and *Trichoderma harzianum* BI against *Meloidogyne javanica* on tomato in the greenhouse and laboratory studies. Arch Phytopathol Plant Prot 46:1632–1640

Jegathambigai V, Wilson-Wijeratnam RS, Wijesundera RLC (2011) Effect of *Trichoderma viride* strain NRRL 6418 and *Trichoderma harzianum* (*Hypocrea lixii* TWC1) on *Livistona rotundifolia* root-knot nematode *Meloidogyne incognita*. J Entomol 8:229–239

Jothi G, Poornima K (2017) Organic amendments for the management of Meloidogyne incognita in tuberose. Pest Manag Horticult Ecosyst 23:72–75

Jyoti S, Singh DP (2017) Fungi as biocontrol agents in sustainable agriculture. In: Singh JS, Singh DP (eds) Microbes and environmental management. Studium Press, Houston

Jonathan, E. I, Barker, K. R, Abdel-Alim, F. F, Vrain, T. C, & Dickson, D. W (2000) Biological control of Moloidogyne incognita on tomato and banana with Rhizobacteria, Actinomycetes, and Pasteuria penetrans. Nematropica, 231–240.

Kalaiarasan, P, Lakhsmanan, P. L, & Samiyappan, R (2010) Induction of oxidative enzyme, peroxidase in groundnut (Arachis hypogaea) by application of Pseudomonas fluorescens, as a defence against the root-knot nematode, Meloidogyne arenaria. Indian Journal of Nematology, 40(1):55

Kavitha, J, Jonathan, E. I, & Umamaheswari, R (2007) Field application of Pseudomonas fluorescens, Bacillus subtilis and Trichoderma viride for the control of Meloidogyne incognita (Kofoid and White) Chitwood on sugarbeet. Journal of Biological Control, 21(2):211–215

Kokalis-Burelle, N (2015) Pasteuria penetrans for control of Meloidogyne incognita on tomato and cucumber, and M. arenaria on snapdragon. Journal of Nematology, 47(3):207

Kumar A, Walia RK, Kapoor A (2005) Field evaluation of Pasteuria penetrans as nursery bed application against Meloidogyne javanica infecting brinjal. Int J Nematol 15:183–186

Khalil MS, Kenawy A, Gohrab MA, Mohammed EE (2012a) Impact of microbial agents on Meloidogyne incognita management and morphogenesis of tomato. J Biopest 5:28–35

Khalil MS, Allam AFG, Barakat AST (2012b) Nematicidal activity of some biopesticide agents and microorganisms against root-knot nematode on tomato plants under greenhouse conditions. J Plant Prot Res 52:47–52

Khan TA, Saxena SK (1997) Effect of root-dip treatment with culture filtrates of on root penetration, development and reproduction of *Meloidogyne javanica* on tomato. Int J Nematol 7:85–88

Khan A, Shaukat SS, Sayed M (2011) Control of nematodes associated with almond using oilcakes in Balochistan. Pak J Nematol 29:171–177

Khattak S, Khattak B (2011) Management of root-knot nematode with *Trichoderma harzianum* and spent mushroom compost. In: Proceedings 46th Croatian and 6th international symposium on agriculture held on 14th–18th February 2011 at Opatija, Croatia, pp 157–160

Kloepper JW, Schroth MN, Miller TD (1980) Effects of rhizosphere colonization by plant growth-promoting rhizobacteria on potato plant development and yield. Phytopathology 70:1078–1082

Koenning SR, Overstreet C, Noling JW, Donald PA, Becker JO, Fortnum BA (1999) Survey of crop losses in response to Phytoparasitic nematodes in the United States for 1994. J Nematol 31:587–618

Koffi MC, Vos C, Draye X, Declerck S (2013) Effects of *Rhizophagus irregularis* MUCL 41833 on the reproduction of *Radopholus similis* in banana plantlets grown under in vitro culture conditions. Mycorrhiza 23:279–288

Kumar P, Chand R (2015) Bioefficacy of *trichoderma harzianum* against root-knot nematode *Meloidogyne incognita* on brinjal. Ann Plant Prot Sci 23:361–364

Kumar V, Jain RK (2010a) Management of root-knot nematode, *Meloidogyne incognita* using *Pochonia chlamydosporium* on okra. Indian J Nematol 40:171–178

Kumar V, Jain RK (2010b) Management of root-knot nematode, *Meloidogyne incognita*, by *Trichoderma viride*, *T. harzianum* and bacterial antagonist, *Pseudomonas fluorescens* as seed treatment on okra. Indian J Nematol 40:226–228

Kumar S, Khanna AS (2008) Role of *Trichoderma harzianum* and neem cake separately and in combination against root-knot nematode on tomato. Indian J Nematol 38:264–266

Kumar V, Singh RV, Singh HS (2011) Management of *Meloidogyne incognita* race 1 and *Rotylenchulus reniformis* by seed treatment with biological agents, organic cakes and pesticides on cowpea. Ann Plant Prot Sci 19:164–167

Kumar D, Bhatt J, Sharma RL (2017) Efficacy of different bio control agents against *Meloidogyne incognita* and *Fusarium oxysporum* on black gram (*Vigna mungo* L). Int J Cur Microbiol Appl Sci 6:2287–2291

Li S, Duan YX, Zhu XF, Chen LJ, Wang YY (2011) The effects of adding secondary metabolites of *Aspergillus niger* on disease resistance to root-knot nematode of tomato. China Vegetables 1:44–49

Li J, Zou C, Xu J, Ji X, Niu X, Yang J, Huang X, Zhang KQ (2015) Molecular mechanisms of nematode-nematophagous microbe interactions, basis for biological control of plant-parasitic nematodes. Annu Rev Phytopathol 53:67–95

Lee, Y. S, & Kim, K. Y (2016) Antagonistic potential of Bacillus pumilus L1 against root-Knot nematode, Meloidogyne arenaria. Journal of Phytopathology, 164(1):29-39.

Luma BAL, Darwish RM, Aburjai T (2003) Effect of plant extracts and essential oils on root-knot nematode. Phytopathol Mediterr 42:123–128

Maheswari, T.ll. & Mani, A (1989) Combined efficacy of Pasteuria penetrans and Paecilomyces lilacinus on the biocontrol ofMiloidogyne javanica ontomato. Int. Nematol. Network Newsl. 5: 10–11

Mankau, R. & Prasad, N (1972) Possibilities and problems in the use of sporozoans endoparasite for biological control of plant parasitic nematodes. Nematropica, 2:7 (Abstr)

Machado V, Berlitz DL, Matsumura ATS, Santin RDCM, Guimarães A, da Silva ME, Fiuza LM (2012) Bactériascomoagentes de controlebiológico de fitonematóides. Oecolog Aust 16:165–182

McGeehan S (2012) Impact of waste materials and organic amendments on soil properties and vegetative performance. Appl Environ Soil Sci 2012:674–684

Mendoza AR, Kiewnick S, Sikora RA (2008) In vitro activity of *Bacillus firmus* against the burrowing nematode *Radopholus similis*, the root-knot nematode Meloidogyne incognita and the stem nematode *Ditylenchus dipsaci*. Biocontrol Sci Technol 18:377–389

Metwally WE, Mostafa FAM, Refaei AR (2015) In vitro study on the antagonistic activity of different native isolates of rhizobacteria against Meloidogyne incognita. Egypt J Agronematol 14:1–9

Meyer SL, Roberts DP (2002) Combinations of biocontrol agents for management of plant-parasitic nematodes and soilborne plant-pathogenic fungi. J Nematol 34:1

Mian IH, Rodriguez-Kabana R (1982) Survey of the nematicidal properties of some organic materials available in Alabama as amendments to soil for control of *Meloidogyne arenaria*. Nematropica 12:235–246

Miller PM, Edgington LV (1962) Effects of paper and sawdust soil amendments on meadow nematodes and subsequent *Verticillium* wilt of tomatoes. Plant Dis Rep 46:745–747

Mohan K (2011) Comparison of inorganic and organic nematicides on the population of soil nematodes in hybrid of *Saccharum* species. J Biopest 4:201–204

Muller R, Gooch PS (1982) Organic amendment in nematode control. An examination of the literature. Nematropica 12:319–326

Muthulakshmi M, Devrajan K (2015) Management of *Meloidogyne incognita* by *Pseudomonas fluorescens* and *Trichoderma viride* in mulberry. Int J Plant Prot 8:1–6

Muthulakshmi M, Devrajan K, Jonathan EI (2010) Biocontrol of root-knot nematode, *Meloidogyne incognita* (Kofoid and white) Chitwood in mulberry (*Morus alba* L.). J Biopest 3:479–482

Nama, C. P, & Sharma, H. K (2017) Bio-management of root-knot nematode, Meloidogyne incognita on cowpea (Vigna unguiculata L.). J Entomo Zool Stud, 5, 50–52

Narasimhamurthy HB, Ravindra H, Sehgal M (2017a) Management of rice root-knot nematode, *Meloidogyne graminicola*. Int J Pure Appl Biosci 5:268–276

Narasimhamurthy HB, Ravindra H, Sehgal M, Ekabote SD, Ganapathi G (2017b) Management of rice root-knot nematode, *Meloidogyne graminicola*. J Entomol Zool Stud 5:1433–1439

Nisha MS, Sheela MS (2016) Effect of fungal egg parasite, *Paecilomyces lilacinus* (Thom.) Samson on *Meloidogyne incognita* in brinjal. Indian J Nematol 46:157–159

Oostendorp M, Dickson DW, Mitchell DJ (1991) Population development of Pasteuria penetrans on Meloidogyne arenaria. J Nematol 23:58–64

Oka Y (2010) Mechanism of nematode suppression by organic soil amendments—a review. Appl Soil Ecol 44:101–115

Ongena M, Jacques P (2008) Bacillus lipopeptides: versatile weapons for plant disease biocontrol. Trends Microbiol 16:115–125

Padgham JL, Sikora RA (2007) Biological control potential and modes of action of *Bacillus megaterium* against *Meloidogyne graminicola* on rice. Crop Prot 26:971–977

Pandey R, Mishra AK, Tiwari S, Kalra A (2011) Nematode inhibiting organic materials and a strain of *Trichoderma harzianum* effectively manages *Meloidogyne incognita* in Withania somnifera fields. Biocontol Sci Technol 21:1495–1499

Parihar K, Rehman B, Ganai MA, Asif M, Siddiqui MA (2015) Role of oil cakes and *Pochonia chlamydosporia* for the management of *Meloidogyne javanica* attacking *Solanum melongena* L. J Plant Pathol Microbiol (special issue) SI:1–5

Priya MS (2015) Biomanagement of rice root-knot nematode, *Meloidogyne graminicola* Golden and Brichfield in aerobic rice. Int J Manag Soc Sci 3:591–598

Ramezani Moghaddam, M, Mahdikhani Moghaddam, E, Baghaee Ravari, S, & Rouhani, H (2014) The nematicidal potential of local Bacillus species against the root-knot nematode infecting greenhouse tomatoes. Biocontrol science and technology, 24(3):279–290.

Ravichandra, N. G, & Reddy, B. M. R (2008) Efficacy ofPasteuria penetrans in the Management ofMeloidogyne incognita InfectingTomato. Indian Journal of Nematology

Rosas-García, N. M (2009) Biopesticide production from Bacillus thuringiensis: An environmentally friendly alternative. Recent Patents On Biotechnology, 3(1):28–36

Raddy HM, Fouad AFA, Montasser SA, Abdel-Lateef MF, El-Samadisy AM (2013) Efficacy of six nematicides and six commercial bioproducts against root-knot nematode, *Meloidogyne incognita* on tomato. J Appl Sci Res 9:4410–4417

Rahman SFSA, Singh E, Pieterse CMJ, Schenk PM (2018) Emerging microbial biocontrol strategies for plant pathogens. Plant Sci 267:102–111

Ramakrishnan S, Nagesh M (2011) Evaluation of beneficial fungi in combination with organics against root-knot nematode, *Meloidogyne incognita*, in FCV tobacco nurseries. J Biol Cont 25:311–315

Ramakrishnan S, Rao CP (2013) Evaluation of *Paecilomyces lilacinus* for the management of root-knot nematode, *Meloidogyne incognita* in flue cured Virginia (FCV) tobacco nursery. Indian J Nematol 43:65–69

Rather MA, Siddiqui MA (2007) Nematicidal effects of some botanicals against root-knot nematode, (*Meloidogyne javanica*) on tomato. Int J Plant Sci 2:49–52

Rehman B, Parihar K, Ganai MA, Usman A, Siddiqui MA (2011) Bio management of root knot nematode through non edible oil seed cakes infesting *Cicer arietinum* L. Int J Appl Biol Pharma Sci 2:411–414

Rodriguez-Kabana R, Morgan-Jones G, Chet I (1987) Biological control of nematodes: soil amendments and microbial antagonists. Plant Soil 100:237–247

Seenivasan, N, David, P. M. M, Vivekanandan, P, & Samiyappan, R (2012) Biological control of rice root-knot nematode, Meloidogyne graminicola through mixture of Pseudomonas fluorescens strains. Biocontrol Science and Technology, 22(6):611–632

Sharma RD (1992) Biocontrol efficacy of Pasteuria penetrans against Meloidogyne javanica. Ciencia Biologica Ecologica e Systematica 12:43–47
Siddiqui, I. A, Atkins, S. D, & Kerry, B. R (2009) Relationship between saprotrophic growth in soil of different biotypes of Pochonia chlamydosporia and the infection of nematode eggs. Annals of Applied Biology, 155(1):131–141
Singh P, Siddiqui ZA (2010) Biocontrol of root-knot nematode Meloidogyne incognita by the isolates of Pseudomonas on tomato. Archiv Phytopath Pl Prot 43:1423–1434
Singh M, Singh J, Gill JS (2008) Impact of Pasteuria penetrans on root-knot nematode (Meloidogyne incognita) infecting tomato (Lycopersicon esculentum). Indian J Agric Sci 78:1092–1094
Spaull, V. W (1984) Observations on Bacillus penetrans infecting Meloidogyne in sugarcane fields in South Africa. Rev. Nematol, 7(3):277–282
Stirling, G. R (1984) Biological control of Meloidogyne javanica with Bacillus penetrans. Phytopathology, 74(1):55–60
Saikia J, Borah A, Bora BC (2007) Efficacy of organic amendments in combination with the nematicide for management of *Meloidogyne incognita* in brinjal. India J Nematol 37(2):179–181
Sansinenea E, Ortiz A (2011) Secondary metabolites of soil Bacillus spp. Biotechnol Lett 33(8):1523–1538
Saraf M, Pandya U, Thakkar A (2014) Role of allelochemicals in plant growth promoting rhizobacteria for biocontrol of phytopathogens. Microbiol Res 169(1):18–29
Seenivasan N (2010) Management of root-knot nematode, *Meloidogyne incognita* with organic amendments in medicinal coleus. Ann Plant Prot Sci 18(2):472–476
Sharf R, Shiekh H, Syed A, Akhtar A, Robab MI (2014) Interaction between *Meloidogyne incognita* and *Pochonia chlamydosporia* and their effects on the growth of *Phaseolus vulgaris*. Arch Phytopathol Plant Prot 47:622–630
Sharon E, Bar EM, Chet I, Herrera EA, Kleifeld O, Spiegel Y (2001) Biological control of the root-knot nematode *M. javanica* by *T. harzianum*. Phytopathology 91:687–693
Siddiqui ZA (2004) Effects of plant growth promoting bacteria and composed organic fertilizers on the reproduction of *Meloidogyne incognita* and tomato growth. Bioresour Technol 95(2):223–227
Siddiqui ZA, Mahmood I (1996) Biological control of *Heterodera cajani* and *Fusarium udum* on pigeonpea by *Glomus mosseae, Trichoderma harzianum*, and *Verticillium chlamydosporium*. Israel J Plant Sci 44:49–56
Sikora RA, Fernandez E (2005) Nematode parasites of vegetables. In: Luc M, Sikora RA, Bridge J (eds) Plant-parasitic nematodes in subtropical and tropical agriculture, 2nd edn. CABI, Wallingford, pp 319–392
Silva JO, Santana MV, Freire LL, Ferreira BS, Rocha MR (2017) Biocontrol agents in the management of *Meloidogyne incognita* in tomato. Ciência Rural 47:1–7. https://doi.org/10.1590/0103-8478cr20161053. Accessed 9 Dec 2017
Simon LS, Pandey A (2010) Antagonistic efficacy of *Paecilomyces lilacinus* and *Verticillium chlamydosporium* against *Meloidogyne incognita* infecting okra. Indian J Nematol 40:113
Singh LM, Mahanta B (2013) Effect of carbosulfan, *Glomus fasciculatum*, *Trichoderma harzianum* and vermicompost alone and combination in management of *Meloidogyne incognita* on green gram. Ann Plant Prot Sci 21:154–156
Singh RS, Sitaramaiah K (1971) Control of root-knot through organic and inorganic amendments of soil: effect of oil cakes and sawdust. Indian J Mycol Plant Pathol 1:20–29
Singh RS, Singh B, Beniwal SES (1967) Observations on the effect of sawdust on the incidence of root knot and yield of okra and tomatoes in nematode infested soil. Plant Disease Rep 51:861–863
Stark JD (2007) Ecotoxicology of neem. ACS Symp Ser 947:275–286
Stirling GR (2014) Nematophagous fungi and oomycetes. Biological control of plant-parasitic nematodes: soil ecosystem management in sustainable agriculture, 2nd edn. CABI, Wallingford

Thakur S, Walia RK (2016) Potential of Bacterial Parasite, Pasteuria penetransApplication as Nursery Soil Treatment and Seed Treatment in Controlling Meloidogyne graminicola Infecting Rice. India J Nematol 46(1):16–19

Tian B, Yang J, Zhang KQ (2007) Bacteria used in the biological control of plant-parasitic nematodes: populations, mechanisms of action, and future prospects. FEMS microbiology ecology, 61(2):197–213

Tiwari S, Pandey S, Singh Chauhan P, Pandey R (2017) Biocontrol agents in co-inoculation manages root-knot nematode [*Meloidogyne incognita* (Kofoid & White) Chitwood] and enhances essential oil content in *Ocimum basilicum* L. Indus Crops Prod 97:292–301

Turatto MF, Dourado FDS, Zilli JE, Botelho GR (2017) Environmental microbiology: control potential of *Meloidogyne javanica* and *Ditylenchus* spp. using fluorescent *Pseudomonas* and *Bacillus* spp. Braz J Microbiol 49(1):54–58

Usta C (2013) Microorganisms in biological pest control—a review (bacterial toxin application and effect of environmental factors). In: Silva-Opps M (ed) Current progress in biological research. InTech, Rijeka, pp 287–317

Vargas R, Acosta N, Moullor A, Betancourt C (1992) Control of Meloidogyne spp. with Pasteuria penetrans (Thorne) Sayre and Starr. J Agric University Puerto Rico 76:63–70

Verdejo-Lucas S (1992) Seasonal population fluctuations of Meloidogyne spp. and the Pasteuria penetrans group in Kiwi orchards. Pl Disease 76:1275–1279

Verdejo-Lucas S (1992) On the occurrence of the" Mediterranean biotype" of Tylenchulus semipenetrans in Spain. Fundamental and applied Nematology, 15(5):475–477

Vetrivelkalai P, Sivakumar M, Jonathan EI (2010) Biocontrol potential of endophytic bacteria on Meloidogyne incognita and its effect on plant growth in bhendi. J Biopest 3(2):452–457

Vikram, Walia RK (2014) Efficacy of bacterial parasite, Pasteuria penetrans application as nursery soil treatment against root-knot nematode, Meloidogyne javanica infecting tomato in different seasons. Indian J Nematol 44:44–49

Vikram, Walia RK (2015) Efficacy of bacterial parasite, Pasteuria penetrans application as seed treatment against root-knot nematode, Meloidogyne javanica. Indian J Nematol 45:1–6

Viggiano JR, Freitas LG, Lopes EA (2015) *Pochonia chlamydosporia* var. chlamydosporia (Goddard) Zare & W. Gams for the management of lettuce infected with *Meloidogyne javanica* (Treub, 1885) Chilean J Agric Res 75:255–258

Vos C, Claerhout S, Mkandawire R, Panis B, de Waele D, Elsen A (2012) Arbuscular mycorrhizal fungi reduce root-knot nematode penetration through altered root exudation of their host. Plant Soil 354:335–345

Wahla, V, Maheshwari, D. K, & Bajpai, V. K (2012) Nematicidal fluorescent pseudomonads for the in vitro and in vivo suppression of root knot (Meloidogyne incognita) of Capsicum annuum L. Pest management science, 68(8):1148–1155.

Wani AH, Bhat MY (2012) Control of root-knot nematode, *Meloidogyne incognita* by urea coated with Nimin or other natural oils on mung, *Vigna radiata* (L) R Wilczek. J Biopest 5:255–258

Wei BQ, Xue QY, Wei LH et al (2009) A novel screening strategy to identify biocontrol fungi using protease production or chitinase activity against *Meloidogyne* root-knot nematodes. Biocontrol Sci Technol 19:859–870

Wu L, Wu HJ, Qiao J, Gao X, Borriss R (2015) Novel routes for improving biocontrol activity of Bacillus based bioinoculants. Front Int Microbiol 6:1395

Ye W, Zeng Y, Kerns J (2015) Molecular characterisation and diagnosis of root-knot nematodes (Meloidogyne spp.) from turfgrasses in North Carolina, USA. PLoS One 10(11):e0143556. https://doi.org/10.1371/journal.pone.0143556

Youssef MMA, El-Nagdi WM (2004) Cellular alteration of root-knot nematode, *Meloidogyne incognita* infected squash plant and intercropping sesame plant or sesame oil seed cake as control measures. Egypt J Phytopathol 32:77–85

Zaki, M. J, & Maqbool, M. A (1990) Effect of Pasteuria penetrans and Paecilomyces lilacinus on the control of root-knot nematodes of brinjal and mung. Pakistan Journal of Phytopathology, 2(1-2):37–42

Chapter 21
Rhizosphere, Rhizosphere Biology, and Rhizospheric Engineering

Pankaj Sharma, Mayur Mukut Murlidhar Sharma, Arvind Malik, Medhavi Vashisth, Dilbag Singh, Rakesh Kumar, Baljinder Singh, Anupam Patra, Sahil Mehta, and Vimal Pandey

Contents

P. Sharma · R. Kumar
Department of Microbiology, CCS Haryana Agricultural University, Hisar, Haryana, India

M. M. M. Sharma
Department of Agriculture and Life Industry, Kangwon National University, Chuncheon, Gangwon, Republic of Korea

A. Malik
Department of Zoology, CCS Haryana Agricultural University, Hisar, Haryana, India

M. Vashisth
Department of Molecular Biology, Biotechnology and Bioinformatics, CCS Haryana Agricultural University, Hisar, Haryana, India

D. Singh
Divison of Microbiology, ICAR-Indian Agricultural Research Institute, New Delhi, India

B. Singh · V. Pandey (✉)
National Institute of Plant Genome Research, New Delhi, India

A. Patra · S. Mehta
International Centre for Genetic Engineering and Biotechnology, New Delhi, India

© The Author(s), under exclusive license to Springer Nature Switzerland AG 2021

H. I. Mohamed et al. (eds.), *Plant Growth-Promoting Microbes for Sustainable Biotic and Abiotic Stress Management*,
https://doi.org/10.1007/978-3-030-66587-6_21

1 Introduction

The global human population is on a perpetual upsurge, however, at the declining rates. The human beings inhabiting this planet are now approaching 7.5 billion, which marks a 100% upturn as compared to that of the early 1960s. This increasing number of human beings, undoubtedly, requires more food resources to proliferate and thrive in the existing environments. Therefore, the major challenge for the agricultural systems is to enhance the food crop production in the upcoming era simultaneously addressing the hazards as well as inconsistency along with the eco-efficiency (Jeranyama et al. 2020). However, some different strategies are already being followed for increasing the food crop production, for instance, use of chemical fertilizers, introduction of genetically modified plants, employment of agrochemicals, as well as the usage of sophisticated machinery. The explicit application of chemical fertilizers has amplified dramatically from 0.5 tons to 23 million tons from 1960 to 2008 correspondingly (Pandey 2018; FAO 2019a, b). The increasing levels of environmental concerns are laying a pressure on the farming community to produce the food crops sustainably (Rani et al. 2019; Singh et al. 2019; Sharma et al. 2019, 2020; Kapoor et al. 2020).

Since the domestication of plants, several strategies have been followed for enhancing the yield of food crops. The advancements in scientific researches and innovation of newer technologies introduced the green revolution which proved to be a milestone in attaining an enhanced food crop production. However, it accounted for a significant enhancement in food crop production persistence of the global monster of hunger coupled with the environmental sustainability concerns requiring the intervention of novel technologies that can fulfill the demands of a higher production along with the preservation of environmental sustainability. The quest to fulfill both these demands puts forward the idea of engineering the rhizospheric portion of plants. The rhizosphere seems to be the most complex habitat of a vast array of microbial population encompassing an intermingled network of plant roots, diverse microbial communities, and soil (Ahkami et al. 2017). This narrow zone of plant-microbe interactions represents the first plant-prejudiced microbial habitat which affects the plant growth in a direct as well as indirect manner. The rhizospheric portion is a complex dynamic and compactly inhabited zone of soil that proves to be an incredible site for the multifaceted set of inter- as well as intraspecies interactions and food web communications which lay a strong effect on the carbon flow as well as transformation (Dessaux et al. 2016; Walker et al. 2011). The plant systems have evolved in a realm of microorganisms. The coevolution of plants with the rhizospheric microbiome has resulted in a state where both these components start affecting each other from the very first day of the dawn of plant life. The roots of plant systems are largely known for altering the physical characteristics of the soil. Plants harbor a vast microbial population by secreting carbon-rich compounds via roots, where such labile substrates are largely favored by the members of microbial communities and they swiftly blend them (Doornbos et al. 2012). The alteration of physical as well as chemical environs of the rhizosphere by the plant systems largely affects the suitability of diverse microbiological clusters and microbial connections and has also encouraged the evolution of novel microbial systems that fit themselves in the rhizospheric life. The gain of fitness sustained by the microbial systems must overshadow the price to the plants in diverted carbon and energy (Vandenkoornhuyse et al. 2015). These plants associated with microorganisms largely assist the plant systems under their plant growth promotion attributes. They not only facilitate the plant systems in the uptake of several key nutrients but also protect them from many biotic as well as abiotic stresses. They are found to enhance the plant productivity directly by fixing the nitrogen, solubilizing the phosphate, producing the siderophore, and indirectly increasing the organic carbon pool of the soil, conferring the plants with the ability to tolerate various biotic as well as abiotic stresses (Mohanram and Kumar 2019). Numerous indications display that plants engineer their rhizospheric microbiome. The most primaeval lines of plants also display a strong capability of altering the comparative richness of different microbial clusters in the soils neighboring their rhizosphere (Chaparro et al. 2014; Valverde et al. 2016) that assists the plant systems in their growth. Apart from this ability of plants to alter their rhizospheric communities, various human practices have also proven to be key drivers in engineering the microbial population of a rhizospheric portion which strongly favors the establishment of advantageous microbial systems on the plant roots which ultimately results in improved plant health and upsurged plant productivity. Therefore, the present chapters strongly target different approaches that are often employed to engineer the plant rhizosphere to bring a qualitative as well as a quantitative upsurge in the productivity of plant systems.

2 Rhizosphere and Root Exudates

The rhizosphere seems to be the most composite microbial territory on the earth, encompassing a cohesive system of plant roots, soil particles, as well as an assorted microbial conglomerate of archaea, bacteria, virus particles, as well as micro-eukaryotes. This fine region of contact amid the soil particles and the plant roots establishes the foremost plant-prompted habitation faced by soil microbiota. The rhizosphere represents an active and compactly inhabited zone of soil upholding a multifarious set of inter- as well as intraspecies communications. In addition to this, it also acts as an active site for the ongoing food web interactions that are known to have a significant influence on the carbon flow and transformation (Ahkami et al. 2017; Dessaux et al. 2016). Adding more to it, the classical description of rhizosphere has described it as a four-dimensional (4D) body: three dimensions for the volume and the fourth dimension representing the time for the rhizospheric functioning (Kuzyakov and Razavi 2019). The assessment of the rhizosphere divulges that it is a habitat for diverse classes of microorganisms. The total volume of microbes inhabited in this zone is represented by some good, by some bad, and by a few ugly microbes. These good, bad, and ugly microorganisms denote at this point the good microbes, plant pathogenic microbes, and opportunistic human pathogenic microbes correspondingly (Dutta and Bora 2019). The microbial dwellers of rhizosphere that have sparked the interest in studies targeting rhizosphere and rhizospheric engineering are the microbes having constructive effects on the plant systems which are largely represented by nitrogen-fixing microorganisms, mycorrhiza, plant growth-promoting rhizobacteria (PGPR), and the microbes possessing antagonistic activity toward plant pathogens. However, the rhizospheric inhabitants that are found to be harmful for the plants take account of the phytopathogenic fungi, oomycetes, bacteria, and nematodes (Mendes et al. 2013).

This natural environment allows different microbial strains to co-occur and to form multifarious microbial populations as well as communities. Therefore, the rhizospheric zone has further been divided into three distinct sub-zones: the endorhizosphere which represents the fragment of the root cortex along with the endodermis where the microorganisms, as well as the mineral ions, exist in the apoplastic space amid the plant cells; the rhizoplane, which denotes the middle zone after the epidermal root cells and mucilage; and the ectorhizosphere, which symbolizes the farthest zone extending from the rhizoplane out into the bulk soil (McNear 2013). The term rhizoplane was denoted the direct exterior surface of plant roots along with any tightly clinging soil particle or debris as well as microbiological populations. The existence of rhizosphere is not under a section of limited extent or shape but should rather be considered as an ascent of physical, chemical, as well as biological possessions alongside the plant root. Therefore, the plant rhizospheric portion is of supreme significance for several valuable ecosystem amenities, for instance, to maintain the nutrient as well as water cycle, seizure of vital nutrients, and the sequestration plus storage of carbon (Adl 2016).

The plant metabolism strongly affects the rhizospheric portion by releasing the carbon dioxide and by emancipating the photosynthates by way of diverse kinds of root exudates predominantly via rhizoplane and ectorhizosphere. The importance of root exudates for plant systems can be understood by the fact that plants discharge approximately 40% of its photosynthates unswervingly into the soil systems primarily as compounds of higher as well lower molecular masses (McNear 2013). The plethora of interactions taking place amid rhizosphere and rhizospheric microbiome governs the plant growth as well as yield in their natural environments. The molecular events taking place in the plant rhizosphere precisely shape the plant rhizospheric microbiome or rhizobiome (Sasse et al. 2018).

The plant roots are the main plant structures that are held accountable for the acquirement of both water and essential nutrients and the secretion of different primary and secondary metabolites called as root exudates. The plant's primary metabolites oozed through roots are predominantly organic acids, carbohydrates, and amino acids. In addition to it, plants also exudate a vast range of secondary metabolites, currently also called plant natural products such as alkaloids, terpenoids, and phenolics. In addition to this, these exudates have also been categorized into two clusters, i.e., low-molecular-weight compounds, for instance, sugars, amino acids, volatile compounds (VOCs), phenolic compounds, organic acids, and other secondary metabolites, and high-molecular-weight compounds, like polysaccharides and proteins. It has also been established that the root exudation is largely responsible for shaping the plant rhizobiome, and these exudates find engrossment in numerous biotic as well as abiotic connections. Several different root exudates have also been found responsible for the initiation of quorum-sensing mechanisms in either the repression or stimulation of quorum-sensing rejoinders of correlated bacterial class. However, the rhizospheric portion has been largely ignored, for its different possible attributes that can enhance the crop yield, predominantly owing to the several confronts allied with the sampling within the rhizospheric soil (Oburger and Schmidt 2016; McCormack et al. 2017; Dutta and Bora 2019). Additionally, the role of plant-allied rhizospheric microbiome has already been unveiled for its different plant growth-promoting attributes. In addition to it, the root exudates, apart from harboring the rhizobiome, are also known for the maintenance of the rhizospheric environment by the possession of several key and unique attributes. The root exudates are also acknowledged for enhancing the accessibility of several key nutrients, for instance, phosphorus, because of the discharge of phosphatases and chelation by the oozed organic acids that are known to concentrate the available phosphorus for the plant uptake (Dakora and Phillips 2002). The exudates are also known for deleteriously affecting the adjoining plants, for instance, via fabrication of allelochemicals (Callaway and Aschehoug 2000) which provides an opportunity for engineering the trait of weed inhibition in the plant systems. The exudates are also known for their possession of root-insect communication trait. The root herbivory by numerous pests like aphids can result in noteworthy reductions in produce as well as the quality of important crops which are known to be inhibited by the root exudates, thereby demonstrating insecticidal activity. The root exudates are also known for altering biochemical and physical properties of the soils inevitably. The root exudates are known to stabilize the soil structure along with an enhancement in the water retention capacity of the soil, thereby indirectly improving the plant growth by managing the soil health. Moreover, they also play an imperative character in the elevation of positive interactions among microbes, for instance, by instigating the colonization with mycorrhizae via releasing strigolactones (Biate et al. 2015). Consequently, an explicit range of traits associated with the rhizosphere are potent enough to be targeted for the improvement in crop yields along with a concomitant reduction in the input of chemical fertilizers and other agrochemicals (Preece and Peñuelas 2020). Therefore, the highly dynamic and potent attitude of plant rhizosphere makes it a suitable area of interest for its manipulation in the quest to obtain improved plant health and enhanced crop productivity.

3 Rhizospheric Microbiome

The rhizosphere acts as a definite hotspot and provides a platform for numerous networks in the interior of the bulk soil. It represents an important biological hotspot where respiration, gaseous altercation, nutrient and moistness usage, and confined provisions of organic matter are deliberated to be most concerned. On the contrary, the bulk soil represents an oligotrophic environ, specifically on the stock of root-instigated organic material. Therefore, the rhizosphere, as affected by root exudation, may encompass up to 10^{11} microbiological cells per gram root and with 10^{12} functional genes per gram soil belonging to over and above 30,000 prokaryotic inhabitants (Mendes et al. 2011; Prosser 2015). The cumulative genome of this rhizospheric microbiome appears to be much greater than the plant genome, and it is, therefore, denoted as the second genome of the plant. The rhizospheric microbiome and its role can be considered similar to the human intestinal microbial populations as they also play a great role in human health maintenance (Berendsen et al. 2012; Bron et al. 2012). The rhizospheric microbiota control diverse biogeochemical cycles along with the various other soil processes by influencing the main rhizosphere progressions, for instance, respiration, nitrification, and denitrification (Breidenbach et al. 2016; Philippot et al. 2013). They are also known to conspicuously influence the iron cycle in soils and have also been demonstrated as the essential drivers of soil organic matter decomposition in the temperate grasslands (Li et al. 2019). Therefore, total characteristics of the agronomic rehearse demand a superior considerate of the different rhizospheric progressions that aid plant progression as well as disease suppression. Consequently, owing to the non-replaceable role of rhizospheric microbiome, the exploration of the complex connection amid crop, soil, and microorganisms in the plant rhizosphere has become the fundamental part for nourishing vigorous as well as high-yielding production structures (Uzoh and Babalola 2018). Therefore, the term rhizosphere diversity is often employed to decrypt a vast array of microorganisms residing in the zone of soil, bordering, and habitually stimulated by plant roots. The intimate interactions of plants with microbial communities in this special zone of soil have made the rhizosphere a place for extraordinary microbial accomplishments (Huang et al. 2014; Nicolitch et al. 2016).

The major proportion of the diverse microbiota harbored by the plant systems is picked up throughout their lifespan from the adjacent environs; thereby, it seems that a considerable part of the plant microbiome finds its origin from the seeds. The seed-allied microbiota is supposed to play an indispensable part in initial phases of the plant development, thereby upsetting the germination as well as the subsistence of the seedling (Pitzschke 2016; Truyens et al. 2015). The soil-based microbes come later into the play and have to contend alongside the previously established microbiota. The microbiota selected in the rhizospheric zone will move to other plant parts and later inhabit diverse plant tissues especially leaves which later represent a major part of the phyllosphere microbiome (Hardoim et al. 2015; Mitter et al. 2016; Sánchez-Cañizares et al. 2017). The plant-originated metabolites known as root exudates play an indispensable role in the root colonization of rhizospheric microbiome. These are usually of low molecular weight and accordingly are straightforwardly easily utilizable, consequently, fashioning an upsurge in the microbiological population thickness of rhizosphere as equated to the bulk soil. The most noticeable and earliest work on the "the rhizosphere effect" was done by Albert Rovira, the research provided detailed views of plant-driven microbial colonization of the rhizosphere at the microscopic scale (Burns 2010; White et al. 2017). This comparative increment in the integer of microbes in plant rhizosphere is usually articulated as the R/S ratio, where R denotes the numbers per gram of soil in the rhizosphere and S in the bulk soil. There is a great variation in these ratios which range between 5 and 50 which may cross 100 also, and this variation is governed by several factors like microbial members, stage of development of plant systems, plant species, as well as the nutritional eminence of plant systems. It should also be taken care of that only a definite percentage of the root surface is shielded by the microbes, for instance, of the total root surface area of maize, the bacteria cover only 4% in apical zones, 7% in the root hair zone, and up to 20% in basal zones. The inhabitation of a root by the rhizospheric microbiome is, however, not limited to rhizoplane only but can also happen in the apoplast of the cortex to varying degrees as indicated by the presence of endophytes (Marschner 2012). The growth of roots into the deeper soil is closely followed by the active colonization of the newer root just behind the meristematic tissues by the microbes attracted toward the root surface. The exudates oozed in the region directly behind the root tip and in the distal zone of elongation zone encourage the growth and proliferation of microorganisms and also appeal additional soil microbes toward the root surface. However, the exudation of metabolites is at a reduced pace and quantity in the root-hair and its neighboring region which furthers marks a decline in the intensity of microbial inhabitants (Marschner 2012). Thus, the fast-growing roots experience an abrupt variation in the microbial community of rhizoplane and rhizosphere from apical to basal regions alongside the root axis (Bowen and Rovira 1991). It is the alteration in category as well as the amount of carbon accessible as exudates in different root zones which stimulates the differences in the community structures (Baudoin et al. 2003; Marschner 2012). However, such differences in the microbial concentration alongside the root axis are vital for the overall nutrient revenue in the interior of the microbial load (Marschner et al. 2011). An upsurge in the microbiological density might lead to an overall nutrient immobilization, while a reduction in microbial load can lead to a net nutrient release.

The plant largely controls the microbial inhabitation of its root environment by secreting highly diverse root exudates. Their diversity and complexity can be taken into account by the fact that the root exudates of even a small plant species may comprehend more than 100 diverse metabolites (van Dam and Bouwmeester 2016). Furthermore, the attitude and class of root exudates only happen to be decisive for the dispersal of bionetworks and niche exactness of definite plant systems (Dakora and Phillips 2002). The release of these composites by the plant roots proceeds by as a minimum of two possible mechanisms, for instance, the exudates may be conveyed crosswise the cell membrane and then discharged into the adjacent rhizosphere, or the plant produces may also be secreted from the root edge cells and root edge-alike cells, which are known to discrete from the root structures as they mature (Hawes et al. 2000; Vicré et al. 2005). The root exudates may contain every possible plant-originated compound excluding some definite composites that find their key involvements in the process of photosynthesis. The rhizospheric microbiome is deliberated as a conglomerate of key engineers that have the potential to be employed to reconstruct the biodiversity and purposes in the tarnished environments. These microorganisms owe an imperative part in the management of growth, health, as well as ecological aptness of their host plant (Buee et al. 2009; Dutta and Bora 2019). Furthermore, these microbial systems have engrossed much attention and have become a subject for rhizospheric engineering due to their possession of key role in the management of both natural and accomplished agriculture soil ecosystems as they find involvement in diverse and significant progressions referring to soil structure formation, organic matter disintegration, toxin exclusion, xenobiotic deterioration, bioremediation, rhizoremediation, nutrient cycling, etc. A plethora of microbes inhabiting the rhizosphere has the capability of doing these jobs for their host plants. However, all the microbes inhabiting rhizosphere are not culturable, but the advances in the techniques of molecular biology and biotechnology have expedited the process of considering the role of other 99% microbes that cannot be cultured in laboratory situations. However, the major plant growth-promoting rhizobacteria that have been reported so far belong to the genera *Azotobacter, Burkholderia, Arthrobacter, Chromobacterium, Caulobacter, Xanthomonas, Azospirillum, Enterobacter, Bacillus, Pseudomonas, Serratia, Flavobacterium, Klebsiella, Erwinia, and Micrococcus* (Bal et al. 2013).

The rhizospheric microbiological inhabitants represent a subdivision of the microbiological society inhabiting the bulk soil. The secretion of exudates by plants allows the proliferation of some specific microbes in the rhizospheric zone as equated to the bulk soil. There have been several theories which have tried explaining the relative assembly of microbial communities in the rhizosphere. However, two main theories have emerged for a possible explanation. The first one is referred to as niche theory, which points out the significance of deterministic progressions, and the second one is deliberated as the neutral theory, which focuses on stochastic processes (Dumbrell et al. 2010). The niche-centered theory forecasts that the variations in the species community configuration are allied to the deviations in the ecological variables because species owe distinctive possessions that reward them the exploitation of matchless niches. The species copiousness in this theory will follow pre-emption, broken stick, log-normal, and Zipf-Mandelbrot models. On the other hand, the neutral theory envisages the structure and configuration of species communities to the geographic remoteness amid the samples on the account of their dispersal limitation, since several species are functionally comparable based on their capability to utilize niches. Consequently, their richness will follow a zero-sum multinomial (ZSM) distribution. Both the theories are well associated with ecological aspects, but none can provide any evidence in the favor of the dynamic nature of microbiological community association in rhizosphere (Mendes et al. 2014).

Since all the members of rhizospheric microbiome are not culturable, however, the culture-grounded approaches have advocated the supremacy of gram-negative microbes in rhizosphere. The proper designation of the microbiota to precise groups requires the use of advanced molecular biology techniques. Since microbial influences in the rhizospheric portion are repeatedly synergistic, thereby, the understanding of microbial system at the community level seems to be most ecologically significant. The community-level depiction of several agriculturally important crops like corn, pea, potato, rice, alfalfa, avocado, tomato, and corn has revealed that in most of the studies, but not all, the *Proteobacteria* was found to be the dominating group. However, the results varied among different classes of *Proteobacteria*, but mostly *Gammaproteobacteria* were found to overpass the other classes (Hawkes et al. 2007). Similarly, Uroz et al. (2010) also found the dominance of *Actinobacteria* and *Proteobacteria* in the oak rhizosphere soil. Likewise, the exploration of the rhizospheric community of three different cultivars of potato also revealed the dominance of the phylum *Proteobacteria* (46%), which was followed by *Firmicutes* (18%), *Actinobacteria* (11%), *Bacteroidetes* (7%), and *Acidobacteria* (3%) (Weinert et al. 2011). The rhizospheric community structure of alfalfa and barley as assessed by Kumar et al. (2018) was also largely represented by *Proteobacteria* (45.9%) which was followed by *Bacteroidetes* (21.4%) and *Actinobacteria* (10.4%). Similarly, the rhizospheric community analysis also proved the dominance of *Proteobacteria* with a share of 47% followed by *Actinobacteria* (23%), *Firmicutes* (6%), and *Acidobacteria* (5%). It also displayed the presence of eukaryote (3%) and archaea and virus (1%). The comparative analysis of rhizospheric soil as compared to the bulk soil confirmed the overexpression of phyla *Actinobacteria*, *Acidobacteria*, *Chloroflexi*, *Cyanobacteria*, *Chlamydiae*, *Tenericutes*, *Deferribacteres*, *Chlorobi*, *Verrucomicrobia*, and *Aquificae* in the rhizospheric soil (Mendes et al. 2014). The rhizospheric microbiome of any particular plant is known to be affected by different factors, and the microbial populations are known to react and acclimatize themselves to such factor, for instance, the loss of nitrogen-fixing symbiosis in *L. japonicus* modifies the assembly of the community accumulations in the roots as well as rhizospheric compartments (Zgadzaj et al. 2016; Sánchez-Cañizares et al. 2017). The patterns of exudates also vary a lot due to plant age, for instance, the GC-MS analysis of root exudates secreted by gnotobiotically nurtured *A. thaliana* displayed that the intensities of sugars and sugar alcohol secretion diminished during the plant development, although the degrees of amino acid and phenolic secretion augmented with time. The exudates comprising of sugars, organic acids, and amino acids intensely shake the configuration of microbiological plant populations, where the members of *Actinobacteria* and *Proteobacteria* represent the principal consumers of such compounds (Chaparro et al. 2014). The effect of exudates on shaping the rhizospheric diversity can be taken into consideration by the fact that a mutation of an ABC transporter, which finds active involvement in the process of exudation, altered the fungal as well as the populations in the rhizosphere of *A. thaliana*. Nevertheless, the incorporation of organic acids rather than sugars, even in the absence of plant systems, encourages bacterial fruitfulness and diversity. Therefore, the procurement of nutrient in any form acts as a strong driver for the microbial assemblage (Badri et al. 2009; Shi et al. 2011). The rhizospheric microbiome of a plant species is also affected by the presence of other plants. Interestingly, the microbial populations of plant systems cultivated in a mixed field are found to contain an enhanced level of microbial biodiversity, which in turn rewards the plant with an enlarged plant height and leaf surface area as equated to the plant cultivated in a monoculture (Lebeis 2015).

4 Plant-Microbe Rhizosphere Interactions

The plant systems have evolved in a realm of tiny microorganisms. The plants started influencing their rhizospheric microbiome from the very first day. The plant roots brought out numerous changes in the soils which ultimately resulted in the alteration of the physical configuration of the soil. Plant systems dug out the key nutrients from the soils, thereby giving a tough competition to the already inhabiting microorganisms. They also took out water from the soils, thereby modifying the soil moisture that too was faced by microorganisms. The plant debris resulted in the accretion of organic carbon that was later handled by the heterotrophic microorganisms, which resulted in the materialization of soil organic matter. The beginning of the process by which plants started releasing their photosynthates via roots favored the quick assimilation of microorganisms (Cotrufo et al. 2013; Lehmann and Kleber 2015; Doornbos et al. 2012). This further lead to the alterations in the physical as well as chemical environs of the rhizosphere, which in turn influenced the fitness of diverse microbial assemblies and communications amid microorganisms and thereby incited the evolution of new microorganisms that were better suited to the life in this thin zone of rhizosphere (Lambers et al. 2009). The sum of genotypic as well as phenotypic deviations in the plant attributes that support the plant-allied microbiomes responsible for upsurging the plant nutrient accessibility, precluding pathogenic microbes, or else refining plant aptness coupled with the plant performance sustains a fitness benefit. Therefore, the aptitude of plant systems toward the sustenance of a constructive microbiota is an attribute under selection. This close relationship of plant systems with the microorganisms is often regarded as an assimilated ecological entity acknowledged as a holobiont (Vandenkoornhuyse et al. 2015). This holobiont has been the unit under selection for several billions of years, thereby supervising the evolutionary pathway headed for plant traits supporting constructive microbiomes.

There is a vast array of microbial systems inhabiting the plant rhizospheric zone, and they are expected to interact with the plant systems in numerous ways. But most frequently only three distinctive classes of such host-microorganism associations are taken into consideration for the activities of the plant-allied microbiome: parasitic, which deleteriously affects the health of plant systems; mutualistic, which aids the plant growth by its growth promotion attributes; and the commensalism, which does not have any effect on the plant systems. However, these descriptions only take into account of the direct influence of the microbial systems on the plant systems and not the indirect belongings on the other community associates, consequently, exclusive of the influence of microbe-microbe communications happening in plant microbiomes. The microorganisms inhabiting the interior of plant tissues are capable of producing numerous growth-prompting molecules, improving nutrient procurement, or persuading defense from several biotic and abiotic stresses. While the beneficial and deleterious communications amid hosts and microbial species can be specifically elaborated, the notion targeting commensalism is not defined with much clarity. A true commensal certainly does not affect the plant health in any form, therefore, it is discreetly impossible to quantity, since it necessitates witnessing the absenteeism of a phenotype. In conclusion, the microbial systems can be deceitfully considered as commensals owing to their transient occurrence, provisional dormancy, or their performance of some formerly uncharacterized roles. Such kind of perceptions necessitates the performance of community-level investigation at the multiple time points and ecological situations (Berendsen et al. 2012; Lebeis 2015; Zapalski 2011). The interactions among numerous microbes inhabiting the rhizosphere also affect the composition of the rhizospheric microbiome. For example, diverse bacterial and fungal rhizospheric inhabitants act as antagonists for numerous soil-dwelling fungal or nematode phytopathogens by the possession of diverse mechanisms. These mechanisms may encompass antibiosis, competition, aptitude of parasitizing the plant pathogens, damage in the phytopathogenic activity via quorum sensing, and initiation of the systemic resistance in plant systems (Ali et al. 2017). However, here, only the account of plant-microbe interactions is taken into consideration.

4.1 Beneficial Interactions: The Good Microbiome

A major proportion of microbiological populations residing the rhizospheric zone have a vital part to perform in enhancing the configuration as well as production of the natural plant systems via safeguarding the persistence and forbearance against diverse biotic as well as abiotic stresses. This job is done by numerous tools, such as bio-fertilization, encouragement of root progression, management of stresses, rhizoremediation, and disease suppression. A large proportion of rhizospheric microbiomes behave synergistically, promote plant growth as well as development, expand the nutrient acquirement, enhance their tolerance, and induce different defense mechanisms in the plant systems. Therefore, these are deliberated as "the good" of rhizospheric microbiomes (Ali et al. 2017). The bacterial members of rhizosphere actively engaged in plant health elevation activities are designated as plant growth-promoting rhizobacteria (PGPR). The plant health and growth promotion trait of rhizosphere-residing bacteria is brought out by maintaining an active supply of numerous vital nutrients that otherwise are either inaccessible or narrowly obtainable by the plant systems, for instance, nitrogen, iron, phosphorus, and zinc. The mechanisms underlying the superior nutrient endorsement encompass phosphate solubilization, nitrogen obsession, solubilization of zinc, and iron chelation via fabrication of siderophores. Additionally, the PGPR also produces several plant hormones, such as indole acetic acid, cytokinin, and gibberellins. Furthermore, the other mechanisms may comprehend the possession of ACC deaminase activity, biofilm materialization, and production of various exopolysaccharides. The active involvement of rhizospheric dwellers in various nutrient cycles results in recovering vital nutrients like N, P, K, Zn, and Fe, thus enhancing their bio-obtainability to the plant systems (Ali et al. 2017; Sharma and Chauhan 2017; Backer et al. 2018). Broadly, such microbes are classified into three major classes according to their possession of plant growth promotion trait. First are the microbes that upsurge the accessibility of the nutrients to plant systems and are designated to be biofertilizers. The second type of microorganisms is responsible for increasing the plant growth by various indirect means such as by protecting from different plant pathogenic attacks. Such organisms are known to be biocontrol agents. The third class comprises microbes that are responsible for stimulating plant growth through secretion of different phytohormones as well as growth regulators, for instance, auxins, gibberellins, cytokinins, etc. Such microorganism is best regarded as biostimulants (Ali et al. 2017).

The PGPRs are also recognized to bring out the accession and assimilation of nitrogen to the plants which is considered as the succeeding most significant occurrence afterward photosynthesis in the plant systems. The process of biological dinitrogen fixation is extremely important to the global agricultural systems. In this process, the inactive dinitrogen from the atmosphere is reduced to ammonia in the occurrence of nitrogenase enzymes and is a doing of diazotrophic microbes (Sulieman 2011; Dixon and Kahn 2004; Franche et al. 2009). The nitrogen fixative microbial systems are commonly classified as (1) symbiotic nitrogen-fixing microbial systems (e.g., rhizobia and *Frankia*) (Zahran 2001; Ahemad and Khan 2012) and (2) nonsymbiotic (free-living, associative, and endophytes) nitrogen-fixing microbial systems like *Cyanobacteria* (*Anabaena*, *Nostoc*), *Azotobacter*, *Azospirillum*, *Azocarus*, etc. The symbiotic association necessitates a multifaceted communication amid the host microbial partners which may result in creation of some specialized structures like nodule formation for the intracellular colonization of bacteria (Bhattacharyya and Jha 2012; Giordano and Hirsch 2004).

PGPR also assist the plant by enhancing the availability of several vital and key nutrients. The method usually employed is the solubilization of the nutrients followed by their enhanced uptake. The solubilization of key nutrients takes place by secretion of some mild organic acids by the microorganism where the enhanced uptake proceeds by the secretion of some chelator molecules like iron. The plant systems usually face a problem which is low phosphate obtainability due to the occurrence of phosphate in insoluble forms. The phosphate-solubilizing bacterial strains convert the insoluble phosphate into its monobasic diabasic forms which are easily available to the plant systems. The phosphate-solubilizing bacteria dwelling the rhizosphere discharge some mild organic acids and enzymes called as phosphatases which facilitate the transformation of inexplicable forms of phosphate to the plant-accessible forms. The major phosphate-solubilizing bacterial strains are represented by *Azotobacter chroococcum*, *Bacillus circulans*, *Cladosporium herbarum*, *Enterobacter agglomerans*, *Pseudomonas chlororaphis*, *P. putida*, *Rhizobium* sp.,

Bradyrhizobium japonicum, *Beijerinckia*, *Burkholderia*, *Pantoea*, *Flavobacterium*, and *Microbacterium* (Ali et al. 2017; Vessey 2003; Lugtenberg and Kamilova 2009).

Iron is another essential nutrient required by the plant systems; however, its comparative insolubility in the soils restricts its accessibility to the plants. It plays a key role by aiding as a cofactor in different enzymes which catalyze numerous biological progressions such as nitrogen fixation, respiration, and photosynthesis. Plant roots favor iron absorption in the form of reduced ferrous ion, but the availability of ferric ion is much common in finely ventilated soils. Several rhizosphere-inhabiting bacteria have the attribute of siderophore production which functions to bind the ferric form of iron, and it is evident that plant species have the capability of absorbing bacterial Fe^{3+}-siderophore complexes (Stein et al. 2009; Andrews et al. 2003; Lemanceau et al. 2009). The siderophores represent some lower molecular mass complexes possessing excessive empathy toward the chelation of ferric ions which is shadowed by the shift and its accretion in the bacterial cells. There can be different types of siderophores like phenol catecholates, hydroxamates, rhizobactin, and pyoverdine siderophores which differ in their structure as well as activity. In addition to this, several fungi are known to produce siderophores which include the rhodotorulic acids which are di- or tri-hydroxamates, the ferrichrome-type siderophores, and the fusarinines. The siderophore production not only provides the iron to the plants, but it also restricts the growth of various bacterial and fungal plant pathogens by restricting the iron availability to those microorganisms. A vast array of microorganisms have been reported for siderophore production that are largely represented by *Agrobacterium tumefaciens*, *Erwinia*, *Bacillus subtilis*, *Pseudomonas stutzeri*, *Mycobacterium*, *Nocardia*, *Rhodococcus*, *Arthrobacter*, *Azotobacter*, *Penicillium*, and *Aspergillus* (Osman et al. 2018; Sheng et al. 2020).

The rhizobacterial members of genera *Bacillus* and *Pseudomonas* have been reported to produce diverse plant growth regulators which further result in the development of fine root fibers by the plant systems, thereby amassing the entire surface area resulting in enhanced nutrient and water uptake. The different types of plant growth hormones secreted by microbes are found to be auxins, mainly indole-3-acetic acid, cytokinin, and gibberellins. These growth regulators are acknowledged to enhance the increase in root length, cell division process, seed and tuber sprouting, movement of water and nutrients, and secondary root development. Additionally, they also mediate geotropic as well as phototropic reactions and thereby confer resistance to different stresses. The microbes are also known to secrete inhibitors like ethylene which influence the hormonal equilibrium in plant systems. Ethylene is considered as a senescence hormone acknowledged for inhibiting plant growth during usual circumstances; however, at lower levels (0.05 ml/l), it is known for stimulating plant growth. This gaseous hormone is called as "stress hormone," and its level is known to upsurge during the plant exposure to different stresses. The rhizobacterial members are also known to produce 1-aminocycloprop ane-1-carboxylase (ACC) deaminase enzyme which cuts the ethylene production in plant, thereby assisting the plant systems in stress recovery (Backer et al. 2018; Ahemad and Kibret 2014).

Plants being immobile living systems have to confront some abiotic stresses like drought stress, temperature stress, salinity stress, etc. These stresses cause a considerable decline in plant fitness and overall crop produce. The plant-allied valuable microbes are known to play an important role in stress abatement along with the expansion of such agricultural systems that are found to be resilient toward the climatic changes. Innumerable studies have proven that numerous rhizospheric microbes like *Rhizobium* and *Azospirillum* possess the trait of plant stress alleviation. The PGPRs are known to secrete several compounds that behave as osmolytes, for instance, the secretion of glycine-betaine, proline, ectoine, trehalose, polyols, and sucrose by PGPR actions in harmonization with the composites secreted by roots in response to various biotic as well as abiotic strains. The bacteria *Pseudomonas pseudoalcaligenes*, *Bacillus pumilus*, *Pseudomonas putida*, *Enterobacter cloacae*, *Serratia ficaria*, *Pseudomonas fluorescence*, *Dietzianatro nolimnaea*, *Bacillus amyloliquefaciens*, etc. are reportedly known for alleviating the salinity stress (Khan and Bano 2019). Similarly, on exposure to drought strain, plants experience the deposit of numerous stress-induced composites, like proline, polysugars, abscisic acid, and glycine betaine, along with an increment in the production of enzymatic as well as nonenzymatic antioxidants. The soil microbiota initiate diverse biological contrivances like accrual of compatible solutes, EPS fabrication, and spore formation. These mechanisms employed by the microorganisms assist the plant systems to cope with the drought stress. Similarly microorganisms employ a variety of stratagems to assist the plant systems in coping with different abiotic and biotic stresses (Priyanka et al. 2019).

The beneficial rhizospheric microflora also assists the plant systems to get rid of different recalcitrant and xenobiotic compounds, which have accreted in soil systems owing to the rapid pace of anthropogenic activities which further results in the soil humiliation and sterility. The coevolution of plant and their allied microbiota has effectively resulted in the reclamation and restoration of the degraded soils without instigating any detrimental by-products, unlike conventional methods. This process is often said to be rhizoremediation. Several root exudates secreted by plants, like linoleic acid, behave as surfactants which enhance the availability of pollutants to the microbial systems by forming a layer on soil particles which also upshot improved attachment of bacteria on the pollutant. The bacteria then secrete several compounds including enzymes and metabolites which function to breakdown the toxic pollutants into their nontoxic forms. The bacteria, namely, *Bacillus licheniformis*, *Bacillus mojavensis*, *Achromobacter xylosoxidans*, *P. aeruginosa*, *Ochrobactrum* sp., *P. fluorescence*, *Microbacterium* sp., *Microbacterium* sp., *Rhizobium* sp., *Rhizobium*, *Pseudomonas*, *Stenotrophomonas*, and *Rhodococcus*, have been reported to degrade various pollutants (Mishra and Arora 2019). Therefore, the possession of numerous and multidisciplinary beneficial attributes of plant-allied rhizospheric microbiota has projected them as an effective substrate for engineering the plant rhizosphere.

4.2 Harmful Interactions: The Bad Microbiome

The plant systems secrete root exudates for attracting beneficial microflora, but some pathogenic microbiota also gets attracted toward plant roots. These microorganisms parasitize the plant systems and result in several severe infections, therefore executing damaging effects on various crops of economic importance. This part of rhizospheric microbiome which affects the health of plant systems and thereby results in a considerable drop in the plant yield as well as economy represents "the bad" rhizosphere microbiome. The soil that endured pathogenic microbiota significantly deteriorates the crops, and among these fungal members of the rhizobiome are found to be most distressing. Consequently, this portion of rhizobiome seems to be a notable chronic menace toward global food production as well as economic steadiness. A vast variety of phytopathogenic fungi finding their origin from the rhizosphere have been reported; however, the most common pathogenic fungi take account of members of genera *Phytophthora*, *Aspergillus*, *Verticillium*, *Fusarium*, *Mucor*, *Pythium*, and *Rhizopus*. On the other hand, several bacteria have also been reported as pathogenic which largely belong to the genera *Pseudomonas*, *Ralstonia*, *Erwinia*, and *Xanthomonas*. The population and a variety of destructive and constructive microbes are interconnected to the measure and eminence of the rhizodeposits and to the aftermath of the microbiological communications happening in the rhizospheric zone (Somers et al. 2004; Tournas and Katsoudas 2005).

There are four major classes of phytopathogens, namely, virus, bacteria, fungi, and nematodes (Agrios 2005); however, only two of these are considered to be key performers in the soils, namely, fungi and nematodes. Nevertheless, bacterial pathogens on a narrow scale are also deliberated to be soil-borne, possibly for the reason that nonspore formers are not able to endure well in soils for longer times. In addition to this, bacterial pathogens also necessitate an injury or an indigenous breach for their penetration into the plants and thereby initiate the infection process. However, some bacterial pathogens are still able to infect the plant systems, for instance, *Ralstonia solanacearum* is responsible for bacterial wilt of tomato and *Agrobacterium tumefaciens* for the crown gall disease. A fewer filamentous bacterial pathogens also exist and infect the plant systems and are better adapted for their survival in soils. However, only fewer viruses are capable of infecting the roots. Their chances of infection are restricted by their requirement of vector and wound in the plant tissues for the initiation of infection. However, nematodes and fungi like *Olpidium* and *Polymyxa* act as the vehicles for viral particles (Campbell 1996; Nester et al. 2005; Raaijmakers et al. 2009). The pathogenic fungal species are causing major harms to crops in the form of various diseases, thereby affecting the overall economy of the field. The major sinks of the crop economy find their origin from several genera like *Pythium*, *Fusarium*, *Verticillium*, *Rhizoctonia*, and *Armillaria* (Ali et al. 2017).

The microbiota inhabiting the rhizosphere is also composed of many nematode species that are found to be parasitic to the plant systems. While a major proportion of the nematodes inhabiting the soils is free-living, 7% of the overall soil-lodging nematodes are found to be pathogenic to diverse plant species. The plant-parasitic nematodes have been found to affect different crops of much economic importance such as wheat, soybean, potato, tomato, and sugar beet. The nematode parasitism produces different signs in plant systems like leaf chlorosis and patchy, wilting, arrested growth coupled with the defenselessness against other major pathogens. The most pathogenic of all these nematodes are said to be root-knot nematodes and cyst nematodes which belong to the Heteroderidae family due to their broad range of host plants. The other major category of parasitic nematodes is migratory endoparasitic nematodes which migrate through roots and detrimentally feed on the plant cells, thereby causing substantial necrosis in the plant tissues. These are largely represented by the rice root nematode (*Hirschmanniella*), lesion nematode (*Pratylenchus*), and burrowing nematodes (*Radopholus*). These nematodes are attracted toward the plant roots by several of the root exudates like alcohols, ketones, organic acids, terpenoids, thiazoles/pyrazidines, cyclic adenosine monophosphate, esters, ions, amines, amino acids, and other aromatic compounds (Moens and Perry 2009; Jones et al. 2013; Ali et al. 2015; Rasmann et al. 2012).

These soil-originated pathogenic microbes have evolved in very hard situations, and therefore these are well fitted to the rhizospheric zone as equated to other microorganisms. They have invented several methodologies in their evolutionary journey to have hard edifices like resting spores, which aid their survival for longer periods in the nonappearance of the host crop.

The rhizospheric soil encompasses numerous microorganisms, somewhat lesser in statistics, which are found to be human pathogens. Such unscrupulous microbial pathogens are "the ugly" ones owing to their most damaging nature by unswervingly infecting the humans. These ugly microbes may either be native to the soils and also be dropped by human deeds, for instance, carried by animal as well as the bird fecal material, manure solicitations, by agricultural machineries, use of slaughterhouse wastes, sewage water, and medical wastes. The major human opportunistic pathogens dwelling the plant rhizosphere are of dermatological significance affecting the skin, hair, nails, etc. The opportunistic human pathogens are mainly represented by fungi like *Microsporum canis*, *Trichophyton mentagrophytes*, *Aspergillus* spp., *Coccidioides*, *Blastomyces dermatitidis*, and *Trichophyton rubrum*. However, the human pathogenic bacterial members especially the spore formers also inhabit the rhizosphere, for instance, *Clostridium tetani*, *C. botulinum*, *Bacillus anthracis*, *Actinomyces israelii*, and *Clostridium perfringens*, and some nonspore formers like enterotoxigenic strains of *E. coli* also inhabit rhizosphere (Berg et al. 2005; Chapman 2005; Baumgardner et al. 2011; Blackburn et al. 2007; Ali et al. 2017). The presence of numerous plant pathogenic microbial systems and unscrupulous human pathogens in the rhizospheric zone has prompted a need to engineer the rhizosphere where only beneficial microbiota can thrive by kicking out the plant and human pathogens so that the release of plant photosynthates via roots can be properly utilized by the plant systems.

5 Rhizospheric Engineering

The plant systems regulate the occurrence of microbial populations in the rhizospheric zone. Plants have also advanced several functions and stratagems for the alteration of rhizosphere and rhizobiome. It has also been proven that both beneficial and pathogenic (plant, human) microbes inhabit the rhizosphere. The configuration, comparative copiousness, and spatial and chronological dynamics of the rhizospheric microbial inhabitants not only affect the plant health and growth but also lay a strong influence on the health of human beings (Ryan et al. 2009; Mendes et al. 2013). The domestication of plant systems was mainly done using artificial selection by selecting crops based on traits excluding reproductive fitness, thereby deviating the whole process from the natural selection. The food crops were mainly selected based on huge seed size, condensed bitterness which is a principal defense mechanism, and some other traits, which unintentionally altered the plant traits regulating the microbiome. Therefore, the domestication process of crops has resulted in the alteration of the microbiomes conscripted by the plant systems (Leff et al. 2016; Pérez-Jaramillo et al. 2016). The advent of employing nitrogen-based fertilizers has also resulted in a paramount deviation from the natural selection. The application of nitrogen-based fertilizers made it sure that the yield of crops was not unswervingly associated with a plants' capability of supporting microbial nutrient cycling. The N fertilization leads to a sharp reduction in the microbial biomass as well as their variety (Treseder 2008; Ramirez et al. 2010), concomitantly leading to the promotion of copiotrophs above oligotrophs (Fierer et al. 2012). The plant selection following explicit fertilizer establishments has promoted the unlinking of soil microbiota from the plant health. The application of ammonium-grounded fertilizers tends to condense the rhizospheric pH, whereas the application of nitrate-based fertilizers leads to an increase in the pH, thereby resulting in an alkaline rhizosphere. It is evident that alterations in soil pH can modify the soil chemistry in the zone surrounding the roots and thus impact the progression along with the configuration of microbial societies (Ryan et al. 2009). The selection of plant systems facing extraordinary fertilization management has resulted in the selection of genotypes supporting microbial N mineralization (Schmidt et al. 2016). Consequently, the present varieties may have experienced a loss in their aptitude of supporting microbiota responsible for degrading the organic forms of nitrogen and solubilizing the mineral nutrients like phosphorus (Wallenstein 2017).

Therefore, the major research interest in this field is precisely leaning toward the development of different approaches that could reshape the rhizospheric microbiota in favor of those microbial systems that have the potential of improving plant health as well as productivity and can also avert the propagation of different plant and human pathogenic microbiota already inhabiting the rhizosphere. Several research programs have already proven that plant's genetic makeup along with soil variety is an important driver for shaping the rhizospheric microbiota (Berg and Smalla 2009; Bakker et al. 2012). Moreover, the fascinating roles played by microorganisms in various natural processes like soil organic materialization, nutrient proclamation, and pathogen burden have projected them for manipulating the microbiome as key for the rhizosphere engineering (Wallenstein 2017). The impact of soils on the rhizospheric microbiota has already been validated for different plant species (Berg and Smalla 2009). The soil systems are composed of extremely multifaceted and assorted environs that considerably affect the physiology of plant systems, a configuration of root exudation, and concurrently the rhizospheric microbiome. The pH of soil systems has also a significant part to play in determining the rhizospheric microbiome. The abundance along with a variety of bacterial populations has been found to fluctuate by the ecosystem type where the soil pH is the key driver. The bacterial variety is utmost in the neutral soils and subordinate in the soils having an acidic pH (Fierer and Jackson 2006; Mendes et al. 2013). Based on the genetic configuration of plant systems also, innumerable methodologies have been suggested for reshaping the microbial configuration of rhizosphere in the quest to redirect the microbial movement. The term "rhizosphere engineering" thereby denotes the alteration of plant's root and adjoining environment in the quest to generate a "biased" milieu that will unambiguously improve the crop yield as well as the plant endurance. Root exudates play an essential role in enticing different plant pathogenic microbes and activation of their virulence factors. Therefore, altering the amount of root exudates through plant breeding experiments or by genetic alteration seems to be an apparent methodology for redirecting rhizospheric microbiome. The other strategy for reshaping the rhizosphere involves various soil amendments like the addition of compost and biochar which favor the colonization by beneficial microbial communities. Other strategies include the introduction of beneficial microbes in soil onto seeds and planting materials (Bhattacharyya and Jha 2012; Mendes et al. 2013). The understanding of the actions involved can help propose the different techniques which can allow the modification of the rhizosphere for an improved plant fitness and enhanced soil output. The different methodologies and representations of rhizospheric engineering are discussed under.

5.1 Soil Amendments

The alteration of the rhizospheric soil, and in turn its microbial constitution which has remained the most involuntary concern of the human activities, such as the frequent farming of some definite crops, may bring about the appearance of disease-oppressive soil systems, and several soil pollutants have also been reported for radically distressing the configuration of soil as well as plant-allied microbiota. The expansion of various novel practices in the field of microbiology and microbial ecology has delivered several prospects for modifying the soil microflora in a way analogous to the discerning "rhizosphere engineering" that happens in nature (Ryan et al. 2009). The amendments in soils seem to be the easiest way of engineering the rhizosphere. A vast array of soil amendments is employed for upsurging the plant productivity which also proves to be an important tool for shaping the rhizospheric

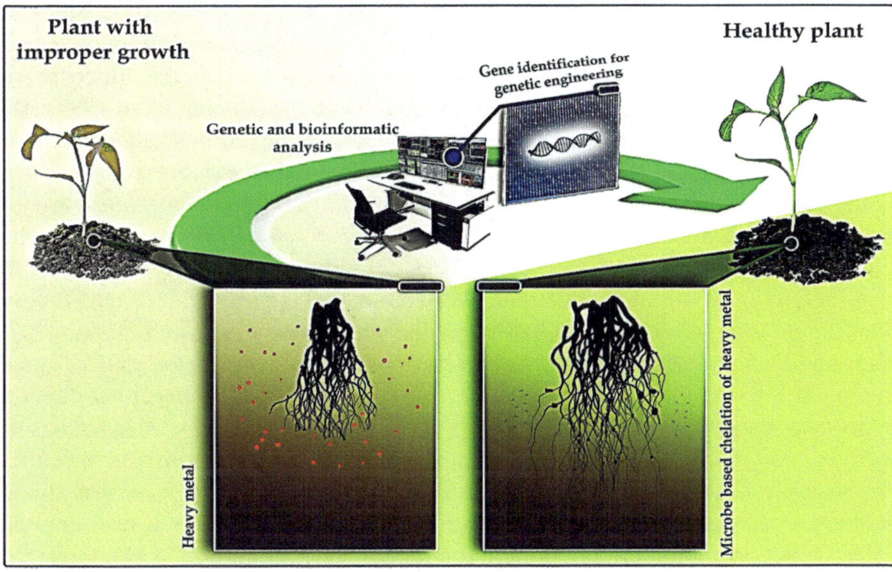

Fig. 21.1 Diagram depicting the different types of soil amendments employed for shaping the rhizospheric microbiome

microbiome (Fig. 21.1). This section takes account of the different types of soil amendments that are often employed for getting a biased rhizosphere.

5.1.1 Soil Amendments with Compost

The addition of compost to the soils is also known for altering the microbial composition of rhizosphere. It increases the soil suppressiveness toward the soil-borne pathogens. However, the soil suppressiveness is dependent on the type of compost added. It also enhances the number of antagonists in the rhizosphere (De Brito et al. 1995). It further improves the physical as well as biochemical belongings of the soil, upsurges the soil water balance, and enhances the nutrient supply to plants, thus altering the soil properties and making it fit for microbial inhabitation. The short-term application of composts increases the rhizosphere soil carbon mineralization and microbial biomass, and this carbon mineralization increases the progression of roots and thin root hairs (Zhang et al. 2014) which further allow the plant systems to harbor beneficial microbiota. The compost brings a source of carbon for the existing rhizospheric microbiota in the form of soil organic matter, and it also acts as a source of diverse classes of microorganism which later inhabit the plant rhizosphere. It also alters the soil chemistry as well as soil structure in a substantial manner and thereby significantly affects the configuration of plant-allied microbial communities (Green et al. 2007). The soil organic matter represents a noteworthy basis of utilizable carbon for different rhizospheric inhabitants (Toal et al. 2000), and it has also been advocated that the incorporation of composts to the soil can upkeep microbes that are not even endured by exudates. This capability for compost-originated organic matter to endure some microbes advises that the "rhizosphere effect" does not act similarly on all microbial inhabitants (Boehm et al. 1997). De Brito et al. (1995) noticed that the compost incorporation to soil augmented the occurrence of bacteria in the rhizosphere of tomato that exhibited antagonism against various soil-borne pathogens like *Rhizoctonia solani*, *Pyrenochaeta lycopersici*, *Fusarium oxysporum* f. sp. *radicis-lycopersici*, and *Pythium ultimum*. The suppression of various pathogenic microbes by addition of compost is known to bring about the recruitment of definite microbes as the suppressive soils tend to lose their suppressive activity on their pasteurization and sterilization (Weller et al. 2002; Haas and Défago 2005). The addition of compost and organic matter enhances the microbial activity in the soil which inhibits the growth of pathogens either directly by its antagonistic activity or indirectly by the possession of competitive actions of recruited soil microorganisms. The suppression incurred to the soil systems either can be general or may also be specific. In case of general suppression, a basal shield contrary to an extensive collection of pathogenic microbes is established, and the defeat is not accredited to any precise microbe (Weller et al. 2002). However, the possession of specific suppression is attributable to the accomplishments of precise microbes that act contrary to specific pathogens and is found to be more operative than general suppression. The compost amendments in the soils not only redesign the structures of a microbial community but also lead to the establishments of new equilibria (Hadar and Papadopoulou 2012). The composts are also known to contain various bacterial and fungal biocontrol agents that later inhabit the plant rhizosphere and are known to advance the regularity of disease control. Antoniou et al. (2017) assessed the consequence of compost addition on the rhizospheric community of tomato along with its effect on the suppression of fungal pathogens. The compost added to the plant was able to suppress the fungus, namely, *Fusarium oxysporum* f. sp. *lycopersici* and *Verticillium dahliae*. It was also observed that the compost lost its disease suppression ability upon sterilization. Furthermore, it was found that the phyla *Firmicutes* and *Ascomycota* were dominating the compost, whereas the phyla *Actinobacteria*, *Proteobacteria*, *Bacteroidetes*, and *Mucoromycota* were rarely isolated. The addition of compost significantly altered the microbiological configuration of the rhizospheric zone as experienced by a reduction in the *Ascomycota* and *Firmicutes*, while *Actinobacteria*, *Bacteroidetes*, and *Proteobacteria* were augmented. Surprisingly, the number of *Proteobacteria* was found to be augmented by 57 times in the rhizosphere samples, while *Actinobacteria* by 6.1 times as equated to the unplanted compost sample. Innumerable studies have evidenced that the incorporation of compost in the agricultural soils protects the plant systems from some pathogenic microbes such as *Pythium ultimum*, *Pythium irregular*, *Phytophthora nicotianae*, *Sclerotinia minor*, and *Sclerotinia sclerotiorum*. The mechanisms may include the direct suppression of the pathogens or activation of the disease resistance genes in plant systems (De Corato 2020). Countless studies have testified a relative increment in the members of *Proteobacteria* and *Actinobacteria* upon compost addition, thus making it the most dominant group in the rhizosphere.

Proteobacteria are also acknowledged for playing a serious role in the global cycling of carbon, nitrogen, iron, and sulfur, whereas *Actinobacteria* are supposed to subsidize the global carbon cycle by degrading the plant biomass, and because of their aptitude of decomposing organic matter in the soils, they are also proficient for fabricating several key enzymes like cellulases, hemicellulases, chitinases, glucanases, and amylases (Mickan et al. 2018; Yang et al. 2019). Conclusively, the amendments of compost in the soils prove to be an effective tool for reshaping the rhizosphere biology and, in turn, the beneficial rhizospheric inhabitants for improved plant health and yield.

5.1.2 Soil Amendments with Biochar

Biochar is a very steady product of thermal deterioration of organic materials in the lack of air (pyrolysis) and is distinguished from charcoal by its use as a soil amendment. The temperature of pyrolysis lies in the range from 300 to 1000 °C. The biomass employed for pyrolysis is principally composed by organic composites like cellulose, hemicellulose, and lignin (Kavitha et al. 2018). It has also been designated as a promising measure to upgrade the soil fertility besides other environmental amenities such as carbon sequestration for the extenuation of climate changes. The addition of biochar is acknowledged for the enhancement of the fertility of soil systems predominantly by uplifting the pH of acidic soils or by enhanced nutrient retention via cation adsorption and by uplifting the water retention capacity of the soil. The desired depth for the application of biochar lies in the range of 4–6 cm (Lehmann et al. 2011; Yu et al. 2019). The biochar amendments in the soils are known to alter the diversity as well as an abundance of the biological community. The alterations induced by the biochar amendment in the microbial community configuration may not only distress nutrient cycling and plant progression but also the dynamics of organic matter present in the soil systems (Wardle et al. 2008; Kuzyakov et al. 2009; Liang et al. 2010). The biochar apertures function as a microenvironment for the proliferation of microbial systems. The microorganisms utilize carbon, nutrients, gases, and water offered by the biochar for growth as well as reproduction. The soil application of biochar at a proportion of 10 t per hectare has resulted in a noteworthy upsurge in the biological nitrogen fixation by red clove as equated to the control. Its amalgamation in the soil is also known to affect the arbuscular mycorrhizal fungi in a positive manner (Jaafar 2014; Mia et al. 2014). Biochar also reduces the tensile strength of the soil, therefore making the root as well as mycorrhizal nutrient mining extra operative. The reduced tensile strength also facilitates the easy seed germination and also simplifies the movement of invertebrates through the soil, thereby modifying the predator/prey dynamic (Lehmann et al. 2011). The biochar addition supports the growth of PGPRs like *Bacillus insolitus*, *Aeromonas hydrophila*, and *A. caviae* which are known to mitigate the salinity stress by the secretion of exopolysaccharide responsible for binding sodium ion that results in a reduced uptake by the plants along with the production of an enzyme called 1-amin ocyclopropane1-carboxylate deaminase which also relieves the salinity stress

(Ashraf and Harris 2004; Ali et al. 2014). In addition to it, the microbial copiousness has also been confirmed in the biochar-amended soils by different methods, like total genomic DNA extraction, plate count, substrate-induced respiration, fumigation-extraction, phospholipid fatty acid extraction, and staining and direct surveillance of discrete biochar particles. Furthermore, it also enhances the rate of reproduction of microbial populations (Lehmann et al. 2011). The microbial communities associated with the nitrogen transformations are known to be altered upon biochar incorporation indicating a reduced soil nitrogen loss and improved nitrogen utilization as indicated by a reduction in the number of *Nitrososphaera* in the rice fields upon biochar amendment (Liu et al. 2017). Moreover, the biochar addition is also known to uplift the network of beneficial fungi in the rhizospheric zone (Wang et al. 2019). Win et al. (2020) evaluated the effect of biochar on the rhizospheric communities using the next-generation sequencing methods and observed that biochar augmented the copiousness of *Proteobacteria* as well as *Actinobacteria* in the rhizoplane particularly after 2 weeks of transplantation. On the contrary, there was a decrease in the number of *Acidobacteria* and *Bacteroidetes*. The members of *Xanthomonadaceae* experienced an increment of 2.8-folds in their numbers after 2 weeks of transplantation followed by *Desulfuromonadales* (1.8-fold), *Burkholderiales* (1.8-fold), and *Actinomycetales* (1.4-fold) along with a concomitant decline in the relative abundance of *Sapropirales* (1.8-fold) and *Nitrososphaerales* (2-fold). Similarly, Cheng et al. (2018a, b) also observed that the supplementation of the soils with the biochar augmented the diversity as well as an abundance of bacteria. The comparative copiousness of *Adhaeribacter*, *Rhodoplanes*, *Pseudoxanthomonas*, and *Candidatus Xiphinematobacter* augmented in the biochar-amended soil; however those of *Lacibacter*, *Pirellula*, and *Kaistobacter* faced a decline. The addition of biochar is also acknowledged for influencing the root metabolome and is known to alter the levels of some amino acids as well as organic acids. Therefore, it is not only the rhizosphere microbiome that is altered upon soil amendments with biochar, but the rhizosphere metabolome is also reshaped. Chen et al. (2017) observed that the biochar addition along with a simultaneous nitrogen reduction caused a 1.75-fold increase in the levels of isoleucine, a 2.16-fold surge in malonate, and a 2.15-fold rise in acetate in exudates. Similarly, Bornø et al. (2018) also observed that the exudates of particularly glucose and fructose were intensely altered by the biochar application, specifying that the plant reaction to biochar application can modify the configuration of root exudates discharged into the rhizosphere. This altered exudation process in turn plays a key role in engineering the rhizospheric microbiome (Fig. 21.2).

5.1.3 Other Soil Amendments

A large number of human practices are known to alter the rhizospheric microbiome in an unintentional way, for instance, addition of fertilizers, addition of substrates for fueling bioremediation processes, use of pesticides and other agrochemicals, etc. The application of glyphosate has been shown to alter the denitrification process

Fig. 21.2 A portrayal depicting a GM plant engineered for the secretion of specific root exudates which later harbors definite microbial populations and alleviate the heavy metal stress

in the grass sward along with a surprising increment of 20- to 30-fold in the denitrification process as equated to the herbicide-untouched grass. The denitrification process in the soil is predominantly attributable to the facultative anaerobic bacteria; thereby, any increment in the process suggests a possible alteration in the diversity and number of accountable microbes in the rhizospheric zone (Tenuta and Beauchamp 1996; Qian et al. 2018). The application of diclofop-methyl leads to a reduction in the nitrification of urea nitrogen in soils. This weedicide is potent enough to inhibit the enzyme acetyl-CoA carboxylase activity and thereby can lead to a reduction in the fatty acid synthesis in the crop. In addition to it, the persistence of residual DM particles in the soil systems is known to affect an extensive range of plant metabolic pathways and thus can lead to an augmented exudation of organic acid (Rensink and Buell 2004; Qian et al. 2012; Chen et al. 2017). The plant root exudates are the crucial influencers of rhizospheric microbiota configuration; therefore, the testified impact of diclofop-methyl on the exudation nurtures the probability that multifaceted plant-microbiome communications could restrain the DM poisonousness and could also alter the copiousness of specific microbes distressing the biogeochemical cycles of nutrients. Qian et al. (2018) reported that the application of DM on rice altered the levels of 28 different exudates in the rice rhizosphere. The altered exudation also affected the rhizospheric microbiome and resulted in an increase in the fraction of *Proteobacteria* from 42.1% in the control to 55.4% after 5 days of DM exposure. However, the comparative richness of phyla, *Firmicutes* and *Acidobacteria*, faced a decline from 22.0 and 16.9% in the control to only 8.9 and 13.9%. Additionally, the comparative richness of the genera *Azospira*, *Clostridiales*, and *Rhodocyclaceae* increased from 7.1, 0.3, and 1.1% in the control to 21.0, 2.4, and 2.3% of total rhizospheric microbes.

The wastewater-borne pollutants are also known to alter the rhizospheric configuration of the holobiont. The wastewater-borne sulfonamides are known to alter the microbiome composition in the constructed wetlands planted with *Cyperus alternifolius*, *Cyperus papyrus*, or *Juncus effusus*. A noteworthy decline in the microbial diversity has been testified along with a precise inhibition of microbes involved in the nitrogen and sulfur cycle. However, the microbes like *Methylosinus*, *Methylotenera*, *Methylocaldum*, and *Methylomonas* which are potent for degradation of sulfonamides are found to be increasing in the rhizospheric zones of the plants (Man et al. 2020). The irrigation with treated wastewater is also known to alter the composition of rhizobiome. The soil ammonia-oxidizing bacterial populations are altered irrespective of the ammonium concentration or the presence of plants. The treated wastewater brings a reduction in the comparative richness of *Actinobacteria* along with a simultaneous upsurge in the comparative copiousness of *Gammaproteobacteria* (Oved et al. 2011; Frenk et al. 2014). Zolti et al. (2019) also reported an upsurge in relative copiousness of *Gammaproteobacteria* and a decline in *Actinobacteria*, in the root microbiome receiving irrigation with treated wastewater. The assessment on more precise levels revealed the abundance of *Pseudomonadales* and a reduction in *Streptomycetales* and *Pseudonocardiales*. Similarly, the wastewater effluent containing aged nanoparticles has also been acknowledged for influencing rhizospheric microbiota. In a study by Liu et al. (2018), it has been claimed that the copiousness of cyanobacteria was amplified by 12.5% as demonstrated predominantly by an upsurge of *Trichodesmium* spp., and the lavishness of unknown archaea was heightened from 26.7% in the control to 40.5% in the soil watered with wastewater effluent containing aged nanoparticles.

Several other organic amendments, such as seed meal for the control of fungal pathogens, also alter the rhizospheric microbiome. The soil amendments with *Brassicaceae* seed meal preparations for the suppression of apple replant disease altered the rhizobiome in a significant way. The amendment not only suppressed the pathogen *Pratylenchus penetrans* but also elevated the level of *Proteobacteria* and *Acidobacteria* in the rhizosphere. In addition to it, the microbial genera engaged in numerous nitrogen-cycling progressions, like *Bradyrhizobium*, *Rhodopseudomonas*, and *Nitrospira*, were found to exhibit more abundance. Similarly, the fungus *Basidiomycota* got reduced in abundance in the apple rhizosphere after the treatment, whereas the abundance of *Zygomycota* got increased (Mazzola et al. 2015).

The addition of fertilizers also changes the structure of rhizosphere microbiome. The soil amendments with high levels of nitrogen fertilizers negatively affect the soil diazotrophs. The discharge of root exudates is reliant on the plant physiological status along with the nutrient obtainability. For instance, maize has been reported to discharge subordinate amounts of amino acids via roots during nitrogen scarcity (Carvalhais et al. 2011, 2013). Therefore, the application of nitrogenous fertilizers alters the nutrient status of the soil and thus affects the rhizospheric microbiome. The analysis of root exudates of maize during nitrogen fertilization has revealed a tremendous increment of 30-folds in the sugar alcohols, 11-folds in sugars, and 7-folds in phenolics. This altered exudation process affected the rhizospheric microbiome by elevating the levels of *Bacillales*, *Nitrosomonadales*, and *Rhodocyclales* and by reducing the abundance of *Chloroflexales*, *Gemmatimonadetes*, and *Phycisphaerae* (Zhu et al. 2016).

6 Engineering the Plant

The plant systems happen to be the strategic elements for shaping the microbial populations in the rhizospheric zone. The plant's ability to employ a diversity of occupations and stratagems to alter its rhizosphere in the quest to circumvent environment-associated stresses has attracted the interest of researchers for modifying the rhizosphere by engineering the plant systems. The understanding of the actions taking place assists in the development of techniques for modifying the rhizosphere for attaining improved plant healthiness and enhanced soil output efficiency. The plants can be genetically engineered for altering the soil organic anion efflux along with its transference from root cells by altering plants with an inordinate aptitude to produce organic anions coupled with their conveyance outside the cell. The plants are also potent enough to be genetically amended for the fabrication of several recombinant proteinaceous molecules, root exudates, and several other metabolites which target a biased rhizospheric colonization (Ryan et al. 2009; Mohanram and Kumar 2019). Nevertheless, the engineering of plant systems drives beyond the presently extensively nurtured, genetically altered plant systems that are resistant to a few pests or resilient to some herbicides.

The role of root exudates in shaping plant microbiome has attracted the attention of plant breeders and plant biotechnologists on a global basis for engineering the plant systems in the quest to get definite root exudates in higher concentration. As early as 1978, Petit et al. recommended to harness the benefit of the close connection prevailing amid the plants and their accompanying microbiota for framing the exudation process. This would offer a selective benefit to certain microbes which would help them in their establishment in the rhizospheric zone, a stratagem later designated "biased rhizosphere" or "artificial symbiosis" (Savka et al. 2002). The earlier reports on engineering plant systems for a biased rhizosphere mainly target the engineering of plant systems to produce opines. The presence of opines in the rhizospheric zone powerfully shakes the native microflora. To be sure, such opine-secreting transgenic plants lead to an increment in the population of opine-consuming associates that may range from 100 to more than 10,000-folds in the non-sterile soils (Mansouri et al. 2002). This phenomenon can result in alterations of the bacterial members that persist evident even in the nonexistence of the selective pressure of opines (Oger et al. 2000) which further validate the excellence of opines as discerning substrates for microbial inhabitants in the rhizospheric zone. For instance, the transgenic lotus plants genetically altered for the production of two opines, namely, mannopine and nopaline, altered the composition of rhizospheric microbiome along with a specific increment in the bacterial communities able to exploit these molecules as sole carbon source (Oger et al. 2004).

The plant metabolism is redesigned for engineering the plant systems for desirable root exudates. The genes directing the synthesis of root exudates are firstly recognized in the plant systems, and then their expression levels are altered for redesigning the rhizosphere for upgraded features. For instance, the GM rice and tomato engineered with the vacuolar H+-pyrophosphatase gene AVP1 from the *Arabidopsis* plant displayed almost 50% more citrate as well as malate efflux as compared to their wild types after their treatment with aluminum phosphate. This was later deduced as a probable mechanism for enhancing resilience toward aluminum-ion-induced strain and to advance the plant aptitude to consume the unsolvable phosphorus (Ahkami et al. 2017; Yang et al. 2007). Similarly, a gene encoding for citrate synthase from *Citrus junos* plant when cloned and overexpressed in *Nicotiana benthamiana* led to a threefold increment in the enzyme activity which further supported the accumulation of citrate in a concentration that was found to be twofolds higher as equated to the wild-type plant systems. Certainly, the root systems of genetically altered plants were found to be more tolerant to aluminum toxicity, and, surprisingly, their roots sustained to lengthen at levels of 100 mM Al, which were enough to constrain growth in wild-type plants (Deng et al. 2009). Likewise, the citrate synthase gene originating from *Pseudomonas aeruginosa* when transferred into papaya also led to an augmented accrual of citrate in the cytoplasm (Rengel 2002) which was further complemented by enlarged efflux of citrate into the vicinity of roots along with an improved forbearance of transformed plants to Al. The secretion of specific root exudates has also been reported for increased plant tolerance toward the deficiency of nutrients. For instance, the transferring of rye chromosome 5R or only a minor segment of chromatin from the long arm of the chromosome 5R to wheat upsurges its lenience toward the copper paucity (Schlegel et al. 1997). The plant's increased tolerance toward copper deficiency after the chromosome transference is also coupled by the fact that genes for mugineic acid synthase and 3-hydroxymugineic acid synthase, the enzymes involved in biosynthesis of common phytosiderophores, are located on the rye chromosome 5R (Rengel 2002). Furthermore, the root exudates are also supposed to play a significant role in the abovementioned process.

The plant systems are evolved with different mechanisms to discharge the exudates into the rhizospheric zone, comprising diverse kinds of passive as well as active transport systems. Conventionally, the exudation has been deliberated to be a passive progression, arbitrated via different pathways: the conveyance over the root membrane by diffusion, ionic channels, and vesicles transport (Baetz and Martinoia 2014). The pitch shaped by their dissimilar levels amid the cytoplasm of root cells and the rhizosphere is a major factor in shaping the exudation process which is also a subject to be affected by the permeability of root membrane, the veracity of root cells, and the polarity of the compounds to be exuded (Badri and Vivanco 2009). The presence of ion channels for secretion of several root exudates also provides a selective prospect for engineering the plants. The ionic channels are held accountable for discharging the carbohydrates along with some precise carboxylates like malate and oxalate, which are oozed not by diffusion, but via a transport machinery facilitated by proteins. Two different anionic channels have been described: SLow Anion Channels (SLACs), originally named S-type (Slow-type), which need several seconds to be activated, and QUick Anion Channels (QUACs), originally named R-type (Rapid type), which can be activated in a few milliseconds (Dreyer et al. 2012). The aluminum-activated malate transporters (ALMT) and multidrug and toxic compound extrusion (MATE) membrane transporters are extensively studied among all the transporters (Sharma et al. 2016; Kang et al. 2011; Vives-Peris et al. 2020). The two approaches that have been tried to upsurge the discharge of organic ions from the roots are engineering the plant systems with an improved ability to synthesize organic ions and genetically altering the plant systems with a heightened aptitude to convey organic ions outside the cell (Ryan et al. 2009). The first approach targets the expression of genes concerned with the synthesis of particular ions, whereas the second approach targets the genes encoding proteins facilitating the movement of organic ions through the plasma membrane. The genetic engineering of plants grounded based on the second approach takes account of genes encoding the transport proteins. The foremost gene that was recognized to translate a transport protein facilitating the efflux of organic anions from plants is *TaALMT1* from *Triticum aestivum* (Sasaki et al. 2004). This gene codes for the first fellow of an innovative membrane protein family that functions as an anion channel to mediate Al^{3+}-activated malate efflux from roots. Thus, it represents an important tool for altering the malate release in the plant rhizosphere. Similarly, the MATE genes are found to efflux a vast array of small organic composites comprising secondary metabolites like flavonoids and alkaloids (Omote et al. 2006). They have also been found to enable citrate efflux from the plant cells. The *Arabidopsis* and tobacco plants transformed with SbMATE1 and HvMATE genes, respectively, have been reported to deliberate Al^{3+}-stimulated citrate efflux along with an augmented tolerance of Al^{3+} stress (Magalhaes et al. 2007; Furukawa et al. 2007). The examples have exhibited the key part of transport proteins in engineering the plant systems for getting a biased rhizospheric zone. Similarly, the plant systems can also be engineered for altering the rhizospheric pH as the plant systems are known to back the rhizospheric acidification by engendering electrochemical gradient potential crosswise the cell membrane of root cells after the efflux of H^+. This acidification assists in the augmentation of the plant's contact to Fe^{3+} and P which are otherwise not accessible to plants (Hinsinger et al. 2003). The efflux of H^+ ions from the plant cells is principally under the control of a large family of H^+-ATPase. Therefore, the manipulation of plant systems for the overexpression of these genes in the quest to amend the rhizospheric pH also seems to be an open opportunity. The expression of the AVP1 pyrophosphatase in *Arabidopsis* beyond the normal levels persuaded a highly acidified rhizospheric environ, speciously by increasing the action of the cell membrane H^+-ATPase (Yang et al. 2007). Therefore, the involvement of diverse biotechnological approaches can be utilized to engineer the plant systems for getting a biased rhizosphere owing to the ability of the engineered plants to produce the desired root exudates, acidify the rhizospheric zone, and therefore harbor the desired set of microbial systems.

7 Engineering of Microbial Partners

The particular aim of microbiome engineering is to influence the microbiota in the direction of an assured type of microbial community that owes the potential of optimizing plant functions of interest. Furthermore, the engineering of microbial partners is always motivated to harnessing the advantage of naturally evolved plant-microbiome communication networks (Quiza et al. 2015). The directing force toward the alteration of rhizospheric microbiome in the quest to upsurge the plant functioning and productivity is the plenty of evidence that has unveiled the critical role of plant-microorganism connection to the healthiness, output-efficiency, and the complete situation of plant systems. Therefore, the only objective of modifying the plant microbiome is to drive the plethora of rhizospheric interactions in the direction of enhanced constructive aftermaths for the plant systems. The plant root exudation-mediated microbial colonization of rhizospheric microbiome is largely explored, but what is of more interest is that the presence of specific microbes in the rhizosphere is also identified to amend and shape the exudation process, for instance, antimicrobial-resistant *Pseudomonas* is potent enough to block the fabrication of plant antimicrobial compounds (Bais et al. 2008; Hartmann et al. 2009; Oburger et al. 2013). Thus, the parameter dealing with the engineering of microbial partners requires a prompt knowledge of rhizospheric interactions. However, the efforts for revealing rhizospheric communications are predominantly focused toward the aptitude of a single plant root exudate to touch the single bacterial or fungal rhizospheric inhabitant. The unblemished constraint tackling this kind of attitude is the removal of the microorganism from any environment that would surely pot the existence of interspecies interactions into ignorance (Ziegler et al. 2013). The other major restrain in this approach is the inability of several rhizospheric microbes to grow in the laboratory and the inadequacy of the culture-dependent approaches for the qualitative scrutiny of rhizosphere microbiome. Interestingly, in spite of these several methodologies, targeting rhizosphere microbiome engineering necessitates the involvement of microbial isolates at hand, thereby pointing the requirement for the escalation of cultivability of rhizospheric microbes. Therefore, the possession of a distinct functional capacity by several microbial isolates puts forward the approach of inoculating these microbial cultures in the plant rhizosphere in the quest to engineer the plant microbiome for improved plant well-being and output (Ryan et al. 2009; Quiza et al. 2015). However, the perseverance, as well as the serviceability of the inoculated isolates, needs to be further measured to ascertain positive influences when used as a definite stratagem for manipulating the rhizospheric microbiome

(Stefani et al. 2015). In addition to this, the inoculation with genetically altered microbial strains also represents an important strategy for manipulating the rhizospheric microbiome. The recombinant strains are genetically altered for any particular desired trait, and in several circumstances, the recombinant strains have the potential to address complications allied with the swift diminution of the population density coupled with their undersized persistence. The recombinant strains may bring out the augmentation of several inhabitants of the endogenous community by the transferal of genetic material via horizontal gene transfer. However, the release of GM strains in the environs necessitates a thorough assessment to appraise the impending risks associated (Ryan et al. 2009). However, the disruption of existing microbial communities of the rhizosphere before the inoculation favors the establishment of biological functions in the rhizosphere. The different approaches for altering the rhizosphere by targeting the microbial partner of the holobiont are explained in detail in the subsequent paragraphs.

7.1 Rhizosphere Engineering by Microbiome Manipulation

The manipulation of rhizospheric microbiome in a direct manner seems to be an easy and more feasible method for engineering the rhizosphere. The inoculation of potent microbial strains seems to be an imperative choice for altering the rhizospheric microbiota. The existence of several novel tactics is potent enough to augment the competence as well as perseverance of the newly introduced microorganism into the soil systems (Bakker et al. 2012). The inoculation process follows some screens and selection perimeters along with a precise evaluation of the different plant health elevation attributes of the retrieved microbial isolates. Furthermore, their survival and growth in the carrier and their efficacy to perform in the natural environments are also assessed before the inoculation (Okafor 2016). The colonization followed by dominance in the rhizospheric zone by the microorganisms is very critical for both beneficial and pathogenic microbes (Bakker et al. 2012). The aptitude of PGPR is being harnessed from several decades as amendments in the form of attributable to their employment as eco-friendly substitute to chemicals, thereby acting as protecting shield against the long-lasting negative impact on different chemicals on the environmental health. However, the employment of this technique has not picked up the anticipated pace regardless of having numerous proven benefits. Therefore, the farming community has lost interest in this technology and thus still relies on the usage of chemical fertilizers (Dubey and Sharma 2019). Several limitations in the abovementioned process came across either with the monoinoculation or even with a consortium assembled with a group of two or more bioinoculants. The direct inoculation of any microbial culture in the rhizosphere is estimated to tackle a substantial degree of competition from the surroundings. It may also alter the already prevailing equipoise in the rhizospheric zone and, thus, can upset the plethora of valuable natural connections (plant-microbe and microbe-microbe interactions) prevailing in the soils. However, some strategies for enhancing the rhizosphere microbiome focusing on the co-inoculation with numerous microbial strains or mixed cultures of arbuscular mycorrhizal fungi (AMF), ectomycorrhizal fungi (ECM), PGPR, and endophytes, enabling combined niche exploitation, cross-feeding, enhancement of one organism's colonization ability, modulating plant growth, and achieving niche saturation and competitive exclusion of pathogens have become successful also (Satyanarayana et al. 2019). The inoculation of microbial culture along with some organic amendment like compost has also proven to be successful and has produced desirable results. The microbial strains that are to be inoculated are the result of the study of any particular plant's microbiome as the plant microbiome consists of several energetic microorganisms that have the potential to alter the plant physiology as well as development and can also prompt the resistance systems against pathogenic microbes along with the elicitation of diverse tolerance mechanisms against numerous plant stresses (Santoyo et al. 2017; Yaish et al. 2017; Yuan et al. 2016). The whole plant microbiome is not capable of assisting plant growth as only a few microbial strains possess these beneficial attributes and the synergistic effects between two strains or more have also been reported for their plant growth supportive attributes (Rojas-Solís et al. 2018; Timm et al. 2016). Therefore, the desired microbial strains are maintained in the form of bioformulations for preserving their viability by shielding them from hostile environmental situations. There are different modes of applications of bioformulations in the field such as biopriming of seeds, foliar spray, seedling dip, and soil drenching. However, the inoculation of the desired microorganisms in the rhizosphere not only increases the number of the inoculated microbes but results in the alteration in the rhizospheric environmental conditions, and therefore the change in the diverse array of communications taking place in the rhizosphere brings out an overall change in the rhizospheric microbiome. For instance, Wan et al. (2017) reported that the inoculation of tomato rhizosphere with the biocontrol agent *Bacillus amyloliquefaciens* altered the rhizospheric composition and increased the abundance of *Pseudomonas* and *Massilia*. Similarly, *Bacillus amyloliquefaciens* when inoculated in the sorghum rhizosphere significantly enhanced the yield and also affected the rhizosphere microbiology as the proportion of *Tremellomycetes* was reduced by 8.87% in the continuous cropping soil (Wu et al. 2019). Likewise, the inoculation of *Pseudomonas putida* Rs-198 in the pepper rhizosphere increased the abundance of *Blastococcus*, *AKYG587*, *Pseudomonas*, *Cyanobacteria*, and *Chloroflexi* (He et al. 2019). The PGPR *Paenibacillus mucilaginosus* when co-inoculated with the rhizobia *Sinorhizobium meliloti* in the rhizosphere of *Medicago sativa* also altered the rhizobiome as displayed by a relative increment in the abundance of *Firmicutes* as well as *Acidobacteria* (Ju et al. 2020). The inoculation with AMF also changes the profiles of rhizospheric microbial community, for instance, the rhizosphere of *Prosopis juliflora* when inoculated with *Glomus intraradices* and a mix of *G. intraradices* and *G. deserticola* also significantly affected the bacterial and fungal community structure (Solís-Domínguez et al. 2011). Similarly, the inoculation of the AMF in the rhizospheres of *Salvia officinalis* L., *Lavandula dentata* L., *Thymus vulgaris* L., and *Santolina chamaecyparissus* also altered the bacterial and fungal communities of rhizosphere. Moreover, the ability of the AM fungus to shape the rhizosphere bacterial community structure was independent of the host plant species (Rodríguez-Caballero et al. 2017). Similarly the inoculation of maize with the phosphate-solubilizing fungi, namely, *Aspergillus niger* P39 and *Penicdlium ozalzcum* P66, also lead to an increased bacterial diversity in the rhizospheric zone as assessed using DGGE fingerprinting (Guang-Hua et al. 2007). Therefore, it can be concluded that the members of rhizospheric microbiota which are often selected from the core microbiome on the basis of their several growth promotion attributes not only directly benefit the plant systems by their valuable possessions but also serve the plant systems by creating a unique environment in the plant rhizosphere. The inoculated microbes assist the growth of plant systems by reshaping the microbial community of the rhizosphere where some genera face a relative increment in their proportion, while the others have to bear a concomitant decline. Thus, this approach inoculating desirable microbes proves to be an important tool for engineering the rhizosphere.

7.2 *Rhizospheric Engineering by Genetic Manipulation of Microbes*

The microbial strains used for inoculation in the quest to engineer the rhizosphere must be established in the rhizosphere and should uphold biologically active populations to outcompete the already adapted occupant microbial systems. However, microbial systems employ a lot of stratagems for successfully inhabiting the new environment, for instance, synthesis of cell surface molecules; at various times the colonization process is not found to be much effective (Ryan et al. 2009). Therefore, the genetic engineering of several microbial strains for various desired traits seems to be a viable option for enhancing their fitness before their inoculation (Fig. 21.3). The genes responsible for the growth promotion attributes of microbial systems have demonstrated to be effective targets for strain enhancement, either by amending the timing or degree of their expression or by transferring and expressing them in alternate hosts with other desirable attributes (Ryan et al. 2009). However, the early efforts comprise the insertion of a heterologous gene encoding a siderophore receptor into a *Pseudomonas fluorescens* strain to render it more competitive in soil (Dessaux et al. 2016). This methodology targets the gene insertion tactic for increasing the number of outer membrane siderophore receptors in microbial strains for making them more efficient on iron acquisition and therefore inhabiting the rhizosphere, for instance, the insertion of the siderophore receptor for ferric pseudobactin 358 into *P. fluorescens* WCS374 resulted in a strain that was found to be more competitive than the WCS374 parental strain for the occupation of the radish rhizosphere (Geetha and Joshi 2013; Raaijmakers et al. 1995). The rhizobacteria are also genetically engineered for the production of several key enzymes and have demonstrated improved plant growth promotion attributes, for instance, *Pseudomonas fluorescens* CHA0 altered with the acdS gene coding for the enzyme ACC

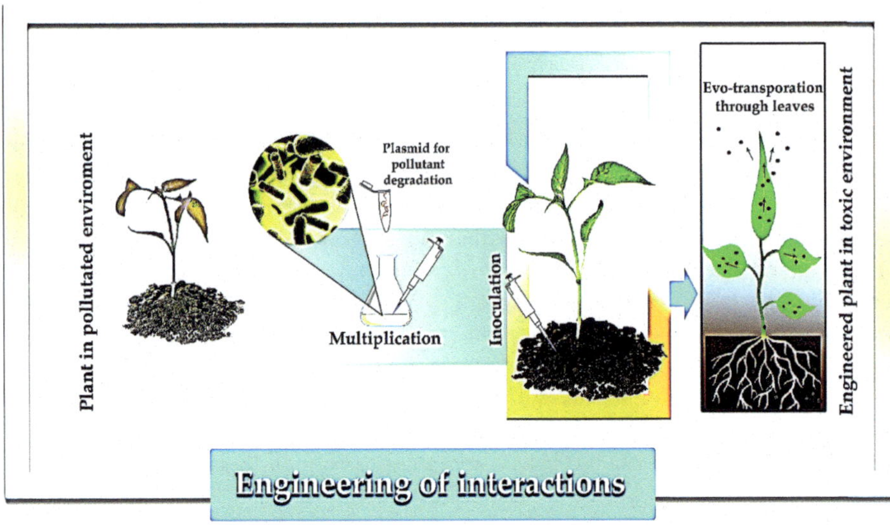

Fig. 21.3 Effect of inoculating plants with GM microorganisms altered for various traits on the plant health

deaminase significantly improved the root length in canola seedlings and also provided enhanced defense against the phytopathogen *Pythium* (Wang et al. 2000). Similarly, the genetically altered *B. subtilis* OKBHF significantly increased the height, fresh weight, and flower along with the fruit number in tomato plants along with a concomitant reduction in the disease rigorousness due to *Cucumber mosaic virus*. The *Bacillus* strain was genetically engineered for the gene coding for the HpaGXooc which is a member of the harpin group of proteins and is responsible for the biocontrol activity (Wang et al. 2011).

The plant systems also face several abiotic stresses, and it is an unhidden fact that several PGPR strains have got unique abilities to aid plant systems during their exposure to different stresses. The competent microbial strains which prove to be effective in coping with the abiotic stresses are isolated and identified, and the molecular cascade of events taking place during the microbial elimination of plant stress is unveiled in the quest to engineer microbial strains with an improved capability of assuaging the plant stress responses. A cadmium-resistant *Pseudomonas aeruginosa* transformed with metallothionein gene has been validated for its tremendous capability of adsorbing cadmium ions via extracellular accrual and was also found to owe an improved aptitude for the immobilization of cadmium divalent ions from the external source. The inoculation of this genetically altered microorganism in cadmium-polluted soil considerably heightened the plant biomass as well as the chlorophyll content in leaf (Huang et al. 2016; Jishma et al. 2019).

The colonization of plant root by the inoculated microorganism represents an important parameter to be considered for genetically altering the microbial systems. The colonization of root surfaces is driven by a molecular cascade of events and also depends on various factors like phenomenon of chemotaxis and biofilm formation (Yaryura et al. 2008). The disruption of gene abrB created a genetically altered strain of *Bacillus amyloliquefaciens* SQR9 which resulted in enhanced root colonization therefore with enhanced biocontrol ability (Weng et al. 2013).

The plants facing insect attacks can also be inoculated with the genetically engineered endophytic microbes transformed with the genes coding for precise insecticidal proteins. Such endophytes are also designated as living vectors meant for the expression of anti-pest proteins in plant systems. The first attempt to insert a heterologous gene into an endophytic microbe was made by Fahey (1988). The other endophyte *Clavibacter xyli* subsp. *cynodontis* was also genetically manipulated with an endotoxin gene originating from *Bacillus thuringiensis*. The genetically improved bacterium was capable of secreting toxin inside the plant that protected the plant systems from insect attacks with a specific reduction in the attacks of *Ostrinia nubilalis* (Tomasino et al. 1995; Lampel et al. 1994). The nitrogen-fixing bacterium *Bradyrhizobium* has also been transformed with the endotoxin gene from *B. thuringiensis* and was later inoculated into the roots of *Cajanus cajan*, where it not only upgraded the nitrogen fixation process but also provided protection to the plant systems against *Rivelia angulata* larvae (Nambiar et al. 1990). Similarly, the endophytic *Bacillus subtilis* WH2 which was genetically engineered to express anti-pest *Pinellia ternata* agglutinin by insertion of PTA gene into plasmid pP43NMK displayed insecticidal activity against white-backed planthopper *Sogatella furcifera* when inoculated in the rice rhizosphere (Qi et al. 2013). Thus, the genetically altered microbes represent an important candidature to be considered for engineering the plant rhizosphere owing to their enhanced performance as compared to their wild relatives. They can be genetically altered for improved colonization of the plant roots as well as for other plant-growth-aiding traits. Moreover, the employment of GM microorganisms could result in the enhancement of many members of the endogenous population by the transmission of genetic information via horizontal gene transfer.

8 Engineering of Interactions

The involvement of root-associated microbiome makes the holobiont a single and complete unit. The association of microorganisms to the plant tissues is a complex process which happens in the soil by way of chemical interactions that takes place with the active involvement of both the partners (Farrar et al. 2014). Taking into account the complication of these communications, a fine understanding of these chemical networks amid all members is indispensable to untangle how microbial inhabitants harmonize their activities and intermingle with the plant roots. Therefore, the portrayal of these interactions is an essential step for understanding the connotations as well as occupations of microbial populations (Kumar et al. 2016). However, many molecules along with the mechanisms involved that synchronize the foundation of precise rhizospheric interactions have already been unveiled and explored in literature. The understanding of such interactions is staggering as the signaling molecules owe the aptitude of upsurging plant functions of interest and provide a unique methodology to access control over the microbial inhabitants if properly understood and harnessed (Guttman et al. 2014; Quiza et al. 2015). The plant's sole purpose of shaping the rhizospheric microbiome is to fascinate favored microbial associates and to deter the pathogens along with the undesirable contestants. These activities happen as a result of different signaling molecules secreted by the plant systems in the form of root exudates. In addition to plant systems, numerous microbes also discharge different signaling compounds in the rhizosphere. These signaling molecules play important roles not only in the life cycles of these organisms but also in their evolution as well as complexity of life (Cornforth et al. 2014; Parks et al. 2014; West et al. 2015). Furthermore, the successful colonization of plant roots by the competent rhizobacteria is possible only due to this bidirectional signaling. Consequently, the collective interests of both the donor and the recipient in the quest to disseminate the unswerving information prompt an operative signaling arrangement to procure numerous health benefits (Kumar et al. 2016). Thus, this bidirectional signaling which accounts for ecological interaction between plant and microbial systems also provides a platform for rhizospheric engineering by manipulating the interaction taking place in the rhizospheric zone. The plant-allied microbial partners yield and exploit diffusible quorum-sensing molecules (e.g., N-acyl-homoserine lactones, AHLs) for signaling each other and thus to order their gene expression (Berendsen et al. 2012). The AHLs of bacterial origin have also been reported to affect root development in the plant systems (Ortíz-Castro et al. 2008) along with the elicitation of the phenomenon acknowledged as induced systemic resistance (ISR) which permits the plant systems to withstand the pathogenic attacks that possibly will be disastrous without the occurrence of such factors of bacterial origin. The plant systems have also developed the ability to utilize the microbial communication systems for manipulating the gene expression in their accompanying microbial populations, such as various plant-allied bacterial members, which owe some LuxR-like proteinaceous molecules which are motivated from different signals originating from plant systems (Ferluga and Venturi 2009). A small proportion of bacterial communities is diverse owing to their ability to quench the signaling process by deteriorating numerous compounds of plant as well as microbial origin in the rhizosphere, thereby leading to the disruption of quorum-sensing process (Tarkka et al. 2009), and other members have also been reported for degrading the compounds, like ethylene, that negatively affect the plant health (Bais et al. 2008). Such members of microbiological community provide an ostensible opportunity for engineering the rhizospheric interactions in the hunt to shape a perfect rhizosphere supporting healthy plant systems. For instance, the members of genus *Pectobacterium* are highly plant pathogenic, and their pathogenicity depends on the fabrication of enzymes that degrade the plant cell wall and are popularly known as macerating enzymes (Liu et al. 2008). The microbe produces these enzymes at great cell density via quorum-sensing mechanisms. The bacterial cell synthesizes a signal molecule, and the concentration of that molecule upturns with the cell density. The quorum-sensing signal is professed after attaining a threshold cell concentration which further prompts the production of the macerating enzymes and in turn the humiliation of the plant tissues. The biocontrol of this plant pathogen is usually based on the alteration of the interactions, i.e., by inhibiting the quorum-sensing mechanism (Faure and Dessaux 2007). Several soil microbes having the potential to degrade the QS signal, for instance, *Bacillus cereus*, *Bacillus thuringiensis*, and *Rhodococcus erythropolis*, have been reported to condense the maceration signs under laboratory conditions (Uroz et al. 2003). It has been found that the bacterium *R. erythropolis* does not hinder the progression of the pathogen, but proficiently averts the accretion of the QS signal and henceforth the deliquescence of the plant tissues (Cirou et al. 2007, 2011, 2012).

Another example of successful engineering of interactions is the successful transformation of soil bacterium *Burkholderia cepacia* with a plasmid encoding toluene degradation (Fig. 21.4). The reinoculation of yellow lupine plants with the transformed bacterial strain sustained the plant growth that too without the appearance of any symptoms of phytotoxicity even at the elevated levels (1000 mg/l) of toluene, contrary to the control plants that displayed symptoms of phytotoxicity at the toluene intensities above 100 mg/l. Some PGPRs are known to aid the plant growth by forming a biofilm around the plant root cells. This biofilm formation happens as a result of microbial response toward the plant root exudates. The addition of root exudates responsible for prompting biofilm formation along with the inoculation of microbial culture is known to enhance plant-microbe interactions and therefore also encourage the biofilm formation (Zhang et al. 2015). Furthermore,

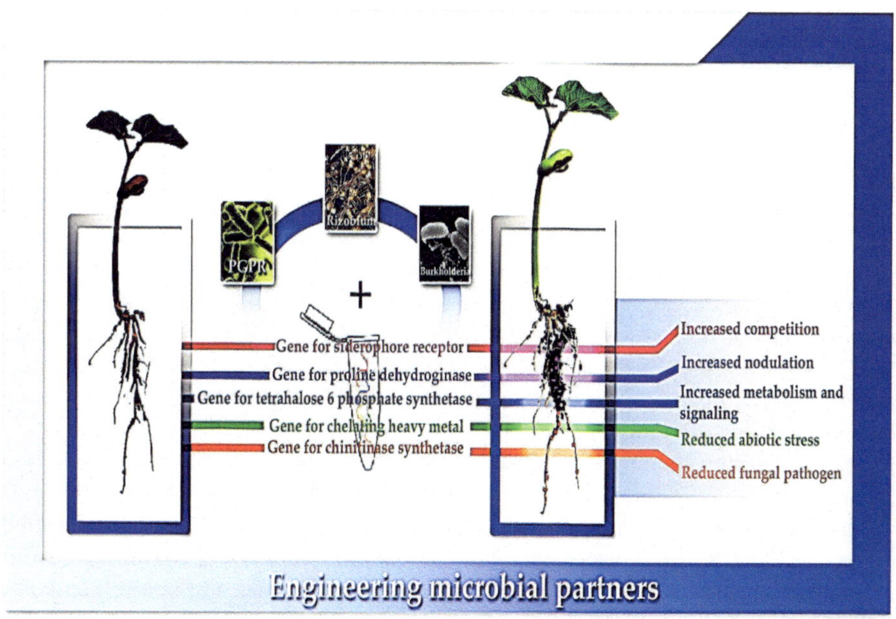

Fig. 21.4 Inoculation of a stressed plant with the genetically engineered microbial partners of holobiont for improved plant-microbe interactions

the combinatorial addition of several microbial strains has also been reported for their improved efficacy as well as improved plant growth assessment parameters. In addition to it, the combinatorial addition has also been proven for supporting greater microbial diversity in plant rhizosphere (Gupta et al. 2019) which probably has happened due to reshaping of the biotic interactions happening in the rhizospheric hotspot. The plant-microbe interactions especially the symbiotic association between plant systems and the rhizospheric microbiota are also engineered for in situ bioremediation of an extensive array of organic pollutants like parathion, trichloroethylene, toluene, and PCBs using genetically altered rhizobacteria or endophytic bacteria (Wu et al. 2006). In a study, the *Arabidopsis thaliana* phytochelatin synthase gene (PCS_{AT}) was expressed in a micro-symbiont, *Mesorhizobium huakuii* subsp. *rengei*, which lives in the nodules of *Astragalus sinicus*. The symbiont expressing the PC synthase possessed the ability to upsurge the cadmium accretion by 1.5-fold in the nodules (Sriprang et al. 2003). Similarly, an antifungal bacterium *Pseudomonas putida* 06909 engineered for plant-microbe symbiotic relationship also exhibited enhanced cadmium-binding properties. The genetic engineering-mediated expression of a metal-binding peptide (EC20) not only upgraded cadmium binding but also alleviated the cellular toxicity of cadmium (Wu et al. 2006). Thus it can be concluded that the interval of interactions between plants and microbes happens to be very critical as it is the process of interaction only which kicks the plant systems as well as microbial systems toward a state of interdependence where both the members can harness the beneficial attributes of each other. Therefore, the engineering of interactions can reshape the plant-microbe interactions for enhanced plant productivity as well as superior plant health.

9 Conclusion and Future Prospects

The rhizosphere is one among the most complex microbial habitats. Plants have evolved into a microbial world where they extended their fine network of roots into the soil already inhabited by a diverse community of microbes. The rapid colonization of the plant roots by the microbes followed by the plant-mediated release of photosynthates via its roots has put both the life forms in a state of interdependence where both these survive as a single unit called as holobiont. Plants are largely known for engineering their rhizospheric microbiomes which differ by the cultivar, age, and variety of plants. However, a large proportion of the rhizospheric microbiome is still represented by the *Proteobacteria* and *Actinobacteria*, and the microbial population varies at the genus and species levels. Plants secrete root exudates to harbor a great diversity of microorganisms. The rhizospheric microbiota responds to these exudates by the phenomenon of chemotaxis and actively colonizes the plant roots. But the prevalence of bad and ugly microbiome proves to be problematic at different times and puts the plant systems in a state of stress. However, the valuable possessions of the beneficial rhizospheric microbiota, for instance, their ability to own plant growth promotion traits and xenobiotic degradation, improve soil structure, and sustain the plant health and productivity, have attracted the attention of researchers to create a "rhizosphere bias". Where only the microbiota beneficial to the plant systems can thrive and aid the plant growth. The rhizosphere can be engineered for the beneficial microbiota by several soil amendments and by direct inoculation of the selected PGPR isolates. However, only a little proportion of rhizospheric microbiome is culturable; therefore, the development of novel processes which can study the valuable microbial possessions in its natural habitat should be a point of major concern. The amendments should be decided after unveiling the requirements of unculturable microbiota. The artificial addition of root exudates is also known to be the important soil amendment, but on the flip side, all the root exudates secreted by the plants at different times haven't been unveiled yet. The interactive effect of all the root exudates should be worked out along with their precise effect on both culturable and non-culturable rhizospheric microbiota. The plant systems are genetically engineered for the production of the desired root exudates, ion efflux, and other metabolites. The advancement in techniques for cheaper production of such metabolites is the need of the hour. Moreover, the artificial production of root exudates at an industrial scale could save a lot of money in the agricultural sector by boosting the overall production. The identification of different biotic and abiotic parts of rhizosphere can also unveil some hidden rhizospheric interactions which can further prove to be an important asset for the agricultural sector. The genetic engineering experiments in the plants have proven to be of only a little success; therefore, the development of robust methodologies which can reveal some novel pathways for metabolic engineering of the plant systems should be addressed. Ultimately, the rhizosphere is a highly dynamic habitat where predictions work the least; thus, this dynamic microbial habitat is a subject to dynamic research.

References

Adl S (2016) Rhizosphere, food security, and climate change: a critical role for plant-soil research. Rhizosphere 1:1–3

Agrios GN (2005) Plant pathology, 5th edn. Elsevier, New York

Ahemad M, Khan M (2012) Effects of pesticides on plant growth-promoting traits of Mesorhizobium stain MRC4. J Saudi Soc Agric Sci 11:63–71

Ahemad M, Kibret M (2014) Mechanisms and applications of plant growth-promoting rhizobacteria: current perspective. JKSUS 26:1–20

Ahkami AH, White RA III, Handakumbura PP, Jansson C (2017) Rhizosphere engineering: enhancing sustainable plant ecosystem productivity. Rhizosphere 3:233–243

Ali S, Charles TC, Glick BR (2014) Amelioration of high salinity stress damage by plant growth-promoting bacterial endophytes that contain ACC deaminase. Plant Physiol Biochem 80:160–167

Ali MA, Abbas A, Azeem F, Javed N, Bohlmann H (2015) Plant-nematode interactions: from genomics to metabolomics. Int J Agric Biol 17:1071–1082

Ali MA, Naveed M, Mustafa A, Abbas A (2017) The good, the bad, and the ugly of rhizosphere microbiome. In: Kumar V, Kumar M, Parsad R, Choudhary DK (eds) Probiotics and plant health 7. Springer, Singapore, pp 253–290

Andrews SC, Robinson AK, & Rodríguez-Quiñones F (2003) Bacterial iron homeostasis. FEMS microbiology reviews, 27(2-3):215–237

Antoniou A, Tsolakidou MD, Stringlis IA, Pantelides IS (2017) Rhizosphere microbiome recruited from a suppressive compost improves plant fitness and increases protection against vascular wilt pathogens of tomato. Front Plant Sci 8:2022

Ashraf MP, Harris PJ (2004) Potential biochemical indicators of salinity tolerance in plants. Plant Sci 166(1):3–16

Backer R, Rokem JS, Ilangumaran G, Lamont J, Praslickova D, Ricci E, Subramanian S, Smith DL (2018) Plant growth-promoting rhizobacteria: context, mechanisms of action, and roadmap to commercialization of biostimulants for sustainable agriculture. Front Plant Sci 9:1473

Badri DV, Vivanco JM (2009) Regulation and function of root exudates. Plant Cell Environ 32:666–681

Badri DV, Quintana N, El Kassis EG, Kim HK, Choi YH, Sugiyama A, Verpoorte R, Martinoia E, Manter DK, Vivanco JM (2009) An ABC transporter mutation alters root exudation of phytochemicals that provoke an overhaul of natural soil microbiota. Plant Physiol 151(4):2006–2017

Baetz U, Martinoia E (2014) Root exudates: the hidden part of plant defense. Trends Plant Sci 19:90–98

Bais HP, Broeckling CD, Vivanco JM (2008) Root exudates modulate plant-microbe interactions in the rhizosphere. In: Karlovesky P (ed) Secondary metabolites in soil ecology. Springer, Berlin, pp 241–252

Bakker MG, Manter DK, Sheflin AM, Weir TL, Vivanco JM (2012) Harnessing the rhizosphere microbiome through plant breeding and agricultural management. Plant Soil 360:1–13

Bal HB, Nayak L, Das S, Adhya TK (2013) Isolation of ACC deaminase producing PGPR from rice rhizosphere and evaluating their plant growth-promoting activity under salt stress. Plant Soil 366(1–2):93–105

Baudin E, Benizri E, & Guckert A (2003) Impact of artificial root exudates on the bacterial community structure in bulk soil and maize rhizosphere. Soil Biol Biochem, 35(9):1183–1192

Baumgardner DJ, Temte JL, Gutowski E et al (2011) The differential diagnosis of pulmonary blastomycosis in Wisconsin: a Wisconsin network for Health Research (WiNHR) study. Wis Med J 110:68–73

Berendsen RL, Pieterse CMJ, Bakker PAHM (2012) The rhizosphere microbiome and plant health. Trends Plant Sci 17(8):478–486

Berg G, Smalla K (2009) Plant species and soil type cooperatively shape the structure and function of microbial communities in the rhizosphere. FEMS Microbiol Ecol 68:1–13

Berg G, Eberl L, Hartmann A (2005) The rhizosphere as a reservoir for opportunistic human pathogenic bacteria. Environ Microbiol 7(11):1673–1685

Bhattacharyya PN, Jha DK (2012) Plant-growth promoting rhizobacteria (PGPR): emergence in agriculture. World J Microbiol Biotechnol 28:1327–1350

Biate DL, Kumari A, Annapurna K, Kumar LV, Ramadoss D, Reddy KK, Naik S (2015) Legume root exudates: their role in symbiotic interactions. In: Arora NK (ed) Plant microbes symbiosis: applied facets. Springer, New Delhi, pp 259–271

Blackburn JK, McNyset KM, Curtis A, Hugh-Jones ME (2007) Modeling the geographic distribution of Bacillus anthracis, the causative agent of anthrax disease, for the contiguous United States using predictive ecologic niche modeling. Am J Trop Med Hyg 77:1103–1110

Buee M, De Boer W, Martin F, Van Overbeek L, Jurkevitch E (2009) The rhizosphere zoo: an overview of plant-associated communities of microorganisms, including phages, bacteria, archaea, and fungi, and of some of their structuring factors. Plant and Soil, 321(1):189–212

Boehm MJ, Wu T, Stone AG, Kraakman B, Iannotti DA, Wilson GE, Madden LV, Hoitink H (1997) Cross-polarized magic-angle spinning (sup13) C nuclear magnetic resonance spectroscopic characterization of soil organic matter relative to culturable bacterial species composition and sustained biological control of pythium root rot. Appl Environ Microbiol 63(1):162–168

Bornø ML, Eduah JO, Müller-Stöver DS, Liu F (2018) Effect of different biochars on phosphorus (P) dynamics in the rhizosphere of Zea mays L. (maize). Plant Soil 431(1–2):257–272

Bowen GD, Rovira AD (1991) The rhizosphere, the hidden half of the hidden half. In: Waisel Y, Eshel A, Kafkafi U (eds) The plant roots, the hidden half. Marcel Dekker, New York, pp 641–669

Breidenbach B, Pump J, Dumont MG (2016) Microbial community structure in the rhizosphere of rice plants. Front Microbiol 6:1537

Bron PA, Van Baarlen P, Kleerebezem M (2012) Emerging molecular insights into the interaction between probiotics and the host intestinal mucosa. Nat Rev Microbiol 10(1):66–78

Burns RG (2010) Albert Rovira and a half-century of rhizosphere research. In: VVSR G, Ryder M, Radcliffe J (eds) The Rovira rhizosphere symposium. The ATSE Crawford Fund, Adelaide, pp 1–10

Callaway RM, Aschehoug ET (2000) Invasive plants versus their new and old neighbors: a mechanism for exotic invasion. Science 290(5491):521–523

Campbell RN (1996) Fungal transmission of plant viruses. Annu Rev Phytopathol 34:87–108

Carvalhais LC, Dennis PG, Fedoseyenko D, Hajirezaei MR, Borriss R, von Wirén N (2011) Root exudation of sugars, amino acids, and organic acids by maize as affected by nitrogen, phosphorus, potassium, and iron deficiency. J Plant Nutr Soil Sci 174(1):3–11

Carvalhais LC, Dennis PG, Fan B, Fedoseyenko D, Kierul K, Becker A, von Wiren N, Borriss R (2013) Linking plant nutritional status to plant-microbe interactions. PLoS One 8(7)

Chaparro JM, Badri DV, Vivanco JM (2014) Rhizosphere microbiome assemblage is affected by plant development. ISME J 8(4):790–803

Chapman SW (2005) Blastomyces dermatitidis. In: Mandell GL, Bennett JE, Dolin R (eds) Mandell, Douglas, and Bennett's principles and practice of infectious diseases, 6th edn. Elsevier, Philadelphia, pp 3026–3040

Chen S, Li X, Lavoie M, Jin Y, Xu J, Fu Z, Qian H (2017) Diclofop-methyl affects microbial rhizosphere community and induces systemic acquired resistance in rice. J Environ Sci 51:352–360

Cheng J, Li Y, Gao W, Chen Y, Pan W, Lee X, Tang Y (2018a) Effects of biochar on Cd and Pb mobility and microbial community composition in a calcareous soil planted with tobacco. Biol Fertil Soils 54(3):373–383

Cheng N, Peng Y, Kong Y, Li J, Sun C (2018b) Combined effects of biochar addition and nitrogen fertilizer reduction on the rhizosphere metabolomics of maize (Zea mays L.) seedlings. Plant Soil 433(1–2):19–35

Cirou A, Diallo S, Kurt C, Latour X, Faure D (2007) Growth promotion of quorum-quenching bacteria in the rhizosphere of Solanum tuberosum. Environ Microbiol 9(6):1511–1522

Cirou A, Raffoux A, Diallo S, Latour X, Dessaux Y, Faure D (2011) Gamma-caprolactone stimulates growth of quorum-quenching Rhodococcus populations in a large-scale hydroponic system for culturing Solanum tuberosum. Res Microbiol 162(9):945–950

Cirou A, Mondy S, An S, Charrier A, Sarrazin A, Thoison O, DuBow M, Faure D (2012) Efficient biostimulation of native and introduced quorum-quenching Rhodococcus erythropolis populations is revealed by a combination of analytical chemistry, microbiology, and pyrosequencing. Appl Environ Microbiol 78(2):481–492

Cornforth DM, Popat R, McNally L, Gurney J, Scott-Phillips TC, Ivens A, Diggle SP, Brown SP (2014) Combinatorial quorum sensing allows bacteria to resolve their social and physical environment. Proc Natl Acad Sci 111(11):4280–4284

Cotrufo MF, Wallenstein MD, Boot CM, Denef K, Paul E (2013) The Microbial Efficiency-Matrix Stabilization (MEMS) framework integrates plant litter decomposition with soil organic matter stabilization: do labile plant inputs form stable soil organic matter? Glob Chang Biol 19(4):988–995

Dakora FD, Phillips DA (2002) Root exudates as mediators of mineral acquisition in low-nutrient environments. In: Adu-Gyamfi JJ (ed) Food security in nutrient-stressed environments: exploiting plants' genetic capabilities. Springer, Dordrecht, pp 201–213

van Dam NM, Bouwmeester HJ (2016) Metabolomics in the rhizosphere: tapping into belowground chemical communication. Trends Plant Sci 21(3):256–265

De Brito AM, Gagne S, Antoun H (1995) Effect of compost on rhizosphere microflora of the tomato and on the incidence of plant growth-promoting rhizobacteria. Appl Environ Microbiol 61(1):194–199

De Corato U (2020) Disease-suppressive compost enhances natural soil suppressiveness against soil-borne plant pathogens: a critical review. Rhizosphere 5:100192
Deng W, Luo K, Li Z, Yang Y, Hu N, Wu Y (2009) Overexpression of Citrus junos mitochondrial citrate synthase gene in Nicotiana benthamiana confers aluminum tolerance. Planta 230(2):355–365
Dessaux Y, Grandclément C, Faure D (2016) Engineering the rhizosphere. Trends Plant Sci 21(3):266–278
Dixon R, Kahn D (2004) Genetic regulation of biological nitrogen fixation. Nat Rev Microbiol 2:621–631
Doornbos RF, van Loon LC, Bakker PA (2012) Impact of root exudates and plant defense signaling on bacterial communities in the rhizosphere. A review. Agron Sustain Dev 32(1):227–243
Dreyer I, Gomez-Porras JL, Riaño-Pachón DM, Hedrich R, Geiger D (2012) Molecular evolution of slow and quick anion channels (SLACs and QUACs/ALMTs). Front Plant Sci 3:263
Dubey S, Sharma S (2019) Rhizospheric microbiome engineering as a sustainable tool in agriculture: approaches and challenges. In: Satyanarayana T, Das SK, Johri BN (eds) Microbial diversity in ecosystem sustainability and biotechnological applications. Springer, Singapore, pp 257–272
Dumbrell AJ, Nelson M, Helgason T, Dytham C, Fitter AH (2010) Relative roles of niche and neutral processes in structuring a soil microbial community. ISME J 4(3):337–345
Dutta J, & Bora U (2019) Rhizosphere microbiome and plant probiotics. In New and Future Developments in Microbial Biotechnology and Bioengineering. Elsevier, 273–281
Fahey JW (1988) Endophytic bacteria for the delivery of agrochemicals to plants. In: Cutler HO (ed) Biologically active natural products. American Chemical Society, Washington DC, pp 120–128
FAO (2019a) The State of Food and Agriculture. Moving forward on food loss and waste reduction. Rome, FAO. Licence: CC BY-NC-SA 3.0 IGO. http://www.fao.org/3/ca6030en/ca6030en.pdf
FAO (2019b) World Food and Agriculture—statistical pocketbook 2019. Rome, FAO. http://www.fao.org/3/ca6463en/ca6463en.pdf
Farrar K, Bryant D, Cope-Selby N (2014) Understanding and engineering beneficial plant–microbe interactions: plant growth promotion in energy crops. Plant Biotechnol J 12(9):1193–1206
Faure D, Dessaux Y (2007) Quorum sensing as a target for developing control strategies for the plant pathogen Pectobacterium. Eur J Plant Pathol 119(3):353–365
Ferluga S, & Venturi V (2009) OryR is a LuxR-family protein involved in interkingdom signaling between pathogenic Xanthomonas oryzae pv. oryzae and rice. Journal of bacteriology, 191(3):890–897
Fierer N, Jackson RB (2006) The diversity and biogeography of soil bacterial communities. Proc Natl Acad Sci U S A 103:626–631
Fierer N, Lauber CL, Ramirez KS, Zaneveld J, Bradford MA, Knight R (2012) Comparative metagenomic, phylogenetic and physiological analyses of soil microbial communities across nitrogen gradients. ISME J 6(5):1007–1017
Franche C, Lindström K, Elmerich C (2009) Nitrogen-fixing bacteria associated with leguminous and non-leguminous plants. Plant Soil 321:35–59
Frenk S, Hadar Y, Minz D (2014) Resilience of soil bacterial community to irrigation with water of different qualities under Mediterranean climate. Environ Microbiol 16(2):559–569
Furukawa J, Yamaji N, Wang H, Mitani N, Murata Y, Sato K, Katsuhara M, Takeda K, Ma JF (2007) An aluminum-activated citrate transporter in barley. Plant Cell Physiol 48:1081–1091
Geetha SJ, Joshi SJ (2013) Engineering rhizobial bioinoculants: a strategy to improve iron nutrition. Sci World J 2013:315890
Giordano W, Hirsch AM (2004) The expression of MaEXP1,a Melilotus alba expansin gene, is upregulated during the sweet clover-Sinorhizobium meliloti interaction. Mol Plant-Microbe Interact 17:613–622
Green SJ, Michel FC, Hadar Y, Minz D (2007) Contrasting patterns of seed and root colonization by bacteria from the genus Chryseobacterium and from the family Oxalobacteraceae. ISME J 1(4):291–299

Guang-Hua WA, Jian J, Mei-Na XU, Xiang-Wen PA, Tang C (2007) Inoculation with phosphate-solubilizing fungi diversifies the bacterial community in rhizospheres of maize and soybean. Pedosphere 17(2):191–199

Gupta R, Singh A, Srivastava M, Shanker K, Pandey R (2019) Plant-microbe interactions endorse growth by uplifting microbial community structure of Bacopa monnieri rhizosphere under nematode stress. Microbiol Res 218:87–96

Guttman DS, McHardy AC, Schulze-Lefert P (2014) Microbial genome-enabled insights into plant–microorganism interactions. Nat Rev Genet 15(12):797–813

Haas D, Défago G (2005) Biological control of soil-borne pathogens by fluorescent pseudomonads. Nat Rev Microbiol 3(4):307–319

Hadar Y, Papadopoulou KK (2012) Suppressive composts: microbial ecology links between abiotic environments and healthy plants. Annu Rev Phytopathol 50:133–153

Hardoim PR, Van Overbeek LS, Berg G, Pirttilä AM, Compant S, Campisano A, Döring M, Sessitsch A (2015) The hidden world within plants: ecological and evolutionary considerations for defining functioning of microbial endophytes. Microbiol Mol Biol Rev 79(3):293–320

Hartmann A, Schmid M, Van Tuinen D, Berg G (2009) Plant-driven selection of microbes. Plant Soil 321(1–2):235–257

Hawes MC, Gunawardena U, Miyasaka S, Zhao X (2000) The role of root border cells in plant defense. Trends Plant Sci 5(3):128–133

Hawkes CV, DeAngelis KM, Firestone MK (2007) Root interactions with soil microbial communities and processes. In: Cardon Z, Whitbeck J (eds) The rhizosphere. Academic, New York, pp 1–29

He Y, Wu Z, Wang W, Liu X, Ye BC (2019) Bacterial community and phosphorus species changes in pepper rhizosphere soils after Pseudomonas putida Rs-198 inoculation. Rhizosphere 11:100164

Hinsinger P, Plassard C, Tang CX, Jaillard B (2003) Origins of root-mediated pH changes in the rhizosphere and their responses to environmental constraints: a review. Plant Soil 248:43–59

Huang XF, Chaparro JM, Reardon KF, Zhang R, Shen Q, Vivanco JM (2014) Rhizosphere interactions: root exudates, microbes, and microbial communities. Botany 92(4):267–275

Huang J, Liu Z, Li S, Xu B, Gong Y, Yang Y, Sun H (2016) Isolation and engineering of plant growth promoting rhizobacteria Pseudomonas aeruginosa for enhanced cadmium bioremediation. J Gen Appl Microbiol 62(5):258–265

Jaafar NM (2014) Biochar as a habitat for arbuscular mycorrhizal fungi. In: Jaafar NM (ed) Mycorrhizal fungi: use in sustainable agriculture and land restoration. Springer, Berlin, pp 297–311

Jeranyama P, Shrestha A, Neupane N (2020) Sustainable food systems: diversity, scope and challenges. In: Rusinamhodzi L (ed) The role of ecosystem services in sustainable food systems. Academic, New York, pp 1–16

Jishma P, Remakanthan A, Radhakrishnan EK (2019) Engineering rhizobacterial functions for the improvement of plant growth and disease resistance. In: Gupta VK, Prabha R (eds) Microbial interventions in agriculture and environment. Springer, Singapore

Jones JT, Haegeman A, Danchin EG, Gaur HS, Helder J, Jones MG, Kikuchi T, Manzanilla-Lopez R, Palomares-Rius JE, Wesemael WM, Perry RN (2013) Top 10 plant-parasitic nematodes in molecular plant pathology. Mol Plant Pathol. https://doi.org/10.1111/mpp.12057

Ju W, Jin X, Liu L, Shen G, Zhao W, Duan C, Fang L (2020) Rhizobacteria inoculation benefits nutrient availability for phytostabilization in copper contaminated soil: drivers from bacterial community structures in rhizosphere. Appl Soil Ecol 150:103450

Kang J, Park J, Choi H, Burla B, Kretzschmar T, Lee Y, Martinoia E (2011) Plant ABC transporters. Arab B 9:e0153

Kapoor D, Sharma P, Sharma MM, Kumari A, Kumar R (2020) Microbes in pharmaceutical industry. In: Sharma SG, Sharma NR, Sharma M (eds) Microbial diversity, interventions and scope. Springer, Singapore, pp 259–299

Kavitha B, Reddy PV, Kim B, Lee SS, Pandey SK, Kim KH (2018) Benefits and limitations of biochar amendment in agricultural soils: a review. J Environ Manag 227:146–154

Khan N, Bano A (2019) Rhizobacteria and abiotic stress management. In: Sayyed RZ (ed) Plant growth promoting rhizobacteria for sustainable stress management. Springer, Singapore, pp 65–80

Kumar V, Baweja M, Singh PK, Shukla P (2016) Recent developments in systems biology and metabolic engineering of plant–microbe interactions. Front Plant Sci 7:1421

Kumar V, AlMomin S, Al-Aqeel H, Al-Salameen F, Nair S, Shajan A (2018) Metagenomic analysis of rhizosphere microflora of oil-contaminated soil planted with barley and alfalfa. PLoS One 13(8)

Kuzyakov Y, Razavi BS (2019) Rhizosphere size and shape: temporal dynamics and spatial stationarity. Soil Biol Biochem 135:343–360

Kuzyakov Y, Subbotina I, Chen H, Bogomolova I, Xu X (2009) Black carbon decomposition and incorporation into soil microbial biomass estimated by 14C labeling. Soil Biol Biochem 41(2):210–219

Lambers H, Mougel C, Jaillard B, Hinsinger P (2009) Plant-microbe-soil interactions in the rhizosphere: an evolutionary perspective. Plant Soil 321(1–2):83–115

Lampel JS, Canter GL, Dimock MB, Kelly JL, Anderson JJ, Uratani BB, Foulke JS Jr, Turner JT (1994) Integrative cloning, expression, and stability of the cryIA gene from Bacillus thuringiensis subsp. kurstaki in a recombinant strain of Clavibacter xyli subsp. cynodontis. Appl Environ Microb 60:501–508

Lebeis SL (2015) Greater than the sum of their parts: characterizing plant microbiomes at the community-level. Curr Opin Plant Biol 24:82–86

Leff JW, Lynch RC, Kane NC, Fierer N (2016) Plant domestication and the assembly of bacterial and fungal communities associated with strains of the common sunflower, Helianthus annuus. New Phytol 214:412–423

Lehmann J, Kleber M (2015) The contentious nature of soil organic matter. Nature 528(7580):60–68

Lehmann J, Rillig MC, Thies J, Masiello CA, Hockaday WC, Crowley D (2011) Biochar effects on soil biota—a review. Soil Biol Biochem 43(9):1812–1836

Lemanceau P, Bauer P, Kraemer S, & Briat JF (2009) Iron dynamics in the rhizosphere as a case study for analyzing interactions between soils, plants and microbes. Plant and Soil, 321(1):513–535

Li H, Su JQ, Yang XR, Zhu YG (2019) Distinct rhizosphere effect on active and total bacterial communities in paddy soils. Sci Total Environ 649:422–430

Liang B, Lehmann J, Sohi SP, Thies JE, O'Neill B, Trujillo L, Gaunt J, Solomon D, Grossman J, Neves EG, Luizão FJ (2010) Black carbon affects the cycling of non-black carbon in soil. Org Geochem 41(2):206–213

Liu H, Coulthurst SJ, Pritchard L, Hedley PE, Ravensdale M, Humphris S, Burr T, Takle G, Brurberg MB, Birch PR, Salmond GP (2008) Quorum sensing coordinates brute force and stealth modes of infection in the plant pathogen Pectobacterium atrosepticum. PLoS Pathog 4(6):e1000093

Liu S, Lu Y, Yang C, Liu C, Ma L, Dang Z (2017) Effects of modified biochar on rhizosphere microecology of rice (Oryza sativa L.) grown in As-contaminated soil. Environ Sci Pollut Res 24(30):23815–23824

Liu J, Williams PC, Geisler-Lee J, Goodson BM, Fakharifar M, Peiravi M, Chen D, Lightfoot DA, Gemeinhardt ME (2018) Impact of wastewater effluent containing aged nanoparticles and other components on biological activities of the soil microbiome, Arabidopsis plants, and earthworms. Environ Res 164:197–203

Lugtenberg B, Kamilova F (2009) Plant-growth-promoting rhizobacteria. Annu Rev Microbiol 63:541–556

Magalhaes JV, Liu J, Guimaraes CT, Lana UGP, Alves VMC, Wang YH, Schaffert RE, Hoekenga OA, Pineros MA, Shaff JE, Klein PE, Carneiro NP, Coelho CM, Trick HN, Kochian LV (2007) A gene in the multidrug and toxic compound extrusion (MATE) family confers aluminum tolerance in sorghum. Nat Genet 39:1156–1161

Man Y, Wang J, Tam NF, Wan X, Huang W, Zheng Y, Tang J, Tao R, Yang Y (2020) Responses of rhizosphere and bulk substrate microbiome to wastewater-borne sulfonamides in constructed wetlands with different plant species. Sci Total Environ 706:135955

Mansouri H, Petit A, Oger P, Dessaux Y (2002) Engineered rhizosphere: the trophic bias generated by opine-producing plants is independent of the opine type, the soil origin, and the plant species. Appl Environ Microbiol 68(5):2562–2566

Marschner P (2012) Rhizosphere biology. Marschner's Miner Nutr Higher Plants 2:369–388

Marschner P, Crowley D, Rengel Z (2011) Rhizosphere interactions between microorganisms and plants govern iron and phosphorus acquisition along the root axis–model and research methods. Soil Biol Biochem 43(5):883–894

Mazzola M, Hewavitharana SS, Strauss SL (2015) Brassica seed meal soil amendments transform the rhizosphere microbiome and improve apple production through resistance to pathogen reinfestation. Phytopathology 105(4):460–469

McCormack ML, Guo D, Iversen CM, Chen W, Eissenstat DM, Fernandez CW, Li L, Ma C, Ma Z, Poorter H, Reich PB (2017) Building a better foundation: improving root-trait measurements to understand and model plant and ecosystem processes. New Phytol 215(1):27–37

McNear DH Jr (2013) The rhizosphere-roots, soil and everything in between. Nat Educat Knowled 4(3):1

Mendes R, Kruijt M, De Bruijn I, Dekkers E, van der Voort M, Schneider JH, Piceno YM, DeSantis TZ, Andersen GL, Bakker PA, Raaijmakers JM (2011) Deciphering the rhizosphere microbiome for disease-suppressive bacteria. Science 332(6033):1097–1100

Mendes R, Garbeva P, Raaijmakers JM (2013) The rhizosphere microbiome: significance of plant beneficial, plant pathogenic, and human pathogenic microorganisms. FEMS Microbiol Rev 37(5):634–663

Mendes LW, Kuramae EE, Navarrete AA, Van Veen JA, Tsai SM (2014) Taxonomical and functional microbial community selection in soybean rhizosphere. ISME J 8(8):1577–1587

Mia S, Van Groenigen JW, Van de Voorde TF, Oram NJ, Bezemer TM, Mommer L, Jeffery S (2014) Biochar application rate affects biological nitrogen fixation in red clover conditional on potassium availability. Agric Ecosyst Environ 191:83–91

Mickan BS, Abbott LK, Fan J, Hart MM, Siddique KH, Solaiman ZM, Jenkins SN (2018) Application of compost and clay under water-stressed conditions influences functional diversity of rhizosphere bacteria. Biol Fertil Soils 54(1):55–70

Mishra I, Arora NK (2019) Rhizoremediation: a sustainable approach to improve the quality and productivity of polluted soils. In: Arora NK, Kumar N (eds) Phyto and rhizo remediation. Springer, Singapore, pp 33–66

Mitter B, Pfaffenbichler N, Sessitsch A (2016) Plant–microbe partnerships in 2020. Microb Biotechnol 9(5):635–640

Moens M, Perry RN (2009) Migratory plant endoparasitic nematodes: a group rich in contrasts and divergence. Annu Rev Phytopathol 47:313–332

Mohanram S, Kumar P (2019) Rhizosphere microbiome: revisiting the synergy of plant-microbe interactions. Ann Microbiol 69(4):307–320

Nambiar PTC, Ma SW, Iyer VN (1990) Limiting and insect infestation of nitrogen-fixing root nodules of the Pigeon pea (Cajanus cajan) by engineering the expression of an entomocidal gene in its root nodules. Appl Environ Microbiol 56:2866–2869

Nester E, Gordon MP, Kerr A (2005) Agrobacterium tumefaciens: from plant pathology to biotechnology. APS, St. Paul, MN

Nicolitch O, Colin Y, Turpault MP, Uroz S (2016) Soil type determines the distribution of nutrient mobilizing bacterial communities in the rhizosphere of beech trees. Soil Biol Biochem 103:429–445

Oburger E, Schmidt H (2016) New methods to unravel rhizosphere processes. Trends Plant Sci 21(3):243–255

Oburger E, Dell'mour M, Hann S, Wieshammer G, Puschenreiter M, Wenzel WW (2013) Evaluation of a novel tool for sampling root exudates from soil-grown plants compared to conventional techniques. Environ Exp Bot 87:235–247

Oger P, Mansouri H, Dessaux Y (2000) Effect of crop rotation and soil cover on alteration of the soil microflora generated by the culture of transgenic plants producing opines. Mol Ecol 9(7):881–890

Oger PM, Mansouri H, Nesme X, Dessaux Y (2004) Engineering root exudation of Lotus toward the production of two novel carbon compounds leads to the selection of distinct microbial populations in the rhizosphere. Microb Ecol 47(1):96–103

Okafor N (2016) Modern industrial microbiology and biotechnology. CRC Press, Boca Raton

Omote H, Hiasa M, Matsumoto T, Otsuka M, Moriyama Y (2006) The MATE proteins as fundamental transporters of metabolic and xenobiotic organic cations. Trends Pharmacol Sci 27:587–593

Ortíz-Castro RANDY, Martínez-Trujillo MIGUEL, López-Bucio JOSE (2008) N-acyl-L-homoserine lactones: a class of bacterial quorum-sensing signals alter post-embryonic root development in Arabidopsis thaliana. Plant Cell Environ 31(10):1497–1509

Osman Y, Mowafy AM, Gebreil AS, Hamed SM (2018) Siderophore production by rhizosphere inhabiting bacteria and fungi. J Plant Product 9(9):717–721

Oved T, Shaviv A, Goldrath T, Mandelbaum RT, Minz D (2011) Influence of effluent irrigation on community composition and function of ammonia-oxidizing bacteria in soil. Appl Environ Microbiol 67(8):3426–3433

Pandey G (2018) Challenges and future prospects of agri-nanotechnology for sustainable agriculture in India. Environ Technol Innov 11:299–307. https://doi.org/10.1016/j.eti.2018.06.012

Parks SE, Cusano DA, Stimpert AK, Weinrich MT, Friedlaender AS, Wiley DN (2014) Evidence for acoustic communication among bottom foraging humpback whales. Sci Rep 4:7508

Pérez-Jaramillo JE, Mendes R, Raaijmakers JM (2016) Impact of plant domestication on rhizosphere microbiome assembly and functions. Plant Mol Biol 90(6):635–644

Petit A, Dessaux Y, Tempé J (1978) The biological significance of opines. I. A study of opine catabolism by Agrobacterium tumefaciens. In: Proceedings of the 4th international conference on plant pathogenic bacteria, vol. 1. Gilbert-Clarey, Angers, pp 143–151

Philippot L, Raaijmakers JM, Lemanceau P, Van Der Putten WH (2013) Going back to the roots: the microbial ecology of the rhizosphere. Nat Rev Microbiol 11(11):789–799

Pitzschke A (2016) Developmental peculiarities and seed-borne endophytes in quinoa: omnipresent, robust bacilli contribute to plant fitness. Front Microbiol 7:2

Preece C, Peñuelas J (2020) A return to the wild: root exudates and food security. Trends Plant Sci 25(1):14–21

Priyanka JP, Goral RT, Rupal KS, Saraf M (2019) Rhizospheric microflora: a natural alleviator of drought stress in agricultural crops. In: Sayyed RZ (ed) Plant growth promoting rhizobacteria for sustainable stress management. Springer, Singapore, pp 103–115

Prosser JI (2015) Dispersing misconceptions and identifying opportunities for the use of'omics' in soil microbial ecology. Nat Rev Microbiol 13(7):439–446

Qi G, Zhang X, Zhao X (2013) Endophytic Bacillus subtilis WH2 containing Pinellia ternata agglutinin showed insecticidal activity against white-backed planthopper Sogatella furcifera. BioControl 58(2):233–246

Qian H, Wang R, Chen J, Ding H, Yong W, Songlin R, Fu Z (2012) Analysis of enantioselective biochemical, physiological, and transcriptional effects of the chiral herbicide diclofop methyl on rice seedlings. J Agric Food Chem 60(22):5515–5523

Qian H, Zhu Y, Chen S, Jin Y, Lavoie M, Ke M, Fu Z (2018) Interacting effect of diclofop-methyl on the rice rhizosphere microbiome and denitrification. Pestic Biochem Physiol 146:90–96

Quiza L, St-Arnaud M, Yergeau E (2015) Harnessing phytomicrobiome signaling for rhizosphere microbiome engineering. Front Plant Sci 6:507

Raaijmakers JM, Sluis LV, Bakker PA, Schippers B, Koster M, Weisbeek PJ (1995) Utilization of heterologous siderophores and rhizosphere competence of fluorescent Pseudomonas spp. Can J Microbiol 41(2):126–135

Raaijmakers JM, Paulitz TC, Steinberg C, Alabouvette C, Moënne-Loccoz Y (2009) The rhizosphere: a playground and battlefield for soilborne pathogens and beneficial microorganisms. Plant Soil 321(1–2):341–361

Ramirez KS, Lauber CL, Knight R, Bradford MA, Fierer N (2010) Consistent effects of nitrogen fertilization on soil bacterial communities in contrasting systems. Ecology 91(12):3463–3470
Rani K, Sharma P, Kumar S, Wati L, Kumar R, Gurjar DS, Kumar D (2019) Legumes for sustainable soil and crop management. In: Meena RS, Kumar S, Singh Bohra J, Jat ML (eds) Sustainable management of soil and environment. Springer, Singapore, pp 193–215
Rasmann S, Ali JG, Helder J, van der Putten WH (2012) Ecology and evolution of soil nematode chemotaxis. J Chem Ecol 38:615–628
Rengel Z (2002) Genetic control of root exudation. Plant Soil 245(1):59–70
Rensink WA, Buell CR (2004) Arabidopsis to rice. Applying knowledge from a weed to enhance our understanding of a crop species. Plant Physiol 135(2):622–629
Rodríguez-Caballero G, Caravaca F, Fernández-González AJ, Alguacil MM, Fernández-López M, Roldán A (2017) Arbuscular mycorrhizal fungi inoculation mediated changes in rhizosphere bacterial community structure while promoting revegetation in a semiarid ecosystem. Sci Total Environ 584:838–848
Rojas-Solís D, Zetter-Salmón E, Contreras-Pérez M, del Carmen R-GM, Macías-Rodríguez L, Santoyo G (2018) Pseudomonas stutzeri E25 and Stenotrophomonas maltophilia CR71 endophytes produce antifungal volatile organic compounds and exhibit additive plant growth-promoting effects. Biocatal Agric Biotechnol 13:46–52
Ryan PR, Dessaux Y, Thomashow LS, Weller DM (2009) Rhizosphere engineering and management for sustainable agriculture. Plant Soil 321:363–383
Sánchez-Cañizares C, Jorrín B, Poole PS, Tkacz A (2017) Understanding the holobiont: the interdependence of plants and their microbiome. Curr Opin Microbiol 38:188–196
Santoyo G, Pacheco CH, Salmerón JH, León RH (2017) The role of abiotic factors modulating the plant-microbe-soil interactions: toward sustainable agriculture. A review. Span J Agric Res 15(1):13
Sasaki T, Yamamoto Y, Ezaki B, Katsuhara M, Ahn SJ, Ryan PR, Delhaize E, Matsumoto H (2004) A wheat gene encoding an aluminum-activated malate transporter. Plant J 37:645–653
Sasse J, Martinoia E, Northen T (2018) Feed your friends: do plant exudates shape the root microbiome? Trends Plant Sci 23(1):25–41
Satyanarayana T, Johri BN, Das SK (2019) Microbial diversity in ecosystem sustainability and biotechnological applications: volume 1. Microbial diversity in normal and extreme environments. Springer, Singapore
Savka MA, Dessaux Y, Oger P, Rossbach S (2002) Engineering bacterial competitiveness and persistence in the phytosphere. Mol Plant-Microbe Interact 15(9):866–874
Schlegel R, Cakmak I, Torun B, Eker S, Köleli N (1997) The effect of rye genetic information on zinc, copper, manganese and iron concentration of wheat shoots in zinc deficient soil. Cereal Res Commun 1:177–184
Schmidt JE, Bowles TM, Gaudin A (2016) Using ancient traits to convert soil health into crop yield: impact of selection on maize root and rhizosphere function. Front Plant Sci 7:373
Sharma R, Chauhan A (2017) Rhizosphere microbiome and its role in plant growth promotion. In: Kalia V, Shouche P, Purohit H, Rahi P, Kalia VC (eds) Mining of microbial wealth and metagenomics. Springer, Singapore, pp 29–56
Sharma T, Dreyer I, Kochian L, Pineros MA (2016) The ALMT family of organic acid transporters in plants and their involvement in detoxification and nutrient security. Front Plant Sci 7:1488. https://doi.org/10.3389/fpls.2016.01488
Sharma P, Sangwan S, Kaur H (2019) Process parameters for biosurfactant production using yeast Meyerozyma guilliermondii YK32. Environ Monit Assess 191(9):531
Sharma P et al (2020) Role of microbes for attaining enhanced food crop production. In: Singh J, Vyas A, Wang S, Prasad R (eds) Microbial biotechnology: basic research and applications. Environmental and microbial biotechnology. Springer, Singapore
Sheng MM, Jia HK, Zhang GY, Zeng LN, Zhang TT, Long YH, Lan J, Hu ZQ, Zeng Z, Wang B, Liu HM (2020) Siderophore production by Rhizosphere biological control bacteria Brevibacillus brevis GZDF3 of Pinellia ternata and its antifungal effects on Candida albicans. J Microbiol Biotechnol 30(5):689–699

Shi S, Richardson AE, O'Callaghan M, DeAngelis KM, Jones EE, Stewart A, Firestone MK, Condron LM (2011) Effects of selected root exudate components on soil bacterial communities. FEMS Microbiol Ecol 77(3):600–610

Singh A, Sharma P, Kumari A, Kumar R, Pathak DV (2019) Management of root-knot nematode in different crops using microorganisms. In: Varma A, Tripathi S, Prasad R (eds) Plant biotic interactions. Springer, Cham, pp 85–99

Solís-Domínguez FA, Valentín-Vargas A, Chorover J, Maier RM (2011) Effect of arbuscular mycorrhizal fungi on plant biomass and the rhizosphere microbial community structure of mesquite grown in acidic lead/zinc mine tailings. Sci Total Environ 409(6):1009–1016

Somers E, Vanderleyden J, Srinivasan M (2004) Rhizosphere bacterial signalling: a love parade beneath our feet. Crit Rev Microbiol 30(4):205–240

Sriprang R, Hayashi M, Ono H, Takagi M, Hirata K, Murooka Y (2003) Enhanced accumulation of Cd2+ by a Mesorhizobium sp. transformed with a gene from Arabidopsis thaliana coding for phytochelatin synthase. Appl Environ Microbiol 69(3):1791–1796

Stefani FO, Bell TH, Marchand C, de la Providencia IE, El Yassimi A, St-Arnaud M, Hijri M (2015) Culture-dependent and-independent methods capture different microbial community fractions in hydrocarbon-contaminated soils. PLoS One 10(6)

Stein RJ, Duarte GL, Spohr MG, Lopes SIG, & Fett JP (2009) Distinct physiological responses of two rice cultivars subjected to iron toxicity under field conditions. Annals Appl Biol, 154(2):269–277

Sulieman S (2011) Does GABA increase the efficiency of symbiotic N2 fixation in legumes? Plant Signal Behav 6:32–36

Tarkka MT, Sarniguet A, & Frey-Klett P (2009) Inter-kingdom encounters: recent advances in molecular bacterium–fungus interactions. Current genetics, 55(3):233–243

Tenuta M, Beauchamp EG (1996) Denitrification following herbicide application to a grass sward. Can J Soil Sci 76(1):15–22

Timm CM, Pelletier DA, Jawdy SS, Gunter LE, Henning JA, Engle N, Aufrecht J, Gee E, Nookaew I, Yang Z, Lu TY (2016) Two poplar-associated bacterial isolates induce additive favorable responses in a constructed plant-microbiome system. Front Plant Sci 7:497

Toal ME, Yeomans C, Killham K, Meharg AA (2000) A review of rhizosphere carbon flow modelling. Plant Soil 222:263–281

Tomasino SF, Leister RT, Dimock MB, Beach RM, Kelly JL (1995) Field performance of Clavibacter xyli subsp. cynodontis expressing the insecticidal protein gene cryIA of Bacillus thuringiensis against European corn borer in field corn. Biol Control 5:442–448

Tournas VH, Katsoudas E (2005) Mould and yeast flora in fresh berries, grapes and citrus fruits. Int J Food Microbiol 105(1):11–17

Treseder KK (2008) Nitrogen additions and microbial biomass: a meta-analysis of ecosystem studies. Ecol Lett 11(10):1111–1120

Truyens S, Weyens N, Cuypers A, Vangronsveld J (2015) Bacterial seed endophytes: genera, vertical transmission and interaction with plants. Environ Microbiol Rep 7(1):40–50

Uroz S, D'Angelo-Picard C, Carlier A, Elasri M, Sicot C, Petit A, Oger P, Faure D, Dessaux Y (2003) Novel bacteria degrading N-acylhomoserine lactones and their use as quenchers of quorum-sensing-regulated functions of plant-pathogenic bacteria. Microbiology 149(8):1981–1989

Uroz S, Buée M, Murat C, Frey-Klett P, Martin F (2010) Pyrosequencing reveals a contrasted bacterial diversity between oak rhizosphere and surrounding soil. Environ Microbiol Rep 2(2):281–288

Uzoh IM, Babalola OO (2018) Rhizosphere biodiversity as a premise for application in bioeconomy. Agric Ecosyst Environ 265:524–534

Valverde A, De Maayer P, Oberholster T, Henschel J, Louw MK, Cowan D (2016) Specific microbial communities associate with the rhizosphere of Welwitschia mirabilis, a living fossil. PLoS One 11(4):e0153353

Vandenkoornhuyse P, Quaiser A, Duhamel M, Le Van A, Dufresne A (2015) The importance of the microbiome of the plant holobiont. New Phytol 206(4):1196–1206

Vessey JK (2003) Plant growth promoting rhizobacteria as biofertilizers. Plant Soil 255(2):571–586
Vicré M, Santaella C, Blanchet S, Gateau A, Driouich A (2005) Root border-like cells of Arabidopsis. Microscopical characterization and role in the interaction with rhizobacteria. Plant Physiol 138(2):998–1008
Vives-Peris V, de Ollas C, Gómez-Cadenas A, Pérez-Clemente RM (2020) Root exudates: from plant to rhizosphere and beyond. Plant Cell Rep 25:1–5
Walker V, Bertrand C, Bellvert F, Moënne-Loccoz Y, Bally R, Comte G (2011) Host plant secondary metabolite profiling shows a complex, strain-dependent response of maize to plant growth-promoting rhizobacteria of the genus Azospirillum. New Phytol 189(2):494–506
Wallenstein MD (2017) Managing and manipulating the rhizosphere microbiome for plant health: a systems approach. Rhizosphere 3:230–232
Wan T, Zhao H, Wang W (2017) Effect of biocontrol agent Bacillus amyloliquefaciens SN16-1 and plant pathogen Fusarium oxysporum on tomato rhizosphere bacterial community composition. Biol Control 112:1–9
Wang CX, Knill E, Glick BR, Defago G (2000) Effect of transferring 1-aminocyclopropane-1-carboxylic acid (ACC) deaminase genes into Pseudomonas fluorescens strain CHA0 and its gacA derivative CHA96 on their growth-promoting and disease-suppressive capacities. Can J Microbiol 46:898–890
Wang S, Wu H, Zhan J, Xia Y, Gao S, Wang W, Xue P, Gao X (2011) The role of synergistic action and molecular mechanism in the effect of genetically engineered strain Bacillus subtilis OKBHF in enhancing tomato growth and Cucumber mosaic virus resistance. BioControl 56(1):113–121
Wang D, Zhang N, Tang H, Adams JM, Sun B, Liang Y (2019) Straw biochar strengthens the life strategies and network of rhizosphere fungi in manure fertilized soils. Soil Ecol Lett 1(1–2):72–84
Wardle DA, Nilsson MC, Zackrisson O (2008) Fire-derived charcoal causes loss of forest humus. Science 320(5876):629
Weinert N, Piceno Y, Ding GC, Meincke R, Heuer H, Berg G, ... & Smalla K (2011) PhyloChip hybridization uncovered an enormous bacterial diversity in the rhizosphere of different potato cultivars: many common and few cultivar-dependent taxa. FEMS microbiology ecology, 75(3):497–506
Weller DM, Raaijmakers JM, Gardener BB, Thomashow LS (2002) Microbial populations responsible for specific soil suppressiveness to plant pathogens. Annu Rev Phytopathol 40(1):309–348
Weng J, Wang Y, Li J, Shen Q, Zhang R (2013) Enhanced root colonization and biocontrol activity of Bacillus amyloliquefaciens SQR9 by abrB gene disruption. Appl Microbiol Biotechnol 97(19):8823–8830
West SA, Fisher RM, Gardner A, Kiers ET (2015) Major evolutionary transitions in individuality. Proc Natl Acad Sci 112(33):10112–10119
White RA III, Rivas-Ubach A, Borkum MI, Köberl M, Bilbao A, Colby SM, Hoyt DW, Bingol K, Kim YM, Wendler JP, Hixson KK (2017) The state of rhizospheric science in the era of multi-omics: a practical guide to omics technologies. Rhizosphere 3:212–221
Win KT, Okazaki K, Ohkama-Ohtsu N, Yokoyama T, Ohwaki Y (2020) Short-term effects of biochar and Bacillus pumilus TUAT-1 on the growth of forage rice and its associated soil microbial community and soil properties. Biol Fertil Soils 56:1–7
Wu CH, Wood TK, Mulchandani A, Chen W (2006) Engineering plant-microbe symbiosis for rhizoremediation of heavy metals. Appl Environ Microbiol 72(2):1129–1134
Wu AL, Jiao XY, Fan FF, Wang JS, Guo J, Dong EW, Wang LG, Shen XM (2019) Effect of continuous sorghum cropping on the rhizosphere microbial community and the role of Bacillus amyloliquefaciens in altering the microbial composition. Plant Growth Regul 89(3):299–308
Yaish MW, Al-Harrasi I, Alansari AS, Al-Yahyai R, Glick BR (2017) The use of high throughput DNA sequence analysis to assess the endophytic microbiome of date palm roots grown under different levels of salt stress. Int Microbiol 19:143–155

Yang H, Knapp J, Koirala P, Rajagopal D, Peer WA, Silbart LK, Murphy A, Gaxiola RA (2007) Enhanced phosphorus nutrition in monocots and dicots over-expressing a phosphorus-responsive type I H+-pyrophosphatase. Plant Biotechnol J 5(6):735–745

Yang Y, Zhang S, Li N, Chen H, Jia H, Song X, Liu G, Ni C, Wang Z, Shao H, Zhang S (2019) Metagenomic insights into effects of wheat straw compost fertiliser application on microbial community composition and function in tobacco rhizosphere soil. Sci Rep 9(1):1–1

Yaryura P, Leon M, Correa O, Kerber N, Pucheu N, Garcia A (2008) Assessment of the role of chemotaxis and biofilm formation as requirements for colonization of roots and seeds of soybean plants by Bacillus amyloliquefaciens BNM339. Curr Microbiol 56:625–632

Yu H, Zou W, Chen J, Chen H, Yu Z, Huang J, Tang H, Wei X, Gao B (2019) Biochar amendment improves crop production in problem soils: a review. J Environ Manag 232:8–21

Yuan Z, Druzhinina IS, Labbé J, Redman R, Qin Y, Rodriguez R, Zhang C, Tuskan GA, Lin F (2016) Specialized microbiome of a halophyte and its role in helping non-host plants to withstand salinity. Sci Rep 6(1):1–3

Zahran HH (2001) Rhizobia from wild legumes: diversity, taxonomy. Ecology, nitrogen fixation and biotechnology. J Biotechnol 91:143–153

Zapalski MK (2011) Is absence of proof a proof of absence? Comments on commensalism. Palaeogeogr Palaeoclimatol Palaeoecol 302(3–4):484–488

Zgadzaj R, Garrido-Oter R, Jensen DB, Koprivova A, Schulze-Lefert P, Radutoiu S (2016) Root nodule symbiosis in Lotus japonicus drives the establishment of distinctive rhizosphere, root, and nodule bacterial communities. Proc Natl Acad Sci 113(49):E7996–E8005

Zhang X, Cao Y, Tian Y, Li J (2014) Short-term compost application increases rhizosphere soil carbon mineralization and stimulates root growth in long-term continuously cropped cucumber. Sci Hortic 175:269–277

Zhang N, Yang D, Wang D, Miao Y, Shao J, Zhou X, Xu Z, Li Q, Feng H, Li S, Shen Q (2015) Whole transcriptomic analysis of the plant-beneficial rhizobacterium Bacillus amyloliquefaciens SQR9 during enhanced biofilm formation regulated by maize root exudates. BMC Genomics 16(1):685

Zhu S, Vivanco JM, Manter DK (2016) Nitrogen fertilizer rate affects root exudation, the rhizosphere microbiome and nitrogen-use-efficiency of maize. Appl Soil Ecol 107:324–333

Ziegler M, Engel M, Welzl G, Schloter M (2013) Development of a simple root model to study the effects of single exudates on the development of bacterial community structure. J Microbiol Methods 94(1):30–36

Zolti A, Green SJ, Mordechay EB, Hadar Y, Minz D (2019) Root microbiome response to treated wastewater irrigation. Sci Total Environ 655:899–907

Chapter 22
Microbial Enzymes and Their Role in Phytoremediation

Kazi Sabnam Siddiqua, Shahabab Ahmad Farooqui, Touseef Hussain, and Heba I. Mohamed

Contents

K. S. Siddiqua · S. A. Farooqui
Department of Zoology, Aligarh Muslim University, Aligarh, Uttar Pradesh, India

T. Hussain (✉)
Department of Botany, Aligarh Muslim University, Aligarh, Uttar Pradesh, India

H. I. Mohamed
Department of Biological and Geological Sciences, Faculty of Education, Ain Shams University, Cairo, Egypt

© The Author(s), under exclusive license to Springer Nature Switzerland AG 2021
H. I. Mohamed et al. (eds.), *Plant Growth-Promoting Microbes for Sustainable Biotic and Abiotic Stress Management*,
https://doi.org/10.1007/978-3-030-66587-6_22

1 Introduction

Due to the dramatic increase in toxic products from various human activities, it has become an important challenge to control environmental pollution. Among them, the major increase in recent years has been soil pollution which might harm human health, crop quality, agriculture, and the climate (Conesa et al. 2012). One of the prominent reasons for soil pollution is due to human activities. The common strategies to remove toxic pollutants from contaminated soils and groundwater are often expensive, labor exhaustive, and not cost-efficient. There are several strategies to remove toxic substances from soil and groundwater. Phytoremediation can be one of these strategies to remove toxic substances from our environment. The plant organism and related microbial networks can be viewed as a daylight-driven hotspot for the turnover of natural, synthetic substances. In such conditions, the destiny of a compound won't just rely upon its inborn auxiliary soundness toward biochemical responses and its bioavailability yet additionally on the practical viability and solidness of common microbial networks as fundamental drivers of characteristic weakening of synthetic concoctions. Late research exhibits that collaborations among plants and microorganisms are significant for the biotransformation of natural, synthetic concoctions, for different procedures influencing the bioavailability of such mixes, and for the dependability of the affected biological system. Persistent natural poisons (POPs) and overwhelming metals, are considered as the most significant compound families that result in soil contamination (Belden et al. 2004; Xia et al. 2009). Due to the usage of insecticides against pests and mosquitoes, DDT has been collected in soil and river sediments (Lunney et al. 2004). The most common heavy metal pollution in soils is cadmium which is toxic to organisms. Low amount of Cd and DDT may influence the thickness of bone and increase the danger of vertebral breakage (Rignell-Hydbom et al. 2009). Bioremediation can convert pollutants to nonhazardous components enzymatically. However, the contaminant detoxification cycle can only continue if the conditions are suitable for the microorganism's growth and movement. Several bacteria complicate the process of eliminating organic contaminants, which rely mainly on the intracellular and extracellular enzymes (Madadi and Abbas 2017). Agricultural drainage and industrial release can be managed by rhizofiltration (Yadav et al. 2011; Yan-de et al. 2007). There can be approximately 275 hazardous substances that cause a threat to human health (Bernard 2010). The top 10 most "priority substances" are presented in Table 22.1. To circumvent the harmful effect of these hazardous compounds, several methodologies have been proposed to lower them from the soil. These techniques mainly incorporate the expulsion of soil to landfill locales or mainly physical methods. Such methods are quick but not cost-effective and may pose a danger to physical, chemical, and biological properties of soil. Moreover, the elimination of toxic substances from the atmosphere may be classified by the various groups and forms of these chemicals. The soil can, for example, be polluted with metals, toxic inorganic compounds, or various organic compounds. Metals include cadmium, cobalt, copper, chromium, lead, zinc, selenium, nickel, or mercury, among others. Other

Table 22.1 The top 10 most toxic metal components are mentioned below

Rank	Substance
1.	Arsenic
2.	Lead
3.	Mercury
4.	Vinyl chloride
5.	Polychlorinated biphenyls
6.	Benzene
7.	Cadmium
8.	Polycyclic aromatic hydrocarbons
9.	Benzo[a]pyrene
10.	Benzo[b]fluoranthene

inorganic mixtures could include nitrate, arsenic, sodium, alkali, or phosphate. Uranium, cesium, or strontium can be radioactive compounds. Chlorinated solvents such as trichloroethylene may form organic compounds: explosives like trinitrotoluene (TNT) and 1,3,5-trinitro-1,3,5-hexahydrotriazine (RDX). Certain constituents include numerous petroleum hydrocarbons such as benzene, toluene, and xylene (BTX), polycyclic aromatic hydrocarbons (PAHs), and pesticides such as atrazine and bentazone.

2 Importance of Phytoremediation

Phytoremediation, a system that utilizes plants to corrupt, balance out, and additionally expel soil pollutants, has been broadly explored. Rhizoremediation, a specific kind of phytoremediation which includes the plants and their related rhizosphere microorganisms, can happen normally/generally or can be impelled through intentionally presenting explicit organisms. In stress condition, such microbes can act as degraders and encourage plant growth (Gerhardt et al. 2009; Ahamd et al. 2019). Whereas certain natural compounds may be metabolized (*i.e.*, remediated) by bacteria that can be contained in or adjacent to the soil, without plants, this technique is usually moderate and incompetent due to the relatively limited number of decaying microorganisms throughout the soil (Brookes and McGrath 1984). In another way, the use of plants for the remediation of polluted soils, *i.e.*, phytoremediation, is a technically safe, effective, and moderately modest technology that is likely to be readily adopted by the applicable accessible. Soil microorganisms which are in close contact with plant roots may often promote metal phytoextraction (Shilev et al. 2001).

Phytoremediation has improved plant biotechnological approaches. The transgenic plants have more potential for productivity and are perfect and modest with economic bioremediation innovations which are highly encouraging; with few difficulties remain. Phytoremediation is a promising innovation that utilizes plants to debase, absorb, use, or detoxify metals, hydrocarbons, pesticides, and chlorinated solvents.

3 Merits and Demerits

The various merits of bioremediation are enlisted below:

1. It is conceivable as well as freely acknowledged (Marmiroli and McCutcheon 2004; Watt 2007).
2. Can be moderated by solar energy (Ali et al. 2013).
3. It can work together with organic compounds (Cofield et al. 2007).
4. Not expensive (Cornish et al. 1995).
5. On the plantation side, it reduces soil erosion by wind and water (Cunningham et al. 1995).
6. The metal-rich plant residue is reusable.
7. Water and airborne secondary diseases can be eliminated (Lili and Hui 2007).

Although some demerits are listed here below:

1. Due to the short root system of plants, only sub-surface contaminants can be cleaned up (Padmavathiamma and Li 2007).
2. Trees with longer root system can tidy up somewhat more profound pollution than plants, regularly 10–15 ft., yet fail to clean up intense springs moving forward without any more structure work.
3. These plants which have absorbed toxic pollutants can be a threat to the food chain (Arthur et al. 2000).
4. It requires large space and intense care.
5. Some volatile compound from groundwater can be a problem for air pollution too (Sakakibara et al. 2010).
6. Plants used in the remedy become inedible (Mejáre and Bülow 2001).
7. It takes a lot of time to clean up a small space (Stomp et al. 1994).

4 Mechanism of Phytoremediation

Rhizoremediation is a kind of phytoremediation which helps clean up pollutants from the low to moderate pollution level suitable mainly for both small and large sites (Zhuang et al. 2007) (Fig. 22.1).

The rhizosphere is identified with the root system and encompassing the surface and sub-surface soil. The three zones of rhizosphere are as follows:

1. *Endorhizosphere*: Some root tissue part (endodermis and cortical layers).
2. *Rhizoplane*: The root surface area where microorganisms associate with soil. It consists of three layers (epidermis, cortex core, and layer of polysaccharides).
3. *Ectorhizosphere*: Zone in which the roots adjoin the soil surface.
4. For expulsion of corruption forms, plants are engaged with several instruments to evacuate both natural and chemical toxic materials from contaminated situations (Rao et al. 2010).

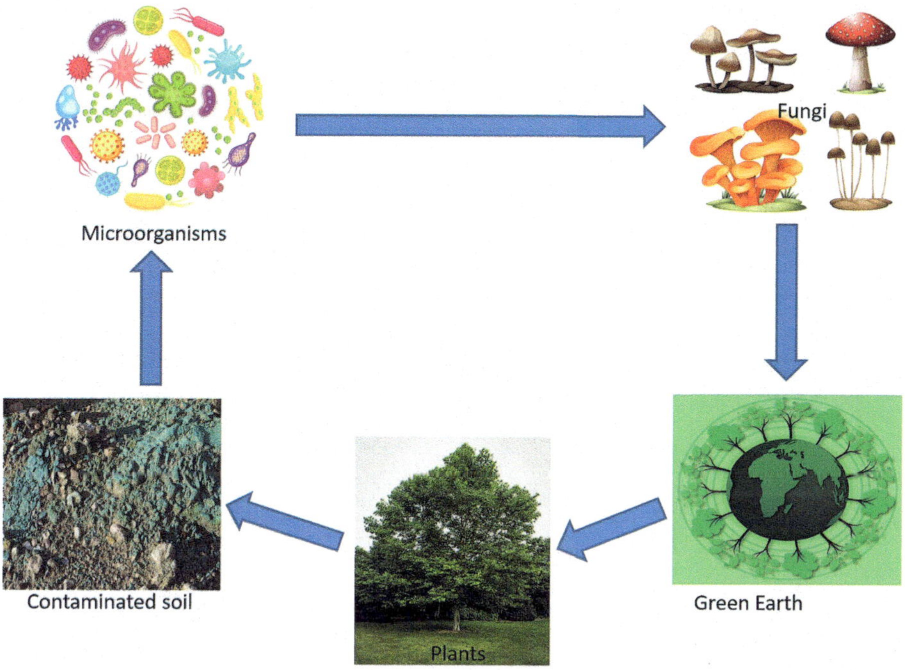

Fig. 22.1 A simple diagrammatic presentation of phytoremediation

Heavy metals pose a grave danger to human and animal health. Heavy metal accumulation in bodies of plants and animals happens after it enters the food chain (Haris et al. 2021; Dhankar et al. 2020; Hussain et al. 2021). They pose a threat because of the mutagenic ability of some heavy metals as it damages the DNA (Mohamed 2011; Mohamed et al. 2016; Akladious and Mohamed 2017). That is why the removal of these heavy metals for soil and several in situ and ex situ technologies that are used for this purpose is required. Phytoremediation is an environmentally sustainable technique, cost-effective for cleaning metal-polluted soils. In their growth, plants embrace various processes to lower the metal in soils without any antagonistic impacts (Table 22.2).

Phytostabilization, phytoextraction, and phytovolatilization are the main mechanisms, but here we are giving a brief explanation of phytovolatilization.

4.1 Phytovolatilization

Changing of toxic heavy metals such as Hg, Se, and As into less dangerous, unforeseeable structures released into the atmosphere by plants is called phytovolatilization (Malik and Biswas 2012). The reasonable utilization of phytovolatilization is

Table 22.2 Various plants used as phytoremediation

Plant	Metal	Reference(s)
Sedum alfredii H.	Pb, Cd	Anjum et al. (2012)
Pteris vittata	As	Datta et al. (2017)
Thlaspi goesingense	Ni	Puschenreiter et al. (2003)
Sedum alfredii	Zn	Yang et al. (2006)
Arabidopsis thaliana	Cd	Kiyono et al. (2012)
Pistia stratiotes	Cd, Pb, Zn	Vesely et al. (2012)
Eichhornia crassipes	As	Theeta et al. (2018)
Pistia stratiotes L.	Cd, Zn	Vidal et al. (2019)
Alyssum species, *Brassica juncea*	Ni	Kerkeb and Krämer (2003)
Oryza longistaminata, *Sorghum arundinaceum*, *Tithonia diversifolia*, and *Hyparrhenia rufa*	Hydrocarbon-contaminated soils	Ruley et al. (2020)
Athyrium wardii	Cd, Pb	Zhang et al. (2012); Zou et al. (2011)
Brassica juncea	Cd	Seth et al. (2008)

addressed because of the arrival of harmful unstable mixes to the environment with a hazard evaluation ought to be finished (Marques et al. 2009). Although some reported that these volatile compounds pose no threat to the environment, they mostly become diluted and dispersed (Meagher et al. 2000). Arsenic effectively volatilized into a mixture of arsenic mixes, arsenite, and arsenate (Sakakibara et al. 2010).

4.2 Phytoextraction

This is the mechanism in which foliage plants remove heavy metals from soil. The heavy metals in the soils are absorbed, transported, and accumulated in the plant's parts above the ground. These plant parts are then collected and safely handled to either dispose of the heavy metals or recycle them. These plants must have the capability of both metal tolerance and fast-growing to produce high biomass (Fig. 22.2).

5 Role of Microbial Enzyme in Phytoremediation

Table 22.3 shows the role of the plant and microbial enzymes in the biodegradation of organic compounds. Microbial sources are identified as (B) the bacterium or (F) the fungus.

Microbial enzymes play an essential role in the removal of environmentally toxic substances that are dispersed in the environment due to human activities. Various catalysts, e.g., oxygenases, are significant chemicals as they are fundamentally associated with the underlying procedure of corruption and reduce and debase the

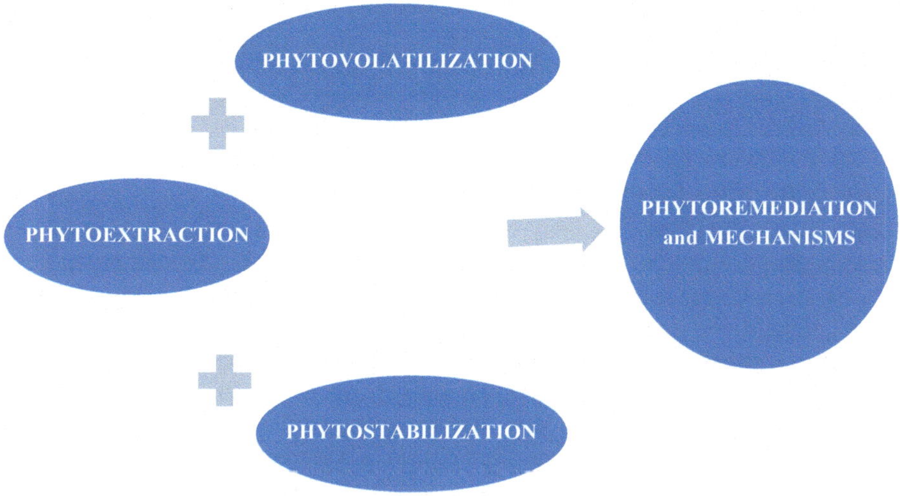

Fig. 22.2 Phytoremediation and its mechanisms

Table 22.3 List of different plant and microbial enzymes which function in organic compound biodegradation

Enzyme family	Examples of known source(s)	References
Various plant enzymes for uptake, transport, sequestration, and degradation	All plants	Pilon-Smits (2005)
Dehalogenase	*Xanthobacter autotrophicus* (B) Hybrid poplar (*Populus* spp.) *Sphingobium chlorophenolicum* (B)	Mena-Benitez et al. (2008) Susarla et al. (2002) Cai and Xun (2002)
Laccase	Alfalfa (*Medicago sativa*) *Trametes versicolor* (F)	Gramss et al. (2013) Novotny et al. (1997)
Dioxygenase	*Pseudomonas* sp. (B) *Mycobacterium* sp. (B)	Pieper et al. (2004) Pieper et al. (2004)
Peroxidase	Horseradish (*Armoracia rusticana*)	Susarla et al. (2002)
Nitrilase	Willow (*Salix* spp.) *Aspergillus niger* (F)	Susarla et al. (2002) Kaplan et al. (2006)
Nitroreductase	*Comamonas* ssp. (B) *Pseudomonas putida* (B) Hybrid poplar (*Populus* spp.)	Liu et al. (2007) Caballero et al. (2005) Susarla et al. (2002)
Phosphatase	Giant duckweed (*Spirodela polyrhiza*)	Susarla et al. (2002)
Cytochrome P450 monooxygenase	Most aerobic bacteria, all fungi, and all plants	McLean et al. (2005)
Oxidoreductases	*Flavobacterium* sp., *Phanerochaete chrysosporium*	Fierer (2017)
Oxygenases	*Bacillus subtilis* (B)	Muthukamalam et al. (2017)
Esterase	*Bacillus subtilis* (B)	Gangola et al. (2018)
Oxidoreductase	Fungi	Barber et al. (2020)

fragrant mixes. They reduce the toxic substances into the substrates. Two major oxygenases are monooxygenases (add one molecule of oxygen) and dioxygenases (add two molecules of oxygen) (Arora et al. 2010; Karigar and Rao 2011).

5.1 Microbial Oxidoreductases

Oxidoreductases used to remove the harmful effect of organic compounds by various bacteria, fungi, and plants (Husain 2006; Karigar and Rao 2011) by oxidative association. Microbes derive energy using biochemical reactions mediated by these enzymes in order to cleave chemical bonds and assist in electron transfer from a reduced organic (donor) substrate to another chemical (acceptor) compound. The pollutants are gradually oxidized to harmless compounds during these oxidation-reduction reactions (Karigar and Rao 2011). Oxidoreductases are involved in humidifying various phenolic substances which are formed in a soil environment from the decomposition of lignin. In the same way, oxidoreductases can also detoxify toxic xenobiotics by polymerization, such as phenolic or anilinic compounds, copolymerization, or binding of humic substances with certain substrates (Park et al. 2006). Microbial enzymes were used to decolorate and degrade azo dyes (Husain 2006). In the energy production process, bacteria consume electrons from organic compounds and use radioactive metal as the final electron acceptor. Eventually, the precipitant can be the product of bacterial redox reactions that reduce metals (Leung 2004).

The most common recalcitrant waste are chlorinated phenolic compounds that are present in the paper and pulp-processed effluents. Such compounds are formed during the process of pulp bleaching upon partial degradation of lignin. Most fungal organisms are considered appropriate for the removal from polluted habitats of chlorinated phenolic compounds. The filamentous fungal mycelia produce extracellular oxidoreductase enzymes which are released into the natural environment and are more effective in penetration of soil pollution than bacteria (Rubilar et al. 2008). Plants can decontaminate water polluted with phenolic compounds using enzymes which are produced and released from their roots. Phytoremediation of chemical contaminants has generally concentrated on three groups of compounds: chlorinated solvents, explosives, and hydrocarbons for petroleum (Duran and Esposito 2000).

5.1.1 Microbial Oxygenases

Oxygenases are a member of the enzyme class called oxidoreductase, FAD/NADH/NADPH used as cosubstrate to transfer oxygen from O_2. Oxygenases are classified into two classes, depending on the number of oxygen atoms used for oxygenation: monooxygenases and dioxygenases. They play a vital position in the chemical process of an organic compound by increasing their reactivity or water solubility or by causing cleavage of the aromatic ring. O_2 atoms are normally incorporated by oxygenase into the organic molecule, leading to cleavage of the aromatic ring (Arora et al. 2009).

5.1.2 Microbial Monooxygenases

The addition of a singlet oxygen molecule is achieved in the substrate by using monooxygenase enzyme. The cofactors used can be divided into two subgroups: (1) monooxygenases based on flavin and (2) monooxygenases P450 (*Bacillus megaterium*). The first subgroup prothetic group is flavin that is activated by using the coenzymes (NADP or NADPH), and the second subgroup includes heme. Monooxygenases are initiated and increase the rate of a chemical reaction activity in the phytoremediation. The other enzymes are cofactor-autonomous that play out their action with the subatomic oxygen as it were. Numerous procedures including desulfurization, denitrification, nitrification, ammonization, dehalogenation, shift, hydroxylation, and fragrant and aliphatic biodegradation are regulated by catalyst monooxygenases (Lock et al. 2017; Sirajuddin and Rosenzweig 2017; Syed et al. 2013).

5.1.3 Microbial Dioxygenases

Those are farraginous systems of enzymes which add molecular oxygen into the substrate. They degenerate the aromatic complex which raises a serious damage to the environment. This can be divided into two subclasses, depending on the enzyme's mode of activity: hydroxylation and cleavage dioxygenases. The hydroxylation enzyme catalyzes the expansion into the substrate of two oxygen atoms, while the cleavage enzyme catalyzes an aromatic ring usually carrying at least two or more groups of hydroxyls. The dioxygenase cleavage is further divided into two groups: intradiol and an extradiol. Such enzymes are concerned with environmental degradation of aromatic molecules. They are soil bacteria that are involved in the transformation process by converting aromatic precursors into aliphatic products (Al-Hawash et al. 2018; Fulekar 2017; Muthukamalam et al. 2017; Xenia and Refugio 2016).

5.2 Microbial Peroxidases

Peroxidases (EC 1.11.1.7) are disseminated widely in the environment. Plants and microorganisms are different sources that produce peroxidase enzymes. These microbial enzymes include degradation of pollution, raw materials, food and paper industries, degradation of textile dyes, lignin degradation paper/pulp industry, decoloration of the dye, sewage treatment, and animal feedstock and as biosensors.

For plants, they help in the production of lignin, the formation of cell walls, auxin metabolism, cell elongation, and channel protection. Also, they are subdivided into both heme and nonheme proteins. Furthermore, heme peroxidases in the prokaryotes and the eukaryotes are classified into three groups based on contrast (Bansal and Kanwar 2013; Falade et al. 2016).

5.2.1 Microbial Lignin Peroxidases (Lip)

During secondary metabolism, the white-rot fungus produces lignin peroxidases. Having the existence of H_2O_2 and mediator like veratryl alcohol LiP, lignin and other phenolic compounds are depleted. During the reaction, H_2O_2 is reduced to H_2O by obtaining electron from LiP (which is oxidized by itself) (Ten Have and Teunissen 2001). Lignin peroxidase (LiP) plays an essential function in the biodegradation of plant cell walls' lignin constituents (Piontek et al. 2011).

5.2.2 Microbial Manganese Peroxidases (MnP)

MnP is produced from basidiomycete fungus that caused lignin-degrading and oxidation of different phenolic compounds (Ten Have and Teunissen 2001), in which a multistep reaction oxidizes Mn^{2+} to the oxidant Mn^{3+}. Mn^{2+} stirs up the output of MnP and plays an important role as a substrate for MnP.

5.2.3 Microbial Versatile Peroxidases (VP)

VP enzymes are capable of oxidizing Mn^{2+} and phenolic aromatic substrates (Ruiz-Duenas et al. 2007). In the absence of manganese, VP has an unusually high specificity of substrates and a tendency to oxidize substrates compared to other peroxidases and plays important role in the bioremediation (Tsukihara et al. 2006).

5.3 *Microbial Laccases*

Laccases belong to multicopper oxidase family that are produced by certain plants and microorganisms which cause oxidation of phenolic and aromatic compounds while at the same time convert the molecular oxygen to water (Nigam 2013). Most microorganisms contain intracellular and extracellular laccases capable of catalyzing the oxidation of polyphenols, polyamines, and lignins (Rodrıguez Couto and Toca Herrera 2006) and repolymerization to humic materials (Viswanath et al. 2014). The production of laccase is depending on the concentrations of nitrogen in the fungi. Typically, the high concentrations of nitrogen are required to obtain large quantities of laccase (Viswanath et al. 2014).

5.4 *Microbial Lipases*

Lipase breaks down lipids which are produced by a wide array of microorganisms, bacteria, actinomycetes, and plants. Recent research has found that lipase is strongly related to the soil's organic pollutants. These microbial lipases are more flexible due to their active industrial use. Lipase enzymes can catalyze different reactions, including hydrolysis, interesterification, esterification, alcoholysis, and aminolysis (Prasad and Manjunath 2011). The lipase activity controlled the dramatic reduction of the total hydrocarbons of polluted soils and plays an important role as bioremediation of oil spills (Riffaldi et al. 2006; Sharma et al. 2011; Okino-Delgado et al., 2017). Lipases cause hydrolysis of triacylglycerol into glycerol and free fatty acids. Lipases were categorized into two groups based on criteria such as (a) enhanced enzyme activity once the triglycerides form an emulsion and (b) protein (lid)-looped lipases that cover the active site (Sharma et al. 2011).

5.5 *Microbial Cellulases*

Cellulases now provide the ability to turn cellulose waste materials into foods to overcome the increase in the population (Bennet et al. 2002). Some organisms formed a bound cell, associated cell envelope, and some extracellular cellulases. Some bacteria and fungi have shown that extracellular cellulases, hemicellulases, and pectinases are expressed constitutively at very low levels (Adriano-Anaya et al. 2005). Cellulose is broken down by cellulases during enzymatic hydrolysis to reduce the amount of sugar that can be fermented to ethanol by yeasts or bacteria (Sun and Cheng 2002). Cellulases extract microfibrils of cellulose that form during washing and the use of cotton-based clothes. This is often known in the textile industry as the brightening of colors and softening of fabrics. *Bacillus* strains produced alkaline cellulases, and *Trichoderma* and *Humicola* fungi produced neutral and acidic cellulases (Leisola et al. 2006).

5.6 *Microbial Proteases*

Proteases cause protein material degradation entering the atmosphere like animal mortality and a by-product in other industries such as livestock, fishing, and clothing, as a result of shedding and molting appendages (Beena and Geevarghese 2010). A varied and unique protease is used in the pharmaceutical industry to grow effective medicinal agents. Clostridial collagenase or subtilisin is used for the treatment of burns and wounds in conjunction with wide-spectrum antibiotics (Beena and Geevarghese 2010; Bhunia and Basak 2014).

5.7 Microbial Pullulanase

Several microorganisms such as *Klebsiella* spp., *Bacillus* spp., and *Geobacillus stearothermophilus* are used to produce pullulanases. It is very common in industrial uses due to its specific enzymatic action on pullulan, particularly in the specific connections (α-1,6 linkages), and starch is very essential as bioprocessor for its action (Karigar and Rao 2011; Lee et al. 2017).

5.8 Microbial Amylases

Alpha-amylases are extracellular enzyme that breaks in starch molecules, the α-1,4-glycosidic bond, and produce oligosaccharides, β-amylase, which also breaks the second maltose α-1,4-glycosidic bond and is synthesized in plants and bacteria. Amylases are important enzymes for their specific application in the process of conversion of industrial starch. Such enzymes are especially active on disaccharides (sucrose) and polysaccharides (starch) and are grouped into the glycoside hydrolase community (Singh et al. 2016; Gopinath et al. 2017).

6 Role of Plant Growth-Promoting Rhizobacteria (PGPR) Under Stress

PGPR is used to improve the execution of plants through different components, such as the production of precious hormones, the upgrading of plant nutrition status, and the decrease of the harm associated with the environment. The association among plants and PGPR happens to specific enthusiasm for situations that are described by imperfect developing conditions like high or low temperatures, dry spell, soil saltiness, and supplement shortage (plant development under stress) (Hussain et al. 2020a–c; Mandal et al. 2021). Primary expects to discuss the fundamental mechanisms of interaction between PGPR and plants and will focus on how PGPR can reduce abiotic stress damage in plants, which are essential crops for human diet (Hussain et al. 2020).

Abiotic stress thusly influences numerous plants like vegetables. In any case, vegetables, which are plants developed for their vegetative parts, are gradually affected by abiotic stress when compared with the family of grasses. The abiotic stress reduces the climate for the vegetable ranch and thus results in reduced crop yields. PGPR are beneficial to soil microscopic organisms suitable for stimulating plant physical substance and natural changes (Mohamed and Gomaa 2012).

Wholesome status, physical and biological properties of the soil, continuously changing environment, and other abiotic stresses are important drivers for reduced output in agriculture (Gopalakrishnan et al. 2015). Abiotic stresses are the fundamental reason for losses in crop yields and hiking food prices in the world with an increasing population. Attempts are being made to create stress-tolerant vegetables through traditional breeding or transgenic approaches, as multiple genes and metabolic procedures are stress-resilient (Ashraf and Akram 2009). The use of useful has recently become a possible new approach for protecting crops from damage caused by abiotic stress (Palaniyandi et al. 2014; Fatnassi et al. 2015; Wang et al. 2016; Hussain et al. 2020a–c).

6.1 Plant Growth-Promoting Bacteria (Subheading)

Natural exudates discharged through the roots are correlated with PGPR into plants and colonize the root surface and soil in direct contact with the root. The rhizosphere is the region of soil in the vicinity of plant roots in which chemistry and microbiology are influenced by their growth, respiration, and nutrient exchange which is illustrated in Fig. 22.3 (Smalla et al. 2006; Martino 2019), whereas the extracellular root surface has called been the rhizoplane (Foster 1986). Exudates discharged from plant roots pull microorganisms in the soil that can colonize rhizosphere or potentially plant tissue. Here, they offer the plant various helpful mixes in the supplement trade, primarily photosynthesis (Kawasaki et al. 2016).

Remarkably, through alternating environmental factors, plants may indirectly influence rhizosphere colonization. For example, increases in pH levels are through the absorption of ions and reduction of O_2 and H_2O levels caused by root respiration and water supply (Philippot et al. 2013). Two different studies (Bouffaud et al. 2012; Peiffer et al. 2013) showed how various genotypes of related plant species can be linked with different bacterial communities of the rhizosphere. Exudates differ in the different parts of the roots, the formative phases of the plant, and the conditions for growth (Zahar Haichar et al. 2008). This implies that after some time and space, a similar plant will link with a large number of different soil bacterial strains (Compant et al. 2010). Several bacterial species can spread from the endodermis of roots, enter, and colonize other stem organs (Compant et al. 2005; Dimkpa et al. 2009).

6.2 Plant Growth Promotion is driven by Rhizobacteria (Subheading)

Interactions with PGPR can lead to increased plant productivity, mineral contents, and plant growth. A portion of the primary benefits obtained by plants due to treatment with PGPB are increased root development, offering better protection against temperature and osmotic pressure, soil poisons, vermin, and pathogens (Lugtenberg and Kamilova 2009).

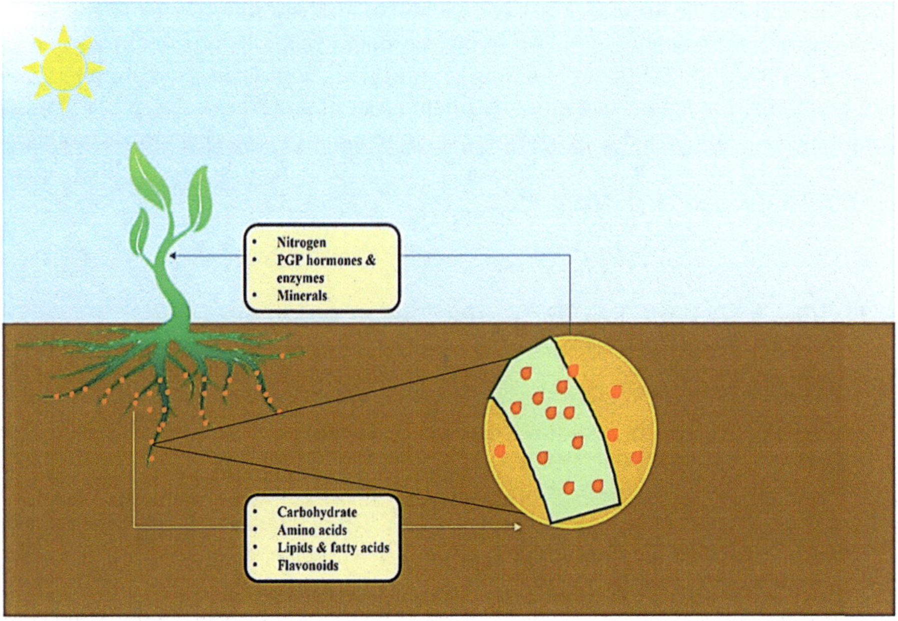

Fig. 22.3 A diagrammatic representation of plants and PGPR

6.3 Hormone-Related Mechanisms (Subheading)

PGPB produced indole-3-acetic acid (IAA) which caused enhancement of plant growth, cell elongation and differentiation, and stimulating lateral root growth (Dimkpa et al. 2009). IAA will roundly boost the plant's dietary status by extending root progression (explicitly sidelong roots), allowing the plant to reach a higher soil substratum, a main feature of nutrients with low mobility such as phosphorus (Wittenmayer and Merbach 2005). Gibberellins (GAs) are considered to play an important role in the promotion of plant development and produced by PGPR (Bastian et al. 1998). These diterpene hormones are present in plants, directing key procedures, for example, germination of the seed, elongation of the stem, expansion of the leaves, root growth, and fullness of root hair (Bottini et al. 2004; Yamaguchi 2008). The function of gibberellins in the reaction of grains to stresses fluctuates relying upon the stress type (Iqbal et al. 2011). The ethylene biosynthetic precursor is ACC, a hormone that is usually found in plants and increased under environmental stress. Ethylene is required for critical procedures such as tissue differentiation, root growth, flowering, grain production, senescence, and abscission; but it may suppress plant performance in case of overproduction (Saleem et al. 2007; Hays et al. 2007). Abscisic acid (ABA) is a plant hormone and increased under abiotic stress (Fahad et al. 2015). ABA is naturally engaged with seeds and bud's torpidity, and ABA imparts the primary biosynthetic strides to cytokinins, a phytohormone class that regularly assumes an adversarial role to ABA. Under salt stress condition, the plant biosynthesis of ABA which moved to leaves and caused stomatal closure, reduced transpiration and water loss (Xing et al. 2004), and reduced photosynthesis due to the CO_2 emission into the leaves (Yang et al. 2009; Barnawal et al. 2017; Shahzad et al. 2017).

6.4 Role of PGPB in Nutrient Stress (Subheading)

Comparatively, the use of PGPB as a biofertilizer has been found to improve plant nutrient usage and promote plant production (Calvo et al. 2015; Çakmakçi 2016). Once added, these inoculants improve plant growth and development or protect plants against pests and diseases (Ramjegathesh et al. 2013). Several microbial inoculants have been used as biofertilizers in this consideration which supply plants with nutrients such as N, P, K, S, and Fe. The more widely used genera as biocontrol agents are *Pseudomonas* (Tewari and Arora 2015), *Bacillus* (Alavo et al. 2015; Hussain and Khan 2020a, b), *Burkholderia* (Pinedo et al. 2015), *Agrobacterium* (Bazzi et al. 2015), and *Streptomyces* (Viaene et al. 2016). By production of antibiotics (Prasannakumar et al. 2015) and siderophores (Patel et al. 2016), by induction of systemic resistance (Zebelo et al. 2016), or any other mechanism, these organisms reduced plant disease.

7 Role of Biotechnology in Phytoremediation

Heavy metal pollution poses a global threat. Pollution from heavy metals remains a global threat. Contamination of heavy metals is an effect on the quality of soil and water as well as to human and animal health since they will pile up in the food chain (El-Beltagi et al. 2020; Moustafa-Farag et al. 2020; Sofy et al. 2020). Phytoremediation is a particular method of bioremediation. It is a characteristic natural procedure of corruption of xenobiotic and stubborn mixes liable for ecological contamination. In this, genetically engineered plants are used which directly uptake the pollutants from the soil (Macek et al. 2000). The word phyto means "plant"; that's why the remediation is mediated by the plant system (Sonali 2011). Phytoremediation includes numerous procedures which are done by the plant during their development on the sullied site. Thus, the pollutants are treated by plants utilizing of these responses like phytoextraction, phytostabilization, phytotransformation, phytostimulation, and phytovolatilization (Sonali 2011). Various contaminations have various destinies in plant-substrate frameworks, so they have diverse rate-restricting variables for phytoremediation that may focus on utilizing hereditary designing. Biotechnology shows us the chance to move hyper-aggregator phenotypes into quickly developing large biomass plants that can be exceptionally successful in phytoremediation (Rupali and Dibyengi 2004; Maurya et al. 2020).

A perfect phytoremediator characterizes more resistance for contamination, the capacity to either debase or assemble the impurities at an elevated amount in the biomass, broad root frameworks, the ability to assimilate a lot of water from the soil, and also quick development rates and significant levels of biomass (Cherian and Oliveira 2005). Albeit a few species can endure and develop in some defiled destinations, these species regularly become gradual, produce extremely low degrees of biomass, and are adjusted to quite certain natural conditions. What's more, trees which have broad root frameworks, high biomass, and low horticultural sources of info necessities endure poisons ineffectively and don't gather them. Traditional plants neglect the requirements for fortunate phytoremediators (Gratão and Braz 2005). The healing limit of plants can be essentially improved by hereditary manipulation and plant transformation technologies (Kraomer 2005). Presentation of novel qualities for the take-up and aggregation of contaminations into high biomass plants is demonstrating a fruitful procedure for the advancement of improved phytoremediators (Martanez et al. 2006). This reviews a portion of the exploration endeavors in this field and highlights future difficulties.

8 Phytoremediation Mechanism of Cd Adopted by Soil Plants

Remediation of Cd-sullied soil is a considerable issue far and wide, and it turned out to be progressively huge because of the exchange of Cd in higher trophic degrees in a natural way of life. Cd hyperaccumulators are exceptionally compelling a direct result of their capacity to endure and take up noteworthy measures of overwhelming metal from soils. Plants of various species have various capacities to hyperaccumulate Cd. Cd has low affinities with soil ligands due to its versatile nature and henceforth is effortlessly extricated by attaches and further shipped to other flying bits of the plant. The factors responsible for plant-based remediation of Cd are pH, temperature, media concentration, and concentration of other than Cd components (Mahajan and Kausha 2018; Dhankar et al. 2020). The phytoremediation process for extracting Cd in soil plants is shown in Fig. 22.4.

9 Conclusion and Future Prospectus

Metal pollution of soils is a common issue in various regions across the globe with varying intensities and magnitudes. Several remediation techniques for each bearing a broad variety of benefits and demerits have already been explored in depth elsewhere. Phytoremediation across all types of remediation is considered environmentally friendly and low cost. Around the same time, the introduction of commercial-scale phytoremediation technology requires careful consideration of the costly

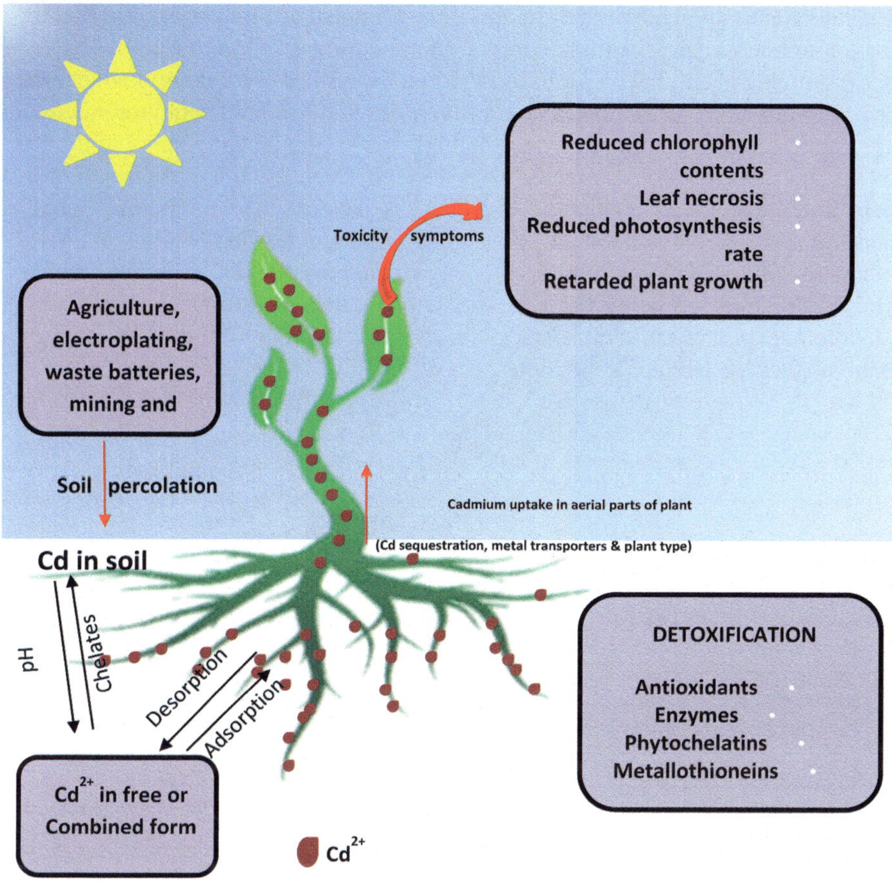

Fig. 22.4 A mechanism for phytoremediation of Cd accepted by soil plants

and time-consuming problems and the fate of the plants being used. It has been recognized that a variety of plants are prepared to accumulate high metal centralizations in their ethereal parts, keep the metals in roots or balance the metals in soils, eventually restrict their translocation to the shoots, and remove the metals from the dirt by amalgamating volatile mixtures. Growing of the above technologies includes different mechanisms that are already discussed in depth. The decision to use innovation in phytoremediation to remediate metal-defiled premises is based on soil type, metal content, degree of tainting, and natural upsetting effect. An understanding of the different processes involved will enhance decision-making when implementing a specific technology. Phytoextraction is commonly used by various advancements in phytoremediation, and a wide variety of hyperaccumulator plants fit for gathering high metal centralizations have been described. Distinguishing evidence and accepting qualities responsible for hyperaccumulation in hyperaccumulator plants into those plants fit for metal accumulation, and high biomass production may disturb the progress in phytoremediation. It requires a deeper understanding of the molecular basis of the pathways involved in pollutant degradation. Further analysis and disclosure of qualities appropriate for phytoremediation are important. Innovation in phytoremediation is still at an early stage of development, and field trials of transgenic plants for phytoremediation are unusually limited. Biosafety concerns should be properly answered, and protocols should be developed to avoid quality streams becoming wild species. Innovations in phytoremediation are currently accessible for only a limited subset of pollutants, and several destinations are being debased with a few synthetic substances. In this way, phytoremediators with various stacked qualities should be designed to satisfy the prerequisites of specific destinations.

Conflict of Interest: None of the Authors have any conflict of interest.

References

Adriano-Anaya M, Salvador-Figueroa M, Ocampo JA, Garcıa-Romera I (2005) Plant cell-wall degrading hydrolytic enzymes of *Gluconacetobacter diazotrophicus*. Symbiosis 40(3):151–156

Ahamd G, Nishat Y, Haris M, Danish M, Hussain T (2019) Efficiency of soil, plant and microbes for the healthy plant immunity and sustainable agricultural system. In: Varma A, Tripathi S, Prasad R (eds) Plant-microbe interface edit. Springer, Cham. https://doi.org/10.1007/978-3-030-19831-2-15

Akladious SA, Mohamed HI (2017) Physiological role of exogenous nitric oxide in improving performance, yield and some biochemical aspects of sunflower plant under zinc stress. Acta Biol Hungarica 68(1):101–114

Alavo TBC, Boukari S, Fayalo DG, Bochow H (2015) Cotton fertilization using PGPR Bacillus amyloliquefaciensFZB42 and compost: impact on insect density and cotton yield in NorthBenin, West Africa. Cogent Food Agric 1(1):1063829

Al-Hawash AB, Alkooranee JT, Abbood HA, Zhang J, Sun J, Zhang X, Ma F (2018) Isolation and characterization of two crude oil-degrading fungi strains from Rumaila oil field Iraq. Biotechnol Rep 17:104–109

Ali H, Khan E, Sajad MA (2013) Phytoremediation of heavy metals—concepts and applications. Chemosphere 91:869–881

Anjum NA, Ahmad I, Mohmood I, Pacheco M, Duarte AC, Pereira E, Umar S, Ahmad A, Khan NA, Iqbal M (2012) Modulation of glutathione and its related enzymes in plants' responses to toxic metals and metalloids—a review. Environ Exp Bot 75:307–324

Arora PK, Kumar M, Chauhan A, Raghava GP, Jain RK (2009) OxDBase: a database of oxygenases involved in Biodegradation. BMC Res Notes 2:67

Arora PK, Srivastava A, Singh VP (2010) Application of monooxygenases in dehalogenation, desulphurization, denitrification and hydroxylation of aromatic compounds. J Bioremed Biodegr 1:1–8

Arthur E, Crews H, Morgan C (2000) Optimizing plant genetic strategies for minimizing environmental contamination in the food chain: report on the MAFF funded joint JIC/CSL workshop held at the John Innes Centre, October 21–23, 1998. Int J Phytoremediation 2:1–21

Ashraf M, Akram NA (2009) Improving salinity tolerance of plants through conventional breeding and genetic engineering: an analytical comparison. Biotechnol Adv 27:744–752

Bansal N, Kanwar SS (2013) Peroxidase(s) in environment protection. Sci World J. https://doi.org/10.1155/2013/714639

Barber EA, Liu Z, Smith SR (2020) Organic contaminant biodegradation by oxidoreductase enzymes in wastewater treatment. Microorganisms 8:122. https://doi.org/10.3390/microorganisms8010122

Barnawal D, Bharti N, Pandey SS, Pandey A, Chanotiya CS, Kalra A (2017) Plant growth-promoting rhizobacteria enhance wheat salt and drought stress tolerance by altering endogenous phytohormone levels and TaCTR1/TaDREB2 expression. Physiol Plant 161(4):502–514

Bastian F, Cohen A, Piccoli P, Luna V, Baraldi R, Bottini R (1998) Production of indole-3-acetic acid and gibberellins A_1 and A_3 by *Acetobacter diazotrophicus* and *Herbaspirillum seropedicae* in chemically-defined culture media. Plant Growth Regul 24(1):7–11

Bazzi C, Alexandrova M, Stefani E, Anaclerio F, Burr TJ (2015) Biological control of Agrobacterium vitis using non-tumorigenic agrobacteria. VITIS-J Grapevine Res 38:31

Beena AK, Geevarghese PI (2010) A solvent tolerant thermostable protease from a psychrotrophic isolate obtained from pasteurized milk. Devel Microbiol Mol Biol 1:113–119

Belden JB, Clark BW, Phillips TA, Henderson KL, Arthur EL, Coats JR (2004) Detoxification of pesticide residues in soil using phytoremediation. Pestic Decontam Detox 863:155–167

Bennet JW, Wunch KG, Faison BD (2002) Use of fungi biodegradation. ASM Press, Washington, DC

Bernard RG (2010) Using soil bacteria to facilitate phytoremediation. Biotechnol Adv 28:367–374

Bhunia B, Basak B (2014) A review on application of microbial protease in bioremediation. In: Pramanik K (ed) Industrial and environmental biotechnology. Studium Press (India) Pvt. Ltd., Darya Ganj, pp 217–228

Bottini R, Cassan F, Piccoli P (2004) Gibberellin production by bacteria and its involvement in plant growth promotion and yield increase. Appl Microbiol Biotechnol 65(5):497–503

Bouffaud ML, Kyselková M, Gouesnard B, Grundmann G, Muller D, Moënne-Loccoz Y (2012) Is diversification history of maize influencing selection of soil bacteria by roots? Mol Ecol 21(1):195–206

Brookes PC, McGrath SP (1984) Effect of metal toxicity on the size of the soil microbial biomass. Soil Sci 35:341–346

Caballero A, Lazaro JJ, Ramos JL, Esteve-Nunez A (2005) PnrA, a new nitroreductase family enzyme in the TNT-degrading strain Pseudomonas putida JLR11. Environ Microbiol 7:1211–1219

Cai M, Xun L (2002) Organization and regulation of pentachlorophenol-degrading genes in Sphingobium chlorophenolicum ATCC 39723. J Bacteriol 184:4672–4680

Çakmakçi, Ramazan & Dönmez, Figen & Aydin, Adil & Sahin, Fikrettin (2016) Growth promotion of plants by plant growth-promoting rhizobacteria under greenhouse and two different field soil conditions. Soil Biology and Biochemistry. 38:1482–1487. https://doi.org/10.1016/j.soilbio.2005.09.019

Calvo P, Watts D, Torbert H, Kloepper J (2015) Application of microbial inoculants promote plant growth, increased nutrient uptake and improve root morphology of corn plants. In: American society of agronomy meetings, November, USA

Cherian S, Oliveira MM (2005) Transgenic plants in phytoremediation: recent advances and new possibilities. Environ Sci Technol 39:9377–9390

Cofield N, Banks MK, Schwab AP (2007) Evaluation of hydrophobicity in PAH-contaminated soils during phytoremediation. Environ Pollut 145:60–67

Compant S, Duffy B, Nowak J, Clément C, Barka EA (2005) Use of plant growth-promoting bacteria for biocontrol of plant diseases: principles, mechanisms of action, and future prospects. Appl Environ Microbiol 71(9):4951–4959

Compant S, Clément C, Sessitsch A (2010) Plant growth-promoting bacteria in the rhizo- and endosphere of plants: their role, colonization, mechanisms involved and prospects for utilization. Soil Biol Biochem 42(5):669–678

Conesa HM, Evangelou MW, Robinson BH, Schulin R (2012) A critical view of current state of phytotechnologies to remediate soils: still a promising tool. Sci World J:173–829

Cornish JE, Goldberg WC, Levine RS, Benemann JR (1995) Phytoremediation of soils contaminated with toxic elements and radionuclides. Battelle Press, Columbus
Cunningham SD, Berti WR, Huang JW (1995) Phytoremediation of contaminated soils. Trends Biotechnol 13:393–397
Datta R, Das P, Tappero R, Punamiya P, Elzinga E, Sahi S, Feng H, Kiiskila J, Sarka D (2017) Evidence for exocellular arsenic in fronds of Pteris vittata. Sci Rep 7:2839. https://doi.org/10.1038/s41598-017-03194-x
Dhankar R, Tyagi P, Kamble SS, Gupta D, Hussain T (2020) Advances in fungi: Rejuvenation of polluted sites. Fungi Bio-Prospects in Sustainable Agriculture, Environment and Nano-Technology, Vol-2, Edit by Sharma VK, Shah MP, Parmar S and Kumar A. Elsevier AP U.K. pg 251–275, https://doi.org/10.1016/B978-0-12-821925-6.00012-5. ISBN: 978-0-12-821925-6
Dimkpa C, Wein T, Asch F (2009) Plant-rhizobacteria interactions alleviate abiotic stress conditions. Plant Cell Environ 32(12):1682–1694
Duran N, Esposito E (2000) Potential applications of oxidative enzymes and phenoloxidase-like compounds in wastewater and soil treatment: a review. Appl Catal B 28(2):83–99
El-Beltagi HS, Sofy MR, Aldaej MI, Mohamed HI (2020) Silicon alleviates copper toxicity in flax plants by up-regulating antioxidant defense and secondary metabolites and decreasing oxidative damage. Sustainability 12:4732. https://doi.org/10.3390/su12114732
Fahad S, Hussain S, Bano A, Saud S, Hassan S, Shan D (2015) Potential role of phytohormones and plant growth-promoting rhizobacteria in abiotic stresses: consequences for changing environment. Environ Sci Pollut Res 22(7):4907–4921
Falade AO, Nwodo UU, Iweriebor BC, Green E, Mabinya LV, Okoh AI (2016) Lignin peroxidase functionalities and prospective applications. Microbiol Open. https://doi.org/10.1002/mbo3.394
Fatnassi IC, Chiboub M, Saadani O, Jebara M, Jebara SH (2015) Impact of dual inoculation with *Rhizobium* and PGPR on growth and antioxidant status of *Vicia faba* L. under copper stress. C R Biol 338(4):241–254
Fierer N (2017) Embracing the unknown: disentangling the complexities of the soil microbiome. Nat Rev Biotechnol 15(10):1–12
Foster R (1986) The ultrastructure of the rhizoplane and rhizosphere. Annu Rev Phytopathol 24(1):211–234
Fulekar MH (2017) Microbial degradation of petrochemical waste-polycyclic aromatic hydrocarbons. Bioresour Bioprocess 4:28
Gangola S, Sharma A, Bhatt P, Khati P, Chaudhary P (2018) Presence of esterase and laccase in *Bacillus subtilis* facilitates biodegradation and detoxification of cypermethrin. Sci Rep 8(1):1–11. 481038/s41598-018-31082-5
Gerhardt KE, Huang XD, Glick BR, Greenberg BM (2009) Phytoremediation and rhizoremediation of organic soil contaminants. Potent Chall Plant Sci 176:20–30
Gopalakrishnan S, Sathya A, Vijayabharathi R, Varshney RK, Gowda CL, Krishnamurthy L (2015) Plant growth promoting rhizobia: challenges and opportunities. 3 Biotech 5(4):355–377
Gopinath SCB, Anbu P, Arshad MKM, Lakshmipriya T, Voon CH, Hashim U, Chinni SV (2017) Biotechnological processes in microbial amylase production. BioMed Res Inter 2017:1272193
Gramss G (2013) Potential contributions of oxidoreductases from alfalfa plants to soil enzymology and biotechnology: A review. Nature Science. 169–224
Gratão LP, Braz J (2005) Phytoremediation: green technology for the clean-up of toxic metals in the environment. Plant Physiol 17:53–64
Haris M, Shakeel A, Hussain T, Ahmad G, Ansari MA, Khan A (2021) New trends in removing heavy metals from industrial wastewater through microbes. In: Shah M.P. (eds) Removal of emerging contaminants through microbial processes. Springer, Singapore. https://doi.org/10.1007/978-981-15-5901-3_9
Hays DB, Do JH, Mason RE, Morgan G, Finlayson SA (2007) Heat stress induced ethylene production in developing wheat grains induces kernel abortion and increased maturation in a susceptible cultivar. Plant Sci 172(6):1113–1123

Husain Q (2006) Potential applications of the oxidoreductive enzymes in the decolorization and detoxification of textile and other synthetic dyes from polluted water: a review. Crit Rev Biotechnol 26(4):201–221

Hussain, T, Akhtar, N, Aminedi, R, Danish M, Nishat Y, Patel S (2020) Role of the potent microbial based bioagent and their emerging strategies for the eco-friendly management of Agricultural Phytopathogens. Natural Bioactive Products in Sustainable Agriculture, Edit. by Singh J and Yadav AN. Springer Singapore, pg 45–66, ISBN-978-981-15-3023-4

Hussain K. et al. (2021) Bioremediation of waste gases and polluted soils. In: Panpatte D.G., Jhala Y.K. (eds) Microbial rejuvenation of polluted environment. Microorganisms for sustainability, vol 26. Springer, Singapore. https://doi.org/10.1007/978-981-15-7455-9_5

Hussain T, Khan AA (2020a) Bacillus subtilis T-AMU and its antifungal activity against Potato black scurf caused by Rhizoctonia solani. Biocatal Agri Biotechnol 23:101433 DOI: 10.1016/j.bcab.2019.101443

Hussain T, Khan AA (2020b) Determining the antifungal activity and characterization of *Bacillus siamensis AMU03* against *Macrophomina phaseolina* (Tassi) Goid. Indian Phytopathol. https://doi.org/10.1007/s42360-020-00239-6

Hussain T, Singh S, Danish, M, Pervez, R, Hussain K, Husain, R (2020a) Natural metabolites an eco-friendly approach to manage plant diseases and for better agricultural farming. Natural bioactive products in sustainable agriculture, Edit. by Singh J and Yadav AN. Springer Singapore, pp 1–13. https://doi.org/10.1007/978-981-15-3024-1_1

Hussain T, Akhtar N, Aminedi R, Danish M, Nishat Y, Patel S (2020b) Role of the potent microbial based bio agent and their emerging strategies for the eco-friendly management of agricultural phytopathogens. In: Singh J, Yadav AN (eds) Natural bioactive products in sustainable agriculture. Springer, Singapore, pp 45–66. https://doi.org/10.1007/978-981-15-3024-1_14

Hussain T, Haris M, Shakeel A, Khan AA, Khan MA (2020c) Bio-nematicidal activities by culture filtrate of Bacillus subtilis HussainT-AMU: new promising biosurfactant bioagent for the management of root galling caused by Meloidogyne incognita. Vegetos 33:229–238 https://doi.org/10.1007/s42535-020-00099-5

Iqbal N, Nazar R, Khan MIR, Masood A, Khan NA (2011) Role of gibberellins in regulation of source-sink relations under optimal and limiting environmental conditions. Curr Sci 100(7):998–1007

Kaplan O, Vejvoda V, Plíhal O, Pompach P, Kavan D, Bojarová P, Bezouska K, Macková M, Cantarella M, Jirků V, Kren V, Martínková L (2006) Purification and characterization of a nitrilase from Aspergillus niger K10. Appl Microbiol Biotechnol 73:567–575

Karigar CS, Rao SS (2011) Role of microbial enzymes in the bioremediation of pollutants: a review. Enzyme Rese 2011. https://doi.org/10.4061/2011/805187

Kawasaki A, Donn S, Ryan PR, Mathesius U, Devilla R, Jones A (2016) Microbiome and exudates of the root and rhizosphere of Brachypodium distachyon, a model for wheat. PLoS One 11(10):e0164533

Kerkeb L, Krämer U (2003) The role of free histidine in xylem loading of nickel in *Alyssum lesbiacum* and *Brassica juncea*. Plant Physiol 131:716–724

Kiyono M, Oka Y, Sone Y, Tanaka M, Nakamura R, Sato MH, Pan-Hou H, Sakabe K, Inoue K (2012) Expression of bacterial heavy metal transporter MerC fused with a plant SNARE, SYP121 in Arabidopsis thaliana increases cadmium accumulation and tolerance. Planta 235:841–850

Kraomer U (2005) Phytoremediation: novel approaches to cleaning up polluted soils. Curr Opin Biotechnol 16:133–141

Lee CW, Jang SH, Chung HS (2017) Improving the stability of cold-adapted enzymes by immobilization. Catalysts 7:112

Leisola M, Jokela J, Pastinen O, Turunen O (2006) Industrial use of enzymes—essay, laboratory of bioprocess engineering. Helsinki University of Technology, Helsinki

Leung M (2004) Bioremediation: techniques for cleaning up a mess. J Biotechnol 2:18–22

Lili L, Hui S (2007) Advance of research on phytoremediation of petroleum-polluted soil. Environ Prot Chem Ind 3:11

Liu L, Jiang CY, Liu XY, Wu JF, Han JG, Liu SJ (2007) Plant–microbe association for rhizoremediation of chloronitroaromatic pollutants with Comamonas sp. strain CNB-1. Environ Microbiol 9:465–473

Lock M, Nichol T, Murrell JC, Smith TJ (2017) Mutagenesis and expression of methane monooxygenase to alter regioselectivity with aromatic substrates. FEMS Microbiol Lett 13:364. https://doi.org/10.1093/femsle/fnx137

Lugtenberg B, Kamilova F (2009) Plant growth-promoting rhizobacteria. Ann Rev Microbiol 63:541–556

Lunney AI, Zeeb BA, Reimer KJ (2004) Uptake of weathered DDT in vascular plants: potential for phytoremediation. Environ Sci Technol 38(22):6147–6154

Macek T, Mackov M, Kas J (2000) Exploitation of plants for the removal of organics in environmental remediation. Biotechnol Adv 18:23–34

Madadi M, Abbas A (2017) Lignin degradation by fungal pretreatment: a review. J Plant Pathol Microbiol 8:398. https://doi.org/10.4172/2157-7471.1000398

Mahajan P, Kausha J (2018) Role of phytoremediation in reducing cadmium toxicity in soil and water. J Toxicol 2018:4864365

Malik N, Biswas A (2012) Role of higher plants in remediation of metal contaminated sites. Sci Rev Chem Commun 2:141–146

Marmiroli N, McCutcheon SC (2004) Making phytoremediation a successful technology. Phytoremediation 1:85–119. https://doi.org/10.1002/047127304X.ch3

Mandal SD, Sonali, Singh S, Hussain K , Hussain T (2021) Plant Microbe Association for the mutual benefits for plant growth and soil health. A. N. Yadav et al. (eds.), Current Trends in Microbial Biotechnology for Sustainable Agriculture, Environmental and Microbial Biotechnology, Springer Nature Singapore, Pg 95–121, https://doi.org/10.1007/978-981-15-6949-4_5

Maurya DK, Kumar A, Chaurasiya U, Hussain T, Singh SK (2020)Modern era of microbial biotechnology: opportunities and future prospects. Microbiomes and Plant Health. Elsevier, U.K ppg 317–343, https://doi.org/10.1016/B978-0-12-819715-8.00011-2

Marques AP, Rangel AO, Castro PM (2009) Remediation of heavy metal contaminated soils: phytoremediation as a potentially promising clean-up technology. Crit Rev Environ Sci Technol 39:622–654

Martanez M, Bernal P, Almela C, Vaolez D, Garca- Agustan P (2006) An engineered plant that accumulates higher levels of heavy metals than Thlaspi caerulescens, with yields of 100 times more biomass in mine soils. Chemosphere 64:478–485

Martino S, Gupta, Sneha and Walker, Robert (2019) The Role of plant growth-promoting bacteria in the growth of cereals under abiotic stresses. https://doi.org/10.5772/intechopen.87083

McLean KJ, Sabri M, Marshall KR, Lawson RJ, Lewis DG, Clift Balding DPR, Dunford AJ, Warman AJ, McVey JP, Quinn AM, Sutcliffe MJ, Scrutton NS, Munro AW (2005) Biodiversity of cytochrome P450 redox systems. Biochem Soc Trans 33:796–801

Meagher R, Rugh C, Kandasamy M, Gragson G, Wang N (2000) Engineered phytoremediation of mercury pollution in soil and water using bacterial genes. In: Terry N, Bañuelos G (eds) Phytoremediation of contaminated soil and water. Lewis Publishers, Boca Raton, FL, pp 201–219

Mejáre M, Bülow L (2001) Metal-binding proteins and peptides in bioremediation and phytoremediation of heavy metals. Trends Biotechnol 19(2):67–73

Mena-Benitez GL, Gandia-Herrero F, Graham S, Larson TR, McQueen-Mason SJ, French CE, Rylott EL, Bruce NC (2008) Engineering a catabolic pathway in plants for the degradation of 1,2-dichloroethane. Plant Phys 147:1192–1198

Mohamed HI (2011) Molecular and biochemical studies on the effect of gamma rays on lead toxicity in cowpea (*Vigna sinensis*) plants. Biol Trace Element Res 144:1205–1218

Mohamed HI, Gomaa EZ (2012) Effect of plant growth promoting Bacillus subtilis and Pseudomonas fluorescens on growth and pigment composition of radish plants (*Raphanus sativus*) under NaCl stress. Photosynthetica 50:263–272

Mohamed HI, Elsherbiny EA, Abdelhamid MT (2016) Physiological and biochemical responses of *Vicia faba* plants to foliar application with zinc and iron. Gesunde Pflanzen 68:201–212

Moustafa-Farag M, Mohamed HI, Mahmoud A, Elkelish A, Misra AN, Guy KM, Kamran M, Ai S, Zhang M (2020) Salicylic acid stimulates antioxidant defense and osmolyte metabolism to alleviate oxidative stress in watermelons under excess boron. Plan Theory 9:724. https://doi.org/10.3390/plants9060724

Muthukamalam SK, Sivagangavathi S, Dhrishya D, Rani SS (2017) Characterization of dioxygenases and biosurfactants produced by crude oil degrading soil bacteria. Brazil J Microbiol 48(4):637–647

Nigam PS (2013) Microbial enzymes with special characteristics for biotechnological applications. Biomol Ther 3:597–611

Novotny C, Vyas BRM, Erbanova P, Kubatova A, Sasek V (1997) Removal of various PCBs by various white-rot fungi in liquid cultures. Folia Microbiol 42:136–140

Okino-Delgado CH, Prado DZ, Facanali R, Marques MMO, Nascimento AS, Fernandes JC (2017) Bioremediation of cooking oil waste using lipases from wastes. PLoS One 12(10):e0186246

Padmavathiamma PK, Li LY (2007) Phytoremediation technology: hyper-accumulation metals in plants. Water Air Soil Pollut 184:105–126

Palaniyandi SA, Damodharan K, Yang SH, Suh JW (2014) Streptomyces sp. strain PGPA39 alleviates salt stress and promotes growth of 'Micro Tom' tomato plants. J Appl Microbiol 117(3):766–773

Park JW, Park BK, Kim JE (2006) Remediation of soil contaminated with 2,4-dichlorophenol by treatment of minced shepherd's purse roots. Archiv Environ Contam Toxicol 50(2):191–195

Patel PR, Shaikh SS, Sayyed RZ (2016) Dynamism of PGPR in bioremediation and plant growth promotion in heavy metal contaminated soil. Indian J Exp Biol 54:286–290

Peiffer JA, Spor A, Koren O, Jin Z, Tringe SG, Dangl JL (2013) Diversity and heritability of the maize rhizosphere microbiome under field conditions. Proc Natl Acad Sci 110(16):6548–6553

Philippot L, Raaijmakers JM, Lemanceau P, Van Der Putten WH (2013) Going back to the roots: the microbial ecology of the rhizosphere. Nat Rev Microbiol 11(11):789

Pieper DH, Martins dos Santos VAP, Golyshin PN (2004) Genomic and mechanistic insights into the biodegradation of organic pollutants. Curr Opin Biotechnol 15:215–224

Pilon-Smits E (2005) Phytoremediation. Annu Rev Plant Biol 56:15–39

Pinedo I, Ledger T, Greve M, Poupin MJ (2015) *Burkholderia phytofirmans* PsJN induces long-term metabolic and transcriptional changes involved in Arabidopsis thaliana salt tolerance. Front Plant Sci 6:466

Piontek K, Smith AT, Blodig W (2011) Lignin peroxidase structure and function. Biochem Soc Transactions 29(2):111–116

Prasad MP, Manjunath K (2011) Comparative study on biodegradation of lipid-rich wastewater using lipase producing bacterial species. Indian J Biotechnol 10(1):121–124

Prasannakumar SP, Gowtham HG, Hariprasad P, Shivaprasad K, Niranjana SR (2015) Delftiatsuruhatensis WGR–UOM–BT1, a novel rhizobacterium with PGPR properties from *Rauwolfia serpentina* (L.) Benth. ex Kurz also suppresses fungal phytopathogens by producing a new antibiotic—AMTM. Lett Appl Microbiol 61(5):460–468

Puschenreiter M, Wieczorek S, Horak O, Wenzel WW (2003) Chemical changes in the rhizosphere of metal hyperaccumulator and excluder Thlaspi species. J Plant Nutr Soil Sci 166:579–584

Ramjegathesh R, Samiyappan R, Raguchander T, Prabakar K, Saravanakumar D (2013) Plant–PGPR interactions for pest and disease resistance in sustainable agriculture. In: Maheshwari DK (ed) Bacteria in agrobiology: disease management. Springer, Berlin, pp 293–320

Rao MA, Scelza R, Scotti R, Gianfreda L (2010) Role of enzymes in the remediation of polluted environments. J Soil Sci Plant Nutri 10(3):333–353

Riffaldi R, Levi-Minzi R, Cardelli R, Palumbo S, Saviozzi A (2006) Soil biological activities in monitoring the bioremediation of diesel oil-contaminated soil. Water Air Soil Pollut 170(1–4):3–15

Rignell-Hydbom A, Skerfving S, Lundh T, Lindh CH, Elmstahl S, Bjellerup P, Jonsson BAG, Stromberg U, Akesson A (2009) Exposure to cadmium and persistent organochlorine pollutants and its association with bone mineral density and markers of bone metabolism on post-menopausal women. Environ Res 109(8):991–996

Rodrıguez Couto S, Toca Herrera JL (2006) Industrial and biotechnological applications of laccases: a review. Biotechnol Advan 24(5):500–513

Rubilar O, Diez MC, Gianfreda L (2008) Transformation of chlorinated phenolic compounds by white rot fungi. Critical Rev Environ Sci Technol 38(4):227–268

Ruiz-Duenas FJ, Morales M, Perez-Boada M et al (2007) Manganese oxidation site in *Pleurotus eryngii* versatile peroxidase: a site-directed mutagenesis, kinetic, and crystallographic study. Biochemistry 46(1):66–77

Ruley JA, Amoding A, Tumuhairwe JB, Basamba TA, Opolot E, Oryem-Origa H (2020) Enhancing the phytoremediation of hydrocarbon-contaminated soils in the sudd wetlands, south sudan, using organic manure. Appl Environ Soil Sci 2020:4614286

Rupali D, Dibyengi S (2004) Biotechnology in phytoremediation of metal-contaminated soils. Proc Indian Natn Sci Acad B 701:99–108

Sakakibara M, Watanabe A, Inoue M (2010) Phytoextraction and phytovolatilization of arsenic from As-contaminated soils by *Pteris vittata.* In: Proceedings of the annual international conference on soils, sediments, water and energy. p 26

Saleem M, Arshad M, Hussain S, Bhatti AS (2007) Perspective of plant growth promoting rhizobacteria (PGPR) containing ACC deaminase in stress agriculture. J Industrial Microbiol Biotechnol 34(10):635–648

Seth CS, Kumar Chaturvedi P, Misra V (2008) The role of phytochelatins and antioxidants in tolerance to Cd accumulation in *Brassica juncea* L. Ecotoxicol Environ Saf 71:76–85

Shahzad R, Khan AL, Bilal S, Waqas M, Kang SM, Lee IJ (2017) Inoculation of abscisic acid-producing endophytic bacteria enhances salinity stress tolerance in *Oryza sativa*. Environ Exp Bot 136:68–77

Sharma D, Sharma B, Shukla AK (2011) Biotechnological approach of microbial lipase: a review. Biotechnology 10(1):23–40

Shilev SI, Ruso J, Puig A, Benlloch M, Jorrin J, Sancho E (2001) Rhizospheric bacteria promote sunflower (Helianthus annuus L.) plant growth and tolerance to heavy metals. Minerva Biotechnol 13:37–39

Singh R, Mittal A, Kumar M, Mehta PK (2016) Amylases: a note on current applications. Inter Sci Comm 5(11):27–32

Sirajuddin S, Rosenzweig AC (2017) Enzymatic oxidation of methane. Biochemistry 54(14):2283–2294

Smalla K, Sessitsch A, Hartmann A (2006) The Rhizosphere: 'soil compartment influenced by the root'. Blackwell, Oxford

Sofy MR, Seleiman MF, Alhammad BA, Alharbi BM, Mohamed HI (2020) Minimizing adverse effects of Pb on maize plants by combined treatment with jasmonic, salicylic acids and proline. Agronomy 10:699. https://doi.org/10.3390/agronomy10050699

Sonali B (2011) Importance of phytoremediation. Biotech Articles. https://www.biotecharticles.com/Applications-Article/Importance-of-Phytoremediation-613.html, (accessed on 8-02-2021)

Stomp AM, Han KH, Wilbert S (1994) Genetic strategies for enhancing phytoremediation. Ann N Y Acad Sci 721:481–491

Sun Y, Cheng J (2002) Hydrolysis of lignocellulosic materials for ethanol production: a review. Bioresour Technol 83(1):1–11

Susarla S, Medina V, Mccutcheon S (2002) Phytoremediation: An Ecological Solution to Organic Chemical Contamination. Ecological Engineering. 18:647–658. https://doi.org/10.1016/S0925-8574(02)00026-5

Syed K, Porollo A, Miller D, Yadav JS (2013) Rational engineering of the fungal P450 monooxygenase CYP5136A3 to improve its oxidizing activity toward polycyclic aromatic hydrocarbons. Protein Eng Des Sel 26(9):553–557

Ten Have R, Teunissen PJM (2001) Oxidative mechanisms involved in lignin degradation by white-rot fungi. Chem Revi 101(11):3397–3413

Tewari S, Arora NK (2015) Plant growth promoting fluorescent *Pseudomonas* enhancing growth of sunflower crop. Int J Sci Technol Soc 1(1):51–53

Theeta S, Meeinkuirt W, Saengwilai P, Pichtel J, Taeprayoon P (2018) Aquatic plants for phytostabilization of cadmium and zinc in hydroponic experiments. Environ Sci Pollut Res 25(15):14964–14976

Tsukihara T, Honda Y, Sakai R, Watanabe T, Watanabe T (2006) Exclusive overproduction of recombinant versatile peroxidase MnP2 by genetically modified white rot fungus, *Pleurotus ostreatus*. J Biotechnol 126(4):431–439

Vesely T, Tlustos P, Szakova J (2012) Organic acid enhanced soil risk element (Cd, Pb and Zn) leaching and secondary bioconcentration in water lettuce (Pistia stratiotes L) in the rhizofiltration process. Int J Phytoremediation 14(4):335–349

Viaene T, Langendries S, Beirinckx S, Maes M, Goormachtig S (2016) Streptomyces as a plant's best friend? FEMS Microbiol Ecol 92(8):1–10.fiw119. https://doi.org/10.1093/femsec/fiw119

Vidal CF, Oliveira JA, da Silva AA, Ribeiro C, Farnese FDS (2019) Phytoremediation of arsenite-contaminated environments: is *Pistia stratiotes* L. a useful tool? Ecol Indic 104:794–801

Viswanath B, Rajesh B, Janardhan A, Kumar AP, Narasimha G (2014) Fungal laccases and their applications in bioremediation. Enzyme Res 2014:163242. https://doi.org/10.1155/2014/163242

Wang C, Wang C, Gao YL, Wang YP, Guo JH (2016) A consortium of three plant growth-promoting rhizobacterium strains acclimates *Lycopersicon esculentum* and confers a better tolerance to chilling stress. J Plant Growth Regul 35(1):54–64

Watt NR (2007) Testing amendments for increasing soil availability of radionuclides. Phytoremediat Methods Rev 23:131–137

Wittenmayer L, Merbach W (2005) Plant responses to drought and phosphorus deficiency: contribution of phytohormones in root-related processes. J Plant Nutri Soil Sci 168(4):531–540

Xenia ME, Refugio RV (2016) Microorganisms metabolism during bioremediation of oil contaminated soils. J Bioremed Biodegr 7:2. https://doi.org/10.4172/2155-6199.1000340

Xia HL, Chi XY, Yan ZJ, Cheng WW (2009) Enhancing plant uptake of polychlorinated biphenyls and cadmium using tea saponin. Bioresour Technol 100(20):4649–4653

Xing H, Tan L, An L, Zhao Z, Wang S, Zhang C (2004) Evidence for the involvement of nitric oxide and reactive oxygen species in osmotic stress tolerance of wheat seedlings: inverse correlation between leaf abscisic acid accumulation and leaf water loss. Plant Growth Regul 42(1):61–68

Yadav BK, Siebel MA, Bruggen JAV (2011) Rhizofiltration of a heavy metal (lead) containing wastewater using the wetland plant Carex pendula. CLEAN—Soil, Air Water 39(5):467–474

Yamaguchi S (2008) Gibberellin metabolism and its regulation. Ann Rev Plant Biol 59:225–251

Yan-de J, Zhen-li H, Xiao-e Y (2007) Role of soil rhizobacteria in phytoremediation of heavy metal contaminated soils. J Zhejiang University Sci B 8(3):192_207

Yang X, Li T, Yang J, He Z, Lu L, Meng F (2006) Zinc compartmentation in root, transport into xylem, and absorption into leaf cells in the hyperaccumulating species of Sedum alfredii Hance. Planta 224:185–195

Yang J, Kloepper JW, Ryu CM (2009) Rhizosphere bacteria help plants tolerate abiotic stress. Trends Plant Sci 14(1):1–4

Zahar Haichar F, Marol C, Berge O, Rangel-Castro JI, Prosser JI, Balesdent JM (2008) Plant host habitat and root exudates shape soil bacterial community structure. ISME J 2(12):1221

Zebelo S, Song Y, Kloepper JW, Fadamiro H (2016) Rhizobacteria activates (+)-δ-cadinene synthase genes and induces systemic resistance in cotton against beet armyworm (Spodopteraexigua). Plant Cell Environ 39(4):935–943

Zhang S, Li T, Huang H, Zou T, Zhang X, Yu H, Zheng Z, Wang Y (2012) Cd accumulation and phytostabilization potential of dominant plants surrounding mining tailings. Environ Sci Pollut Res 19:3879–3888

Zhuang X, Chen J, Shim H, Bai Z (2007) New advances in plant growth promoting rhizobacteria for bioremediation. Environ Inter 33:406–413

Zou T, Li T, Zhang X, Yu H, Luo H (2011) Lead accumulation and tolerance characteristics of Athyrium wardii (Hook.) as a potential phytostabilizer. J Hazard Mater 186:683–689

Index

© The Author(s), under exclusive license to Springer Nature Switzerland AG 2021
H. I. Mohamed et al. (eds.), *Plant Growth-Promoting Microbes for Sustainable Biotic and Abiotic Stress Management*,
https://doi.org/10.1007/978-3-030-66587-6

D

E

F

I

Q

R

T

Printed by Printforce, the Netherlands